Fundamentals of
Structural Mechanics

The book *Fundamentals of Structural Mechanics* is written with a view to classify various concepts necessary for all engineers and designers. The clarity of these concepts are essentially required for application of engineering in design and construction of **safe structures** and **machine** components. No engineering design is possible without understanding the concept of structural mechanics. The fundamentals of structural mechanics clarifies **behaviour of materials** under different types of loads. No structure can be **designed or built** without considering the behaviour of its components. The behaviour and strength of materials and components of structures subjected to different loadings is studied under structural mechanics. Thus this book on fundamentals of structural mechanics forms the foundation course for theory of structures, design of structures, theory of machines, etc.

The text material is presented considering learning theories and heterogeneity of learners. The material is arranged in such a way that first the concepts are described and clarified to develop principles for application in design and analysis of structural elements. This approach of presentation makes even a weak student to learn effectively the fundamentals essential in their engineering professional career. The presentation of new knowledge to the learners is based on the principles of **'known to unknown'**, **'easy to difficult'**, and **'simple to complex'**.

The presentation of new knowledge is immediately followed by solved examples sequenced in increasing difficulty levels so as to enthuse and maintain the motivation of different types of learners. The book will facilitate self-learning also.

The unsolved practice exercises for students shall facilitate **mastery learning** to excel in the examinations and professional career. The focus of the book is on **learning fundamentals and developing problem-solving skills** of learners by application of fundamental knowledge.

The **learning focus** makes the book equally suited to **degree and diploma** students of civil, mechanical, electrical and other engineering disciplines. Use of simple language makes the book **student-friendly** in different regions to prepare them for the gaining advance knowledge. Some difficult and challenging problems shall facilitate **top students to excel** in their studies and profession.

Fundamentals of
Structural Mechanics

PS Gahlot ME (Str. Engg.)
Former Director General
Yagyavalkya Institute of Technology
Sitapura, Jaipur, Rajasthan
Former Professor in Civil Engineering
National Institute of Technical Teacher's Training and Research
Chandigarh 160 019

Deep Gehlot BE, BSc
Senior Engineering and Construction Manager
TCS

CBS

CBS Publishers & Distributors Pvt Ltd

New Delhi • Bengaluru • Chennai • Kochi • Kolkata • Mumbai
Hyderabad • Nagpur • Patna • Pune • Vijayawada

Fundamentals of Structural Mechanics

ISBN: 978-81-239-2192-1

Copyright © Authors and Publisher

First Edition: 2012
  Reprint: 2017

Published by Satish Kumar Jain and Produced by Varun Jain for

CBS Publishers & Distributors Pvt Ltd
4819/XI Prahlad Street, 24 Ansari Road, Daryaganj, New Delhi 110 002, India.
Ph: 23289259, 23266861, 23266867   Website: www.cbspd.com
Fax: 011-23243014   e-mail: delhi@cbspd.com; cbspubs@airtelmail.in.
Corporate Office: 204 FIE, Industrial Area, Patparganj, Delhi 110 092
Ph: 4934 4934   Fax: 4934 4935   e-mail: publishing@cbspd.com; publicity@cbspd.com

Branches

- **Bengaluru:** Seema House 2975, 17th Cross, K.R. Road,
  Banasankari 2nd Stage, Bengaluru 560 070, Karnataka
  Ph: +91-80-26771678/79   Fax: +91-80-26771680   e-mail: bangalore@cbspd.com
- **Chennai:** 7, Subbaraya Street, Shenoy Nagar, Chennai 600 030, Tamil Nadu
  Ph: +91-44-26680620, 26681266   Fax: +91-44-42032115   e-mail: chennai@cbspd.com
- **Kochi:** Ashana House, No. 39/1904, AM Thomas Road, Valanjambalam,
  Ernakulam 682 016, Kochi, Kerala
  Ph: +91-484-4059061-65   Fax: +91-484-4059065   e-mail: kochi@cbspd.com
- **Kolkata:** 6/B, Ground Floor, Rameswar Shaw Road, Kolkata-700 014, West Bengal
  Ph: +91-33-22891126, 22891127, 22891128   e-mail: kolkata@cbspd.com
- **Mumbai:** 83-C, Dr E Moses Road, Worli, Mumbai-400018, Maharashtra
  Ph: +91-22-24902340/41   Fax: +91-22-24902342   e-mail: mumbai@cbspd.com

Representatives

- **Hyderabad** 0-9885175004   - **Nagpur**   0-9021734563   - **Patna** 0-9334159340
- **Pune**   0-9623451994   - **Vijayawada** 0-9000660880

Printed at India Binding House, Noida, UP, India

Preface

Structural mechanics forms the basic foundation course for all structural and machine element designs. Engineering students and designers cannot design efficiently without fully understanding the fundamentals of structural mechanics. Keeping this need of engineering students and professional engineers in mind, this book *Fundamentals of Structural Mechanics* is written for the benefit of engineering students desirous of excelling in the profession. The authors have long experience in teaching strength of materials, mechanics of solids, theory of structures, applied mechanics, design of structures, etc. and in the profession to derive the basic needs of the subject. Keeping this need in mind, the book has been brought out to benefit the engineering students desirous of excelling in their examinations as well as in the profession.

The contents are presented in a graded manner incorporating the principles of 'known to unknown', 'easy to difficult' and 'simple to complex' so that the knowledge is easier to comprehend. The variety of solved and unsolved practice exercises makes the text material learner-friendly so as to facilitate independent learning by the students.

The text material covers engineering degree as well as diploma syllabi of various technical boards and universities. The presentation of the text material suits heterogeneous student masses with variable IQ. The solved and unsolved problems prepare the learners for professional competitive examinations as well as professional career in design, construction and manufacture.

The book is highly suited for the basic course on strength of materials, mechanics of structures, and mechanics of solids, specially engineering students of civil, mechanical, automobile, aeronautics, and other engineering courses both at degree and diploma levels.

The book is written in simple language easily understood by national and other regional language-speaking students.

PS Gahlot
Deep Gehlot

Acknowledgements

The authors have developed this book on the fundamentals of structural mechanics based on learning theories, vast experience of teaching the subject at different levels including the teachers, slow and fast learning students, and working professionals, viz. structural designers. The authors express their gratitude to Dr MM Malhotra (former Director, National Institute of Technical Teacher's Training and Research, Chandigarh) for providing learning modules of strength of materials submitted for the award of PhD for structuring and sequencing different topics. Authors are also indebted to the authors of many other books consulted during preparation of this book. The authors are also thankful to all those teachers and students who provided feedback on learning of strength of materials for refining and developing this book on the fundamentals of structural mechanics.

The authors also express their thanks to Mr Yogesh Mahawar for his untiring work of typing manuscript. The authors also wish to express their thanks to Mrs Manorama Gahlot for providing logistic support during the preparation of the book.

PS Gahlot
Deep Gehlot

Contents

Unit II

AXIAL FORCE STRUCTURES

Unit III
BENDING STRUCTURES (BEAMS)

Unit IV
BENDING STRUCTURES: DEFORMATIONS

Unit V
TORSION IN SHAFTS

Unit VI

LONG AND SHORT COLUMNS: LOAD CAPACITY ANALYSIS

12 Long and Short Columns in Structures   515

Unit VII

COMBINED STRESSES AND PRINCIPAL STRESSES

13 Principal Planes and Principal Stresses   561

14 Unsymmetrical Bending   593

Unit I

Fundamental Concepts

1

Force System and Equilibrium

LEARNING OBJECTIVES

After studying this chapter the learner will be able to:

1.1 **Understand** basic concepts of **force system**.

1.2 **Draw free body diagram** of a body subjected to certain forces.

1.3 **Understand** the conditions of **equilibrium** of forces.

1.4 **Differentiate** between **vector** and **scalar** quantities.

1.5 **Know** basic **principles** of analyzing coplanar force system.

1.6 **Understand resultant** of coplanar force system.

1.7 **Determine orthographic components** of a force system.

1.8 **Understand parallelogram law** of forces for resultant.

1.9 **State Lami's theorem** for **concurrent coplanar** force system.

1.10 **Understand** drawing of **free body diagram** to solve **problems** of mechanisms in **equilibrium**.

1.11 **Understand triangle law** of forces.

1.12 **Understand polygon law of forces** for graphical method.

1.1 INTRODUCTION

Engineering Mechanics

During your school days, you have studied mechanics as one of the topics in physics. Engineering mechanics is a branch of *physical science*, which provides understanding of the *actions and reactions* of bodies at rest or in motion.

In engineering, be it civil, mechanical, electrical or electronics, there are innumerable problems that are related to the determination of the *state of motion* of bodies under the *action of* a set of forces. Examine various bodies and structures under the action of a set of forces, e.g.

- Whether a dam is stable or it tilts due to action of water pressure on it?
- What is the power of motor (action) required for raising a particular mass of water through a certain height in a certain period of time?

3

For solving all such problems, an engineer needs the application of the *principles of mechanics*. A clear understanding of these principles will be essential for understanding the subjects such as *hydraulics, strength of materials, structural analysis* and *design* that you would study in subsequent semesters. It is necessary to understand certain *basic concepts* before it is possible to understand the meaning of *"mechanics"*, the *problems* covered under it and the principles upon which this branch of science is based.

1.2 BASIC CONCEPTS

Space: Space is region, which extends *in all directions* and contains everything in it such as solar system with sun and its planets, stars, etc. The *position of a body* in space is determined with respect to a *reference system*. The position of an aircraft in space is determined with respect to *earth*. Similarly the position of an object on the earth's surface is determined with reference to a fixed or immovable station on the earth by taking *linear and angular measurements* or *three-dimensional coordinates*.

Time: Time is a measure of *duration* between *successive events*.

Motion: When a body occupies a fixed position for any length of time, it is said to be *static or at rest* during that interval of time. But, when the *position of a body* is *varying*, it is said to be in *motion*. The *change of position* of the moving body is called its *displacement*.

Let the body moves from O along a straight line OX (Fig. 1.1). The body reaches P in an interval of t seconds. OP is the *displacement* of the body. Suppose at time t' the body is at Q, the *displacement* in the *interval* $(t' - t)$ is PQ.

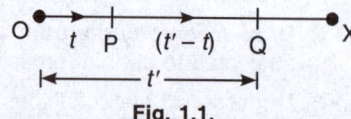

Fig. 1.1.

Mass: Mass is a measure of the *quantum of material* in an object. Mass *does not change* with a body's position, movement or alteration of its shape unless material is *added or removed*.

Weight: The *force* with which a body is *attracted towards the centre of earth* is called as *"weight"*. Weight is a gravitational *force* acting on a body mass.

Particle: The *point mass without measurable dimension* but containing a very small amount of *matter* is called a *"particle"*.

Body: Any *matter* that is *bounded* by a *closed surface* is called a *body*. An object having *definite mass* occupying the *definite space* is called a body. The *total mass* of a body acts at a *certain point* and this point is called the *centre of mass*.

Force: Force is that which *changes*, or *tends to change*, the *state of rest* or *motion* of a body. *Newton* (N) is the SI unit of force. One *Newton of force* is capable of causing an *acceleration of 1 ms^{-2}* in a body of *mass 1 kg* in ideal conditions.

$$1000 \text{ N} = 1 \text{ kN} \quad \text{(one kilo Newton)}$$

Equilibrium: A body acted upon by a *system of forces* is said to be in *equilibrium*, if it either remains in a *state of rest* or *continues to move in a straight line* with *uniform velocity*.

Rigid body: A rigid body is that which *does not undergo deformation* on the application of forces. The body, which does not undergo any change in its dimensions or shape, even after application of force, is called a *rigid body*. It is impossible to find a body that does not deform under the effect of forces. However, if the *deformations are relatively small*, the body is *considered rigid.* In this chapter we shall be *assuming bodies as rigid* to understand the basic concepts.

Mechanics: Mechanics is the science dealing bodies *at rest* as well as *in motion.* It describes the conditions when a body is said to be *at rest* or *in motion* under the effect of various forces. Mechanics, therefore, forms the basis of analysis of all engineering structures and machines. All modern developments and research in the field of *strength and stability* of *soils, structures, machines, fluid flow, vibrations, molecular, atomic* and *subatomic* behavior are dependent upon the principles of mechanics. Much of modern engineering mechanics is based on *Isaac Newton's laws of motion.*

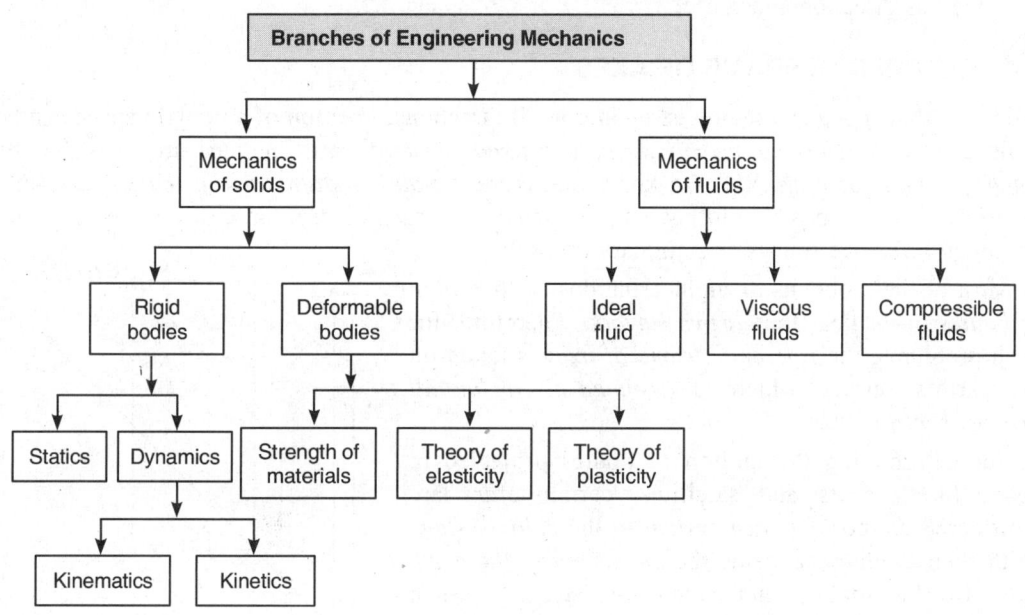

Fig. 1.2.

Dynamics: It deals with the problems concerning the *motion of bodies and effect of forces* acting on these bodies. Dynamics may further be divided in *kinematics* and *kinetics.*

Kinematics: Kinematics deal with the problems related to the determination of the *nature of motion* of bodies i.e. their *displacement, velocity* and *acceleration* without any regard to the forces acting on them, or the masses involved. It is *study of geometry of motion* of bodies without consideration of forces causing the motion.

Kinetics: Kinetics deals with the problems, which require the determination of the *effect of forces* on the *motion of a body* or conversely the forces causing a certain motion. Kinetics is the *study of effect of forces on the motion* of bodies.

Statics: Statics deal with the problems concerning the *equilibrium of bodies* under the effect of *forces.* It is the *study of body in a state of rest* or *uniform motion* caused by the action of forces.

1.3 QUANTITIES

1. **Scalar quantity:** Scalar quantities are those, which can be completely defined by their *quantum/magnitude* or numerical value. *Speed* of vehicle, say 40 km/hr, *mass* of a body, say 30 kg, the *temperature* of a body, say 50° C are all *scalar* quantities.

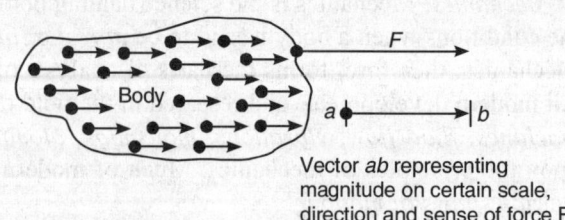

2. **Vector quantity:** Vectors are those quantities which require magnitude as well as direction and sense.

Vector *ab* representing magnitude on certain scale, direction and sense of force F.

Fig. 1.3.

1.4 REFERENCE FRAME OF AXES

We know that *space* is a region extending in all directions. *Position* of a body in space can be defined only by *referring measurements* to a *fixed frame of axes*. The measurements may be *linear, angular* or *both*. We also know that there is *nothing stationary in the universe* and therefore, it is not possible to have a *fixed frame of reference*. Certain cases can be dealt by assuming reference frames as stationary or fixed.

Most of the problems in engineering deal with *motion of bodies* on or near the *earth's surface*. Therefore, for such problems, the *reference frame of axes* is taken on the earth's surface which is *assumed fixed for all practical purposes*.

For determining the motion of bodies projected in space like rockets and satellites, certain *stars* are considered as the *reference system* in the *solar system*. With such a reference frame the *movement of the earth* can also be studied and taken into account. Such problems will not be dealt here.

Let us now consider methods to *define the position* of a particle with respect to a reference frame. To define the

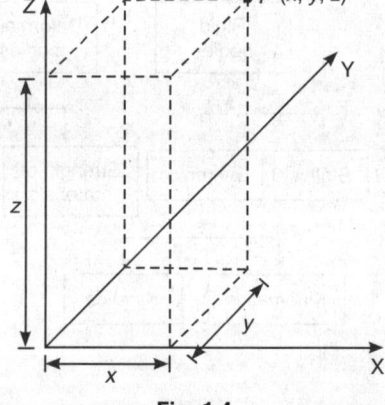

Fig. 1.4.

position of a particle in space, three measurements are necessary to be made; e.g. the position 'P' of a particle with respect to the reference frame *X-Y-Z* is defined by *three coordinates x, y, z,* as shown in Fig. 1.4.

In case the motion of a particle is taking place in a *plane*, its *position can be defined by only two measurements*. This can be done in two ways, viz. *Rectangular* and *Polar Coordinate Systems*.

Rectangular Coordinate System

In this system the position of any particle 'P' is determined by the rectangular coordinates (x, y) where, x is the linear distance of the particle 'P' from the origin O measured *parallel to the X-axis* and

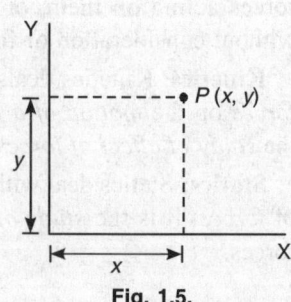

Fig. 1.5.

y is the linear distance of the particle P from the origin O measured *parallel to the Y-axis* (Fig. 1.5).

Polar Coordinate System

In polar co-ordinate system the position of a particle 'P' is defined (r, θ), where r is the radial distance of the particle 'P' from the origin, and θ is the angle made by the *radial line OP* with the *base line OX* (Fig. 1.6).

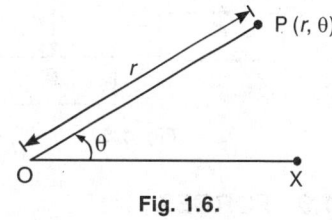

Fig. 1.6.

1.5 PRINCIPLES OF MECHANICS

The science of mechanics is based on the following principles, which are *empirical* (i.e. these are established through *experimentation* and *observation*), and have *no analytical proof.*

Newton's First Law of Motion

Any particle will continue to move in a straight line with uniform velocity or remain in a state of rest, if there is no unbalanced force acting on it. For example, let there be two *collinear forces F_1 and F_2* acting on a particle. If F_1 is equal and opposite to F_2 and the unbalanced force will be zero, the particle will be in *equilibrium.* In case they are not equal, the difference between the two will be the *unbalanced force* and the particle will *not remain in equilibrium* and will move in the *direction of greater force.*

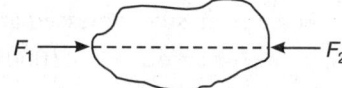

F_1, F_2 are collinear forces

Fig. 1.7.

Newton's Second Law of Motion

The *rate of change of momentum* of a body or particle is equal to the unbalanced force acting on it and takes place in the direction of the unbalanced force.

The momentum of a particle is the *product of its mass and velocity*, with the mass remaining constant, the rate of change of momentum will be the product of *mass and the rate of change of velocity*, which in turn will mean the *product of mass and acceleration* of the particle.

$$F = \frac{mv_1 - mv_2}{t} = \frac{m(v_1 - v_2)}{t}$$

$$F = \frac{d(mv)}{dt} = m\frac{dv}{dt}$$

or

$$F = ma, \qquad \text{where } a = \frac{dv}{dt}, \ m = \text{constant mass}$$

i.e.

$$F \propto a$$

Newton's Third Law of Motion

To *every action* there is *an equal and opposite reaction,* e.g. the action of a body A on body B is followed by *an equal* and *opposite reaction* of body B on body A.

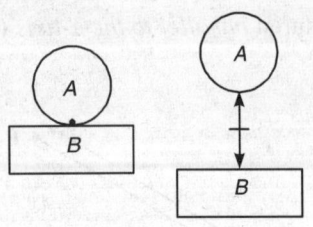

$R = ma$ (reaction of body B on body A)

ma (action of a body A on body B),
where m = mass of body 'A' responsible for action
and a = acceleration of body A at the time of action

Fig. 1.8.

1.6 FORCES

An external agency which *tends to change the state of rest or of uniform motion* of a body is called *"force"*. A force has *magnitude,* sense as well as *direction* and it is a *vector quantity.*

SI Unit of Force

Newton (N) is the S.I. unit of force.

Newton is a force in an ideal condition causing an *acceleration of* 1m/sec^2 when acting on a body of *mass* 1kg.

Force is mostly expressed and measured in Newton (N)

$$1000 \text{ N} = 1 \text{ kN}$$

Body of 1 kg *mass* subjected to an acceleration due to gravitation (9.81 m/sec^2) exerts a force of 9.81 N and known as *weight of* 1 kgf. Therefore 1 kg mass will have weight of 9.81 N or 1 kgf.

Characteristics of Force

A force is completely defined by its (i) *point of application,* (ii) *magnitude,* and (iii) *its direction*

(Arrowhead indicates direction or sense of force while length may represent magnitude to certain scale)

Effect of Forces

When a *force* acts on a *body*, it causes *changes* in the body as follows:

    i. *Change the state of rest* or *motion* of the body

    ii. *Accelerate* or *retard the motion* of the body

    iii. *Turn or rotate* the body

    iv. Keep the *body in equilibrium*

    v. *Change the shape and size* of the body

1.7 SYSTEM OF FORCES

When a number of forces act on a body or group of bodies they form a force system. Depending upon the *orientation of the forces acting on a body*, the system of forces is classified as (i) *coplanar* force system and (ii) *non coplanar* force system.

Coplanar force system consists of a set of forces with their *lines of action lying in the same plane*.

Non coplanar force system consists of a *set of forces* whose *lines of action do not lie in the same plane*.

Coplanar and non-coplanar forces can be classified into various types as explained in subsequent paras.

i. **Coplanar, Concurrent force system:** In this system all the forces are in *the same plane and their lines of action meet at a point O*, which is known as the *point of concurrency* of the force system (Fig. 1.9).

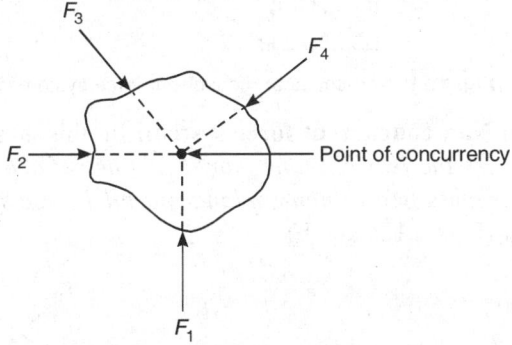

Fig. 1.9: Coplanar concurrent force system

ii. **Coplanar, Non-concurrent force system:** In this system, the forces acting on a body are in the same plane but their *lines of action do not meet at the same point*. This system of forces can further be classified as (a) *coplanar parallel force system*, in which the lines of action of all the forces are parallel (Figs 1.10a and b) *Coplanar, non-parallel force system* in which the lines of action of the forces are *neither parallel nor concurrent* (Fig. 1.10b).

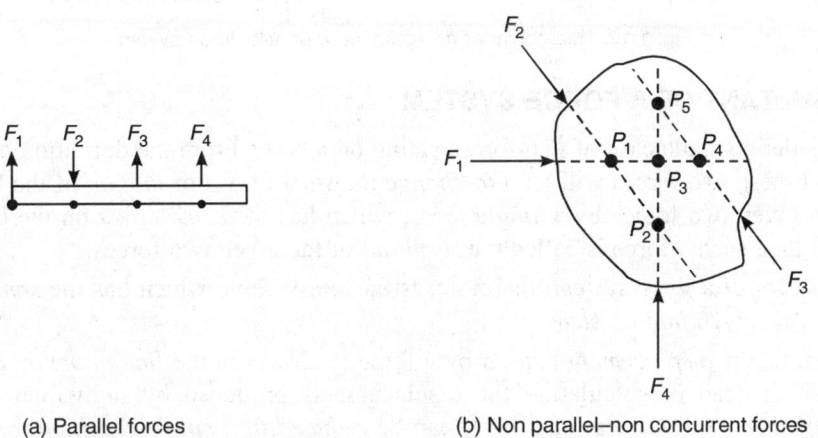

(a) Parallel forces   (b) Non parallel–non concurrent forces

Fig. 1.10: Coplanar, non-concurrent force system

iii. **Non coplanar, concurrent force system:** In this system the forces acting on a body *do not lie in the same plane* but their *lines of action do meet at a point* (P) in space (Fig. 1.11).

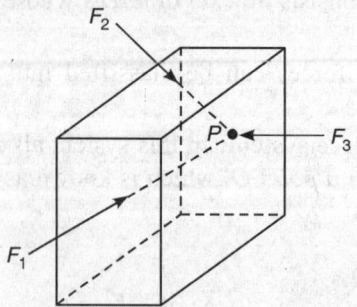

Fig. 1.11: Non-coplanar, concurrent force system

iv. **Non-coplanar and Non-concurrent force system:** In this system, the forces acting on a body *neither lie in the same plane, nor their lines of action meet at a point*. This system of forces is further classified as *non-coplanar parallel force system* and *non-coplanar non-parallel force system* (Figs 1.12a and b).

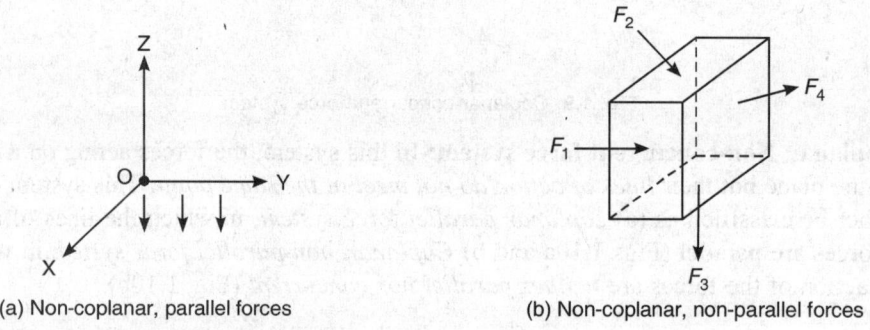

(a) Non-coplanar, parallel forces   (b) Non-coplanar, non-parallel forces

Fig. 1.12: Non-coplanar non-parallel or parallel force sysem

1.8 RESULTANT OF A FORCE SYSTEM

Let us consider a simple case of two forces acting on a body. From the definition of force, we can say that these two forces will *tend to change* the *state of rest* or *motion* of the body. If we *replace* the given two forces by a *single force*, which has the *same effect* on the body as the two forces, then such a force is called the *resultant* of the given two forces.

The *resultant* of a *force system,* therefore, is the *single force* which has the *same effect* on the body as the given *force system.*

To illustrate, the *displacement caused* by a force is always in the *line of action of the force* in its sense. Instead of calculating the displacements produced by individual forces and summing them up, the *force system* itself can be *reduced to a single resultant force* and the effect of the *resultant force* will then give the *effect* of the total forces of the system (Fig. 1.13).

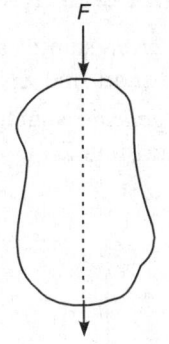

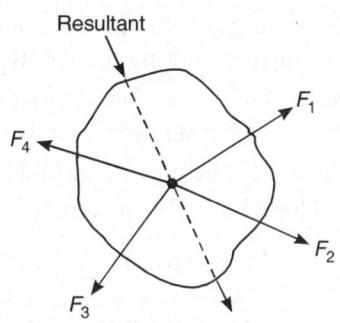

(a) Direction of displacement   (b) Resultant displacement

Fig. 1.13.

1.8.1   Resolution of Force

The method of *splitting a force* into *components* without changing the effect on the body is called as *resolution* of force or method of resolution. This method is used to find resultant in magnitude and direction of more than two concurrent forces. A force may be resolved in *two mutually perpendicular* directions, i.e. X and Y.

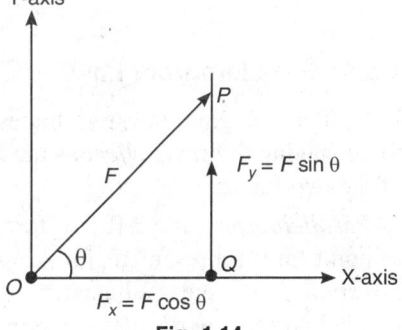

Fig. 1.14.

Let force F makes an angle θ with horizontal or 'X' axis. This force can be resolved into two rectangular components, i.e. *horizontal component* (F_x) and vertical component (F_y) as shown in Fig. 1.14. If length $OP = F$, $OQ = F_x$, and $QP = F_y$.

F_x = horizontal component, F_y = vertical component

$$\sin\theta = \frac{F_y}{F}, \qquad F_y = F\sin\theta,$$

$$\cos\theta = \frac{F_x}{F}, \qquad F_x = F\cos\theta,$$

Horizontal component F_x and vertical component F_y are as shown in Fig. 1.14.
F_x and F_y are also called *orthogonal* components.

Sign Convention

For horizontal components i.e. F_x
Rightward as + *ve*, i.e. $\rightarrow$ + *ve*
and
Leftward as – *ve*, i.e. $\leftarrow$ – *ve*

For vertical components i.e. F_y
Upward as + *ve* $\uparrow$
and
Downward as – *ve* $\downarrow$

1.8.2 Determine Resultant (R) of Coplanar Concurrent Force System

a. First *resolve* the given forces *horizontally* and find the *algebraic* sum of all the horizontal components by considering proper sign convention of F_x. In short find ΣF_x.

b. Then *resolve* the given forces *vertically* and find the *algebraic* sum of all the vertical components considering proper sign convention of F_y. In short find ΣF_y.

c. Then using Pythagoras theorem, find the magnitude of resultant.

Let the resultant be in first quadrant, as shown in Fig. 1.15.

$$R^2 = (\Sigma F_x)^2 + (\Sigma F_y)^2$$

$$R = \sqrt{(\Sigma F_x)^2 + (\Sigma F_y)^2}$$

$$\tan \theta = \frac{\Sigma F_y}{\Sigma F_x}$$

$$\theta = \tan^{-1}\left(\frac{\Sigma F_y}{\Sigma F_x}\right)$$

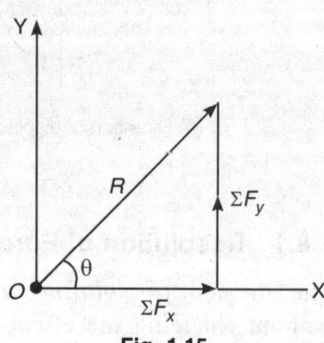

Fig. 1.15.

1.8.3 Parallelogram Law

Using Parallelogram law, two forces acting on a particle may be replaced by a single diagonal force, having the *same effect* as the original forces. This *equivalent force* is called the *resultant* of the two forces.

Parallelogram Law: "If two forces acting at a point are represented in *magnitude and direction* by the adjacent sides of a Parallelogram then the *diagonal* passing through their *point of intersection represents the resultant,* both in magnitude and direction."

Let the forces P $(\overrightarrow{OA})$ and Q $(\overrightarrow{OB})$ act at O (Fig. 1.16). Let OA and OB represent the forces P and Q in magnitude and direction acting at an

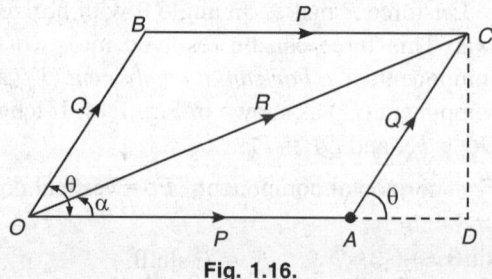

Fig. 1.16.

angle θ with each other. Complete the Parallelogram $OA\ CB$. Draw $CD \perp^{ar}$ to OA. Let $\angle COA = \alpha$.

Let R denote the *magnitude* of the *resultant*

$$OD = OA + AD = OA + AC \cos \theta = P + Q \cos \theta \qquad \text{(Since } AC = OB = Q\text{)}$$

Also $\quad DC = AC \sin \theta = Q \sin \theta$

$$R^2 = OC^2 = OD^2 + DC^2 = (P + Q \cos \theta)^2 + (Q \sin \theta)^2$$

$$= P^2 + Q^2 \cos^2 \theta + 2PQ \cos \theta + Q^2 \sin^2 \theta$$

or $\quad R^2 = P^2 + Q^2 (\cos^2 \theta + \sin^2 \theta) + 2PQ \cos \theta = P^2 + Q^2 + 2PQ \cos \theta$

$$R = \sqrt{P^2 + Q^2 + 2PQ \cos \theta}$$

$$\tan \alpha = \frac{DC}{OD} = \frac{Q \sin \theta}{(P + Q \cos \theta)}$$

1.8.4   Lami's Theorem

If *three forces* acting on a particle or a body keep it in *equilibrium,* then each force is proportional to the *sine of the angle between the other two forces* (Fig. 1.17).

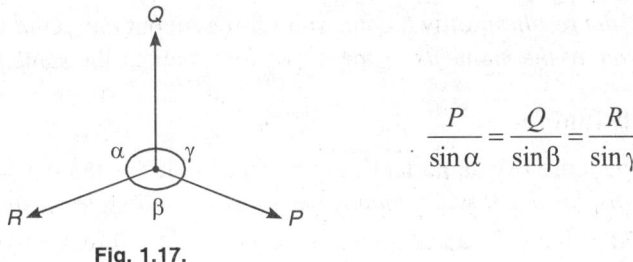

$$\frac{P}{\sin \alpha} = \frac{Q}{\sin \beta} = \frac{R}{\sin \gamma}$$

Fig. 1.17.

Limitations of Lami's Theorem

  i. There should be *three non-parallel* forces acting on a body.
 ii. These three forces acting on a body should keep the *body in equilibrium.*
iii. *Three forces should be concurrent,* i.e. acting at a point on a body.
 iv. The *forces should act away* from the point of *concurrence.*
  v. The angle between *any 2 forces* should be *less than 180°* for equilibrium.

1.8.5   Turning Effect of Forces

It is a common experience that when we push, or pull, a door by holding at the handle, the door opens by *rotating about the hinges* which connect the door shutter to the door frame. The shutter is prevented from moving forward as a whole because of its connection with the frame through the hinges. You might have also noticed that the handle is placed as far away from the hinges as possible. You can easily find out for yourself that if you have to open a door, it *requires a larger force when applied near the hinges* then when the force is applied far away from the hinges.

Now think of a bar, firmly embedded in a wall and projecting out as shown in Fig. 1.18. This is commonly known as *cantilever.* If you hold the bar at its free end and hang yourself, you can see the bar taking up the shape shown by dotted lines. This curving or rotation of the bar is more if the bar is thin. Your weight essentially provides the turning effect on the bar because of which it bends. The *turning effect* of a *force* is measured by a quantity known as the *'moment'.*

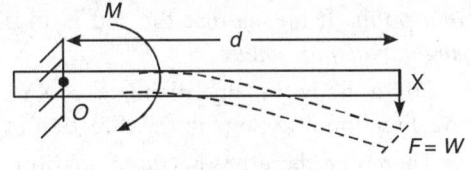

Fig. 1.18.

Moment

Moment of a force is a measure of the *turning effect* of the force about an axis. Moment '*M*' of a force *F* about an axis is defined as the *product of the force 'F' and the perpendicular distance 'd'* from the *axis of rotation* to the *line of action* of the force.

$$M = F \times d$$

Varignon's Theorem

The *moment of the resultant* of two concurrent forces about *any point* in their plane is *equal to the algebraic sum of the moments* of these two forces about the *same point*.

1.8.6 Equilibrium

Force and *Moment* are two important agents responsible for the *motion* of a body. *Force* will cause a *linear displacement* while *moment* will cause an *angular displacement*.

Hence, when a body is acted upon by a system of forces on it and if it remains in equilibrium, it means the body undergoes *neither a linear nor an angular displacement*. It means the *resultant force* should be equal to *zero* or the total *X-components* and the total *Y-components* should each be equal to *zero*. The *sum of the moments* of the forces about any point in the plane of the forces *must also be zero*.

Now consider the rod *ABC* subjected to downward vertical force of 20 N at *A* and 10 N at *C* and an upward force of 30 N at *B*, the length *AB* = 1.5 m and the length *BC* = 3 m (Fig. 1.19).

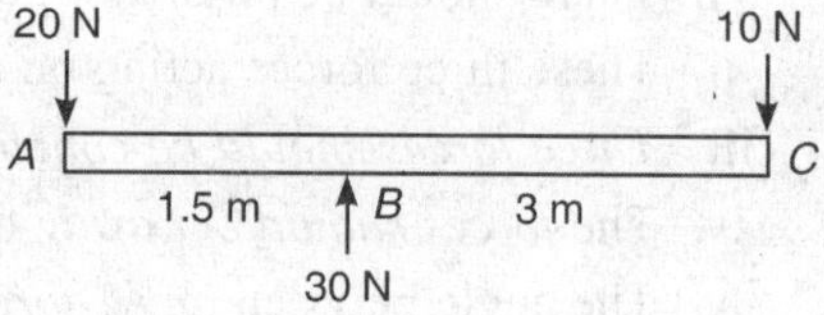

We may conclude that the rod will be in *equilibrium*, because

$\Sigma F_y = 0$

$30 - 20 - 10 = 0,$

Taking moments about *B*

$(10 \times 3 - 20 \times 1.5) = 30 \text{ Nm} - 30 \text{ Nm} = 0,$

Taking moments about *A*

$10 \times 4.5 - 30 \times 1.5 = 0$

Fig. 1.19.

$(\uparrow + \text{ve}, \downarrow - \text{ve})$

(Clockwise moment + ve, $\curvearrowright$
anticlockwise moment − ve $\curvearrowleft$)

If the net moment about *B* is not zero the rod is liable to rotate about *B*. In general if the net *moment about any point* in the plane of the forces is *not zero* the rod is liable to *rotate about that point*. It means that the rod is in perfect *equilibrium if it has neither a linear nor an angular displacement*.

For a body in any plane, say XY plane, it can have three possible motions: one in the direction of X; one in the direction of Y and one *rotation* about the axis Z (Fig. 1.20).

There can therefore be *three* possibilities of *unbalanced forces:* two unbalanced forces in the direction of *axes* X *and* Y and one *unbalanced moment* of the couple about Z-axis. For any planar body to be in *equilibrium*, the unbalanced forces in *directions* X *and* Y, as also the *unbalanced moment* about the Z-axis *must be zero*.

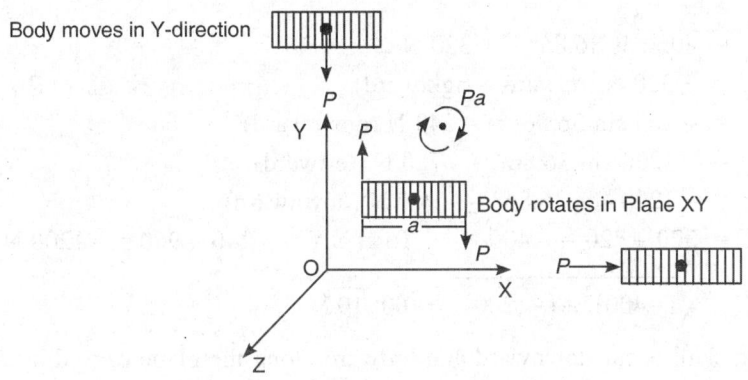

Fig. 1.20.

The unbalanced force or moment in any direction is given by the algebraic sum of all the forces in that direction. Accordingly the *three essential conditions* of *equilibrium* for a body in a plane are:

$$\Sigma F_x = 0, \quad \Sigma F_y = 0, \quad \Sigma M_z = 0$$

EXAMPLE 1.1: Determine the X and Y components of each of the forces shown in Fig.1.21 and also determine the resultant.

Solution: $\theta_1 = \tan^{-1} (12/5) = 67.38°$, $\cos \theta_1 = 5/13$, $\sin \theta_1 = 12/13 = 0.923077$

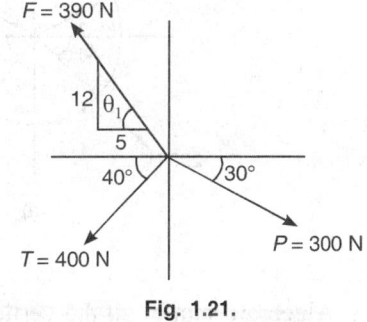

$$F_x = F \cos \theta_1 = -390 \times \cos 67.38° = -150 \text{ N}$$
$$F_y = F \sin \theta_1 = 390 \times \sin 67.38° = 360 \text{ N}$$
$$P_x = P \cos 30° = +300 \times \cos 30° = 260 \text{ N}$$
$$P_y = -P \sin 30° = -300 \sin 30° = -150 \text{ N}$$
$$T_x = -T \cos 40° = -400 \cos 40° = -306.4 \text{ N}$$
$$T_y = -T \sin 40° = -400 \sin 40° = -257 \text{ N}$$
$$R_x = -150 + 260 - 306.4 = -196.4 \text{ N}$$
$$R_y = +360 - 150 - 257 = -47 \text{ N}$$
$$R = \sqrt{(-150)^2 + (-47)^2} = \textbf{201.94 N}$$

Fig. 1.21.

Thus the resultant motion will occur in opposite direction of X and Y (leftward and downward).

EXAMPLE 1.2: The body on the inclined plane as shown in Fig. 1.22 is subjected to the vertical and horizontal forces shown. Find the components of each force along X-Y axes which are oriented parallel and perpendicular to the incline.

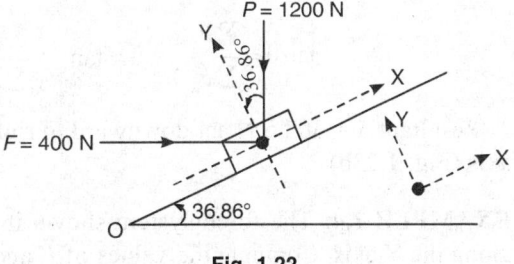

Fig. 1.22.

Solution:

$$F_x = 400 \cos 36.86° = +320 \text{ N}$$
$$F_x = +320 \text{ N (upward – rightward)}$$
$$F_y = -400 \sin 36.86° = -240 \text{ N (downward)}$$
$$P_x = -1200 \sin 36.86° = -720 \text{ N (leftward)}$$
$$P_y = -1200 \cos 36.86° = -960 \text{ N (downward)}$$

Total $\Sigma X = 320 - 720 = -400$ N,   Total $\Sigma Y = -240 - 960 = -1200$ N

$$R = \sqrt{(-400)^2 + (-1200)^2} = 400\sqrt{10} \text{ N}$$

Thus motion shall occur downward and leftward along the plane caused by resultant force of $400\sqrt{10}$ N.

EXAMPLE 1.3: Determine the *resultant* of the four forces acting on the body shown in Fig. 1.23a:

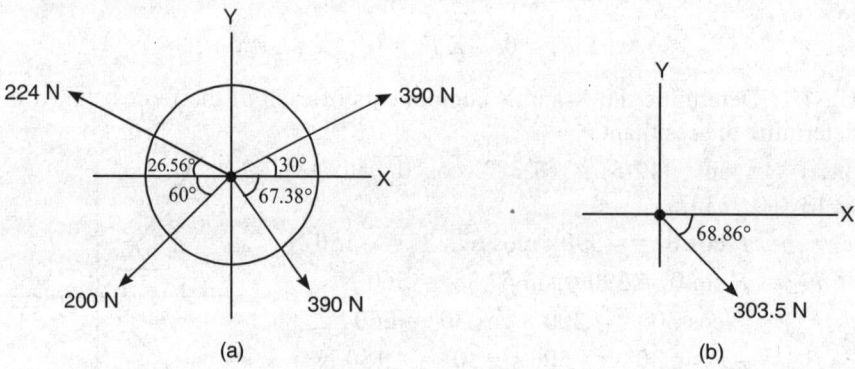

Fig. 1.23.

Algebraic sum of all the vertical components

$$\Sigma F_y = +300 \sin 30° + 224 \sin 26.56° - 200 \sin 60° - 390 \sin 67.38° = -283 \text{ N } (\downarrow)$$

Algebraic sum of all the horizontal components

$$\Sigma F_x = 300 \cos 30 - 224 \cos 26.50 - 200 \cos 60 + 390 \cos 67.38 = +109.4 \text{ N } (\rightarrow)$$

Resultant $R = \sqrt{(\Sigma F_x)^2 + (\Sigma F_y)^2} = \sqrt{(109.4)^2 + (-283)^2} = 303.5$ N (rightward and downward)

$$\tan \theta = \frac{\Sigma F_y}{\Sigma F_x}, \quad \theta = \tan^{-1}\left(\frac{\Sigma F_y}{\Sigma F_x}\right) = \tan^{-1}\frac{283}{109.4} = 68.86°$$

Resultant $R = 303.5$ N act downward to right side at an angle $\theta = 68.86°$ with the horizontal axis (Fig. 1.23b).

EXAMPLE 1.4: The force system shown in Fig. 1.24 has a resultant of 200 N pointing up along the Y-axis. Compute the values of F and θ required to give this resultant.

Solution: Here the resultant is 200 N pointing up along the Y-axis. That means,

$$Rh = 0 \text{ and } Rv = + 200 \text{ N}$$

$$\Sigma V = R_v = 200 = F \sin \theta - 240 \sin 30° - 0$$

$$F \sin \theta = 320 \text{ N}$$

$$\Sigma H = R_h = F \cos \theta + 240 \cos 30° - 500 = 0$$

$$F \cos \theta = + 292.15 \text{ N}$$

$$\tan \theta = 320 / 292.15,$$
$$\theta = \tan^{-1} (1.095) = 47.6°,$$

$$F = \frac{320}{\sin 47.6} = 433.33 \text{ N}$$

$$F = 433.3 \text{ N}$$

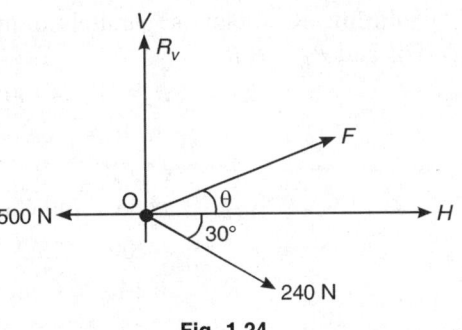

Fig. 1.24.

EXAMPLE 1.5: Repeat example 1.4 if the resultant is 300 N down to the right and 60° with the X-axis.

Solution: Resultant is 300 N down to the right at 60° with X-X, that means,

$$Rv = -300 \sin 60° = -259.8 \text{ N}$$

$$Rh = +300 \cos 60° = 150 \text{ N}$$

Considering forces given in example 1.4:

$$\Sigma F_y = F \sin \theta - 240 \sin 30° = -259.8$$

$$F \sin \theta = -259.8 + 240 (0.5) = -139.8 \text{ N}$$

$$\Sigma F_x = F \cos \theta + 240 \cos 30° - 500$$

$$= 150 = R_h$$

$$F \cos \theta = 150 + 500 - 240 (0.866) = +442.1 \text{ N}$$

$$\tan \theta = -139.8/442.1, \quad \theta = 17.54,$$

$$F = 463.6 \text{ N} \quad \text{(Downward and to rightward)}$$

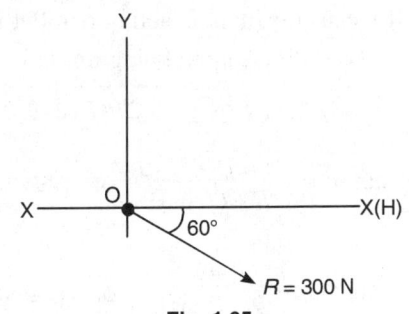

Fig. 1.25.

EXAMPLE 1.6: The block shown in Fig. 1.26 is acted on by its weight $W = 400$ N, a horizontal force $F = 600$ N and the pressure P exerted by the inclined plane. The resultant R of these forces is parallel to the incline. Determine P and R. Does the block move up or down the incline.

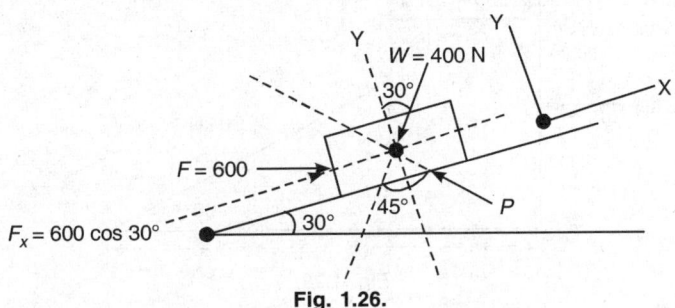

Fig. 1.26.

Solution: Resultant is parallel to inclined plane, i.e. X-axis, that means $R_y = 0$ along Y-axis and $R_H = \pm R$

$$\Sigma V = R_y = 0 = -400 \cos 30° + P \cos 15° - 600 \sin 30°$$
$$0 = -346.4 + 0.966 P - 300 = 0, \qquad \therefore \quad P = 646.4/0.966 = 669.1 \text{ N}$$
$$P = 669.1 \text{ N}$$
$$\Sigma H = R_H = -400 \sin 30° - P \sin 15° + 600 \cos 30°$$
$$= -200 - 173.2 + 519.6 = +146.4$$
$$R = +146.4 \text{ N} \quad \text{means block } moves \, up \text{ along the line.}$$

EXAMPLE 1.7: Two locomotives on opposite banks of a canal pull a vessel moving parallel to the banks by means of two horizontal ropes. The tensions in these ropes are 2000 N and 2400 N while the angle between them is 60°. Find the resultant pull on the vessel and the angle between each of the ropes and the sides of the canal.

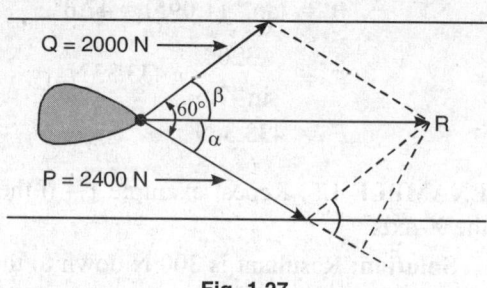

Fig. 1.27.

Solution: Vessel is moving parallel to banks. It means resultant is acting parallel to canal.

According to parallelogram law

$$R = \sqrt{P^2 + Q^2 + 2PQ \cos\theta} = \sqrt{(2400)^2 + (2000)^2 + 2 \times 2400 \times 2000 \cos 60°} = \mathbf{3815.7 \text{ N}}$$

$$\tan \alpha = \frac{Q \sin\theta}{P + Q\cos\theta} \quad \text{or} \quad \alpha = \tan^{-1} \frac{2000 \sin 60°}{2400 + 2000 \cos 60°} = 27°$$

$$\tan \beta = \frac{P \sin\theta}{Q + P\cos\theta} \quad \text{or} \quad \beta = \tan^{-1} \frac{2400 \sin 60°}{2000 + 2400 \cos 60°} = 33°$$

$$\alpha = \mathbf{27°}, \qquad \alpha = \mathbf{33°}$$

EXAMPLE 1.8: The force acting on 1 m in length of the dam are shown in Fig. 1.28. Determine the resultant force acting on the dam. Calculate the point of intersection of the resultant with the base.

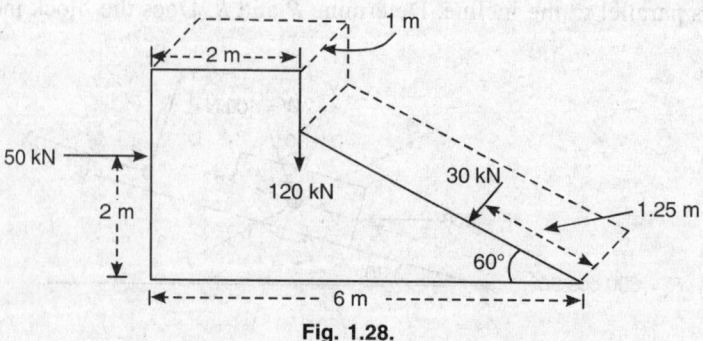

Fig. 1.28.

Solution:

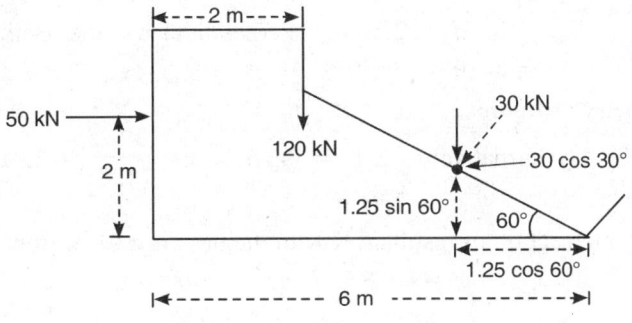

Fig. 1.29.

Resolving the forces acting on the dam horizontally

$$\Sigma F_x = 50 - 30 \cos 30° = \mathbf{24.02 \ kN}$$

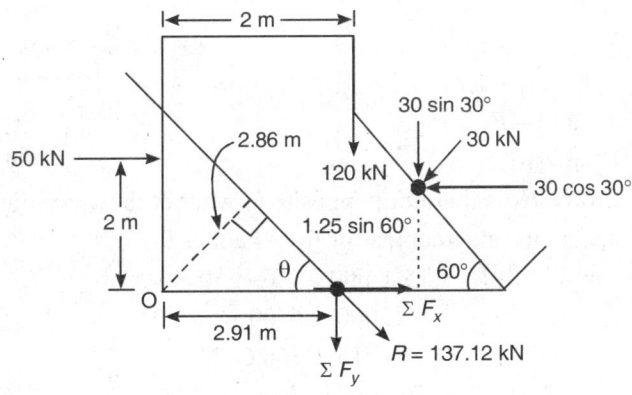

Fig. 1.30.

Resolving the forces vertically

$$\Sigma F_y = -120 - 30 \sin 30 = -\mathbf{135 \ N}$$

∴ $$R = \sqrt{(24.02)^2 + (-135)^2} = \mathbf{137.12 \ kN} \ \text{(rightward and downward)}$$

$$\theta = \tan^{-1}\left[\frac{-135}{24.02}\right] = \tan^{-1}(5.6032) = 79.91°, \ \text{with the horizontal}$$

To find the point of intersection of the resultant with the base.

Let '*d*' be the perpendicular distance between resultant *R* and point *O* (Face of Base)

Taking moments of forces about *O*, we get

$$\Sigma M_0 = 50 \times 2 + 120 \times 2 + 30 \sin 30° \times (6 - 1.25 \cos 60°) - 30 \cos 30° \times (1.25 \sin 60)$$

$$= 392.5 \ \text{kNm}$$

Applying *Varignon's theorem of moment about 'O'*, we get

$$R \times d = \Sigma \, M_0$$
$$137.12 \, d = 392.5$$
$$d = 2.86 \text{ m, Perpendicular to the resultant '}R\text{'}.$$
$$\theta = 79.91°$$

From the geometry of figure

$$\sin 79.91° = 2.86/x \text{ (Alternatively } x.\Sigma \, F_y = 392.5, \therefore x = \frac{392.5}{135} = \textbf{2.91 m})$$

$$x = \textbf{2.91 m}$$

Hence point of intersection of resultant R with the base is 2.91 m from the face point 'O' as shown (Fig. 1.30).

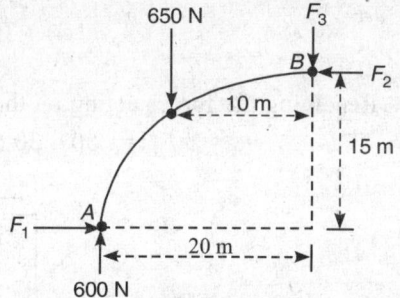

EXAMPLE 1.9: A coplanar system of forces as shown in Fig. 1.31 is in *equilibrium*. Determine the unknown forces F_1, F_2 and F_3.

Solution: The system of forces is in equilibrium

$$\therefore \quad \Sigma \, F_x = 0, \quad \Sigma \, F_y = 0, \quad \Sigma \, M_z = 0$$

Unknown forces F_1, F_2, F_3

i. $\Sigma \, F_x = 0 = F_1 - F_2, \; F_1 = F_2$

ii. $\Sigma \, F_y = 0 = 600 - 650 - F_3$

$\quad \therefore \; F_3 = -50$ N (upward)

Fig. 1.31.

The sense of force F_3 is therefore opposite to what is shown in Fig. 1.31.

iii. For $\Sigma \, M_z = 0$, Let us take moment of forces about B

$$-F_1 \times 15 + 600 \times 20 - 650 \times 10 = 0$$
$$-15 \, F_1 = -12000 + 6500 = -5500$$
$$F_1 = 5500/15 = 366.66 \text{ N}$$
$$F_1 = F_2 = 366.66 \text{ N}, \; F_3 = -50 \text{ N} \qquad (\text{i.e. } F_3 \text{ is 50 N upward}).$$
$$F_1 = 366.66 \text{ N (rightward)}, F_2 = 366.66 \text{ N (leftward)}$$

EXAMPLE 1.10: A coplanar system of forces as shown in Fig. 1.32 is in *equilibrium*. Determine the unknown forces F_1, F_2 and F_3.

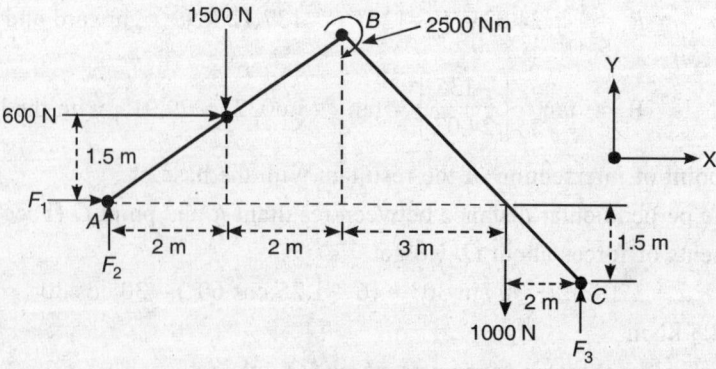

Fig. 1.32.

Solution:

The system of forces is in equilibrium, $\therefore$ $\Sigma F_x = 0$, $\Sigma F_y = 0$, $\Sigma M_z = 0$

Unknown forces are F_1, F_2 and F_3.

(i) For $\Sigma F_x = 0 = F_1 + 600 = 0$, $\therefore$ $F_1 = -600$ N (i.e. leftward)

The sense of the force F_1 is opposite to that shown in Fig. 1.32.

(ii) $\Sigma F_y = 0 = F_2 - 1500 - 1000 + F_3$, $\therefore$ $F_2 + F3 = 2500$ N

(iii) Taking moments of all forces about A, we have

$\Sigma M_z = 0 = 600 \times 1.5 + 1500 \times 2 + 2500 + 1000 \times 7 - F_3 \times 9$

$9 F_3 = 900 + 3000 + 2500 + 7000$

$9 F_3 = 13400$

$F_3 = 13400/9 = 1488.8$ N (upward)

Substituting for F_3 in equation $F_2 + F_3 = 2500$,

$\therefore$ $F_2 = 2500 - 1488.8 = F_2 = 1011.2$ N (upward)

$F_2 = 1011.2$ N

EXAMPLE 1.11: A weight of 400 N is hung with the help of two strings as shown in Fig. 1.33a, compute the forces in the strings AO and BO.

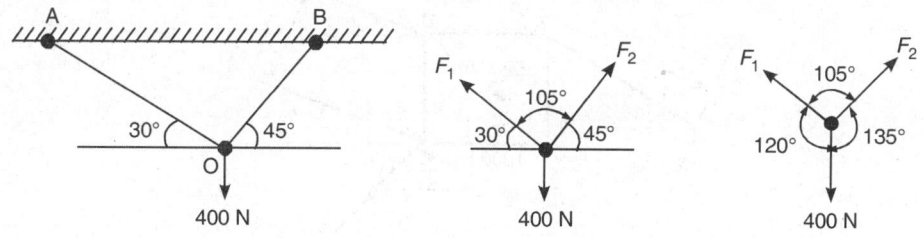

Fig. 1.33.

Solution: Applying Lami's theorem (Fig. 1.33b and c).

$$\frac{F_1}{\sin 135°} = \frac{F_2}{\sin 120°} = \frac{F_3}{\sin 105°}$$

$$F_1 = \frac{400}{\sin 105°} \times \sin 135° = 292.8 \text{ N}$$

and $$F_2 = \frac{400}{\sin 105°} \times \sin 120° = 358.6 \text{ N}$$

PRACTICE EXAMPLE 1.12: Determine completely the resultant of the four forces shown in Fig. 1.34a.

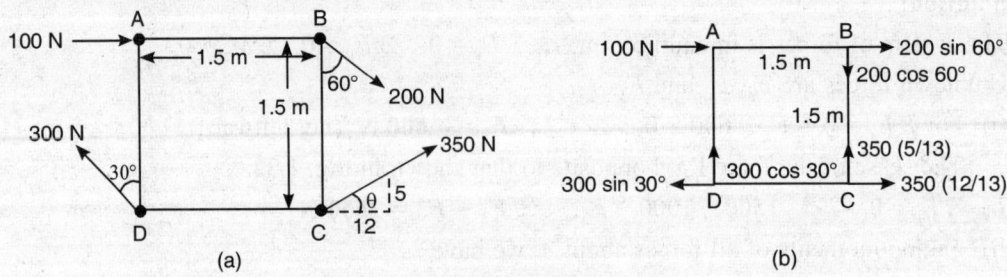

Fig. 1.34.

Ans. 534.6 N, 33.4°, 0.67 m from O.

PRACTICE EXAMPLE 1.13: A system of *coplanar* forces as shown in Fig. 1.35 is in equilibrium. Determine the unknown forces F_1, F_2 and F_3.

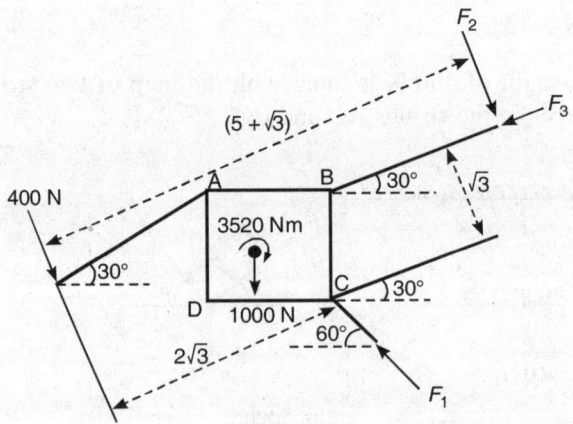

Fig. 1.35.

PRACTICE EXAMPLE 1.14: A 12 m bar AB of negligible weight rests in a horizontal position on the smooth incline. Compute the distance x at which a load $T = 100$ kN should be placed from point B to keep the bar horizontal with force $P = 200$ kN.

Solution: $\Sigma V = 0$

$200 + T = N_A \cos 30° + N_B \cos 45°$

$0.866 N_A + 0.707 N_B = 300$

$\Sigma H = 0$

$N_A \sin 30° - N_B \sin 45° = 0$

$0.5 N_A = 0.707 N_B$

$N_A = 1.414 N_B$

$N_B = 155.3$ kN

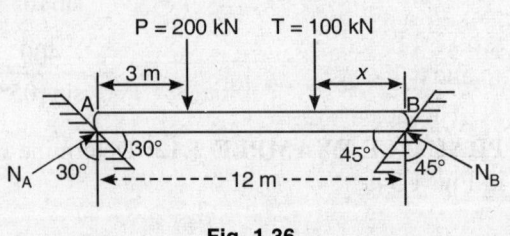

Fig. 1.36.

$$N_A = 219.6 \text{ kN}$$
$$\Sigma M_A = 0$$
$$200 \times 3 + 100 \times (12 - x) - N_B \cos 45° \times 12 = 0,$$
$$\underline{1800} - 100x - 155.3 \times 12 \times 0.707 = 0, \ 1800 - 100x - 1317.56 = 0$$
$$x = \textbf{4.824 m}$$

PRACTICE EXAMPLE 1.15: The roof truss in Fig. 1.37 is supported by a hinge at A and a horizontal roller support at B. The wind loads are perpendicular to BC. Compute the total reactions at A and B.

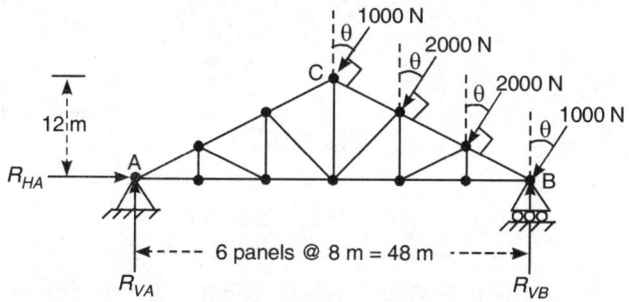

Fig. 1.37.

Solution: $\Sigma V = 0$, $\tan \theta = \dfrac{12}{24}$, $\tan \theta = \dfrac{1}{2}$,

$$\theta = 26.56°, \ BC = \sqrt{(12)^2 + (24)^2} = 12\sqrt{5} \text{ m} = 26.833 \text{ m}$$
$$R_{VA} + R_{VB} = (1000 \cos 26.56°) \times 2 + (2000 \cos 26.56°) \times 2$$
$$R_{VA} + R_{VB} = 1788.93 + 3577.86 = 5366.80 \text{ N}$$

$\Sigma H = 0$

$R_{HA} = 1000 \sin 26.56° \times 2 + 2000 \sin 26.56° \times 2 = 894.27 + 1788.5 = 2683.28 \text{ N}$

$\Sigma M_B = 0$

$R_{VA} \times 48 = 1000 \times 26.83 + 2000 \times 17.88 + 2000 \times 8.94 + 1000 \times 0$

$R_{VA} = 1677 \text{ N}$

$R_{VB} = 3690 \text{ N}$

PRACTICE EXAMPLE 1.16: Two cylinders A and B rest in a box as shown in Fig. 1.38. *A* has a diameter of 300 mm and weighs 1200 N. *B* has a diameter of 200 mm and weighs 360 N. The box is 450 mm wide at the bottom. Assume that all *surfaces are smooth*. Find the reactions at the supporting surfaces.

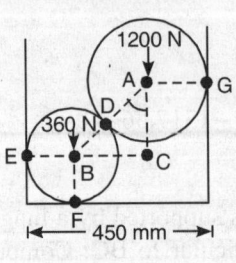

(a) Bodies in equilibrium

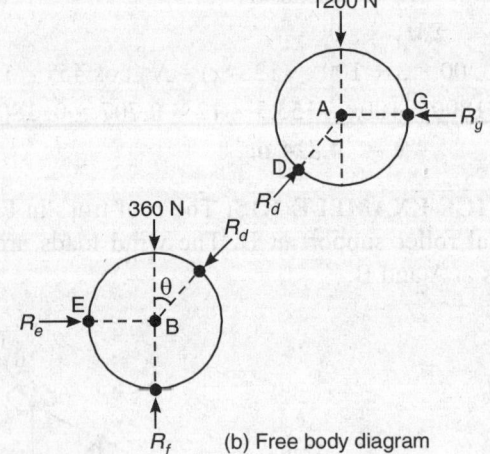

(b) Free body diagram

Fig. 1.38.

Solution:

$$BA = 150 + 100 = 250 \text{ mm}$$
$$BC = 450 - 150 - 100 = 200 \text{ mm}$$
$$\sin \theta = 0.80, \quad \cos \theta = 0.60, \quad \theta = 53.13°$$

Consider the equilibrium of the sphere A

Resolving the forces acting on the sphere A vertically

$$R_d \cos \theta = 1200 \qquad \therefore R_d = 1200/\cos \theta = 1200 \times 5/3 = 2000 \text{ N}$$

Resolving horizontally

$$R_g = R_d \sin \theta = 2000 \times (4/5) = 1600 \text{ N}$$

Now consider the equilibrium of sphere B

Resolving horizontally: $\quad R_e = R_d \sin \theta = 2000 \times (4/5) = 1600 \text{ N}$

Resolving vertically: $\quad R_f = 360 + R_d \cos \theta = 360 + 2000 \times (3/5) = 1560 \text{ N}$

PRACTICE EXAMPLE 1.17: A frame consisting of three members is supported and loaded as shown in Fig. 1.39. Find the rectangular components of the forces transmitted from one member to another through the connecting pins E, F and G. Support B rests on smooth surface.

Solution: Considering the equilibrium and the free body diagram of the whole frame, taking moment about A for external forces

$$V_b \times 3 = 250 \times 1.5 + 250 \times 4.5$$
$$V_b = 500 \text{ N} \downarrow$$
$$V_a = -750 \text{ N} \uparrow \text{ (opposite to that shown in Fig. 1.39b)}$$
$$\Sigma H = 0$$
$$H_a + 250 = 0, \qquad H_a = -250 \text{ (opposite to that shown in Fig. 1.39b)}$$

Step 2: Now consider the equilibrium of the horizontal member DEF.

Considering reactions at E and F. All reactions developed by links are equal and opposite on each other at joints.

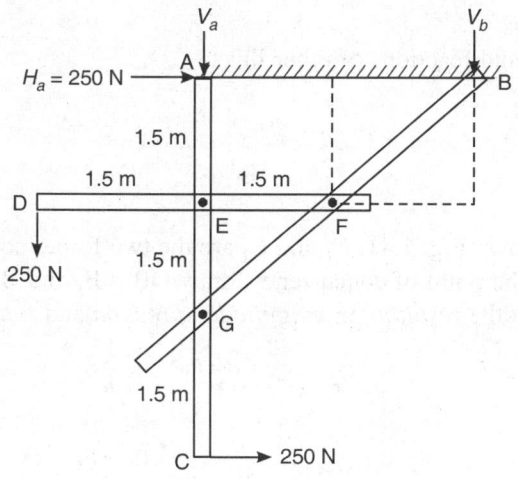

(a) Mechanism with forces

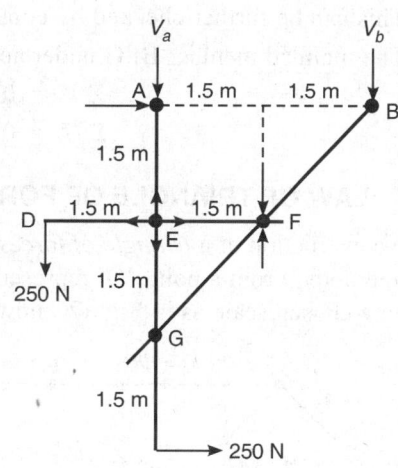

(b) Free body diagram of mechanism

Fig. 1.39.

$$250 \times 1.5 = V_r \times 1.5, \qquad \therefore V_r = 250 \text{ N} \downarrow \text{ (downward)}$$
$$\Sigma V = 0, \qquad \therefore V_e = 500 \text{ N} \uparrow \text{ (upward)}$$
$$\Sigma H_f + H_e = 0$$

Step 3: Now consider the equilibrium

For the vertical member AEGC.

Taking moments *about E*

$$250 \times 3 + 250 \times 1.5 = H_g \times 1.5$$
$$H_g = 750 \text{ N} \leftarrow$$
$$H_e = 750 \text{N} \rightarrow$$
$$H_f + H_e = 0, \quad H_f - 750 = 0, \quad H_f = + 750$$

Resolving vertically

$$V_g + 500 = 750$$
$$V_g = 250 \text{ N} \downarrow$$

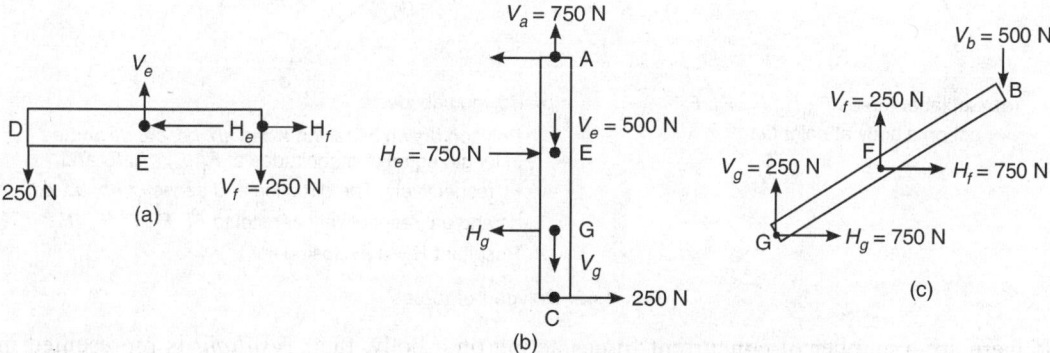

Fig. 1.40: Free body diagram of elements

This can be further checked by considering:

The inclined member BFG under actions and reactions of other links.

$$\Sigma V = 0$$
$$\Sigma H = 0$$

1.9 LAW OF TRIANGLE OF FORCES

The construction of a *triangle of forces* is shown Fig. 1.41. F_1 and F_2 are the two forces acting on the body. From a point 'O' representing the point of concurrency, draw OB = F_1 and BC = F_2 to a chosen scale as in Fig. OC now gives the *resultant* in *magnitude, direction* and *sense*.

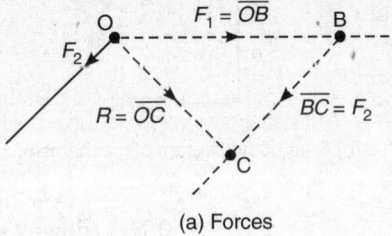

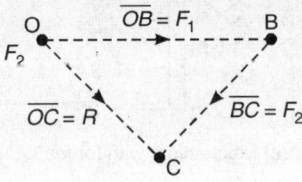

(a) Forces

Forces F_1 and F_2 act on a body at O.
The resultant 'R' of forces F_1 and F_2 can be found by triangle law of forces.

(b) Vector diagram

$\overline{OB}$ represents F_1 in magnitude and direction.
$\overline{BC}$ represents F_2 in magnitude and direction.
$\overline{OC}$ represents R in magnitude while CO represents equilibrium.

Fig. 1.41: Triangle law

1.10 FORCE POLYGON

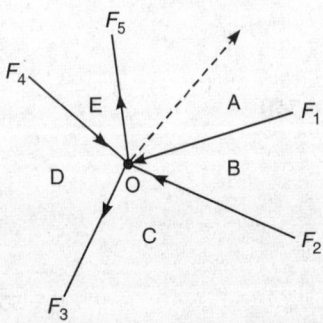

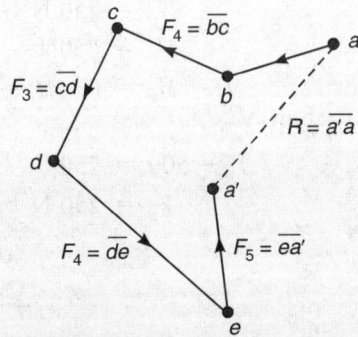

(a) Coplanar forces F_1, F_2, F_3, F_4, F_5
act on a body at point O.

(b) Polygon abcdea'a

Polygon drawn by taking sides *ab, bc, cd, de* and *ea'* in length equal to magnitudes of F_1, F_2, F_3, F_4 and F_5 respectively. The direction and sense of *ab, bc, cd, de* and *ea'* respectively parallel to F_1, F_2, F_3, F_4, F_5. Resultant $R = \overline{a'a}$ (closing line)

Fig. 1.42: Polygon of forces

If there are a number of concurrent forces acting on a body, their *resultant* is represented in magnitude, direction and sense by the *closing line, joining the first point to the last point,* of the force polygon. The remaining sides of the polygon represent the forces on the body in *magnitude, direction* and *sense*. An important point to be noticed about the force polygon is that the order of taking the forces of the system for drawing the polygon does not effect the resultant in any way.

Before studying next chapter, please ensure whether you know the following things or not. If not then study this chapter again to develop your competence for proper understanding.

1. Define the term '*force*', and state clearly the *effects of force*.
2. State and prove *parallelogram law* of forces.
3. Explain clearly the procedure for finding out the *resultant force* analytically.
4. State *triangle law* of forces and *polygon law of forces*.
5. Explain the meaning of *moment of a force*. Explain it mathematically.
6. Explain and define *equilibrium*. State the *conditions of equilibrium*.

OBJECTIVE TYPE QUESTIONS

Select and write the correct responses of following statements.

1. In a *clockwise moment*, we actually use wall clock in order to know the time for which the moment is applied.
   - (a) True
   - (b) False

2. If a *number of coplanar forces* are acting simultaneously on a particle, the algebraic sum of the moments of all forces about any point is equal to the moment of their resultant force about the same point. The principle is known as
   - (a) Principle of moments
   - (b) Principle of Levers
   - (c) Varignon's theorem
   - (d) None of them

3. Which of the following group of statements facilitate to define the *force*:
   - (a) A force is an agent which produces or tends to produce motion.
     A force is an agent which destroys or tends to destroy motion.
   - (b) Air force of a country.
     Armed police force.
   - (c) Forcep to pickup or hold things.
     Student forces the teacher to leave the class early.

4. In order to determine the effects of a force acting on a body, we must know
   - (a) its magnitude.
     direction of the line along which it acts and point of action.
     the nature (whether push or pull).
   - (b) the material of the body
     the shape of the body
     the smoothness of the body

5. The resultant R of two forces P and Q is acting at an angle (α) with P. the correct formula for a will be:

(a) $\tan \alpha = \dfrac{P \sin \theta}{P + Q \cos \theta}$ (b) $\tan \alpha = \dfrac{P \cos \theta}{P + Q \cos \theta}$ (c) $\tan \alpha = \dfrac{Q \sin \theta}{P + Q \cos \theta}$

6. The resultant of two forces P and Q acting at an angle θ is equal to:

(a) $\sqrt{P^2 + Q^2 + 2PQ \sin \theta}$ (b) $\sqrt{P^2 + Q^2 + 2PQ \cos \theta}$

(c) $\sqrt{P^2 + Q^2 - 2PQ \sin \theta}$

7. The difference between *mass* and *weight* is explained by:

(a) using physical and electronic balances.

(b) defining scalar and vector quantities.

(c) finding the nature of material of a body.

8. One *Newton* (N) force is equal to a *force* causing:

(a) an acceleration of 1 metre per $\sec^2$ in a body of 1 kg mass.

(b) an acceleration of 'g' in a body of 1000 gm mass.

(c) a speed of 1 metre per sec in a body of 1000 Newton weight.

9. Two forces *equal* in *magnitude* and acting on a body can be in *equilibrium* if:

(a) two forces are parallel and at equal distance from the center of the body.

(b) two forces are parallel and opposite in direction.

(c) two forces are opposite in direction and pass through the same point in the body.

10. Three *concurrent coplanar* forces P, Q and R keeps a body in *equilibrium*. The angles between P-Q, Q-R and R-P are respectively A, B and C. The relation between P, Q and R will be:

(a) $\dfrac{P}{\sin A} = \dfrac{Q}{\sin B} = \dfrac{R}{\sin C}$ (b) $\dfrac{P}{\cos A} = \dfrac{Q}{\cos B} = \dfrac{R}{\cos C}$

(c) $\dfrac{P}{\sin B} = \dfrac{Q}{\sin C} = \dfrac{R}{\sin A}$ (d) $\dfrac{P}{\cos B} = \dfrac{Q}{\cos C} = \dfrac{R}{\cos A}$

Response Sheet

S. No.	1	2	3	4	5	6	7	8	9	10
Response										

EXERCISE I

Q.1.1. The following forces act at a point:

(i) 20 N inclined at $30°$ towards North of East

(ii) 25 N towards North

(iii) 30 N towards North West, and

(iv) 35 N inclined at $40°$ towards South of West.

Find the magnitude and direction of the resultant force.

$\Sigma V = 0, \quad \Sigma H = 0, \quad R = \sqrt{F_x^2 + F_y^2}$

Q.1.2. A system of forces are acting at the corners of a rectangular block ABCD as shown in Fig. Ex. 1.2. Determine the magnitude and direction of the resultant force.

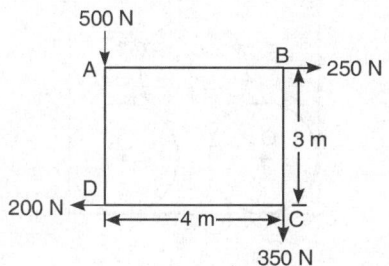

Fig. Ex. 1.2.

Q.1.3. A machine shaft BC 1.5 m long and of mass 100 kg is supported by two ropes AB and CD as shown in Fig. given below:

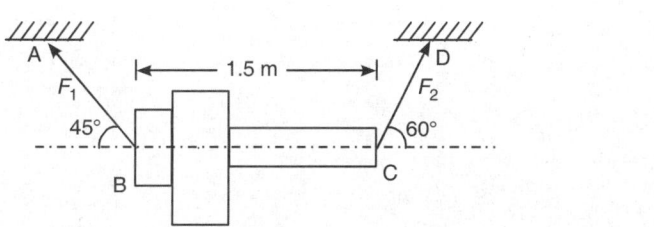

Fig. Ex. 1.3.   **FBD Ex. 1.3.**

Calculate the tensions F_1 and F_2 in the rope AB and CD.

Q.1.4. A horizontal line PQRS is 12 m long, where $PQ = QR = RS = 4$ m. Forces of 1000, 1500, 1000 and 500 N act at P, Q, R and S respectively with downward direction. The lines of action of these forces make angles of 90°, 60°, 45° and 30° respectively with PS. Find the magnitude, direction and position of the resultant force.

Q.1.5. An electric light fixture weighing 15 N hangs from a point C, by two strings AC and BC. The string AC *is inclined at* 60° to the horizontal and BC *at* 45° to the vertical as shown in Fig. Ex.1.5. Using Lami's theorem, or otherwise, determine the forces in the strings AC and BC.

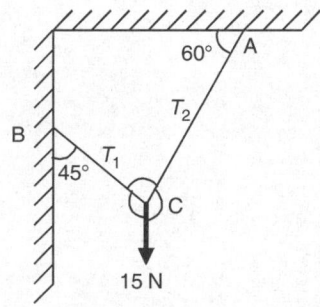

Fig. Ex. 1.5.

Q.1.6. Three Cylinders weighing 100 N each and of 80 mm diameter are placed in a channel of 180 mm width as shown in Fig. Ex. 1.6.

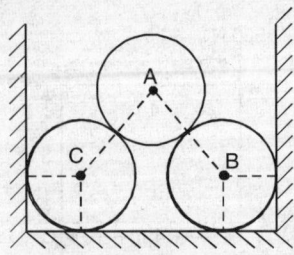

Fig. Ex. 1.6.

Determine the pressure exerted by (i) the cylinder A on B at the point of contact; (ii) the cylinder B on the base and (iii) the cylinder B on the wall.

2

Engineering Properties of Materials

LEARNING OBJECTIVES

After studying this chapter the learner will be able to:

2.1 **Know** the meaning of the mechanical properties.

2.2 **List** the **engineering properties** of materials.

2.3 **Explain** the **structural behaviour** with respect to mechanical properties.

2.4 **Identify design parameters** from the study of mechanical properties.

2.5 **Differentiate** in design approaches based on mechanical properties.

2.6 Describe various **engineering properties.**

2.1 INTRODUCTION

In determining strength of a structure or for designing a structure, a designer or construction supervisor must know the properties of construction materials. By properties we mean specific characteristics, which help to identify a particular material. Some of these properties are: unit-weight, toughness, strength, malleability, ductility, heat and electrical conductivity, to mention just a few. A combination of certain properties in material makes it useful for engineering applications. Properties of materials can be grouped under mechanical, physical, chemical, electrical and thermal properties. Without the knowledge of properties of materials, it will be difficult for an engineer/technologist to make proper selection of materials used in specific construction. Inspection with regard to the correct use of materials in construction work will also necessitate knowledge of properties. Engineers employed in the design office shall not be able to carry out structural design or check the strength of structural member without the knowledge of **mechanical properties** of materials.

You must be familiar with the various construction materials prior to studying this subject. Higher is the degree of self-motivation in you, greater are the chances of your learning this chapter to make you fully competent construction engineer.

The contents of this chapter are important before taking up problems on determining strength of structure or structural elements. The knowledge gained from this chapter will be used in studying **Strength of materials** and **Design of structures**.

Upon completion of study of this chapter, you will be able to:

- List the various properties of materials.
- Classify properties under various groups.
- Explain the meaning and significance of properties.
- Relate properties with specific application.
- Explain the meaning and significance of working stress and factor of safety.

2.2 CLASSIFICATION OF PROPERTIES

General properties of materials can be classified into seven basic groups for the purpose of study and understanding. These seven groups of properties are:

i. **Physical Properties:** Dimensions, shapes, density or specific gravity, macro structure and microstructure, unit-weight, etc. are **physical** properties.

ii. **Chemical Properties:** Oxide or compound composition, acidity, alkalinity, resistance to corrosion or weathering, etc are **chemical** properties.

iii. **Mechanical Properties:** Strength (tension, compression, shear, flexure, impact), endurance, stiffness, toughness, elasticity, plasticity, malleability, ductility, brittleness, hardness, etc are properties referred as **mechanical**.

iv. **Thermal Properties:** Specific heat, coefficient of expansion, conductivity, etc are referred as **thermal** properties.

v. **Electromagnetic Properties:** Conductivity of electrons, magnetic permeability, galvanic action, etc are considered as **electromagnetic** properties.

vi. **Optical Properties:** Colour, light transmission, light reflection, light refraction, etc are referred as **optical** properties.

vii. **Acoustical Properties:** Sound transmission, sound reflection, sound absorption, etc are considered as **acoustical** properties.

We shall study here in detail only the **mechanical properties** of materials, which are of use to us in **structural engineering analysis**.

Recall and classify the various properties of engineering materials.

2.3 IMPORTANT MECHANICAL PROPERTIES OF METALS

i. Elasticity

Whenever a member made of an elastic material is subjected to loading, it deforms and changes in its shape and size. On removal of the load, the member returns to its **original size and shape**. **Elasticity is the ability of materials to return to its original shape and size after the removal of the loads.** The elasticity is the property of materials by virtue of which the material deforms on loading within limits and returns to original shape and size on removal of the loads.

State examples of elasticity

When a member is loaded beyond **certain limit** it does not recover its original shape and size even after removal of loads. This load limit beyond which the material **does not return to its original dimensions** is called its **elastic limit**.

When a material is loaded beyond the elastic limit, the difference between its **original dimensions** and **final dimensions** each at **no load condition** is called the **permanent set**.

All properties may or may not be present simultaneously in the same material. The criteria of suitability of materials depend on the usage and possession of one or more of the above properties. The suitability of these materials is assessed through mechanical tests in laboratory.

When an engineering component is subjected to external force, the component tends to deform while the internal molecular forces acting between the molecules offer resistance against deformation. The deformation continues till full resistance (R) to external force (F) is setup and equilibrium is established between **external force** and **internal resistance**.

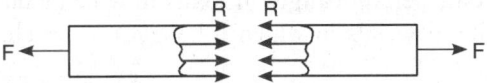

If the force F is removed gradually, the deformation also gets reduced and the component **reaches its original state** on full removal of the external force if the load F is within certain limits.

This property of material to undergo deformation under the force and return to its original shape and size on removal of the force is called "**ELASTICITY**".

Most of the engineering materials are **elastic** only within **certain limits of force**. In case of **complete regain** of original shape and size of a component on removal of load, the material is known as **perfectly elastic**. For each material there is critical value of the load upto which it exhibits **elasticity**. Engineering materials can be perfectly elastic or plastic in certain range of loading. Steel, Aluminium, Copper, Concrete, etc. are considered **perfectly elastic within certain limits**.

Stress-Strain Relation

The load per unit area, normal to the applied load is known as **Stress (P)**. The linear deformation per unit length is known as **Strain (e)**. Elastic properties of engineering materials are determined in laboratories using small lengths. Measured loads are applied gradually and elongations are measured over a certain **gauge length**. This process of increase in load and measurement of linear increase in the direction of load is plotted as Load-deformation curve. Load point is specially noted when the deformation becomes excessively large for **almost constant load or minor increment in load**. Such a point is called **Yield Point**. Generally mild steel displays distinct yield point characteristics. Mild steel exhibits elastic properties slightly below yield point.

Elastic limit is maximum stress on the material specimen upto which the material behaves as elastic i.e. on unloading the material specimen returns to its original shape and size. If the loading is applied beyond this elastic limit, the material specimen undergoes **much larger deformation** for small increase in stress. This behaviour of material is known as **plasticity**. If the specimen, **unloaded** beyond elastic limit there will be some residual strain or deformation.

Homogeneity and Isotropy

A material is homogeneous if it has same composition throughout the body. For such a material, the elastic properties are the same at each and every point in the body. It is interesting to note that for a homogeneous material, the elastic properties need not be the same **in all the directions**. If a material is equally **elastic in all the directions**, it is said to be an **isotropic**. If, however, it is **not equally elastic** in all directions, i.e. it possesses different elastic properties in different directions, it is called an **anistropic**. A theoretically ideal materials meet the requirements of homogeneity and isotropy. We shall be dealing with only the **homogeneous and isotropic** material in this discussion.

ii. Plasticity

A plastic material is that which when deformed **does not return to its original shape and size** even after removal of loads. Plasticity is thus opposite to elasticity. No material is fully elastic or fully plastic. There are only certain **ranges of loads** in which a material is **plastic** or **elastic**. The range of loads in some materials in which it behaves as an elastic material can be more than those in others.

List some materials, which are more plastic.

Elastic materials when deformed beyond their **elastic limits** undergo **plastic deformations** resulting into a **permanent set**. Many engineering materials display **elastic** as well as **plastic** deformations when loaded beyond certain range of loading.

A material in plastic state is permanently deformed by the application of load, and it has no tendency to recover. Every elastic material also possesses the property of plasticity. Under the action of large forces, most engineering materials become plastic and behave in a manner similar to a viscous liquid. The characteristic of the material by which it **undergoes inelastic strains** beyond those at the elastic limit is known as **plasticity**. When large deformations occur in a ductile material loaded in the plastic region, the material is said to undergo **plastic flow**. The property is particularly useful in the operations of pressing and forging. **Plasticity** is also useful in the design of structural members, utilizing its ultimate strength.

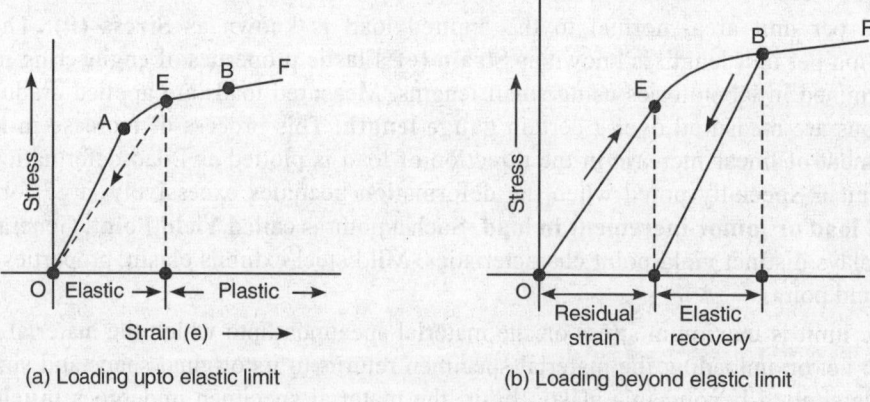

Fig. 2.1: Stress-Strain Diagram (Enlarged)

iii. Strength

Strength is the ability of any material to **resist applied forces** or to resist deformation. Strength of a material may refer to the resistance offered by it against a number of applied type of forces. Accordingly the strength is measured as the **tensile strength, compressive strength** or **shear strength** of materials.

This is the most important property of a material from design point of view. The specific **type of strength** of a material enable it to resist fracture under specific **type of loading**. The **load** required to cause **fracture**, divided by the specific area of the test specimen, is termed as the **ultimate strength** of the material, and is expressed as the unit stress. An important consideration in engineering design is the **capacity of the object** (such as building structure, machine component, air craft, vehicle, ship, etc.) to support or transmit loads. If structural **failure** is to be avoided, the load **capacity of structure** must be greater than the maximum loads it will be required to sustain during its service period. Since the ability of a structure to **resist load is called the strength**, the governing criterion is that the actual strength of a structure must exceed the required working strength. The ratio of the **actual strength** to the **required working strength** is called the factor of safety. However, failure may occur under the action of **tensile, compressive** or **shear** loads. It is essential to know the **ultimate strength** of material in each of these three conditions and the three **ultimate strengths** are separately determined experimentally.

Tensile Strength

To understand tensile strength of a material, let us take the case of a **mild steel** bar subjected to axial tension (Fig.2.2). For certain range of axial tension, the extension of the bar is proportional to the load applied. Thus stress is proportional to strain up to a certain limit. This limit is called the **proportional limit**. The shape of the graph depicting relationship between stress and strain up to this point is thus a straight line. As the load is further increased, a stage comes when small increase in load causes a **permanent set**. This **limiting value** of the load beyond which the bar **does not return to its original shape and size** after removal of the load is called the **elastic limit**.

Differentiate between proportional limit and elastic limit.

Points P and E refer to the **proportional** and **elastic limit** on the stress strain diagram (Fig. 2.2). It is not necessary that proportional limit and elastic limit have the same values. Therefore, a material can behave elastically without having a linear stress strain relationship. The slope of the stress strain diagram up to the **limit of proportionality** is called the **Young's Modulus of Elasticity**.

As the bar is loaded beyond the **elastic limit, yielding** starts and the material comes in the **plastic stage**. There is sudden drop in the load with further increase in strain. This point is the **yield point**. The material **deform without any further increase in load**. There are two yield points-upper and lower.

After the yield region, an increase in **load** is required to **cause further strain**. This behaviour is called **work hardening** or **strain hardening**, and the material does in fact become harder and requires increase in load for further deformations.

As the load is increased, **the bar reduces** in section. At the point where the bar **ultimately breaks**, load has dropped down. In fact the stress is the **maximum at the breaking of the bar** although the load is not maximum, **area** of the bar **has reduced** from its original value. The relation of stress-strain is shown in the graph (Fig. 2.2). The **ultimate stress** denoted by 'T' is worked out for metals by taking the **maximum load** reached **divided** by the **original cross-sectional area**. Tension test is performed on a universal testing machine, where a bar of standard gauge length is fixed between the wedge grips and then loaded gradually.

Differentiate between yield stress, proportional limit and elastic limit.

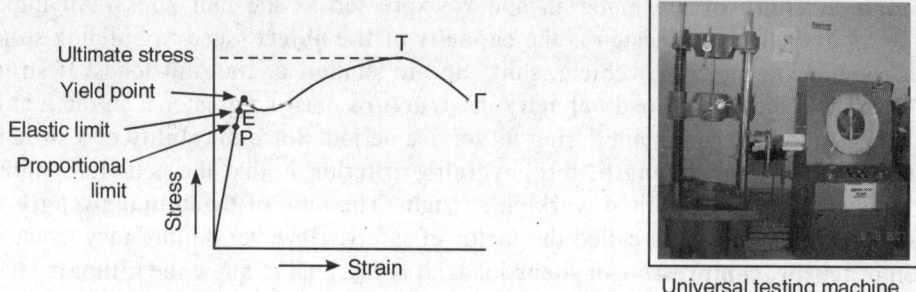

Fig. 2.2: Stress-strain graph of mild steel

Strength in Compression

The mechanical properties of a **ductile material** are generally obtained from a **test in tension**. However, **compression test** is of importance for materials, which are **brittle** and primarily used to **resist compressive load** such as **concrete** and **cast iron**.

Materials exhibit similar behaviour in both compression and tension in the elastic range. In other words, the elastic **modulus, proportional limit** and **yield point** are approximately equal in both tension and compression. Brittle materials when tested in compression fail on plane of maximum shear as illustrated in Fig. 2.3.

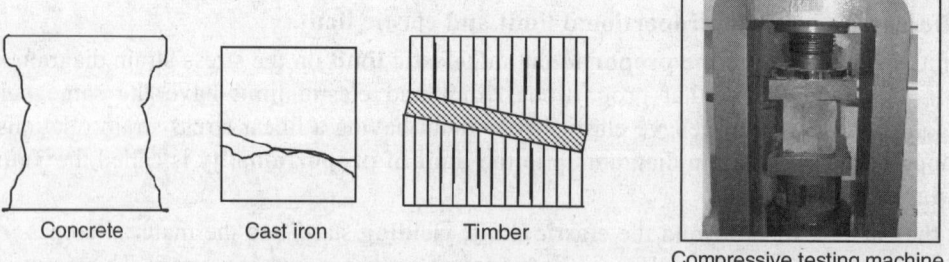

Fig. 2.3: Failure of materials under compression

Strength in Shear

The commonly used test for obtaining a relationship between shear stress and shear strain is the torsion test. This test is conducted on a solid or a hollow circular bar of the material under investigation. Equal and opposite torque 'T' is applied to each end of the bar and the **angle of**

twist θ is measured for a specified gauge length. Thus a relationship between 'T' and angle of twist 'θ' is plotted on a graph.

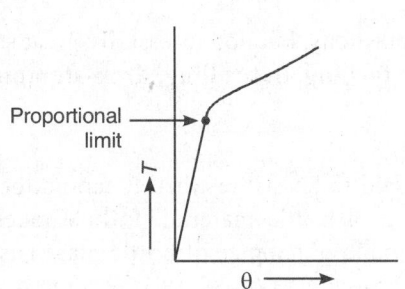

Torsion Testing Machine

Fig. 2.4: Torque '*T*' v/s angle of twist 'θ'

Sketch the shape of the T-θ diagram up to elastic limit.

The shape of the *T*-θ diagram plotted experimentally is like the load-extension diagram of the tension test (Fig. 2.2). It is a straight line up to the elastic limit. Thus the relationship between ***T*** and **θ** is linear up to elastic limit. The shear stress is proportional to the shear strain. From the ***T*-θ** diagram shown in Fig. 2.4, we can find the value of '*C*', the modulus of rigidity.

In torsion test, the shear stress generated varies over the cross-section of the bar. It is **maximum** at the **outer fibres** where it starts to exceed the limiting values while the material in the core of the bar is still **elastic**. With continued twisting, more and more of the cross section yields. No neck of the bar is formed, as in the case of tensile test. Fracture of ductile metals in torsion occurs in the plane of maximum shear perpendicular to the axis of the bar, whereas brittle materials fail along a 45° helix to the axis of the bar due to tensile stress across that plane.

iv. Ductility

A material, which undergoes **considerable deformation before breaking,** is called **ductile**. Materials used for structural purposes need to be ductile. In case they are not ductile, then the failure is sudden and the occupants of structures do not get any warning of its failure. All codes of practice prescribe a **minimum ductility** of steels used for structural purposes. In a tension test, ductility is expressed in two ways: one by the **percentage elongation** of the gauge length after **fracture** and second by the **percentage reduction in cross-sectional area** referred to the neck or minimum section at fracture.

Ductility is the characteristic which permits a material to be **drawn out longitudinally** to a reduced section, under the action of **tensile force**. The property of ductility is utilized in wire drawing. In a ductile material, therefore, **large deformation** is possible before absolute failure or rupture takes place. A ductile material must possess a **high degree of plasticity** along with strength. A ductile material shows a **certain degree of elasticity**, together with a **considerable degree of plasticity**. Ductility is measured in the tensile test specimen of the material, either in terms of **percentage elongation** or in terms of **percentage reduction in the cross-sectional area of the test specimen before failure or breaking**.

Explain why is the ductility considered an important property of structural steel.

v. Malleability

Malleability is the property by virtue of which it permits the material to be extended in all directions under loading without rupture.

A malleable material possesses a high degree of plasticity, but not necessarily great strength. This property is utilized in many operations such as **forging, hot rolling, drop-stamping**, etc.

vi. Brittleness

Brittleness implies lack of ductility. A material is said to be brittle when it can not be drawn out to smaller section by application of tension. In a brittle material, **failure** takes place **without warning** and the property is highly undesirable. Examples of brittle materials are (i) Cast Iron (ii) High carbon steel, (iii) Concrete, (iv) Stone, (v) Glass, (vi) Ceramic materials, and (vii) Many common metallic alloys.

vii. Stiffness

Stiffness is the quality of a structural member of certain material to **resist deformation**. Under the same load rubber deforms more than wood, thus wood is stiffer than rubber. Stiffness depends on elastic properties of material and cross-sectional dimensions of the member. Stiffness property, helps in structural analysis of indeterminate structures.

viii. Hardness

Hardness of any material is primarily understood as a resistance to some kind of deformation. This resistance could be against **indentation, abrasion, scratching** or machining. Generally hardness is measured as the **resistance offered against indentation**.

Many tests for hardness are used by measuring resistance against indentation. Most popular among these are: **Brinell test, Vicker test** and **Rockwell test**. In all these tests pressing a hard element on the material whose hardness is to be tested, with application of some force to make an impression is adopted. Both the applied force and the geometry of the impression made are measured to ascertain the value of hardness. The **hardness** of a material can be **related to its tensile strength**. In other words, the **tensile strength** of a material can be approximated from its **hardness number**.

Hardness is the ability of a material to **resist indentation** or **surface abrasion**. Since these resistances are not necessarily synonymous, it is usual to estimate the hardness of a material on the basis of resistance to indentation only. Tests on hardness may be classified into (i) scratch test, and (ii) indentation test. The scratch test consists of pressing a loaded diamond into the surface of the specimen, and then pulling the diamond so as to make a scratch. The hardness number is then determined on the basis of (i) load required to make a scratch of a given width, or (ii) the width of the scratch made with a given load. The **indentation test** consists of pressing a body of standard shape into the surface of the test specimen. In the commonly used Brinell hardness test a hardened steel ball of a given diameter is pressed into the surface of test specimen, under a fixed standard load 'P' and then surface area of the indent is measured. **Brinell's** hardness number (BHN) is then given by:

$$\text{BHN} = \frac{P}{\frac{\pi D}{2}\left[D - \sqrt{D^2 - d^2}\right]}$$

Where

P = Standard Load (N)

D = Diameter of steel ball (mm)

d = Diameter of the **indent** (mm)

ix. Toughness or Tenacity

Toughness or Tenacity of a material is defined as **its capacity to store energy** by the material upto the point of fracture. Fig. 2.5 shows the **load extension** diagram of a material. Toughness or Tenacity represents the strength with which the material opposes rupture or tearing apart. It is due to the attraction of molecules for each other.

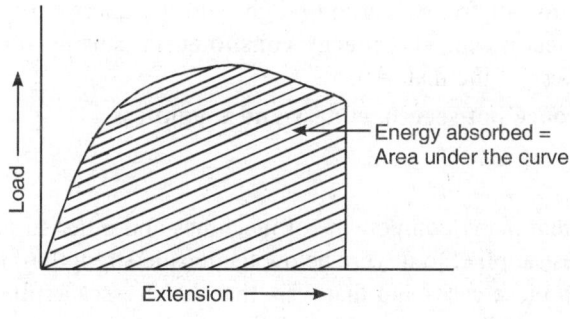

Fig. 2.5.

According to the energy principle, work done externally on a system will be equal to the energy stored internally in the system.

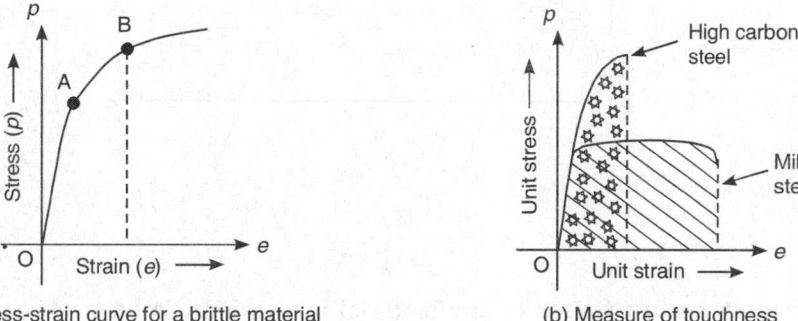

(a) Stress-strain curve for a brittle material

(b) Measure of toughness

Fig. 2.6.

Find the work done by varying force when the graph of load versus extension is given.

Toughness is the property of a material which enables it to **absorb energy without fracture**. This property is desirable in components subjected to **cyclic** or **shock loading**.

Toughness is measured in terms of **energy required per unit volume** of the material to cause **rupture** under the action of gradually increasing tensile load. This energy includes the work done upto the elastic limit which is small in comparison with the energy subsequently stored before rupture or failure. **Toughness** is expressed as energy absorbed by the material per unit volume of the material and is calculated as the area under stress-strain diagram.

The work done externally on the bar will be equal to the **area under the load extension graph**. This area (shown shaded in the figure) will than be equal to internal energy stored in the material upto fracture. In other words it gives its capacity of **absorbing energy** before failure.

Toughness is an extremely important property in structure that is used to absorb **elastic strain energy**. Think of the case of structures subjected to vibrations due to some shock say an earthquake. Due to vibrations the structure deforms. If it does not posses ductility and is brittle, then it cannot absorb energy and will fail. That is the reason why structures constructed in bricks fail during earthquakes.

Impact tests are used to measure **toughness** of materials. In these tests, (Charpy and Izod) a hammer is allowed to fall from a certain height and the energy possessed by it is used to cause fracture in the test piece. The **energy consumed** in causing fracture of the test piece **measures the toughness** of the material.

Explain the difference between hardness and toughness.

x. Fatigue

It has been observed that many components of machines and other structures have failed even when the nominal stress applied to it were below the tensile strength of the material. One thing common was found in these cases and that was 'that the stresses applied to these components were not steady in magnitude and nature but varied in a cyclic manner'. This cyclic variation of stresses is shown in Fig. 2.7. The failure of a material when subjected to a number of variable stress cycles is known as failure due to fatigue.

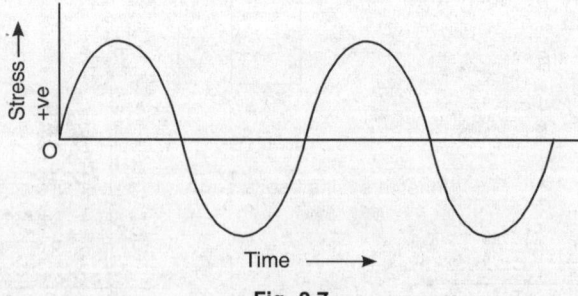

Fig. 2.7.

The strength against fatigue of any material is measured in terms of endurance limit. Endurance is measured in terms of the number of loading cycles applied without failure at any stress level.

Endurance limit is the stress range which gives a specific large number of cycles usually 50×10^6. Materials can, therefore, be stressed safely below the endurance limit.

xi. Working Stresses

In order to design safe structure or machine component, we must assure that the stresses produced in them due to applied loads are kept less than those causing failure, **or fracture of materials used in structures or machine components**.

In designing structures, the maximum stresses resulting from applied loads are limited to values much smaller than the ultimate stresses of the materials.

These limiting values of stresses **are called the** working **or** permissible **stresses**.

State the reasons for working stresses much smaller than the failure stresses.

The reasons to keep stresses in structures much lower than failure stresses are that there are many uncertainties in the designing of structures. Important of these are:-

- **We do not know the** exact loads **to which the structure is likely to be subjected to during its lifetime. Even if the loads are known, we are not sure of the nature and combination of these loads accurately**.
- **We are not certain about the exact** stress distribution **in the members**.
- **Our assumptions about** support conditions **for the members may not be exact**.
- **We are not sure about the** absolute properties **of materials being used by us**.
- Unknown defects may be produced by poor workmanship, poor assembly, poor treatment, poor welding and by imperfections in material itself.

We therefore, see that there are large factors, many of them beyond our control that make it necessary for us to keep the allowable stress well below **the value that produces** failure.

For materials that are brittle **the criteria of failure is generally taken to be** ultimate strength. **For materials that are** ductile, **the criterion of failure is generally taken to be the** yield stress.

The term factor of safety is used to express the ratio between the failure **stress and the allowable/working or** permissible stress. **For example, if mild steel has a yield stress of 260 N/mm^2 and a working stress of 150 N/mm^2 for design purposes, then the** factor of safety **will be 260/150 = 1.733 and the margin of safety will be 1.733**.

In designing and analyzing structures, we make so many idealizing assumptions, which are difficult to realize in practice. It is therefore, not possible to state truly the safety margin kept in designing a structure until or unless it is actually loaded.

SUMMARY

In this chapter you have learned the engineering or mechanical properties of materials.

Elasticity

A member made of an elastic material deforms when subjected to loading. On removal of the load, the member returns to its original **size and shape. This property of returning to the** original shape and size **by a material after unloading is called**

---

elasticity. **Different materials behave as** elastic **only upto** certain limit of loading **known as** elastic limit. **The** stress **remains** proportional to strains **upto a** stress limit **known as** proportional limit.

Plasticity

When a material is loaded, it deforms. On unloading if the material does not resume its original position, it is said to have undergone a plastic deformation. Plasticity **is a property of a material, that** refers to permanent set **or partial recovery from the** final deformed shape **even after** unloading. **The** ductile **materials undergo** large elongation **before failure. This property of large elongation and** non resuming **of** original shape and size **even after** full unloading **is known as** plasticity.

Ductility

A material is said to be ductile**, when it** undergoes large deformation before failure **or** rupture. **Thus ductile materials fain gradually and take time before collapse of the element**. Ductility **plays important role in design of structures.**

Stiffness

It is the quality of the material and the size of the section to resist deformation under loading.

Tensile Strength

A ductile material when subjected to tensile loading, it undergoes deformation such that its extension is proportional **to the load applied or** stress is proportional to strain **upto a certain limit called the** proportional limit. **The ratio between stress and strain upto this limit is called the** modulus of elasticity. **A ductile material has a distinct** yield point **under tensile loading.** Yield point **stress is also considered as** tensile strength.

Compressive Strength

The properties of elastic material in tension and compression are approximately the same. For example the proportional limit, elastic limit, modulus of elasticity are the same for ductile materials. Compressive strength becomes important for brittle materials. Compression failure takes place along the plane of maximum shear. The compressive stress **just** before failure **represents the** compressive strength.

Shear Strength

The relationship between torsion (T) and shear strain θ is determined by torsion test. In the torsion test the T-θ diagram is similar to the load extension diagram in the tension test. The ratio between the shear stress **and** shear strain **is determined from the T-θ curve drawn for solid or hollow rod. From the curve T-θ and dimensions of the rod, the** shear strength **can be determined. Shear strength is also measured within the elastic limit of the material.**

Hardness

Hardness **is the property of the material that denotes its** resistance against indentation. **It can be related to the tensile strength of a material. It is expressed as Brinell Hardness Number (BHN). BHN** $= \dfrac{2P}{\pi D\left(D - \sqrt{D^2 - d^2}\right)}$, **D and d are** respectively **diameter of ball and indent.**

Toughness

Toughness **is the capacity of a material to** absorb energy per unit volume **of the material upto the point of** rupture or failure. **It represents the strength of material to** oppose rupture **or** tearing.

Fatigue

The failure of a structure subjected to cyclic stresses varying in magnitude from negative to positive is called fatigue. **The** strength **of a material** in fatigue **is indicated by its** endurance limit. **Endurance limit is the** stress limit **which when develops very large number of cycles (50 × 10^6)** without rupture **or failure in a material.**

Working Stresses and Factor of Safety

Working stresses **or allowable stresses are the** limiting stresses **that are** allowed to occur **while designing a structure. These must be** less than **the** failure stresses. **The** ratio **between the** failure stress **and the** working stress **is called the** factor of safety.

Working or **permissible stresses** are considered in **design of structure** to account for many type of **uncertainties** in material **properties** and **type** and **quantum of loadings**. **Ultimate strength** is based on the **maximum load prior to failure** and the original section.

Understanding of **basic concepts and principles of equilibrium** and **engineering properties** will play most critical role in understanding of all subjects related to **structural analysis, design** and **construction**.

EXERCISE 2

Q.2.1. For ductile materials what kind of stress is taken to be failure stress?

Q.2.2. Define working stress and factor of safety.

Q.2.3. Why do we keep working stresses less than the failure stresses? State any three reasons.

Q.2.4. List eight common properties of construction materials.

Q.2.5. Classify properties of materials listed by you in Q.4 under the seven basic groups (Physical, Chemical, Mechanical, Thermal, Electromagnetic, Optical and acoustical properties).

Q.2.6. Read each of the following statements. Encircle T if a statement is true and encircle F if a statement is false.

T	F	i	Strength is the only property that needs to be considered for materials and used in structures.
T	F	ii	Strength relates to the ability of a material to offer resistance against applied loads.
T	F	iii	Ability of material to resist penetration or scratching is called toughness.
T	F	iv	Ability of material to be deformed without breaking is called the property of hardness.
T	F	v	Brittleness refers to a place of material cracking or breaking.
T	F	vi	Mild steel is more ductile than cast iron.
T	F	vii	Elasticity refers to the ability of a material to deform extensively without rupture.

Q.2.7. Write the letter of the relevant property in column B against each specific application in column A. Each term in column B may be used once, more than once or not at all.

Column A

i.   Use of designing a column in a building.

ii. Use in structures subjected to seismic loads.

iii. Use in structures subjected to reversal of stresses.

iv. Use for wearing surface of steps in a stair.

v.   Use as a flexural member of any residential building.

Column B

A. Hardness

B. Endurance Limit

C. Compressive Strength

D. Strength and Ductility

E. Weight, Strength, Ductility

F. Tensile, Compressive and Shear Strengths

G. Plasticity

H. Elasticity

Q.2.8. Define following properties in not more than 80 words.

i.	Elasticity	ii.	plasticity	iii.	yield point
iv.	ultimate strength	v.	Factor of safety	vi.	Ductility
vii.	I sotropy	viii.	working strength	ix.	proportional limit
x.	Elastic limit	xi.	Young's Modulus of elasticity	xii.	Tensile strength
xiii.	Compressive Strength	xiv.	Malleability	xv.	Hardness
xvi.	Toughness	xvii.	Fatigue	xviii.	Endurance
xix.	shear strength	xx.	Brittleness.		

3

Concepts in Structural Analysis, Loads, Supports and Free Body Diagrams (FBD)

LEARNING OBJECTIVES

After studying this chapter the learner will be able to:

3.1 **Know** the meaning of the term **structure.**

3.2 **Know** the meaning of **plane** and **space structure.**

3.3 Know the meaning of determinate and indeterminate structure.

3.4 **Know** the different types of **loads** on a **structure.**

3.5 **Understand** the different kinds of **structural supports** and the **reactive forces** offered by each.

3.6 **Draw free body diagram** of **structures** and their components.

3.7 **Understand** the meaning of the term **internal forces** in **structures.**

3.8 **Understand** the term **stress: tensile stress, compressive stress** and **shear stress.**

3.9 **Understand** the meaning of the term **deformation: Axial (Linear) deformation, shear deformation** and **bending deformation.**

3.0 INTRODUCTION

In order that we can understand the analysis of loaded structural elements, it is important to know what the term **structure** means. What different type of structures are there and what do we understand by internal forces, stresses and deformations in structures.

In this chapter, the learner will learn: (a) definition of structure, (b) different type of **loads** on structures, (c) different type of supports used in structures, (d) different type of **internal forces, stresses** and **deformations** caused in structures as a result of external loads.

Understanding of these concepts will greatly facilitate in learning and solving problems in the analysis of structures.

3.1 TYPE OF STRUCTURES

You have seen buildings, dams, bridges, ships, aeroplanes, lathe machines, table and chairs. All these are the examples of **structures**. All of these are subjected to some kinds of **external**

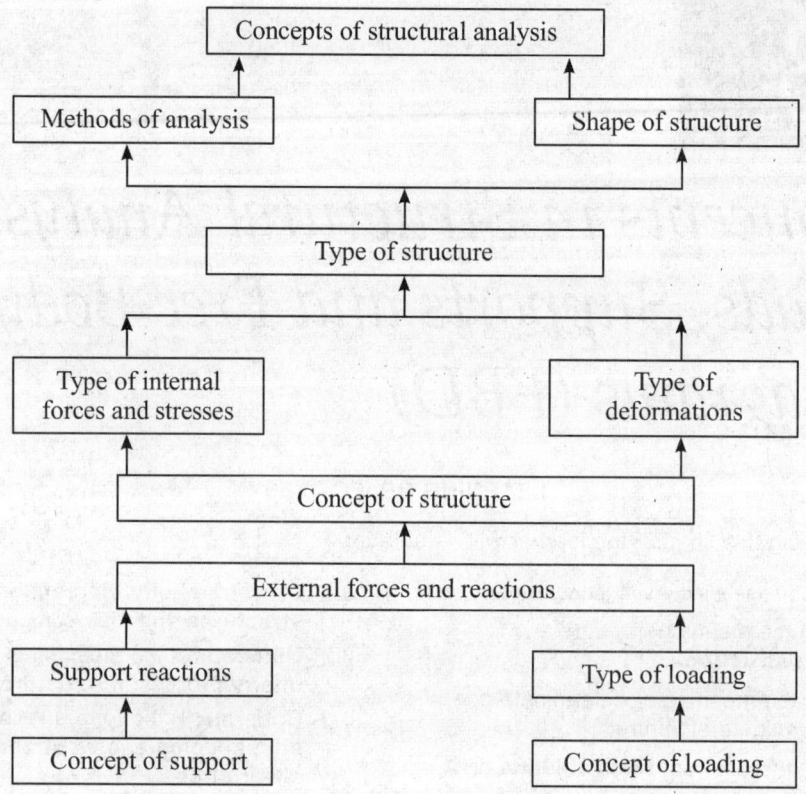

Fig. 3.1: Learning structure "Structural analysis concepts"

forces called **loads**, for example a building may be subjected to **forces** caused by **weight** of its **own components**, **weight of the occupants**, **weight of the materials** stored, **wind pressure** and the environmental effects such as **temperature changes**. As a result of **external forces resisting forces** are set up in the members of structure. These resisting forces are called internal forces, also known as stress. The **loads on a structure** are held in **equilibrium** by the **internal forces** without appreciable **deformation** of the parts of a structure. The structure is thus **stable** and helps in **transferring imposed loads to the bearing medium,** say **foundation soil** in the case of a building or a bridge. The internal resisting forces are developed to resist deformations of structural elements

Structure

A **structure** is a body/device composed of **one** or **more component** parts. It is acted upon by certain **loads**, which are held in **equilibrium by internal forces** developed in its members without any appreciable **deformation of one part relative to another**. It is **stable** under the effect of imposed loads and helps in **transferring** these **loads to the bearing media**, through its **supports**.

Which of the following can be classified as a structure?

a. A rod of 10 N weight hanging from the ceiling of a building and free to rotate like a pendulum.

b. A horizontal bar resting on **two rollers** at its ends and subjected to loads inclined to vertical.

c. A rectangular frame made of four members joined together by pins.

d. A bar supported at one end by **hinge** and at other by a **roller** and subjected to loads inclined to vertical.

3.2 LOADS ON STRUCTURE

Figure 3.2 shows the picture of a highway bridge beam. Let us identify the various types of external forces (loads), acting on the bridge. Firstly there is the **weight** of bridge itself. For example, the deck of the bridge has a uniform thickness and its weight is therefore **uniformly distributed**. The total weight of the bridge deck acts at the centre of the span, being the C.G. of the weight. Secondly loads due to the **vehicular traffic** such as a bus, truck, road roller, tank, etc. act upon the bridge at the points of location. In the case of a bus/truck or a road roller the load is transferred through the wheels at their **surface of contact**. Considering the area of contact surface to be very small, such a load is taken to be **concentrated at the point** of the **contact area**. When we look to a bridge pier, it is subjected to loads from the deck plus **water pressure** of stream. The **water pressure** acting on the nose of the pier is zero at the free water surface and increases at a uniform rate with the depth of water. In addition to these loads, bridge is also subjected to **wind loads** and **seismic loads**. The information makes you to think of the different ways in which **external forces** (loads) act on a structure. You have observed three different ways in which loads can act on a structure. These are: (a) **Uniformly Distributed**, (b) **Concentrated**, (c) **Uniformly Varying**.

a. Uniformly Distributed Loads

Load acting on a structure is said to be **uniformly distributed,** if its intensity (weight per unit area or per unit length of the member) is the same throughout. It is represented by either of the symbols shown in Figs 3.2 or 3.3. The weight of deck slab of uniform thickness in the case of

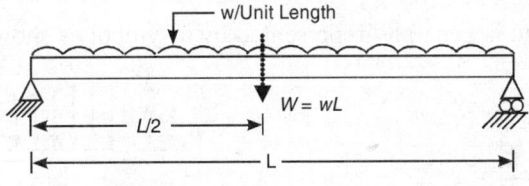

Fig. 3.2: Symbol for uniformly distributed loads

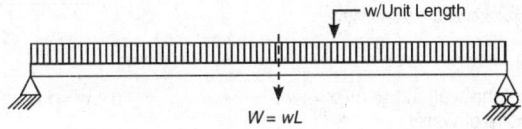

Fig. 3.3: Symbol for uniformly distributed loads

a bridge is a uniformly distributed load. If **w** is taken as the load per unit length of the member, then the total load on the whole length of member will be **wL**, where **L** is the length of the member. The total load **wL** will act at its **C.G.** i.e. at a distance of **L/2** from either end of the member.

State three examples of uniformly distributed loads on structures.

b. Concentrated or Point Loads

Actually no load can be called a point load **concentrated at a point,** because the area of contact between the imposed load and the structure has always a certain value. However, for all practical purposes if the **contact area** is very small as compared to the surface area of the structure, then the load is assumed to be concentrated at a point (centre of contact). It is generally denoted by an arrow, which indicates the line of action of the load. The load of the bus/truck or road roller was assumed to be concentrated at the **points of contact of their wheels** (Fig. 3.4).

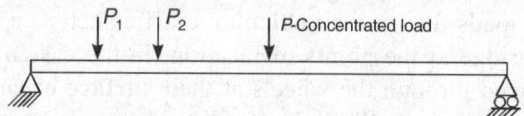

Fig. 3.4: Symbol for concentrated loads

State the three examples of concentrated loads on structures.

c. Uniformly Varying Loads

An imposed load is said to be varying uniformly, if its int[ensity is varying uniformly from one end to another, e.g. the earth pressure acting on the back of a wall holding earth or a beam supporting a triangular load as shown in Figs 3.5a and b respectively. The total uniformly distributed load 'W' acts at its C.G., i.e. at $\dfrac{h}{3}\left(\text{or } \dfrac{l}{3}\right)$ from the triangular load base (from the maximum intensity side) and is equal to $\dfrac{wL}{2}\left(\text{or in case of dam } \dfrac{ph}{2} = \dfrac{wh^2}{2} = P\right)$ shown in

Figs 3.5a and b. The load is generally represented by a symbol as shown in Fig. 3.6.

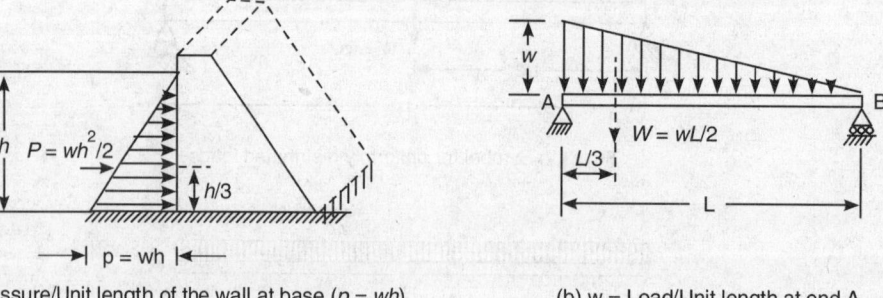

(a) Pressure/Unit length of the wall at base (p = wh)
(w = density of water)

(b) w = Load/Unit length at end A

Fig. 3.5: Uniformly varying distributed load

State two situations other than those mentioned earlier, where uniformly varying distributed loads act on structures.

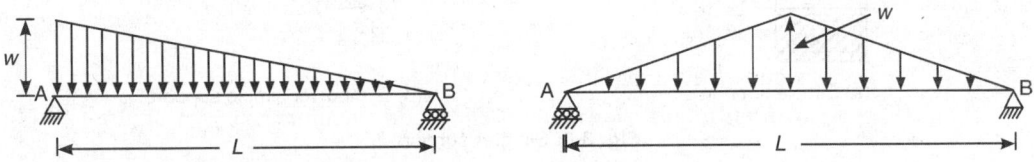

Fig. 3.6: Symbol of uniformly varying distributed load (max. intensity w)

3.3 STRUCTURAL SUPPORTS

The loads applied to a structure are transmitted to its supports after **deformation**. The supports in turn transmit these loads to the **bearing media.** There are three types of structural supports, which are in common use. These are **roller, hinged** (or pin) and **fixed** (or encastre) **supports.** All these supports **offer reactive forces**, which hold the loads acting on the structure in **equilibrium.** Each of the supports used for a plane structure offers reactions in the plane of the structure to establish **equilibrium** with external loads on the structure. Support reactions, thus, keep the structure in **equilibrium** against the **external loads** (or forces).

a. Roller Support

Figure 3.7a shows a **roller** support. You generally find this kind of support used to support bridge girder. This support does not offer any **resistance to the rotation** of structure about the axis perpendicular to the plane in which the structure lies. It also allows **free movement** of the structure in a direction tangential to the roller support **without any reaction**. It thus **offers only the reactive force perpendicular** to the base of the support. The common symbols used to represent this support is shown in Fig. 3.7b. Another common support, which behaves approximately like a **roller support**, is a **simple support**, which is a smooth flat surface. You find the use of this support in building where **beams are supported over bed plates on walls**. Such a support is shown in Fig. 3.8. This support is also similar to a roller support and **does not offer resistance to the rotation** of structure about an axis perpendicular to the plane of the structure. It also **does not offer resistance to movement** in a direction tangential to the base of the support. It only offers **one reactive force** perpendicular to the base of the support.

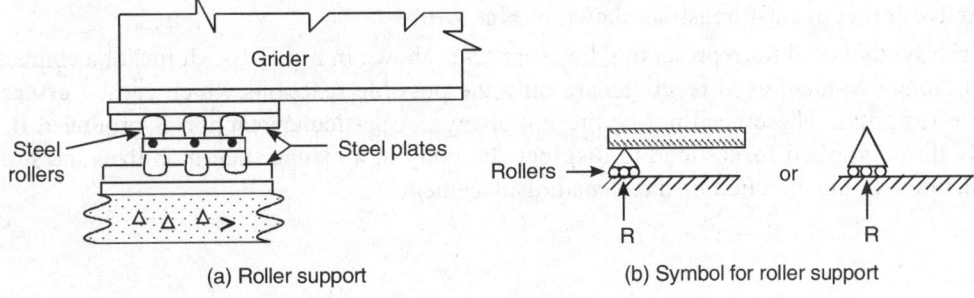

(a) Roller support   (b) Symbol for roller support

Fig. 3.7: Roller support

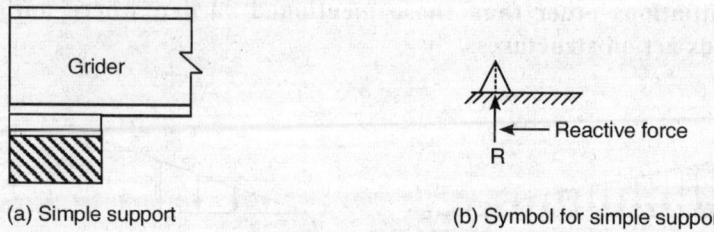

(a) Simple support   (b) Symbol for simple support

Fig. 3.8: Simpler support

b. Hinged (or Pin) Support

Figure 3.9a shows a **hinged** (or **rocker**) support. Figure 3.9b shows another type of a **rocker hinge** base. This support **cannot offer resistance to rotation** of a structure about an axis perpendicular to the plane of the structure. However, it **does not permit displacement** of the

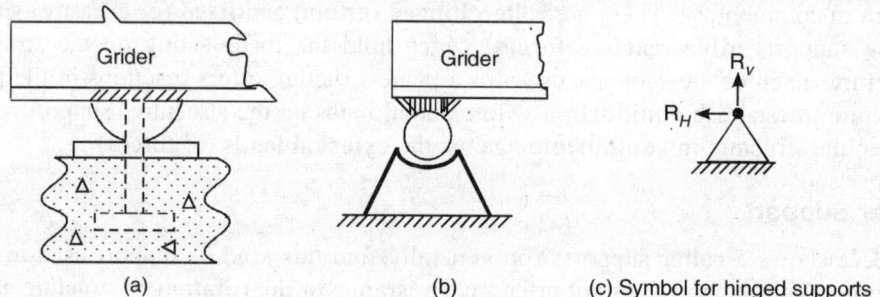

(a)   (b)   (c) Symbol for hinged supports

Fig. 3.9: Hinged supports

structure either **along** or **perpendicular** to its base. A hinge (or pin) support, can therefore, offer **two reactive forces,** one **along** and the other **perpendicular** to its base. The symbol used for representing a hinged support is given in Fig. 3.9c.

c. Fixed (or Encastre) Support

A **fixed, encastre** or **a built in** support is shown in Fig. 3.10. It is very common to see the use of this support in all kinds of structures. This support fixes the **built in** end of the structure and offers **resistance** against all kinds of displacements—**rotational** or **transitional** in any direction. The support therefore, offers **three reactions, one moment** against rotation and two **reactive forces** against transit as shown in Fig. 3.10b.

The symbol used for representing this support is shown in Fig. 3.10c. It must be emphasized that the above-mentioned reactions are only the possible reactions which can be **offered** by these **supports**. These need not be present always. A particular reaction is produced **if, and only if,** the **applied forces** tend to **displace** the body in a manner that mobilizes the **support resistance** in the direction of threatened displacement.

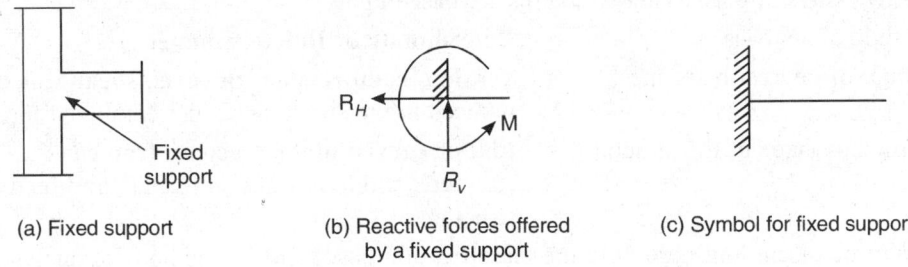

(a) Fixed support

(b) Reactive forces offered
by a fixed support

(c) Symbol for fixed support

Fig. 3.10: Fixed support

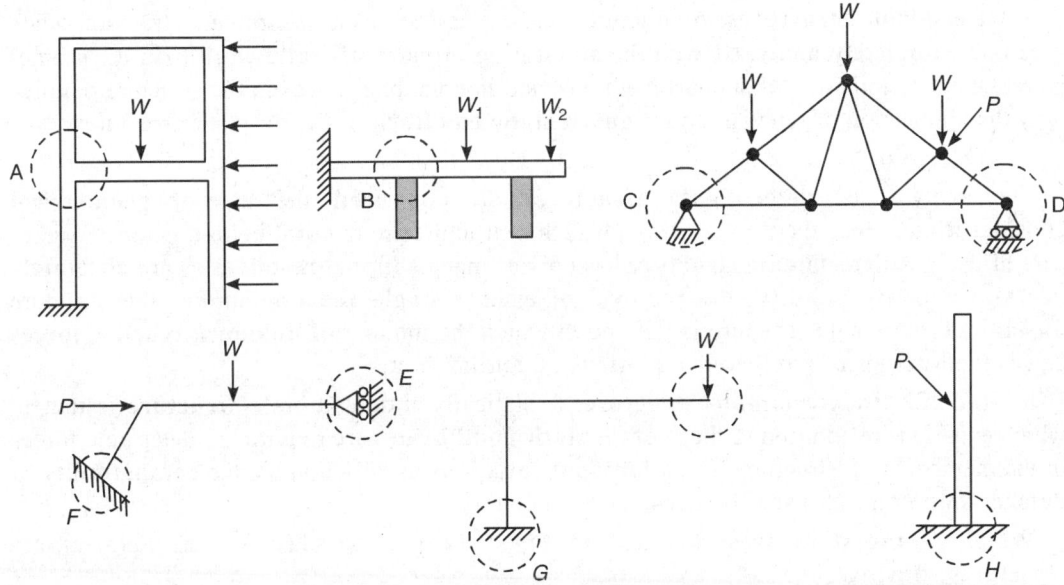

Fig. 3.11: Structures with different supports

Name the type of supports *A, B, C,* D, *E, F* and *G* given in Fig. 3.11 and the number of reactive forces offered by each.

Draw the directions of reactive forces offered by supports *A, B, C, D, E, F, G,* and *H* in the case of structures shown in Fig. 3.11.

3.4 CLASSIFICATION OF STRUCTURES

There are different ways of classifying structures. One of the ways is to classify them as **plane** or **space** structures. **Plane structure** is the one whose component members lie in the same plane e.g. a pole, roof truss or a frame. Also the loads acting on the structure also lies in the plane. **Space structure** is the one in which the loads acting on the structures **do not lie in one plane** (e.g. a dome, a shell, etc) and the components also lie in space (i.e. in all the three perpendicular directions).

Other two ways of classifying structures are based upon

- Method of analysis   –   **Determinate** or **Indeterminate**.
- Nature of internal resistances   –   **Tension, compression, flexural, shear** which are induced in the structures under combined forces.
- Form and shape of the structure –   **Flat** or **curved plate** structure, **framed** structure, **single bar** structure and **combined form** structure.

We shall be discussing here only the classification based on the methods of analysis. For other classifications of structures you are recommended to read the references on **theory of structures**.

a. **Determinate Structures:** A structure can be classified as a determinate structure, when it can be **completely analyzed** with the application of **rules of static equilibrium**. In other words, the determination of support reactions, and internal resistances in the structure requires only the application of relevant **equations of static equilibrium** i.e. for plane structures

$$\Sigma F_x = 0, \qquad \Sigma F_y = 0, \qquad \Sigma M_z = 0,$$

The analysis of **determinate structures** do **not need** the use of **geometrical considerations**. Once the forces acting on such a structure are in **equilibrium**, geometry takes care of itself. A **determinate structure** has only as many **support reactions** as are **absolutely necessary** for its **stability**, the **removal of even a single reaction makes** the structure **unstable**. Determinate structure is the one in which the number of **unknown reactive forces** equals the **number of applicable equations** of **equilibrium**.

b. **Statically Indeterminate Structure:** A statically **indeterminate structure** is that in which equations in addition to those of the **static equilibrium** are required to determine forces in the members of a structure. The **additional equations** usually involve the **compatibility of deformation** in the structure by considering geometry.

Which of the structures sketched in Fig. 3.12 can be classified as determinate structures? Justify.

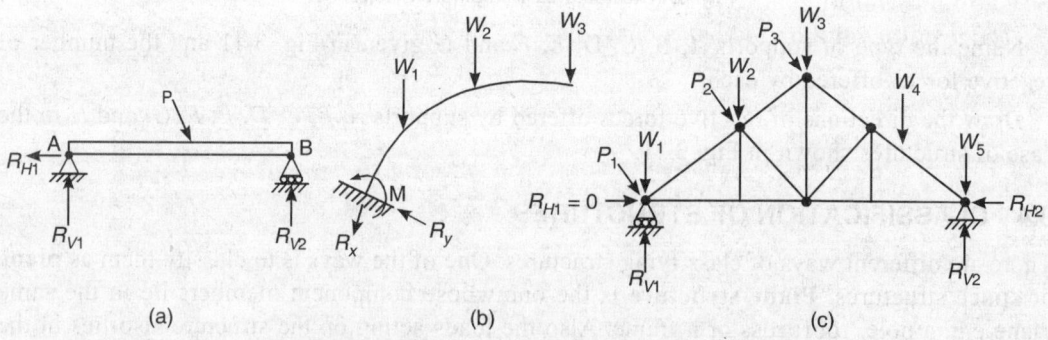

Fig. 3.12: Different structures with loads and supports reactions

3.5 INTERNAL RESISTANCE, STRESS AND DEFORMATION

It was explained earlier that when a structure is subjected to loads, the structure undergoes deformation and as a result **internal resistances (or stresses)** are set up in the structure. Different type of **internal forces (or stresses)** and **deformations** are produced in the structure due to external loads which are explained in the following paragraphs. Before dealing with these, it is important to understand the concept of **free body diagram (FBD)**. It is not only useful in understanding the concept of **internal forces** but also helps in solving problems in mechanics of structures by use of equilibrium conditions.

3.5.1 Free Body Diagram

A structure as a whole is a body, which is in **equilibrium externally** under the action of **applied forces** and **reactive forces** offered by its supports. Similarly, any portion of a structure taken out **free** from the main structure will be in **equilibrium** under the action of **applied forces** and **internal resistances** in the members **which are cut out**. A diagram which shows the structure or a portion of it in **equilibrium** under the **action of applied forces, support reactions** and **internal resistances** in the member that are **cut out**, is called a **free body diagram (FBD)**. The free body diagram of the truss (Fig. 3.13a) and one of its shaded parts are shown in Fig. 3.13b and Fig. 3.13c respectively. In drawing the free body diagram of the truss, its **supports are removed and instead the reactive forces offered** by these supports are introduced. Support A is a hinge and will offer only a **vertical reaction R_A** as horizontal reaction will be zero because the **loads acting on the truss are vertical**. Support B, being a roller, offers only one reactive force R_B **normal** to its base **(vertical)**. The FBD is shown in Fig. 3.13b. To draw free body diagram of the shaded portion of the truss and draw it separately as shown in Fig. 3.13c. The members 1, 2, 3 have been cut out by the section mn. The only forces that can exist in these members are the axial forces. Therefore, at the points of cut the three internal forces S_1, S_2, S_3 in members 1, 2, 3 are introduced. At the end A, the support is removed and the **vertical reaction R_A** offered by the support is introduced. This part of the structure is in **equilibrium** under external forces and internal forces S_1, S_2, S_3 in cut members.

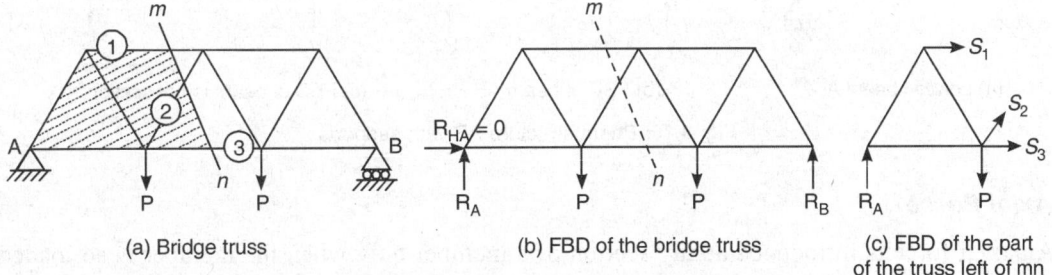

(a) Bridge truss (b) FBD of the bridge truss (c) FBD of the part of the truss left of mn

Fig. 3.13: Bridge truss with loads, support reaction and FBD

A **free body diagram** is thus a diagram of a structure or the part of structure, which is made **free from all contacts** with other structural supports or its surroundings showing all **external** and **internal forces** in cut members keeping the part in **equilibrium condition**. At the points where it is made free from its contacts with original main structure, forces and/or couples are introduced to keep it in **static equilibrium** as necessary.

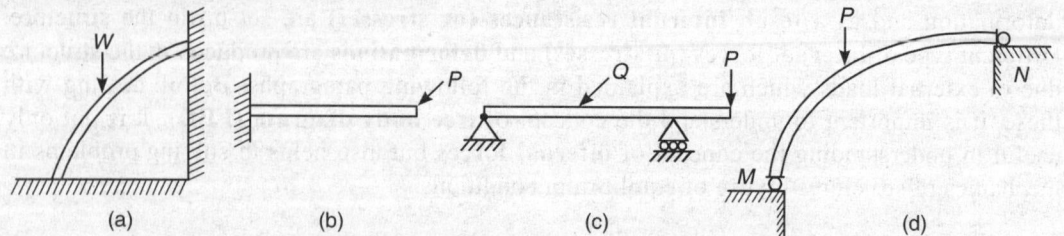

Fig. 3.14: Different structures with loading and supports

3.5.2 Internal Resistances

Let us consider a beam AB simply supported at its ends (one end A hinged and the other end B supported on rollers) and subjected to loads as shown in Fig. 3.15a. Supports at the ends of the beam hold it in **equilibrium**. The **loads** on the beam as well as the **reactive forces** offered by the supports (P, V_A, V_B, H_A) are called the **external forces** on the beam (Fig. 3.15b).

The **structure** (in this case beam) offers **internal resistance** at each of its sections to **counteract** the effects (deformations) of **external forces**. One can visualize these **internal resistances** (internal forces) by cutting the structure at any section X-X. For example, at section XX of the beam internal forces H_i, V_i and M_i must act to maintain the part of beam to the **left of section** XX in **equilibrium** (Fig. 3.15c). In the actual beam, these internal forces are supplied by the portion of the beam on the **right** of the section XX. These **internal forces** at the cut section of the beam are so determined that they keep each part of the beam in **equilibrium**. The various types of internal **resisting** forces can be **axial force, moment, shear force** and **torque**.

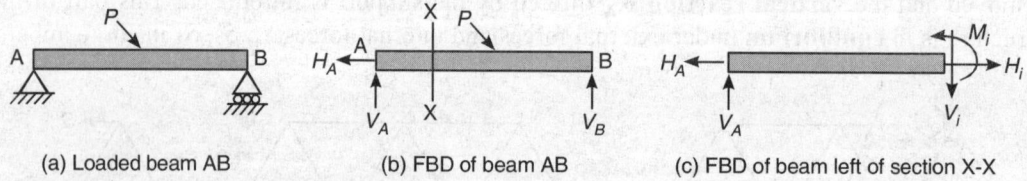

(a) Loaded beam AB   (b) FBD of beam AB   (c) FBD of beam left of section X-X

Fig. 3.15: Beam structure AB with supports

Axial Force

An axial force is introduced at any section of a member only when the member is so loaded that there is a component of the **load parallel to the axis** of the member (Fig. 3.16). Axial force acting at any section of a member could be tensile or compressive. Axial force is **tensile** when the member **tends to elongate** under its effect as shown in Fig. 3.16b. Axial force is **compressive** when the member tends to **shorten** under its effect as shown in Figs 3.17a and b.

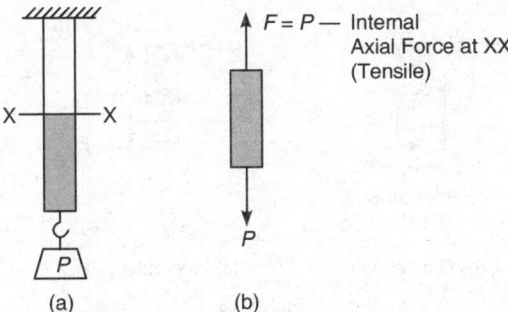

Fig. 3.16: Hanging bar structure and FBD

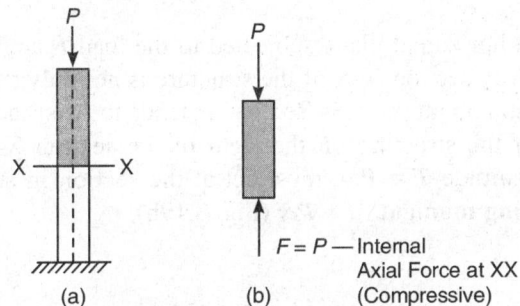

Fig. 3.17: Column structure and FBD

Shear Force

Let us consider a vertical post subjected to a horizontal load at its free end as shown in Fig. 3.18. Consider any section XX of the post. If the post is assumed as cut at this section then the effect of the load (which is **transverse to the longitudinal axis** of the structure and is tangential to section XX) is to **slide or shear** the cut portion of the post over the remaining portion. If the post is in **equilibrium**, then **every part** of the post will also be in **equilibrium**. There exist an internal force P, tangential to the horizontal section XX, to maintain the **shaded post in equilibrium**. Internal force P is the **resisting shear force** at section XX. Resisting shear force acting at any section of a member is, therefore, tangential to the sectional plane and **resists the shearing** or separation of one portion from another portion of the structure.

Resisting moment

Referring to Fig. 3.18a, the effect of force P at any section XX of the post has tendency not only to separate or shear the shaded portion from the un-shaded portion but also has tendency to **bend the member**. Looking at the shaded portion of the post, it can be said that it has tendency to **rotate** under the effect of the **couple formed by the parallel and opposite forces** P (Fig. 3.18b). To maintain this portion in **equilibrium** there must exist an opposite moment $M = P.x$, which must be **equal and opposite** to the moment of the couple at the section XX (Fig. 3.18d). The internal moment $M = P.x$, is thus the **resisting** moment acting at the section as shown in FBD in Fig. 3.18d.

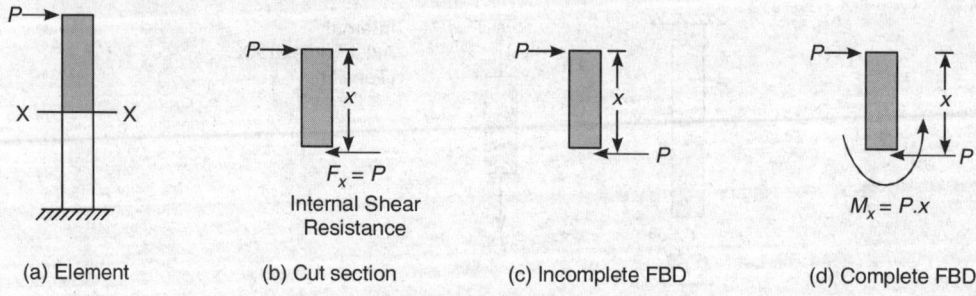

| (a) Element | (b) Cut section | (c) Incomplete FBD | (d) Complete FBD |

Fig. 3.18.

Resisting torque

Consider a structure in a horizontal plane subjected to the load P acting vertically downward as shown in Fig. 3.19a. Any section **X-X** of the structure is not only subjected to shearing and bending, but also a rotation about the **axis Z**. Load P tends to twist the structure about axis Z. In order that the part of the structure on the right of the section X-X is in **equilibrium** a **resisting torque** of magnitude $T = P.d$, must act at the section in addition to the **resisting shear force P** and **resisting moment** $M = P.x$ (Fig. 3.19b).

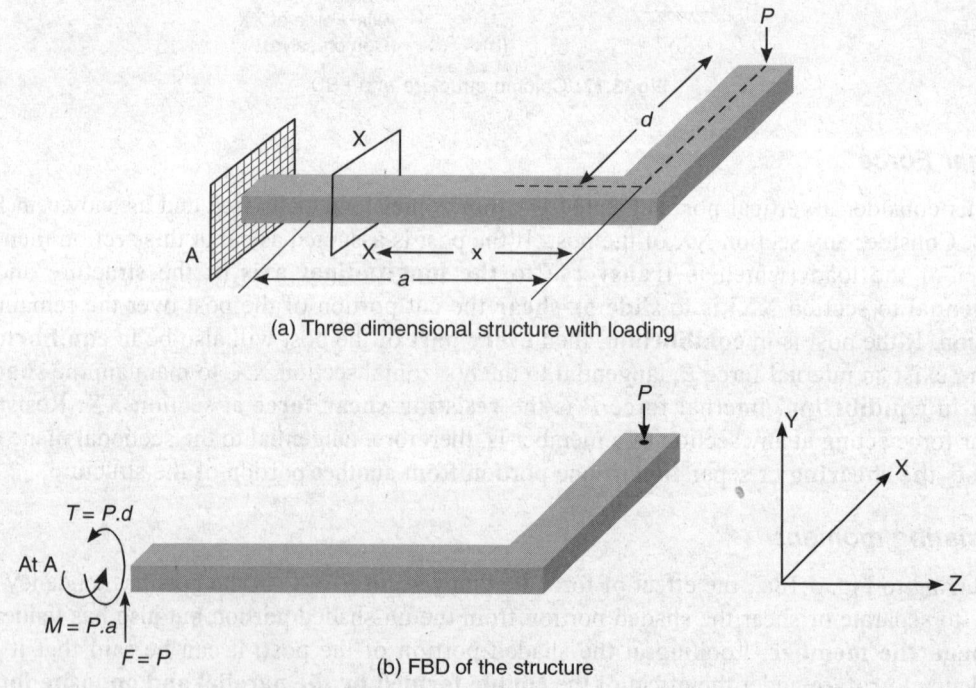

(a) Three dimensional structure with loading

(b) FBD of the structure

Fig. 3.19: Three dimensional structure and FBD

Name the type of force and moments that will act at section XX of the structures shown in Fig. 3.20.

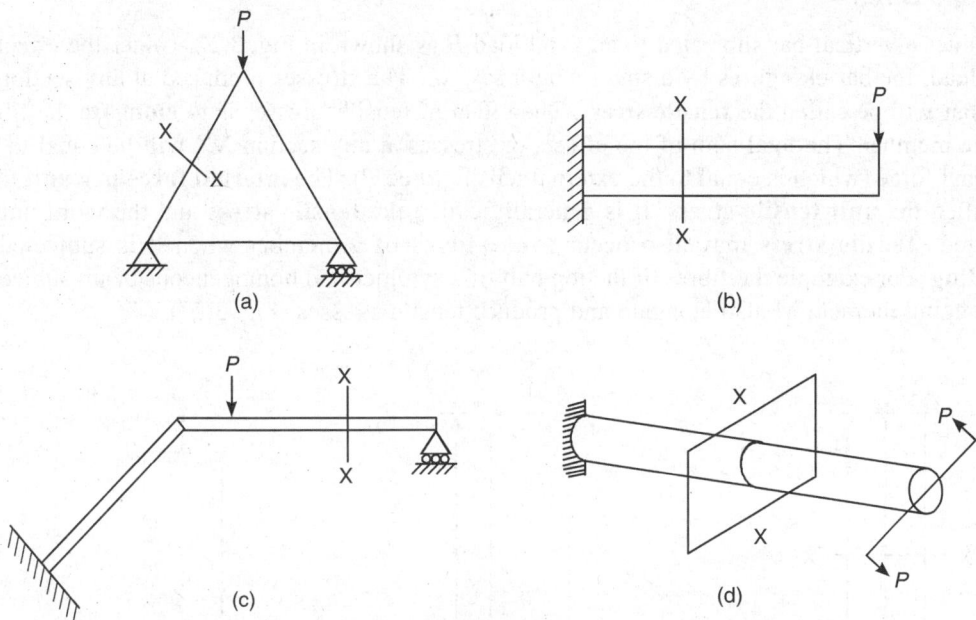

Fig. 3.20: Different structures with load and supports

3.5.3 Stress

Consider a structure subjected to a variety of loads as shown in Fig. 3.21. The structure though in equilibrium tends to deform under the effect of these loads. As a result of this, **internal resistances** are set up at **each point or section** of the structure. These internal resistances thus set up are the **internal forces and hold the external forces** on the structure in **equilibrium** as has been said earlier. These internal forces are in fact the total effects of the deformations and stresses actually produced in the fibres of the structural member shown. **Stress is the internal resistance per unit area**, set up in the fibres of a structural member to **oppose** the tendency to deform due to the external loads. There are basically three types of stresses: **Tensile** stress, **compressive** stress and **shear** stress. These stresses are explained subsequently.

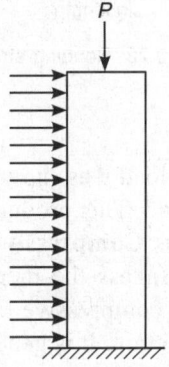

Fig. 3.21: Column with axial and lateral load

Tensile Stress

Consider a vertical bar subjected to an axial load P as shown in Fig. 3.22. Under the effect of this load, the bar **elongates** by a small length say ΔL. The stresses produced at any section of this bar will be called the **tensile stress**. The effect of tensile stresses is to **elongate** the fibres of the member. The total sum of the effects or stresses at any section XX will be equal to the internal force (which is equal to the external tensile force P). The **internal force per unit area** is called the **unit tensile stress**. It is generally called the **tensile stress** and the word unit is omitted. **Tensile stress** may also occur in the fibres of a member when it is subjected to **bending**. For example the fibres of the top half of a symmetrical homogeneous beam subjected to hogging moment M also elongate and produce tensile stresses (Fig. 3.23).

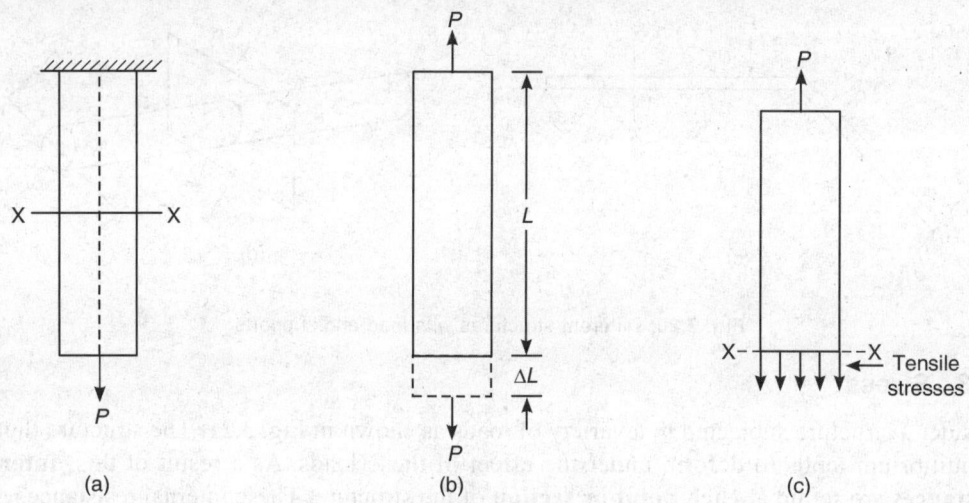

Fig. 3.22: Tensile stress

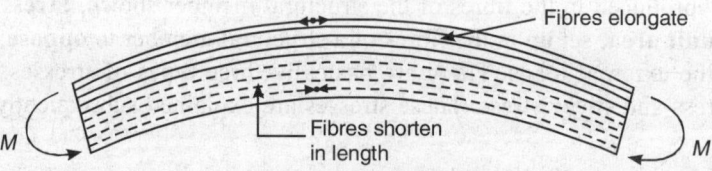

Fig. 3.23: Bending stress

Compressive Stresses

Consider a column subjected to an axial load P as shown in Fig. 3.24. The column shortens say by a length ΔL under the effect of the load. This produces stresses in the fibres of the member, which are called compressive stresses. **Compressive stresses** thus push the particles of material **against one another** and **compress** the member. The member shortens under the effect of compressive stresses. Unit compressive stress is equal to the internal axial compressive force per unit area of the section. It is generally, called compressive stress and the term unit is omitted.

Bending of a member also results in the creation of compressive stresses. For example in Fig. 3.23 the fibres in the **lower half** of the beam will be **subjected to compressive stresses**.

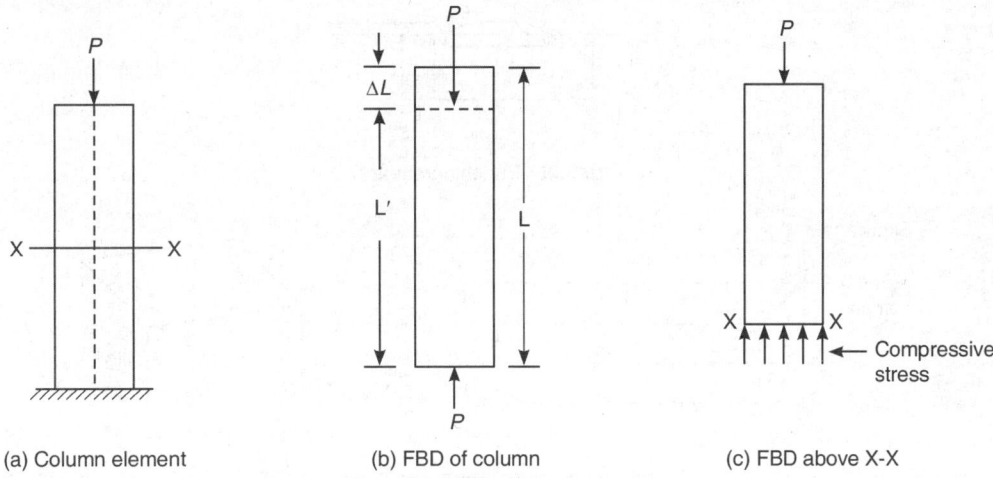

(a) Column element   (b) FBD of column   (c) FBD above X-X

Fig. 3.24: Axial compressive stress

Shear Stress

Figure 3.25a shows a riveted joint. Rivets tend to shear under the effect of loads on the joint. Fig. 3.26 shows **punching** of a hole in sheets of paper. The hole punch applies a load, which causes shearing of paper. Take a beam built into a wall at one of its end. The load at the end of the beam **tends to shear off** the beam from the wall at its end. Shear introduces deformations, which change the shape of a rectangular element into a skewed parallelogram. The forces producing the deformation act on planes along which sliding takes place. These forces measured per unit surface area are called unit shear stress or simply shear stress (Fig. 3.27).

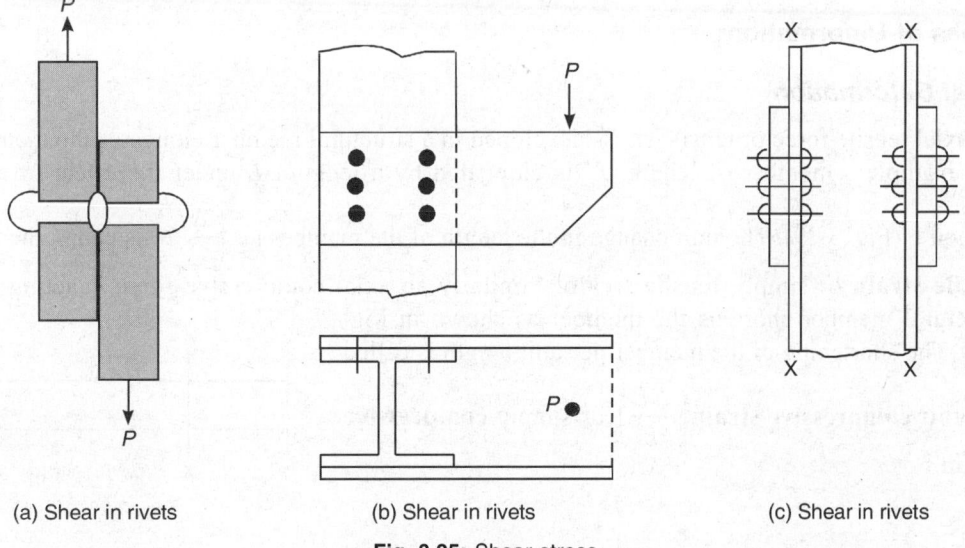

(a) Shear in rivets   (b) Shear in rivets   (c) Shear in rivets

Fig. 3.25: Shear stress

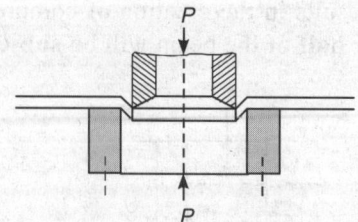

Fig. 3.26: Punching shear

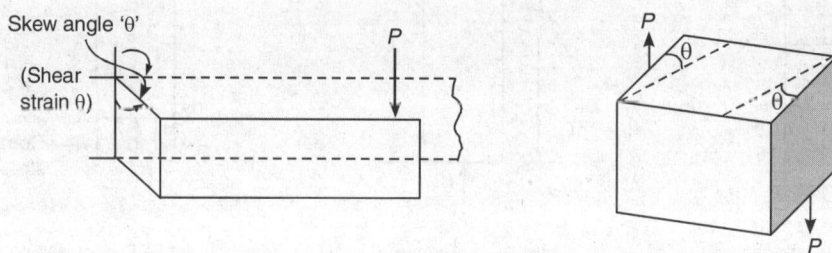

Fig. 3.27: Shear strain in beams

Torque also produces shear stresses in the fibres of the material. The state of stress is also called torsion.

What kind of stresses will be produced at section XX of the structures shown in Fig. 3.28.

3.5.4 Deformation

When a structure is subjected to a set of loads, it does not stay in its original form. It changes its **shape and size**. This is so because no structure is rigid. The deformation of the structure or its member depends upon the nature of the **internal forces** developed in it due to the action of external forces.

Types of Deformation

Axial Deformation

An axial tensile force or tensile stress developed in a structural member elongates the member. For example a member of length 'L' is elongated by a length ΔL under the effect of axial tension P (Fig. 3.29). The unit change in the length of the member i.e. $\left(\dfrac{\Delta L}{L}\right)$ is called the **unit tensile strain** or simply **tensile strain**. Similarly an axial compressive force P acting in a structural member shortens the member as shown in Fig. 3.30. The shortening of the member per unit length is called the **unit compressive strain** $\left(\dfrac{\Delta L}{L}\right)$ or simply **compressive strain**.

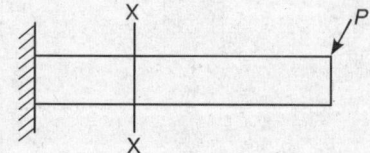

Fig. 3.28.

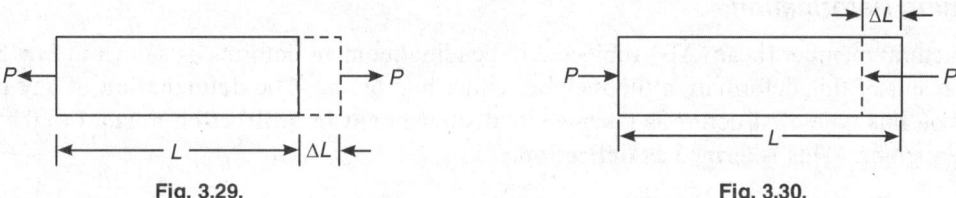

Fig. 3.29.   Fig. 3.30.

Shear Deformation

Shear Stress tends to change the **shape of a rectangular element ABCD** into a skewed parallelogram AB'C'D. For example, a rectangular block ABCD, fixed at bottom and subjected to shear P as shown in Fig. 3.31a deforms into a parallelogram AB'C'D as shown in Fig. 3.31b. A rectangular element 'A' of beam in Fig. 3.32a will deform as shown in Fig. 3.32b under the effect of shear forces. Shear deformation is the **change in the angle (θ)** of the rectangular element from 90° rather than the change in its unit length.

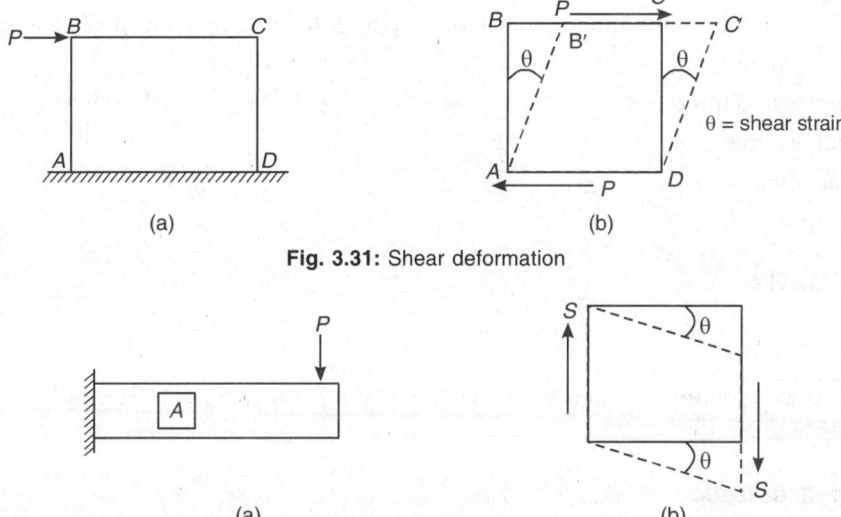

(a)   (b)

Fig. 3.31: Shear deformation

(a)   (b)

Fig. 3.32: Shear deformation in beam element

It is also called the **shear strain.** As a result of torsion, the member also undergoes shear deformation. The member is twisted under the effect of torque. The resulting deformation is a shear deformation. The **shear strain 'è'** in a shaft of certain length and subjected to a **torque** *T* is shown in Fig. 3.33.

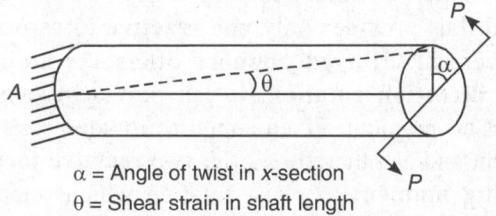

α = Angle of twist in x-section
θ = Shear strain in shaft length

Fig. 3.33: Shear deformation due to torsion in the shaft

Bending Deformation

A structural member (beam AB) subjected to bending moment deforms as shown in Fig. 3.34. As a result of this deformation the member either hog or sag. The **deformation** of any point say P on this type of structure is taken as its **displacement in a direction normal** to the axis of the member. This is termed as **deflection.**

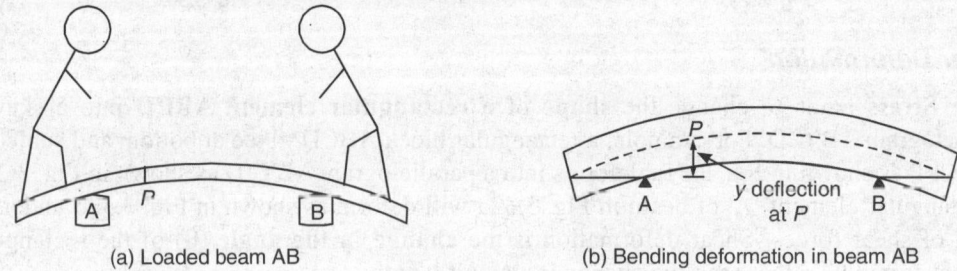

(a) Loaded beam AB   (b) Bending deformation in beam AB

Fig. 3.34: Bending of beam AB under loading

Name the kinds of deformation associated with the following applied forces and couples.

 a. Tensile force

 b. Compressive force

 c. Bending force

 d. Shear force

 e. Torque

3.6 SUMMARY

Structure

A structure is an **assembly of members** which is **stable, safe** and in **static equilibrium** under the effect of **loads.**

Loads on structures

The external forces acting on a structure are called loads. These could be **concentrated, uniformly distributed** or **uniformly varying distributed** loads.

Structural Supports

There are ideally **three types of supports**, which are used, in supporting structures. These are **roller, hinged** and **fixed** supports. **Roller support** offers no constraint against movement in longitudinal direction and thus provides only **one reactive** force perpendicular to its axis to resist perpendicular movement. **Hinged support** offers constraint against movement in position but not against **direction rotation**. It can provide **two reactive forces** to resist movement in position but no moment. Fixed support provides constraint against position as well as direction movement and can therefore, offer **two reactive forces** to resist movement in position and **one opposing moment** to resist rotation which remains zero. Thus **reactive forces** are developed if the position **movement is not allowed** and **reactive moment** is developed if the **rotational movement is not allowed. Movement or rotational freedom develops no resistance. When movement is not allowed, reactive forces** are developed.

Types of Structures

Structures can be classified on different basis. According to one basis they are classified as **determinate** or **indeterminate** structures. **Determinate** structures are those whose **analysis** can be carried out merely with the **application of principles of static equilibrium**. Analysis of these determinate structures does not need **geometrical considerations**.

Internal Resistances

There are **resistances** offered at different sections of structural members to **balance the effects** of the **applied loads** on the structure. The different kinds of internal **resisting** forces are: **Axial force, moment, shear force and torque. The internal resisting forces** are **equal and opposite** to the **externally applied forces** and these are: **Axial force, bending moment, shear force and torque.**

Stresses

The **internal resistances per unit area** set up in each fibre of a member of a loaded structure to **balance the externally applied loads** are called **stresses**. There are basically three types of stresses: **Tensile stress, compressive stress** and **shear stress**. Unit stress is defined as resistance developed per unit area.

Deformation

All elastic bodies **change their shape and size** when subjected to **loads**. This **change in the shape and size** of the bodies is called **deformation**. Depending on the type of forces acting in the member of a structure it can undergo **axial, shear, bending** or **torsional deformation**. The **unit change** in size and shape are called **strains**. Change in shape occurs due to shear stresses and is measured as slope (θ).

EXERCISE 3

Q.3.1. Classify structures according to (a) method of analysis (b) nature of internal stresses, (c) form and shape.

Q.3.2. Explain: external loads, internal resistance and unit axial stress.

Q.3.3. Explain with sketch: compressive, tensile and shear stresses.

Q.3.4. Differentiate between loads and resistances.

Q.3.5. Explain type of external loads.

Q.3.6. Explain type of different supports and reactions offered with sketches.

Q.3.7. Explain the concept of freebody diagram with sketches.

Q.3.8. Draw freebody diagrams of

(a)   Simply supported beam with central point load W,

(b)   Short column carrying 20 kN axial load,

(c)   Short column subjected to 10 kN lateral load at the top.

Q.3.9. Complete the conceptual freebody diagram of:

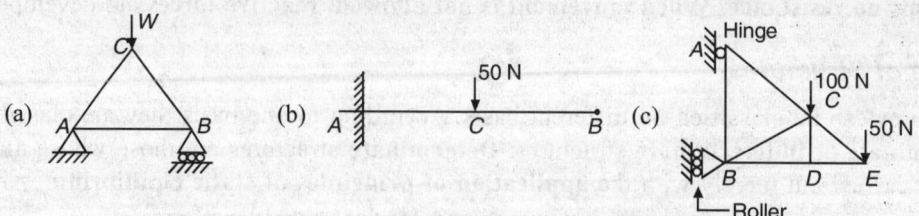

Q.3.10. Explain with sketch: (a) shear stress, (b) compressive stress in a punching machine.

Q.3.11. Explain axial deformation in an axial element.

Q.3.12. Explain bending of a beam with sketch showing various deformations.

Unit II

Axial Force Structures

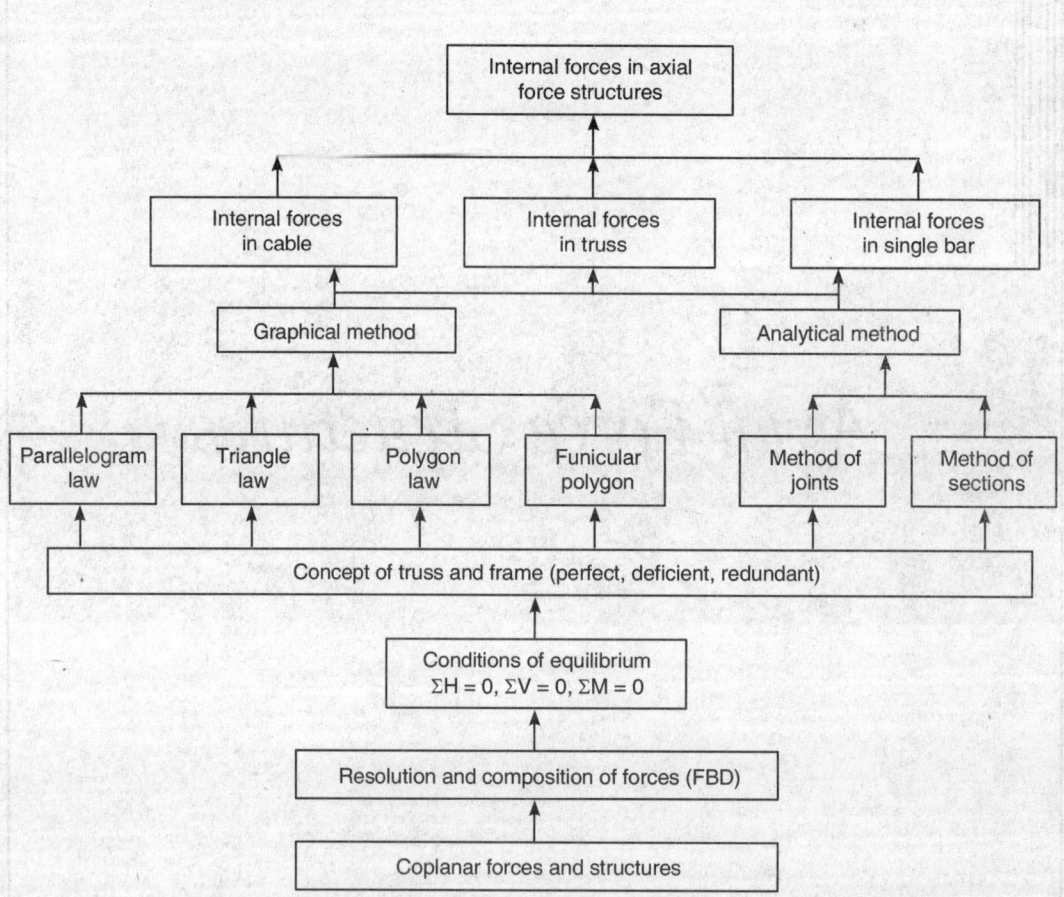

Fig. 4.0: Learning structure for axial force structures

4

Internal Forces in Axial Force Structures: Plane Trusses

LEARNING OBJECTIVES

After studying this chapter the learner will be able to:

4.1 **Understand** the meaning of **axial-force structures** (single bar and multi member axial-force structures).

4.2 **Draw free body diagram** of various axial-force members.

4.3 **Understand** the meaning of the term: **Truss, perfect truss, deficient truss** and **redundant truss**.

4.4 **Know** the assumptions made in analyzing trusses.

4.5 **State the conditions of static equilibrium** applied in the method of joints and the method of sections.

4.6 **Know** the graphical condition of static equilibrium.

4.7 **Compute internal forces** in the member of a perfect truss by using the method of joints.

4.8 **Compute forces** in the members of a truss by using the method of sections.

4.9 **Determine graphically** the **two unknown** forces in a system of coplanar forces, which keep a **body in equilibrium**.

4.10 **Determine graphically** the internal forces in the members of a truss by the use of funicular polygon and polygon of forces.

4.1 INTRODUCTION

There are a large variety of planar structures, which behave as axial-force structures. An axial force structure is that which is **subjected to axial forces** at all of its cross-sections. Some examples of axial-force structures are **columns** in buildings, **bridge piers, posts, tie** in trusses, cable structures, funicular arches, and trusses (Fig. 4.1). In all these structures the members are either subjected to axial **compression** or axial **tension**. These structures could therefore, be single bar structures (Figs 4.1a and b), curved bar structures (Figs 4.1c and d) and trusses (Fig. 4.1e). Fig. 4.2a, shows a member subjected to two equal and opposite forces P whose lines of actions are collinear with the axis of member. These forces tend to **stretch** (elongate) the member causing **tension** in it. The internal forces developed tend to pull the joint.

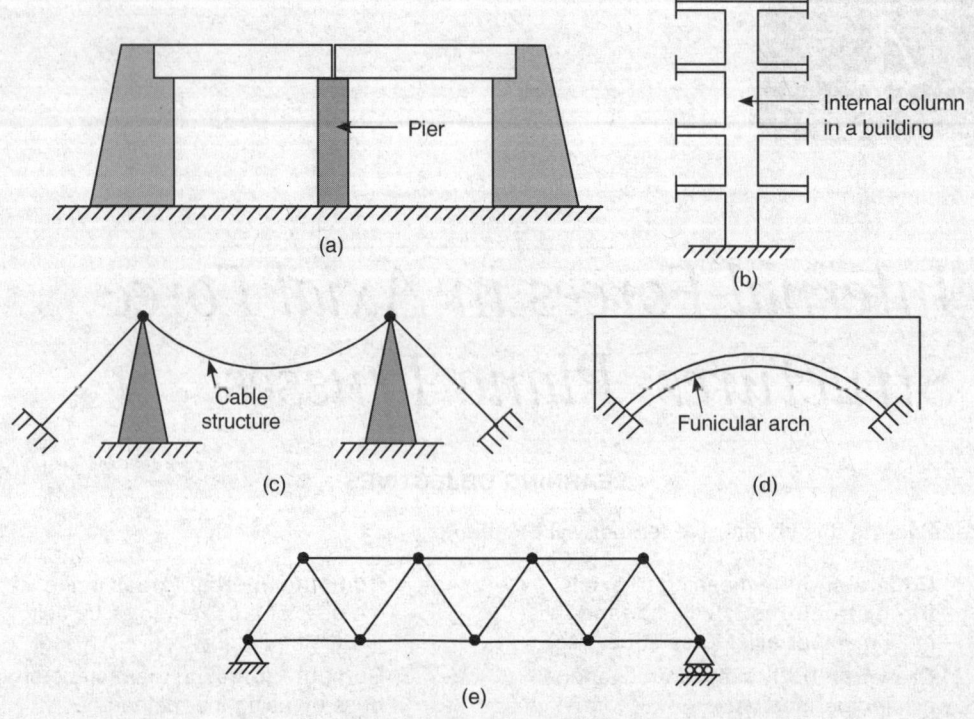

Fig. 4.1: Axial force structure

Figure 4.2b, shows a member subjected to axial compression. The member is subjected to two equal and opposite forces at its ends collinear with the axis of the member and moving towards each other. In the process, the forces tend to **shorten** the member causing **compression** in it. The force arrow tends to push the joint.

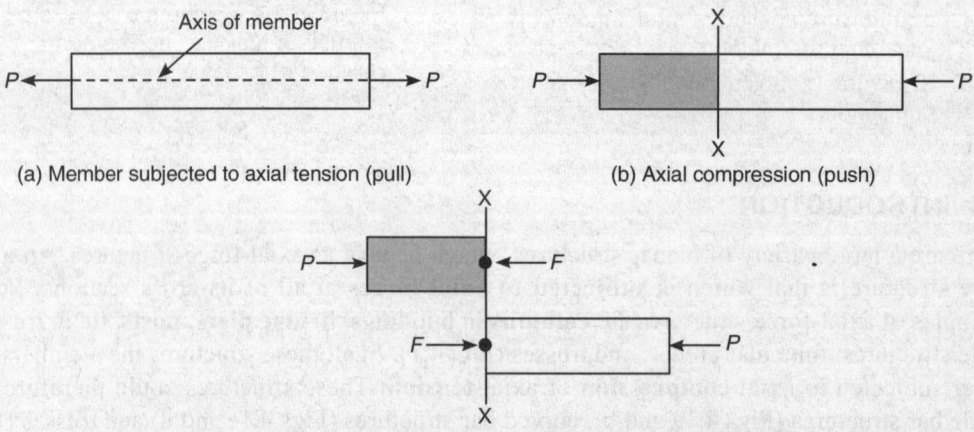

Fig. 4.2: Single bar axial members

A member subjected to either **axial tension** or **axial compression** is also called a two-force member.

Consider any section XX across any of these members say, the member in Fig. 4.2b. You will find that at the face XX, an axial force F equal to P must act to keep the part of the bar on the left or right of section XX in **equilibrium.** This force F is called the **internal resisting axial force** and is offered by the member at its cross-section XX. This is thus the only type of internal force acting in axially loaded structures (Fig. 4.2c).

It is equal and opposite to the externally applied axial force or direct force P acting at the section. Direct force acting at a section is thus equal to the algebraic sum of the axial force acting on one side of the section (left or right). We shall now calculate axial force at different sections of axially loaded structures.

4.2   AXIAL FORCES IN SINGLE BAR STRUCTURE

A **single bar axial-force structure** is a structure made up of a single bar and subjected to axial loads. For example a column in a building, guy rope, a bridge pier, etc. You will now determine forces in a straight single bar axial-force structure.

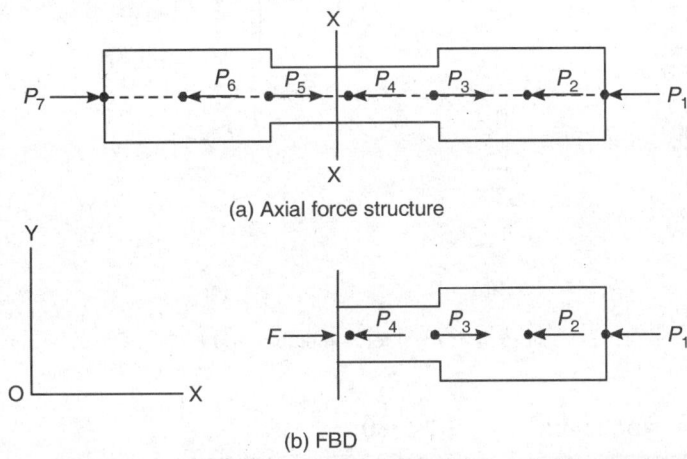

(a) Axial force structure

(b) FBD

Fig. 4.3: Axial force structure

Let us consider a member AB loaded as shown in Fig. 4.3a. All the loads $P_1, P_2, P_3, P_4, P_5, P_6$, and P_7 are axial loads and the member is in **equilibrium.**

To determine forces acting at any cross–section say XX of the bar, draw free body diagram **(FBD)** of the bar on the left or right of the section (Fig. 4.3b). Let F be the internal resisting axial force at XX.

Then for equilibrium we have,

$\Sigma F_x = 0$, i.e. $(F - P_4 + P_3 - P_2 - P_1) = 0$ or $F = P_4 - P_3 + P_2 + P_1$

The axial force or direct force applied at section XX which is equal and opposite to the internal resisting axial force F is therefore equal to the **algebraic sum** of the axial forces acting on one side of the section.

If the sense of F is towards the section (as is the case here) then the force acting on the section is **compressive**. In case the sense of F is away from the section, then it is **tensile**.

We will now solve a number of examples to illustrate computation of axial or direct force in axially loaded single bar structures.

SOLVED PROBLEMS

EXAMPLE 4.1: An internal column in a building is subjected to loads as shown in Fig. 4.4a. Determine the internal forces acting at the base of the column C-C and also at section A-A and B-B of the column.

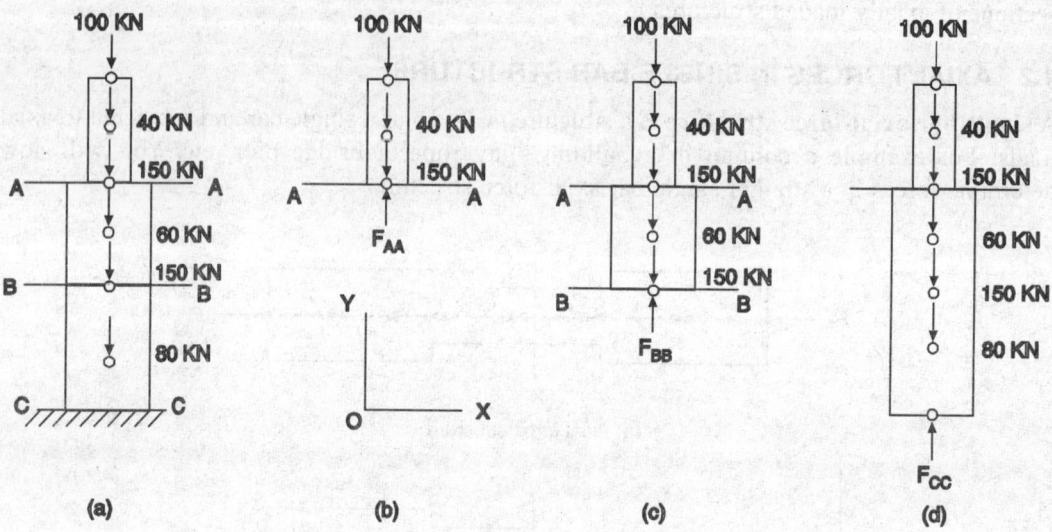

Fig. 4.4: Axial force member and FBD

Solution:

1. **Equations applicable** : $\Sigma F_y = 0$
2. **Unknown quantities** : Axial forces at AA, BB and CC
3. **Data given** : As in Fig. 4.4a.
4. **Evaluation of** : Force at AA (Just above AA)

i. Draw FBD of the column **just below** section AA (Fig. 4.4b).

Apply conditions of static equilibrium
$$\Sigma F_y = 0$$
$$-100 - 40 - 150 + F_{AA} = 0$$
∴
$$F_{AA} = \textbf{290 kN (Compressive)}$$

ii. Force at BB (Just **above** BB)

Draw F.B.D. of the column **above** BB.

Apply conditions of static equilibrium (Fig. 4.4c), $\Sigma F_y = 0$
$$-100 - 40 - 150 - 60 + F_{BB} = 0$$
∴
$$F_{BB} = \textbf{350 kN (Compressive)}$$

Just below *BB*,

$F'_{BB} = 100 + 40 + 150 + 60 + 150 = $ **500 kN (Compressive)**

(iii) Force at the base *CC*

Draw F.B.D. of the column just **above** the base *CC*.

Apply the **conditions** of static **equilibrium** (Fig. 4.4d), $\Sigma F_y = 0$

$- 100 - 40 - 150 - 60 - 150 - 80 + F_{CC} = 0$

$\therefore F_{CC} = $ **580 kN (Compressive)**

EXAMPLE 4.2: A 4 m steel member of uniform cross-section hang vertically from a fixed end at top. The weight of the member is 2000 N/m. Calculate axial force acting at the fixed end and at mid length (Fig. 4.5a).

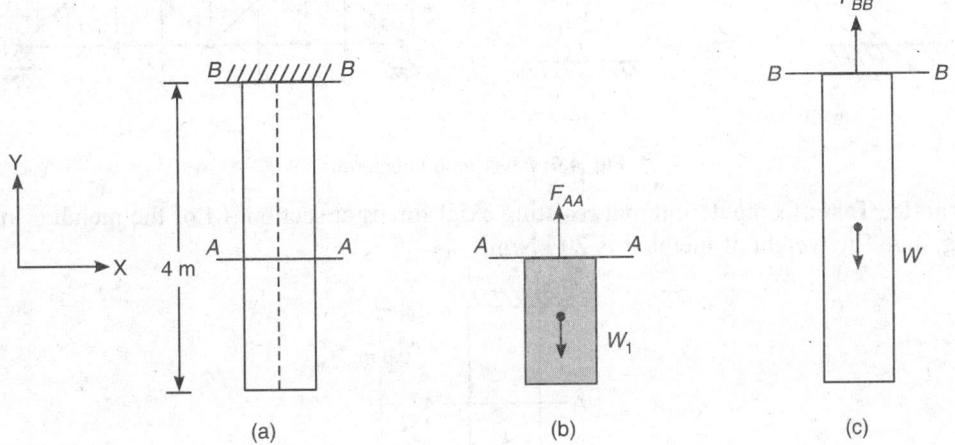

Fig. 4.5: Axial force member

Solution:

i.   **Relations**   : $\Sigma F_y = 0$

ii. **Unknown**   : Axial forces at *AA* and *BB*

iii. **Data given**   : As in Fig. 4.5a.

iv. **Evaluation**

a.   Internal resisting axial force at mid length

Draw F.B.D. of the lower half-length of the bar. The force acting on this bar is W_1 equal to its weight. Weight of half-length

$$W_1 = 2 \times 2000 = 4000 \text{ N}$$

For equilibrium, $F_y = 0$, $F_{AA} - W_1 = 0$, (Fig. 4.5b)

$\therefore$ $F_{AA} = $ **4000 N (Tensile)**

b.   Internal resisting axial force at the fixed end **BB**.

Draw the F.B.D. of the bar. The only force acting on the bar is its total weight W.

$$W = 4 \times 2000 = \mathbf{8000 \ N}$$

For equilibrium, $\quad \Sigma F_y = 0, \quad F_{BB} - W = 0,$ (Fig. 4.5c)

$\therefore \qquad\qquad F_{BB} = \mathbf{8000 \ N \ (Tensile)}$

Practice Task: Identify from the structures given in Fig. 4.6 those that can be classified as axial force structures.

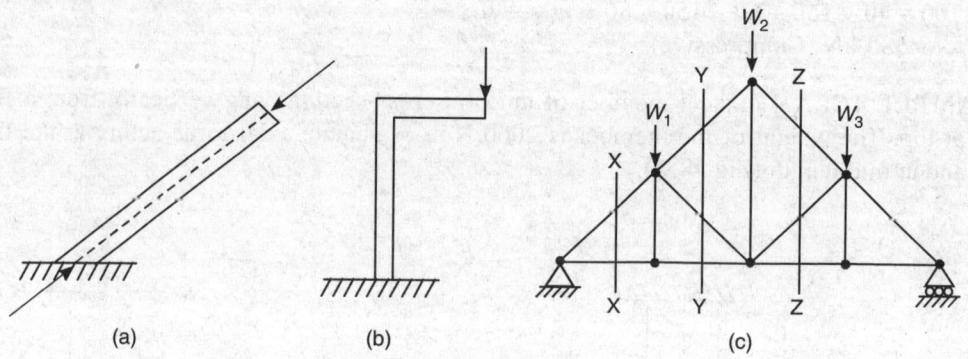

Fig. 4.6: Axial force structures

Practice Task: Compute internal resisting axial forces at section AA of the member loaded in Fig. 4.7. The weight of member is 20 kN/m.

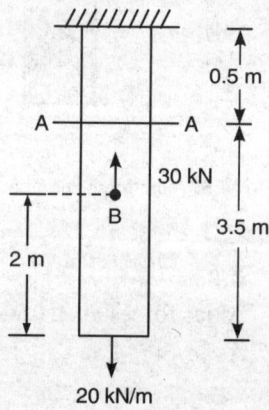

Fig. 4.7: Loaded axial element

4.3 DETERMINING INTERNAL FORCES IN MEMBERS OF TRUSSED STRUCTURES

It is important to understand what we mean by a trussed structure before attempting to determine forces in them.

4.3.1 Truss

You must have seen trusses used in bridges, cinema halls, industrial sheds or workshops. Could you observe some common features in all these trusses? On keen observation you will find (a) the member of all the trusses are so connected as to produce a **form** which consists of **one** or **more number of triangles** connected with one another, (b) the members are connected through **pin joints** or **single rivet**, which permit **free rotation** of members at joints, (c) the **loads are applied** at the panel **joints**. As a result of the above characteristics, truss members act as **axial-force structure** subjected to either **axial compression** or **tension**.

Consider extension of single member by taking two members connected by a pin joint. This structure of two members remain unstable under loads (Fig. 4.8a). Add a third member by connections with a pin at unconnected joints to form a triangle which becomes stable under the actions of loads (Fig. 4.8b). Such a structure can further be extended by two members to form additional triangles for stability (Fig. 4.8c).

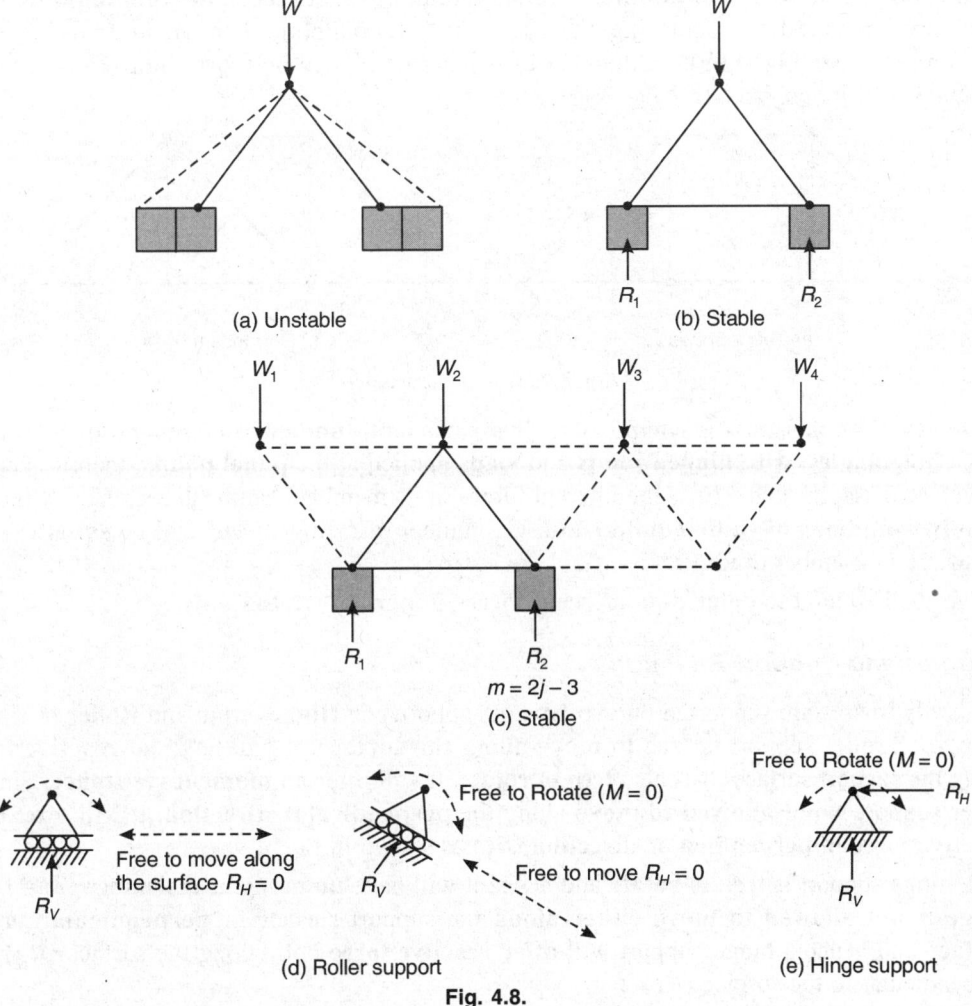

Fig. 4.8.

In case the members of a truss are connected at their ends by means of a number of rivets, bolts or welds, the **rotation** of such members **at their ends** is **constrained** and the members are subjected to additional forces due to **bending**.

When the **loads** are also applied in **between the panel points**, the members would develop **bending** in addition to **axial forces** and the truss members required will be relatively heavy.

We can now say that a **truss is a structure** formed out of **straight members** connected at their ends so as to **form a shape consisting of a triangle** or **a number of triangles** connected with one another. The members of a truss act as **tension** or **compression** members, provided these are connected by means of **hinged joints** and the truss is **loaded at its panel points.** Joints with single rivet or bolt can be considered as hinged joint.

In case the **number of members** in a truss is such that those fail to enclose the space in a **triangle(s)**, then the truss will **not be stable** and is called a **deficient truss** (Fig. 4.9a), i.e. number of members '**m**' will be **less than** $(2j - 3)$, where j = number of joints.

In another case when the **number of members** in a truss **exceeds the minimum number** than those required for enclosing the space in a **triangle(s)**, then the truss is called a **redundant truss** (Fig. 4.9b), i.e. number of members '**m**' will be **more than** $(2j - 3)$, where j = number of joints.

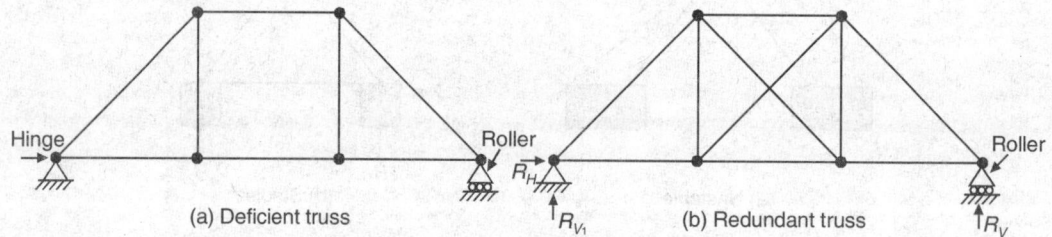

(a) Deficient truss   (b) Redundant truss

Fig. 4.9: Trussed structures

However, when a truss is composed of **just sufficient number of members** to enclose the space in **triangles, with hinged joints** and **loads** applied at the **panel points,** then it is called a **perfect truss**. In such a truss the internal forces in its members can be determined by the use of **only conditions of static equilibrium.** i.e. number of members 'm' will be **equal** to $(2j - 3)$, where j = number of joints.

We shall attempt to **determine internal forces in perfect trusses only.**

Support and Support Reactions

Generally trusses are supported on two type of supports viz **Hinge** or pin and **Roller** or simple support. A roller support is **free to move along the surface**, it will have **no reactive force** along the support surface. It is also **free to rotate**, it will offer **no moment resistance**. Since a roller support is **not allowed to move** along the **perpendicular direction**, it will offer only **reactive force in perpendicular direction** (R_V) as shown in Fig. 4.9b.

A **hinge** support is **free to rotate** and hence it will have **no moment resistance** ($M = 0$). A **hinge** is **not allowed to move either along** the support surface or **perpendicular to the surface** and hence a hinge support will **offer reactive force** both along the surface (R_H) and perpendicular to the surface (R_V).

Practice Task: Name which of the trusses given in Fig. 4.10 are perfect, deficient and redundant trusses.

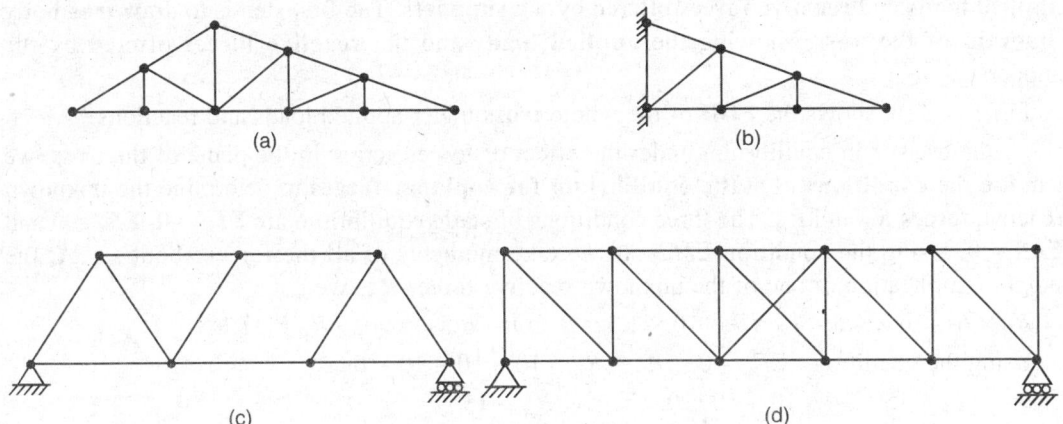

Fig. 4.10: Trussed structures

4.3.2 Method of Determining Forces In Trusses

Figure 4.11a shows the line diagram of one of the plane trusses used in the roof of a building. Let this truss be subjected to loads at its panel points as shown in the diagram. The problem is to determine internal forces in the members of the truss.

Determination of internal forces in the members of a truss will enable us: (a) to know **safe load** the truss can carry or (b) to **design the size** and shape of the members so as to **safely resist internal forces**.

There are two different ways in which we can solve the problems of determining forces in the members of a truss. One is **analytical** and the other **graphical**.

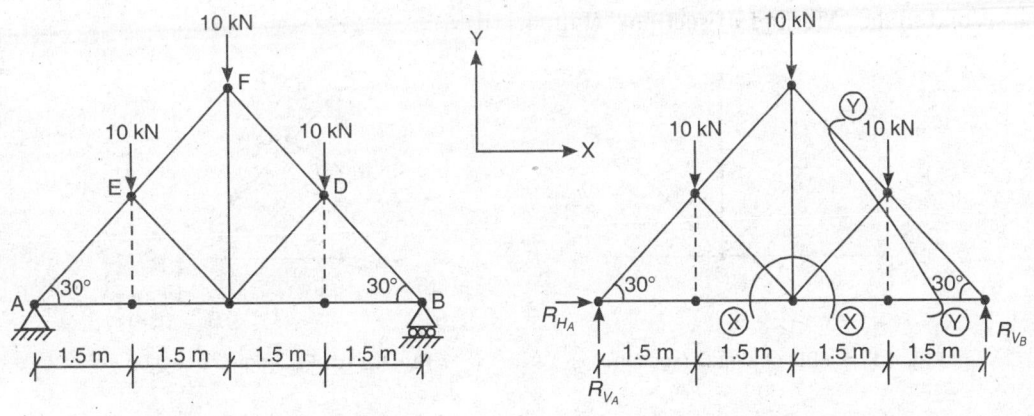

Fig. 4.11a: Truss with loads   **Fig. 4.11b:** FBD of truss

4.3.3 Analytical Method

We know that a truss is **supported** at its two ends and is in **equilibrium** under the effect of **applied loads** and **reactive forces** offered by the **supports**. The first step is to draw **free body diagram** of the truss, showing the **applied loads** and the **reactive forces** offered by the supports.

Figure 4.11b, shows the **FBD** of the whole **truss** under applied loads and reactions.

As the truss is in equilibrium under the effect of forces acting in the plane of the truss, we can use the **conditions of static equilibrium for coplanar forces** to determine the unknown reactive forces R_A and R_B. The three conditions of static equilibrium are $\Sigma F_x = 0$, $\Sigma F_y = 0$ and $\Sigma M_z = 0$. Using the condition $\Sigma M_z = 0$, we take moments of all the forces about say **A**, the point of application of one of the unknown reactive forces R_A. We get,

$R_B \times 6 - 10 \times 4.5 - 10 \times 3 - 10 \times 1.5 = 0$ Or $6\,R_B = 90$, $R_B = 15$ kN

Using the condition $\Sigma F_y = 0$, $R_A + R_B - 10 - 10 - 10 = 0$

or $R_A + 15 - 10 - 10 - 10 = 0$, $\therefore$ $R_A = 15$ N

In this particular case, we could have found the values of R_A and R_B without making any calculation, because the truss is **symmetrical** and R_A and R_B will be each **equal to half of the total load** on the truss (i.e. $R_A = R_B = \frac{1}{2} \times (10 + 10 + 10) = $ **15 kN**).

As the truss is in **equilibrium** under the effect of all applied **loads** and all the reactions. We can say that each of its **part** must also be in **equilibrium**. A portion of a structure can (in imagination) be cut free from the whole. This portion will be in **equilibrium** under the action of **applied loads** and **internal forces** in the members, which are **cut**. We can thus draw the free body diagram of such parts of the truss. Two typical free body diagrams for the parts of a truss are drawn in Fig. 4.11c, (Joint C) and Fig. 4.11d (**section y-y**). Free body diagram of the part in Fig. 4.11c, is obtained by making a cut around a joint C (**section x-x**). The other shown in Fig. 4.11d, is obtained by making a complete cut across the truss along **section y-y**. Using either of the two types of FBD, we can determine the unknown internal forces in the members of the truss. The first type of free body diagram (Fig. 4.11c) refers to '**Method of joints**' and the second to the '**Method of sections**' (Fig. 4.11d).

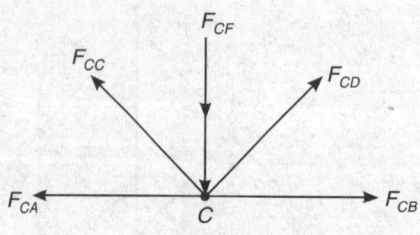

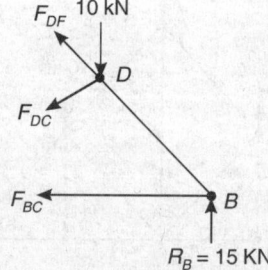

Fig. 4.11c: FBD of joint C (section x-x)   **Fig. 4.11d:** FBD right of section y-y

a. METHOD OF JOINTS

In this method, the free body diagrams of the first type are used. In other words **cut** is made **around the joint**. The free body diagram of part of the truss at **joint A** is shown in Fig. 4.11e.

F_1 and F_2 indicate the **internal forces** in members AC and AE respectively. This is the case of **concurrent coplanar forces meeting at point A**, which are in **equilibrium**. We can therefore use the following conditions of static equilibrium in determining the unknown forces F_1 and F_2.

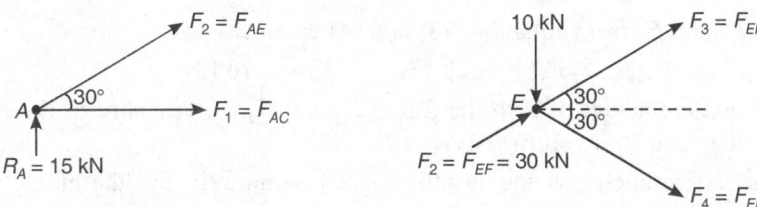

Fig. 4.11e: FBD of joint A   **Fig. 4.11f:** FBD of joint E

$$\Sigma F_x = 0, \qquad \Sigma F_y = 0$$

For $\Sigma F_x = 0$, we have, $F_1 + F_2 \cos 30^\circ = 0$

or $F_1 + (\sqrt{3}/2)F_2 = 0$ ... (1)

For $\Sigma F_y = 0$, we have,

$$15 + 0 + F_2 \sin 30^\circ = 0$$

or $15 + 0.5\, F_2 = 0$

$\therefore$ $F_2 = -\,\mathbf{30\ kN}$ (Compressive), (i.e. opposite of assumed) ... (2)

This means that the nature of force F_2 **is opposite to the one assumed**. It is not tensile as assumed, but is compressive.

Substituting for F_2 in equation (1), we have

$F_1 - (\sqrt{3}/2) \times 30 = 0,$ $\therefore\ F_1 = +\,\mathbf{26\ kN}$ **(Tensile, as assumed)**

JOINT E

Consider **FBD** of the part of the truss **around joint E** (Fig. 4.11f).

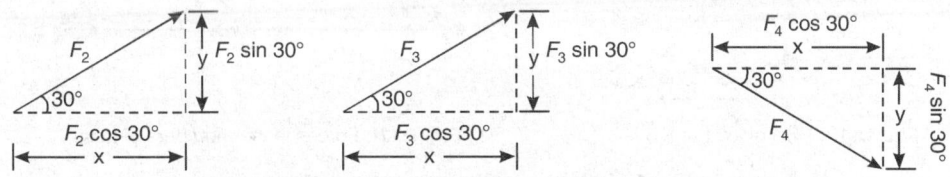

Fig. 4.11g: Rectangular components of forces at joint E

F_3 and F_4 are the internal forces in members EF and EC of the truss and **assumed to be tensile**.

F_2 is the force in member AE and we have already found it equal to **30 kN compressive**. Because it is **compressive**, therefore, the **arrow points towards the joint E** (An action of member to push the joint).

Applying the conditions of equilibrium $\Sigma F_x = 0$, $\Sigma F_y = 0$.

For $\Sigma F_x = 0$, we have, $F_2 \cos 30^\circ + F_4 \cos 30^\circ + F_3 \cos 30^\circ + 0 = 0$,

or $30 \times (\sqrt{3}/2) + F_4 \times (\sqrt{3}/2) + F_3 \times (\sqrt{3}/2) = 0$

or $\qquad\qquad\qquad F_4 + F_3 = -30$ $\qquad\qquad\qquad\qquad$ ... (3)

For $\Sigma F_y = 0$, we have, $\quad F_2 \sin 30^\circ - F_4 \sin 30^\circ + F_3 \sin 30^\circ - 10 = 0$

or $\quad 30 \times (1/2) - F_4 \times (1/2) + F_3 \times (1/2) - 10 = 0$

or $\qquad\qquad\qquad F_3 - F_4 = -10$ $\qquad\qquad\qquad\qquad$ ... (4)

Solving for F_3 and F_4 from equations (3) and (4) we have

$$F_3 = -20 \text{ kN}, \qquad F_4 = -10 \text{ kN}$$

This means that the nature of both the forces F_3 and F_4 are **opposite to the one assumed**. F_3 and F_4 are therefore both **compressive**.

As the truss is symmetrical and is also loaded symmetrically, therefore, the forces in symmetrical members will be the same.

Force in member (Magnitude)		Force in member (Nature)
DB Same as in	AE = 30 kN	Compressive
BC Same as in	AC = 26 kN	Tensile
CD Same as in	CE = 10 kN	Compressive
DF Same as in	EF = 20 kN	Compressive

JOINT F

FBD of the part of the truss around joint F is shown in Fig. 4.11h.

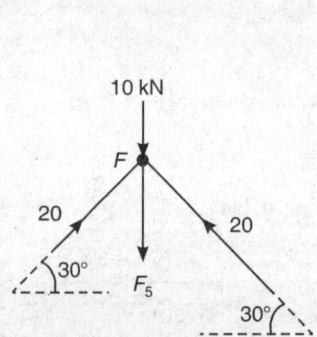

Fig. 4.11h: FBD of joint at F

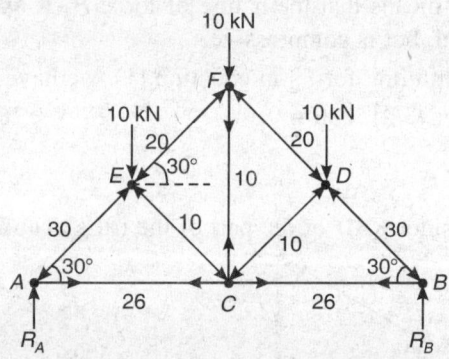

Fig. 4.11i: Forces in the members of truss

Fig. 4.11: Analysis of internal forces in truss

F_5 is the force in member FC and is assumed to be tensile.

Applying the conditions of static **equilibrium**, $\Sigma F_x = 0$, $\Sigma F_y = 0$

For $\Sigma F_y = 0$, $\qquad\qquad 20 \sin 30^\circ - 10 + 20 \sin 30^\circ - F_5 = 0$

$\therefore \qquad\qquad\qquad\qquad F_5 = +10 \text{ kN}$ (Tensile)

The forces in various members of the truss are indicated in Table 4.1 and shown in Fig. 4.11i.

Table 4.1: Forces in the members of the truss

Member	Force	Nature
AE and BD	30 kN	Compressive
EF and DF	20 kN	Compressive
AC and BC	26 kN	Tensile
CE and CD	10 kN	Compressive
CF	10 kN	Tensile

In using this method, you must have observed that we:

1. Start by drawing **FBD** of the joint, where there are only **two members** meeting at the joint. In other words there are **not more than two unknown forces** to be determined. **Two unknowns** can be found by using **two equations** of equilibrium.

2. Select the next joint for analysis with the consideration that **not more than two unknown member forces** exist which need to be determined at that joint. For example after analyzing **joint A**, we took up **joint E** and not joint F. Had we taken up joint F, there were **three unknown forces** (in members **EF, DF, CF**) and therefore it was not possible to solve for these **three unknowns with only two equations of equilibrium** $\Sigma F_x = 0$, $\Sigma F_y = 0$.

3. In this method of joints, we could not determine forces in some members (say member **EF**) **directly** in one step. We had to solve for the members at the **joint A**, before taking up the **joint E** to solve for the member EF since number of unknowns at the joint E are more than two.

Practice Task: Using method of joints, determine internal forces in the members of a truss loaded as shown in Fig. 4.12.

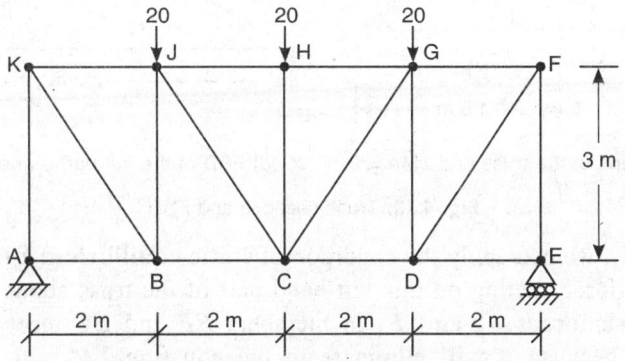

Fig. 4.12: Loaded truss (ABCDEFGHJHKA)

b. *Method of sections*

This method involves the application of the conditions of static **equilibrium** to the **free body diagram** of the type given in Fig. 4.11d. This free body diagram was obtained by a **cut** made completely across the truss through section Y-Y.

Consider the same truss as used in the earlier example (Fig. 4.11). To determine forces in members **EF**, **EC** or **AC** we can **cut the truss** by a **section x-x** passing through these members and dividing the truss in two parts (Fig. 4.13a). We can then consider **equilibrium of the left hand** α right hand part of the truss as per convenience. Here we are considering the equilibrium of the **left hand part,** for which the free body diagram is drawn as shown in Fig. 4.13. F_3, F_4, and F_1 are the three unknown **internal forces** in the members **EF**, **EC** and **AC** respectively.

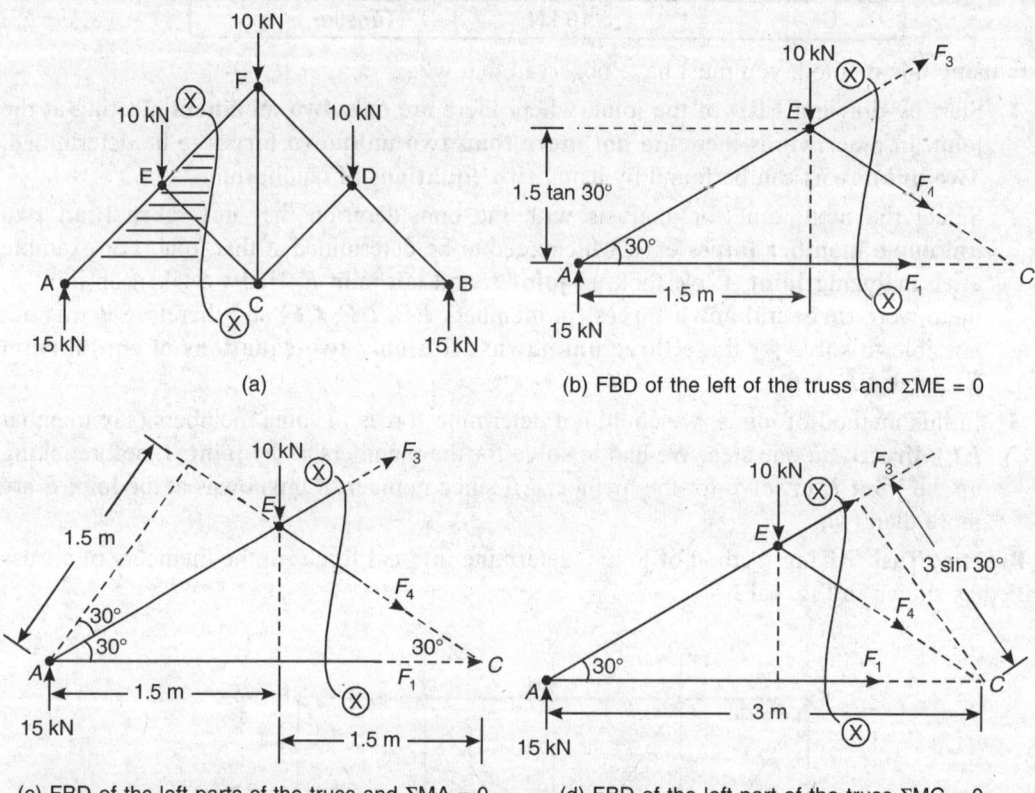

(a)

(b) FBD of the left of the truss and $\Sigma M_E = 0$

(c) FBD of the left parts of the truss and $\Sigma M_A = 0$

(d) FBD of the left part of the truss $\Sigma M_C = 0$

Fig. 4.13: Truss sections and FBD

To determinate F_1, we can apply the condition of **static equilibrium** $\Sigma M = 0$ about E. Take moments of all the forces acting on this left hand part of the truss about point **E where the other two unknown forces** F_3 and F_4 in members **EF** and **EC** meet (point E has been especially chosen, because it will eliminate in the equation $\Sigma M = 0$, the two unknown quantities F_3 and F_4, because their moment about the point E is zero as these unknown forces are passing through E).

Taking moments about E (for $\Sigma M = 0$), we have,

$$F_4 \times 0 + F_3 \times 0 + F_1 \times 1.5 \tan 30^\circ + 10 \times 0 - 15 \times 1.5 = 0$$

or $\qquad F_1 \times 1.5 \times (1/\sqrt{3}) = 22.5, \quad \therefore \quad F_1 = 26 \text{ kN (Tensile)}$

To determine F_4, take moments of all the forces about **point** A, where the two other unknown forces $\boldsymbol{F_3}$ **and** $\boldsymbol{F_1}$ **meet** (Fig. 4.13c).

Taking moments about A. For $\Sigma M = 0$, we have,

$$F_1 \times 0 + F_3 \times 0 - F_4 \times 3 \sin 30^\circ - 10 \times 1.5 + 15 \times 0 = 0$$

or $\qquad 0 - F_4 \times 1.5 - 15 = 0,$

$\therefore \qquad\qquad\qquad\qquad F_4 = -10$ kN (i.e. Compressive as opposite of assumed)

To determine F_3, take moments about the point **where the other two unknown forces** $\boldsymbol{F_4}$ **and** $\boldsymbol{F_1}$ **meet** i.e. at point C (Fig. 4.13d)

Taking moments about point C. For $\Sigma M = 0$, we have,

$$F_4 \times 0 + F_1 \times 0 - F_3 \times 3 \sin 30^\circ - 15 \times 3 + 10 \times 1.5 = 0$$

or $\qquad -1.5\, F_3 - 45 + 15 = 0,$

$\therefore \qquad\qquad\qquad\qquad F_3 = -\,\mathbf{20}$ **kN** (**Compressive** as opposite of assumed)

In applying this method, we have observed that for the condition of static equilibrium $\Sigma M = \mathbf{0}$, yield results, **it must cut the section across the complete truss. In doing so the section shall cut neither more nor less than three members of the truss.** Secondly to determine the unknown force in one of the cut members, **take moments of the forces about the point, where the other two unknown forces meet.** Thirdly when two of the three cut members are parallel (say members **a** and **b** in Fig. 4.14), then force in member **c** can not be determined by taking moments about a meeting point which does not exist (as the forces in members **a** and **b** are parallel). In other words the condition of equilibrium $\Sigma M = 0$, will not help in determining the force in **c.** In such a case, we need to apply the **other two conditions of equilibrium, i.e.**

$$\boldsymbol{\Sigma F_x = 0, \quad \Sigma F_y = 0}.$$

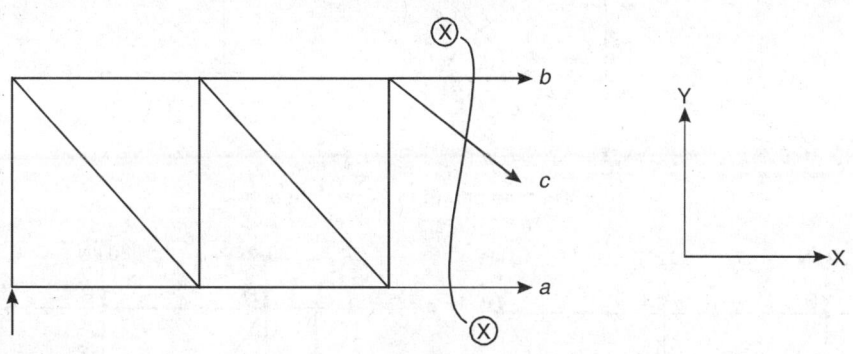

Fig. 4.14.

This method of sections is very useful, when we are required to **determine forces in only selected members** of a truss. It is however, **lengthy** if forces in **all the members** are to be determined when the loading is also unsymmetrical and at variable inclinations.

SOLVED EXAMPLES

EXAMPLE 4.3: Determine forces in members (1), (2) and (3) of the truss loaded as shown in Fig. 4.15.

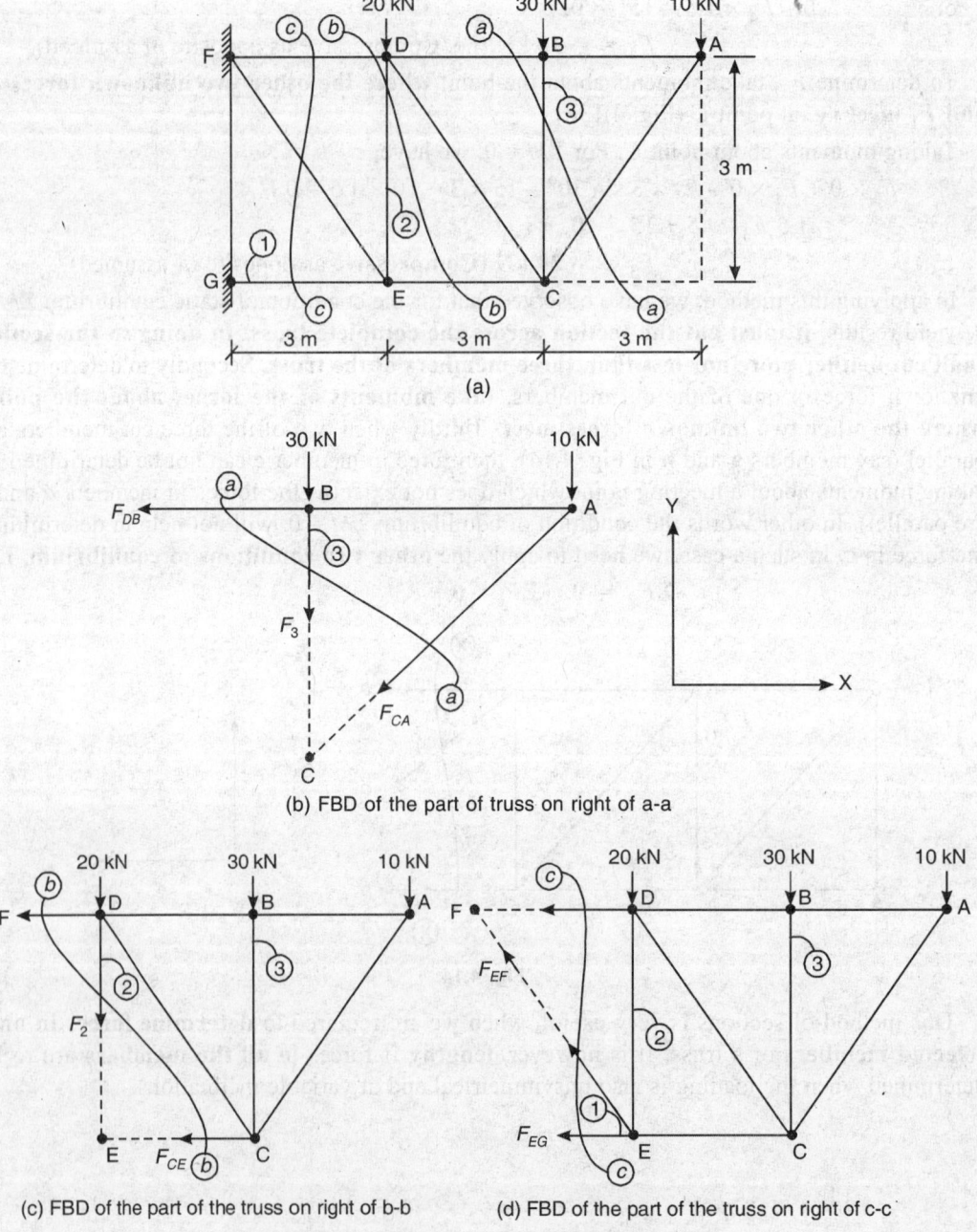

(a)

(b) FBD of the part of truss on right of a-a

(c) FBD of the part of the truss on right of b-b (d) FBD of the part of the truss on right of c-c

Fig. 4.15: Cantilever truss with loadings and sections

Solution:

1. **Relation**   : $\acute{O}M = 0$
2. **Unknowns**   :   Forces in members 1, 2, 3?
3. **Data given**   :   As shown in Fig. 4.15a.
4. **Evaluation**

i. As it is a cantilever truss, we need not calculate the reactive forces, we can always consider **equilibrium of the free end part** of the truss.

ii. To determine force in vertical member **BC** (3), cut the truss by section **a-a**. Draw **FBD** of the part of the truss on the right of section a-a (Fig. 4.15b).

Let the force in member **BC** (3) be F_3. Take moments about **A**, where the other two unknown forces in cut members **BD** and **AC** meet.

For $\acute{O}M = 0$, we have,

$F_3 \times 3 + 10 \times 0 + 30 \times 3 = 0$

∴ $F_3 = -30$ kN (i.e. Compressive as opposite of assumed one)

iii. To determine forces in vertical **member DE** (2), cut the truss by section b-b and draw **FBD** of the part of truss on the right of b-b (Fig. 4.15c). The condition of static equilibrium $\acute{O}M = 0$, alone will not be adequate to determine directly, because the other two unknown forces in members **DF** and **CE** are parallel and **do not meet.** We can, therefore, use the **other conditions of equilibrium** $\acute{O}F_x = 0, \acute{O}F_y = 0$.

For $\acute{O}F_y = 0,$ $-10 - 30 - 20 - F_2 = 0$

∴ $F_2 = -60$ kN (Compressive as opposite to that assumed)

iv. To determine force in member EG (1), **cut** the truss by section **c-c** and draw FBD of the part of truss on the **right side of c-c** (Fig. 4.15d).

Take moments of forces about *F*.

For $\acute{O}M = 0$, we have, $0 - F_1 \times 3 - 20 \times 3 - 30 \times 6 - 10 \times 9 = 0$

or $3F_1 + 60 + 180 + 90 = 0,$

∴ $F_1 = -110$ kN (Compressive i.e. opposite of assumed one)

Practice Task: Determine forces in members DE (F_1), EF (F_2) and AB (F_3) of the truss shown in Fig. 4.16.

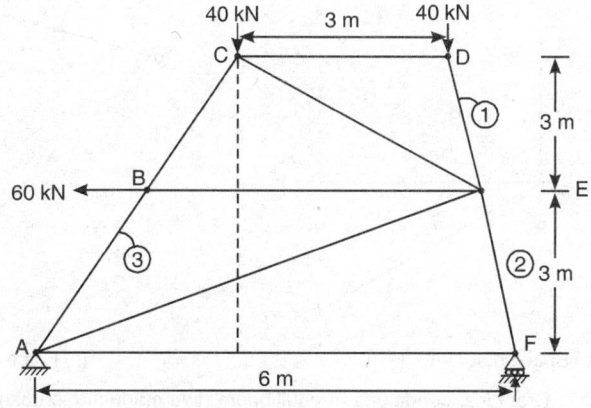

Fig. 4.16: Tower truss with loading

4.4 GRAPHICAL METHOD

In section 4.3.3 you have learned two alternative **analytical methods** (method of **joints** and method of **sections**) for determining internal forces in trusses. Now you will learn to determine internal forces in these trusses **graphically**. To do so, you must understand the **graphical conditions of static equilibrium** and possess the skill of **drawing triangle of forces, polygon of forces** and **funicular polygon**.

4.4.1 Graphical Conditions of Static Equilibrium of Coplanar Forces

We have already studied in the subject 'applied mechanics' that for a system of coplanar forces, if the **force polygon closes**, the **resultant cannot be a force** but may be a **couple**. If, however, the **funicular polygon also closes**, then the **resultant couple also vanishes**. There are thus two graphical conditions, which a system of coplanar force must satisfy, if the **forces are to be in equilibrium**. These are:

1. The **force polygon must close** (if this condition is satisfied, **the resultant force will be zero**).

2. The **funicular polygon must close** (if this condition is satisfied the **resultant couple will be zero**).

We shall review here our knowledge of the **force polygon** and **funicular polygon.** Also we shall understand the sufficiency of the conditions of equilibrium for different systems of **coplanar forces** to be in **equilibrium.** In other words, the **graphical conditions** required for different systems of coplanar **forces to be in equilibrium.**

4.4.2 Force Polygon

A **force polygon must close** is a **sufficient condition** of static **equilibrium** for any problem that involves a system of **coplanar concurrent forces.** For example, forces acting on a joint in a plane truss, can be solved **graphically by the use of force polygon** provided the number of unknowns is **not more than two**. Those two **unknown forces** may be **acting in a known direction** (Fig. 4.17) or **one force**, which is **unknown (R)** in **both magnitude and direction** (Fig. 4.18). Let us consider joint (1) in Fig. 4.17i. It is subjected to **two known** forces of

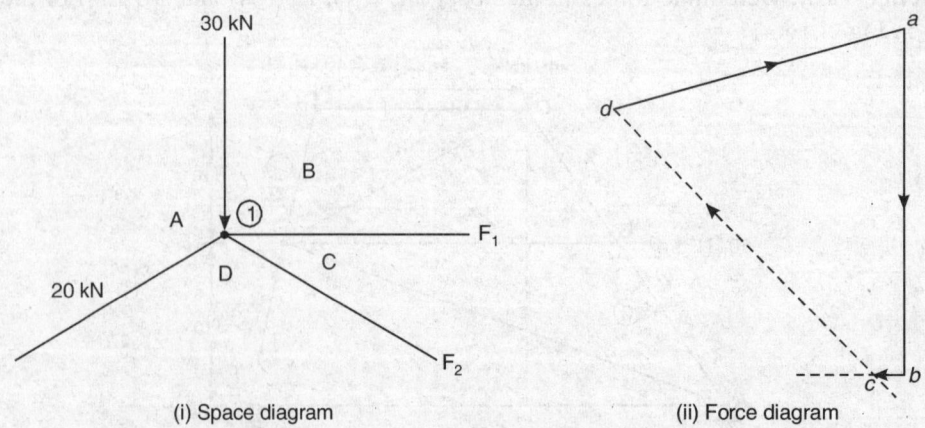

(i) Space diagram   (ii) Force diagram

Fig. 4.17: Graphical conditions of equilibrium (two magnitudes unknown).

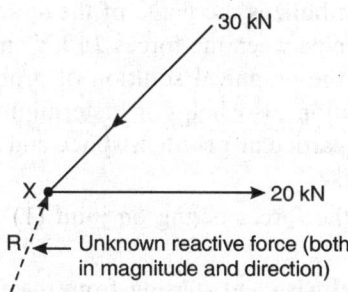

Fig. 4.18: One totally unknown force (magnitude and direction unknown)

30 kN and 20 kN (in **magnitude** and **direction**) and **two unknown forces** F_1 and F_2 (with known **directions** but **magnitudes** not known). This diagram is also called the **space diagram**.

4.4.3   Graphical Method to Determine Unknown Reactions

Graphical method is quite suitable when **all** the members of the truss are required to be **analyzed** and also the applied **loads** and members are placed along **various directions** with variable inclinations. Resolved components of such forces require a large number of calculations.

In graphical method first the truss is drawn on white paper sheet with certain assumed convenient linear scale for length (span and height). All the members and loads are marked correctly according to the direction and points of actions. Reactive forces are also shown depending on the type of support.

Having drawn the truss and applied loads, mark the **spaces** by **capital letters A, B, C,** .... etc. between two **consecutive applied loads** either in **clockwise** or **anticlockwise manner** on a continuous basis till all the **external spaces** between the loads are marked. This is done by using **Bow's notations**. According to **Bow's notation** all these forces will be known by the name of spaces on either side (in space diagram) and marked by **small case letters** (a, b, c, ...) in the **force** or **vector** diagram. For example, a force P lies between spaces **A** and **B** in space diagram, force P will be marked as *a-b* in vector diagram. The length of *a-b* will be taken by assuming **certain scale** for force e.g. **1 kN = 10 mm** and drawn **parallel** to the **direction of load** P (*a-b*) in the space diagram.

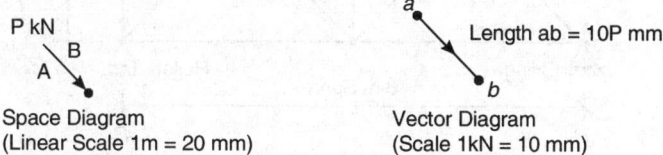

Bow's Notation

To determine magnitude of F_1 and F_2 (Fig. 4.17), let us first name all the forces in the space diagram using **Bow's notation.** According to this notation, each force in the **space diagram is designated by two letters between which it lies**. Thus the force of 20 kN in Fig. 4.17i is designated by **AD**. **Letter A symbolizes** the name of the **space** between the forces of **20 kN** and **30 kN** and letter **D** the **space** between the forces **20 kN** and F_2. By adapting this standard notation of designating forces, the graphical solution of problems related to coplanar force system is simplified. This notation also helps in determining the **direction** in which any unknown force will act. **In one** particular **problem space** and **force** designation is carried out **either clockwise** or **anticlockwise**.

After having designated all the forces acting on joint (1) (Fig. 4.17i) we can proceed to draw the polygon of forces.

Choosing any order, say **clockwise** and starting from the **force DA,** draw the vector '*da*' representing **force DA** in direction and magnitude on some scale (Fig. 4.17ii). From *a,* draw the **vector '*a – b*'** representing the **force AB** in direction and magnitude on the assumed scale. From *b* draw a line *bc* **parallel** to the **line of action** of the **force BC.** From *d* **draw** a line **parallel to the force DC,** meeting the line *bc* in *c.* Thus the force polygon representing forces **DA, AB, BC** and **CD** acting at joint (1) is *d-a-b-c-d* and is in **equilibrium** (Fig. 4.17ii) as the force polygon closes.

To determine the nature of the unknown forces (Compressive or Tension) Bow's notation is of great help. Referring to the force polygon in Fig. 4.17ii, we know that it has been completed by drawing **vectors da, ab, bc** and **cd** in a **clockwise order.** The arrows representing the sense of direction in the vectors *bc* and *cd* can therefore be marked by **maintaining the chosen order** (in this case **clockwise**). The sense of direction in the forces **BC** and **CD** are then marked on the **space diagram**, so that it is consistent with that of the **force polygon.** Force BC and CD are both forces of **compression** because they **push at the joint (1)**.

Practice Task: Designate the various forces acting on the truss (spaces) shown in Fig. 4.19 by using Bow's notation.

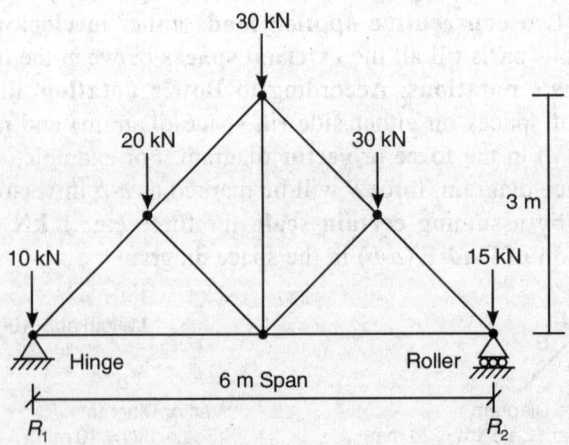

Fig. 4.19: Loaded truss with hinge and roller supports

Practice Task: The free body diagram of a joint of a plane-truss is shown in Fig. 4.20. Determine the unknown forces F_1 and F_2 using polygon of Forces. (Nature and Magnitude).

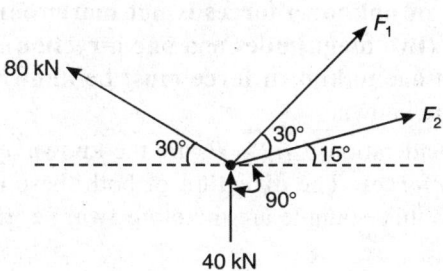

Fig. 4.20: Two unknown forces (both magnitudes)

4.4.4 Funicular Polygon

In problems related to coplanar force systems, which are not concurrent, sufficient conditions for the system to be in equilibrium are not only that the **force polygon must close** but also the **funicular polygon must close**. When the force polygon for a system of **coplanar** concurrent forces closes, it ensures that the **resultant force** is **zero**, but this **does not ensure non existence of a resultant couple in a coplanar non concurrent force system**. To ensure that the **resultant couple is also zero, the funicular polygon must also close**.

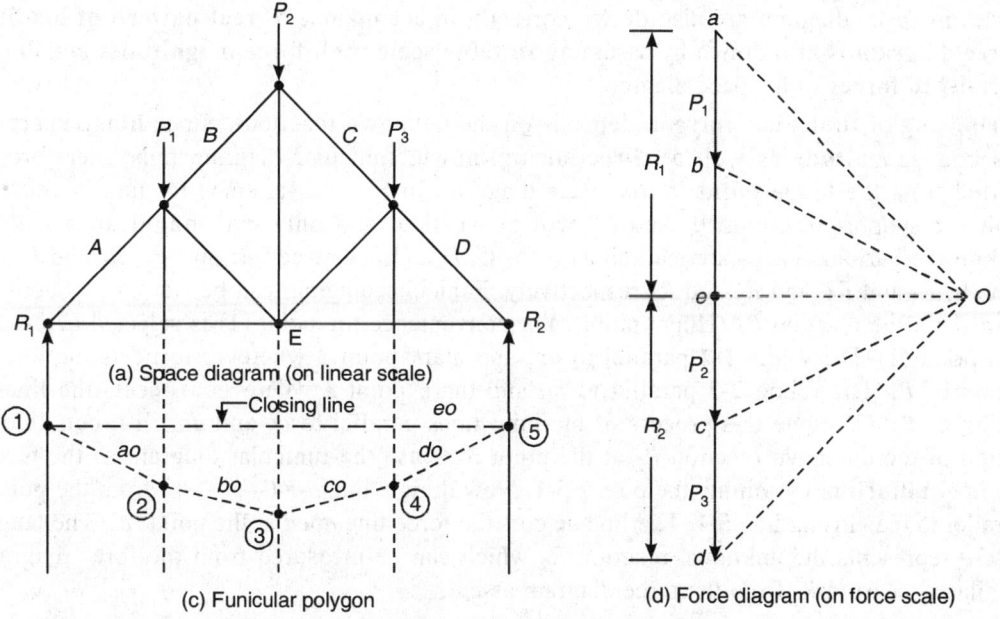

Fig. 4.21: Graphical determination of R_1 and R_2

Consider a plane truss subjected to loads as shown in Fig. 4.21. Loads P_1, P_2, P_3 are applied loads. R_1 and R_2 are the reactive forces offered by the end supports. This system of forces (P_1, P_2, P_3, R_1 and R_2) is an example of a **coplanar non-concurrent system**. To determine **unknown** (say R_1 and R_2) in this system, use of force polygon and funicular polygon can be made, provided the number of **unknown forces is not more than two** with a **maximum of three unknown quantities** (**two magnitudes** and **one direction**). This would mean that the **line of action** (direction) of one **unknown force must be known** and for the other **a point on its line of action** must be known.

In the example under consideration, P_1, P_2, P_3 are the **known** applied loads. R_1 and R_2 are the two **unknown reactive forces.** The **direction** of both these reactive forces are **known**. The number of unknowns in this example are therefore **two**, i.e. **the magnitudes of reactive forces R_1 and R_2.**

To determine these, first **name all the forces** using **Bow's notation** in the **space diagram** (Fig. 4.21a). Draw **vector ab** to represent force **AB** (P_1) as shown in space diagram) in **magnitude** (on some scale) and direction parallel to the force. From **b** draw **bc** to represent force **BC** (P_2) in **magnitude** (on the same assumed scale) and **direction** parallel to the force. From **c** draw **cd** to represent force **CD** (P_3) shown in space diagram in **magnitude** (on the same assumed scale) and **direction** parallel to the force. Choose a point **O** as a **pole** in **force polygon** (vector diagram) and draw lines **ao, bo, co** and **do** by joining points **a, b, c,** and **d** with the **pole O** in vector diagram. Each force in the vector diagram is now represented by its two components e.g. force **vector ab** is represented by its components **ao** and **bo** (Triangle law of forces). It may be noted that in **graphical** method the **space diagram** is drawn with some **suitable linear scale** in reference to **length** and **height** of the truss. **Directions** of forces in space diagram are also drawn correctly in accordance to **real pattern of loading**. **Force diagram** is also drawn by assuming **suitable scale** w.r.t. **force magnitudes** and drawn **parallel to forces** as in space diagram.

Drawing of **funicular polygon** depends on the unknown reactions. Since **hinge reaction** has both **magnitude** as well as **direction unknown**, funicular diagram must therefore be started from the **hinge point** in the space diagram. In Fig. 4.21a, since loading is vertical, both the support reactions R_1 and R_2 will be **vertical** and only two **magnitudes** will be **unknown.** Various spaces are marked as A, B, C, D and E between R_1 and P_1, P_1 and P_2, P_2 and P_3, P_3 and R_2, and R_2 and R_1 respectively. Funicular diagram can be started by choosing point **1** on the reaction R_1 (Hinge point). For convenience the point (1) is selected below the real point (1). Draw line **1-2** parallel to **ao**, and mark point **2** wherever it meets the line of action of P_1. Draw line **2-3** parallel to **bo** and mark point **3** wherever it meets the line of action of P_2. Continue this process of marking lines parallel to **co** and **do** till it cuts line of action of the unknown reaction R_2 at the point **5**. **Close** the funicular diagram as the forces are in **equilibrium** by joining the points **5-1**. Now draw a line **o-e** (——·——·——) from the **pole O** parallel to the closing line **5-1**. The line **oe** cuts the force line **abcd** at the point 'e'. The **length** of **d-e** represents the unknown reaction R_2 which can be measured from the force diagram. Similarly measure R_1 from the force diagram as **e-a**.

Practice Task: A system of coplanar forces as shown in Fig. 4.22 is in static equilibrium. Determine the unknown force F_1 and F_2 graphically.

Practice Task: A system of coplanar forces as shown in Fig. 4.22 is in static equilibrium. Determine the unknown force F_1 and F_2 graphically.

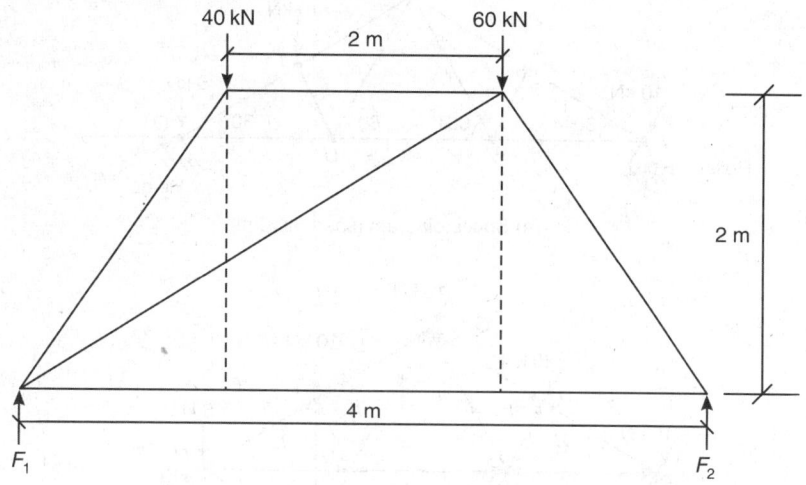

Fig. 4.22

4.5 DETERMINATION OF INTERNAL FORCES IN TRUSS MEMBERS

Having learned the concepts of Bow's notation in space diagram, force polygon and funicular polygon, we shall now apply these concepts for the determination of **internal forces** in the members of a determinate truss. Let a truss be loaded as shown in Fig. 4.23a.

i. The first step is to draw the **free body diagram** of the truss showing on it the **applied forces** and the **unknown reactive forces** at the supports (Fig. 4.23a). The **direction** of the reactive force R_P at support P (a roller) is **known** and is therefore, marked as vertical (**perpendicular** to roller) in the space diagram. However, the direction of R_q the **unknown reactive force** at the hinge support Q is **not known** and is therefore left as unmarked. Tentative reaction R_q may be marked inclined at an angle of θ with horizontal.

ii. Next name each of the applied and reactive force using Bow's notation (Fig. 4.23b) by marking spaces A, B, C, D, E, F and G on either side of the forces.

iii. Determination of reactive forces is done by drawing the funicular polygon by choosing some scale for force. Fig. 4.23c, shows **vectors ab, bc, cd, de, ef** representing the applied forces **AB, BC, CD, DE** and **EF** to selected scale. Choose a **pole O** in force or **vector diagram** and draw rays ao, bo, co, do, eo, and fo by joining a, b, c, d, e & f points with the pole O.

Start the drawing of funicular polygon from the hinge support point Q (Fig. 4.23b), the known point on the line of action of the **unknown reactive force** (with unknown magnitude and direction) at Q. Ray fo is between the forces **FG** (R_q) and **EF** (5 kN) and therefore point Q represents the point of intersection of these two forces. From Q, **draw Q-1′** in funicular diagram parallel to ray eo of vector diagram between the forces **EF** and **DE**, which meets the force line **DE** at the point **1′**. From **1′**, draw line **1′-2′** parallel to ray do between the force **DE** and **DC** meeting force line **DC** at the point **2′**. From **2′**

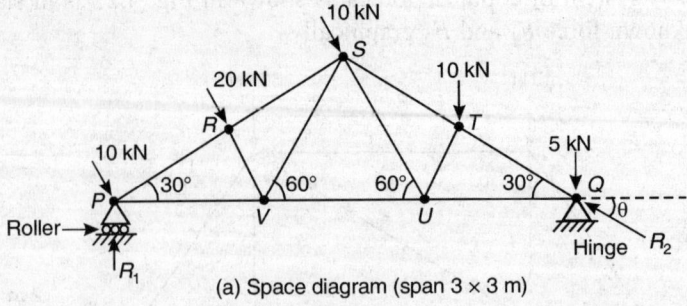

(a) Space diagram (span 3 × 3 m)

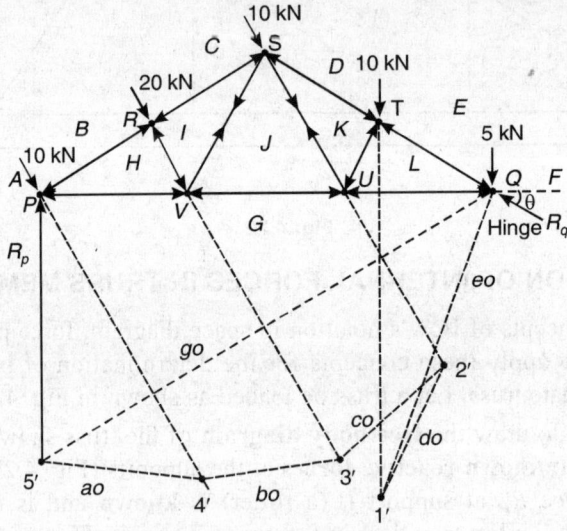

(b) FBD and funicular diagram

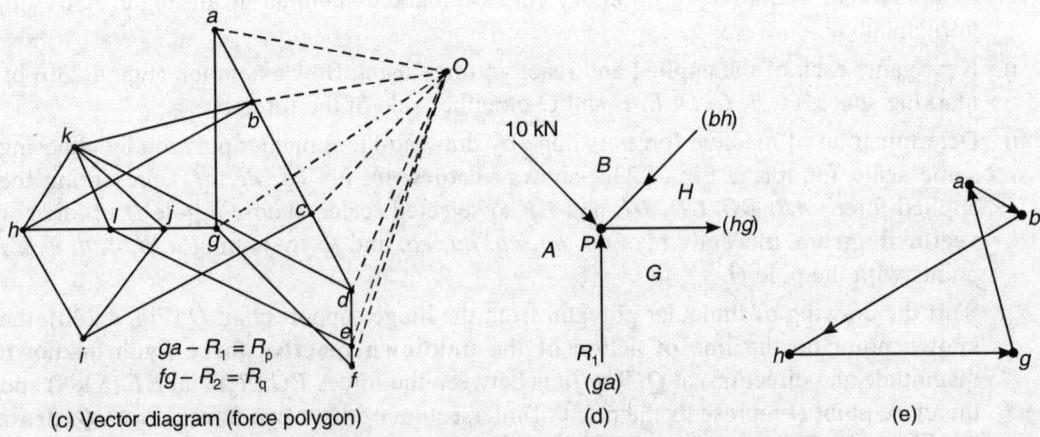

(c) Vector diagram (force polygon)

$ga - R_1 = R_p$
$fg - R_2 = R_q$

(d)

(e)

Fig. 4.23: Loaded truss, FBD, and vector diagram (force polygon)

draw **2'-3'** parallel to ray **co** between the forces **CD** and **BC** meeting force line *BC* at the point **3'**. From point **3'** draw line **3'-4'** parallel to ray **bo** between the forces **BC** and **AB** meeting force line AB at the point **4'**. From point **4'** draw line **4'-5'** parallel to ray **ao** between the forces **AB** and **AG** (R_P) meeting the line of action R_p at the point **5'**. As these set of forces are in **perfect equilibrium close the funicular polygon** by joining the starting point **Q and 5'** (Fig. 4.23b). From the point **O** in vector diagram draw a ray *og* parallel to **Q-5'** meeting line *ag* drawn parallel to the direction of reaction R_p cutting the line *ag* at the point *g* (Fig. 4.23c). Join *gf* to complete the force polygon (vector diagram) *abcdefga*. *fg* and *ga* in vector diagram represent the **magnitudes** and **directions** of the **reactive forces FG** and **GA** of space diagram respectively.

iv. The next step is to determine internal forces in each member of the truss. Name the forces in each member of the truss by using **Bow's notation**. As the part of the truss at each joint is also in **equilibrium**, we can separately draw the **FBD** of each joint and apply the polygon of forces to determine the **unknown internal force** in the members of the truss. We must however, be careful in selecting the joints for this purpose. The order followed by us in selecting these joints should be such that the number of **unknown member forces** is **not more than two** at any selected joint. Keeping this rule in mind we can start either at the support joint *P* or *Q*. Let us start at joint *P*. The **FBD** of the joint *P* is shown in Fig. 4.23d. *AB* and *GA* are the **two known forces** and *BH*, *HG* are the **two unknown internal forces** acting in the respective members of the truss. These unknown internal forces can be determined by completing the force polygon *abhga*. *ab* represents force *AB* in direction and certain assumed scale for magnitude. From *b* draw *bh* **parallel to BH**. From *g* draw *gh* **parallel to GH** cutting *bh* at *h*. Determine the nature of the internal forces *BH* and *HG*, by drawing vector diagram and measuring bh and hg and converting with the selected force scale. Mark the direction of the vector '*ab*' then *bh*, *hg* and finally *ga* in cyclic order respectively (Fig. 4.23e).

Force **BH** is thus a **compressive** force (**pressing** the joint) and **HG** is **tensile** force (**pulling** the joint).

Next we can select another joint where the unknowns are not more than two. Accordingly we shall select the **joint R** and draw the polygon of forces for the forces **BC (20 kN)**, **CI, IH** and **HB (known from the joint P)** at the **joint R**. Now select the **joints V, S, T, U** and complete the **vector diagram** for the whole of the truss (Fig. 4.23c).

v. Scale the values of internal forces from the vector diagram and write down internal forces in each member in a tabular form **(Table 4.2)**.

Table 4.2: Internal forces in the truss

Member	Name of Force Bow's Notation	Magnitude (kN)	Nature
PR	BH (*bh*)	36	Compressive
RS	CI (*ci*)	36	Compressive
ST	DK (*dk*)	33	Compressive
TQ	EL (*el*)	38	Compressive
UQ	GL (*gl*)	13	Tensile
UV	GJ (*gj*)	5.2	Tensile
PV	GH (*gh*)	26.5	Tensile
VR	HI (*hi*)	20	Compressive
TU	KL (*kl*)	8	Compressive
SV	IJ (*ij*)	20	Tensile
SU	JK (*jk*)	8	Tensile

EXAMPLE 4.4: Determine graphically internal forces in the members of a trussed structure loaded as shown in Fig. 4.24a.

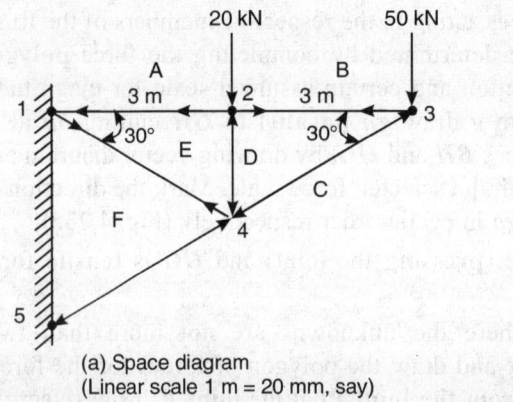

(a) Space diagram
(Linear scale 1 m = 20 mm, say)

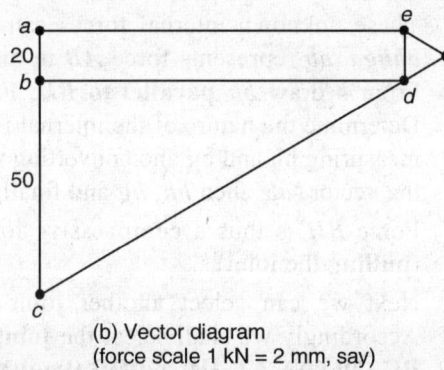

(b) Vector diagram
(force scale 1 kN = 2 mm, say)

Fig. 4.24.

Solution:

i. This is a cantilever truss with one end free and the other supported on hinges. In such a case we need not first determine the unknown reactive forces at hinge supports **1** and **5**. We can directly determine the unknown forces in the members by starting drawing of the **vector diagram from the free end** of the truss.

ii. Name forces in each member of the truss **in space diagram** using Bow's notation.

iii. Next select **joint 3** and draw the vector diagram of the forces **bc**, **cd** and **db** acting at this joint 3 parallel to the force *BC*, member *CD* and member *DB* respectively selecting suitable scale. The lines *bc*, *cd* and *db* represent respective forces in magnitude with assumed scale and direction. Next select joint **2** and draw the force polygon for the forces *AB*, *BD*, *DE* and *EA*. Next take up joint **4** and draw **force polygon** for the forces **ED**, **DC**, **CF** and **EF**. Measure lengths of vectors *ed*, *dc*, *cf* and *ef* and convert by the assumed scale.

Forces in the members of the truss are **measured** from the vector diagram and their magnitude and nature are shown in a tabular form in Table 4.3.

Table 4.3: Internal forces in the truss

Member	Name of force	Magnitude	Nature
1 – 2	AE	86.6 kN	Tensile
2 – 3	BD	86.6 kN	Tensile
3 – 4	DC	100 kN	Compressive
4 – 5	FC	120 kN	Compressive
2 – 4	ED	20 kN	Compressive
4 – 1	EF	20 kN	Tensile

Practice Task: Determine graphically the forces in the members of a truss loaded as shown in Figs 4.25 and 4.26.

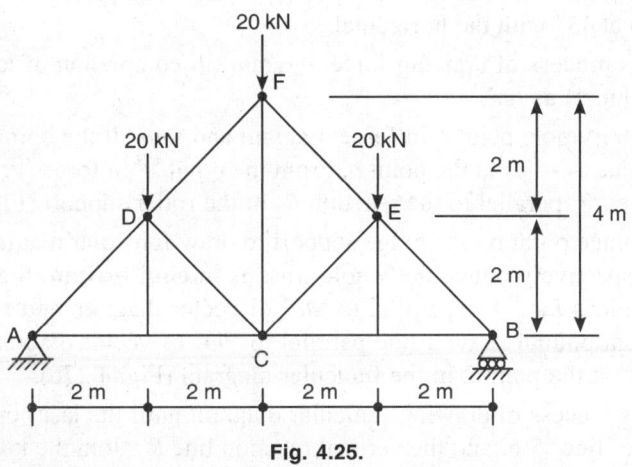

Fig. 4.25.

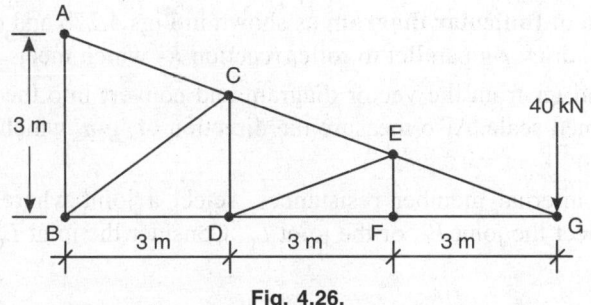

Fig. 4.26.

EXAMPLE 4.5: A fink truss of 10 m span and 3m height carrying resultant loads acting at various panel points as shown in Fig. 4.27a. Determine support reactions and internal resistances in members: L_2L_3, L_3L_4, U_2L_3, U_2U_3, U_3L_3, U_3U_4 and L_3U_4.

Solution: Hints – Graphical Method

   i. Draw the truss space diagram selecting suitable linear scale, (say 1 m = 10 mm) for graphical method. Mark the support reactive forces R_1 and R_2 according to the nature of support (Hinge and roller respectively).

  ii. Mark the spaces between various forces and reactions as indicated in space diagram (draw with linear scale) shown in Fig. 4.27a, as **A, B, C, ...G** in clockwise direction. Also mark the internal spaces as **H, I, J, K, L, M, N, P,** and **Q** in clockwise order (Fig. 4.27b).

 iii. Assume a suitable force scale for drawing vector or **force diagram** (1kN = 4mm, say).

  iv. Draw force diagram taking **force AB** (15 kN) as $a - b = 15 \times 4 = 60$ mm and parallel to **AB** of space diagram i.e. horizontal line 'a' to 'b' measuring 60 mm (Fig. 4.27c).

   v. From 'b' draw a line 'bc' of **50 mm** (12.5 × 4) length and parallel to **force BC** (12.5 kN) inclined at **30°** with the horizontal **line 'a-b'**.

  vi. From 'c' draw a line 'cd' of 70 mm (17.5 × 4) length and parallel to the force **CD** (17.5 kN) inclined at **45°** with the horizontal.

 vii. Continue this process of drawing force diagram till completion of last known force EF of 10 kN (40 mm) as 'ef'.

viii. Select a arbitrary pole point o in force diagram and join all the points a, b, c, d, e and f with dotted line (-------) to the point o. From the point 'f' in force (vector) diagram draw a vertical line 'fg' parallel to the reaction R_2 at the roller support (Fig. 4.27c).

  ix. Choose the hinge point on the hinge support to draw funicular diagram $L_0 - 1 - 2 - 3 - 4 - 5 - 6$ respectively. Since the whole truss is in **equilibrium**, the funicular diagram must **close**. Lines $L_0 - 1$ is parallel to 'ao' of vector diagram and meets the force line AB at 1. From point 1 draw a line parallel to 'bo' of vector diagram which meets the force line BC at the point 2 in the funicular diagram (Fig. 4.27b).

   x. Continue this process of drawing funicular diagram until the last point '6' is arrived by intersection of line '5-6' and the vertical reaction line R_2. Join the last point '6' with the starting point 'L_0' by a **chain** line. Now draw a line 'og' from pole o in force diagram parallel to L_0-6 of **funicular diagram** as shown in Figs 4.27b and c. From the point f in vector diagram draw *f-g* parallel to roller reaction R_2 which meets o-g at 'g'.

  xi. Measure *fg* and **ga** from the vector diagram and convert into the values of R_2 and R_1 from the assumed scale. Also measure the direction of 'g-a' which represents the slope of R_1.

 xii. To determine internal member resistances, select a joint where there are only two unknowns. Select the joint L_0, or the joint L_6. Consider the joint L_6.

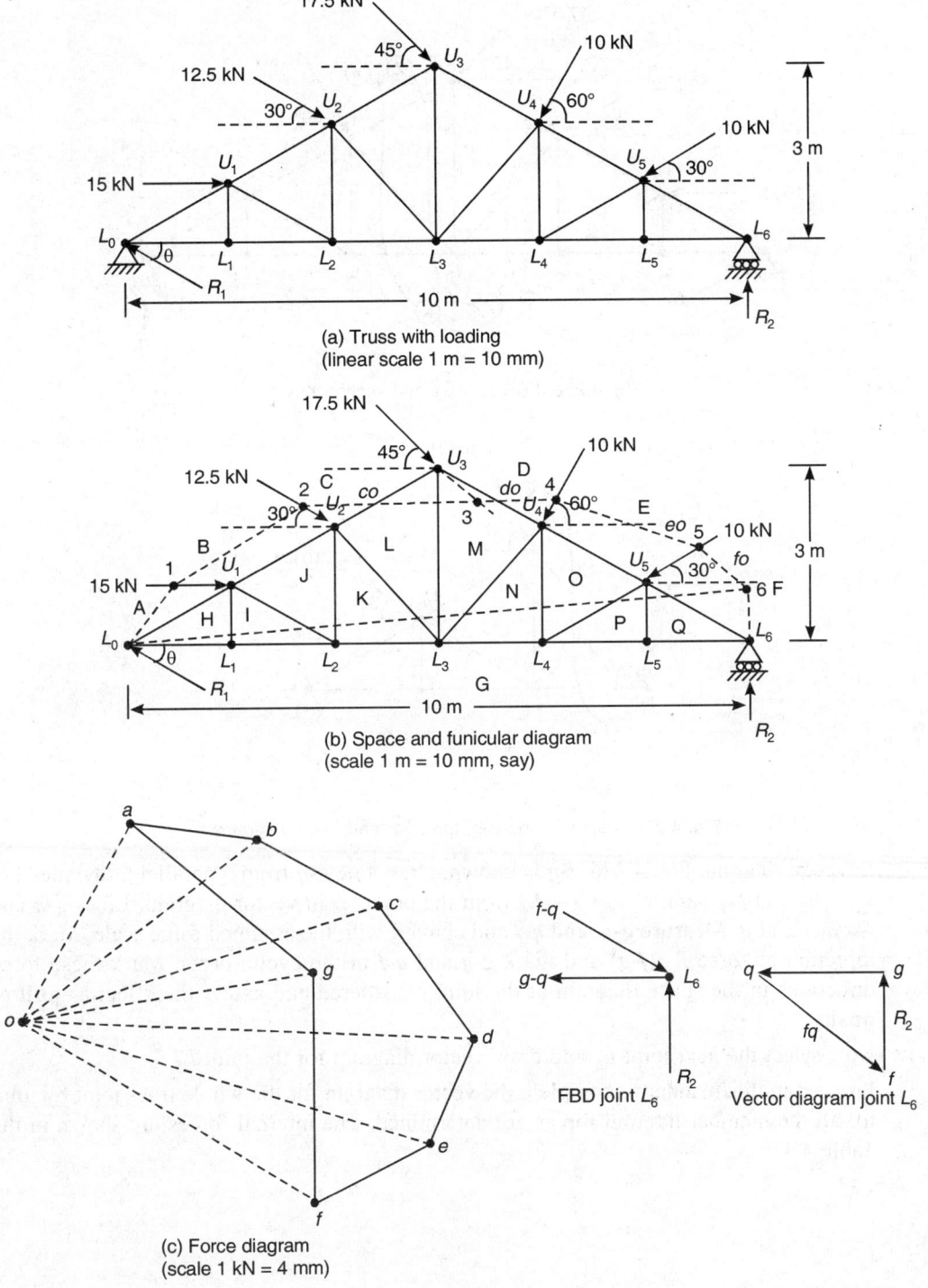

(a) Truss with loading
(linear scale 1 m = 10 mm)

(b) Space and funicular diagram
(scale 1 m = 10 mm, say)

(c) Force diagram
(scale 1 kN = 4 mm)

FBD joint L_6

Vector diagram joint L_6

Fig. 4.27d.

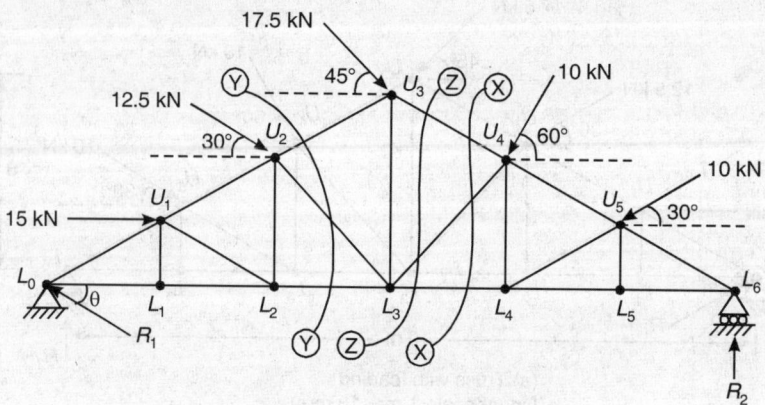

Fig. 4.27e: FBD of truss and section X-X

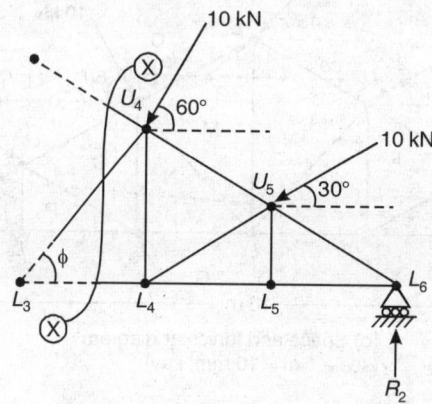

Fig. 4.27: Truss with loading, funicular and force diagrams

xiii. In vector diagram Fig. 4.27d, *f-g* is known. Draw line **g-q** from g parallel to the member $L_5 - L_6$ and *f-q* parallel to $U_5 - L_6$ from the point '*f*' in vector diagram. Lines **g-q** and *f-q* meets at *q*. Measure **g-q**, and *q-f* and convert with the assumed force scale. Mark the direction of force R_2 (*f-g*) and mark **g-q** and *q-f* in the cyclic order. Mark these force directions in the space diagram at the joint considered and assess the effect as **pull** or **push**.

xiv. Now select the next joint L_5 and draw vector diagram for the joint L_5.

xv. Proceed in this manner to complete the vector diagram for the whole truss joint by joint till all the member internal forces are determined. The internal forces are shown in the Table 4.4.

Table 4.4: Internal forces in truss members

S. No.	Member	Internal Force		Nature
		Symbol	Magnitude	
1	L_0U_1	a-h	17 kN	Compressive
2	U_1U_2	b-j	25.5 kN	Compressive
3	U_2U_3	c-l	27.5 kN	Compressive
4	U_3U_4	d-m	41.5 kN	Compressive
5	U_4U_5	e-o	46.5 kN	Compressive
6	U_5L_6	f-q	46.0 kN	Compressive
7	L_0L_1	g-h	39.0 kN	Tensile
8	L_1L_2	g-i	39.0 kN	Tensile
9	L_2L_3	g-k	31.5 kN	Tensile
10	L_3L_4	g-n	31.0 kN	Tensile
11	L_4L_5	g-l	39.5 kN	Tensile
12	L_5L_6	g-q	39.5 kN	Tensile
13	L_1U_1	h-i	0 kN	-----
14	L_2U_1	i-j	9.0 kN	Compressive
15	L_2U_2	j-k	4.5 kN	Tensile
16	L_3U_2	k-l	15.0 kN	Compressive
17	L_3U_3	l-m	23.0 kN	Tensile
18	L_3U_4	m-n	15.0 kN	Compressive
19	L_4U_4	n-o	5.0 kN	Tensile
20	L_4U_5	o-p	10.0 kN	Compressive
21	L_5U_5	p-q	0 kN	-----

METHOD OF SECTIONS

Consider imaginary cut sections X-X, Y-Y and Z-Z for the analysis of internal forces in desired members (Fig. 4.27e). First determine reactions by considering FBD of the truss. Resolve applied forces **horizontally** and **vertically** as applied on various panel points.

Joint U_1

Horizontal Component = 15.0 kN, Vertical Component = 0,

Joint U_2

Horizontal Component = 12.5 cos 30° = 12.5 × 0.866 = **10.83 kN**

Vertical Component = 12.5 sin 30° = 12.5 × 0.50 = **6.25 kN**

Joint U_3

Horizontal Component = 17.5 cos 45° = 17.5 × 0.7071 = **12.37 kN**

Vertical Component = 17.5 sin 45° = 17.5 × 0.707 = **12.37 kN**

Joint U_4

Horizontal Component = 10 cos 60° = 10 × 0.5 = **5.0 kN**

Vertical Component = 10 sin 60° = 10 × 0.866 = **8.66 kN**

Joint U_5

Horizontal Component = 10 cos 30° = 10 × 0.866 = **8.66 kN**

Vertical Component = 10 sin 30° = 10 × 0.5 = **5.0 kN**

Each panel span = 10/6 = 1.666 m

Vertical height of panel points = 3/3 = 1 m, 2 m & 3 m.

Slope of Truss: $\tan \theta_1 = 3/5, \sin \theta_1 = 3/\sqrt{34} = 0.5145,   \cos \theta_1 = 0.8575$

Reaction Determination

Take moment about the joint L_0

$$
\begin{aligned}
R_2 \times 10 &= 15 \times 1 + 10.83 \times 2 + 6.25 \times 3.333 + 12.37 \times 3 + 12.37 \\
&\quad \times 5 - 5 \times 2 + 8.66 \times 6.333 - 8.66 \times 1 + 5.0 \times 8.667 \\
&= 15 + 21.66 + 20.83 + 37.11 + 61.65 - 10 + 54.82 - 8.66 \\
&\quad + 43.335 \\
&= 23.5945 \\
R_2 &= 23.60 \text{ kN} \\
R_{1H} &= 15 + 10.83 + 12.37 - 5 - 8.66 = \textbf{24.54 kN} \ (\leftarrow) \\
R_{1V} &= 0 + 6.25 + 12.37 + 8.66 + 5 - 23.60 = \textbf{8.68 kN} \ (\uparrow) \\[4pt]
R_1 &= \sqrt{(8.68)^2 + (24.54)^2} = \sqrt{75.34 + 602.21} = \textbf{26.03 kN}
\end{aligned}
$$

$$
\tan \theta = \frac{8.68}{24.54}, \quad \boldsymbol{\theta = 19.48^\circ}, \quad \sin \theta = 0.3335, \quad \cos \theta = 0.9428
$$

$$
\tan \phi = \frac{2.0}{\dfrac{5}{3}} = 1.2, \quad \phi = \textbf{50.19}^\circ, \quad \sin \phi = 0.7682, \quad \cos \phi = 0.6402
$$

Consider section X-X and FBD of RHS to determine internal forces in cut members (Fig. 4.27e).

Take **moment** of all forces **about** L_6 for RHS FBD

$$
P_{U_4U_3} \times 0 + P_{L_3L_4} \times 0 + P_{U_4U_3} \times 5 \sin \phi + \left(10 \sin 60 \times \frac{2}{3} \times 5 + 10 \cos 60 \times 2\right) + 10 \sin 30 \times \frac{5}{3}
$$

$$
+ 10 \cos 30 \times 1 = 0
$$

$$
P_{U_4U_3} = -\frac{1}{5 \times 0.7682}\left[\frac{8.66 \times 10}{3} + 5 \times 2 + \frac{5 \times 5}{3} + 8.66\right] = \frac{-1}{3.8410}[28.87 + 10 + 8.33 + 8.66]
$$

$$
= [28.87 + 10 + 8.33 + 8.66]
$$

$$
= -14.5 \text{ kN (Compressive)}
$$

Take **moments** about U_4

$$
P_{L_3L_4} \times 2 - 23.6 \times \frac{2 \times 5}{3} + 10 \cos 30 \ (2 - 1) + 10 \sin 30 \times \frac{5}{3} = 0,
$$

$$P_{L_3L_4} = (78.667 - 8.66 - 8.33)\frac{1}{2} = \frac{61.7}{2} = + \textbf{30.85 kN (Tensile)}$$

Take **moment about** L_3

$$P_{U_3U_4} \times 5 \sin \theta_1 + 10 \cos 60 \times 2 - 10 \sin 60 \times \frac{5}{3} + 23.6 \times 5 - 10 \sin 30 \times \frac{2 \times 5}{3} + 10 \cos 30 \times 1 = 0$$

$$P_{U_3U_4} = \left\{ \frac{50}{3} \times 0.866 + \frac{100}{3} \times 0.5 - 20 \times 0.5 - 23.6 \times 5 - 10 \times 0.866 \right\} \frac{1}{5 \times 0.5145}$$

$$= -41.03 \text{ kN (Compressive)}$$

Similarly proceed with other sections to determine internal forces in other members. The internal forces in these members are given in Table 4.5 as calculated by method of sections.

Table 4.5: Internal forces in selected members

S. No.	Member	Internal Force		Remarks
		Magnitude (kN)	Nature	
1	L_2L_3	+ 31.5	Tensile	Same as in graphical Method
2	L_3L_4	+30.85	Tensile	Approximately as Table 4.4
3	U_2L_3	− 14.50	Compressive	Approximately as Table 4.4
4	U_2U_3	− 28.0	Compressive	Approximately as Table 4.4
5	U_3L_3	23.0	Tensile	Approximately as Table 4.4
6	U_3U_4	− 41.10	Compressive	Approximately as Table 4.4
7	L_3U_4	− 14.50	Compressive	Approximately as Table 4.4

EXAMPLE 4.6: A 6 m length truss shown in Fig. 4.28a, carries vertical loads and supported on a hinge and a roller support. Calculate internal forces in all the members.

Solution: Hints: Since internal forces are required to be determined in all the members, any method of analysis can be employed due to symmetry of truss and loading.

Support Reactions: Length of each panel = 2m

Symmetrical : $\qquad R_1 = R_2 = \dfrac{20 \times 2 + 10 \times 3}{2} = 35$ kN each.

Also taking moments about L_0,

$$R_2 \times 6 = 10 \times 2 + 10 \times 4 + 20 \times \frac{3}{2} + 10 \times 3 + 20 \times (6 - 1.5)$$

$$R_2 = \frac{210}{6} = \textbf{35 kN} = R_1$$

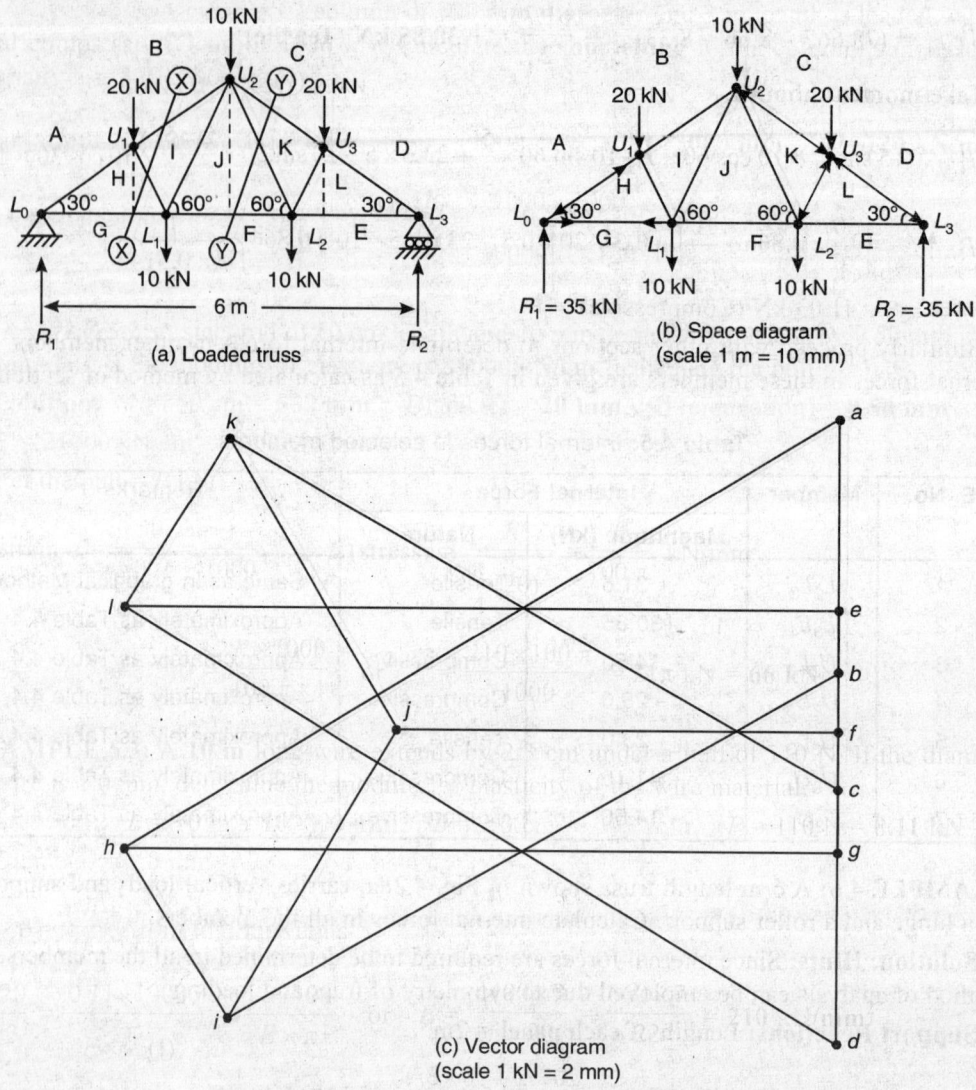

(a) Loaded truss

(b) Space diagram
(scale 1 m = 10 mm)

(c) Vector diagram
(scale 1 kN = 2 mm)

Fig. 4.28: Fink truss with loads, and vector diagrams

Consider Joint L_3, using **method of joints**

$$P_{U_3L_3} \sin 30 + 35 = 0, P_{U_3L_3} = -35 \times 2 = -\textbf{70 kN} \text{ (Compressive)}$$

$$P_{U_2L_3} - 70 \cos 30 = 0, P_{U_2L_3} = 70 \times 0.866 = \textbf{60.62 kN} \text{ (Tensile)}$$

Continuing the process, the internal forces can be found as given in Table 4.6.

Consider the section X-X. Take **moment about Joint U_1**, considering FBD of LHS section X-X.

$$P_{L_0L_1} \times 1.5 \tan 30 - R_1 \times 1.5 \text{ m} = 0,$$

$$P_{L_0L_1} = \frac{35 \times 1.5}{1.5 \tan 30} = 35\sqrt{3} = \textbf{60.62 kN (Tensile)}$$

Take **moment about the joint L_0** consider LHS FBD of section X-X.

$$P_{U_1L_1} \times 2 \sin 60 + 20 \times 1.5 = 0,$$

$$P_{U_1L_1} = \frac{30}{2 \times 0.866} = -17.32 \text{ kN (Compressive)}$$

Take **moment about L_1** considering **LHS FBD** (Section X-X)

$$P_{U_1U_2} \times 2 \sin 30 - 20(2 - 1.5) + 35 \times 2 = 0,$$

$$P_{U_1U_2} = \frac{-70 + 10}{1} = -\textbf{60 kN (Compressive)}$$

Process can be continued and internal forces can be found as in Table 4.6.

Table 4.6: Internal forces in truss members (Fig. 4.28)

S. No.	Member	Internal force		Remarks
		Magnitude (kN)	**Nature**	
1	$L_0L_1 = L_2L_3$	60.6	Tensile	
2	L_1L_2	37.5	Tensile	
3	$L_0U_1 = L_3U_3$	70.0	Compressive	
4	$L_1U_1 = L_2U_3$	17.32	Compressive	
5	$U_1U_2 = U_2U_3$	60	Compressive	
6	$L_1U_2 = L_2U_2$	28.87	Tensile	

These values may also be found by **graphical method** by drawing **vector diagram** by marking spaces in space diagram and assuming **suitable scales** for space diagram (1m = 10 mm) vector diagrams (1kN = 2mm). Complete the vector diagram considering joint by joint and starting from the joint L_0 or L_3. The values can be scaled from the vector diagram as shown in Figs 4.28b and c. Check the values from the Table 4.6.

EXAMPLE 4.7: A parallel member pratt truss of 6 equal panels of 18 m total span and 4m height carries loads on lower panel points as shown in Fig. 4.29a.

Solution: Hints:

Since loading and the truss are symmetrical and the loads are only vertical,

$$R_1 = R_2 = \frac{5 \times 20}{2} = \textbf{50 kN}$$

Slope of panels: $\tan \phi = \dfrac{4}{3}$, $\sin \phi = \dfrac{4}{5}$, $\cos \phi = \dfrac{3}{5}$

Internal forces in truss members can be found by **method of joints** starting from the joints L_0 or L_6. This can also be done by considering FBD of **LHS of the section X-X or Y-Y**.

Consider the joint L_0:

$$\Sigma V = 0, \quad P_{L_0 U_1} \sin \phi = 0, \quad P_{L_0 U_1} = -\dfrac{50}{\dfrac{4}{5}} = -\ 62.5\ \text{kN (Compressive)}$$

$$\Sigma H = 0, \quad P_{L_0 L_1} - 62.5 \cos \phi = 0, \quad P_{L_0 L_1} = 62.5 \times \dfrac{3}{5} = +\ 37.5\ \text{kN (Tensile)}$$

Consider the joint L_1

$$\Sigma V = 0, \quad P_{L_1 U_1} - 20 = 0, \quad P_{L_1 U_1} = +\ 20\ \text{kN (Tensile)}$$

$$\Sigma H = 0, \quad P_{L_1 L_2} - 37.5 = 0, \quad P_{L_1 L_2} = +\ 37.5\ \text{kN (Tensile)}$$

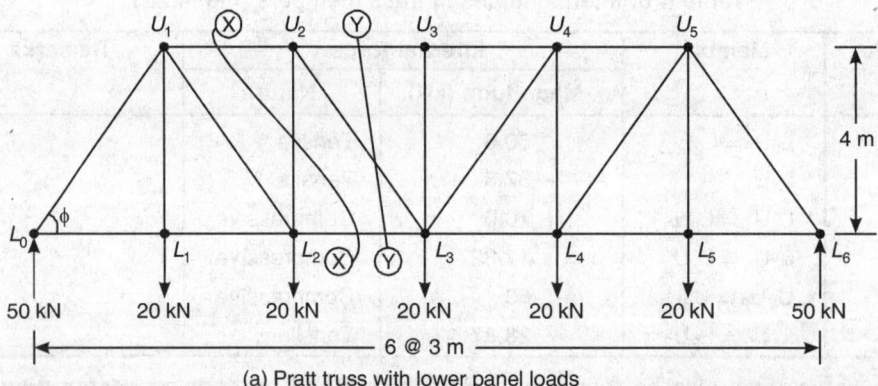

(a) Pratt truss with lower panel loads

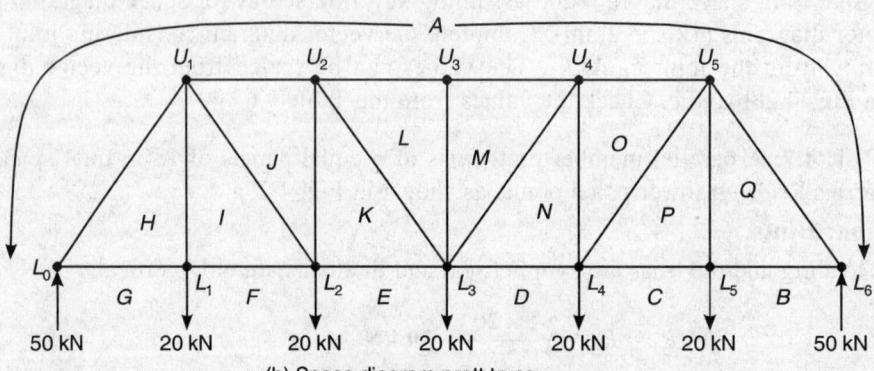

(b) Space diagram-pratt truss
(scale 1m = 5 mm)

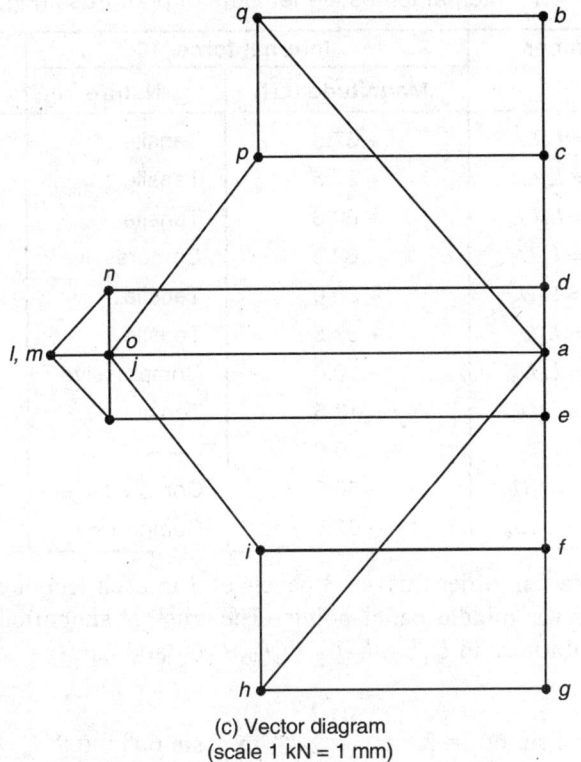

(c) Vector diagram
(scale 1 kN = 1 mm)

Fig. 4.29: Pratt truss, FBD and vector diagram

Consider section X-X

Moments about L_2 for LHS of X-X FBD:

$\Sigma M = 0, \quad P_{U_1 U_2} \times 4 + 50 \times 6m - 20 \times 3 - 50 \times 6 = 0,$

$$P_{U_1 U_2} = -\frac{240}{4} = -60 \text{ kN (Compressive)}$$

$\Sigma M = 0$, about U_2: $\quad P_{L_2 L_3} \times 4 + 20 \times 3 - 50 \times 6 = 0, \quad P_{L_2 L_3} = +60 \text{ kN (Tensile)}$

$\Sigma V = 0$, Left of X-X: $\quad P_{L_2 U_2} + 50 - 20 - 20 = 0, \quad P_{L_2 U_2} = -10 \text{ kN (Compressive)}$

Consider section Y-Y and LHS FBD

ΣM about $L_3 : 0 = 20 \times 3 + 20 \times 6 - 50 \times 9 - P_{U_2 U_3} \times 4, \qquad \boldsymbol{P_{U_2 U_3} = -\dfrac{270}{4} = -67.5 \text{ kN}}$

ΣM about $U_2 : 0 = 50 \times 6 - 20 \times 3 - P_{L_2 L_3} \times 4, \qquad \boldsymbol{P_{L_2 L_3} = \dfrac{240}{4} = +60 \text{ kN}}$

Graphical method can also be applied by drawing **space** and **vector diagrams** as shown in Figs 4.29b and c.

Table 4.7: Internal forces in members of pratt truss (Fig. 4.29)

S. No.	Member	Internal force		Remarks
		Magnitude (kN)	**Nature**	
1	$L_0L_1 = L_5L_6$	+ 37.5	Tensile	
2	$L_1L_2 = L_4L_5$	+ 37.5	Tensile	
3	$L_2L_3 = L_3L_4$	+ 60.0	Tensile	
4	$L_0U_1 = L_6U_5$	− 62.5	Compressive	
5	$L_1U_1 = L_5U_5$	+ 20.0	Tensile	
6	$L_2U_1 = L_4U_5$	+ 37.5	Tensile	
7	$L_2U_2 = L_4U_4$	− 10.0	Compressive	
8	$L_3U_2 = L_3U_4$	12.5	Tensile	
9	L_3U_3	0.0	——-	
10	$U_1U_2 = U_4U_5$	− 60.0	Compressive	
11	$U_2U_3 = U_3U_4$	− 67.5	Compressive	

EXAMPLE 4.8: A warren girder truss of 3 panels of 4 m each length carries loads of 75 kN and 45 kN at two lower middle panel points. The truss is supported on simple supports. Determine the internal forces in U_1U_2, U_2U_3, L_1L_2, L_1U_2 and L_2U_2.

Solution: Hints:

Vertical distance = $4 \sin 60^\circ = 2 \times \sqrt{3} = 2\sqrt{3}$ **m**, $\sin 60^\circ = 0.866$, $\cos 60^\circ = 0.50$

Determine reactions analytically or graphically.

Moment about L_0, $R_2 \times 12 - 75 \times 4 - 45 \times 8 = 0$, $R_2 = \dfrac{660}{12} = \textbf{55 kN}$

Moment about L_3, $R_1 \times 4 \times 3 - 75 \times 8 - 45 \times 4 = 0$, $R_1 = \dfrac{780}{12} = \textbf{65 kN}$

Method of sections: Consider section X-X, FBD of LHS,

Moment about U_2: $0 = P_{L_1L_2} \times 2\sqrt{3} - 65(4 + 2) + 75 \times 2$,

$$P_{L_1L_2} = = 40\sqrt{3} = + \textbf{69.3 kN (Tensile)}$$

Moment about L_1: $65 \times 4 + P_{U1U_2} \times 2\sqrt{3} = 0$,

$$P_{U_1U_2} = -\frac{260}{2\sqrt{3}} = -\textbf{75.05 kN (Compressive)}$$

Resolving Vertically: $P_{L_1U_2} \sin 60 - 75 + 65 = 0$,

$$P_{L_1U_2} = +\frac{10 \times 2}{\sqrt{3}} = +\textbf{11.55 kN (Tensile)}$$

Consider section Y-Y and analyze other members.

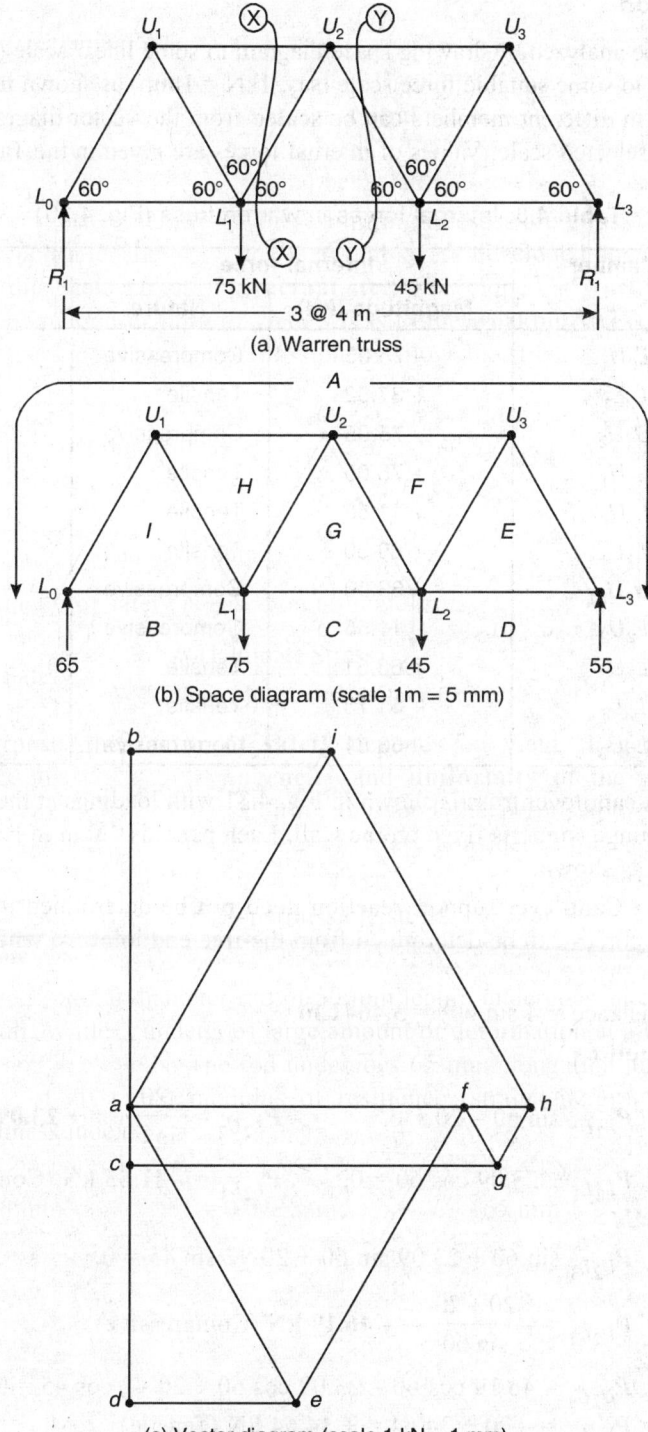

(a) Warren truss

(b) Space diagram (scale 1m = 5 mm)

(c) Vector diagram (scale 1 kN = 1 mm)

Fig. 4.30: Warren truss, space and vector diagram

Graphical Method

The truss can also be analyzed by drawing space diagram to some linear scale (say, 1m = 10 mm) and vector diagram to some suitable force scale (say, 1kN = 1mm) as shown in Figs 4.30b and c. The internal forces in different members can be **scaled** from the vector diagram Fig. 4.30c, and converted with the selected scale. Values of internal forces are given in the Table 4.8.

Table 4.8: Internal forces in warren truss (Fig. 4.30)

S. No.	Member	Internal force		Remarks
		Magnitude (kN)	**Nature**	
1	L_0U_1	− 75.05	Compressive	
2	L_0L_1	+ 37.52	Tensile	
3	U_1U_2	− 75.05	Compressive	
4	L_1U_1	+ 75.05	Tensile	
5	L_1U_2	+ 11.55	Tensile	
6	L_1L_2	+ 69.30	Tensile	
7	U_2U_3	− 63.50	Compressive	
8	L_2U_2	− 11.55	Compressive	
9	L_2U_3	+ 63.51	Tensile	
10	L_2L_3	+ 31.75	Tensile	
11	L_3U_3	− 63.51	Compressive	

EXAMPLE 4.9: A cantilever truss is shown in Fig. 4.31 with loadings at the panel points. The truss rests on two hinge supports fixed on the wall. Each panel is 4.0 m in length and members are inclined at $60°$ as shown.

Solution: Hints: Cantilever support reaction **need not** be determined in the beginning as **unknown** member forces can be determined from the free end joint L_2, **where the unknowns are only two.**

Perpendicular distance = $4 \sin 60° = $ **3.4641 m**

Consider the joint L_2

$\Sigma V = 0$, $\qquad P_{L_2U_2} \sin 60 - 20 = 0$, $\qquad P_{L_2U_2} = \dfrac{20}{\sin 60} = $ **+ 23.09 kN** (Tensile)

$\Sigma H = 0$, $\qquad P_{L_2L_1} + 23.09 \cos 60 = 0$, $\qquad P_{L_2L_1} = $ **− 11.55 kN** (Compressive)

Consider joint U_2

$\Sigma V = 0$, $\qquad P_{U_2L_1} \sin 60 + 23.09 \sin 60 + 20 \sqrt{2} \sin 45 = 0$,

$$P_{U_2L_1} = -\dfrac{20 + 20}{\sin 60} = \text{− 46.19 kN (Compressive)}$$

$\Sigma H = 0$, $\qquad P_{U_2U_1} - 46.19 \cos 60 - 23.09 \cos 60 + 20 \sqrt{2} \cos 45 = 0$,

$\qquad\qquad P_{U_2U_1} = - 20 + 34.64 = $ **+ 14.64 kN** (Tensile)

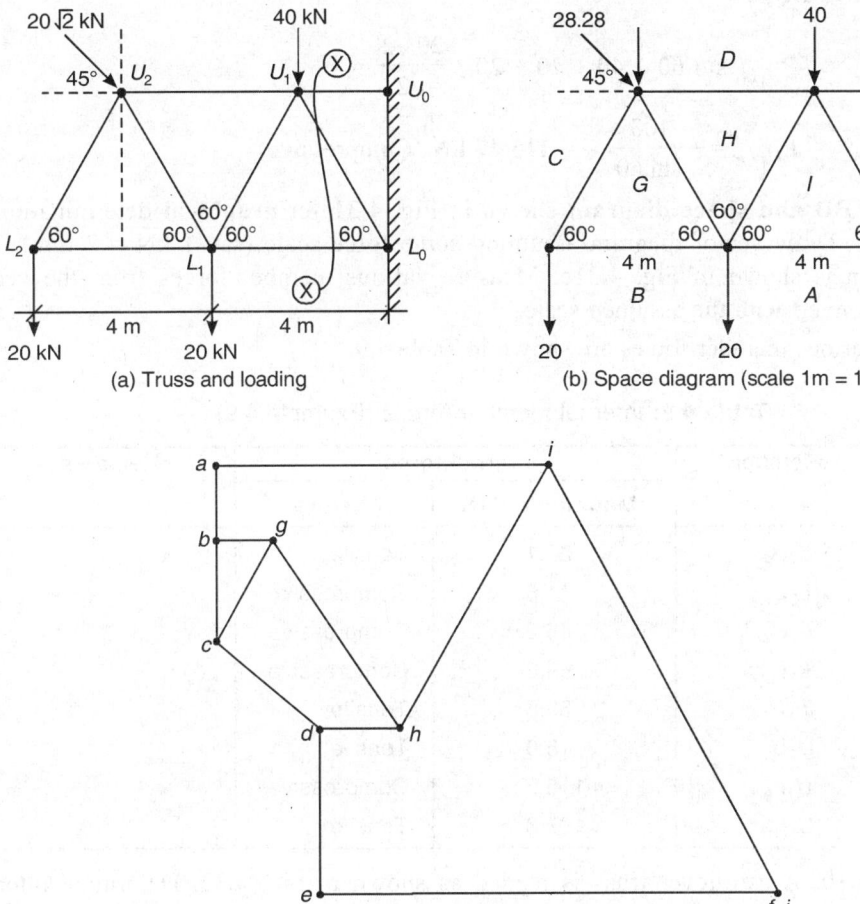

Fig. 4.31: Cantilever truss with loads, FBD and vector diagram

Consider LHS of section X-X FBD

$\Sigma M = 0$ at U_1, $P_{L_1 L_0} \times 4 \sin 60 + 20 (2 + 6) + 20 \sqrt{2} \sin 45 \times 4 = 0$,

$$P_{L_1 L_0} = -\frac{240}{4 \sin 60} = -\textbf{69.28 kN (Compressive)}$$

$\Sigma M = 0$ at L_0,

$$P_{U_1 U_0} \times 4 \sin 60 - 40 \times 2 - 20 \times 4 - 20 \times 8 - 20 \times 6 + 20 \times 4 \sin 60 = 0$$

$$P_{U_1 U_0} = \frac{(440 - 69.3)}{4 \sin 60} = +\textbf{107.80 kN (Tensile)}$$

$\Sigma V = 0$, **LHS FBD**,

$$P_{U_1 L_0} \sin 60 + 40 + 20 + 20 + \frac{20\sqrt{2}}{\sqrt{2}} = 0,$$

$$P_{U_1 L_0} = -\frac{100}{\sin 60} = -115.47 \text{ kN (Compressive)}$$

Consider FBD and space diagram shown in Fig. 4.31 for **graphical** determination of internal forces. Draw vector diagram assuming some force scale (say 1 kN = 2 mm). The vector diagram is shown in Fig. 4.31c. Measure various member forces from the vector diagram and convert with the assumed scale.

Values of various member forces are shown in Table 4.9.

Table 4.9: Internal member forces (Example 4.9)

S. No.	Member	Internal force		Remarks
		Magnitude (kN)	Nature	
1	$L_2 U_2$	23.1	Tensile	
2	$L_2 L_1$	11.6	Compressive	
3	$L_1 U_2$	46.2	Compressive	
4	$L_1 L_0$	69.3	Compressive	
5	$L_1 U_1$	69.6	Tensile	
6	$U_2 U_1$	15.0	Tensile	
7	$U_1 L_0$	115.5	Compressive	
8	$U_1 U_0$	107.8	Tensile	

EXAMPLE 4.10: A cantilever truss is loaded as shown in Fig. 4.32. Determine internal resistance in members $U_2 U_3$, $U_2 L_2$ and $L_1 L_2$.

Solution: Hints:

$$\tan \theta = \frac{2}{4}, \quad \sin \theta = \frac{1}{\sqrt{5}}, \quad \cos \theta = \frac{2}{\sqrt{5}}$$

For finding internal stresses in members $U_2 U_3$, $U_2 L_2$ and $L_1 L_2$ the shortest and simplest approach is **method of sections**.

Consider LHS FBD of section X-X (Fig. 4.32a):

Moment about U_2:

$\Sigma M = 0$, $\qquad 30 \times 4 + P_{L_1 L_2} \times 2 = 0$, $\qquad P_{L_1 L_2} = -60 \text{ kN (Compressive)}$

Moment about L_2:

$\Sigma M = 0$, $\qquad (30 + 10) 4 + 30 \times 8 - P_{U_2 U_3} \times 4 \sin (90 - \theta) = 0$

$$P_{U_2 U_3} = +\frac{400\sqrt{5}}{4 \times 2} = 50\sqrt{5} = +111.8 \text{ kN}$$

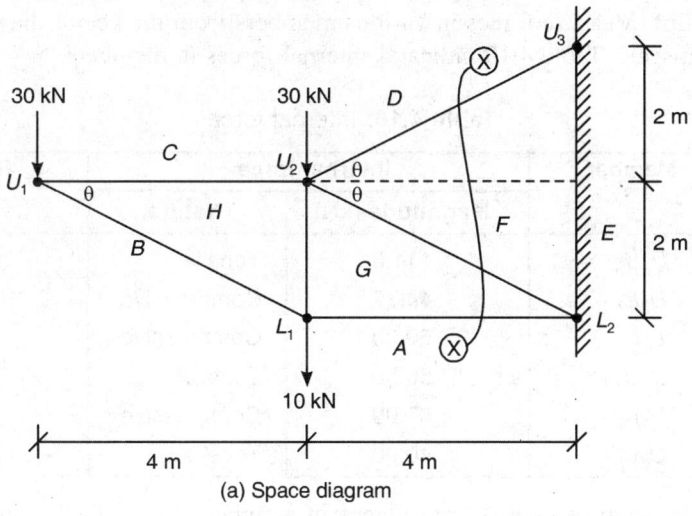

(a) Space diagram

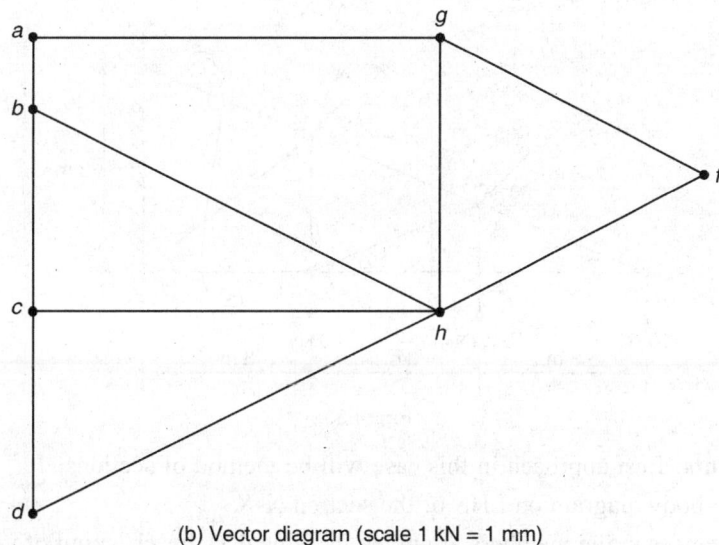

(b) Vector diagram (scale 1 kN = 1 mm)

Fig. 4.32: Cantilever truss and vector diagram

Resolving along horizontal all forces on LHS of section X-X:

$$P_{U_2U_3} \cos \theta + P_{U_2L_2} \cos \theta + P_{L_1L_2} = 0,$$

$$P_{U_2L_2} = -111.8 + \frac{60}{\cos \theta} = -111.8 + 30 \sqrt{5} = -44.72 \text{ kN}$$

$$P_{U_2L_2} = -44.72 \text{ kN (Compressive)}$$

Graphical determination can also be done by starting from the free end side joint (vector diagram Fig. 4.32b). Measure forces in various members from the vector diagram and convert with the assumed scale. Table 4.10 indicates internal forces in members.

Table 4.10: Internal forces

S. No.	Member	Internal force		Remarks
		Magnitude (kN)	Nature	
1	U_2U_3	111.8	Tensile	
2	U_2L_2	44.72	Compressive	
3	L_1L_2	60.00	Compressive	
4	U_1U_2	60.00	Tensile	
5	U_1L_1	67.00	Compressive	
6	U_2L_1	40.00	Tensile	

EXAMPLE 4.11: Analyze internal resistances in members U_2U_3, U_2L_4 and L_3L_4 in case of cantilever truss shown in Fig. 4.33.

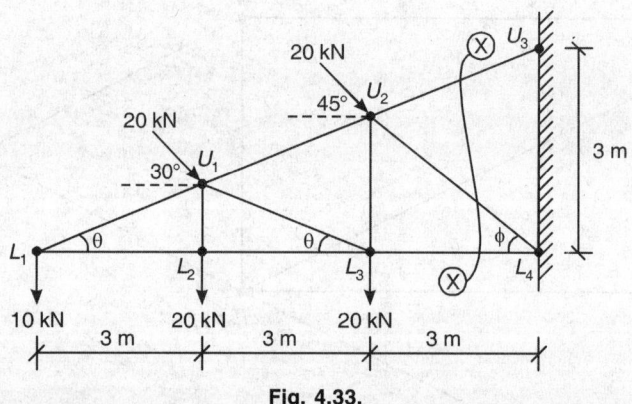

Fig. 4.33.

Solution: Hints: Best approach in this case will be method of sections.

Consider free body diagram on LHS of the section X-X

Internal resistances in the members are given in Table 4.11 for checking of your analysis by any method.

4.6 SUMMARY

This chapter deals with the computation of **internal forces** in **axial force structures**, only single bar tension or compression members connected by pin joints in plane trusses have been considered for computation of internal forces.

Axial Force Structure

An **axial force structure** is the one, which is acted upon by only direct normal forces at centroid of its section.

Table 4.11: Approximate internal stresses (Fig. 4.33)

S. No.	Member	Internal force		Remarks
		Magnitude (kN)	Nature	
1	U_2U_3	114.8	Tensile	
2	U_2U_4	68.0	Compressive	
3	L_3L_4	83.66	Compressive	
4	L_3U_2	37.9	Tensile	
5	U_1U_2	70.0	Tensile	
6	L_1U_1	31.6	Tensile	
7	L_1L_2	30.0	Compressive	
8	L_2U_1	20.0	Tensile	
9	L_2L_3	30.0	Compressive	
10	U_1L_3	56.0	Compressive	
11	L_0U_1	35.0	Compressive	
12	L_0L_1	26.6	Tensile	
13	U_1U_2	20.0	Compressive	
14	U_1L_1	25.2	Compressive	
15	U_2L_1	11.4	Tensile	
16	U_2U_3	25.2	Compressive	
17	U_3L_2	25.2	Compressive	
18	U_3L_1	0.0	——	
19	L_1L_2	5.0	Tensile	

Direct normal force at any section of an **axial force structure** is the algebraic sum of the **axial forces** acting on one side of the section. Direct normal force at a section can be **tensile** or **compressive**.

Truss

Truss is a structure consisting of members joined at their ends with hinge or pin in a manner so as to form a **triangle** or a **number of triangles**. The members of a truss are subjected to an axial centric force (either tension or compression).

In analyzing trusses, it is assumed that members of the truss are connected through pin joints and the loads are applied at the panel joints only.

Perfect Truss

If the number of members in a truss are such that they are **just sufficient to enclose a space in triangle(s)**, then the truss is said to be **perfect**. Also the joints are connected through hinges or pins and the loads act at panel points only. The number of members m are equal to $(2j - 3)$, where j = number of joints and m is the number of members.

Perfect, stable and determinate

Deficient Truss

A truss is said to be **deficient if the number of member are less and fail to enclose the space in triangle(s)**. Such trusses are not stable under general loading. The number of members 'm' in a truss with 'j' joints are such that $m < (2j - 3)$.

Deficient and unstable

Redundant Truss

A truss is said to be **redundant,** when the number of members exceed that required for enclosing the space in triangles. The number of members m in a truss with 'j' joints are such that $m > (2j - 3)$.

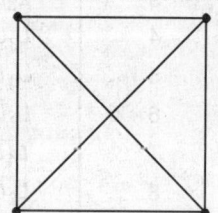

Redundant and statistically indeterminate

Determining Internal Forces in Truss Members

Support **reactions** are first determined graphically or analytically.

There are two methods of determining internal forces. These are analytical and graphical methods. Again, under the analytical method there are two methods viz. **method of joints** and **method of sections**. All these methods are based on the conditions of external or internal **equilibrium**. Internal member forces are determined by considering **FBD** of **joints** or **portion of structure** on one side of the **selected section**.

Method of Joints

In this method, FBD is first drawn for selected joint of a truss with **not more than two unknown quantities (both magnitudes or one magnitude and one direction) for unknown forces** is drawn. Using the **conditions of static equilibrium** viz. $\Sigma F_x = 0$, $\Sigma F_y = 0$, the unknown forces are determined. This procedure is adopted for the rest of the truss to complete the **analysis of joints one by one**. The sequence of joints is so chosen that every time the selected joint has two unknown magnitudes since the directions will be same as that of the members.

Method of Sections

This method is preferred, when it is required to determine **forces in a few members** of a truss. In this method the truss is **cut by a section**, so as to divide it into two parts. The section is so passed that it **cuts the member** in which the force is **to be determined** and at the same time **two other unknown members** of the truss shall be **meeting at a joint**. The condition of static equilibrium used in this method is $\Sigma M = 0$. **FBD** of the two cut parts is drawn and equilibrium equations are analyzed to find internal forces in the members.

Graphical Conditions of Static Equilibrium

For a truss subjected to a **system of coplanar forces**, the conditions of static equilibrium are that (a) **the force polygon must close**, (b) **the funicular polygon must also close** to satisfy $\Sigma M = 0$, condition.

Method of Determining Forces in Trusses Graphically

The steps involved in determining internal forces in the members of a truss graphically are: (a) Draw **FBD** of the truss. (b) Name all the forces-internal and external by using **Bow's notation.** (c) Draw **force diagram** for the known forces acting on the system. (d) Draw a **polar diagram**. (e) Draw the **funicular polygon**. (f) **Complete the force polygon** for the **external forces** acting on the truss (assuming suitable force scale). (g) Complete the force **(vector) diagram** for all the internal forces in the members. (h) **Scale** the value of the **forces in different members** from the vector diagram. Considering these in a cyclic order for the same joint. **Scale** of drawing vector diagram must be the same throughout the same problem. The parallel lines must be drawn accurately by taking the dimensions and angles of truss members correctly.

EXERCISE 4

Q.4.1. A king post truss with panel loads is shown in Fig. Q4.1. Determine internal resistances offered by the members of the truss. Truss is simply supported.

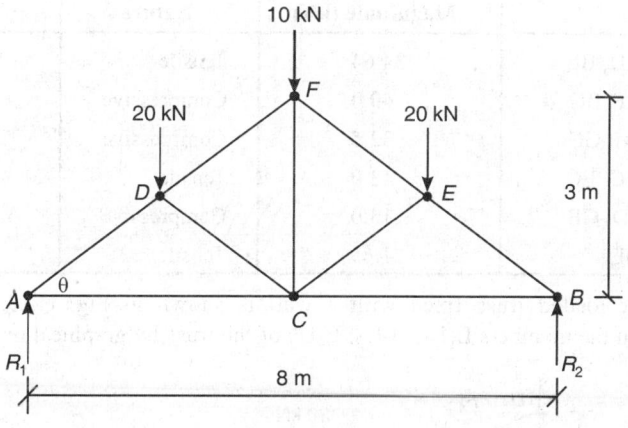

Fig. Q4.1.

Ans. Q.4.1.

S. No.	Member	Internal Force		Remarks
		Magnitude (kN)	**Nature**	
1	AD, BE	41.66	Compressive	
2	AC, BC	33.3	Tensile	
3	DF, EF	25	Compressive	
4	DC, EC	16.7	Compressive	
5	FC	20.0	Tensile	

Q.4.2. A simply supported loaded truss is shown in Fig. Q4.2. Determine internal stresses in the members using graphical or analytical approach.

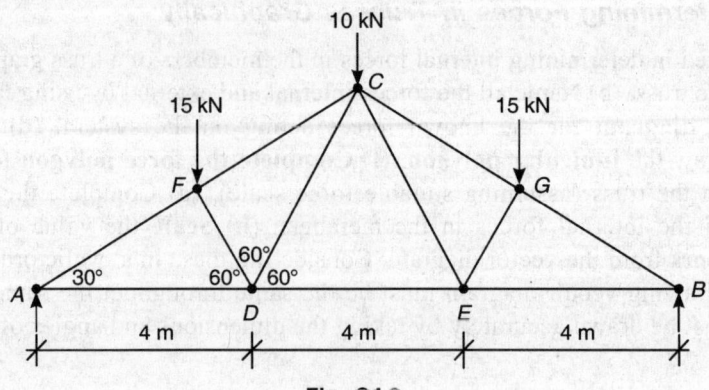

Fig. Q4.2.

Ans. Q.4.2

S. No.	Member	Internal Force		Remarks
		Magnitude (kN)	**Nature**	
1	AD, BE	34.64	Tensile	
2	AF, BG	40.0	Compressive	
3	FC, GC	32.5	Compressive	
4	DC, EC	13.0	Tensile	
5	FD, GE	13.0	Compressive	
6	DE	21.65	Tensile	

Q.4.3. A cantilever loaded truss fixed with a wall is shown in Fig. Q4.3. Determine internal resistances in the members L_1L_2, U_1U_2, L_1U_2 of the truss by graphical or analytical method.

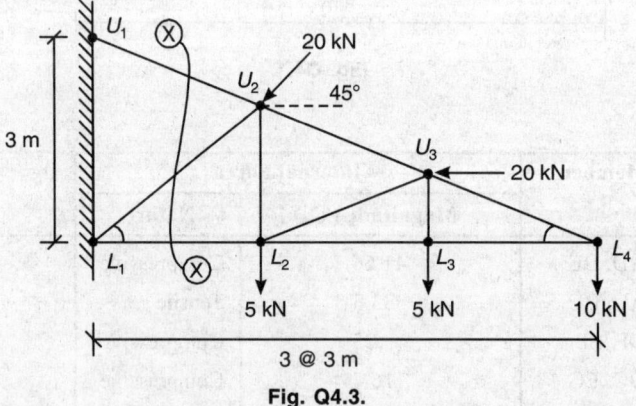

Fig. Q4.3.

Ans. Q.4.3

S. No.	Member	Internal Force		Remarks
		Magnitude (kN)	Nature	
1	L_3L_4	30.0	Compressive	
2	L_4U_3	31.62	Tensile	
3	L_2L_3	30.0	Compressive	
4	L_3U_3	5.0	Tensile	
5	U_2U_3	29.0	Tensile	
6	L_2U_2	10.83	Tensile	
7	L_2U_3	18.43	Compressive	
8	L_1L_2	**47.5**	Compressive	
9	L_1U_2	**16.51**	Compressive	
10	U_1U_2	**27.82**	Tensile	

Q.4.4 A cantilever loaded truss fixed in a wall at A and B is shown in Fig. Q4.4. Determine internal member stress in members **AC**, **BC** and **BD**.

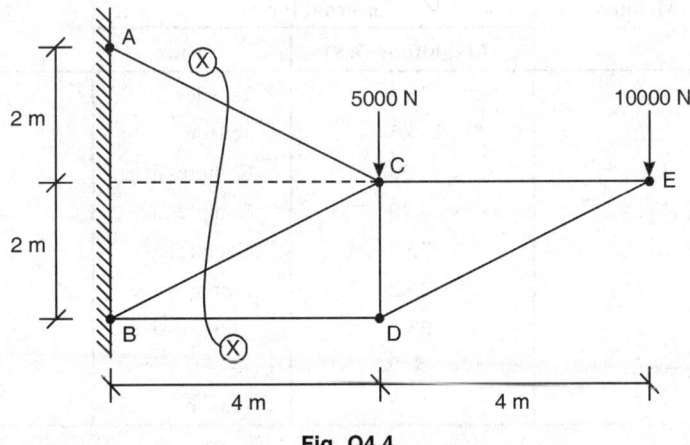

Fig. Q4.4.

Most efficient and quick method is to consider FBD on RHS of section X-X.

Ans. Q.4.4

S. No.	Member	Internal force		Remarks
		Magnitude (kN)	Nature	
1	AC	$+ 12500\sqrt{5} + 27950$	Tensile	
2	BC	$- 2500\sqrt{5} - 5590$	Compressive	
3	BD	$- 20000$	Compressive	
4	CD	$+ 10000$	Tensile	
5	CE	$+ 20000$	Tensile	
6	DE	$- 10000\sqrt{5}$	Compressive	

Q.4.5. Determine the internal resistances graphically or analytically in members of a king post truss subjected to wind loadings as shown in Fig. Q4.5. The truss is supported on roller on left end and hinge on right end.

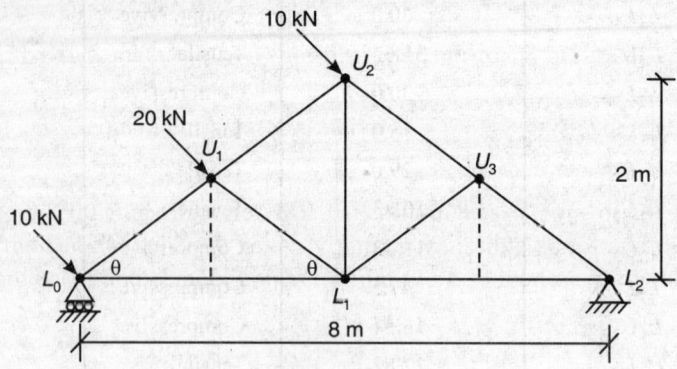

Fig. Q4.5.

Ans. Q.4.5

S. No.	Member	Internal force		Remarks
		Magnitude (kN)	**Nature**	
1	L_0L_1	26.8	Tensile	
2	L_1L_2	4.5	Tensile	
3	L_0U_1	35	Compressive	
4	U_1U_2	20	Compressive	
5	U_2U_3	25.0	Compressive	
6	U_3L_2	25.2	Compressive	
7	U_1L_1	25.0	Compressive	
8	U_3L_1	0.0	——	
9	U_2L_1	11.2	Tensile	

Q.4.6. Analyze the internal resistances in members of a simply supported truss shown in Fig. Q4.6.

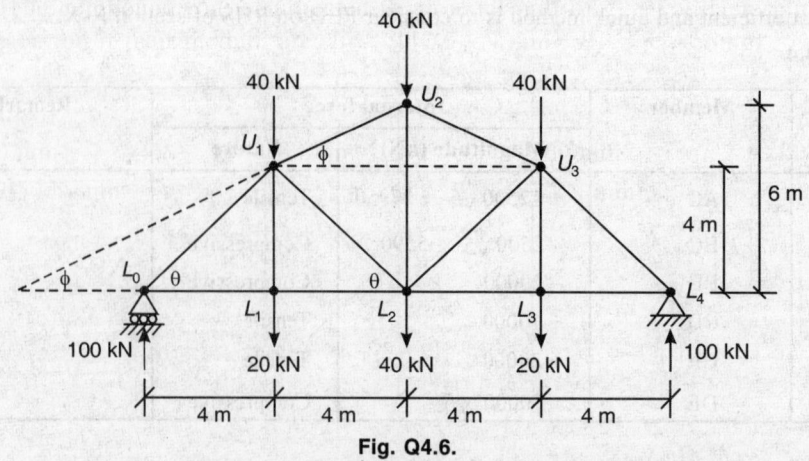

Fig. Q4.6.

Ans. Q.4.6

S. No.	Member	Internal Force		Remarks
		Magnitude (kN)	Nature	
1	L_0U_1, L_4U_3	141.4	Compressive	
2	L_0L_1, L_3L_4	100.0	Tensile	
3	L_1U_1, L_3U_3	20.0	Tensile	
4	L_1L_2, L_2L_3	100.0	Tensile	
5	U_1L_2, U_3L_2	9.6	Compressive	
6	U_1U_2, U_2U_3	104.0	Compressive	
7	U_2L_2	53.2	Tensile	

Q.4.7. A K-truss loaded vertically as shown in Fig. Q4.7 and supported on simple supports. Each panel span is **4m** and height is **6m**. Determine internal resistances graphically or by method of section in members U_3U_4, M_3U_3, M_3L_3, L_3L_4, M_3L_4 and M_3U_4.

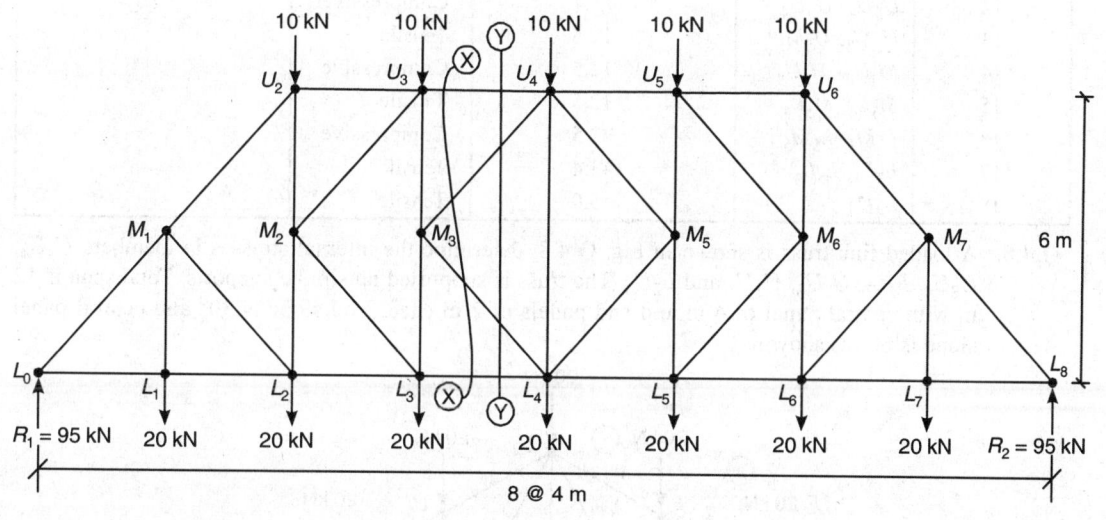

Fig. Q4.7.

Ans. Q.4.7

S. No.	Member	Internal force		Remarks
		Magnitude (kN)	Nature	
1	L_0M_1, L_8M_7	158.4	Compressive	
2	L_0L_1, L_6L_7, L_7L_8, L_1L_2	127.0	Tensile	
3	L_1M_1, L_7M_7	20.0	Tensile	
4	M_1L_2, M_7L_6	16.8	Compressive	
5	M_1U_2, M_7U_6	142	Compressive	
6	U_2U_3, U_6U_5	113.4	Compressive	
7	U_2M_2, U_6M_6	75.4	Tensile	
8	M_2U_3, M_6U_5	37.6	Compressive	
9	M_2L_3, M_6L_5	37.6	Tensile	
10	L_2M_2, L_6M_6	30.0	Tensile	
11	L_2L_3, L_6L_5	113.4	Tensile	
12	U_3U_4, U_5U_4	113.4	Compressive	
13	U_3M_3, U_5M_5	12.5	Tensile	
14	M_3U_4, M_5U_4	12.5	Compressive	
15	M_3L_4, M_5L_4	12.5	Tensile	
16	L_3M_3, L_5M_5	2.5	Compressive	
17	L_3L_4, L_5L_4	143.4	Tensile	
18	L_4U_4	5.0	Tensile	

Q.4.8. A loaded fink truss is shown in Fig. Q 4.8, determine the internal stresses in members U_3U_4, U_8U_4, L_2L_3, U_2U_8, U_2U_3 and L_2U_8. The truss is supported on simple supports. Total span is 12 m with central panel of 4 m and end panels of 2 m each. End slope is $30°$ and central panel slope is $60°$ as shown.

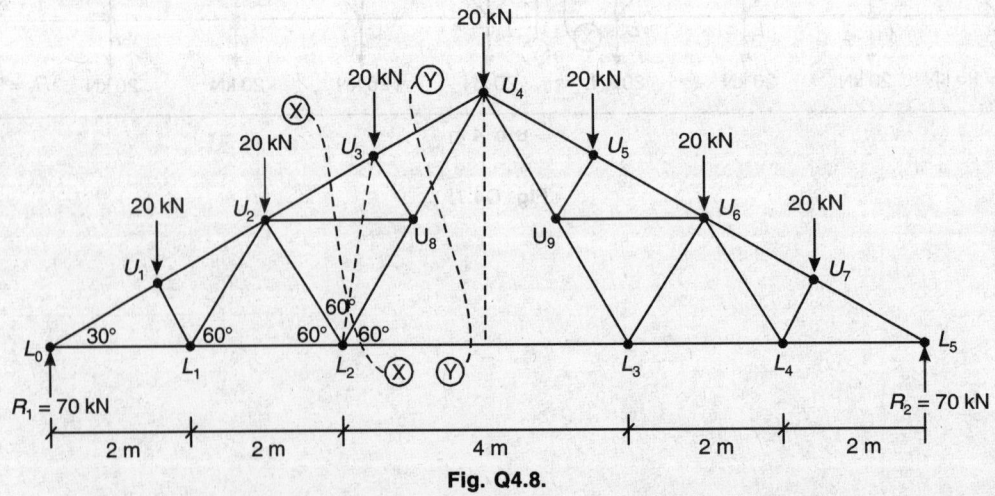

Fig. Q4.8.

Values of internal forces can be found solving joint by joint. Initially members U_3U_8 and U_2U_8 may be replaced by an **imaginary** member U_3L_2 and this member is again replaced by the original members after analysis. Original replaced members be analyzed again.

Ans. Q.4.8

S. No.	Member	Internal Force		Remarks
		Magnitude (kN)	Nature	
1	U_3U_4, U_4U_5	110	Compressive	
2	U_8U_4, U_9U_4	52.0	Tensile	
3	L_2L_3	70.0	Tensile	
4	U_2U_8, U_6U_9	17.32	Tensile	
5	U_2U_3, U_5U_6	120.0	Compressive	
6	L_2U_8, L_3U_9	34.64	Tensile	
7	L_0U_1, L_5U_7	140	Compressive	
8	L_0L_1, L_4L_5	121.3	Tensile	

Q.4.9. A truss shown in Fig. Q4.9 carries horizontal and vertical loads as shown. The truss is supported on rollers at L_2 and hinge at L_4. Determine nature & magnitude of member in U_3U_2, L_3U_3, L_4U_3, L_3U_2, L_2L_3 & L_3L_4.

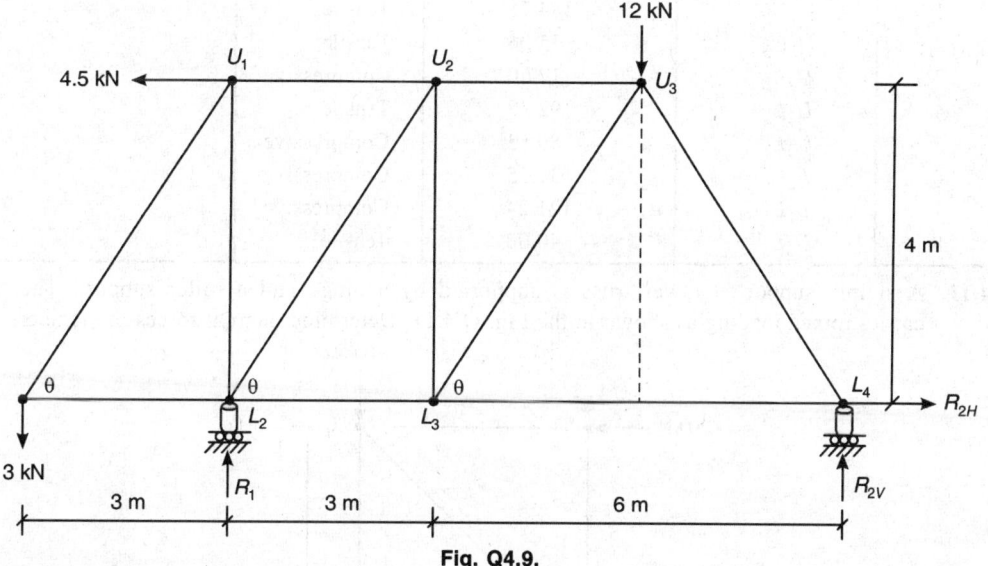

Fig. Q4.9.

Ans. Q.4.9

S. No.	Member	Internal Force		Remarks
		Magnitude (kN)	Nature	
1	U_3U_2	1.50	Tensile	
2	U_3L_3	8.75	Compressive	
3	U_3L_4	6.25	Compressive	
4	L_3L_2	3.00	Tensile	
5	L_3U_2	7.0	Tensile	
6	L_3L_4	8.25	Tensile	

Q.4.10. A cantilever loaded truss is shown in Fig. Q4.10. The truss is fixed at U_1 & L_1. Determine internal member forces in U_1U_2, U_2U_3, U_2L_1, U_2L_2, U_3L_2 & L_2L_3.

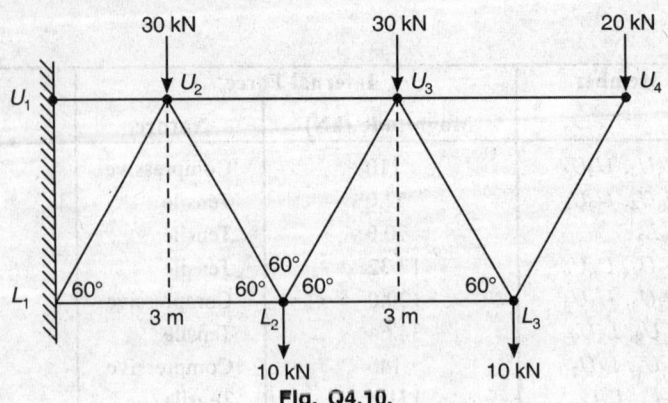

Fig. Q4.10.

Ans. Q. 4.10

S. No.	Member	Internal Force		Remarks
		Magnitude (kN)	Nature	
1	U_1U_2	184.75	Tensile	
2	U_2U_3	75.06	Tensile	
3	U_2L_1	127.0	Compressive	
4	U_2L_2	92.38	Tensile	
5	U_3L_2	80.83	Compressive	
6	L_2L_3	31.55	Compressive	
7	L_1L_2	121.24	Compressive	
8	U_3L_3	40.00	Tensile	

Q.4.11. A simply supported tower truss is supported by a hinge and a roller support. The truss carries mixed loading as shown in the Fig. Q 4.11. Determine internal forces in members.

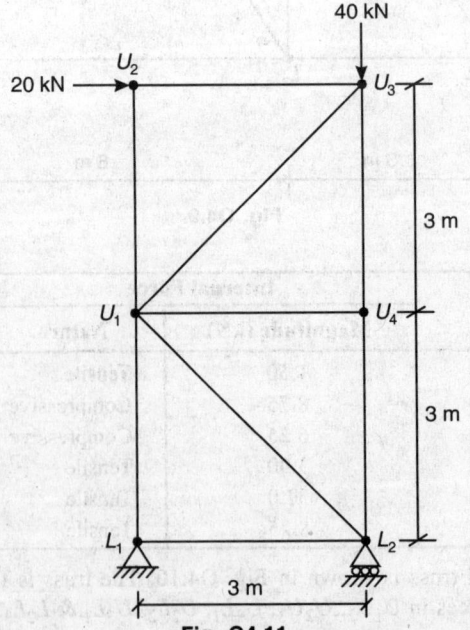

Fig. Q4.11.

Ans. Q. 4.11

S. No.	Member	Internal Force		Remarks
		Magnitude (kN)	**Nature**	
1	U_2U_3	20	Compressive	
2	U_1U_3	$20\sqrt{2}$	Tensile	
3	U_1L_1	40	Tensile	
4	U_3U_4	60	Compressive	
5	U_1L_2	$20\sqrt{2}$	Compressive	
6	U_4L_2	20	Tensile	

Q.4.12. A trauss supported on a hinge and inclined roller is shown in Fig. Q4.12. Calculate member forces around the joint U_3 and the joint L_2.

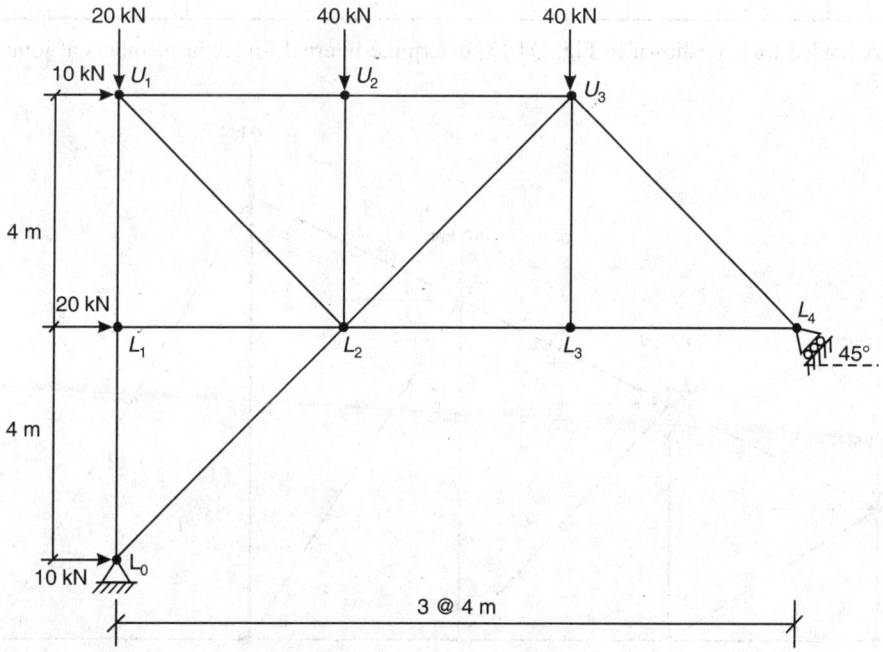

Fig. Q4.12.

Ans. Q.4.12

S. No.	Member	Internal Force		Remarks
		Magnitude (kN)	**Nature**	
1	U_2U_3	40	Compressive	
2	L_2U_3	0	——	
3	L_3U_3	0	——	
4	L_4U_3	$40\sqrt{2}$	Compressive	
5	L_0L_2	$10\sqrt{2}$	Compressive	
6	L_1L_2	20	Compressive	
7	U_1L_2	$30\sqrt{2}$	Tensile	
8	U_2L_2	40	Compressive	
9	L_2L_3	0	——	

Q.4.13. A loaded truss is shown in Fig. Q4.13, determine internal forces in members at joints U_2 and U_3.

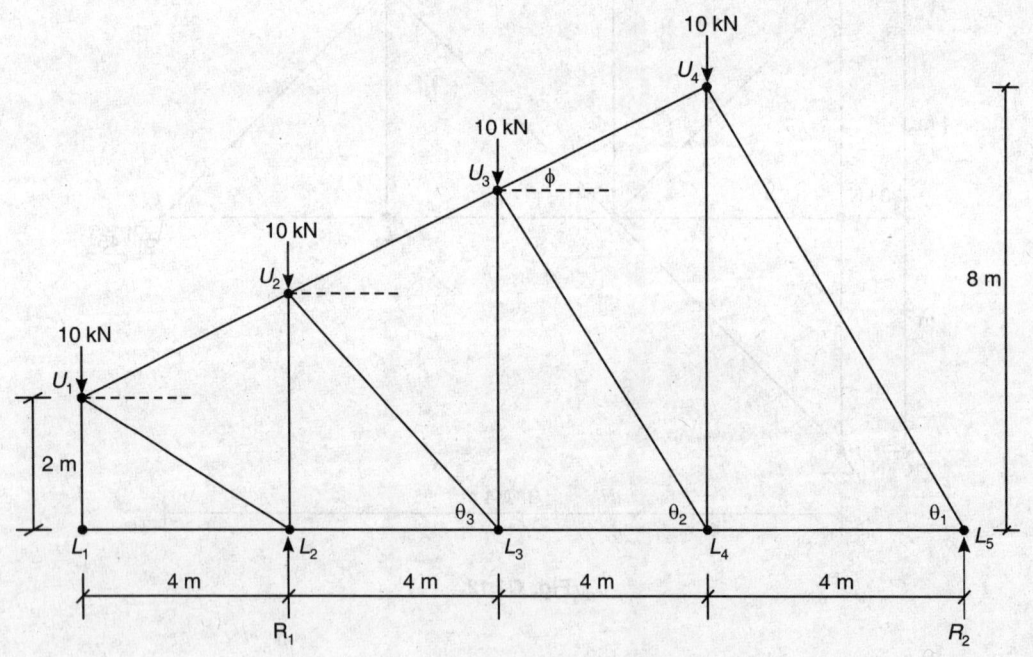

Fig. Q4.13.

Ans. Q.4.13

S. No.	Member	Internal Force		Remarks
		Magnitude (kN)	**Nature**	
1	$U_2 U_1$	$5\sqrt{5}$	Tensile	
2	$U_2 L_2$	$85/3$	Compressive	
3	$U_2 L_3$	$10\sqrt{2}$	Tensile	
4	$U_2 U_3$	0	——	
5	$U_3 L_3$	-10	Compressive	
6	$U_3 L_4$	$5\sqrt{13}/3$	Tensile	
7	$U_3 U_4$	$-5\sqrt{5}/3$	Compressive	
8	$U_4 L_5$	$-10\sqrt{5}/3$	Compressive	
9	$L_4 L_5$	$10/3$	Tensile	
10	$U_4 L_4$	5	Compressive	

Q.4.14. A loaded truss is shown in Fig. Q4.14, determine internal stresses in members $U_1 U_2$, $U_1 L_3$, $L_2 L_3$, $U_3 L_3$, $U_2 U_3$ & $L_3 L_4$.

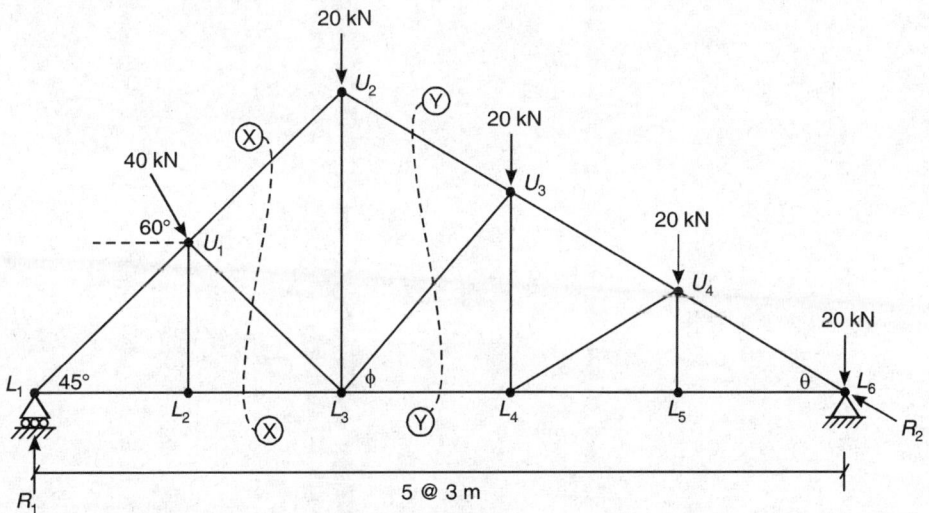

Fig. Q4.14.

Ans. Q.4.14

S. No.	Member	Internal Force		Remarks
		Magnitude (kN)	**Nature**	
1	U_1U_2	57.12	Tensile	
2	U_1L_3	38.64	Compressive	
3	L_2L_3	47.75	Tensile	
4	L_3L_4	35.39	Tensile	
5	U_3L_3	24.29	Compressive	
6	U_2U_3	48.54	Compressive	
7	U_2L_3	33.48	Compressive	

5

Axial Force Members, Stress, Strain and Elasticity

LEARNING OBJECTIVES

After studying this chapter, the learner **understands stress, strain** and **elastic deformations** in axial force members subjected to **linear loads**. The learner will be able to:

5.1 **Explain** different stresses and strains in elastic materials.

5.2 **Explain** stress–strain curve for elastic materials loaded axially.

5.3 **Explain Hooke's Law** for elastic materials.

5.4 **Explain** stress–strain relationship in a elastic material of **prismatic sections** subjected to axial loading.

5.5 **Explain stress–strain relationship** in an elastic material in a variable sections subjected to axial loading.

5.6 **Explain lateral stress-strain** and **Poisson's** ratio in elastic materials.

5.7 **Explain** stress–strain in **composite members** under axial loading.

5.8 **Calculate stress–strain** and different **deformations** under axial loading.

5.9 **Explain thermal stress–strain** in elastic materials.

5.10 **Explain shear stress-strain** relations, complementary shear stress and **modulus of rigidity**.

5.11 **Calculate thermal** stresses and strains in elastic materials.

5.12 **Explain volumetric strain** and **bulk modulus** of elasticity.

5.13 **Explain** relationship in various **modulii of elasticity**.

5.14 **Explain strain energy** for gradually applied and impact loads.

5.15 **Calculate** different **elastic deformations** from the given data and **modulii of elasticity, rigidity and bulk modulus of elasticity**.

5.1 INTRODUCTION

We have already learned the analysis of **axial force structures** by considering **external equilibrium** of the structure (truss) as a whole and **equilibrium** of individual components at joints or FBD of one side of sections. These axial structures were considered **rigid** neglecting any deformation which occur due to loading. These structures were considered in stable condition and in equilibrium. In real life these elements are **not in perfect** or **ideal rigid** condition but undergo some **deformations** when subjected to forces. All elements comprise a large number of particles. When the body as a whole is in equilibrium, all such particles of the body are also in **equilibrium**. All such particles remain connected to each other by certain internal **force of attraction** based on its specific characteristics. When these particles are subjected to **external forces**, these particles **tend to deform**. These deformations are resisted by developing **internal resistance**. The development of this internal resistance continues till it **balances the external force**. The equilibrium of external and internal forces can be understood by removing a portion of body from the main body and marking the balancing internal forces.

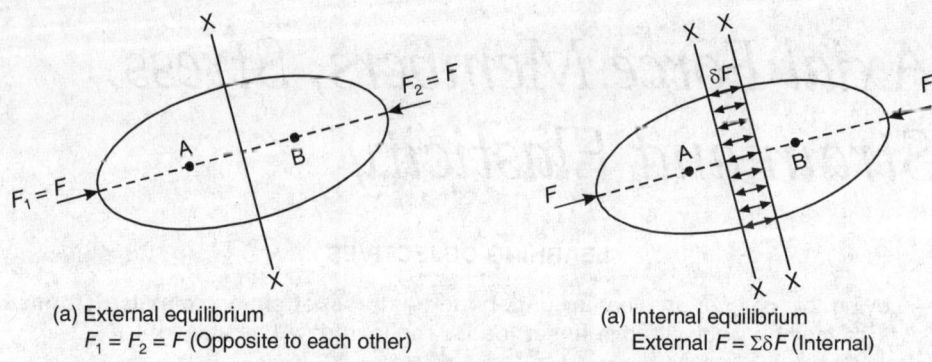

(a) External equilibrium
$F_1 = F_2 = F$ (Opposite to each other)

(a) Internal equilibrium
External $F = \Sigma\delta F$ (Internal)

Fig. 5.1: External and internal equilibrium

5.2 STRESS

Consider portion A or B of the body shown in Fig. 5.1 and draw FBD of the portion A. At a cut section X-X a total force $\Sigma\delta_F$ is developed internally in all the particles to balance the **external resultant force F**. For equilibrium $\Sigma\delta_F = F$.

This internal δF developed in a small portion δA per unit area is known as **stress**. Stress is sometimes called **unit stress**. Stress is equal to $p = \dfrac{\delta F}{\delta A}$. Since the force is **vector quantity**, the **stress** is also a **vector quantity**. Resultant force has **normal** and tangential components. **Normal stress** (axial stress) is defined as the ratio of **normal force** to **cross-sectional area**.

Tangential force per unit area is called **tangential stress** (p_t).

$$\mathbf{p_n} = \frac{\delta F_x}{\delta A}, \text{ as } \delta A \text{ tends to zero.} \quad \mathbf{p_t} = \frac{\delta F_t}{\delta A}, \text{ as } \delta A \text{ tends to zero.}$$

Thus, **stress** may be defined as **internal force per unit area** of the cross-section. When the internal force is in tangential direction, the stress is known as **tangential stress**.

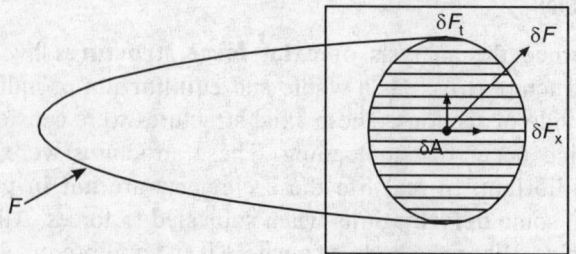

Fig. 5.2: Normal and tangential stress

Average **normal** stress $\qquad p_n = \dfrac{\delta F_x}{\delta A} = \dfrac{F_x}{A}$, Limit as $\delta A \to 0$

Average **tangential** stress $\quad p_t = \dfrac{\delta F_t}{\delta A} = \dfrac{F_t}{A}$, Limit as $\delta A \to 0$

Thus stress $\qquad\qquad\qquad p = \dfrac{\delta F}{\delta A}$, as $\delta A \to 0$ $\qquad\qquad\qquad$ ... (5.1)

Consider prismoidal rods subjected to **axial loads** as shown in Fig. 5.3. The load is acting along the **axis passing through the centroid** of the X-section. Since the load is acting along the centroidal axis, the average stress may be assumed uniform throughout the X-section. The axial force can be **compressive** or **tensile** as shown in Fig. 5.3.

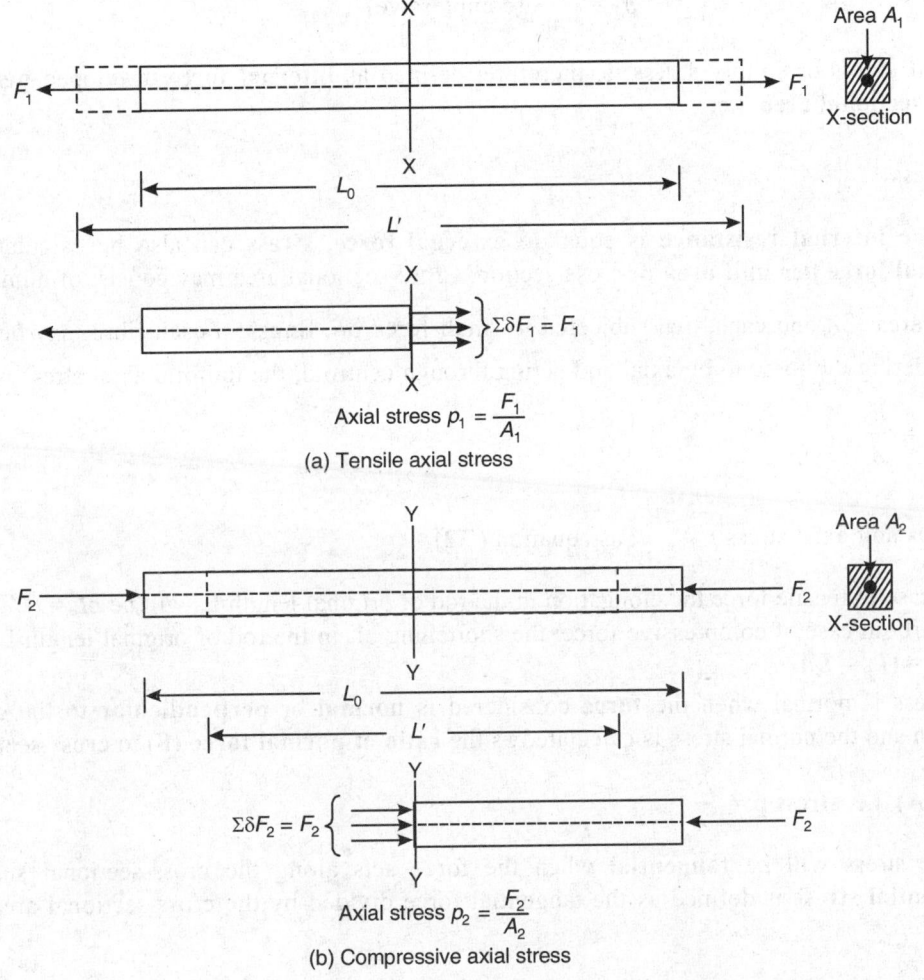

Area A_1

X-section

Axial stress $p_1 = \dfrac{F_1}{A_1}$

(a) Tensile axial stress

Area A_2

X-section

Axial stress $p_2 = \dfrac{F_2}{A_2}$

(b) Compressive axial stress

Fig. 5.3: Axial stress

Considering the rods or axial elements as **elastic**, the dimensions of the rod undergo changes. The axial force when acts on the rod, it causes **elongation** or **shortening** as shown in Fig. 5.3. When the axial force (F_1) causes **elongation** or increase in length (L_0 to L'), the force (F_1) is known **tensile** (pull) as shown in Fig. 5.3a. When the axial force (F_2) tends to cause **shortening** or decrease in length (L_0 to L'), the force (F_2) is called **compressive** (push) as shown in Fig. 5.3b.

If the axial tensile force is F_1 and area of cross-section is A_1, the average tensile stress in the rod will be,

$$p_1 = \frac{F_1}{A_1} \text{ (tensile)}$$

Similarly, the compressive stress will he,

$$p_2 = \frac{F_2}{A_2} \text{ (compressive)}$$

Unit stress or simply **stress** is, therefore, defined as **internal force** developed **per unit cross-sectional area**, i.e.

Stress $$p = \frac{F}{A} \qquad \qquad \dots (5.2)$$

Since **internal resistance** is equal to **external force**, **stress** can also be calculated as **external force per unit area** of cross-section. Cross-sectional area may consist of number of small areas δA and each area subjected to small force δF. Stress in each fibre may be $\frac{\delta F}{\delta A}$. Considering the force to be axial and acting through centroid, the uniform axial stress will be

$$\frac{\Sigma \delta F}{\Sigma \delta A} = \frac{F}{A}.$$

Thus unit axial stress $p = \frac{F}{A}$, as equation (5.2)

In case of tensile force the elongation in the rod of original length L_0 will be $\delta L = (L' - L_0)$. Similarly in case of compressive force, the shortening δL in the rod of original length L_0 will be $\delta L = (L_0 - L')$.

Stress is normal when the **force** considered is **normal** or **perpendicular** to the **cross-section** and the normal stress is calculated as the **ratio of normal force (F)** to **cross-sectional area (A)**, i.e. **stress p** $= \frac{F}{A}$.

The stress will be **tangential** when the force acts along the cross-sectional surface. **Tangential stress** is defined as the tangential force divided by the cross-sectional area, i.e.

$$p_t = \frac{F_t}{A}.$$

Linear strain is defined as **linear deformation per unit length** of the member, i.e.

$$\text{Strain } (e) = \frac{\text{Deformation } (\delta)}{\text{Original Length } (L_0)},$$ ... (5.3)

i.e.
$$e = \frac{\delta}{L_0},$$

it is ratio and hence has no units. But **units** of both δ and L_0 must be the **same**.

Units of stress are generally newton (N) per square millimeter (mm^2) or kilo newton per square metre, i.e. N/mm^2 (MPa), kN/m^2, etc.

It is assumed that the axial forces are within **elastic limits** and the materials of the axial elements are **elastic**. Properties of elastic materials have already been studied and we shall make use of these properties.

5.3   ELASTICITY

When a body is subjected to certain external force, the internal grains or particles undergo **deformation** to develop **internal resistance**. Thus the particles store **internal strain energy**. On removal of the external force, the stored strain energy releases the deformation and brings the particles back to their original position. In an **ideal elastic material**, the **deformation disappears completely** on removal of external load. In case of perfect elastic material, the **work done** by external force gets completely stored as **elastic strain energy**. It has been practically proved that most of **construction materials behave** as elastic up to certain loading limits. This **limiting stress** up to which the **material** behaves as elastic is known as **elastic limit**. Similarly, within certain limits, the **axial stress** developed **remains proportional** to the **linear strain created** in the **elastic material**. This is known as **Hooke's law** which states that in **elastic materials** the **axial stress remains proportional to the linear strain** up to the proportional limit.

$$p \propto \frac{\delta}{L}, \quad \text{or} \quad p \propto e, \quad \text{or} \quad p = E \cdot e$$

Where, E represents the constant, e represents the unit strain, p represents the unit stress. In case of elastic materials loaded axially within the proportional limits.

$$E = \frac{p}{e} \quad \text{or} \quad e = \frac{p}{E}$$ ... (5.4)

$$\frac{\delta}{L} = \frac{p}{AE} \quad \text{or} \quad \delta = \frac{PL}{AE}, \quad \text{(Where } P = pA\text{)}$$ ... (5.5)

Values of modulus of elasticity, also known as **Young's modulus of elasticity 'E'** are given in Table 5.1.

Designers of structures make use of **elastic properties** of construction materials for **simplicity**. Modern designers also make use of **plastic properties** of materials as a result of **research findings**, development of **material technology** and considerations of **economy**.

Table: 5.1: Values of modulus of elasticity

S. No.	Material	E (kN/mm², Gpa)
1	Aluminium Pure	70
2	Aluminium Alloy	70 – 79
3	Brass	96 – 110
4	Bronze	96 – 120
5	Cast Iron	83 – 170
6	Concrete	18 – 30
7	Copper (Pure)	110 – 120
8	Steel	190 – 210
9	Wrought Iron	190
10	Rubber	0.0007 – 0.004

5.4 LINEAR AND LATERAL STRAINS—POISSON'S RATIO

It has been found that the axial elements of elastic material not only undergo **linear deformations** due to application of axial forces but these elements also produce **deformations in lateral directions**. A French mathematician Mr. Poisson proved that for **elastic materials** the **lateral strains (e') are proportional to the linear strains (e)**. The ratio of **lateral strain (e')** to **linear strains (e)** is **constant** in elastic materials and this constant is known as **Poisson's ratio** (μ or $1/m$)

i.e.
$$\mu \text{ or } \frac{1}{m} = \frac{e' (\text{lateral strain})}{e (\text{longitudinal strain})} \qquad \qquad ...(5.6)$$

or
$$e' = \mu e \quad \text{or} \quad \frac{e}{m},$$

where, e' is lateral strain and e is linear strain.

Poisson's constant for common construction materials is given in Table 5.2.

Table: 5.2: Poisson's ratio for engineering materials

S. No.	Material	Poisson's Ratio m
1	Aluminium, Aluminium Alloys	0.33
2	Brass, Bronze	0.34
3	Cast Iron	0.20 to 0.30
4	Cement Concrete (Comp.)	0.10 to 0.20
5	Copper	0.33 to 0.36
6	Steel	0.27 to 0.30
7	Wrought Iron	0.30
8	Rubber	0.45 to 0.50

The length of any rod L will change to $L(1 + e)$, when a **tensile axial force P** is applied while the length will change to $L(1 - e)$, when a compressive axial force P is applied. The length of axial element under axial force changes in the ratio of $(1 \pm e) : 1$, + for tensile and – for compressive forces.

Lateral Strain $\qquad\qquad e' = -\dfrac{\delta t}{t} = -\mu e = -\mu\,(p/E)$

Also Lateral Strain $\qquad e' = -\dfrac{\delta B}{B} = -\mu e = -\mu\,(p/E)$

–ve sign indicates reduction in lateral dimension for axial tensile stress.

i.e. $\qquad\qquad\qquad e' = \dfrac{\delta t}{t} = \dfrac{\delta B}{B} = -\mu\,(p/E)$ $\qquad\qquad$ ... (5.7)

Lateral dimensions change in ratio of $(1 \pm e') : 1$, **–ve for tensile axial force** and **+ve for compressive force.**

There cross-sectional area will change in ratio of $(1 - e')(1 - e') : 1$ or $(1 - 2e') : 1$, neglecting small quantities of higher power (e'^2 etc.).

Thus $\dfrac{A'}{A} = (1 - 2e')$, or $A' = (1 - 2e')A$, or $\delta A = (A - A') = 2e'A$

or $\qquad\qquad\qquad \dfrac{\delta A}{A} = 2e' = 2\mu e = 2\mu\,(p/E)$

$$\delta A = 2\mu\,(p/E)\,.\,A \qquad\qquad\qquad ... (5.8)$$

Thus change in cross-sectional area δA is equal to **twice Poisson's ratio** multiplied by **linear strain** (p/E) **times original area**.

Since length changes in the ratio of $(1 + e) : 1$ and cross-sectional area in the ratio of $(1 - 2e') : 1$, the **volume** will change in the ratio of $(1 + e)(1 - 2e') : 1$

or $\quad (1 + e)(1 - 2\mu e) : 1$ or $(1 + e - 2\mu e - 2\mu e^2) : 1$

Neglecting small quantities, thus,

$$\dfrac{V'}{V} = (1 + e - 2\mu e) = 1 + e\,(1 - 2\mu)$$

$$V' = V + Ve\,(1 - 2\mu) \quad \text{or} \quad \dfrac{V' - V}{V} = e\,(1 - 2\mu)$$

or $\quad \dfrac{\delta V}{V} = e\,(1 - 2\mu)$ or $\delta V = \dfrac{p}{E}\,(1 - 2\mu)\,V$, (under linear loading) $\qquad$ ... (5.9)

Change in volume (δV) of a member under an axial **stress p** will be linear strain (p/E) multiplied by $(1 - 2\mu)$ times the original volume (V).

EXAMPLE 5.1: A circular CI column of **100 mm diameter** and 3 m length is loaded with an axial compressive load of **1000 kN**. Calculate the **shortening** in the column if modulus of elasticity **E = 120 kN/mm²**.

Solution: Load $P = $ **1000 kN**, $A = \dfrac{\pi}{4}(100)^2$ mm², $L = $ **3000 mm**, $E = $ **120 kN/mm²**

Deformation $\delta = e \cdot L = (p/E) \cdot L = \dfrac{PL}{AE} = \dfrac{1000 \times 3000}{10000 \dfrac{\pi}{4} \times 120} = \dfrac{10}{\pi} = $ **3.18 mm**

EXAMPLE 5.2: A steel rod of 50 cm length and 20 mm diameter elongates by 0.50 mm under certain pull. If the modulus of elasticity is 21000 kN/cm², calculate the pull.

Solution: $L = 50$ cm = **500 mm**, Diameter = **20 mm**, δ (elongation) = **0.50 mm**
$E = 21000$ kN/cm² = **210 kN/mm²**,
Let the pull = P kN

Strain $e = \dfrac{0.50}{500} = \dfrac{1}{1000}$, $\quad p$ (stress) $= \dfrac{P}{\dfrac{\pi}{4}(20)^2} = \dfrac{P}{100\,\pi}$ kN/mm²

$E = p/e \quad$ or $\quad 210 = \dfrac{P \times 1000}{100\,\pi \times 1} \quad$ or $\quad P = \dfrac{210 \times 100\,\pi}{1000} = 21\pi$ kN = **66 kN**

EXAMPLE 5.3: A **10 m** long wire extends by 2/3 cm under a pull of 110 N. If the diameter of wire is 1.0 mm, determine the modulus of elasticity of the wire material.

Solution: $L = 10$ m = **10, 000 mm**, $\delta = 2/3$ cm = **20/3 mm**, $P = 110$ N = **0.11 kN**
Diameter = 1.0 mm

$\delta = \dfrac{PL}{AE}$, All in **kN, mm** units

$\dfrac{20}{3} = \dfrac{0.11 \times 10000}{E \times \dfrac{\pi}{4}(1)^2} = \dfrac{4 \times 1100}{E \times \pi} \quad$ or $\quad E = \dfrac{4400 \times 3}{\pi \times 20} = \dfrac{660 \times 7}{22} = $ **210 kN/mm²**

EXAMPLE 5.4: A steel rod of 40 cm length and uniform cross-sectional area of 2.0 cm² is subjected to axial loads as shown in Fig. 5.4. If the material has elastic modulus of 210 kN/mm², determine the total elongation of the rod.

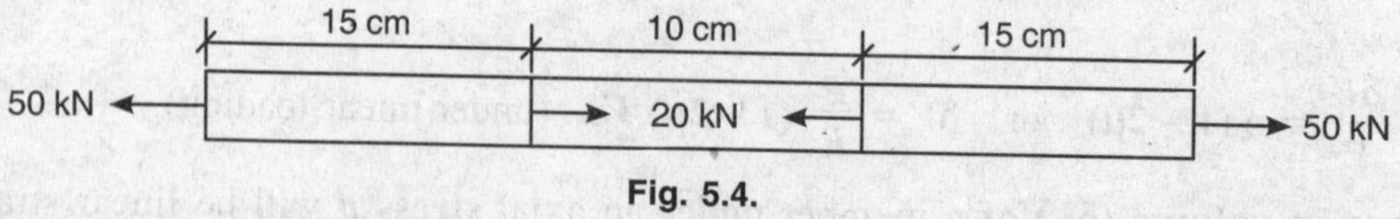

Fig. 5.4.

Solution: $L_1 = 15$ cm = **150 mm**, $L_2 = 10$ cm = **100 mm**, $L_3 = 15$ cm = **150 mm**
$A_1 = A_2 = A_3 = 2.0$ cm² = **200 mm²**, $E = $ **210 kN/mm²**
$P_1 = +\,50$ kN $= P_3$, $\qquad\qquad P_2 = +\,(50 - 20) = +\,30$ kN

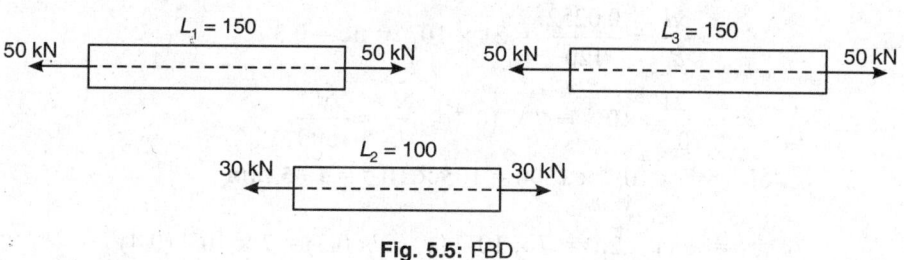

Fig. 5.5: FBD

Deformation $\delta = \dfrac{PL}{AE}$

Total deformation $\delta = \delta_1 + \delta_2 + \delta_3 = \dfrac{P_1 L_1}{A_1 E} + \dfrac{P_2 L_2}{A_2 E} + \dfrac{P_3 L_3}{A_3 E}$

or $\quad \delta = \dfrac{(7500 + 3000 + 7500)}{200 \times 210} = \dfrac{3}{7} = \mathbf{0.4285\ mm}$

EXAMPLE 5.5: Prismoidal rod of 10 mm × 20 mm cross-section and 1 m length is subjected to a pull of 40 kN. If the rod material modulus of elasticity is 200 kN/mm², and Poisson's ratio $\mu = 0.30$, calculate the elongation, change in breadth and depth of section. Also calculate the change in area and volume.

Solution: $\quad P = \mathbf{40\ kN}, \qquad A = 10 \times 20 = \mathbf{200\ mm^2}, \qquad L = \mathbf{1000\ mm},$

$E = 200\ \text{kN/mm}^2, \qquad\qquad \mu = \mathbf{0.3}, \quad D = 20\ \text{mm} \qquad\qquad B = 10\ \text{mm},$

$\delta L = \dfrac{PL}{AE} = \dfrac{40 \times 1000}{200 \times 200} = \mathbf{1\ mm}$

from equation 5.7

$\delta B = e'.B = (\mu e).B = \left(\mu \cdot \dfrac{P}{E} \cdot B \right) = \left(\dfrac{0.30 \times 40}{200 \times 200} \right) \cdot B = \times 10 = \mathbf{0.003\ mm}$

$\delta D = e'.D = (\mu e).D = \left(\mu \cdot \dfrac{P}{E} \cdot D \right) = (\mu.).D = \times 20 = \mathbf{0.006\ mm}$

Area change $\delta A = (2e').A = (2\mu e).A = \left(\dfrac{0.30 \times 40}{200 \times 200} \right) \times 200 = \mathbf{0.12\ mm^2}$

Change in volume $\delta V = (1 - 2\mu).V = (1 - 2\mu).AL = \ = \mathbf{80\ mm^3}$

EXAMPLE 5.6: A circular rod of 12 cm diameter undergoes a reduction of 0.0252 mm in its diameter when an axial pull is applied. Find the pull and change in its length and volume if the length is 1.50 m, Poisson's ratio $\mu = 0.3$ and modulus of elasticity E = 210 kN/mm².

Solution: $\quad L = 1.5\ \text{m} = \mathbf{1500\ mm}, \quad d = 12\ \text{cm} = \mathbf{120\ mm}, \quad \delta d = \mathbf{0.0252\ mm},$

$E = 210\ \text{kN/mm}^2, \qquad\qquad \mu = 0.3$

$$e' = \frac{\delta d}{d} = \frac{0.0252}{120} = 21 \times 10^{-5} = \mu e = 0.3\ e$$

$$e = \frac{21}{0.3} \times 10^{-5} = 7 \times 10^{-4} = \frac{\delta L}{L} = \frac{\delta L}{1500}$$

$$\delta L = 7 \times 10^{-4} \times 1500 = 105 \times 10^{-2} = \mathbf{1.05\ mm}$$

$$\frac{\delta_v}{V} = e\ (1 - 2\mu) = 7 \times 10^{-4}\ (1 - 2 \times 0.3) = 7 \times 10^{-4}\ (0.4)$$

$$\delta V = 2.8 \times 10^{-4} \times \frac{\pi}{4}\ (120)^2 \times 1500 = 0.7 \times 10^{-4} \times \frac{22}{7} \times 14400 \times 1500$$

$$= 10^{-5} \times 10^4 \times 22 \times 144 \times 15 = 10^{-1} \times 330 \times 144 = \mathbf{4752\ mm^3}$$

$$e = \frac{p}{E} \quad \text{or} \quad 7 \times 10^{-4} = \frac{P}{AE} = \frac{P}{210 \times \frac{\pi}{4} \times (120)^2}$$

or $\quad P = 7 \times 10^{-4} \times 210 \times \frac{\pi}{4} \times (120)^2 = \mathbf{1663.2\ kN}$

EXAMPLE 5.7: A circular rod of **12 mm** diameter and **1.50 m** length has modulus of elasticity **E = 210 kN/mm^2**, Poisson's ratio **μ = 0.30**. If pull of 5 kN is applied on the rod, find its change in **length, area** and **volume**.

Solution: $\quad P = 5$ kN, $\qquad E = 210$ kN/mm^2, $\qquad \mu = 0.30$, $\qquad L = 1500$ mm,

Diameter $= 12$ mm

$$\delta L = \frac{PL}{AE} = \frac{5 \times 1500}{\frac{\pi}{4}(12)^2 \times 210} = \frac{20 \times 1500}{\pi(144) \times 210} = \mathbf{0.3156\ mm}$$

$$\delta A = 2\mu \cdot e \cdot A = 2 \times 0.3 \times \frac{5 \times \frac{\pi}{4} \times (12)^2}{\frac{\pi}{4}(12)^2 \times 210} = \frac{0.6 \times 5}{210} = \mathbf{0.014286\ mm^2}.$$

$$\delta V = V \cdot e\ (1 - 2\mu) = A \cdot L \times \frac{p}{E}\ (1 - 0.6) = \frac{PL}{E}\ (0.4) = \frac{5 \times 1500 \times 0.4}{210} = \mathbf{14.286\ mm^3}$$

5.5 TENSILE TEST ON DUCTILE MATERIALS

Many of the mechanical properties have already been explained through axial **tensile** or **compressive** test of ductile or brittle materials in ξ 2.3 Chapter 2. Standard tensile test is shown for mild steel ductile bars. Standard test specimen for tensile test is prepared with **gauge length** markings of **5 times diameter**. Test specimen is fixed in jaws of Universal Testing Machine (UTM) and tensile load is gradually increased. The linear **deformation** is also measured with the help of **extensometer** fixed with the standard **gauge length (5 times diameter)**. Load and corresponding deformations are measured at regular intervals and a

stress-strain diagram is drawn **manually** or **automatically** by an attachment on the roller drum fixed with pen and paper by resistering displacements and load applied.

The stress-strain diagram for mild steel bar is shown in Fig. 5.6 for some important points in **tensile test** curve.

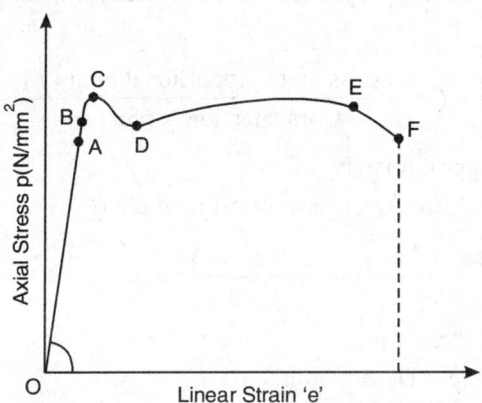

Fig. 5.6: Stress–strain curve in tension for mild steel

The stress-strain or force-deformation curve is shown in Fig. 5.6. The stress-strain curve remains straight line up to **point A** which is known as **proportional limit**. **Proportional limit** is the **maximum axial stress** up to which stress-strain is perfectly straight line and Hooke's law is applicable.

Point B represents **elastic limit** of stress up to which on removal of the load the deformation disappears completely. **Point C** represents **yield point** where the material starts yielding (increase in deformation **without any increase in stress**). The phenomenon of increase in deformation without any increase of stress is known as **yielding** and the material enters in the plastic stage of deformation. **Point C** is also known as **upper yield point**.

Point D is **end of yielding** and the material gets **strain hardened** to develop **further increase in stress-resistance** with increase in deformation. **Point D** is called **lower yield point** as stress at **point D** is lower than that at **point C**.

In plastic stage from **D to E to F**, the strain (deformation) increases with minor increase in stress (force). The **maximum stress** is at **E** and known as **ultimate stress**. The material rod **breaks** at the **point F** which shows **maximum strain** (deformation). Although the **point F** denotes **less stress** than the **point E** but the actual stress will be higher at the point F due to reduction in the cross-sectional area at the time of breaking of the specimen. The stress at this time of breaking is known as **breaking stress**. All the stresses are calculated and represented with reference to the **original cross-sectional area**.

Ductility is measured by **percent elongation** at the time of **breaking**, i.e. ductility

index $= \dfrac{(L' - L_0)}{L_0} \times 100$, where L' is the length of the tensile specimen just before breaking and L_0 is the original gauge length before loading. Generally gauge length of the specimen is taken equal to 5 times the diameter. Ductility index is also $\dfrac{(A_0 - A')}{A_0} \times 100$.

Tensile test provides material properties which are useful in design of element (size and shape).

- Modulus of Elasticity $E = \dfrac{\text{Stress up to proportional limit } (f)}{\text{Corresponding strain } (e)}$.

- Elastic Limit $= f_y$, approximately.
- Ultimate Strength $= f_u$ (max. stress w.r.t original area)

- Ductility index $= \dfrac{(L' - L)}{L} \times 100$ or $\dfrac{(A - A')}{A} \times 100$

- Modulus of strain energy/volume

Energy $\qquad U = \text{Force} \times \text{Deformation}$

or $\qquad U = (\text{Average Stress} \times \text{Area}) \times (\text{Strain} \times \text{Length})$,

or $\qquad U = \text{Average Stress} \times \text{Strain} \times \text{Volume}$, (Since, Area $\times$ Length = volume)

or $\qquad U = \text{Average Stress} \times \dfrac{\text{Stress}}{E} \times \text{Volume}$, Elastic Limit $= f_y$.

or $\qquad U = \dfrac{f_y}{2} \times \dfrac{f_y}{E} \times \text{Volume}$

Thus modulus of strain energy $U_0 = \dfrac{U}{\text{Volume}} = \dfrac{f_y^2}{2E}$ ... (5.10)

- Toughness index

$T_o = f_p \cdot e_p$, $\quad f_p = f_u$ (ultimate), $\quad e_p = \dfrac{(L' - L)}{L}$, ultimate strain at breaking.

$T_o = f_u \cdot \dfrac{(L' - L)}{L}$ kN-m/m^3 or kN-mm/mm^3 or N-mm/mm^3 (Energy/unit volume) at the time of breaking.

For **design** of structural elements these properties will be used appropriately.

5.6 WORKING STRESS AND FACTOR OF SAFETY

The design of structures is based on important factors of:
- Safety of strength
- Stability
- Avoidance of excessive deformation

- Durability
- Appearance (aesthetics), etc.

In design we shall consider strength of materials, stability and deformations of structures or elements. In this context we always keep the **actual stress** less than proportional or **elastic limits**. Generally in ductile materials there is **minor difference** in **proportional limit, elastic limit, upper yield** and **lower yield stress** and hence all these values are assumed as 'f_y'.

For the design purpose generally all these limiting stresses are considered equal to **yield stress** (a distinct point in tensile test) and the **actual stress** developed under **working loads** is kept much below this **yield stress**. The **actual stress** developed at **working loads** is called **working stress**. The margin or ratio of **yield stress** to the **working stress** is called **factor of safety**. **Working stress** is also called permissible or allowable stress.

$$\textbf{Factor of Safety} = \frac{\text{Yield Stress}}{\text{Working Stress}} \qquad\qquad \text{... (5.11)}$$

$$\text{or} \quad \text{Working Stress} = \frac{f_y}{\text{Factor of Safety}}$$

Sometimes the factor of safety is based on ultimate stress.

$$\text{i.e.} \quad \text{Factor of Safety} = \frac{f_u\,(\text{Ultimate Stress})}{\text{Working or Permissible Stress}}$$

For design purpose, the **factor of safety** depends on yield strength, type of loads, calculations of **actual loading, homogeneity** and **uniformity** of the material. In brittle materials such as concrete, stone, cast iron, or wood the permissible stress is based on **ultimate stress** and factor of safety. **Factor of safety** considered for engineering materials may be **1.50 to 3.0**, generally **2.0**. Thus permissible stress $= \dfrac{f_y}{2}$ or $\dfrac{f_u}{2}$.

EXAMPLE 5.8: A 12.8 mm diameter rod of 50 mm length undergoes an elongation of 0.042 mm with 22 kN pull. While it undergoes large amount of deformation at a load of 28 kN. The ultimate load attained is 58.5 kN. The rod undergoes 65 mm elongation just before breaking. Calculate modulus of elasticity, modulus of resilience (strain energy), ultimate strength, ductility and toughness index.

Solution: $F = 22$ kN, $\qquad F_y = 28$ kN, $\qquad F_u = 58.5$ kN, $\qquad L = 50$ mm,

Diameter = 12.8 mm, $\qquad \delta L = 0.042$ mm, $\qquad L' = 65$ mm

$$f = \frac{22000}{\dfrac{\pi}{4}(12.8)^2} = \textbf{171 N/mm}^2, \qquad e = \frac{0.042}{50} = \textbf{0.00084}$$

$$E = \frac{f}{e} = \frac{171}{0.00084} = \textbf{203.5 kN/mm}^2$$

$$f_y = \frac{28000}{\frac{\pi}{4}(12.8)^2} = \textbf{217.6 N/mm}^2$$

Modulus of Elastic Resilience $= \dfrac{f_y^2}{2E} = \dfrac{(217.6)^2}{2 \times 203.5} = \textbf{0.1164 N-mm/mm}^3$

$$f_u = \frac{58500}{\frac{\pi}{4}(12.8)^2} = \textbf{455 N/mm}^2$$

Ductility Index $\qquad D_i = \dfrac{(65-50)}{50} \times 100 = \textbf{30\%}$

T_o (Toughness Index) $= f_u \cdot D_i = 455 \times 0.3 = \textbf{136.5 N-mm/mm}^3$

EXAMPLE 5.9: A steel circular rod carries an axial tensile load of 831.6 kN. The ultimate stress of the material is 441 N/mm². If a factor of safety is 1.50, determine the safe diameter of the rod. If the rod is of 1 m length, determine the change in length, volume and diameter. $\mu = 0.30$, E (modulus of elasticity) = 210 kN/mm².

Solution: $F = 831.6$ kN, $\qquad f_u = 441$ N/mm², $\qquad$ Factor of Safety = 1.50,
$E = 210$ kN/mm², $\qquad\qquad \mu = 0.30$

Working Stress 'f' $= \dfrac{f_u}{\text{Factor of Safety}} = \dfrac{441}{1.5} = \textbf{294 N/mm}^2$

Safe area required $= \dfrac{\text{Load}}{\text{Safe Working Stress}} = \dfrac{831.6}{294} \times 1000 = 2828.5714 \text{ mm}^2$

$A = \dfrac{\pi}{4}(d)^2 = 2828.5714, \qquad d = \textbf{60.0273 mm}$

$\delta L = e \cdot L = \dfrac{f}{E} \cdot L = \dfrac{294 \times 1000}{210000} = \textbf{1.40 mm}$

$\delta d = e' \cdot d = \mu \cdot e \cdot d = 0.3 \times 0.0014 \times 60.027 = \textbf{0.0252 mm}$

$\delta V = e \, (1-2\mu) \cdot V = \dfrac{294 \times 1000}{210000} \times (1-0.6) \times 2828.57 \times 1000 = \textbf{1584.0 mm}^3$

EXAMPLE 5.10: A steel punch has permissible stress of 200 N/mm². Determine the diameter of punch hole which can be done in a steel plate of 10 mm thickness if the **ultimate shear strength** of the plate is 400 N/mm². Also find the load on the punch.

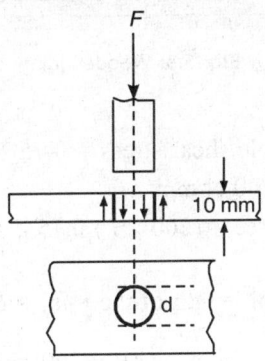

Fig. 5.7: Punching

Solution:

f (permissible stress) = 200 N/mm², $\quad fu$ = 400 N/mm² in shear

Thickness = **10 mm**, $\qquad\qquad$ Diameter of the hole = d

Area of punch hole = $\dfrac{\pi}{4}(d)^2$, **Shear area** of cylindrical hole = $\pi\, d.t$

$A = \dfrac{\pi}{4}(d)^2$, $\quad$ Shear area = $\pi\, d \,.\, t = 10\,\pi d$

Ultimate shear load = $400(10\pi\, d)$

Safe axial punch load = $\dfrac{\pi}{4}(d)^2 \times 200$

For hole to be punched $\qquad \dfrac{\pi}{4}(d)^2 \times 200 \geq 400\,(10\,\pi d)$

$$d \times 50 \geq 4000,\ d \geq \dfrac{4000}{50} = \textbf{80 mm}$$

Load $F = \dfrac{\pi}{4}(80)^2 \times 200$ N = **1004.80 kN**

EXAMPLE 5.11: A wooden joint shown in Fig. 5.8 is subjected to an axial force of 45 kN. Width of the joint is 200 mm. The allowable shear stress in wood is 0.75 N/mm² along the grains and compressive stress of 5.0 N/mm² locally. Determine the depth of the joint AB and length of the joint 2L.

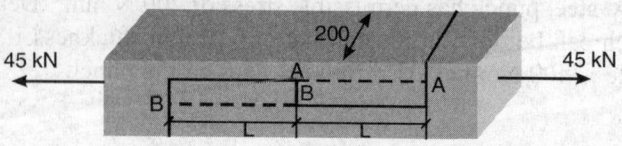

Fig. 5.8: Wooden joint

Solution:

Force $F = 45000$ N,   Allowable shear stress $= 0.75$ N/mm^2

Allowable compressive stress $= 5.0$ N/mm^2

Maximum allowable shearing force $= (200 \times L)\, 0.75 \geq 45000$, for safety.

$$\therefore \quad L = \frac{45000}{150} = \textbf{300 mm}, \text{Total length of the joint} = \textbf{600 mm}$$

Maximum allowable compressive force $= (AB \times 200)\, 5.0 \geq 45000$, for safety.

$$\therefore \quad AB = \frac{45000}{1000} = \textbf{45 mm}$$

5.7 VARIABLE X-SECTION BARS UNDER AXIAL FORCES

The **axial force elements** are of **uniform** cross-section throughout the length, the axial stress remains uniform throughout the length. If the cross-section of axial force element changes along the length of the member, the **axial stress** varies along the length. **Strains** in unit

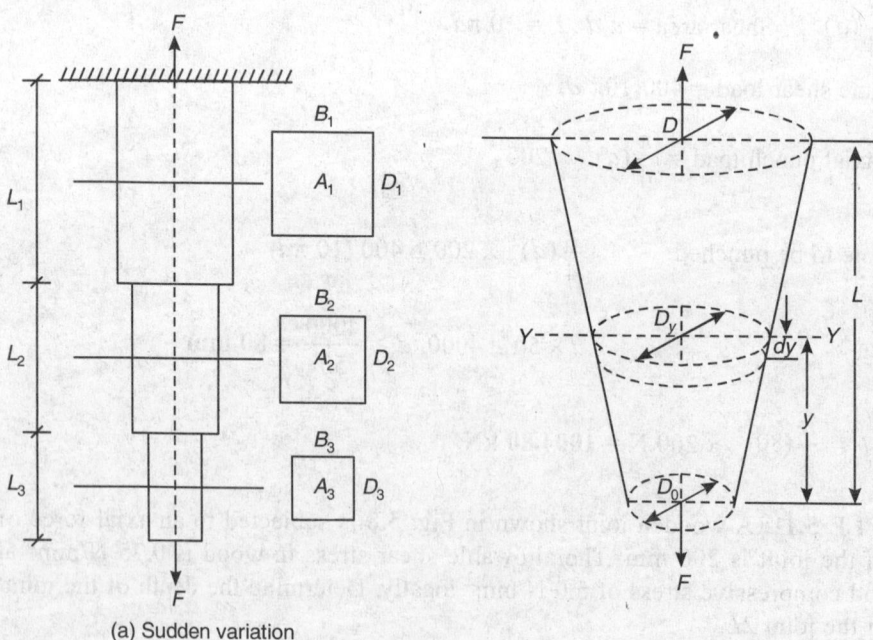

(a) Sudden variation

Fig. 5.9: Variable cross-section axial force elements

length also **varies** along the length. The variation of cross-section may be due to sudden change in diameters or cross-sections at certain points or it may vary gradually along the length (Fig. 5.9)

In this case the axial force 'F' causes variable stress according to x-section and the deformation caused is calculated separately in different sections. Since the material is same in all the sections, the modulus of elasticity 'E' is the same. Thus deformations in different sections will be:

$$\delta_1 = \frac{p_1 L_1}{E}, \qquad \delta_2 = \frac{p_2 L_2}{E}, \qquad \delta_3 = \frac{p_3 L_3}{E}$$

or $\quad \delta_1 = \dfrac{FL_1}{A_1 E}, \qquad \delta_2 = \dfrac{FL_2}{A_2 E}, \qquad \delta_3 = \dfrac{FL_3}{A_3 E}$

Total deformation $\delta = \delta_1 + \delta_2 + \delta_3$

or $\quad \delta = \dfrac{FL_1}{A_1 E} + \dfrac{FL_2}{A_2 E} + \dfrac{FL_3}{A_3 E} = \dfrac{F}{E}\left(\dfrac{L_1}{A_1} + \dfrac{L_2}{A_2} + \dfrac{L_3}{A_3}\right)$ ... (5.12)

(b)   Uniformly Varying X-Section

Let the variable cross-section be changing in diameter from D_0 (minimum) to diameter D (maximum). Let the uniform axial load be 'F'. Diameter D_y at a distance of y above minimum diameter point will be

$$D_y = \frac{(D - D_0)}{L} y + D_0 = (D_0 + Ky), \qquad \text{where } K = \frac{(D - D_0)}{L}$$

Cross-sectional area $A_y = \dfrac{\pi}{4}(Dy)^2 = \dfrac{\pi}{4}(D_0 + Ky)^2$

Stress at the section X-X $= \dfrac{F}{A_y} = \dfrac{F \cdot 4}{\pi(D_0 + Ky)^2}$

Strain $e_y = \dfrac{4F}{\pi E (D_0 + Ky)^2} = \dfrac{4F}{\pi E}(D_0 + Ky)^{-2}$

Deformation δy in length $dy = \dfrac{4F}{\pi E}(D_0 + Ky)^{-2} \, dy$

Total deformation in full length **0 to L** will be

$$\delta L = \int_0^L \frac{4F(D_0 + Ky)^{-2}}{\pi E} \, dy = \frac{4F}{\pi E}\int_0^L (D_0 + Ky)^{-2} \, dy = \frac{4F}{\pi E}\left[\frac{(D_0 + Ky)^{-1}}{-1 \cdot K}\right]_0^L$$

$$= \frac{-4F}{\pi EK}[(D_0 + KL)^{-1} - D_0^{-1}] = \frac{-4F}{\pi EK}\left[\frac{1}{D_0 + KL} - \frac{1}{D_0}\right]$$

$$= \frac{-4F}{\pi EK}\left[\frac{D_0 - D_0 - KL}{D_0(D_0 + KL)}\right] = \frac{-4F}{\pi EK}\left[\frac{-KL}{D_0(D_0 + KL)}\right] = \frac{4FKL}{\pi EK}\left[\frac{1}{D_0 \cdot D}\right]$$

[Since $D = D_0 + KL$]

$$= \frac{4FL}{\pi E}\left[\frac{1}{D_0 \cdot D}\right] = \frac{FL}{\frac{\pi}{4}(D_0 \cdot D)E} \qquad \qquad ...(5.13)$$

If $D = D_0$, i.e. uniform section, then,

$$\delta L = \frac{FL}{\frac{\pi}{4}(D^2)E} = \frac{FL}{A \cdot E} \text{ same as found earlier.} \qquad ...(5.12)$$

EXAMPLE 5.12: A 10 m long steel rod has uniform diameter of 30 mm in first 4 m. Next 3 m length has uniform diameter of 40 mm and for remaining 3 m length has 20 mm diameter. If the rod is subjected to an axial load **50 kN**, find the total elongation of the rod. $E = 200$ kN/mm^2. Also find the strain energy stored in the rod.

Solution:

$$A_1 = \frac{\pi}{4}(30)^2 = \textbf{225 } \boldsymbol{\pi} \textbf{ mm}^2, \qquad L_1 = \textbf{4000 mm}, \qquad\qquad E = \textbf{200 kN/mm}^2$$

$$A_2 = \frac{\pi}{4}(40)^2 = \textbf{400 } \boldsymbol{\pi} \textbf{ mm}^2, \qquad L_2 = \textbf{3000 mm}, \qquad\qquad F = \textbf{50 kN}$$

$$A_3 = \frac{\pi}{4}(20)^2 = \textbf{100 } \boldsymbol{\pi} \textbf{ mm}^2, \qquad L_3 = \textbf{3000 mm}$$

$$\delta L \text{ (Total)} = \frac{F}{E}\left(\frac{L_1}{A_1} + \frac{L_2}{A_2} + \frac{L_3}{A_3}\right) = \frac{50}{200}\left(\frac{4000}{225\pi} + \frac{3000}{400\pi} + \frac{3000}{100\pi}\right)$$

$$= \frac{1000}{4\pi}\left(\frac{4}{225} + \frac{3}{400} + \frac{3}{100}\right) = \frac{552.777}{4\pi} = \textbf{4.4011 mm}$$

Strain energy = Average Resistance × Deformation

= (50/2) × 4.4011 = **110.03 kN-mm**

EXAMPLE 5.13: A 10 m long steel rod has 20 mm diameter at one end and varies uniformly to 10 mm diameter at the other end. If the rod carries an axial force of 20 kN and has material $E = 200$ kN/mm^2 and $\mu = 0.30$, determine elongation and volume change in the rod.

Solution:

$L = 10,000$ mm, $\quad\quad E = 200$ kN/mm^2, $\quad\quad F = 20$ kN, $\quad\quad \mu = 0.30$,

$D_1 = 10$ mm, $\quad\quad\quad\quad\quad\quad\quad\quad\quad\quad D_2 = 20$ mm

$$\delta L = \frac{FL}{\frac{\pi}{4}(D_0 \cdot D)E} = \frac{20 \times 10000}{200 \times \frac{\pi}{4}(10 \times 20)} = \frac{4000}{200\pi} = \frac{20}{\pi} = 6.37 \text{ mm}$$

$$\frac{\delta V}{V} = e_1(1 - 2\mu) = \frac{6.37}{10000}(1 - 0.6) = 2.548 \times 10^{-4}$$

$$V = \frac{\pi}{3}(r_1^2 + r_2^2 + r_1.r_2) \cdot L = \frac{\pi}{3}(5.0^2 + 10.0^2 + 5 \times 10) \times 10000 = \frac{175\pi}{3} \times 10000$$

$$\therefore \quad \delta V = 2.548 \times 10^{-4} \times \frac{175\pi}{3} \times 10^4 = 466.70 \text{ mm}^3$$

5.8 AXIAL STRESS AND STRAIN DUE TO SELF WEIGHT

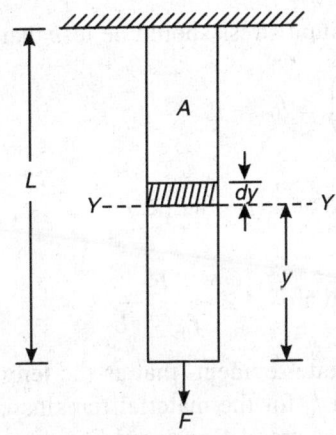

Fig. 5.10: Stress–strain due to self weight

Consider a rod of length L and **cross-sectional area A**. Let the weight **density** of rod material be w N/mm^3. Consider a section Y-Y at a distance of y from the bottom end.

Weight of rod below section **Y-Y**

$F = A \cdot y \cdot w$

Stress at Y-Y $= \dfrac{F}{A} = \dfrac{A \cdot wy}{A} = wy$ ... (5.14)

Strain at Y-Y $= \dfrac{f}{E} = \dfrac{wy}{E}$

Deformation in small strip $dy = \dfrac{wy\,dy}{E}$

Total deformation in the rod length $y = 0$ to $y = L$

$$\delta L = \int_0^L \frac{w \cdot y \cdot dy}{E} = \frac{w}{E}\left(\frac{L^2}{2} - 0\right) = \frac{wL^2}{2E}$$

Total $\qquad\qquad \delta L = \dfrac{wL^2}{2E}$ ... (5.15)

If the rod also carries axial load F in addition to its self load, the stress at any point will be

$$f_y = \left(wy + \frac{F}{A}\right)$$ ... (5.16)

f_y will be maximum when y is maximum, i.e. $y = L$

$\therefore \qquad\qquad f_{y\,(max)} = \left(wL + \dfrac{F}{A}\right)$

For the design purpose, maximum stress should be less than permissible stress f,

i.e. $\qquad\qquad \left(wL + \dfrac{F}{A}\right) \le f_0.$

or $\qquad\qquad \dfrac{F}{A} \le (f_0 - wL)$ or $A\,(f_0 - wL) \ge F$

$\therefore \quad$ Necessary cross-sectional area $A \ge \dfrac{F}{f_0 - wL}$ ... (5.17)

From this equation it is quite evident that as the **length increase, area** required **also increase** for safe stresses. w and f_0 for the material remains constant and as 'L' increases, at some stage $f_0 = wL$. In such a case area 'A' required becomes **infinite** and such a rod must be designed with a **variable cross-sectional area**.

Total deformation = sum of deformation due to constant load F and uniformly varying stress due to self load (wAy)

$$\delta L \text{ (Total)} = \left(\frac{FL}{A \cdot E} + \frac{wL^2}{2E}\right)$$ ... (5.18)

It is evident that **deformation** due to **self weight** is **half** of uniform **axial stress** 'f'.

EXAMPLE 5.14: A vertical steel shaft of a mining pump is 100 m long. Permissible stress is 50 N/mm^2 and density is 7.8×10^{-5} N/mm^3. Stroke length of piston = 200 mm. Find the safe diameter of the rod and radius of the crank shaft if net resistance in upward movement is **10 kN** while in downward movement it will be only **1 kN**.

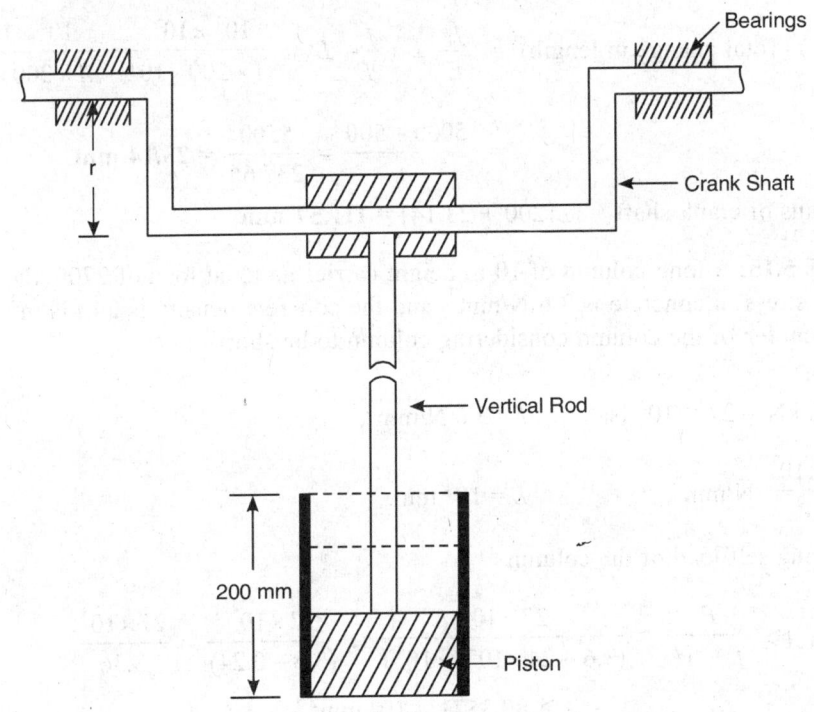

Fig. 5.11: Long vertical rod

Solution:

Density $w = 7.8 \times 10^{-5}$ N/mm^3, $L = 100{,}000$ mm $= 10^5$ mm

f (permissible) $= 50$ N/mm^2,   Piston stroke $= 200$ mm

Resistance in downward stroke $= 1$ kN   Resistance in upward stroke $= 10$ kN

The shaft rod must be safe in maximum force, i.e. for **10 kN**.

Since rod is very long, its self weight will influence the stress and strains.

Safe area of x-section $A \geq \dfrac{F}{f - wL}$.

or $A = \dfrac{\pi}{4}(d^2) \geq \dfrac{10000}{(50 - 7.8 \times 10^{-5} \times 10^5)}$

or $A = \dfrac{\pi}{4}(d^2) \geq \dfrac{10000}{42.2} = 237$, $d^2 \geq \dfrac{10000 \times 4}{\pi (42.2)}$

$d \geq 17.37$ **mm**, Diameter be taken as **17.4 mm** ($A = 237.66$ mm^2).

When the piston moves **upward**, the vertical rod is in **compression** and the **rod shortens** while the rod **elongates** when the piston moves **downward** and the rod is **under**

tension. The total shortening and elongation must be adjusted in the radius of crank shaft, i.e. radius of crank shaft = ½ [stroke + $\delta_1 + \delta_2$].

$$(\delta_1 + \delta_2) \text{ (Total change in length)} = \frac{f_1}{E} \cdot L + \frac{f_2}{E} \cdot L = \frac{10^4 \times 10^5}{A \times 200 \times 10^3} + \frac{10^3 \times 10^5}{A \times 200 \times 10^3}$$

$$= \frac{5000 + 500}{A} = \frac{5500}{237.66} = \textbf{23.14 mm}$$

Thus radius of crank shaft = ½ [200 + 23.14] = **111.57 mm**

EXAMPLE 5.15: A long column of 10 m height carries an axial load of 2700 kN. If the safe permissible stress in concrete is 3.6 N/mm², and the concrete density is 24 kN/m³, determine the safe **diameter** of the column considering column to be short.

Solution:

$F = 2700 \text{ kN} = 27 \times 10^5 \text{ N}, \qquad f = 3.6 \text{ N/mm}^2,$

$w = \dfrac{24 \times 10^3}{10^9} \text{ N/mm}^3, \qquad L = 10^4 \text{ mm}$

Considering self load of the column

$$\text{Safe area } A \geq \frac{F}{f - wL} = \frac{27 \times 10^5}{(3.6 - 24 \times 10^{-6} \times 10^4)} = \frac{27 \times 10^5}{(3.6 - 0.24)} = \frac{27 \times 10^5}{3.36}$$

$$A \geq 80.3571 \times 10^4 \text{ mm}^2,$$

If the column is circular having diameter 'd'.

$$\frac{\pi}{4} (d^2) = 80.3571 \times 10^4, \quad d^2 = \frac{80.3571 \times 4 \times 10^4}{\pi} = 102.366 \times 10^4$$

$$d = 1011.76 \text{ mm}$$

Safe diameter of the concrete column = **1012 mm**

5.9 AXIAL FORCE ELEMENTS OF COMPOSITE MATERIALS

When the axial force element consists of more than one material having different modulii of elasticity, such elements are called **composite** members. Both the materials of **composite** members **resist** external forces jointly as per their elastic properties. **Composite** elements can be formed by annularly joining **two tabular** or one tabular section with other solid cylindrical and fixed concentrically.

When axial force F acts, both the materials of the composite bar jointly put up the resistance F_1 and F_2 ($\therefore$ $F_1 + F_2 = F$). Since both the material undergo the **same deformation** and have the **same** length, strains in both are also equal, i.e.

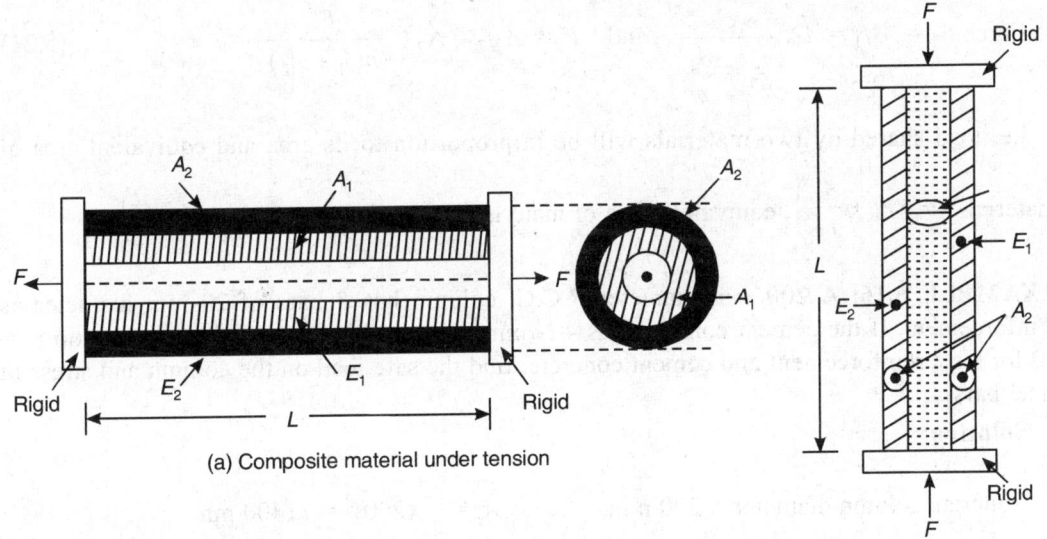

(a) Composite material under tension

(b) Composite material rod under compression

Fig. 5.12: composite section

$$e_1 = e_2 = e \left(\text{Since } \frac{\delta_1}{L_1} = \frac{\delta_2}{L_2} = \frac{\delta}{L} \right)$$

or $$\frac{f_1}{E_1} = \frac{f_2}{E_2} \quad \text{or} \quad f_1 = \frac{E_1}{E_2} \cdot f_2 \qquad \qquad \dots (5.19\ (a))$$

or $$\frac{f_1}{f_2} = \frac{E_1}{E_2} \qquad \qquad \dots (5.19\ (b))$$

i.e. ratio of stresses in two materials are in the ratio of their modulii of elasticity.

If $$\frac{E_1}{E_2} = m, \qquad f_1 = mf_2 \qquad \qquad \dots (5.19\ (c))$$

Thus these two equations $F_1 + F_2 = F$ and $f_1 = mf_2$ can be used to find the strains in the materials of composite bar.

$$\therefore \quad F = (f_1 \cdot A_1) + (f_2 \cdot A_2) = mf_2 \cdot A_1 + f_2 \cdot A_2 = f_2\ (mA_1 + A_2)$$

or $$f_2 = \frac{F}{(mA_1 + A_2)}, \quad \text{Also } f_1 = mf_2 = \frac{mF}{(mA_1 + A_2)} = \frac{F}{A_1 + \dfrac{A_2}{m}} \qquad \dots (5.20)$$

i.e. $$f_1 = \frac{F}{A_1 + \dfrac{A_2}{m}}, \qquad f_2 = \frac{F}{(mA_1 + A_2)}$$

Force $F_1 = A_1 f_1 = A_1 \cdot \dfrac{F}{A_1 + \dfrac{A_2}{m}}$ and $F_2 = A_2 f_2 = A_2 \cdot \dfrac{F}{(mA_1 + A_2)}$ ... (5.21)

i.e. load shared by two materials will be in proportion to its area and equivalent area of

material-1 $= \left(A_1 + \dfrac{A_2}{m} \right)$, equivalent area of material-2 $= (mA_1 + A_2)$, where $m = \dfrac{E_1}{E_2}$.

EXAMPLE 5.16: A 200 mm diameter R.C.C. column has 5 bars of 20 mm diameter as reinforcement. If the cement concrete has 4 N/mm² permissible stress and modular ratio $m = 20$ for steel reinforcement and cement concrete, find the safe load on the column and stress in steel bars.

Solution:

Concrete column diameter = 200 mm, $\qquad A = \dfrac{\pi}{4} (200)^2 = 31400 \text{ mm}^2$

Steel bar diameter = 20 mm, $\qquad A_S = 5 \times \dfrac{\pi}{4} (20)^2 = 1570 \text{ mm}^2$

Concrete Area $A_C = A - A_S = 31400 - 1570 = 29830 \text{ mm}^2$
Both the materials deform together $\therefore e = e_S = e_C$

or $\quad \dfrac{f_S}{E_S} = \dfrac{f_C}{E_C} \quad$ or $\quad f_S = \dfrac{E_S}{E_C} \cdot f_C = 20 \times 4 = 80 \text{ N/mm}^2$

Total safe load $F = F_S + F_C = A_S \cdot f_S + A_C \cdot f_C = 1570 \times 80 + 29830 \times 4$
or $\quad F = 125600 + 119320 = 244920 \text{ N}$
or $\quad F = 244.920 \text{ kN} \approx \textbf{245 kN}$

EXAMPLE 5.17: Three rods of equal length comprise of copper, zinc and aluminium having x-sectional areas of 600, 900 and 1200 mm² respectively. Ends of these rods are rigidly connected. The compound rod is subjected to an axial pull of 228 kN. Determine the load shared by each rod if their modulii are: $E_C = 112.5 \text{ kN/mm}^2$, $E_Z = 85 \text{ kN/mm}^2$, $E_A = 70 \text{ kN/mm}^2$.

Solution:

$A_C = 600 \text{ mm}^2, \qquad A_Z = 900 \text{ mm}^2, \qquad A_A = 1200 \text{ mm}^2, \qquad F = \textbf{228000 N}$
$E_C = 112.5 \text{ kN/mm}^2, \qquad\qquad\qquad E_Z = 85 \text{ kN/mm}^2, \qquad E_A = 70 \text{ kN/mm}^2$

$$ e = e_C = e_Z = e_A $$

i.e. $\qquad\qquad \dfrac{f_C}{E_C} = \dfrac{f_Z}{E_Z} = \dfrac{f_A}{E_A} = e \qquad\qquad\qquad$ ... (1)

Total load $F = F_C + F_Z + F_A$,
i.e. $\qquad\qquad F = f_C \cdot A_C + f_Z \cdot A_Z + f_A \cdot A_A \qquad\qquad$ ... (2)

or $\quad 228 \times 10^3 = f_C \cdot 600 + f_Z \cdot 900 + f_A \cdot 1200$

or $\quad 228 \times 10^3 = e \cdot E_C \cdot 600 + e \cdot E_Z \cdot 900 + e \cdot E_A \cdot 1200$

or $\quad 228000 = e \times 112.5 \times 10^3 \times 600 + e \times 85 \times 10^3 \times 900 + e \times 70 \times 10^3 \times 1200$

or $\quad 228 = 100 \cdot e \, [675.0 + 76.5 + 840] = 100 \cdot e \, [2280]$

or $\quad e = \dfrac{1}{1000} = 10^{-3}$

Substituting the value of e, we have

$F_C = 600 \times 112.5 \times 10^3 \times 10^{-3} = 67.5$ kN, $\quad F_Z = 76.5$ kN, $\quad F_A = 84$ kN

Total $F = $ **228 kN**

EXAMPLE 5.18: A steel bolt of 800 mm length and 800 mm^2 cross-section has thread of pitch 2 mm. The bolt is centrally placed in a copper tube having cross-sectional area of 1600 mm^2 and fixed with the tube by rigid washers at ends.

a. If the nut is given half extra turn after tightening, find the stresses in bolt and tube if initially no stress exists. $E_S = 210$ kN/mm^2, $E_C = 112.5$ kN/mm^2.

b. If the nut is turned to create an axial compression of 200 kN determine the stresses in the bolt and tube and angle of rotation.

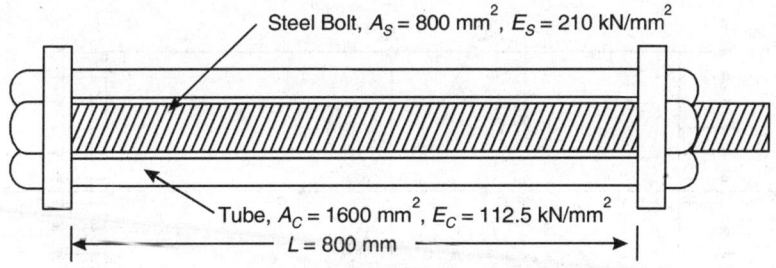

Steel Bolt, $A_S = 800$ mm^2, $E_S = 210$ kN/mm^2

Tube, $A_C = 1600$ mm^2, $E_C = 112.5$ kN/mm^2
$L = 800$ mm

Fig. 5.13.

Solution:

a. Pitch $p = 2$ mm, Total deformation by half turn $\delta = \frac{1}{2}\,(p) = 2/2 = $ **1 mm**

Total deformation $\delta = $ Contraction in copper tube + elongation of steel bolt.

i.e. $\quad \delta = 1$ mm $= \dfrac{f_C}{E_C} \cdot L + \dfrac{f_S}{E_S} \cdot L = \dfrac{FL}{A_C \cdot E_C} + \dfrac{FL}{A_S \cdot E_S}$,

[Since Force on tube = Force on bolt]

or $\quad 1 = FL \left(\dfrac{1}{A_C \cdot E_C} + \dfrac{1}{A_S \cdot E_S} \right) = F \times 800 \left(\dfrac{1}{112.5 \times 10^3 \times 1600} + \dfrac{1}{210 \times 10^3 \times 800} \right)$

or $\quad 1 = F \times \dfrac{800 \times 10^{-5}}{8} \left(\dfrac{1}{225} + \dfrac{1}{210} \right) = F \times 10^{-3} \left(\dfrac{210 + 225}{225 \times 210} \right)$

or $\quad F = \dfrac{1 \times 225 \times 210}{435 \times 10^{-3}} = \dfrac{10^3 \times 15 \times 210}{29} = \mathbf{108.62\ kN}$

$f_C = \dfrac{108.62}{1600} \times 10^3\ \text{N/mm}^2 = \mathbf{67.89\ N/mm^2}$

$f_S = \dfrac{108.62}{800} \times 10^3\ \text{N/mm}^2 = \mathbf{135.78\ N/mm^2}$

b. $\quad F = 200 \times 1000\ \text{N},\quad f_C = \dfrac{200000}{1600} = 125\ \text{N/mm}^2,\quad f_S = \dfrac{200000}{800} = 250\ \text{N/mm}^2$

$\delta = \delta C + \delta S = \dfrac{125 \times 10^{-3}}{112.5} \times 800 + \dfrac{250 \times 10^{-3}}{210} \times 800 = \dfrac{100}{112.5} + \dfrac{200}{210}$

$\quad\quad = 0.889 + 0.9524 = 1.8413$

Angle Turn $\theta = 2\pi \times \dfrac{1.8413}{2.0} = \mathbf{5.7816}$ radian or $\mathbf{0.9206}$ turn $(\mathbf{331.43^0})$

5.10 STRESSES ON INCLINED PLANES DUE TO AXIAL STRESS

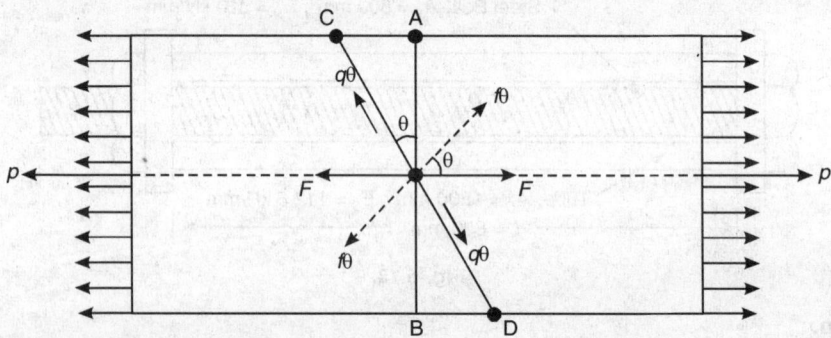

Fig. 5.14: Normal and tangential (shear) stresses on inclined plane due to axial stress

Consider an axial force member having cross-sectional area of A and carries an axial force F. the **normal stress** across the transverse plane AB will be $p = \dfrac{F}{A}$.

Let us find the stress across an **inclined plane CD** inclined at an angle θ (anticlockwise) with the transverse plane **AB** (normal of the inclined plane CD is at angle of θ anticlockwise with the axis of member).

Length CD = AB sec θ,　Uniform thickness of the member = t

Area of the inclined plane $CD\ A' = t$. $CD = t$. AB sec $\theta = A$ sec θ

Component of axial force F along the normal of the inclined plane $CD = F$ cos θ.

∴　Normal stress f_θ along the inclined plane $CD = \dfrac{F \cos \theta}{A \sec \theta} = \dfrac{F}{A} \cos^2 \theta$

or $$f_\theta = p \cdot \cos^2 \theta \qquad \qquad \dots (5.22)$$

Tangential (shear) component along the inclined plane $CD = F \sin \theta$

Tangential stress $$q_\theta = \frac{F \sin \theta}{A \sec \theta} = \frac{F}{A} \cdot \sin \theta \cos \theta = \frac{P}{2} \cdot \sin 2\theta \qquad \dots (5.23)$$

Thus stresses on inclined plane CD inclined at an angle θ will be

Normal Stress $$f_\theta = p.\cos^2 \theta = \frac{p}{2} (1 + \cos 2\theta) \qquad \dots (5.22 \text{ (b)})$$

Tangential Stress $$q_\theta = p.\sin \theta \cos \theta = \frac{p}{2} \sin 2\theta \qquad \dots (5.23 \text{ (b)})$$

f_θ and q_θ will always be **less than** p as values of $\cos \theta$ and $\sin \theta$ will always be less than 1.0 or at the most equal to 1.0. Thus maximum values of f_θ and q_θ are:

$$(f_\theta)_{max} = p, \text{ when } \cos^2 \theta = 1, \text{ or } \boldsymbol{\theta = 0}, (q_\theta)max = \frac{p}{2}, \text{ when } \sin 2\theta = 1, \theta = 45^0 \quad \dots (5.24)$$

5.11 COMPLEMENTARY SHEAR AT ANY POINT IN STRESSED BODY

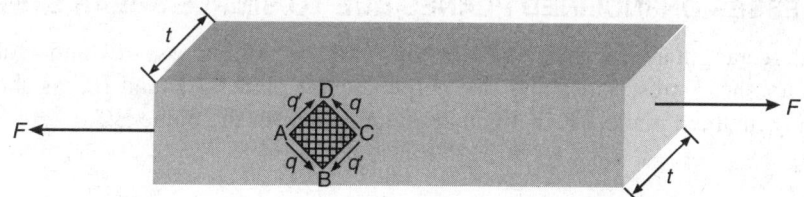

Fig. 5.15: Complementary shear in an element

Consider any member of uniform thickness 't' subjected to certain load F. Let there be small element ABCD at any point in the stressed body. Since the body is in equilibrium the element **ABCD** will also be in **equilibrium**. Let the faces AB and CD has shear stress q and the faces BC and AD has tangential force F' causing shear stress q'.

Force: $$F = q \cdot t \cdot AB, \qquad F' = q' \cdot t \cdot BC$$

Since the element is in equilibrium, the moment of the couple formed along the faces AB and CD will be equal and opposite to the moment of the couple formed along the faces BC and DA.

i.e. $$F \times BC = F'. CD$$

or $$(q \cdot t \cdot AB) BC = (q' \cdot t \cdot BC) CD$$

or $$q \cdot (AB \times BC \times t) = q' (BC \times CD \times t)$$

$$\boldsymbol{AB = CD}$$

$\therefore$ $$q (AB \times BC \times t) = q' (BC \times AB \times t)$$

or $$\boldsymbol{q = q'} \text{ (Complementary Shear)} \qquad \dots (5.25)$$

Hence any element in a stressed body is always subjected to a pair of shear stresses of equal intensity on transverse or perpendicular planes. In other words **shear stress q on any plane is** always accompanied by another shear stress *q* along the perpendicular plane. This **shear stress of equal intensity** and **opposite nature** on the perpendicular plane is called **complementary shear**.

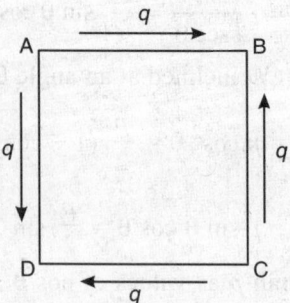

Fig. 5.16: Complementary shear stress

Shear stress on **any plane can not occur without complementary shear** on the **perpendicular plane**, i.e. if any plane in a stressed body is subjected to any **shear stress**, its perpendicular plane will also have **shear stress** of the **same intensity** and **opposite nature**.

5.12 STRESSES ON INCLINED PLANES DUE TO SIMPLE SHEAR STRESSES

Consider a rectangular element **ABCD** of uniform thickness '*t*' and subjected to complementary shear stresses *q* along two perpendicular planes **AB** and **BC** as shown in Fig. 5.17. Take any inclined plane **BE** inclined at an angle *q* with the plane BC.

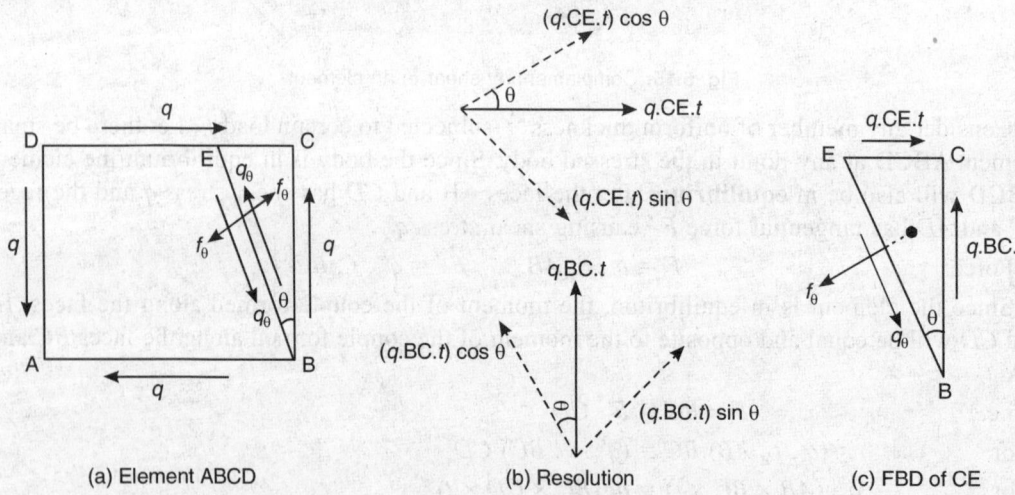

(a) Element ABCD   (b) Resolution   (c) FBD of CE

Fig. 5.17: Stresses on inclined plane BE due to shear stress

Consider **FBD** (Free Body Diagram) of the inclined plane element **BCE**.

Force along $CE = q \cdot CE \cdot t$,   **Force** along $BC = q \cdot BC \cdot t$

Components along the **normal of BE**

$$(f_\theta \cdot BE \cdot t) = (q \cdot BC \cdot t) \sin \theta + (q \cdot CE \cdot t) \cos \theta$$

or $$f_\theta = q \cdot \frac{BC}{BE} \cdot \sin\theta + q \cdot \frac{CE}{BE} \cdot \cos\theta$$

$$= q \{\cos\theta \cdot \sin\theta + \sin\theta \cdot \cos\theta\}$$

or $$f_\theta = q \ (\sin 2\theta) \qquad \qquad ... (5.26)$$

Tangential component along BE

$$(q_\theta \cdot BE \cdot t) = -(q \cdot CE \cdot t) \sin\theta + (q \cdot BC \cdot t) \cos\theta$$

or $$q_\theta = q \left\{ \frac{BC}{BE} \cdot \cos\theta - \frac{CE}{BE} \cdot \sin\theta \right\} = q \{\cos^2\theta - \sin^2\theta\}$$

or $$q_\theta = q \ (\cos 2\theta) \qquad \qquad ... (5.27)$$

f_θ will be maximum when $\sin 2\theta$ is maximum, i.e. $\sin 2\theta = 1$, $2\theta = 90°$ or $270°$, $\theta = \mathbf{45°}$ or **135°**

Normal stress f_θ **will be maximum** along a plane at **45°** and **135°** and of the same magnitude as the shear stress.

$$(f_\theta)_{\text{max}} = q \qquad \qquad ... (5.28)$$

q_θ will be maximum when $\cos 2\theta$ is maximum, i.e. $\cos 2\theta = 1$, $2\theta = 0$ or $180°$, i.e. $\theta = \mathbf{0}$ or **90°**.

$$(q_\theta)_{\text{max}} = q \qquad \qquad ... (5.29)$$

Hence **maximum shear** occurs along the **same plane** where the **simple shear is acting**. It is important to note that along the **plane of maximum shear** there is **no normal stress**. In case of pure shear stress, there exists **two perpendicular** planes along which **normal stresses** are **maximum** and **minimum** and equal to the **pure shear q in magnitude** and these **planes of maximum and minimum stress** is located at **45°** or **(90° + 45°)** with the plane of pure shear. The maximum normal stress may be tensile or compressive depending on the direction (nature) of pure shear stress.

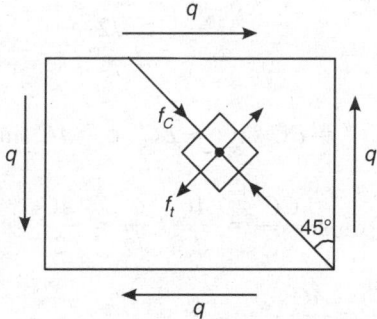

Fig. 5.18: Planes of maximum and minimum normal stresses

5.13 RELATIONS BETWEEN VARIOUS ELASTIC MODULII (E, G AND K)

Elastic modulus (E) is defined as ratio of **linear stress** (p) to the **elastic strain** (e) within **limits of proportionality**.

Shear modulus of elasticity (**G**) also known as **modulus of rigidity** is the ratio of **pure shear stress q** to the corresponding **shear strain** (**φ**).

Bulk modulus of elasticity (**K**) is defined as the **ratio of three dimensional uniform stress** ($p = p_x = p_y = p_z$) to the corresponding **volumetric strain** (e_v).

(a) Consider a square element ABCD subjected to pure **shear stress q** as shown in Fig. 5.19. Assume one edge of the element as fixed and the element **ABCD deforms** to **ABC′D′** causing **shear deformation of φ**.

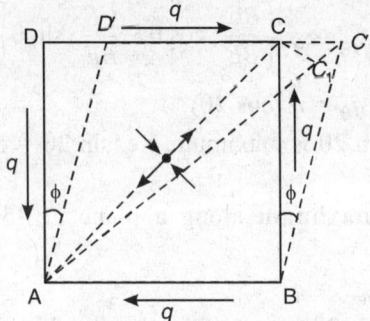

Fig. 5.19: Square element under pure shear

The sides AD and BC deforms to D' and C' respectively. The diagonal AC is subjected to **tensile stress of 'q'** due to the action of pure shear 'q'. It may be noted that the diagonal AC is inclined at **45°** and the diagonal BD is inclined at **135°**. The diagonal BD develops a linear stress of '$- q$' i.e. Compressive stress 'q' transverse to **AC**. Let CC_1 be perpendicular to AC'. **AC ≈ AC_1**.

Elongation in $AC = AC' - AC = AC' - AC_1 = C'C_1$

Angle $CC'C_1 = 45°$ approximately,

∴
$$C_1C' = CC' \cos 45° = \frac{CC'}{\sqrt{2}} \qquad \text{... (1)}$$

(φ being very small angle)

$$CC' = BC \sin \phi \approx BC . \phi = AC \sin 45°.\phi = \frac{AC}{\sqrt{2}} \cdot \phi \qquad \text{... (2)}$$

$$C_1C' = \frac{CC'}{\sqrt{2}} = \frac{AC}{\sqrt{2}} \cdot \frac{\phi}{\sqrt{2}} = \frac{AC}{2} \cdot \phi$$

Linear strain e in $AC = \dfrac{C_1C'}{AC} = \dfrac{AC}{2} \dfrac{\phi}{AC} = \dfrac{\phi}{2}$

$$e = \frac{\phi}{2} \qquad \text{... (3)}$$

Strain in diagonal AC can also be calculated in terms of **tensile stress q** along AC and **lateral compressive stress q**.

$$e = \frac{q}{E} + \mu \cdot \frac{q}{E} = \frac{q}{E}(1 + \mu) \qquad \dots (4)$$

By Hooke's law

$$G = \frac{q}{\phi} \quad \text{or} \quad \phi = \frac{q}{G} \qquad \dots (5)$$

From (3), (4) and (5)

$$\therefore \qquad e = \frac{\phi}{2} = \frac{q}{2G} = \frac{q}{E}(1 + \mu)$$

or

$$\frac{q}{2G} = \frac{q}{E}(1 + \mu)$$

or

$$\boldsymbol{E = 2G\,(1 + \mu)} \qquad \dots (5.30)$$

(b) Volumetric strain is defined as **change in volume per unit** volume due to action of uniform stress (pressure) in **three perpendicular** directions. The ratio of this **uniform stress p** in three perpendicular directions and corresponding **volumetric strain** is known as **Bulk Modulus of elasticity (K).**

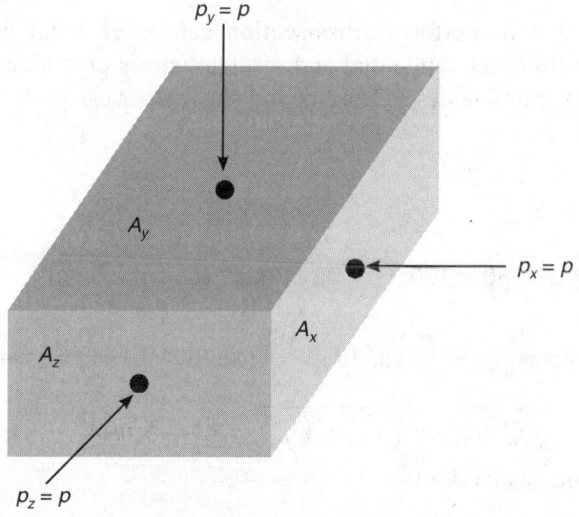

Fig. 5.20: Cuboid under pressure p

Consider a cuboid subjected to three dimensional stress (pressure) p. **Elastic strain** in **each**

direction will be $e_x = \dfrac{p}{E} - \mu\dfrac{p}{E} - \mu\dfrac{p}{E} = \dfrac{p}{E}(1 - 2\mu)$ $\qquad \dots (6)$

If original volume is $V = x^3$, Changed volume $V' = \{x\,(1 + e_x)\}^3$

Volumetric strain $\qquad e_V = \dfrac{\delta V}{V} = \left(\dfrac{V'-V}{V}\right) = \dfrac{x^3(1+e_x)^3 - x^3}{x^3} = (1+e_x)^3 - 1$

or $\qquad\qquad\qquad\qquad e_V = 3e_x + 3e_x^2 + e_x^3$

neglecting higher order terms of small quantities, we have,

$$e_V = 3e_x \qquad\qquad\qquad ... (7)$$

By definition K (Bulk Modulus) $= \dfrac{p}{e_V}$ or $e_V = \dfrac{p}{K}$ $\qquad\qquad ... (8)$

From equations (6), (7) and (8), we have

$$\frac{p}{K} = 3\left\{\frac{p}{E} - 2\mu\frac{p}{E}\right\} - 3\frac{p}{E}(1 - 2\mu)$$

or $\qquad\qquad\qquad \dfrac{1}{K} = \dfrac{3(1-2\mu)}{E}$

or $\qquad\qquad\qquad \boldsymbol{E = 3K(1 - 2\mu)} \qquad\qquad\qquad ... (5.31)$

Equations 5.30 and 5.31 can be combined as

$$\boldsymbol{E = 2G(1 + \mu) = 3K(1 - 2\mu)} \qquad\qquad ... (5.32)$$

Equation **5.32** gives relationship between 3 elastic modulii.

EXAMPLE 5.19: A bar of uniform cross-section carries an axial compressive stress of 100 N/mm^2. Determine normal, tangential and resultant stress on a plane inclined at $30°$ with the axis. Also find maximum values of normal and shear stresses.

Solution:

$p = 100$ N/mm^2, $\qquad \theta = 30°$

Nominal Stress

$$f_\theta = p\cos^2\theta = 100\cos^2 30 = 100\,\frac{3}{4} = \textbf{75 N/mm}^2 \text{ (Compressive)}$$

Tangential (Shear Stress) $q_\theta = \dfrac{p}{2}\sin 2\theta = \dfrac{100}{2}\sin 60° = \dfrac{100\sqrt{3}}{4} = \textbf{43.3 N/mm}^2$

Resultant Stress $f_r = \sqrt{f_\theta^2 + q_\theta^2} = \sqrt{75^2 + 43.3^2} = \textbf{86.6 N/mm}^2$

Max. $f_\theta = 100$ N/mm^2, when $\theta = 0°$.

Max. $q_\theta = \dfrac{p}{2} = \dfrac{100}{2} = 50$ N/mm^2, $\qquad\qquad\qquad\qquad$ when $\theta = 45°$

EXAMPLE 5.20: A rectangular bar is subjected to a pure shear stress of 100 N/mm^2 at a point. Determine maximum normal stress at that point.

Solution:

$q = 100$ N/mm^2

$(f_\theta)_{max} = q$, $\qquad\qquad$ at planes inclined at $45°$ and $135°$.

i.e. $f_\theta = \pm 100$ N/mm^2 (Tensile and Compressive) at $\theta = 45°$ and $135°$ respectively.

EXAMPLE 5.21: A square element ABCD with 50 mm side is made of a material having $E = 70$ kN/mm^2 and $\mu = 0.25$. Determine the relative displacement of CD with AB when subjected to pure shear of 70 N/mm^2.

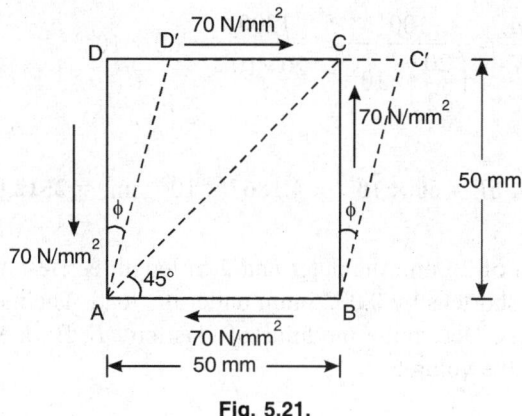

Fig. 5.21.

Solution:

$q = 70$ N/mm^2, $E = 70$ kN/mm^2, $\mu = 0.25$

Shear Strain $\phi = \dfrac{q}{G}$, $E = 2G\,(1 + \mu)$ or $70 \times 10^3 = 2G\,(1 + 0.25)$

$G = \dfrac{70 \times 10^3}{2 \times 1.25} = 28 \times 10^3$ N/mm^2

$\phi = \dfrac{70}{28 \times 10^3} = 2.5 \times 10^{-3} = 0.0025$ radians $(0.1433°)$

Elongation strain of diagonal $AC = e = \dfrac{\phi}{2} = 0.00125.$

$\delta_{AC} = 50\sqrt{2} \times 0.00125 =$ **0.08839 mm**.

Also $CC' = BC\,\phi = 50 \times 0.0025 = 0.1250$ mm

Extension in $AC = CC' \cos 45° = \dfrac{0.1250}{\sqrt{2}} =$ **0.08839 mm**

EXAMPLE 5.22: A steel sphere of 200 mm diameter is subjected to a uniform pressure of 100 N/mm^2 from all the directions. If the value of modulus of elasticity $E = 200$ kN/mm^2, $\mu = 0.30$, determine the change in volume.

Solution:

Volume $V_0 = \dfrac{4}{3}\pi r^3 = \dfrac{4}{3}\pi\,(100)^3 =$ **4.1867 $\times$ 10^6 mm^3**,

$p = 100$ N/mm^2,

$$E = 3K(1 - 2\mu), \quad K = \frac{E}{3(1 - 2\mu)} = \frac{200 \times 10^3}{3(1 - 0.6)} = N/mm^2 = \frac{2 \times 10^5}{1.2} = 1.667 \times 10^5 \ N/mm^2$$

$$\text{Volume Strain } e_V = \frac{p}{K} = \frac{100}{\left(\frac{20}{12}\right) \times 10^5} = \frac{1200}{20 \times 10^5} = 60 \times 10^{-5}$$

$$\frac{\delta V}{V} = e_V = 60 \times 10^{-5}, \ \delta V = \mathbf{60 \times 10^{-5}} \times 4.1867 \times 10^6 \ mm^3 = \mathbf{2512.02 \ mm^3}$$

EXAMPLE 5.23: A rod of 20 mm diameter and 2 m length carries an axial pull of 31.4 kN. The diameter of the rod shortens by 0.0025 mm under the load. The modulus of rigidity of the material $G = 80 \ kN/mm^2$. Determine modulus of elasticity E, Bulk Modulus (K), Poisson's Ratio (μ) and change in the volume.

Solution:

$$A = (20)^2 = 314 \ mm^2, \qquad \text{Load} = 31400 \ N, \qquad \text{Stress } p = \frac{31400}{314} = \mathbf{100 \ N/mm^2}$$

$$\text{Lateral Strain } e' = \mu e = \mu \frac{p}{E} \quad \text{or} \quad \frac{0.0025}{20} = \frac{\mu \times 100}{E},$$

$$\text{or} \qquad\qquad E = \frac{2000 \ \mu}{0.0025} = \mathbf{8 \times 10^5 \ \mu}$$

$E = 2G(1 + \mu) \quad \text{or} \quad 8 \times 10^5 \ \mu = 2 \times 80000 (1 + \mu)$

or $\quad 8 \times 10^5 \ \mu = 160000 + 160000 \ \mu$

$$\text{or} \quad (80 - 16) \times 10^4 \ \mu = 16 \times 10^4, \quad \text{or} \quad \mu = \frac{16 \times 10^4}{64 \times 10^4} = \mathbf{0.25}$$

Substituting value of μ

$E = 8 \times 10^5 \times 0.25 = \mathbf{2 \times 10^5 \ N/mm^2} \ (200 \ kN/mm^2)$

$E = 3K(1 - 2\mu) = 3K(1 - 0.5),$

$$K = \frac{2 \times 10^5}{3 \times 0.5} = \mathbf{1.3333 \times 10^5 \ N/mm^2} \ (\mathbf{133.33 \ kN/mm^2})$$

Change in volume $A(1 - \mu e)(1 - \mu e) L(1 + e) - AL$

$\quad \delta V = AL \{(1 - 2\mu e + \mu^2 e^2)(1 + e)\} - AL$

$\qquad = AL \{1 - 2\mu e + \mu^2 e^2 + e - 2\mu e^2 + \mu^2 e^3\} - AL$

Neglecting higher powers of small quantities

$\quad \delta V = (e - 2\mu e) AL = e(1 - 2\mu) AL$

$$= e\,(1 - 0.5)\,314 \times 2000 = \frac{50 \times 314 \times 2000}{2 \times 10^5}, \left(e = \frac{31400}{314\,E} \right)$$

$$= \frac{50 \times 314}{100} = \textbf{157 mm}^3$$

5.14 TEMPERATURE STRESSES IN SINGLE MATERIAL AXIAL FORCE MEMBERS

When axial force members are heated or cooled, these members **expand** or **contract** freely in proportion to the **change in the temperature** and **coefficient of expansion** of the material. When the **natural expansion** or **contraction** due to **change of temperature** is **forced** not to expand or contract **due to fixity, stress** is developed in the member. The amount of stress depends on the amount of expansion or contraction (**not allowed to occur**) and the modulus of elasticity of the material '*E*'. We must understand very clearly that if the **expansion or contraction occurs freely** due to the change of temperature, **no stress will develop**. Stress develops **only when** free **expansion or contraction is restricted** or resisted by end grips or fixed supports.

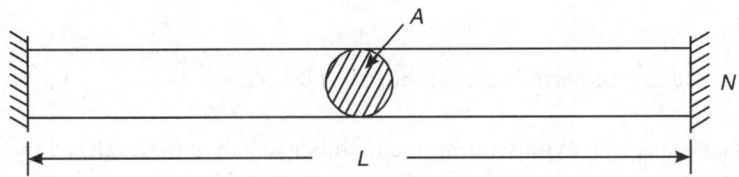

Fig. 5.22: Temperature stress

Consider a rod *MN* of cross-sectional area *A* and length *L* fixed at ends in rigid supports *M* and *N* at **normal temperature t_0 with no stress**. Temperature of the rod *MN* suddenly changes to t_1. The material of the rod has **coefficient of expansion α** and **modulus of elasticity *E*.**

Change in temperature $t = (t_1 - t_0)$

Free expansion (rise) or contraction (fall) = $L \cdot \alpha \cdot t$

(a) When the expansion or contraction is **restricted or neutralized fully** by application of an **axial force** due to end **grips or supports**, the elastic **deformation** due to axial support force $F = L \cdot \alpha \cdot t$

i.e. $\dfrac{FL}{AE} = L \cdot \alpha \cdot t$, or $\dfrac{F}{A} = (\alpha \cdot t \cdot E)$

or **Stress** developed due to **full restrain** = $(\alpha \cdot t \cdot E)$ ... (5.33)

Total force of end support = Area × Stress = $A\,(\alpha \cdot t \cdot E)$

i.e. **Force** $F = A\,(\alpha \cdot t \cdot E)$ ... (5.34)

Let the end supports are **not fully fixed** and **yields by δ** due to the change of temperature '*t*' and the **deformation restricted** by supports = $(L \cdot \alpha \cdot t \cdot - \delta)$, elastic

strain '*e*' = $\dfrac{(L \cdot \alpha \cdot t - \delta)}{L} = \left(\alpha t - \dfrac{\delta}{L} \right)$

$$\therefore \qquad \textbf{Stress} = \textbf{Strain} \times E = \left(\alpha t - \frac{\delta}{L} \right) E \qquad \qquad \dots (5.35)$$

Force $F = (\text{Stress} \times \text{Area}) = E \times e \times A = E \left(\alpha t - \frac{\delta}{L} \right) A$

$$\textbf{Force} \qquad \qquad F = A \cdot E \left(\alpha t - \frac{\delta}{L} \right) \qquad \qquad \dots (5.36)$$

(b) When expansion or contraction due to temperature change is **totally restricted** by supports.

Stress induced $\qquad p = \alpha \cdot t \cdot E \qquad \qquad \dots (5.33)$

Force induced $\qquad F = A \cdot \alpha \cdot t \cdot E \qquad \qquad \dots (5.34)$

(c) When the expansion or contraction due to temperature change is **restricted partially** due to **yielding** of end supports.

Induced **stress due to partial restriction** $p = E \left(\alpha t - \frac{\delta}{L} \right) \qquad \qquad \dots (5.35)$

Induced force due to partial restriction $F = AE \left(\alpha t - \frac{\delta}{L} \right) \qquad \qquad \dots (5.36)$

Temperature rise causes **expansion** in the member which is **neutralized** by pushing by the **rigid end supports**, the member develops **compressive stress**. Similarly, when the **temperature falls it causes contraction** in the member which is **neutralized by pulling by the rigid end supports**, the member develops **tensile stress**.

5.15 TEMPERATURE STRESSES IN COMPOSITE MATERIAL AXIAL FORCE MEMBERS

Composite material comprise of **two or more materials monolithically** attached to each other. These materials have **different coefficients of expansion 'α'** and **modulii of elasticity E**. The **composite materials actually expand or contract together** when subjected to heating or cooling due to rigid connection. Since their **coefficients of expansion are different**, these materials tend to expand differently. Actual expansion or contraction is **forced** to be **equal** due to application of **internal force on each other**. Thus the **actual expansion** is the algebraic sum of two components i.e. **natural** and **forced** due to internal stress generated due to rigidity. Thus the composite material develops **stress** in its constituents due to change of temperature.

Consider a composite rod of two constituent materials having coefficients of expansion α_1, α_2 and modulii of elasticity E_1, E_2 respectively. Areas of cross-sections may be taken as A_1, A_2 respectively. Let α_1 is more than α_2 and ends are rigidly connected.

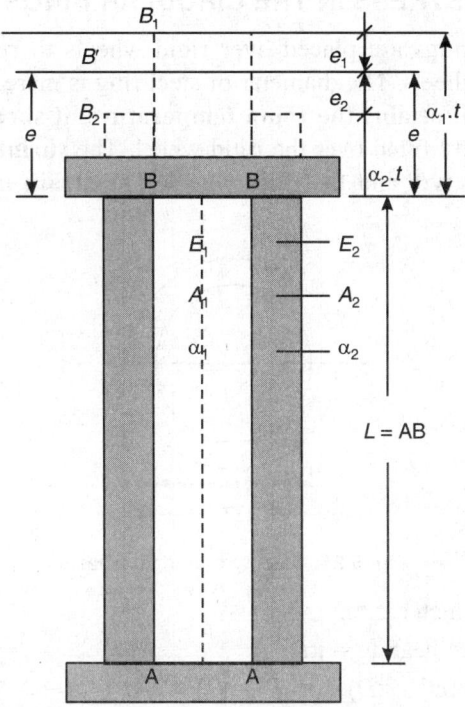

Fig. 5.23: Composite bar with temperature stresses

Let the composite rod **AB** be of **unit length**. When the temperature changes, two materials try to expand differently but due to rigidity of supports or monolithic action, these expand equally (e). This causes reduction (e_1) in expansion of material-1 having more α_1 and extra stretching (e_2) of material-2 having less α_2 (refer Fig. 5.23).

Free expansion material-1 : $BB_1 = \delta_1 = \alpha_1 \cdot t$

Free expansion material-2 : $BB_2 = \delta_2 = \alpha_2 \cdot t$

Actual expansion in composite material $BB' = e$ ... (1)

Strain e_1 in material-1 $e_1 = (\alpha_1 \cdot t - e)$ forced due to composite action   ... (2)

Strain e_2 in material-2 $e_2 = (e - \alpha_2 \cdot t)$, induced due to composite action   ... (3)

Add (2) and (3):

$$e_1 + e_2 = (\alpha_1 \cdot t - \alpha_2 \cdot t) \qquad\qquad \text{... (4)}$$

Internal force developed in two materials will be equal and opposite. Material-1 is compressed with force F while material-2 is stretched with the same tensile force F (since no external force is acting on the rod) as the member is in equilibrium.

$$F = p_1 A_1 = p_2 A_2 \qquad\qquad \text{... (5.37)}$$

From (4)

$$\frac{p_1}{E_1} + \frac{p_2}{E_2} = (\alpha_1 - \alpha_2)\, t \qquad\qquad \text{... (5.38)}$$

5.16 TEMPERATURE STRESS IN THE CIRCULAR RINGS

Some times strong steel rings are placed over rigid wheels to **reduce wear and tear** and increase the life of these wheels. The diameter of steel ring is increased by heating to a higher temperature so that when it **attains the room temperature, it shrinks** to the diameter of the rigid wheel to remain tightly fitted over the rigid wheel. The **shrinking** of the outer steel ring causes **hoop stress** in the tyre and the rigid wheel. The **strain** induced is known as **hoop strain**.

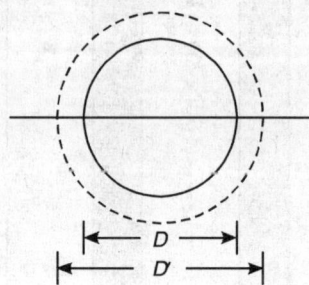

Fig. 5.24: Steel tyre on rigid wheels

On heating the ring perimeter = $\pi D'$

Original perimeter before heating = πD

Change in perimeter = $(\pi D' - \pi D) = \pi (D' - D)$

Strain
$$e = \frac{\pi D' - \pi D}{\pi D} = \frac{D' - D}{D} \qquad \qquad \dots (1)$$

Since the wheel is rigid and does not allow free expansion or contraction, strain e is induced in the ring. Stress in the ring $= e \cdot E$

$$p = eE = \left(\frac{D' - D}{D}\right) . E, \left(\text{Since} = \frac{D' - D}{D} = \alpha \cdot t\right) \quad \dots (5.39)$$

$$D' = D (1 + \alpha . t)$$
$$p = eE = \alpha . t . E$$

i.e.
$$p = (\alpha . t . E) \qquad \qquad \dots (5.40)$$

Where p = Stress (Hoop stress)

   α = Coefficient of expansion

   t = Change in temperature from normal

   E = Modulus of elasticity

   D' = Diameter of the ring after heating.

EXAMPLE 5.24: A copper rod AB of 750 mm length and 120 mm² cross-sectional area is rigidly connected axially with a steel rod BC of 450 mm length and 480 mm² cross-sectional area at the common point B. Points A and C are embedded in fixed supports at A and C. Determine the stresses developed in the two materials and the displacement of B when temperature rises by 60°. Material properties are:

Copper: $E_C = 105$ kN/mm^2, $\alpha_C = 18 \times 10^{-6}$ /°C
Steel: $E_S = 210$ kN/mm^2, $\alpha_S = 12 \times 10^{-6}$ /°C

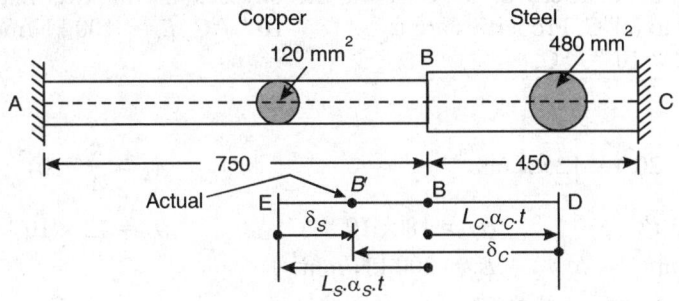

Fig. 5.25: Displacement due to temperature

Solution:

$E_C = 105$ kN/mm^2, $E_S = 210$ kN/mm^2, $A_C = 120$ mm^2
$A_S = 480$ mm^2, $L_C = 750$ mm, $L_S = 450$ mm
$\alpha_C = 18 \times 10^{-6}$ /°C $\alpha_S = 12 \times 10^{-6}$ /°C

Let the point B shifts to B' due to temperature change. Actual $BB' = \delta$.

Free expansion of the copper rod $= BD = L_C.\alpha_C.t = 750 \times 18 \times 10^{-6} \times 60 = \textbf{0.81 mm}$

Free expansion of the steel rod $= BE = L_S.\alpha_S.t = 450 \times 12 \times 10^{-6} \times 60 = \textbf{0.324 mm}$

Copper rod may have induced deformation $\delta_C = BD + BB' = \textbf{(0.81} + \boldsymbol{\delta}\textbf{)}$ ... (1)

Steel rod may have induced deformation $\delta_S = BE - BB' = \textbf{(0.324} - \boldsymbol{\delta}\textbf{)}$ ... (2)

Adding (1) and (2),

$$\delta_C + \delta_S = (0.81 + \delta) + (0.324 - \delta) = \textbf{1.134 mm} \qquad \text{... (3)}$$

Since there is no external load, the load induced by steel on copper rod will be the same as that induced by copper rod on the steel rod.

i.e. $P_S = P_C$ or $p_S . A_S = p_C . A_C$

or $p_S \times 480 = p_C \times 120$ or $\boldsymbol{p_C = 4p_S}$ ... (4)

From (3)

$$\frac{p_C}{E_C} \cdot L_C + \frac{p_S}{E_S} \cdot L_S = 1.134 \quad \text{or} \quad \frac{4p_S \cdot 750}{105000} + \frac{p_S \cdot 450}{210000} = 1.134 \text{ mm},$$

$$p_S = \frac{1.134 \times 210000}{6450} = \textbf{36.92 N/mm}^2$$

$\therefore$ $p_C = \textbf{147.683 N/mm}^2$,

From equation (2)

$$\frac{p_S}{E_S} \cdot L_S = 0.324 - \delta, \quad \delta = 0.324 - \frac{36.92 \times 450}{210000} = 0.324 - 0.07912 = \textbf{0.244 N/mm}^2$$

EXAMPLE 5.25: A copper tube has external diameter 30 mm and internal diameter 20 mm. A steel rod of 20 mm diameter is placed in the tube with ends rigidly connected. The composite bar has no stress at 30°C. Find the stresses in the two materials when the temperature rises to 80°C. Properties are $\alpha_C = 18 \times 10^{-6}$ /°C, $E_C = 100$ kN/mm^2, $E_S = 200$ kN/mm^2 and $\alpha_S = 12 \times 10^{-6}$ / °C.

Solution:

$$A_C = \frac{\pi}{4} (30^2 - 20^2) = 125 \,\pi \text{ mm}^2, \qquad\qquad A_S = \frac{\pi}{4} (20)^2 = 100 \,\pi \text{ mm}^2,$$

$$t = 80 - 30 = 50°C, \qquad \alpha_C = 18 \times 10^{-6}/°C, \qquad \alpha_S = 12 \times 10^{-6}/°C,$$
$$E_C = 100 \text{ kN/mm}^2, \qquad\quad E_S = 200 \text{ kN/mm}^2$$

Since there is no external force acting on the composite bar, the **force** induced in each material will be **equal**, i.e.

$p_C \cdot A_C = p_S \cdot A_S$ or $\mathbf{125\,\pi p_C = 100\,\pi p_S}$ or $p_S = \mathbf{1.25}\,p_C$ ... (1)

For composite bar $e_C + e_S = (\alpha_C - \alpha_S)\, t$

or $\dfrac{p_C}{E_C} + \dfrac{p_S}{E_S} = (18 - 12) \times 10^{-6} \times 50 = 3 \times 10^{-4}$

or $\dfrac{p_C}{100 \times 10^3} + \dfrac{p_S}{200 \times 10^3} = 3 \times 10^{-4}$ ... (2)

From (1) and (2)

$$\frac{p_C}{10^5} + \frac{1.25\,p_C}{2 \times 10^5} = 3 \times 10^{-4} \quad\text{or}\quad \frac{2\,p_C + 1.25\,p_C}{2 \times 10^5} = 3 \times 10^{-4}$$

or $3.25\, p_C = 6 \times 10$ or $p_C = \dfrac{60}{3.25} = \dfrac{240}{13} = \mathbf{18.46 \text{ N/mm}^2}$

$p_S = 1.25\, p_C = 1.25 \times 18.46 = \mathbf{23.015 \text{ N/mm}^2}$

EXAMPLE 5.26: A steel thin circular flat tyre is fixed on a rigid material wheel of 1 m diameter. The tyre steel has maximum permissible stress of 40 N/mm^2, $\alpha = 11 \times 10^{-6}$, $E = 200$ kN/mm^2. Determine the internal diameter of the tyre and the temperature to which the tyre may be heated to slip on to the rigid wheel.

Solution:

$D' = 1000$ mm,   Internal Diameter $= D$, $\alpha = 11 \times 10^{-6}$ /°C,
$E = 200 \times 10^3$ N/mm^2,   Permissible $p = 40$ N/mm^2

Maximum Stress p (in ring) = Strain $\times E = \left(\dfrac{D' - D}{D}\right) \cdot E = \left(\dfrac{D'}{D} - 1\right) \cdot E$

i.e. $40 = \left(\dfrac{D'}{D} - 1\right) \times 2 \times 10^5$ or $\dfrac{D'}{D} = \left(\dfrac{40}{2 \times 10^5} + 1\right) = \dfrac{40 + 2 \times 10^5}{2 \times 10^5}$

or $\dfrac{D'}{D} = \dfrac{200040}{2 \times 10^5}$, (Since, $D' = 1000$ mm given)

$$\therefore \quad D = \frac{1000 \times 2 \times 10^{-5}}{200040} = \frac{1 \times 10^{8} \times 10^{-5}}{1.00020} 1.0002 = 10^{3} = 10^{3}\,(1 - 0.0002) = \textbf{999.8 mm}$$

If the rise in temperature $= t$, then

$$D' = D\,(1 + \alpha \cdot t) \quad \text{or} \quad \frac{D' - D}{D} = \alpha \cdot t$$

$$\frac{1000 - 999.8}{999.8} = 11 \times 10^{-6}\,t \quad \text{or} \quad t = \frac{0.2 \times 10^{6}}{1000 \times 11} = \frac{200}{11} = \textbf{18.18}^{\textbf{o}}\textbf{C}$$

EXAMPLE 5.27: A rectangular flat link shown in Fig. 5.26 carries an axial force of 16 kN. The link has 25 mm diameter hole at the enlarged end to transfer the load through a 25 mm cross pin. If the cross pin is in double shear, find the shear stress in pin, tensile stresses in the link at the sections X-X and Y-Y.

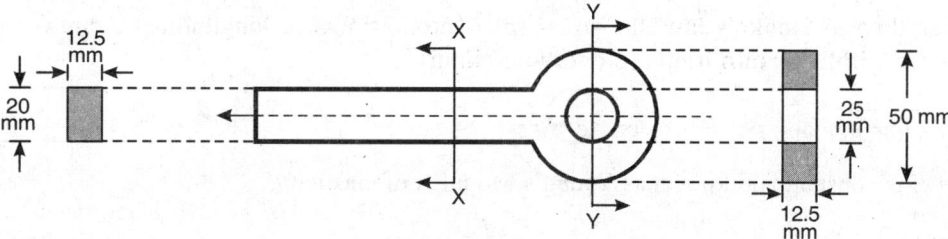

Fig. 5.26: A link

Solution:

$A_X = 20 \times 12.5 = 250$ mm^2, $A_Y = (50 - 25) \times 12.5 = 312.5$ mm^2,

Load $P = \textbf{16000 N}$

Pin area for double shear $= 2 \times \dfrac{\pi}{4}\,(25)^2 = 312.5\,\pi$ mm$^2 = 981.25$

p_X (tensile stress) at X-X $= \dfrac{16000}{250} = \textbf{64 N/mm}^{\textbf{2}}$

p_Y (tensile stress) at Y-Y $= \dfrac{16000}{312.5} = \textbf{51.2 N/mm}^{\textbf{2}}$

q in pin (shear stress) $= \dfrac{16000}{312.5\,\pi} = \textbf{16.306 N/mm}^{\textbf{2}}$

5.17 SUMMARY

When an axial force element of length (L) is subjected to an axial force (F), it undergoes an axial **deformation** (δ). If the force is within **elastic limit**, the deformation disappears completely on removal of the force. The axial elements are in **equilibrium** both externally and under **internal resistances**. Consider an axial element of length (L) and cross-sectional area (A) and subjected to an axial force (F).

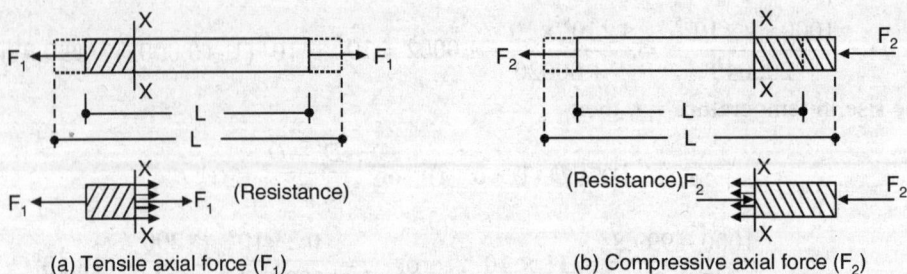

(a) Tensile axial force (F_1) (b) Compressive axial force (F_2)

Fig. 5.27: Axial tension and compression

The axial **resistance per unit area** is known as **stress** (p). The deformation (δ) in axial direction per unit length is known as strain (e).

i.e.
$$p = \frac{F}{A}, \, e = \frac{\delta}{L} \qquad \qquad \qquad ... (i)$$

According to **Hooke's law**, the stress (p) is proportional to longitudinal strain (e) if the forces are **within certain limits** (proportional limit).

i.e. $p \propto e$, or $p = E \cdot e$ or $e = \dfrac{p}{E}$ $\qquad \qquad ... (ii)$

where E is constant and known as Young's Modulus of elasticity.

$$\delta = e \cdot L = \frac{FL}{A \cdot E} \qquad \qquad ... (iii)$$

Axial stress will be **tensile** if the element tends to **elongate** and stress will be **compressive** if the element tends to **shorten** under the action of an **axial force**. Elastic materials undergo **deformations in lateral dimensions** also when subjected to axcial forces. **Change in lateral dimension** per unit lateral dimension is known as the **lateral strain** (e'). The lateral strains (e') are proportional to the longitudinal strain (e) for elastic materials.

i.e. $e' \propto e$, or $e' = \mu e$, where μ is constant and known poisson's ratio.

or
$$\mu = \frac{\text{lateral strain} (e')}{\text{longitudinal strain} (e)} \qquad \qquad ... (iv)$$

Volumetric strain due to axial stress (p) will be

$$\frac{\delta_v}{V} = \frac{p}{E}(1 - 2\mu) \qquad \qquad ... (v)$$

Strain in lateral area $\left(\dfrac{\delta_A}{A}\right)$ under axial stress (p) will be

$$\frac{\delta_A}{A} = 2\mu\left(\frac{p}{E}\right) \qquad \qquad ... (vi)$$

Strain energy under axial stress (p) will be

$$U = \frac{p^2}{2E} \times \text{Volume} \qquad \qquad ... (vii)$$

Tensile test on ductile materials (like mildsteel) provides structural behaviour of engineering materials.

A – Proportional limit

B – Elastic limit

C – Yield point stress

D – Lower yield point stress

E – Ultimate maximum stress

F – Breaking stress with maximum deformation

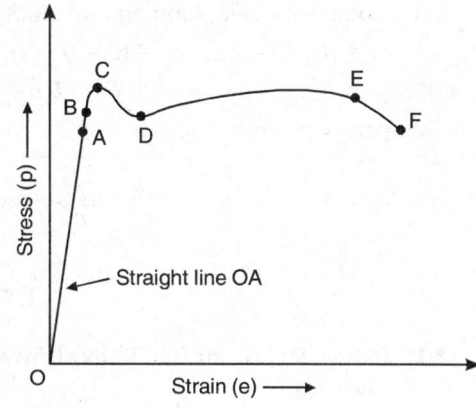

Factor of safety is defined as the ratio of **maximum ultimate** (or **yield**) stress to its **working stress**.

i.e.
$$\text{F.S.} = \frac{\text{Ultimate (or yield) Stress}}{\text{Working Stress}} \qquad \text{... (viii)}$$

Total deformation in variable section under axial force will be

$$\delta = \delta_1 + \delta_2 + \delta_3 = \frac{F}{E}\left(\frac{L_1}{A_1} + \frac{L_2}{A_2} + \frac{L_3}{A_3}\right) = \frac{F \cdot L}{\frac{\pi}{4}(D_0 D)E} \qquad \text{... (ix)}$$

Longitudinal deformation under own self weight

$$\delta_L = \frac{\omega}{2E}\cdot L^2 \text{, where } \omega = \text{weight/unit volume} \qquad \text{... (x)}$$

Composite material axial elements :

$$f_1 = \frac{F}{\left(A_1 + \dfrac{A_2}{m}\right)}, \quad f_2 = \frac{F}{(m A_1 + A_2)}, \text{ where '}m\text{'} = \frac{E_1}{E_2}, f_1 = mf_2 \qquad \text{... (xi)}$$

Stress along inclined plane

$$f_\theta = \frac{p}{2}(1 + \cos 2\theta), \quad q_\theta = \frac{p}{2}\sin 2\theta \qquad \text{... (xii)}$$

Shear stress 'q' on any plane is **always accompanied** by another shear stress 'q' (known as **complementary shear**) along the perpendicular plane.

Relation between various **elastic modulii** (E, G, and K)

$$E = 2G(1 + \mu) = 3K(1 - 2\mu) \qquad \text{... (xiii)}$$

Temperature stress $p = \alpha \cdot t \cdot E$, when axial expansion/contraction is restricted. The temperature stress induced when expansion/contraction is partially restricted

$$p = E\left(\alpha \cdot t - \frac{\delta}{L}\right) \qquad \text{... (xiv)}$$

Axial composite bars, temperature strains e_1, e_2, net strain 'e' are

$$e_1 = (\alpha_1 \cdot t - e), \quad e_2 = (e - \alpha_2 \cdot t), \quad (e_1 + e_2) = (\alpha_1 - \alpha_2)t \qquad \text{... (xv)}$$

Also $$F = A_1 p_1 = A_2 p_2$$

Hoop stress in rings

$$p = e \cdot E = \left(\frac{D' - D}{D} \right) E, \quad \frac{D' - D}{D} = \alpha \cdot t \qquad \text{... (xvi)}$$

EXERCISE 5

Q.5.1. **Define Stress, Strain, Lateral Strain, Poisson's Ratio, Hooke's Law** and **Modulus of Elasticity**.

Q.5.2. A 200 mm long steel rod has rectangular cross-section of 50 mm × 50 mm. If the rod carries 200 kN axial compression, determine the stress, change in volume, Poisson's ratio = 0.3, modulus of elasticity E = 200 kN/mm^2.

Ans. $\delta V = 80$ **mm**^{3}.

Q.5.3. A 6 m long steel rod has rectangular cross-section of 15 mm × 20 mm. The rod carries an axial tension of 300 kN. If Poisson's ratio = 0.26 and E for steel = 200 kN/mm^2, determine change in length, breadth, depth and volume.

Ans. $\delta V = 432$ **mm**^{3}, $\delta L = 30$ **mm**, $\delta B = 0.0195$ **mm**, $\delta D = 0.0260$ **mm**.

Q.5.4. A 5 m long rod of 10 mm diameter elongates 6.67 mm under an axial load of 22 kN. Determine the value of E and change in volume and diameter, if Poisson's ratio = 0.3.

Ans. $E = 210$ **kN/mm**^{2}, $\delta V = 209.3$ **mm**^{3}, $\delta d = 0.004$ **mm**.

Q.5.5. A steel rod of 120 mm diameter undergoes 0.050 mm reduction in its diameter and extension of 2.1 mm in its 1.5 m length under an axial load. Elastic modulus $E = 210$ kN/mm^2, determine Poisson's constant, change in area and volume.

Ans. $\mu = 0.3$, $\delta V = 9504$ **mm**^{3}, $\delta A = 2e'.A = 9.504$ **mm**^{2} **(reduction)**.

Q.5.6. A circular steel rod carries axial load of 1663.2 kN. If the ultimate stress in steel is 441 N/mm^2 and the factor of safety is 3, find the safe diameter of the rod.

Ans. **Diameter = 120 mm**.

Q.5.7. A punch of 80 mm diameter is used to punch a plate of 10 mm thickness. The ultimate shear stress of the plate is 300 N/mm^2. Find the stress in the punch and its nature.

Ans. **Stress in the punch = 150 N/mm**^{2} **compressive**.

Q.5.8. A short column carries an axial load of 704 kN. If the permissible stress is 80 kN/mm^2 and the external diameter is 160 mm, find the thickness 't' of the circular hollow column section.

Ans. $t = 20$ **mm**.

Q.5.9. Copper and steel wires of the same length carries equal tensile load. If both the wires undergo the same extension, determine diameter of copper wire when the steel wire diameter is **1.414 mm**. $E_S = 200$ kN/mm^2, $E_C = 100$ kN/m^2.

Ans. $d_{\text{Copper}} = 20$ **mm**.

Q.5.10. A 4 m long steel bar carries an axial pull of 845 kN. If the safe permissible stress in steel is 130 N/mm^2, and the maximum allowable elongation is 2.5 mm, determine the safe area of cross-section. $E = 200$ kN/mm^2.

Ans. **Safe** $A = 6760$ **mm**^{2}.

Q.5.11. Determine safe axial load W on an aluminium rod of 1000 mm^2 cross-section if permissible stress is 100 N/mm^2 and maximum allowable elongation should not exceed 0.25 mm in a gauge length of 250 mm. $E_a =$ **70 kN/mm^2**.

Ans. $W_{max} =$ **70 kN**.

Q.5.12. Define: Modulus of elasticity, Modulus of rigidity, Bulk modulus of elasticity and state their inter relationship.

Q.5.13. A rod of 50 mm diameter and 200 mm gauge length elongates by 0.33 mm and the diameter shortens by 0.0255 mm when an axial pull of 750 kN is applied. Determine modulus of elasticity, Poisson's ratio, shear modulus and bulk modulus of elasticity.

Ans. $E =$ **77.2 kN/mm^2**, $\mu =$ **0.3091**, $G =$ **29.488 kN/mm^2**,

$K =$ **67.404 kN/mm^2**.

Q.5.14. A concrete column of **250 mm × 250 mm** size is reinforced with **4 bars** of cross-section **1500 mm^2** each bar. If the **column** is short and carries an axial load of 305 kN, find the stresses in the two materials. **Modular ratio** of steel concrete = 15.

Ans. $p_C =$ **2.082 N/mm^2**, $p_S =$ **31.23 N/mm^2**.

Q.5.15. Three rods of 2 m length consists of 3 different materials are rigidly connected at ends. The composite bar system is subjected to a pull of **231 kN**. The cross-sections areas are 700, 1400 and 2000 mm^2 respectively. The elastic modulii are respectively 200, 100 and 70 kN/mm^2. Find the stresses.

Ans. **Load shared by each = 77 kN, $p_1 =$ 110 N/mm^2, $p_2 =$ 55 N/mm^2, $p_3 =$ 38.5 N/mm^2.**

Q.5.16. A copper wire of 60 m length is pulled by an axial stress 100 N/mm^2. Determine elongation in the wire if $E_C =$ 100 kN/mm^2. Also find the change in temperature for the same expansion in the wire if $\alpha_C =$ 18 × 10^{-6} /°C.

Ans. **$\delta L =$ 30 mm, $t =$ 55.5°C**.

Q.5.17. A 6 m long steel rod of 10 mm diameter is hanging vertically from a horizontal ceiling. The rod carries an axial pull of 16.5 kN. If the rod is heated by 40°C. The original length was measured at room temperature of 10°C. If $E_S =$ 210 kN/mm^2 and $\alpha_S =$ 12 × 10^{-6} /°C, find the stress and total elongation in the rod.

Ans. **Stress = 210 N/mm^2, Total $\delta =$ 8.16 mm.**

Q.5.18. A 4 m long steel rod of 1200 mm^2 cross-section is fixed in the rigid supports at both this ends. (i) If the temperature falls by 30°C after fixing in end supports, find the force developed in the rod due to full fixity of the support (ii) If the supports yield by 0.0044 mm, find the force exerted on the supports due to 30°C fall of temperature below the normal temperature. $E_S =$ 210 kN/mm^2, $\alpha =$ 12 × 10^{-6} /°C.

Ans. **i.** **Pull on the supports = 90.72 kN**

ii. **Pull on the supports = 63 kN**

Q.5.19. A steel tape measures exactly 30 m at a room temperature of 20°C and pull of 50 N. If the temperature rises to 40°C and a pull of 250 N is applied to measure. Find the exact length of the tape. $\alpha =$ 11 × 10^{-6} /°C, $E =$ 200 kN/mm^2. Cross-sectional area = 0.0007 mm^2.

Ans. **Tape length = 30.0109 m**.

Q.5.20. Two rods of copper and steel are rigidly connected at ends. Cross-sectional areas are respectively 1000 mm^2 and 1500 mm^2. The stress in the two bars is zero at 20°C. If temperature rises to 100°C, find the stresses in the two bars. $E_C =$ 100 kN/mm^2, $E_S =$ 200 kN/mm^2, $\alpha_C =$ 17 × 10^{-6} /°C, $\alpha_S =$ 11 × 10^{-6} /°C.

Ans. $f_C =$ **36 N/mm^2, $f_S =$ 24 N/mm^2.**

Q.5.21. A short steel column is shown in Fig. Q5.21. The column carries axial loading as shown in the Fig. Q5.21. The column has cross-sectional area of 25000 mm² and modulus of elasticity E = 200 kN/mm², $h = 0.3$. Neglecting bending of column calculated the change in length and volume.

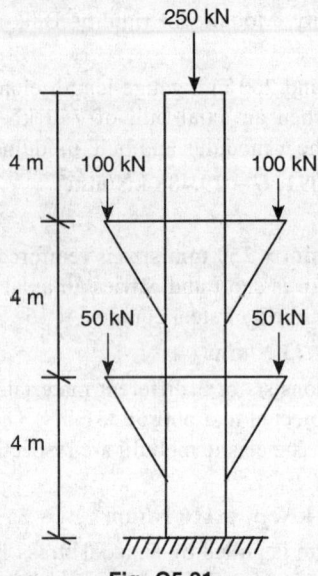

Fig. Q5.21.

Ans. **$\delta L = 1.0$ mm reduction, $\delta V = 10000$ mm³ reduction**.

Q.5.22. A copper rod of 1000 mm² cross-sectional area is shown in Fig. Q5.22. If the modulus of elasticity $E_C = 100$ kN/mm². Determine the change in length.

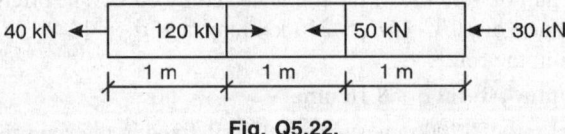

Fig. Q5.22.

Ans. **0.70 mm reduction**.

Q.5.23. A kotter pin lock key $b \times 40$ mm is shown in Fig. Q5.23. Permissible shear stress in the key is 40 N/mm². Determine the safe value of 'b' if the maximum load on the lever is 1 kN at 250 mm from the centre.

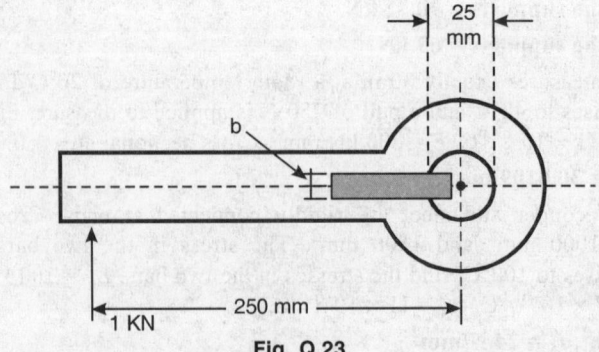

Fig. Q.23.

Ans.: **$b \geq 12.5$ mm**

Q.5.24. A 20 mm bolt with nut resting on washer is shown in Fig. Q5.24. The bolt carries a pull of 50 kN. The nut bolt rests on a hollow circular washer with outer diameter 'd' and inner (hole) diameter of 30 mm. The permissible bearing stress in washer is 1.5 N/mm^2 and shear stress in the nut is 80 N/mm^2. Determine the outer diameter 'd' of washer and thickness 't' of the nut.

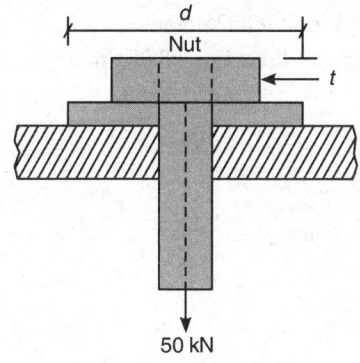

Fig. Q5.24.

Ans. $d \geq 208.24$ **mm,** $t \geq 9.952$ **mm.**

Q.5.25. A piston rod is subjected to 80 kN compressive force and 72 kN tensile force. If permissible compressive stress is 90 N/mm^2 and permissible tensile stress 80 N/mm^2, find the safe cross-sectional area of the rod.

Ans. **A = 900 mm^2.**

Q.5.26. A steel tube of 30 mm outer diameter and 5 mm wall thickness. A solid copper rod of 15 mm diameter is placed centrally in the steel tube and attached rigidly at ends. The composite member carries an axial load of 50 kN. Find the stresses in the steel tube and the copper rod if $E_S = 200$ kN/mm^2, $E_C = 100$ kN/mm^2. Also find δL in 1 m length.

Ans. $p_S = 104$ **N/mm^2,** $p_C = 52.0$ **N/mm^2,** $\delta L = 0.52$ **mm.**

Q.5.27. A steel tube of 30 mm outer diameter and 5 mm wall thickness contains centrally solid copper rod of 15 mm diameter. The tube and the rod is attached rigidly at ends. The composite member is heated through 40°C. (i) Find the stresses in the two materials if $E_S = 200$ kN/mm^2, $E_C = 100$ kN/mm^2, $\alpha_S = 12 \times 10^{-6}$ /°C, $\alpha_C = 18 \times 10^{-6}$ /°C. (ii) If the composite member also carries 50 kN axial compression, determine **total stresses** in the steel tube and solid brass rod.

Ans. i. $p_C = 19.592$ **N/mm^2,** $p_S = 8.816$ **N/mm^2 (tensile).**

ii. p_C **(total) = + 71.592 N/mm^2,** $p_S = + 95.184$ **N/mm^2.**

Q.5.28. A steel rod of 40 m length hangs from top. A load of 24 kN hangs at the bottom. Density of steel = 78 kN/m^3. If the permissible stress is 80 N/mm^2, find the diameter and change in length and change in diameter.

Ans. $d \geq 19.96$ **mm,** $\delta L = 18.468$ **mm.**

Q.5.29. A 500 mm rod of 25 mm uniform thickness and variable breadth of 25 mm at one end to 75 mm at the other end. The rod carries an axial tensile load of 200 kN. Determine the total deformation of the rod. Also determine stress at a distance of x from the smaller end. $E = 200$ kN/mm^2.

Ans. $\delta L = 0.19085$ **mm,** $p_x = \dfrac{8 \times 10^4}{(250 + x)}$ **N/mm^2,** x **from smaller end in mm.**

Q.5.30. In a tensile test of a material following observations are taken:

Load at proportional limit = 28 kN, Ultimate (max.) load = 56 kN,

Original diameter of the specimen = **12.5 mm**

Gause length L = 50 mm

Deformation at proportional limit 0.0525 mm

Total deformation at the time of breaking = 8.075 mm

Determine:

    a.   Proportional limit,

    b.   Modulus of elasticity,

    c.   Modulus of elastic resilience $u = \dfrac{f_y^2}{2E}$,

    d.   Ultimate stress,

    e.   Ductility index

    f.   Toughness index

Ans. **a.** **228.07 N/mm^2,**   **b.** **217.21 kN/mm^2**

      c. **119.74 × 10^{-3} N-mm/mm^3**   **d.** **456.14 N/mm^2**

      e. **DI = 16.15 %**   **f.** **$T_u = f_u \times Di$ = 73.66 N-mm/mm^3.**

Q.5.31. A tie member carries an axial load of 120 kN. The uniform cross-section of the bar is 40 mm × 30 mm. Determine normal and shear stress on a plane inclined at an angle of 45° with the plane of the axial load.

Ans. f_θ = 500 N/mm^2, q_θ = 500 N/mm^2.

Q.5.32. A rigid rectangular slab is supported by three circular bars of 20 mm diameter. Two outer bars are of copper and middle bar is of steel. The heights of bars are so adjusted that each bar shares 15 kN load equally. If the rigid slab supports additional load of 20 kN, calculate the total stress in each material. E_S = 200 kN/mm^2, E_C = 100 kN/mm^2.

Ans. p_C = 31.82 N/mm^2, p_S = 47.73 N/mm^2 (compressive).

Q.5.33. Three rods of 20 mm diameter are connected by rigid slabs at each end. The two outer bars are of copper and middle bar is of steel.

    a.   If the temperature rises by 50°C above normal, determine the stress developed when the rigid slabs remain parallel. E_S = 200 kN/mm^2, E_C = 100 kN/mm^2, α_S = 12 × 10^{-6} /°C, α_C = 18 × 10^{-6} /°C

    b.   If the rigid slab also carries a load of 40 kN, find the total stresses.

Ans. **a.** p_C = 15 N/mm^2, p_S = 30 N/mm^2 (tensile).

      b. p_C = 46.85 N/mm^2, p_S = 33.70 N/mm^2 (compressive).

Q.5.34. A 3 m long rigid concrete slab BDE is supported by 750 mm long copper rod and 1000 mm long steel rod having uniform cross-sections. (a) If the rigid slab BDE remains horizontal before and after loading, find the overhang 'x' (DE) as shown in Fig. Q5.34. b. If the each rod cross-section is 1000 mm^2, find stresses. E_C = 100 kN/mm^2, E_S = 200 kN/mm^2.

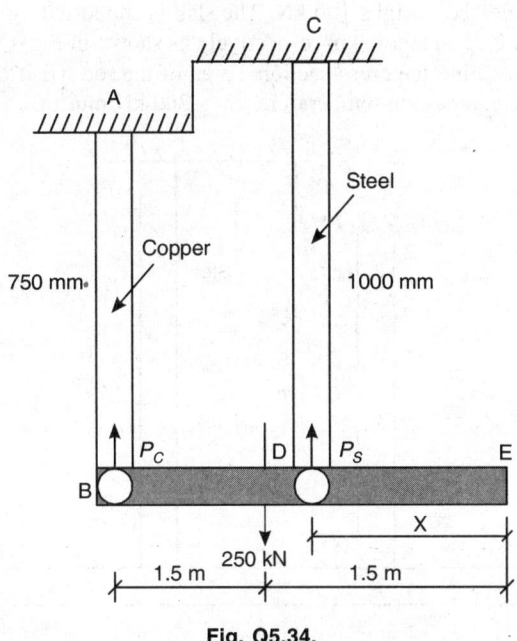

Fig. Q5.34.

Ans. a. $x = 0.5$ m,

 b. $p_C = 100$ N/mm^2, $p_S = 150$ N/mm^2.

Q.5.35. A rigid beam ABC hinged at A supported at B by aluminium wire BD and at C by steel wire CE. Wire BD = CE = 1 m and $A_A = 3$ mm^2, $A_S = 2$ mm^2 respectively. $E_S = 200$ kN/mm^2, $E_A = 70$ kN/mm^2. If a point load W of 20 kN acts at the mid point of BC (AB = BC = 1.0 m), determine the loads shared by steel and aluminium wires.

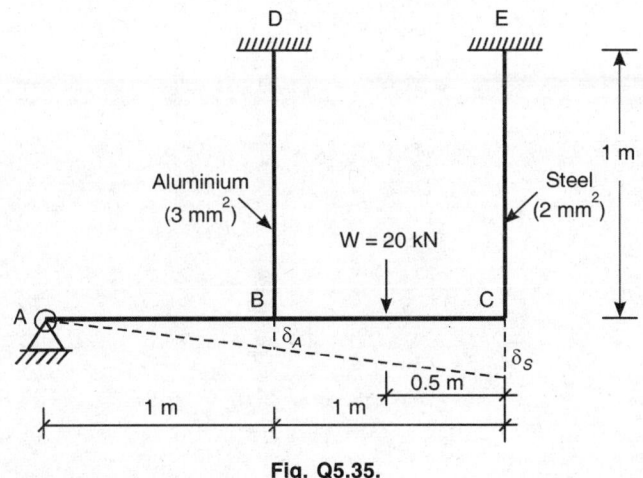

Fig. Q5.35.

Ans. $P_{Steel} = 13.26$ kN and $P_{Al} = 3.48$ kN.

Q.5.36. A 5 m long concrete slab weighs 100 kN. The slab is supported by two steel rods AB and CD of original lengths 0.75 m and 1.0 m respectively as shown in Fig. Q 5.35. Cross-sectional area of CD is 1500 mm². Find the cross-sectional area of the rod AB if the rigid slab has to remain horizontal on 30°C increase in temperature. $E_S = 200$ kN/mm², $\alpha_S = 12 \times 10^{-6}$ /°C.

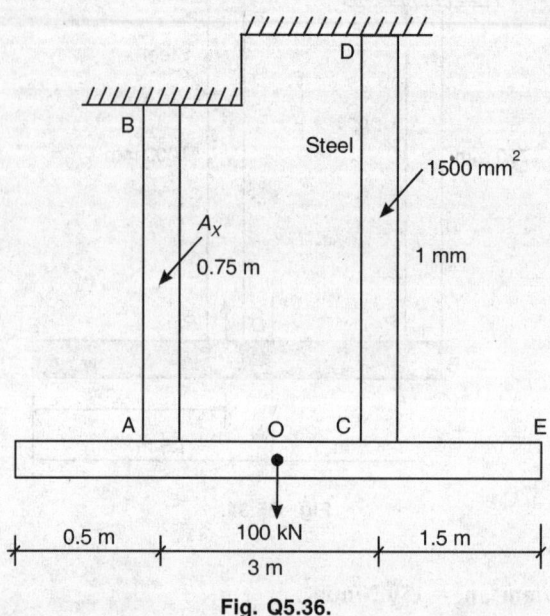

Fig. Q5.36.

Ans. Total $\delta_A = \delta_C$, $A_x = 400$ mm².

Unit III

Bending Structures (Beams)

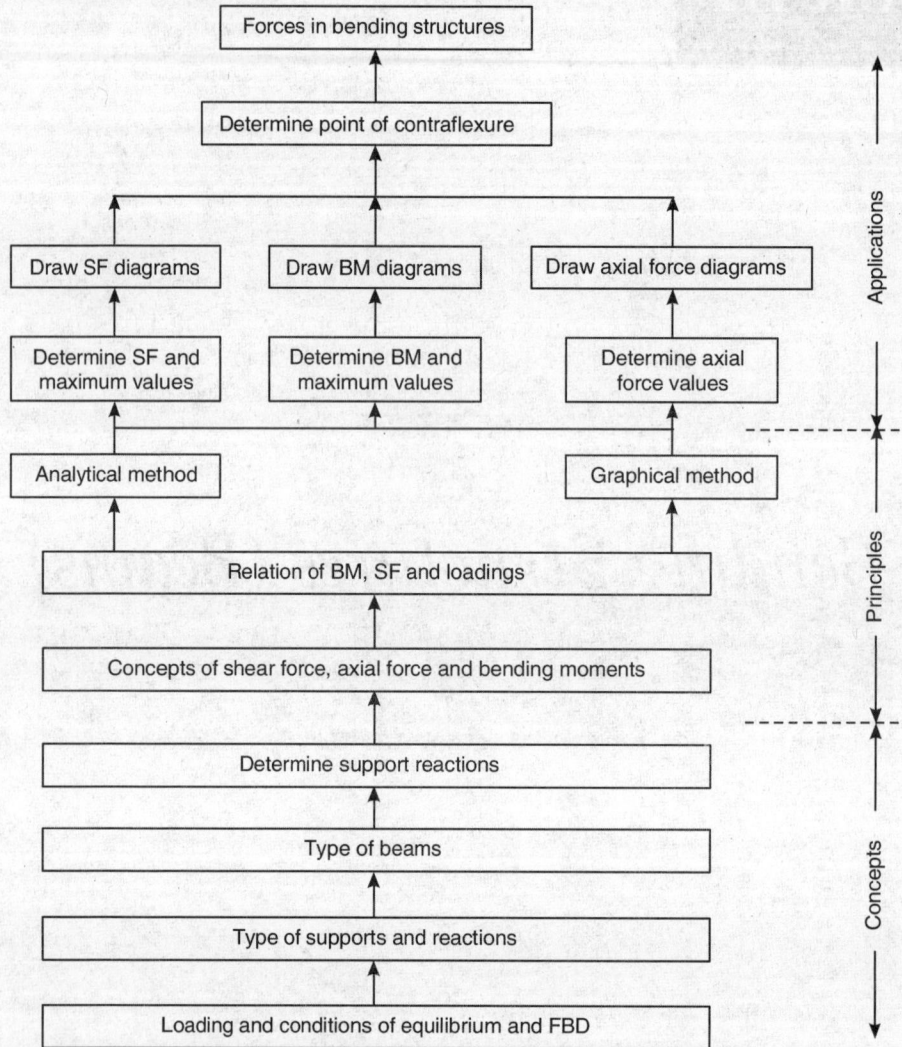

Fig. 6.0: Learning structure for forces in bending structures

6

Shear Force, Axial Force and Bending Moments in Bending Structures

LEARNING OBJECTIVES

After studying this chapter, the learner **understands Shear Force (SF)**, **Axial Force (AF)** and **Bending Moment (BM)** and will be able to:

6.1 **Explain external and internal** equilibrium in a **loaded bending structure.**

6.2 **Explain** the concept of **beam and supports.**

6.3 **Explain** different type of **supports** and **beams.**

6.4 **Draw Free Body Diagram (FBD)** of loaded beam section showing **SF, AF** and **BM (or moment of resistance)** for a given **loading.**

6.5 **Explain Shear Force (SF), Axial Force (AF)** and **Bending Moment (BM).**

6.6 **Calculate SF, AF and BM** in a simply supported beam carrying **point loads.**

6.7 **Calculate SF, AF and BM** in a simply supported beam carrying **UDL.**

6.8 **Calculate SF, AF and BM** in a simply supported beam carrying **mixed loading.**

6.9 **Calculate SF, AF and BM** in a cantilever beam carrying **mixed loading.**

6.10 **Calculate SF, AF and BM** in a overhanging beam carrying **mixed loading.**

6.11 **Calculate SF, AF and BM** in an inclined beam member carrying **mixed loading.**

6.12 **Calculate SF, AF and BM** in a beam **(Simply Supported or Cantilever)** carrying **uniformly varying loading.**

6.13 **Explain** relation between **loading, SF and BM.**

6.14 **Draw SF, AF and BM diagram** for a simply supported beam carrying **mixed loading.**

6.15 **Draw BMD** from a given **SFD.**

6.16 **Calculate** and **draw SFD** from a given **BMD.**

6.17 **Calculate maximum BM** from the given **SF.**

6.18 **Explain** the **point of contraflexure (inflexion).**

6.19 **Calculate** the **point of contra-flexure** in a beam with given **loading.**

6.1 INTRODUCTION

You have learned to determine internal forces in **axially loaded** structural elements and **pin jointed truss** structures in earlier units. In this section you will learn to determine **internal resistances** in structural members subjected to forces having components **transverse** to the axis of the member (called bending element). If the **equilibrium** of any portion of the bending element (**beam**) is considered, it will be found that at each section of the structural element internal resistances will be produced to establish equilibrium of each section. Depending upon the loads, the internal forces produced will be **Bending Moment** (BM), **Shear Force** (SF) and **Axial Force** (AF). These can be easily explained and determined by drawing **Free Body Diagram** (FBD) of the section considered.

As an engineer you will be faced with many situations where you are required to **design** or **check the design** of bending structures or elements subjected to SF, BM, and AF. The activity of **design** or **check of design** requires determination of the internal resisting forces caused by external loads. The design or check of design will be essential before actual construction or production of these members.

Mastering the contents of this chapter will **facilitate** you in **understanding** the rest of the units of strength of materials, theory of structures and theory of machines, etc. This topic will lay a strong foundation for your engineering knowledge and skills. The mastering and understanding of the basic principles of "**equilibrium**" and drawing of "**Free Body Diagram** FBD" are essential before effective understanding of **SF**, **AF** and **BM** in beams.

We shall first understand "**What is bending?**" "**What causes bending?**" "**What is a bending structure or a beam?**" What type of internal forces (**SF** and **BM**) are caused due to bending and how to determine these internal forces?

These bending structures are called **beams** if these members are mainly subjected to **transverse loading** and the **length** of the axis of the member is relatively **large** compared to its **cross-sectional dimensions**. Therefore a **beam** is defined as a structural member subjected to **transverse loading** acting along longitudinal **axial plane symmetrical** and perpendicular to the **cross-section**. The beams may have uniform cross-section throughout the length or sometimes a beam may be designed to have variable section to suit special **loading pattern**. In simple cases of beams, the line of action of the loads lie along the **longitudinal axial plane**. The components of the loads are **transverse** or lie along the longitudinal axis of the beam. The transverse components of the loads will **tend to bend** the longitudinal axis of the member. The axial components will tend to deform (Shorten or elongate) longitudinally which has already been explained earlier in axial structural elements.

The chapter will be devoted to the understanding of **Shear Force** (SF), **Axial Force** (AF) and **Bending Moment** (BM).

6.2 BENDING

Consider a straight bar as shown in Fig. 6.1a. The bar carries transverse loads in the **YZ** plane and acting normal to the axis-**Z** of the beam. As a result of these transverse loads, the bar bends and the Z-axis assumes a **curved shape** in the plane **YZ** as shown in Fig. 6.1b. The bending of the longitudinal axis of bar takes place in the **plane YZ** or about an axis **normal** to the **YZ** plane (i.e. about X-axis). Therefore, this bending of the bar is due to moment taken about **X-axis (M_x)**.

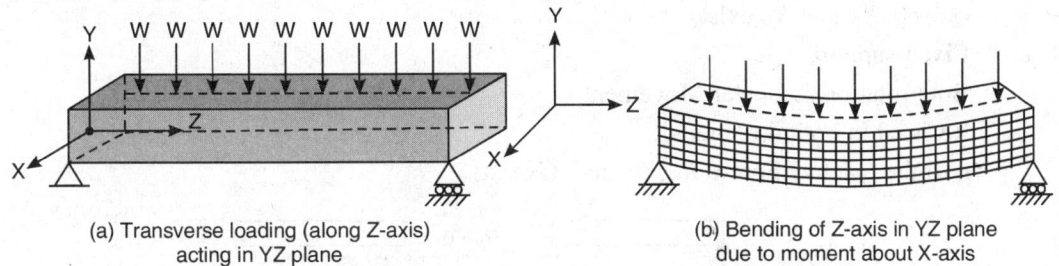

(a) Transverse loading (along Z-axis)
acting in YZ plane

(b) Bending of Z-axis in YZ plane
due to moment about X-axis

Fig. 6.1: Bending of Z-Axis in YZ Plane (about X-Axis)

Consider a case when the bar is subjected to the loads **normal to the longitudinal axis Z**, but the loads are acting in the **X-Z** plane (Fig. 6.2). The bar under these loads in X-Z plane tends to bend Z-axis or takes a curved shape in the X-Z plane due to moment of loads about Y-axis (M_y) and the bending is said to take place about an axis normal to the plane X-Z, i.e. Y-axis.

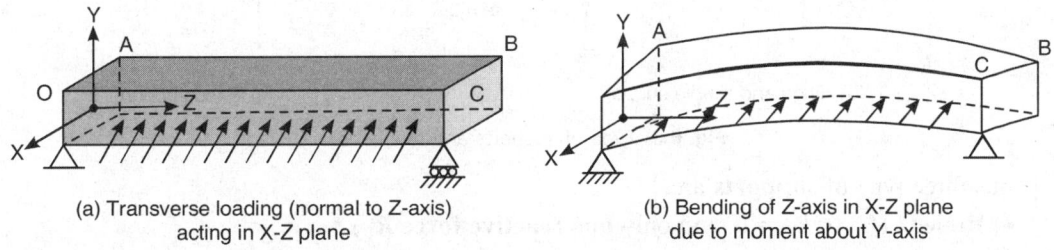

(a) Transverse loading (normal to Z-axis)
acting in X-Z plane

(b) Bending of Z-axis in X-Z plane
due to moment about Y-axis

Fig. 6.2: Bending of Z-axis in XZ plane (about Y-axis)

Bending of a member therefore, occurs whenever it is subjected to loads having a component **normal to its axis** or when the loads are **acting transverse** to the member. Components of the loads which are acting **along the longitudinal axis** of the member cause only **axial deformations** (elongation or shortening). These components are called **Axial Forces** (AF).

6.3 SUPPORT OR BASE OF MEMBERS

Any structural bending member requires to be supported to transfer the loads to the ground or base. These bases are of different types which help in deciding the type of bending structure or beam. A base or support represents the end condition of structure to connect it to the ground or foundation. This connection of the structure with foundation is known as support.

There are different type of supports as given below:

a. **Roller** support:

Rotation & axial

movement **free**

($M = 0$, $R_H = 0$, only R_V exist)

   b.   **Hinge** support and **simple** support:

          Only rotation free

          ($M = 0$, R_H and R_V exist)

   c.   **Fixed** support:

          Both rotation & Linear Movement

          Resisted in both directions

          (All the reactive Forces R_H, R_V and M exist).

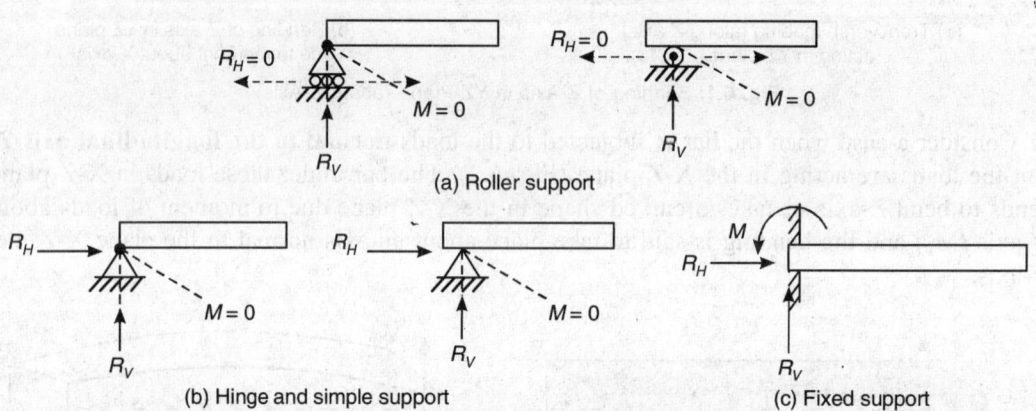

Fig. 6.3: Type of supports and reactions

Thus three type of **supports** are:

- **Roller** ($M = 0$, $R_H = 0$, and only **one reactive force R_V**).
- **Hinge** or Simple ($M = 0$, and only **two** reactive forces R_H and R_V).
- **Fixed** (All the **three** reactive forces M, R_H and R_V).

For analysis of structural element, support reactions are required to be determined by using conditions of **equilibrium** and type of support.

6.4 TYPE OF BEAMS

We have seen that structural elements subjected to **transverse forces** undergo **bending** and the part of the section is in tension or compression. The Structural elements having much larger length compared to its cross-sectional dimensions and subjected to transverse loads, such members **predominantly** undergo **bending**. Thus structural members, having **length much larger** than cross-sectional dimensions and subjected to **transverse loads**, are called **beams**. These beams are classified on the basis of its type and number of **supports**.

Consider beam structures shown in Fig. 6.4. These beam structures are classified in different categories according to the **type** and **number of supports**. We shall first consider plane structures subjected to a loading in a plane.

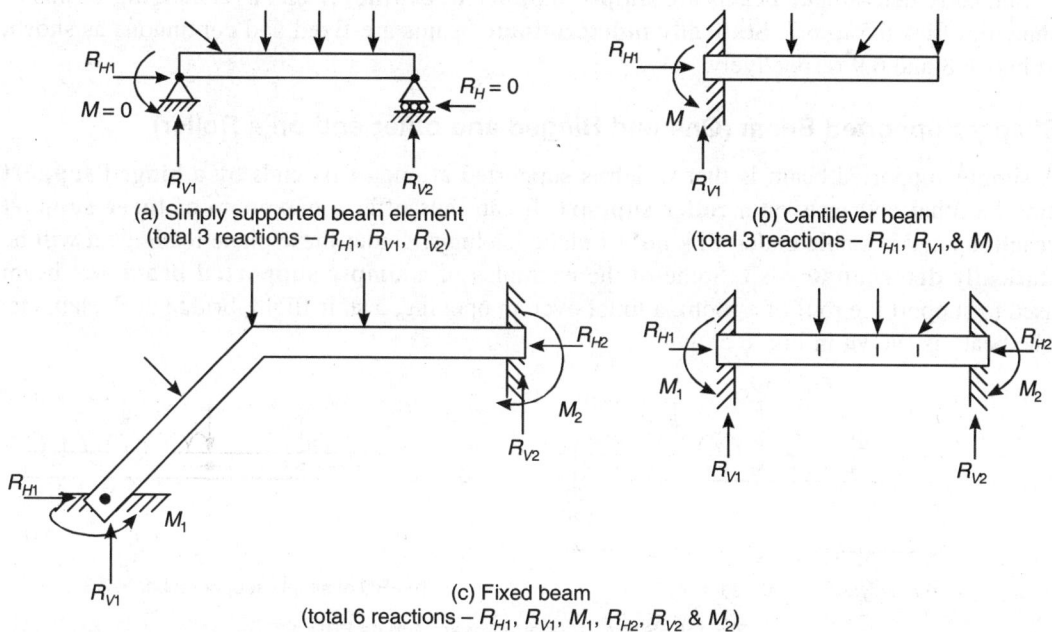

(a) Simply supported beam element
(total 3 reactions – R_{H1}, R_{V1}, R_{V2})

(b) Cantilever beam
(total 3 reactions – R_{H1}, R_{V1},& M)

(c) Fixed beam
(total 6 reactions – R_{H1}, R_{V1}, M_1, R_{H2}, R_{V2} & M_2)

Fig. 6.4: Different beams and support reactions

Bending structures shown in Fig. 6.4 are all examples of **beams** supported at ends. If these beams represent **plane structures** and subjected to the coplanar loads, there will be **at least 3 reactive forces** at the supports. If the support **reactions are 3**, the structure will be in **stable equilibrium**. If reactions are **less than 3**, the structure will be **unstable**. If the reactions are **more than 3**, the structure will be **statically indeterminate** and 3 equations of statics will **not be adequate** to determine internal forces and deformations. Fig. 6.4a, represents **simply supported beam** (SS), and the beam is **determinate**. Fig. 6.4b, represents a **cantilever** beam and is also **determinate**, with 3 unknown reactions. Fig. 6.4c, represents a beam with both ends fixed and having 6 unknown support reactions. The structural element will require **only 3 support reactions** to keep it in stable equilibrium and hence **6 support reactions** will make the beam **statically indeterminate**. The fixed beam shown in Fig. 6.4c, is **statically indeterminate**.

The common attributes of all simple bending structures are:

- All members are connected to its supports or bases.
- Bending structures are subjected to loading in a plane.
- Cross-sections along the longitudinal axis remains uniform in general.

Plane beams with plane loading can also be classified on the basis of support conditions and support reactions as:

- Statically **determinate** beams with **three** total unknown support reactions.
- Statically **indeterminate** with **more than three** total unknown support reactions.

Statically determinate beams are **simply supported, cantilever** and **overhanging** beams as shown in Figs 6.5 to 6.7. **Statically indeterminate** beams are fixed and continuous as shown in Figs 6.8 and 6.9 respectively.

SImply Supported Beam (One end Hinged and other end on a Roller)

A simply supported beam is that which is supported at one of its ends by a **hinged support** and the other end rests on a **roller support**. It can thus offer a maximum of **three support reactions** and is stable under all kinds of plane loadings. Under these plane loadings it will be **statically determinate** also. Some of the examples of a **simply supported beam** are: beam used to support the roof of a room, a lintel over an opening, a stair flight, bridge deck slab, etc. The beam is shown in Fig. 6.5.

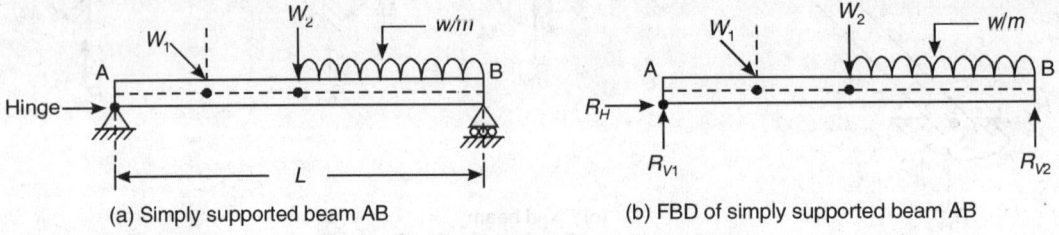

| (a) Simply supported beam AB | (b) FBD of simply supported beam AB |

Fig. 6.5: Simply supported beam and its FBD

Cantilever Beam (One end **Fixed** and other end **Free**)

A cantilever beam or a cantilever (Fig. 6.6) is a beam supported at its **one end** by a **fixed support** and the **other end is free**. It can, thus, offer **three reactive forces** at its fixed end to keep the beam stable. Some examples of a cantilever beam are: Sunshade over openings, a cantilever bridge, an arm of a jib crane, etc. A cantilever beam is a **statically determinate** structure since the number of reactive forces and the number of equilibrium equations are equal (i.e. 3 in this case).

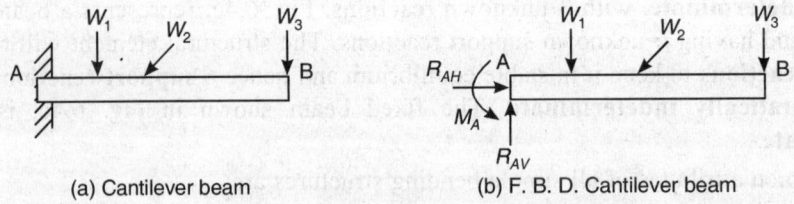

| (a) Cantilever beam | (b) F. B. D. Cantilever beam |

Fig. 6.6: Cantilever beam and its FBD

Overhanging Simply Supported Beam (One or both ends projected)

An **overhanging** beam is **similar** to simply supported beam except that the supports of the beam are so positioned that a part of the **beam overhangs** the base or support. The overhang may be on one or both ends of the beam. Examples are, sunshades over windows, railway sleepers used to support railway track, balanced cantilever beams in bridges, beams used to support room slabs along with balcony slab or verandah and room slabs.

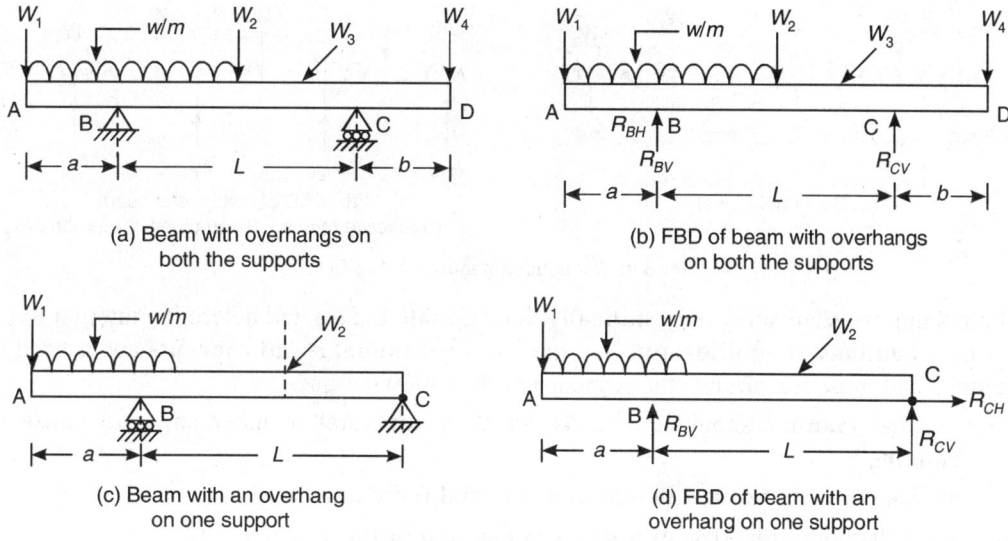

(a) Beam with overhangs on
both the supports

(b) FBD of beam with overhangs
on both the supports

(c) Beam with an overhang
on one support

(d) FBD of beam with an
overhang on one support

Fig. 6.7: Overhanging beam and its FBD

Fixed Beam

A **fixed** beam (Fig. 6.8) is that which is supported at its both ends by fixed supports. Each support is capable of **offering three reactive forces**. The total number of **reactive forces** offered by **two fixed supports** in this beam, could therefore, be **six**. There are only **3 equations** of statics available for the determination of 6 reactive forces. The beam is therefore **statically indeterminate** and requires 3 other conditions of geometry and elastic deformations.

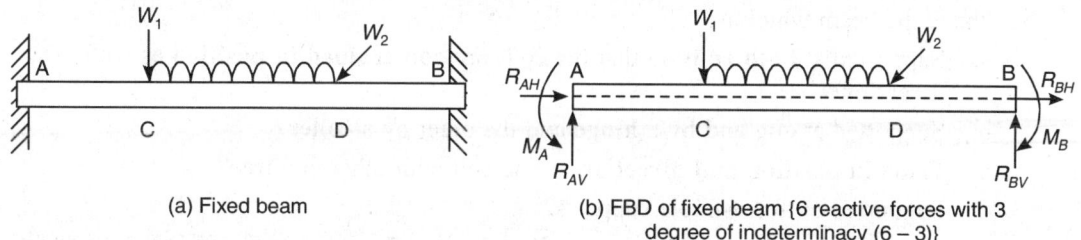

(a) Fixed beam

(b) FBD of fixed beam {6 reactive forces with 3
degree of indeterminacy (6 – 3)}

Fig. 6.8: Fixed Beam and its FBD

Continuous Beam

A **continuous** beam (Fig. 6.9) is that which is supported by **more than two supports**. The continuous supports are simple supports and the end supports could be **fixed, hinged** or **roller** supports. This beam is also **statically indeterminate** due to **more than three** reactive forces offered by the number of supports. The degree of **indeterminacy** depends on the **type** and **number** of supports (Total number of reactive forces minus three equations of equilibrium represents degree of indeterminacy).

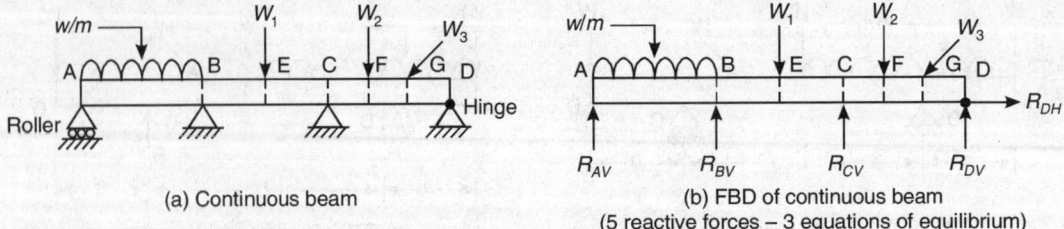

(a) Continuous beam

(b) FBD of continuous beam
(5 reactive forces – 3 equations of equilibrium)

Fig. 6.9: Continuous beam and its FBD

In this unit we shall study only **statically determinate beams** and determine internal forces by using **equations of equilibrium**. You have learned **fundamental concepts** about bending structures and must recapitulate the response to the following questions.

 i. In what **manner** should the loads act on a structural member so as to **cause its bending**?

 ii. In what way is a **beam** different from an **axial force member**?

 iii. The behaviour of a **truss as a whole** is that of a **beam**, explain, how?

 iv. What type of reactive forces are offered by (a) roller support, (b) hinge support, and (c) fixed support?

 v. Name the statically determinate beams showing their support reactions.

 vi. Explain, how a bending structure is required to offer an **axial force** resistance?

 vii. In a beam element which type of load causes bending?

 viii. Which type of support **does not offer resistance to rotation**?

 ix. Which type of support **offer resistance to rotation and displacement**?

 x. Name the beam which is:

    a. Supported at **both ends** so that the ends are **constrained in position** as well as **direction**?

    b. Supported at **one end** by a **hinge** and the other by a **roller**?

    c. **Fixed in position and direction** at one end with other end free?

    d. Supported on four simple supports?

    e. A beam **AB** of 10 m span supported on two simple supports C and D 6 m apart.

6.5 TYPE OF LOADS

A bending structural element is subjected to different type of transverse loads. These transverse loads can be classified by way of its movement along the length (span) or by spread and distribution. From movement point of view, loads are classified as **Static** (stationary) or **Dynamic** (moving with time). Static loads can be applied on a **very small area (point loads)** or the loads are distributed over a large area uniformly or non-uniformly). Loads acting on a very small area (negligible area) are called **point loads** or **concentrated loads** (Fig. 6.10i). Distributed loads over a large span can be **uniform** or **non-uniform**. When the distributed loads are **equally** applied **per unit area** or unit length, it is called **uniformly distributed load** (UDL) and represented by w N/m (Fig. 6.10ii). When the distributed load varies uniformly from a certain minimum value to a maximum value, it is called **uniformly variable distributed load** (UVDL) as shown in Fig. 6.10iii. When the distributed loading varies without any pattern of variation, it is called **non-uniformly distributed variable** loading (Fig. 6.10iv). For analysis of non-uniform distributed loading, loads are considered on a small length and each load is assumed to act at the centre of each small length or area.

These loads are symbolically represented as:

i. Concentrated or Point Load W or P

Example: A beam resting at ends on a column or other supports. The beam transmits concentrated load on the supports.

ii. Uniformly Distributed Load (w/m)

Example: A **beam** supporting a slab carries **uniformly distributed load** per unit length of beam. The load on the beam per unit length is the **same** along the length. **Self weight** of the beam also remain the same per unit length throughout the span if the beam cross-section is uniform.

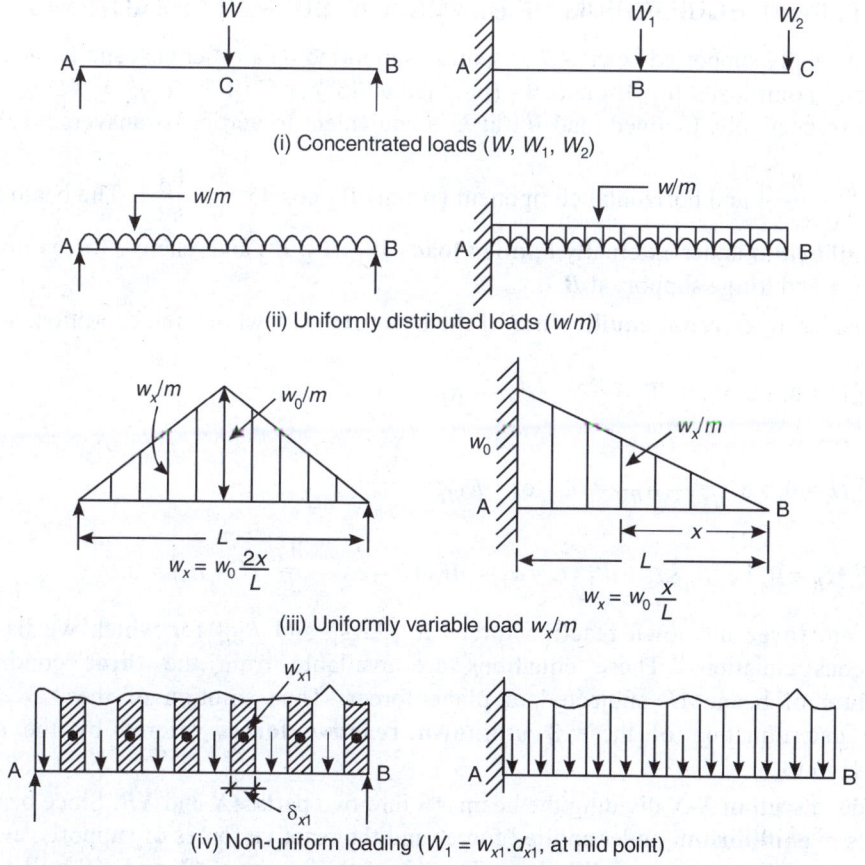

(i) Concentrated loads (W, W_1, W_2)

(ii) Uniformly distributed loads (w/m)

$$w_x = w_0 \frac{2x}{L}$$

$$w_x = w_0 \frac{x}{L}$$

(iii) Uniformly variable load w_x/m

(iv) Non-uniform loading ($W_1 = w_{x1}.d_{x1}$, at mid point)

Fig. 6.10: Type of loading

iii. Uniformly Variable Distributed Loads

Example: Loading exerted by a wall over a **window lintel** or an applied **triangular load** on the beam or cantilever. Such load intensity can be analyzed by using principle of **similar triangles** from the given pattern of loading minimum to maximum.

iv. Non-Uniformly Distributed Loads

Bridge beams carrying variable loads from the slab panels subjected to different movable loads of Buses, Trucks, Scooters, Cars, etc.

Dynamic loads are caused by moving articles/objects such as vehicles on bridges. Based on the pattern of application, dynamic loads can also be **concentrated (point)** loads. Loads applied through **wheels** of a vehicle are point loads. Loads applied through a chain in case of tank are **distributed over the area of contact** (uniform or variable). Thus dynamic loads can also be **point loads, uniformly distributed loads, uniformly variable distributed loads**, or **non-uniform distributed loads**. In this section only **static loadings** on **determinate beams** shall be considered.

6.6   EXTERNAL EQUILIBRIUM OF BEAMS AND SUPPORT REACTIONS

Consider a simply supported beam AB of span L, supported on a roller at A and hinge at B. The beam carries point loads W_1, W_2 and W_3 (inclined at 45°) at C ($AC = a_1$), D ($AD = a_2$) and E ($AE = a_3$) respectively. Inclined load W_3 at E is equivalent to vertical (transverse) component $W_3 \sin 45° \left(= \dfrac{W_3}{\sqrt{2}} \right)$ and horizontal component (Axial) $W_3 \cos 45°$ $\left(= \dfrac{W_3}{\sqrt{2}} \right)$. The beam is **stable** and in **equilibrium** under externally **applied loads** W_1, W_2, W_3 and reactive forces from **roller** support at A and **hinge** support at B.

By considering **external equilibrium** of the beam AB as a whole, the conditions are:

(i) $\Sigma V = 0$, i.e. $W_1 + W_2 + \dfrac{W_3}{\sqrt{2}} - R_{AV} - R_{BV} = 0$.

(ii) $\Sigma H = 0$, i.e. $\dfrac{W_3}{\sqrt{2}} - R_{BH} = 0$,   or $R_{BH} = \dfrac{W_3}{\sqrt{2}}$

(iii) $\Sigma M_B = 0$, i.e. $R_{AV}.L - W_1 (L - a_1) - W_2 (L - a_2) - \dfrac{W_3}{\sqrt{2}} (L - a_3) = 0$.

There are **three** unknown reactive forces R_{AV}, R_{BV} and R_{BH} for which we have **three** simultaneous equations. These equations are available from the **three** conditions of **equilibrium** of beam AB subjected to plane forces. Thus solution of these 3 equations facilitate determination of these **3 unknown reactive forces** offered by the supports. (Figs 6.11i to v).

Consider a **section X-X** dividing the beam AB into two parts AX and XB. Since beam AB as a whole is in **equilibrium** under applied forces and the reactive forces of supports, **every part** of the beam will also be in **equilibrium**. Therefore, the portions AX and XB will also be in **equilibrium**. Considering equilibrium of parts AX and XB **separately**. It becomes quite evident that independent equilibrium of each part AX or XB **necessitates** consideration of

internal forces F_x and M_x at the section X-X to keep each part AX or XB in equilibrium. These **internal forces** required to establish equilibrium under transverse external loads are **shear forces** (F_x) and **bending moments** (M_x). We shall determine these **internal forces** caused at various sections to establish equilibrium under external **transverse loads**.

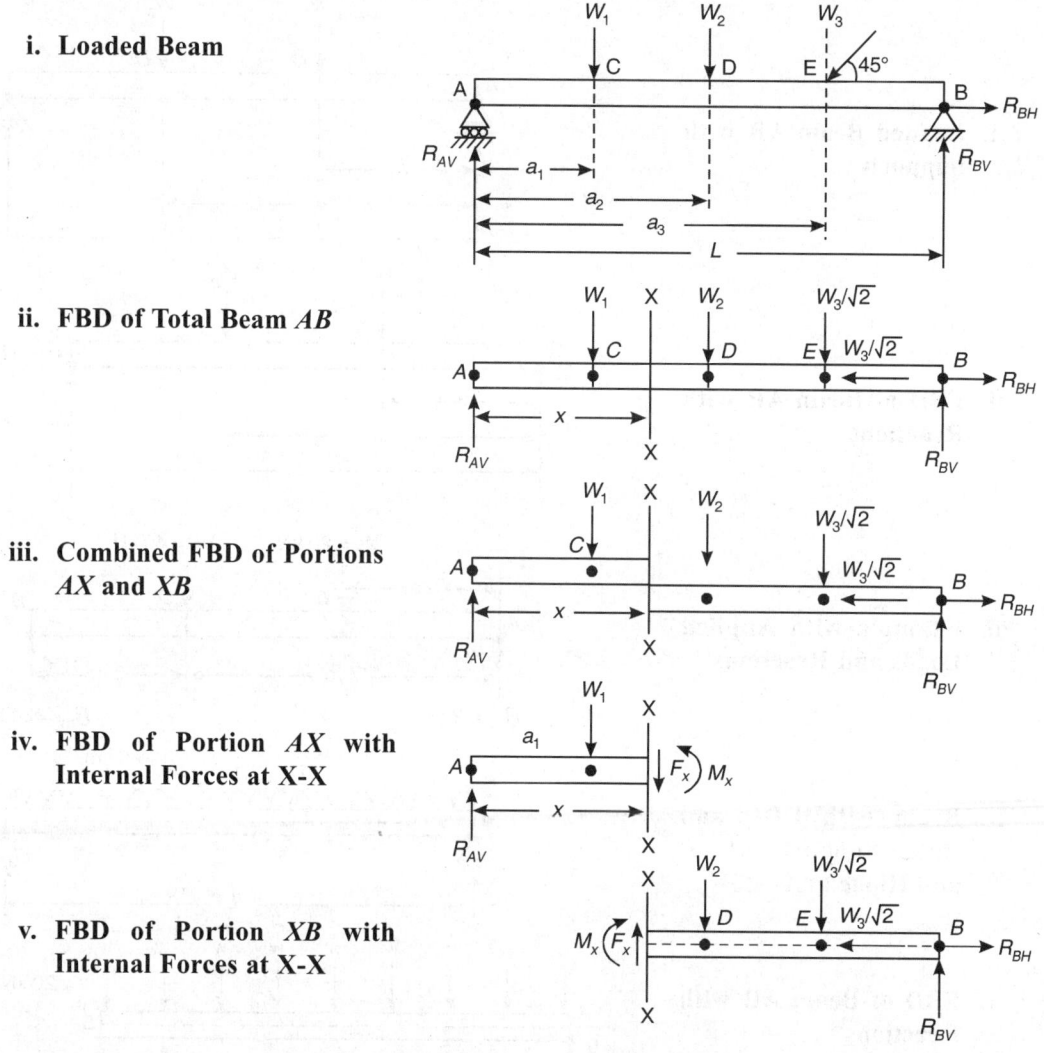

i. Loaded Beam

ii. FBD of Total Beam AB

iii. Combined FBD of Portions AX **and** XB

iv. FBD of Portion AX **with Internal Forces at X-X**

v. FBD of Portion XB **with Internal Forces at X-X**

Fig. 6.11: Beam AB with external loads and FBDs

6.7 SUPPORT REACTIONS

For analyzing any determinate beam (bending structure), it is essential to determine support reactions by considering **external equilibrium** of the structure as a whole. As an example consider a simply supported beam AB of span L carrying concentrated loads W_1 and W_2 at C and D respectively (Fig. 6.12i). Let $AB = L$, $AC = a_1$, $AD = a_2$.

Roller support at A will have only **vertical reaction R_{AV}**. Hinge support at B can have **vertical** and **horizontal reactions** $(R_{BV}$ & $R_{BH})$. In this case since loading is vertical only, the horizontal reaction $R_{BH} = 0$.

Free body diagram with support reactions is shown in Fig. 6.12ii.

i. Loaded Beam AB with Supports

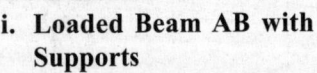

ii. FBD of Beam AB with Reactions

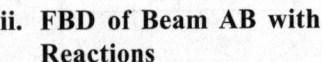

iii. Example with Applied Loads and Reactions

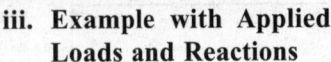

iv. Beam with UDL and Roller Support at B and Hinge at A

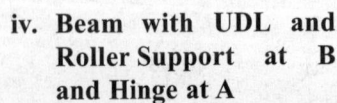

v. FBD of Beam AB with Reactions

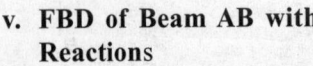

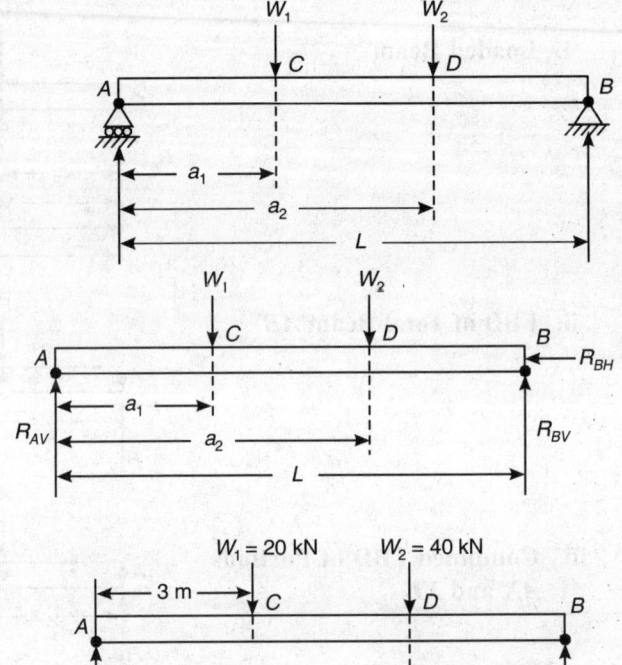

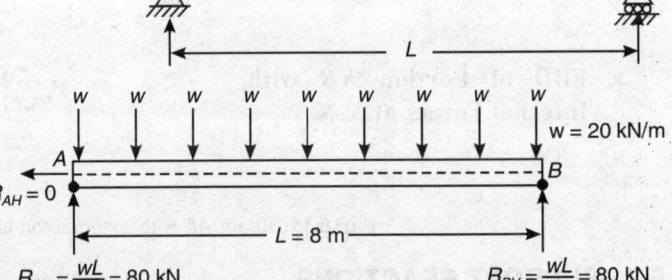

Fig. 6.12: Support reactions (simply supported beams)

Considering **equilibrium** of the beam AB under external loads W_1 and W_2 at C and D respectively we have by resolving vertically and horizontally:

$$R_{AV} + R_{BV} - W_1 - W_2 = 0 \qquad \text{... (1)}$$

$$R_{BH} - 0 = 0, \quad R_{BH} = 0 \qquad \text{... (2)}$$

Moment about the hinge support B:

$$R_{AV} \cdot L - W_1 (L - a_1) - W_2 (L - a_2) = 0$$

or
$$R_{AV} = (W_1 + W_2) - \frac{(W_1 a_1 + W_2 a_2)}{L} \qquad \dots (3)$$

From equation (1) and (3) we have

$$R_{BV} = \frac{(W_1 a_1 + W_2 a_2)}{L} \qquad \dots (4)$$

Consider an **example** (Fig. 6.12iii) with $L = 10$ m, $W_1 = 20$ kN, $W_2 = 40$ kN, $a_1 = 3$ m, $a_2 = 7$ m.

$$R_{BV} = \frac{1}{10} (20 \times 3 + 40 \times 7) = \textbf{34 kN}, \qquad R_{AV} = (20 + 40) - 34 = \textbf{26 kN}$$

Consider another example (Fig. 6.12iv) with *udl w/m* (20 kN/m) on entire simply supported span ($L = 8$ m).

Equilibrium equations reveal:

$$R_{AV} + R_{BV} = wl \qquad \text{or} \qquad R_{AV} + R_{BV} = 20 \times 8 = 160 \qquad \dots (4)$$

$$\Sigma M_A = 0, \qquad R_{BV} \cdot 8 - 20 \times 8 \times \frac{8}{2} = 0, \qquad R_{BV} = \textbf{80 kN}, \qquad R_{AV} = 160 - 80 = \textbf{80 kN}$$

Loads $w = 20$ kN acts at centre of each one metre of span (Fig. 6.12v).

Consider another example of cantilever beam AB (span 2 m) carrying loads of 10 kN at the free end B and 25 kN at C (Fig. 6.13i). The beam is in **equilibrium** and hence conditions of equilibrium provides support reactions. Resolving all the forces,

Vertically: $\qquad\qquad\qquad\qquad R_{AV} - 25 - 10 = 0, \qquad\qquad R_{AV} = \textbf{35 kN} \uparrow$

Horizontally: $\qquad\qquad\qquad\qquad R_{AH} - 0 = 0, \qquad\qquad\qquad R_{AH} = \textbf{0}$

Moment about A: $\qquad\qquad\qquad M_A - 25 \times 1 - 10 \times 2, \qquad M_A = \textbf{45 kNm}$

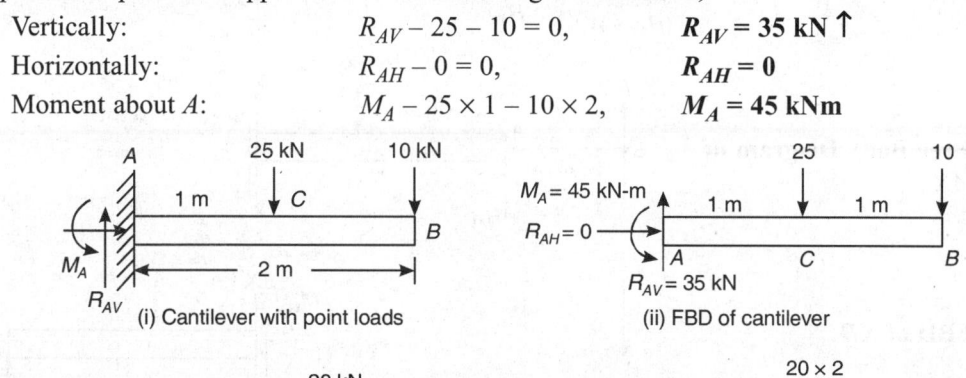

(i) Cantilever with point loads   (ii) FBD of cantilever

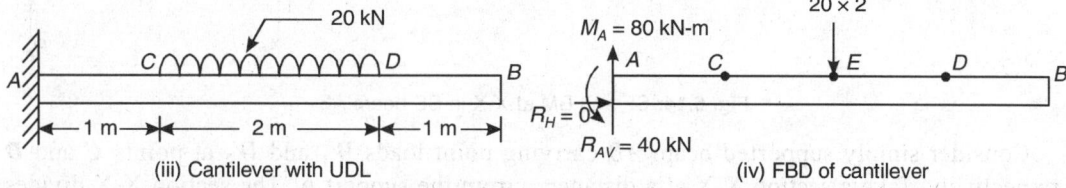

(iii) Cantilever with UDL   (iv) FBD of cantilever

Fig. 6.13: Cantilever beam

Consider another example of cantilever beam AB with udl of 20 kN/m over CD (2 m) as shown in Fig. 6.13iii. From **equilibrium** conditions (Fig. 6.13iv) we have

$$R_{AH} - 0 = 0, \qquad\qquad \boldsymbol{R_{AH} = 0}.$$

Resolving Vertically: $\qquad R_{AV} - 20 \times 2 = 0, \qquad\quad \boldsymbol{R_{AV} = 40 \text{ kN} \uparrow},$

Moment about A: $\qquad\qquad M_A - 20 \times 2\left(1 + \dfrac{2}{2}\right) = 0, \qquad \boldsymbol{M_A = 80 \text{ kNm}}$

Support reactions in determinate beams can easily be determined by applying **conditions of equilibrium** as explained in examples (Figs 6.12 and 6.13).

6.8 SHEAR FORCE AND BENDING MOMENTS

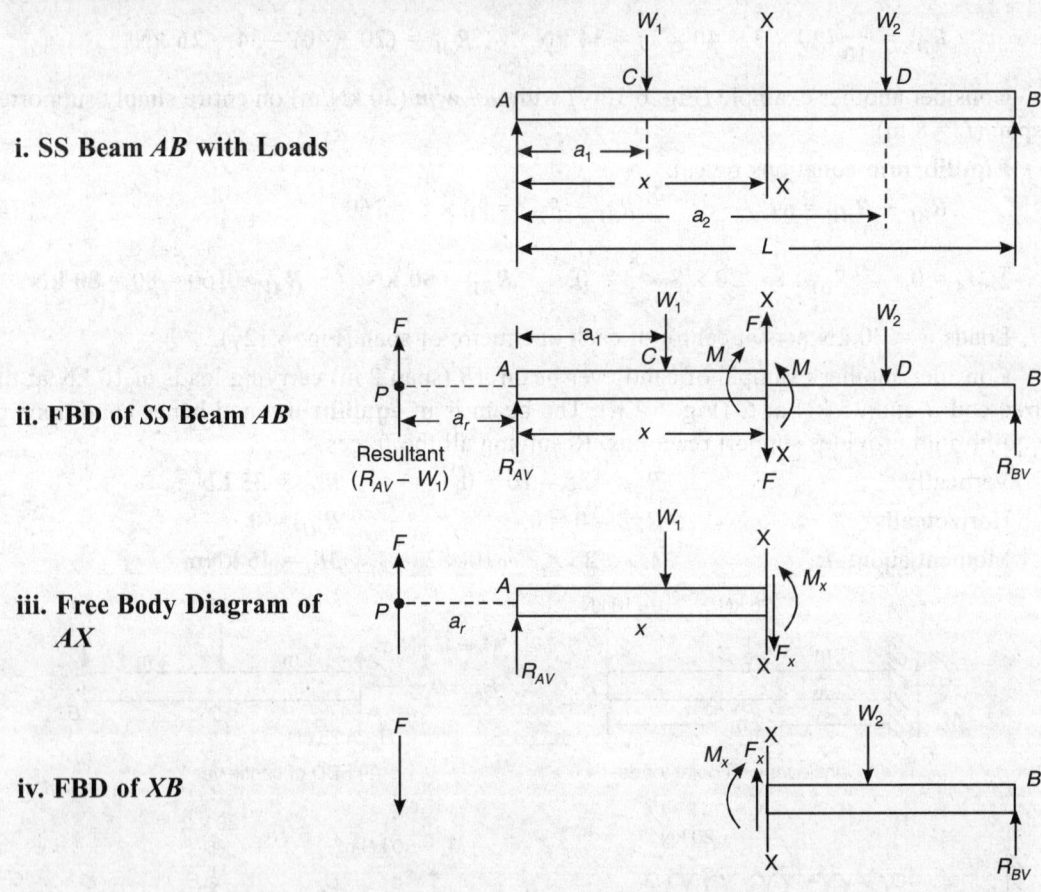

i. SS Beam AB with Loads

ii. FBD of SS Beam AB

iii. Free Body Diagram of AX

iv. FBD of XB

Fig. 6.14: SF and BM at X-X in SS beam AB

Consider simply supported beam AB carrying point loads W_1 and W_2 at points C and D respectively. Take a section X-X at a distance x from the support A. The section X-X divides the beam in two parts AX and XB. Since the whole beam AB is in **equilibrium** under the applied loads W_1 and W_2 alongwith support **reactions at A and B**, the parts AX and XB will also be in **equilibrium**.

Consider the **equilibrium** of the portion AX subjected to vertical load W_1 at C and support **reaction R_{AV}** at A. These two forces are parallel and opposite to each other. The **resultant** of these two parallel forces will be $\qquad F = (R_{AV} - W_1) \uparrow.$ ... (1)

The resultant force 'F' acts at a **point P** along the axis at a distance a_r from the support A. If the **resultant force $F = (R_{AV} - W_1)$** is not zero, then AX can not be in **equilibrium** unless a force F acts internally at the section X-X in opposite direction (i.e. downward). The moment of two forces about the **point P** where resultant acts will be:

$$R_{AV} \cdot a_r - W_1(a_r + a_1) = 0, \qquad \text{or} \qquad a_r = \frac{W_1 a_1}{(R_{AV} - W_1)} \qquad \text{... (2)}$$

This downward **internal transverse force 'F' developed at the section X-X** will neutralize the resultant transverse force F but these two vertical forces are equal and opposite and forms a couple. The portion AX will be in equilibrium only if the couple of moment $F(a_r + x) = (R_{AV} - W_1)(a_r + x)$ acts in opposite direction at the section X-X.

The balancing **transverse internal force 'F_x'** is called **shear force** at X-X. The moment of

the couple is $M_x = F_x(a_r + x) = (R_{AV} - W_1)\left\{ \dfrac{W_1 a_1}{(R_{AV} - W_1)} + x \right\}$

$$\text{or} \quad M_x = (R_{AV} - W_1)\left\{ \frac{W_1 a_1 + R_{AV} \cdot x - W_1 x}{(R_{AV} - W_1)} \right\} = R_{AV} \cdot x - W_1(x - a_1) \qquad \text{... (3)}$$

This **moment** of the internal couple developed at the section X-X is called **bending moment** at the section X-X of the beam and is equal to **algebraic sum** of the moments of external transverse forces on one side of the section. It is evident that for **equilibrium** of any part (say AX) created by considering a section X-X, the section tends to develop **internal transverse force 'F_x'** called **shear force** and **internal moment 'M_x'** is called **bending moment**.

Consider free body diagram (FBD) of the portions AX and XB (Figs 6.14iii and iv). The part AX remains in **equilibrium** due to development of **downward internal transverse force F_x** and **anticlockwise internal moment M_x** at the section X-X by the other part **XB**. Similarly by considering the **equilibrium** of the part XB, the part AX develops a **upward internal transverse force F_x** and **clockwise internal moment M_x** at the section X-X. Thus the **unbalanced external upward transverse** force F_x and the **external clockwise moment M_x** from the left of the section X-X are balanced by the external **downward** transverse force F_x and the external **anticlockwise moment M_x** from the **right** of the section X-X as shown in Fig. 6.15. The beam remains in **equilibrium** under externally applied loads and support reactions.

Considering part AX of the beam, it is clear that **algebraic sum** of all **transverse forces** on the part AX equal to F which is balanced by the **internal upward** resistance equal to F developed at the section X-X. This **internal transverse resistance 'F'** is called **shear force F_x** at the section X-X. This **internal resistance to shear** per unit area is known as **shear stress**.

Thus for numerical calculations, **shear force** F_x at X-X is considered as **algebraic sum** of external **transverse forces** taken on left (*AX*) of the section X-X.

Similarly, the **algebraic sum** of moments **about the section X-X** of all transverse forces acting on the **left of the section X-X** will be considered as **bending moment** (M_x) at the section X-X. This bending moment M_x is created by developing internal resisting moment equal to M_x at the section X-X. The internal resisting moment is known as **moment of resistance** M_r and is numerically equal to bending moment M_x. The longitudinal stresses developed in the cross-section due to bending moment are known as **bending stresses**.

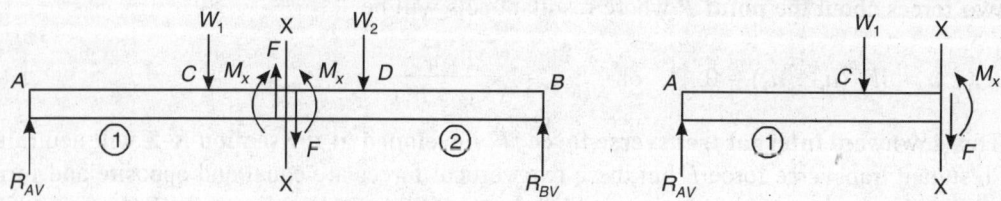

Fig. 6.15: SF and BM

Shear force (SF) at any section is, therefore, **defined as algebraic sum of all transverse forces** considered either on **left** or **right of the section**.

Bending moment (BM) at any section is, therefore, **defined as algebraic sum of moments** taken **about the section** for all **transverse forces** considered **either left** or **right of the section**.

6.9 AXIAL FORCE IN BEAM

If a beam structure is subjected to inclined loading, the inclined load will have two components – **Transverse** (normal to axis) and **Axial** (along the axis). We have already studied the effect of transverse loads to determine SF and BM. The component of load parallel to axis of beam causes **axial stress**. **Axial force** at any section X-X is **algebraic sum of forces parallel to the longitudinal axis taken on either left or right of the section**. Consider a beam *AB* carrying an inclined load W at C (Fig. 6.16). The line of action of the load W makes an angle of θ with the longitudinal axis as shown in Fig. 6.16. The inclined load **W** can be **replaced by transverse load W sin θ** and an **axial load W cos θ**. Axial load **W cos θ** will cause axial stresses (compressive or tensile) depending on the **nature** and **magnitude** of the axial load and **area of cross-section**. **Hinge supports** are capable of **offering axial resistances** while **roller supports** do not offer any **axial resistance**. Axial forces can also be determined by considering equilibrium of forces along the axis on the left or right of the section.

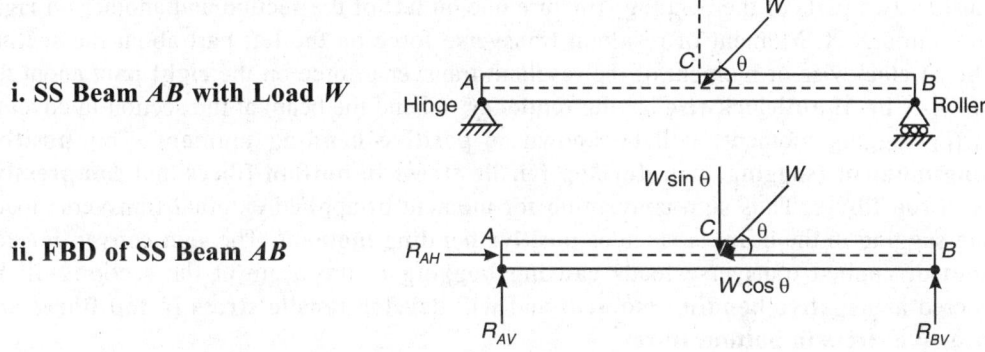

i. SS Beam *AB* with Load *W*

ii. FBD of SS Beam *AB*

Fig. 6.16: Axial force in beam

6.10 SIGN CONVENTIONS OF SF AND BM

We must adopt certain sign conventions for analyzing SF and BM in bending structures. Consider a section X-X in a beam structure dividing it into two parts left and right of the section X-X. These two portions will tend to **separate along the section X-X** due to application of transverse forces (Fig. 6.17).

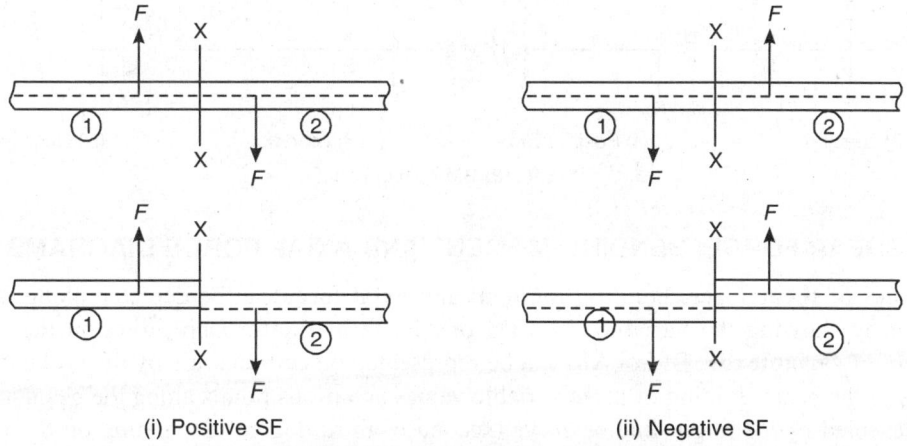

(i) Positive SF   (ii) Negative SF

Fig. 6.17: Sign conventions SF

Let the resultant of applied transverse force be '*F*' on the left part of the section and acts upward. The resultant of applied forces on the right part of the section will be '*F*' acting downward. The tendency to separate the two parts will be to move **left portion upwards** and **right portion downwards** (Fig. 6.17i). Let us adopt the convention to call left upward **force *F*** and **right downward** force *F* as **positive SF**. If the resultant '*F*' of **applied transverse forces be downward on the left** of the section X-X and **upward on the right** of the section X-X, the SF will be called **negative** (Fig. 6.17ii). We shall apply the **same convention throughout** our study. These sign conventions are for convenience of analysis and there are no hard and fast rules. Some books may **adopt opposite of this sign convention** but whatever sign conventions are chosen, the **same convention must be adopted throughout** the study.

Consider two parts of the bending structure one on **left** of the section and another on **right** of the section X-X. Moment of resultant transverse force on the left part about the **section X-X be M clockwise** or moment of the resultant transverse force on the **right part** about the **section X-X be M anticlockwise** having **tendency** to bend the beam at the section in concave shape (or sagging moment) will be known as **positive bending moment**. This **positive bending moment (sagging)** will **develop tensile stress in bottom fibres** and **compressive stress in top fibres**. Thus sign convention for moment of applied external transverse loads causing **sagging** in the beam is taken as **positive bending moment**. The **sign convention for moment** of applied transverse loads **causing hogging** in the beam at the section will be considered as **negative bending moment** and will develop **tensile stress in top fibres** and **compressive stress in bottom fibres**.

There is no **hard and fast rule** to take the same sign convention. Some books may consider opposite of the sign convention adopted but we should take **care that whatever sign convention** is adopted, the same is **followed throughout the analysis** (Fig. 6.18).

Thus **external moment** causing **sagging in the beam** at the section will be considered as **positive** while the external moment causing **hogging** in the beam at the section will be considered as **negative**.

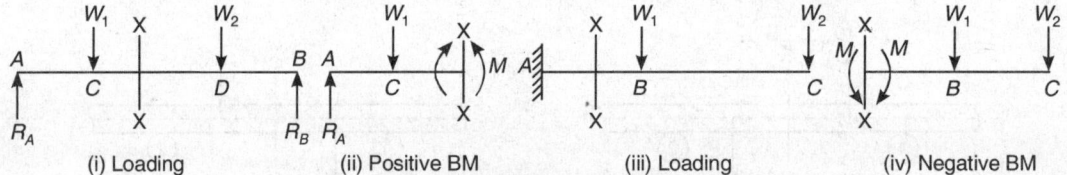

(i) Loading   (ii) Positive BM   (iii) Loading   (iv) Negative BM

Fig. 6.18: BM convention

6.11 SHEAR FORCE, BENDING MOMENT AND AXIAL FORCE DIAGRAMS

The values of **shear force**, **bending moment** and **axial force** can be depicted along the axis of beam by showing the variable (SF, BM or AF) as **ordinate**. Depending on the general **equation of variable** (SF, BM or AF) can be represented on certain scale by the ordinate along the axis of the beam. Joining of these variable values at various points along the beam axis can be represented by a straight line or curve (second or third degree) depending on the **type of loading** and pattern of **distribution** of loading. Axial force will be developed in the beam only if there are **inclined** or **axial force** components. Shear force and bending moments are created due to **transverse** loadings.

SF, BM and **AF** diagrams are drawn by adopting certain linear scale for the axial distances and force or bending moment scale for the variable ordinates. Line joining the ordinate points represent the variable (SF, BM or AF) diagram. These diagrams will be explained by specific examples.

EXAMPLE 1: A simply supported beam *AB* (span *L*) carries a concentrated vertical load *W* at *C* (*AC* = *a*, *BC* = *b*). Draw SF and BM diagrams.

Solution:

Consider equilibrium of beam

Moment about $B = 0$, $\qquad R_A \cdot L - W \cdot b = 0,$ $\qquad \boldsymbol{R_A = \dfrac{Wb}{L}}$, upwards

ΣM about $A = 0$, $\qquad R_B \cdot L - W \cdot a = 0,$ $\qquad \boldsymbol{R_B = \dfrac{Wa}{L}}$, upwards

Check: $\Sigma V = 0$ $\qquad R_A + R_B - W = \dfrac{Wb}{L} + \dfrac{Wa}{L} - W = 0$

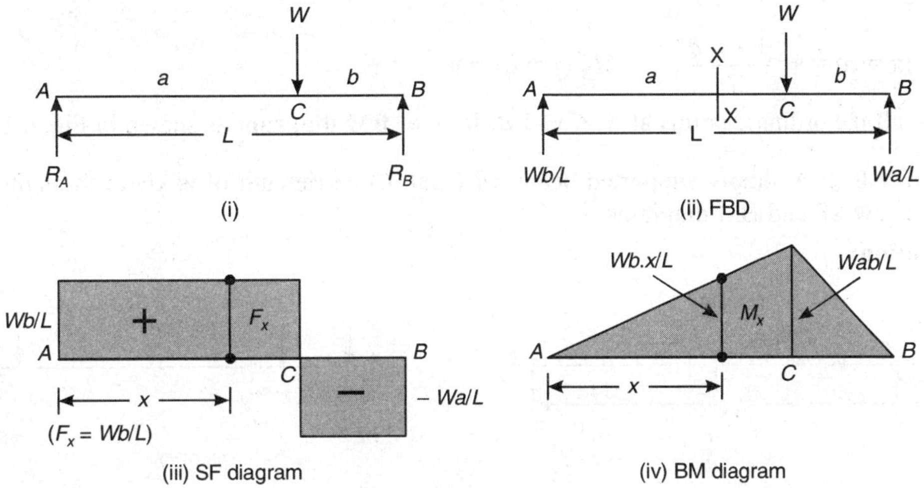

Fig. 6.19: SFD and BMD

Shear Force Diagram

SF at $A = R_A = +\dfrac{Wb}{L}$ $\qquad$ SF at X-X $(AX = x)$

F_x $(A$ to $C) = +\dfrac{Wb}{L}$, constant

$F_C = +\dfrac{Wb}{L}$ (Just left of C), $\quad F_C$ (Just right of C) $= \dfrac{Wb}{L} - W = -\dfrac{W}{L}(L-b) = -\dfrac{Wa}{L}$

F_x $(x > a) = -\dfrac{Wa}{L}$, constant

F_B (Just left of B) $= -\dfrac{Wa}{L}$, $\quad F_B$ (Just right of B) $= -\dfrac{Wa}{L} + \dfrac{Wa}{L} = 0$

Join all the ordinate points at A, C and B. It gives SF diagram as shown in Fig. 6.19iii.

Bending Moment Diagram

BM at A: $\quad M_A = \dfrac{Wb}{L} \times 0 = 0$,

BM at X-X $(x = 0$ to $a)$: $M_x = +\dfrac{Wb}{L} \cdot x$ $\qquad\qquad\qquad$ (Straight line variation)

$$M_C (x = a) = \frac{Wb \cdot a}{L} = +\frac{W \cdot a \cdot b}{L}$$

BM from C to B ($x \geq a$):

$$M_x = +\frac{W \cdot b \cdot x}{L} - W(x - a) = \frac{Wx}{L}(b - L) + W \cdot a = W \cdot a - \frac{Wx}{L}(L - b) = W \cdot a - \frac{W \cdot x \cdot a}{L}$$

or $\quad M_x (x \geq a) = +\dfrac{W \cdot a}{L} (L - x)$ (Straight line variation)

$$M_C (x = a) = +\frac{W \cdot a \cdot b}{L}, \qquad M_B (x = L) = \mathbf{0}.$$

Join all the ordinate points at A, C and B. It gives **BM diagram** as shown in Fig. 6.19iv.

EXAMPLE 2: A simply supported beam AB (span L) carries udl of w kN/m throughout the beam. Draw SF and BM diagrams.

Solution:

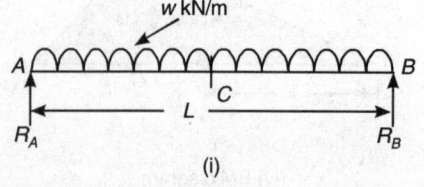

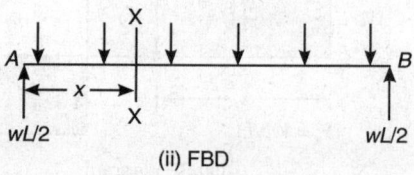

(i)   (ii) FBD

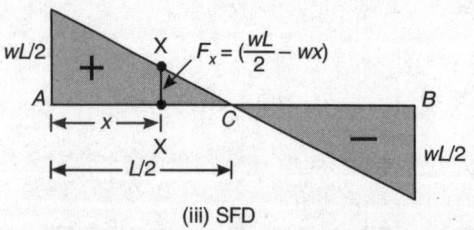

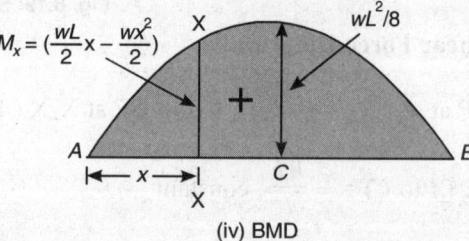

(iii) SFD   (iv) BMD

Fig. 6.20: SFD and BMD

Considering equilibrium of beam AB,

$R_A + R_B - w \cdot L = 0$, as the loading is symmetrical,

$$R_A = R_B = \frac{wL}{2}, \text{ Also } \Sigma M_A = 0, \qquad R_B \cdot L = w \cdot L \cdot \frac{L}{2} \qquad \text{or} \qquad R_B = \frac{wL}{2}$$

Shear Force

$$\text{SF at } A = +R_A = +\frac{wL}{2}, \qquad\qquad \text{SF at X-X } (AX = x), F_x = +\frac{wL}{2} - wx$$

$$F_x = w\left(\frac{L}{2} - x\right), \qquad\qquad\qquad \textbf{Straight Line} \text{ (Fig. 6.20iii)}$$

$$F_C\left(x=\frac{L}{2}\right)=0, \qquad\qquad F_B=-\frac{wL}{2},$$

Bending Moment

BM at A, $M_A =\dfrac{wL}{2}\times 0 = 0,$

BM at X-X, $(AX = x)$ $\qquad\qquad M_x =\dfrac{wL}{2}\cdot x - wx\cdot\dfrac{x}{2}=\dfrac{wx}{2}(L-x)$

$M_x =\dfrac{wx}{2}(L-x),$ $\qquad\qquad$ **Parabolic** (Fig. 6.20iv)

For Max. BM $\dfrac{dM_x}{dx}=0,$

or $\quad \dfrac{d}{dx}\left\{\dfrac{wL\cdot x}{2}-\dfrac{wx^2}{2}\right\}=0 \quad$ or $\quad \dfrac{wL}{2}-\dfrac{w}{2}(2x)=0, x=\dfrac{L}{2}$

$\therefore \quad \mathbf{M_{max.}}\left(x=\dfrac{L}{2}\right)=\dfrac{w}{2}\cdot\dfrac{L}{2}\left(L-\dfrac{L}{2}\right)=\dfrac{w\cdot L^2}{8}$

Max. BM $=\dfrac{w\cdot L^2}{8}$ **at** $x =\dfrac{L}{2}$, **Point of zero SF** (Fig. 6.20iv).

EXAMPLE 3: A simply supported beam AB (span L) carries a triangular load with maximum rate of loading w at the centre.

Solution:

Consider equilibrium of the beam AB.

$$\Sigma M \text{ about } A = R_B \cdot L -\frac{wL}{2}\cdot\frac{L}{2}=0, \quad R_B =\frac{wL}{4},$$

$$\Sigma V = 0, R_A + R_B -\frac{wL}{2}=0, \quad\text{or}\quad R_A +\frac{wL}{4}-\frac{wL}{2}=0, \quad R_A =\frac{wL}{4}$$

A **to** $C\left(x\le\dfrac{L}{2}\right)$

Rate of loading $wx =\dfrac{2w}{L}\cdot x$, straight line variation

Shear Force Diagram

SF at A (Just right of A) $= +\dfrac{wL}{4},$

SF at X-X, Algebraic sum of transverse forces on one side of the section.

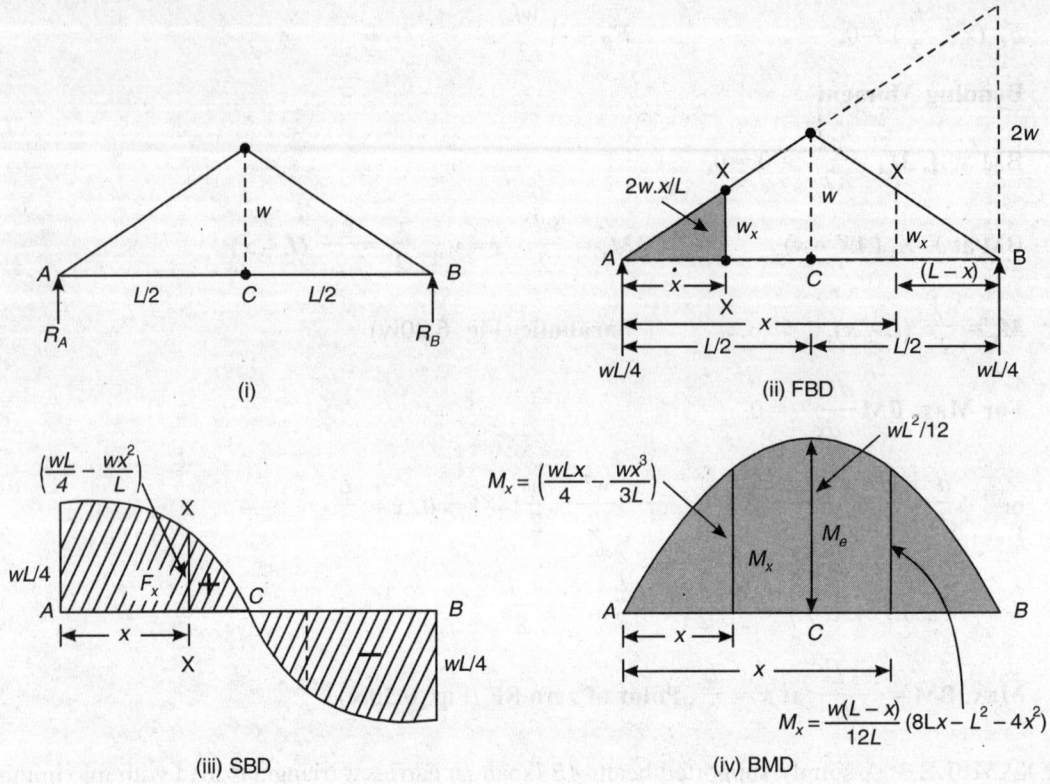

Fig. 6.21: SFD and BMD

$$F_x = +\frac{wL}{4} - \frac{w_x \cdot x}{2} = \frac{wL}{4} - \frac{2w}{L} \cdot x \cdot \frac{x}{2}, \quad F_x = \left(\frac{wL}{4} - \frac{wx^2}{L}\right), \text{ Parabolic variation}$$

$$F_C\left(x = \frac{L}{2}\right) = \frac{wL}{4} - \frac{w}{L}\left(\frac{L^2}{4}\right) = 0$$

$$w_x \text{ (Beyond } C\text{), } x > \frac{L}{2}, \, w_x = \frac{2w}{L} \cdot (L - x)$$

$$F_x \text{ (Beyond } C, x > \frac{L}{2}) = \frac{wL}{4} - \frac{wL}{2 \times 2} - \left(\frac{w + w_x}{2}\right)\left(x - \frac{L}{2}\right)$$

$$\text{or} \quad F_x = -\frac{1}{2}\left\{w + \frac{2w}{L}(L - x)\right\}\left(x - \frac{L}{2}\right)$$

$$F_C\left(x = \frac{L}{2}\right) = 0, \qquad F_B(x = L) = -\frac{L}{4}\{w + 0\} = -\frac{wL}{4}$$

Bending Moment Diagram

BM at $A = 0$, $\qquad\qquad$ $\boldsymbol{M_A = 0}$

$$M_x \left(x \le \frac{L}{2}\right) = \frac{wL}{4} \cdot x - \frac{2w}{L} \cdot x \cdot \frac{x}{2}\left(\frac{x}{3}\right) = \left(\frac{wL}{4} \cdot x - \frac{w \cdot x^3}{3L}\right),$$ $\qquad$ **Cubic parabola**

$$M_C \left(x = \frac{L}{2}\right) = \left(\frac{wL^2}{8} - \frac{wL^2}{24}\right) = \frac{wL^2}{12} \text{ , Max. B.M. at } C \qquad \text{(Point of zero shear force)}$$

$$M_x \left(x \ge \frac{L}{2}\right) = \frac{wL}{4}(L-x) - \frac{w_x \cdot (L-x)}{2}\left(\frac{L-x}{3}\right) = \frac{wL}{4}(L-x) - \frac{2w}{L}(L-x)\frac{(L-x)^2}{6}$$

$$= w\,(L-x)\left[\frac{3L^2 - 4(L^2 - 2Lx + x^2)}{12L}\right]$$

or $\qquad M_x = w\,(L-x)\,[8Lx - L^2 - 4x^2],$

$$= \frac{w(L-x)}{12L}\,[8Lx - L^2 - 4x^2],$$ $\qquad$ **Cubic parabola**

$$M_C \left(x = \frac{L}{2}\right) = \frac{wL}{24L}\,[4L^2 - L^2 - 4\frac{L^2}{4}\,] = \frac{w}{24}\,(2L^2) = \frac{wL^2}{12}$$

$M_B\,(x = L) = 0$

For Max. BM, we have $\dfrac{dM_x}{dx} = 0 = \dfrac{d}{dx}\left\{\dfrac{wLx}{4} - \dfrac{w}{3L} \cdot x^3\right\}, \left\{x \le \dfrac{L}{2}\right\}$

or $\qquad \dfrac{d}{dx}\left\{\dfrac{wL}{4} \cdot x - \dfrac{w}{3L} \cdot x^3\right\} = 0,$ or $\left(\dfrac{wL}{4} - \dfrac{3wx^2}{3L}\right) = 0,$ $\qquad \therefore\ x^2 = \dfrac{L^2}{4}, x = +\dfrac{L}{2}$

$\therefore \qquad M_{max} \text{ (at } x = \dfrac{L}{2}) = \left\{\dfrac{wL}{4} \cdot \dfrac{L}{2} - \dfrac{w}{3L} \cdot \dfrac{L^3}{8}\right\} = \dfrac{wL^2}{8} - \dfrac{wL^2}{24} = \dfrac{wL^2(3-1)}{24} = \boxed{\dfrac{wL^2}{12}}$

EXAMPLE 4: A cantilever AB of span L carries (a) concentrated load W at B (b) *udl* of w/m on the entire span.

Solution:

Consider equilibrium of the cantilever AB of span L.

Take a **section X-X** at a distance of **x from the free end**. We can draw Free Body Diagrams with reference to free end as there are no unknown reactions at the free end.

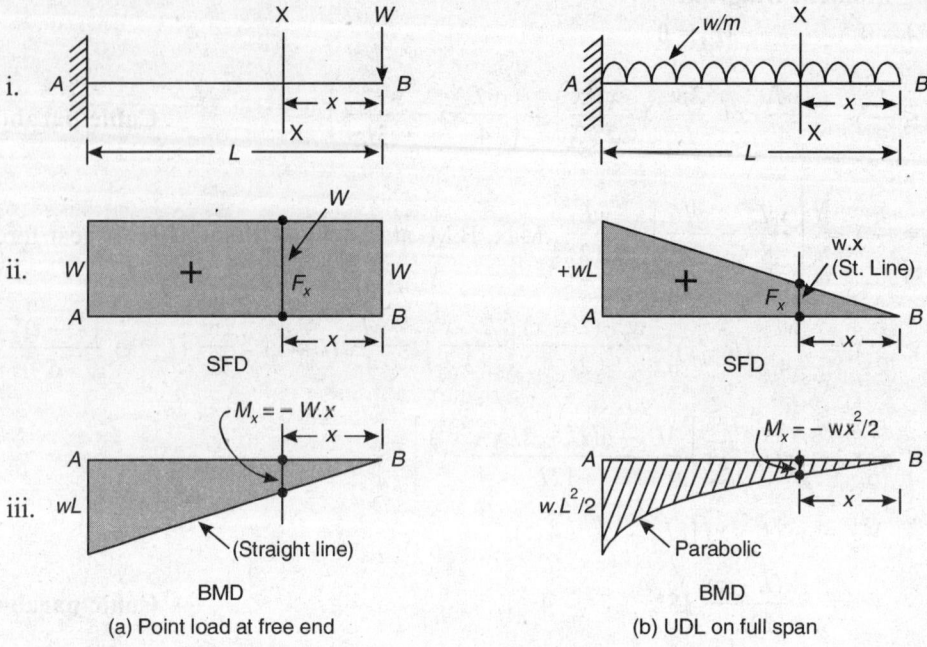

Fig. 6.22: Cantilever beams

a. **Shear Force**

$$F_B = + W, \qquad\qquad F_x\,(x = 0 \text{ to } L) = + W \text{ (constant)}, \qquad\qquad F_A = + W$$

Bending Moment

BM at free end

$$M_B\,(x = 0) = 0,$$

$$M_x\,(x = 0 \text{ to } L) = -W \cdot x \qquad\qquad \text{(straight line Fig. 6.22aiii)}$$

$$M_A\,(x = L) = -W \cdot L \qquad\qquad \text{(Max.)}$$

M_x will be maximum when x is maximum.

b. **Shear Force**

$$F_B = + 0,\ F_x\,(x = 0 \text{ to } L) = + w \cdot x \qquad\qquad \text{(straight line)}$$

$$F_A\,(x = L) = + w \cdot L$$

Bending Moment

BM at free end

$$M_x\,(x = 0 \text{ to } L) = -\frac{w \cdot x^2}{2}, \qquad\qquad \text{(Parabolic Curve Fig. 6.22biii)}$$

$$M_B\,(x = 0) = 0,$$

For maximum or minimum BM: $\dfrac{dM_x}{dx} = 0,$

i.e. $\dfrac{d}{dx}\left(-\dfrac{wx^2}{2}\right) = 0, \quad x = 0$ $\qquad$ (Minimum)

$$M_A\,(x = L) = -\frac{w \cdot L^2}{2} \qquad \text{(Maximum)}$$

Considering the principles of **external** and **internal equilibrium** of determinate beams, the values of **SF** and **BM** can be calculated and diagrams can be drawn as explained in earlier examples 1 to 4. **SF** and **BM** diagrams will further be explained with the help of mixed illustrative numerical examples.

ILLUSTRATIVE EXAMPLES

EXAMPLE 6.1: A simply supported beam AB is of 10 m span and it carries an udl of 20 kN/m on 5 m length CD ($AC = 2$ m and $BD = 3$ m). Determine reactions and draw S.F. and B.M. diagrams.

Solution:

Moment about B:

$R_A \times 10 - 20 \times 5 \times (3 + \dfrac{5}{2}) = 0,$

$R_A = \dfrac{100 \times 5.5}{10} = 55$ kN,

Moment about A:

$R_B \times 10 - 20 \times 5 \times (2 + \dfrac{5}{2}) = 0,$

$R_B = 10\,(4.5) = 45$ kN

Shear Force

$F_A = + 55$ kN,

$F_C = + 55$ kN

$F_x\,(x = 0 \text{ to } 2 \text{ m}) = + 55$ kN $\quad$ (constant)

$F_x\,(x = 2 \text{ m to } 7 \text{ m}) = + 55 - 20\,(x - 2)$

For $F_x = 0, \quad (95 - 20x) = 0,$

$x_0 = = 4.75$ m $\qquad$ (zero S.F. point at E)

$F_D\,(x = 7\text{m}) = (95 - 140) = - 45$ kN

$F_x\,(7\text{m to } 10\text{m}) = - 45$ kN, $\quad$ Constant

$F_B = - 45$ kN

Bending Moment

$M_x\,(x = 0 \text{ to } 2 \text{ m}) = + 55x,$ $\quad$ (Straight Line)

$M_C\,(x = 2 \text{ m}) = + 110$ kN-m

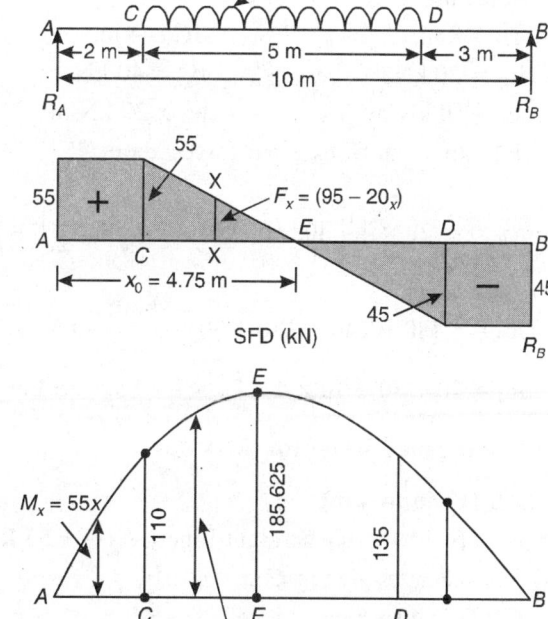

Fig. 6.23: SFD and BMD

x from A

$$M_x \ (x = 2 \text{ to } 7 \text{ m}) = + 55x - \frac{20 \ (x-2)^2}{2} = 55 \ x - 10 \ (x-2)^2, \qquad\qquad \text{Parabolic}$$

$M_C \ (x = 2 \text{ m}) = + \textbf{110 kNm}, \qquad M_D \ (x = 7 \text{ m}) = + \textbf{135 kNm}$

$M_{\max} \ (x = 4.75 \text{ m i.e. point of zero S.F.}) = 55 \ (4.75) - 10 \ (2.75)^2$

(Point E) $M_E = - 261.25 - 75.625 = + \textbf{185.625 kNm}$

$M_x \ (x = 7 \text{ m to } 10 \text{ m}) = 45 \ (10 - x), \quad M_D \ (x = 7 \text{ m}) = 135 \text{ kNm} \qquad \text{(Straight Line)}$

$M_B \ (x = 10) = 0$

Shear force and bending moment diagrams with various values shown in Fig. 6.23.

EXAMPLE 6.2: A simply supported beam AB of 8 m span carries two point loads of 20 kN and 40 kN at 4 m and 6 m respectively from the support A. The beam also carries an udl of 10 kN/m from A for 4 m span and another udl of 20 kN/m for 2 m span from the support B. Determine the values of SF and BM and draw these diagrams. Also find maximum BM and its location.

Solution:

$AB = 8 \text{ m}, \qquad\qquad AC = 4 \text{ m}, \qquad\qquad BD = 2 \text{ m}$

$W_1 = 20 \text{ kN}, \qquad\qquad W_2 = 40 \text{ kN}$

$w_1 = 10 \text{ kN/m}, \qquad\qquad w_2 = 20 \text{ kN/m}$

Take moment of external forces about A:

$$R_B \cdot 8 - 20 \times 4 - 40 \times 6 - 20 \times 2 \times \left(6 + \frac{2}{2}\right) - 10 \times 4 \times 2 = 0$$

$$R_B = \frac{1}{8} \ [80 + 240 + 280 + 80] = \frac{680}{8} = \textbf{85 kN}$$

$$R_A = 20 + 40 + 10 \times 4 + 20 \times 2 - 85 = \textbf{55 kN}$$

Shear Force Diagram (kN)

$AC \ (x = 0 \text{ to } 4 \text{ m})$

$F_x = (+ 55 - 10x), \text{ Straight Line}, \quad F_A = + \textbf{55 kN}, \quad F_C \ (x = 4) = + \textbf{15 kN}$

S. F. changes sign at $C \ (x = 4 \text{ m})$ \hfill Left of C

$CD \ (x = 4 \text{ to } 6 \text{ m})$

$F_x = 55 - 10 \times 4 - 20 = - 5, \text{ Constant}$

$F_C' = - 5, \qquad F_D = - 5, \qquad F_D' = - 5 - 40 = - \textbf{45}$

(Right of C) \quad (Left of D) \quad (Right of D)

$DB \ (x = 6 \text{ to } 8 \text{ m})$

$F_x = 55 - 10 \times 4 - 20 - 40 - 20 \ (x - 6) = (75 - 20x), \text{ Straight Line}$

$F_B \ (x = 8) = - \textbf{85}, \qquad F_B' = 0$

Left of B

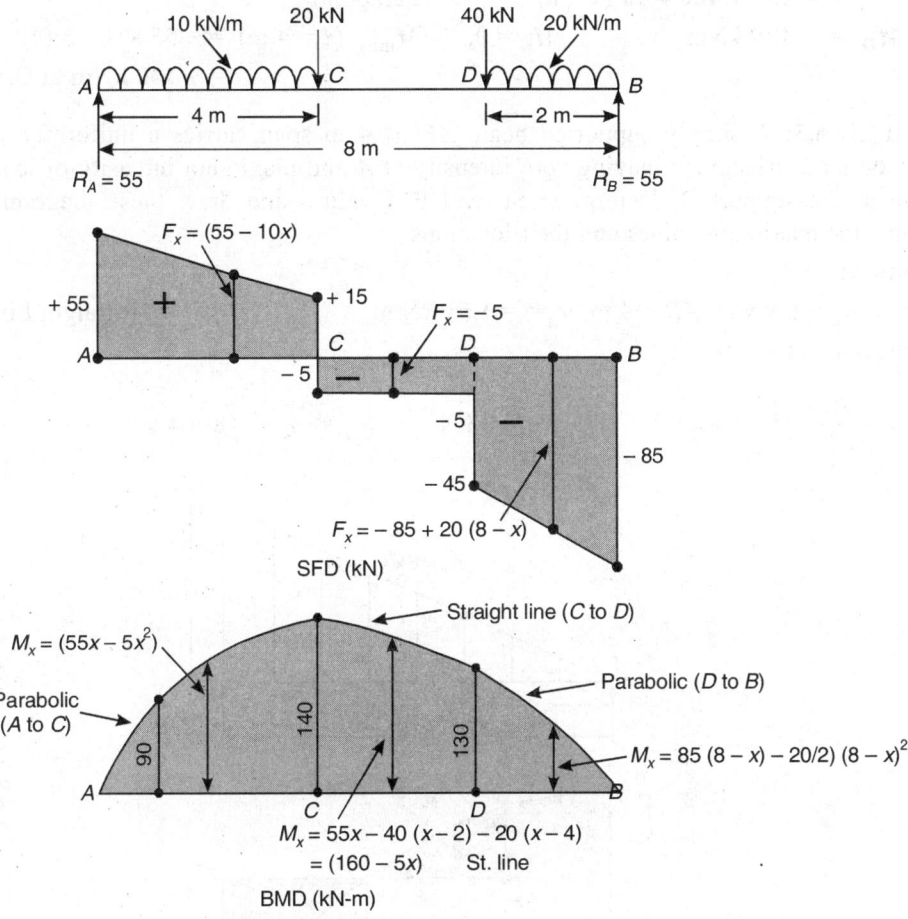

Fig. 6.24: SFD and BMD

Bending Moment

AC (*x* = 0 to 4)

$$M_x = 55x - \frac{10x^2}{2} = (55x - 5x^2) \quad \text{Parabolic,}$$

$M_A = 0$, M_C (*x* = 4 m) = **+ 140 kN-m**

CD (*x* = 4 to 6 m)

$M_x = 55x - 10 \times 4 (x - 2) - 20 (x - 4) = (- 5x + 160)$, Straight Line

M_C (*x* = 4 m) = **+ 140 kNm**, M_D (*x* = 6 m) = **+ 130 kNm**

DB (*x* = 6 to 8 m)

$$M_x = 55x - 40 (x - 2) - 20 (x - 4) - 40 (x - 6) - \frac{20}{2} (x - 6)^2$$

$$= -45x + 400 - 10 (x - 6)^2, \qquad \text{Parabolic}$$

$$M_D = +130 \text{ kNm}, \qquad M_B = 0, \qquad M_{max.} (x = 4 \text{ m}) = +55 \times 4 - 5 (4)^2$$

$$= +140 \text{ kNm at C.}$$

EXAMPLE 6.3: A simply supported beam AB of 4 m span carries a uniformly variable distributed load (triangular) having zero intensity at A and maximum intensity of loading of 6 kN/m at the support B. Determine SF and BM values and draw these diagrams. Also determine the maximum values and their locations.

Solution:

$w_a = 0$, $w_b = 6$ kN/m, $AB = 4$ m, $w_x = = \mathbf{1.5}x$ kN/m   (Straight Line)

Moment about $A = 0$

$$R_B \cdot 4 - \frac{6 \times 4}{2} \times \left(\frac{2}{3} \times 4 \right) = 0, \qquad R_B = 8 \text{ kN}, \qquad R_A = \frac{6 \times 4}{2} - 8 = 4 \text{ kN}$$

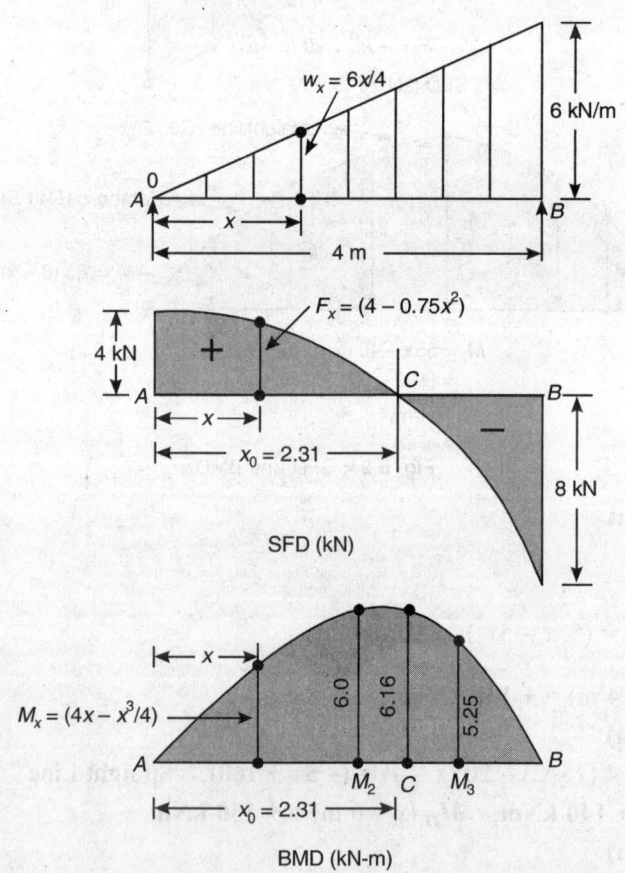

Fig. 6.25: SFD and BMD

Shear Force

$F_A = + 4$ kN

$$F_x \ (x = 0 \text{ to } 4 \text{ m}) = + 4 - \frac{w_x \cdot x}{2} = \left(4 - \frac{1.5x^2}{2} \right) = (4 - 0.75x^2), \qquad \text{Parabolic}$$

$F_B = (4 - 0.75 \times 4^2) = - 8$ kN,

$F_x = 0$, zero S. F. point x_0 at C.

SF (Max.) $= + 4$ kN at A, $- 8$ kN at B.

$$0 = 4 - 0.75x_0^2, \qquad x_0 = 2 \times \frac{2}{\sqrt{3}} = 2.31 \text{ m}$$

$F_{2.31} = 4 - 0.75 \ (2.31)^2 = 0$, B.M. will be maximum at point of zero SF C.

(i.e. $x = 2.31$ m)

Bending Moment

$M_A = 0$,

$$M_x \ (x = 0 \text{ to } 4 \text{ m}) = 4x - \frac{w_x \cdot x}{2} \times \frac{x}{3} = 4x - \frac{1.5x^3}{6} = x \ (4 - 0.25 \ x^2)$$

For maximum value, $\dfrac{dM_x}{dx} = 0$, or the point of zero S.F. $(x = 2.31$ m)

$$\frac{dM_x}{dx} = 0 = 4 - \frac{3 \times 1.5x^2}{6} = \left(4 - \frac{3}{4}x^2 \right), \qquad x = \sqrt{\frac{16}{3}} = \frac{4}{\sqrt{3}} = 2.31 \text{ m}$$

$$M_{\text{max.}} = 2.31 \ (4 - 0.25 \times 2.31^2) = 2.31 \left(4 - \frac{4}{3} \right) = 6.16 \text{ kNm}$$

$$M_1 \ (x = 1 \text{ m}) = 3.75 \text{ kNm}, \qquad M_2 \ (x = 2 \text{ m}) = 4 \times 2 - \frac{1}{4} \ (2)^3 = 6.0 \text{ kNm}$$

$$M_3 \ (x = 3 \text{ m}) = 4 \times 3 - \frac{1}{4} \ (3)^3 = 12 - 6.75 = 5.25 \text{ kNm}$$

$M_{\text{max.}} \ (x = 2.31$ m$) = + 6.16$ kNm

EXAMPLE 6.4: A cantilever beam AB of 3 m span, carries concentrated load of 10 kN at the free end A, 20 kN at C ($AC = 2$ m), and 30 kN at D ($AD = 2.5$ m). It also carries a udl of 20 kN/m on AC. Determine SF and BM at the fixed end B, free end A, and at C. Also draw these diagrams showing maximum values.

Solution:

Cantilever beams can be easily analyzed with reference to its **free end** (A).

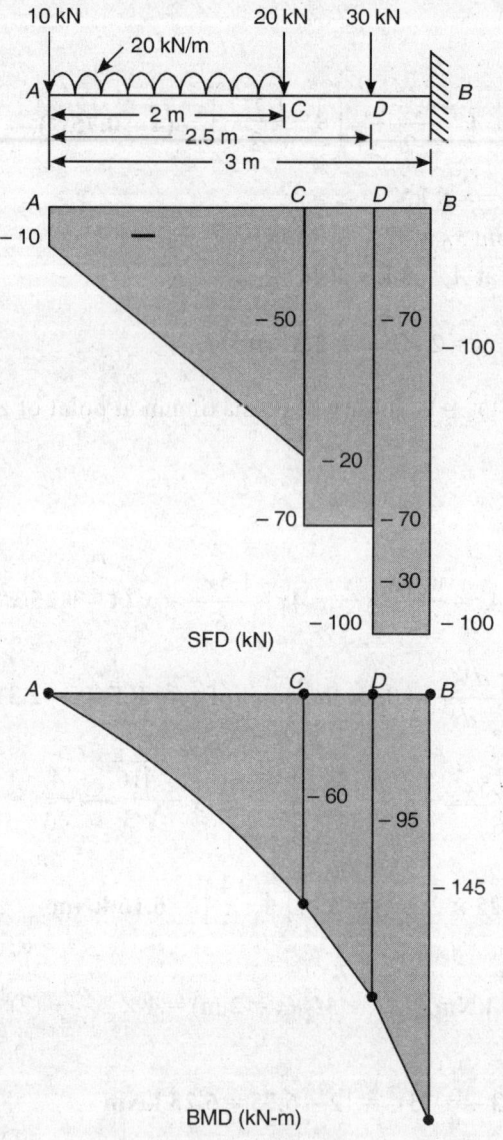

Fig. 6.26: SFD and BMD

Shear Force (kN)

$F_A = -10$ kN,

F_x (A to C) $= -(10 + 20x)$,   **Straight Line**

Left of C

$F_C = -10 - 20 \times 2 = -50$,

Right of C

$F_C' = -50 - 20 = -70$ **kN**

F_x (*C* **to** *D*) = – 70 kN (constant)

Left of *D*

F_D = – 70 kN, F_D' (Right of *D*) = – 70 – 30 = – **100 kN**

F_x = (*D* to *B*) = – 100 kN (**constant**),

F_B = – **100 kN** (SF Max.)

Bending Moment (kN-m)

$$M_x \ (A \ to \ C) = -10x - \frac{20 \cdot x^2}{2}$$

$$= -10x \ (1 + x), \quad \text{Parabolic}$$

M_A (*x* = 0) = 0, M_C (*x* = 2 m) = – **60 kNm**

M_x (*C* to *D*) = – 10x – 40 (x – 1) – 20 (x – 2)

$\qquad = (- \textbf{70}\textbf{\textit{x}} + \textbf{80}), \quad$ Straight Line

M_C (*x* = 2 m) = – 60 kNm,

M_D (*x* = 2.5 m) = – **95 kN-m**

M_x (*D* to *B*) = – 10x – 40 (x – 1) – 20 (x – 2) – 30 (x – 2.5)

or $M_x = (- \textbf{100}\textbf{\textit{x}} + \textbf{155}), \qquad$ Straight Line

M_B (*x* = 3 m) = – **145 kNm** ($M_{max.}$)

SFD and BMD for example 6.4 shown in **Fig. 6.26**.

EXAMPLE 6.5: A simply supported beam *ABC* of 5 m length is supported at *A* and *B* (*AB* = 4 m). The overhanging end *C* carries a concentrated load of **60 kN**. The beam carries a udl of **10 kN/m** on the supported span *AB*. A concentrated load of 100 kN is also applied at a point *D* (1 m from the support *A*). Determine reactions, SF and BM and draw diagrams indicating maximum values along with their locations.

Solution:

$\qquad \Sigma M_A = 0, \quad R_B \times 4 - 100 \times 1 - 60 \times 5 - 10 \times 4 \times 2 = 0$

$$R_B = \frac{480}{4} = 120 \text{ kN}, \quad R_A = 100 + 10 \times 4 + 60 - 120 = \textbf{80 kN}$$

Shear Force (kN)

F_A (Just right of A) = + **80 kN**, F_A (Just left of *A*) = 0

F_x (*A* to *D*) = + 80 – 10x,   Straight Line

F_D (*x* = 1 m) = 80 – 10 = + **70 kN** (Just left of *D*)

F_D' (Just right of *D*) = 80 – 10 – 100 = – **30 kN**

F_x (*D* to *B*) = + 80 – 10x – 100 = – 10 (x + 2)   Straight Line

F_D' (Just right of *D* x = 1 m) = – **30 kN**,

F_B (Just left of *B* x = 4 m) = – **60 kN**

F_B (Just right) = – 60 + 120 = + **60 kN**

F_x (*x* = 4 m to 5 m) = + 80 – 100 – 40 + 120 = + **60 kN** (constant)

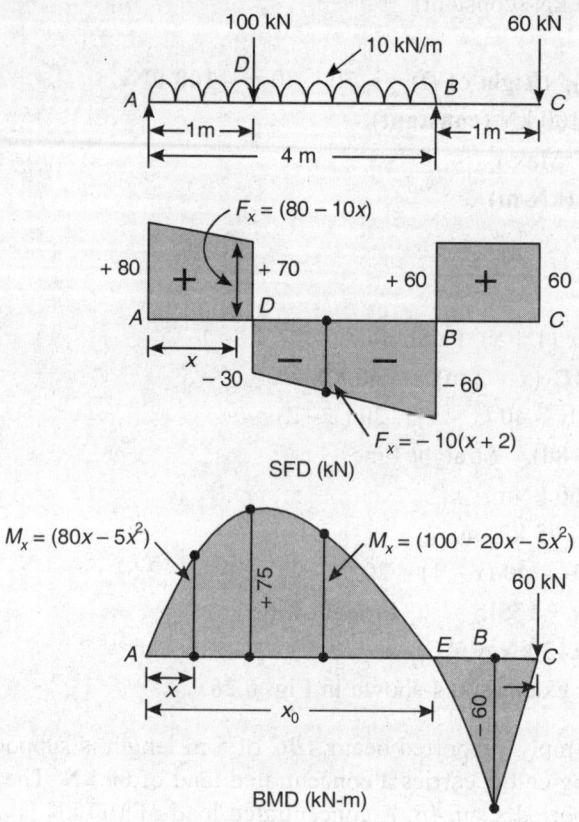

Fig. 6.27: SFD and BMD

F_C (Just left of C, $x = 5$ m) = **+ 60 kN** F_C' (Just right of C) = 0

It may be noted that wherever point loads are acting new equation of F_x is formed.

Bending Moment (kN-m)

$M_A = 0$, M_x ($x = 0$ to 1 m) $= 80x - \dfrac{10x^2}{2} = (80x - 5\,x^2)$,   Parabolic

M_D ($x = 1$ m) $= + 75$ kNm,

M_x (D to B) $= 80\,x - 5x^2 - 100\,(x - 1) = \mathbf{100 - 20x - 5x^2}$,   Parabolic

M_D ($x = 1$ m) $= 100 - 20 - 5 = \mathbf{+ 75}$ **kNm,**   (Max. +ve B.M. at D)

M_B ($x = 4$ m) $= 100 - 80 - 80 = \mathbf{- 60}$ **kNm,**   (Max. –ve B.M. at B)

BM diagram changes sign from +75 kNm at D to – 60 kNm at B. Thus the BM changes sign between D and B and becomes zero at E. This point E is known as point of contra flexure or inflexion.

i.e. $100 - 20x - 5x^2 = 0$,   or $x^2 + 4x - 20 = 0$,

$x = \dfrac{-b \pm \sqrt{b^2 - 4ac}}{2a} = \dfrac{-4 \pm \sqrt{16 + 4 \times 1 \times 20}}{2} = -2 \pm 4.899 = \mathbf{+ 2.899}$ **m**

$M_{2.899} = 100 - 57.98 - 42.02 = 0$

$M_E = 0$, E represents point of **contra flexure** or **inflexion**.

EXAMPLE 6.6: A **6 m** long *ABCD* is supported with a hinged at *A* and a roller at *B*. Supports *A* and *B* are **4 m** apart. The **free end *D*** overhangs support B and carries concentrated load of **50 kN** at *D*. A point load of 40 kN also acts at *C* (*BC* = 0.5 m). Sketch SF and BM diagram showing values at important locations.

Solution:

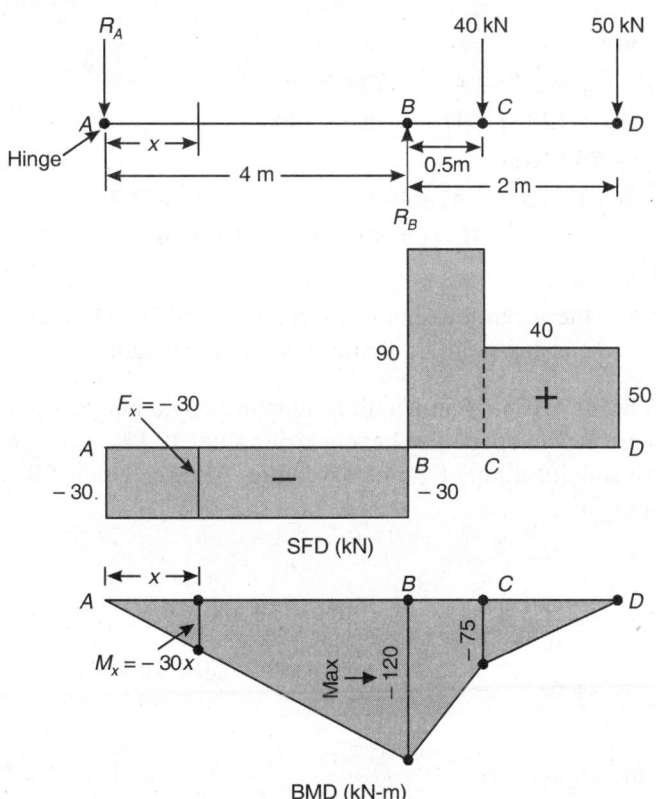

Fig. 6.28: SFD and BMD

$\sum M_A = 0$,   i.e. $R_B \times 4 - 50 \times 6 - 40 \times 4.5 = 0$, $R_B = \dfrac{480}{4} = 120$ kN

$R_A + 120 - 40 - 50 = 0$, $R_A = -30$ kN (i.e. downward)

Shear Force (kN)

$F_A = -\textbf{30 kN}$,   (Just right of *A*)

F_x (*A* to *B*) = **– 30 kN**,   (Constant)

F_B (Just left of *B*) = **– 30 kN**, F_B' (Just right of *B*) = – 30 + 120 = **+ 90 kN**

F_x (*B* to *C*) = **+ 90 kN**,   (Constant)

F_C (*x* = 4.5 m) = **+ 90 kN**,   (Just left of *C*)

F_C' (Just right of C) = $+ 90 - 40 = + 50$ kN

F_x (C to D) = $+ 50$ kN,   (Constant)

F_D (Just left of D) = $+ 50$ kN, F_D' (Just right of D) = $+ 50 - 50 = 0$

Shear force changes sign at B where values pass through **zero** line and hence BM shall be maximum at B.

Bending Moment (kN-m)

x from hinge end A

$M_A = 0$

M_x (A to B) = $- 30x$,   Straight Line

$M_{\text{max.}}$ ($x = 4$ m) = $M_B = - 30 \times 4 = - 120$ kN-m

M_x (B to C) = $- 30x + 120 (x - 4) = + 90x - 480$,   Straight Line

M_C ($x = 4.5$ m) = $- 75$ kN-m

M_x (C to D) = $- 30x + 120 (x - 4) - 40 (x - 4.5) = + 50x - 300$,   Straight Line

M_D ($x = 6$ m) = 0, M_C ($x = 4.5$ m) = $- 75$ kN-m

$M_{\text{max.}}$ = $- 120$ kN-m at B

The point of zero value when the sign changes. SFD and BMD of example 6.6 shown in Fig. 6.28. **Check** if there is any point of contra flexure or inflexion?

EXAMPLE 6.7: A beam ABC of 5 m length is supported by a hinge at A and a roller support at B. Two supports are at 3 m apart. The beam carries a udl 12 kN/m on the entire 5 m length. Determine SF, BM and location of point of contra flexure. Draw SF and BM diagrams indicating maximum values.

Solution:

$$\Sigma M_A = 0 = R_B \times 3 - 12 \times 5 \times \frac{5}{2}, \qquad R_B = 50 \text{ kN}$$

$$R_A + R_B = 12 \times 5, \qquad\qquad R_A = 60 - 50 = 10 \text{ kN}$$

Shear Force (kN)

$F_A = + 10$ kN, (Just right of A)

F_x (A to B) = $(10 - 12x)$,   Straight Line

Point of zero shear: $0 = (10 - 12x_d)$, $x_d = \dfrac{10}{12} = 0.833$ m

F_B (Just left of B, $x = 3$ m) = $10 - 12 \times 3 = - 26$ kN

F_B' (Just right of B) = $- 26 + 50 = + 24$ kN

F_x (B to C) = $10 - 12x + 50 = (60 - 12x)$, Straight Line

F_B' ($x = 3$ m) = $60 - 12 \times 3 = + 24$ kN

F_C ($x = 5$ m) = $60 - 12 \times 5 = 0$.

Shear force **zero** at D ($x_d = 0.833$ m) and **at B.**

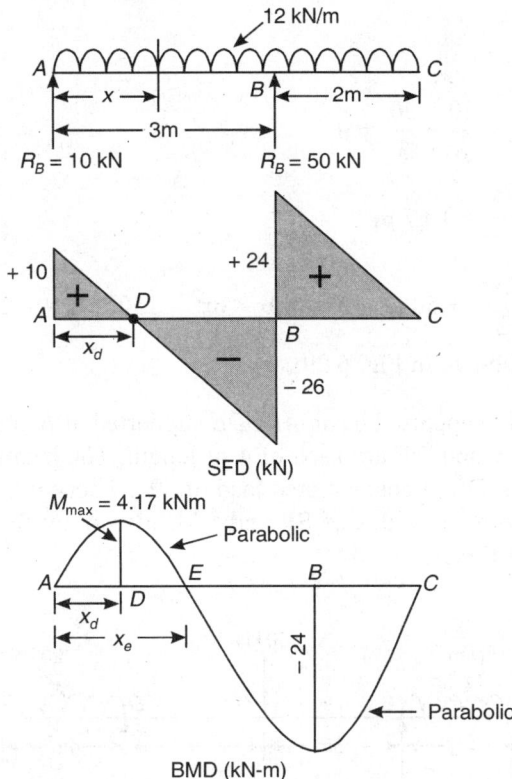

Fig. 6.29: SFD and BMD

Bending Moment

$M_A = 0$,

$$M_x \ (A \ \text{to} \ B) = 10x - \frac{12x^2}{2} = (10x - 6x^2), \ \text{Parabolic}$$

$$M_D \ (x = 0.833 \ \text{m}) = 10\frac{5}{6} - 6\left(\frac{5}{6}\right)^2$$

$$= 8.33 - 4.16 = + \textbf{4.17 kNm}$$

$M_{\text{max.}} = + 4.17$ kNm at D.

$M_B \ (x = 3 \ \text{m}) = + 10 \ (3) - 6 \ (3)^2$

$$= 30 - 54 = - \textbf{24 kNm}$$

BM changes sign D to B (+ 4.17 to – 24.0).

$M_x = 0$, **point of contra flexure** x_e.

$0 = 10x_e - 6 \ (x_e)^2$

or $\quad 5x_e - 3x_e^2 = 0, \quad$ or $\quad x_e \ (5 - 3x_e) = 0,$

$$x_e = 0, \text{ or } x_e = \frac{5}{3} \text{ m}$$

$$M_E = 10\left(\frac{5}{3}\right) - 6\left(\frac{5}{3}\right)^2 = \frac{50}{3} - \frac{50}{3} = 0,$$

point of inflexion $x_e = \frac{5}{3} = 1.67$ m

$$M_x \ (B \text{ to } C) = 10x - \frac{12x^2}{2} + 50 \ (x - 3) = 60x - 6x^2 - 150, \ M_B = -24 \text{ kNm}, \ M_C = 0$$

SF and BM diagrams shown in Fig. 6.29.

EXAMPLE 6.8: A simply supported beam *ABCDE* supported at *B* and *D*. Supported span *BD* is **10 m** and overhangs *AB* and *DE* are each of **4 m** length. The beam carries udl of **20 kN/m** on each overhangs *AB* and *DE*. A concentrated load of 50 kN acts at *C*, the middle of span *BD*. Calculate reactions, maximum values of SF and BM, also draw SF and BM diagrams and determine point of contra flexure.

Solution:

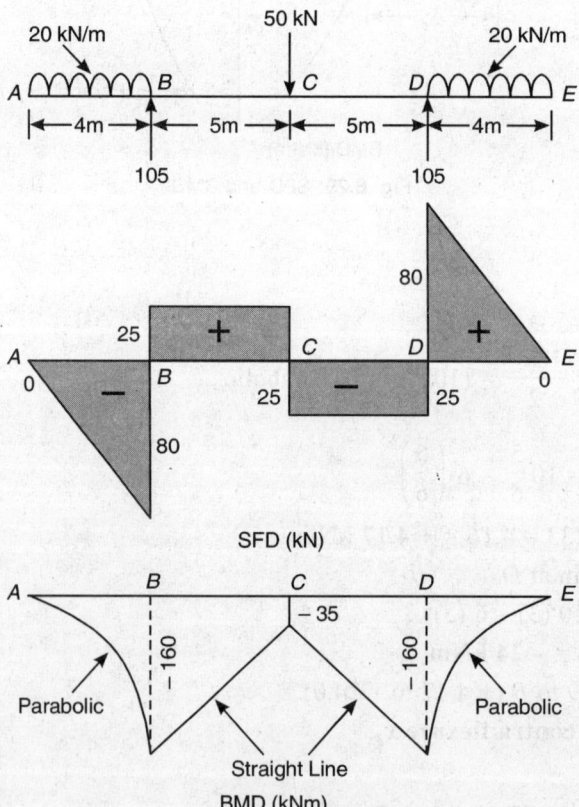

Fig. 6.30: SFD and BMD

$$\Sigma M_B = 0, \quad R_D.10 + 20 \times 4 \times \left(\frac{4}{2}\right) - 50 \times 5 - 20 \times 4 \ (10 + 2) = 0, \quad R_D = 105 \ \text{kN}$$

$R_B + R_D - 80 - 80 - 50 = 0, \quad \boldsymbol{R_B = 105 \ \text{kN}}$

Shear Force

$F_A = 0, \quad F_x \ (A \ \text{to} \ B) = -20x, \qquad$ Straight line

F_B (Just left) $= -20 \times 4 = -\textbf{80 kN},$

F_B' (Just right of B) $= -80 + 105 = +\textbf{25 kN}$

$F_x \ (B \ \text{to} \ C) = -20 \times 4 + 105 = +\textbf{25 kN}, \quad$ (Constant)

F_C (Just left of C) $= +\textbf{25 kN},$

F_C' (Just right of C) $= +25 - 50 = -\textbf{25 kN}$

$F_x \ (C \ \text{to} \ D) = -\textbf{25 kN}, \qquad$ (Constant)

$F_D \ (x = 4 + 10 \ \text{m}) = -25 \ \text{kN},$

F_D' (Just right of D) $= -25 + 105 = +\textbf{80 kN}$

$F_x \ (D \ \text{to} \ E) = +80 - 20x, \qquad$ Straight line between D and E (x from D)

$F_E = +0$

Bending Moment

$$M_A = 0, \quad M_x \ (A \ \text{to} \ B) = -20 \ . \ \left(\frac{x^2}{2}\right) = -10x^2, \qquad \text{Parabolic}$$

$M_B = -10 \ (4)^2 = -\textbf{160 kNm},$

$M_x \ (B \ \text{to} \ C) = -20 \times 4 \ (x - 2) + 105 \ (x - 4) = +25x + 160 - 420$

x from $A = +25x - 260, \qquad$ Straight line

$M_B \ (x = 4 \ \text{m}) = -\textbf{160 kNm},$

$M_C \ (x = 9) = -\textbf{35 kNm}$

$M_x \ (C \ \text{to} \ D) = -80 \ (x - 2) + 105 \ (x - 4) - 50 \ (x - 9) = -25x + 190, \quad$ Straight line

$M_C \ (x = 9 \ \text{m}) = -\textbf{35 kNm},$

$M_D \ (x = 14 \ \text{m}) = -\textbf{160 kNm}$

$$M_x \ (D \ \text{to} \ E) = -80 \ (x - 2) + 105 \ (x - 4) - 50 \ (x - 9) + 105 \ (x - 14) - (x - 14)^2 \ \frac{20}{2}$$

$M_x = +80x - 10 \ (x - 14)^2 - 1280,$

$M_D \ (x = 14) = -\textbf{160 kNm},$

$\boldsymbol{M_E = 0}$

Since BM does not change the sign, there is no point of **contraflexure**.

$M_{\text{max.}} = -160 \ \text{kNm}$ at B and D, $\quad M_{\text{min.}} = -35 \ \text{kNm}$ at C

SF and BM diagrams shown in Fig. 6.30.

EXAMPLE 6.9: A simply supported beam *ABCD* of 10 m length carries udl of 20 KN/m. The beam is supported at B and C with **equal** overhangs AB and CD. Find the value of overhang AB if the absolute maximum BM is to be of minimum value.

Solution:

$L = 10$ m, Overhang $= a$, Supported length $BC = (10 - 2a)$,

$w = 20$ kN/m

$$R_A = R_B = \frac{WL}{2} \text{ (symmetry)} = 10 \times 10 = 100 \text{ kN}$$

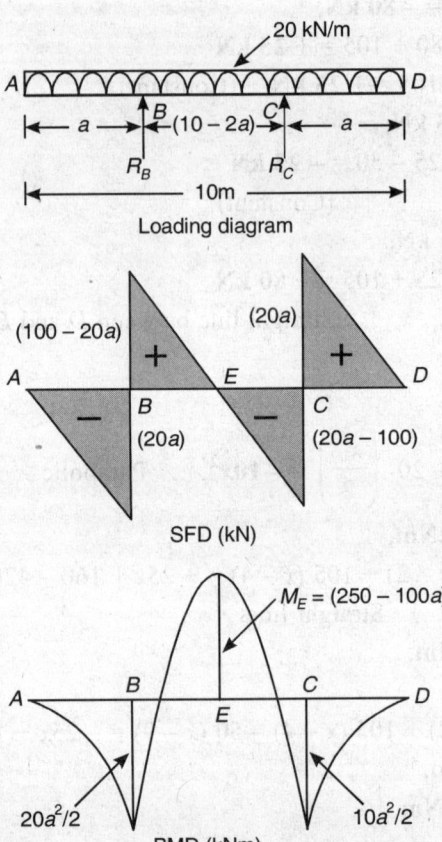

Fig. 6.31: SFD and BMD

Shear Force

$F_A = 0$, $F_B = -20a$, $F_B' = +100 - 20a = 20(5 - a)$

$F = 0$, at the point E at x from A.

i.e. $F_x = -20x + 100 = 0$, $x = \dfrac{100}{20} = 5$ m

$F_C = -20(10 - a) + 100 = (20a - 100)$

$F_C' = (20a - 100) + 100 = +20a$

SF crosses zero line at B, E and C.

Maximum numerical values of BM will be at B, E and C. $M_B = M_E$.

Hence, $M_B = - \{M_E\}$ or $\boldsymbol{M_B + M_E = 0}$

Bending Moment

$$M_A = 0, \quad M_B = \frac{20a^2}{2} = - 10a^2,$$

$$M_x \ (B \ \text{to} \ C) = - 20 \cdot \frac{x^2}{2} + 100 \ (x - a) = - \boldsymbol{10x^2 + 100 \ (x - a)}$$

x from A

$M_E \ (x = 5 \ \text{m}) = - 250 + 100 \ (5 - a) = - 100a + 250$

$M_B + M_E = 0$,   i.e. $- 10a^2 - 100a + 250 = 0$,   or $a^2 + 10a - 25 = 0$

$$a = \frac{-10 \pm \sqrt{100 + 4 \times 25}}{2} = (- 5 \pm 5\sqrt{2}) = 5 \ (\sqrt{2} - 1) = 5 \ (0.4142) = 2.071$$

Thus, when $\boldsymbol{a = 2.071}$ **m**, the maximum BM at B and at E will be numerically equal. Maximum $M_B = -$ **42.89 kNm**, also $M_E = +$ **42.89 kNm**.

SF and BM diagrams are shown in Fig. 6.31.

EXAMPLE 6.10: A beam $ABCD$ is hinged at A and supported on a roller at D. The length AD is 10 m. The beam carries a point load of 100 kN at the point C at 7 m from the hinge A. It also carries an eccentric load (pull) of **20 kN** on a bracket at the point B, 3 m away from A. The pull is 2 m away from the bracket base B as shown.

Solution:

$\Sigma M_A = 0$, $R_D \times 10 - 100 \times 7 + (20 \times 3 + 40) = 0$, $\boldsymbol{R_D = 60 \ \text{kN} \ (\uparrow)}$

$\boldsymbol{R_A = 100 - 20 - 60 = 20 \ \text{kN} \ (\downarrow)}$

Shearing Force

$F_A = +$ 20 kN, $F_x \ (A \ \text{to} \ B) = +$ 20 kN,   (constant)

F_B (Right of B) $= + 20 + 20 = 40$ kN, $F_x \ (B \ \text{to} \ C) = +$ 40 kN, (constant)

F_C (Left of C) $= +$ 40 kN, F_C' (Just right) $= + 40 - 100 = -$ **60 kN**,

F_D (Left of D) $= -$ 60 kN, F_D' (Right of D)$= - 60 + 60 = 0$

Bending Moment

$M_A = 0$, $\boldsymbol{M_x \ (A \ \text{to} \ B) = +} \ 20x$,   Straight line

$\boldsymbol{M_B = 20 \times 3 = +} \ \textbf{60 kNm}$, $\boldsymbol{M_B'}$ (Just right) $= + 60 - 40 = +$ **20 kNm**

$\boldsymbol{M_x}$ (B to C) $= 20x - 40 + 20 \ (x - 3) = \boldsymbol{(40x - 100)}$,   Straight line

$M_C \ (x = 7\text{m}) = +$ 180 kNm

$\boldsymbol{M_x}$ (C to D) $= + 20x + (20) \ (x - 3) - 40 - 100 \ (x - 7)$

$\boldsymbol{M_x = - 60x + 600 = 60 \ (10 - x)}$,   Straight line

$\boldsymbol{M_D \ (x = 10) = 0}$

$M_C (x = 7) = 60 (10 - 7) = + 180 \text{ kNm}$ (Maximum BM)

SF and BM diagrams are shown in Fig. 6.32.

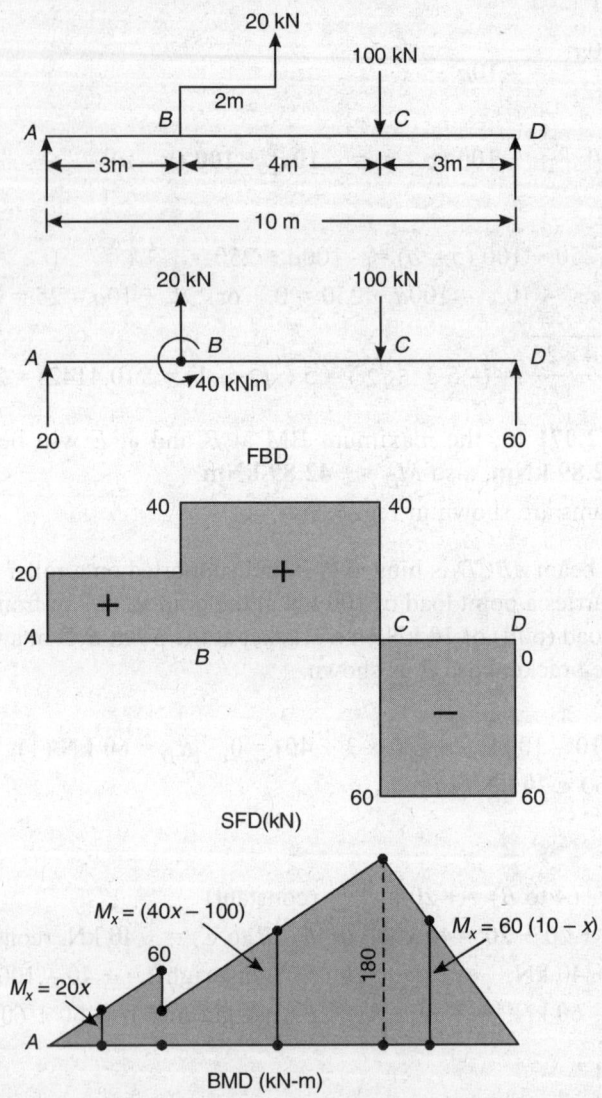

Fig. 6.32: SFD and BMD

EXAMPLE 6.11: An inclined beam ABC of 10 m length is inclined at 30° with the horizontal. The beam is supported on a hinge support at A and roller support at C. A point load of **50 kN** acts vertically at a point B situated 4 m from the hinge support A. Determine the values of axial force, shear force and bending moment and draw these diagrams.

Solution: Convert loads into transverse and tangential components. The beam is inclined at

$30°$, and hence **transverse** component will be $50 \cos 30° = 50 \times \dfrac{\sqrt{3}}{2} = \mathbf{25\sqrt{3}\ kN}$. Tangential components will be $50 \sin 30° = 25$ kN (causing axial force).

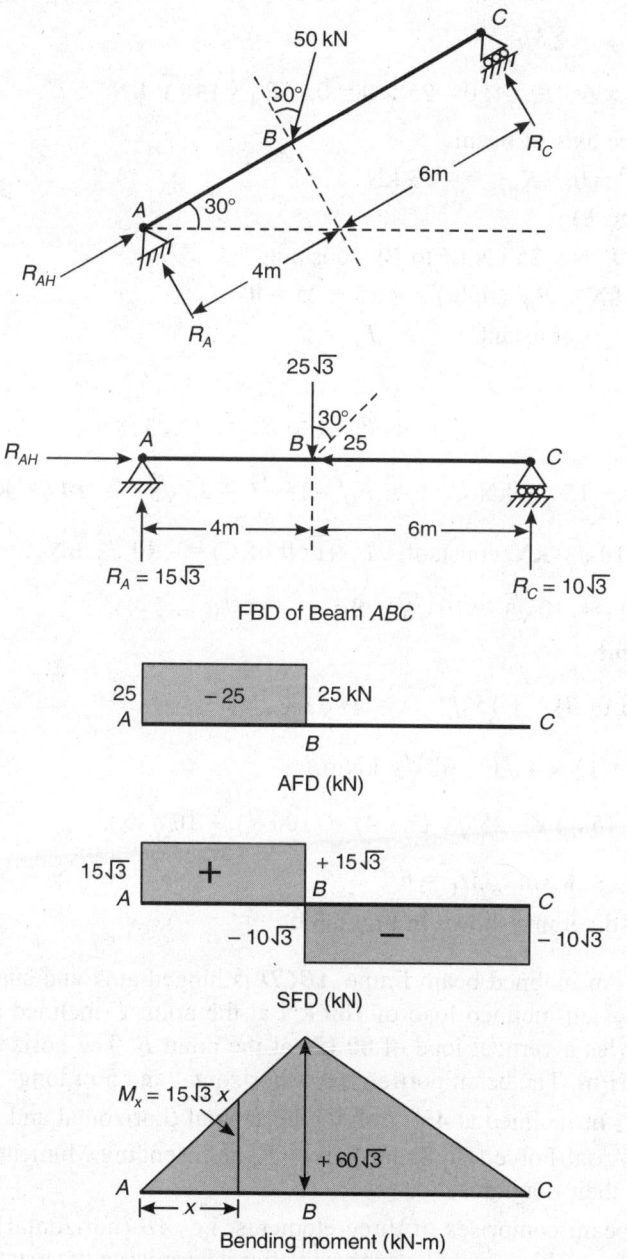

FBD of Beam *ABC*

AFD (kN)

SFD (kN)

Bending moment (kN-m)

Fig. 6.33: AF, SF and BM diagram

Reactions

Moment about A, $\Sigma M_A = 0$

$R_C \times 10 - 25\sqrt{3} \times 4 - 25 \times 0 = 0$, $\boldsymbol{R_C = 10\sqrt{3}}$ **kN**   (Upward)

Moment about C, $\Sigma M_C = 0$

$R_A \times 10 - 25\sqrt{3} \times 6 - R_{AH} \times 0 - 25 \times 0 = 0$, $\boldsymbol{R_A = 15\sqrt{3}}$ **kN**   (Upward)

Resolve along the axis of beam

$R_{AH} - 50 \sin 30° = 0$, $\boldsymbol{R_{AH} = -25}$ **kN**   (Compressive)

Axial Force (A to B)

$P_A = -25$ kN, $P_x = -25$ kN (A to B), constant

$\boldsymbol{P_B}$ (left) $= -\boldsymbol{25}$ **kN**, $\boldsymbol{P_B'}$ (right) $-\ -25 + 25 = \boldsymbol{0}$

P_x (B to C) $= 0$,   constant, $P_C = 0$

Shear Force

$F_A = +15\sqrt{3}$ kN, F_x (A to B) $= +15\sqrt{3}$ kN,

$\boldsymbol{F_B}$ (Left of B) $= +\ \boldsymbol{15\sqrt{3}}$ **kN**, $\boldsymbol{F_B'} = 15\sqrt{3} - 25\sqrt{3} = -\boldsymbol{10\sqrt{3}}$ **kN**

F_x (B to C) $= -\boldsymbol{10\sqrt{3}}$ **kN**, constant, $\boldsymbol{F_C}$ (Left of C) $= -\boldsymbol{10\sqrt{3}}$ **kN**,

$\boldsymbol{F_C'}$ (Right of C) $= -10\sqrt{3} + 10\sqrt{3} = \boldsymbol{0}$

Bending Moment

$\boldsymbol{M_A = 0}$, M_x (A to B) $= +15\sqrt{3} \cdot x = \boldsymbol{15\sqrt{3}\,x}$,   Straight line

M_B ($x = 4$ m) $= +15 \times 4\sqrt{3} = \boldsymbol{60\sqrt{3}}$ kNm,   (Max.);

M_x (B to C) $= +15\sqrt{3}\,x - 25\sqrt{3}\,(x-4) = (100\sqrt{3} - 10\sqrt{3}\,x)$,   Straight line

M_B ($x = 4$) $= \boldsymbol{60\sqrt{3}}$ **kNm**, $\boldsymbol{M_C = 0}$

AF, SF and BM diagrams shown in Fig. 6.33.

EXAMPLE 6.12: An inclined beam frame **ABCD** is hinged at A and supported at roller end D. The beam carries an inclined load of **100 kN** at the point C inclined at $45°$ with vertical. The beam also carries a vertical load of **80 kN** at the point B. The horizontal span **CD** (**4 m**) carries udl of **10 kN/m**. The beam portion AB is horizontal and **5 m** long. The inclined portion **BC** has length $3\sqrt{2}$ **m** inclined at $45°$ with the horizontal (horizontal and vertical components $= 3$ m). Calculate Axial Force (P), Shear Force (F) and Bending Moment (M) at points A, B, C and D and draw their diagrams.

    Solution: The beam comprises of three elements, i.e. AB (horizontal), BC (inclined) and CD (horizontal). The problem must be analyzed after determining the reaction components and drawing free body diagrams for each part separately. Consider joints A, B, B', C, C' and D.

Reactions at A and D

Take moment of all forces about the hinge A.

$$R_{DV}(5+3+4) - 4 \times 10\left(5+3+\frac{4}{2}\right) - \frac{100}{\sqrt{2}} \times 3 - 80 \times 5 - 8 \times \frac{100}{\sqrt{2}} = 0$$

$$R_D = \frac{1}{12}\left[800 + \frac{1100}{\sqrt{2}}\right] = 131.485 \text{ kN}, \qquad \text{(Vertical-Normal to roller)}$$

Resolve all forces in vertical direction, $\quad \Sigma V = 0$

$$R_{DV} + R_{AV} - 80 - \frac{100}{\sqrt{2}} - 10 \times 4 = 0, \quad R_{AV} = 59.23 \text{ kN}$$

Resolve all forces in horizontal direction, $\quad \Sigma H = 0$

$$\left(R_{AH} - \frac{100}{\sqrt{2}}\right) = 0, \quad R_{AH} = 50\sqrt{2} = 70.71 \text{ kN} \ (\leftarrow)$$

Also take moment about the point D, $\quad \Sigma M_D = 0$

$$R_{AH} \cdot 3 - 4 \times 10 \times \frac{4}{2} - 50\sqrt{2} \times 4 - 80 \times 7 + 59.23 \times 12 = 0$$

$R_{AH} = 70.71 \text{ kN}$ (OK)

Portion AB

Axial Force $P_A = + 70.71$ kN, $\qquad P_x$ (A to B) $= + 70.71$ kN, $\qquad$ (constant) $= P_B$

Shear Force $F_A = 59.23$ kN, $\qquad F_x$ (A to B) $= + 59.23$ kN, $\qquad$ (constant) $= F_B$ (left)

F'_B (right) $= 59.23 - 80 = - 20.77$ kN

Bending Moment $M_A = 0$, $\quad M_x$ (A to B) $= 59.23x$, $\qquad$ (Straight line)

M_B ($x = 5$m) $= + 296.15$ kN-m

Portion BC

Axial Force $P'_B = (70.71 + 20.77)\dfrac{1}{\sqrt{2}} = 64.69$ kN,

P_x (BC) $= + 64.69$ kN, $\qquad$ (Constant)

$P_{CB} = + 64.69$ kN

Shear Force $F_{BC} = (70.71 - 20.77) = + 35.31$ kN,

F_x (BC) $= + 35.31$ kN, $\qquad$ (constant)

$F_C = + 35.31 - 100 = - 64.69$ kN

Bending Moment $M_B = 296.15$ kNm,

M_x (B to C) $= 296.15 + 35.31x$, $\quad$ (Straight line)

M_C ($x = 3$ along BC) $= 445.96$ kNm

Portion CD (Horizontal)

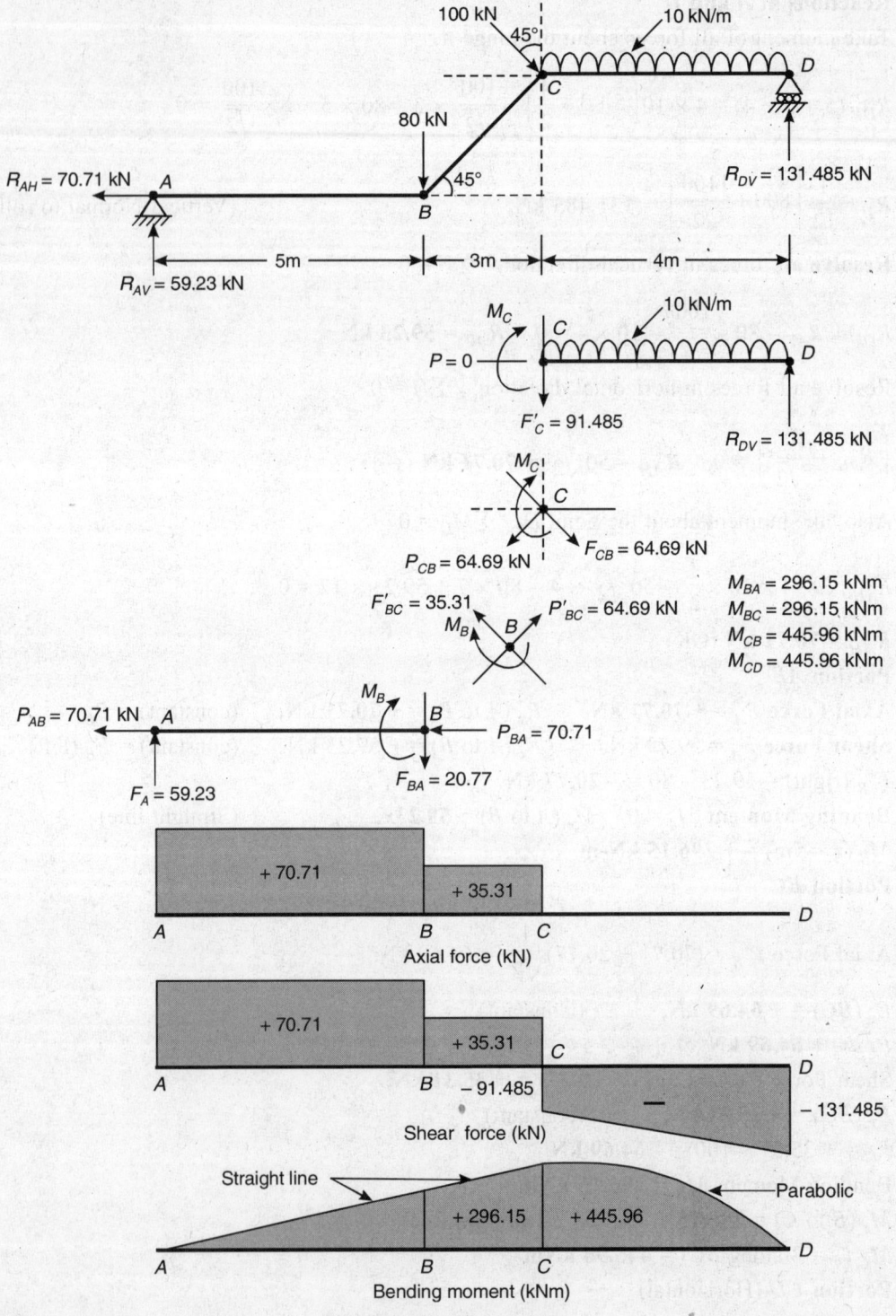

Fig. 6.34: AF, SF and BMD

Axial Force P_C $(CD) = (64.69 \times \dfrac{1}{\sqrt{2}} - 64.69 \times \dfrac{1}{\sqrt{2}}) = 0,$ P_x $(CD) = 0$

$P_D = 0$

Shear Force F_C $(CD) = -(64.69 + 64.69) = -$ **91.485 kN**

F_x $(CD) = -$ **(91.485 + 10x),**   Straight line,

F_D (Just left of D, $x = 4$ m) $= -$ **131.485 kN**

F'_D (Just right) $= -131.485 + 131.485 = $ **0**

Bending Moment $M_C = +445.96$ kNm,

x from C

$$M_x \ (CD) = 445.96 - 91.485x - \frac{10x^2}{2} = (445.96 - 91.485x - 5x^2) \text{ Parabolic}$$

M_D $(x = 4$ m$) = 0$

AFD, SFD and BMDs are shown in Fig. 6.34.

6.12   RELATIONSHIP OF LOADING, SHEAR FORCE AND BENDING MOMENT

Bending moment, shear force and loadings on any beam are related to each other with a definite relationship. Consider a **small element 'δx'** of a beam between section mn and $m_1 n_1$ having different situations of loading on the element δx.

(a) Let the small element 'δx' carries **no load** and the section *mn* has positive shear force 'F' and positive moment 'M' say. Since there is no load on the small element δx (mn to $m_1 n_1$), the section $m_1 n_1$ will also have positive shear force F and bending moment ($M + \delta M$).

Consider equilibrium of small element δx, we have

$F - F = 0,$ $M + F \cdot \delta x - (M + \delta M) = 0,$ or $F \cdot \delta x - \delta M = 0,$

or $\underset{\delta x \to 0}{\text{Lim}} \dfrac{\delta M}{\delta x} = F$ i.e. $\dfrac{dM}{dx} = F$ ... (1)

The rate of **change of moment with respect to distance** $x \left(\dfrac{dM}{dx} \right)$ represents **shear force** F at the section **X-X**. It can also be stated that the **slope of bending moment curve at any section X-X represents the shear force 'F'** at that section X-X.

(b) Consider another case where the small element δx carries uniformly distributed load $w/$ *m*. Total load on the element δx will be ($w \cdot \delta x$). Considering equilibrium of the element δx, we have

$F - w \cdot \delta x - (F + \delta F) = 0,$ or $- w \cdot \delta x - \delta F = 0,$ $\delta F = - w \cdot \delta x$

or $\underset{\delta x \to 0}{\text{Lim}} \dfrac{\delta F}{\delta x} = - w,$ or $\dfrac{dF}{dx} = - w$ ... (2)

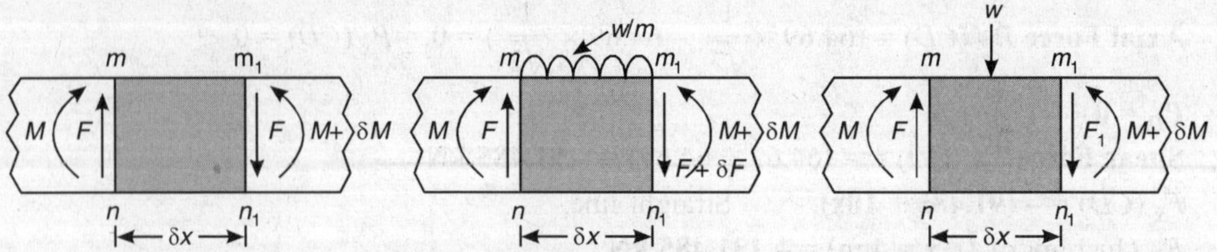

(a) No load on δx (b) u.d.l. w on δx (c) Point load W on δx

Fig. 6.35.

It can be stated that the **rate of change of shear force** with respect to distance 'dx' will be equal to opposite of the **rate of loading $(-w)$** at the section X-X. It can also be stated that the **slope of the shear force diagram** represents the **rate of distributed load** with opposite sense $(-w)$ at the section X-X.

Taking moment of forces on the small element δx (mn to m_1n_1).

$$\Sigma M = 0, \quad \text{or} \quad M + F \cdot \delta x - w \cdot \delta x \cdot \frac{\delta x}{2} - (M + \delta M) = 0$$

or $F \cdot \delta x - w \dfrac{\delta x^2}{2} - \delta M = 0$, neglecting square of small quantities.

or $F\delta x - \delta M = 0$,   Or $F = \underset{\delta x \to 0}{\text{Lim}} \dfrac{\delta M}{\delta x}$

or $\dfrac{dM}{dx} = F$,   Same as in earlier case   ... (1)

Hence it can be said that the equation $\dfrac{dM_x}{dx} = F$, also holds good in case of udl on the section. It can be generally stated that the **slope of BM curve** at any **section X-X** **represents the shear force** at that point.

(c) Consider a case when the small element δx (mn to m_1n_1) carries a point load W resulting in **sudden change** in shear force. Thus considering equilibrium of the small element δx, we have

$F - W - F_1 = 0,$ $\mathbf{F_1 = (F - W)}$ ... (3)

The **shear force** on the left of the point where W acts will be F and on the **right** of the load W, it will be $(F_1 = F - W)$. The equation $\dfrac{dM}{dx} = F$ is also applicable at the point of application of W, where the slope of BM curve is infinite due to sudden change of SF

Equation $\dfrac{dM_x}{dx} = F$ can also be written as $\mathbf{dM = F\ dx}$ or $M = \int F\ dx$, (or $\Sigma F_1 \delta x$), i.e.

moment M at any section is equal to the area of shear force diagram between the two limiting points.

The equation $\dfrac{dF}{dx} = -w$, can also be written as $dF = -w\,dx$ or $\int dF = F = -\int w\,dx$ $= -\sum w_1\,dx$.

The shear force at any point 'F' will be equal to the **area of load distribution curve** between the limits of summation. Above equations can be used to determine the values of **shear force** or **bending moment** from the given **bending moment** or **shear force** **diagrams**.

EXAMPLE 6.13: A shear force diagram in a simply supported beam ABC of 10 m span is shown in Fig. 6.36. Determine its bending moment diagram and loading diagrams.

Solution:

$$M_x = \int_A^X F\,dx$$

$$= \text{Area of '}F\text{' Diagram}$$

$$M_B = 36 \times 4 = 144 \text{ kNm}$$

$$M_D \text{ (Max)} = 36 \times 4 + = 176.4 \text{ kNm}$$

$$M_C = 176.4 - 84 \times = 0$$

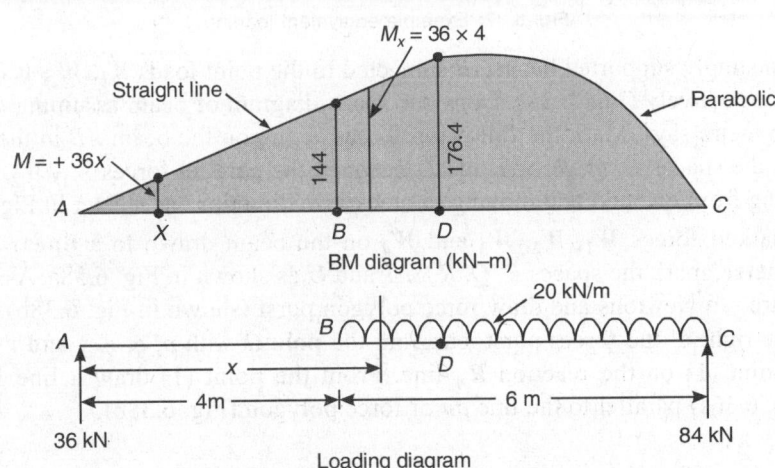

$$F_x = 36 - \frac{36}{1.8}(x-4) = 36 - 20(x-4) \text{ kN}$$

SF diagram (kN)

BM diagram (kN–m)

Loading diagram

Fig. 6.36: BMD and loading from SFD

Loading Diagram

$$\frac{dF}{dx} = -w$$

Constant A to B equal to $\quad R_A = 36$ kN

$$R_C = 84 \text{ kN}$$

$$\frac{dF_x}{dx} = -20 = -w \ (B \text{ to } C)$$

6.13 GRAPHICAL METHOD FOR SF AND BMD

Shear force and bending moment diagrams in beams can also be drawn with the help of graphical approach using the principles of polygon of coplanar forces. Space diagram and force polygon are drawn assuming suitable **linear** and **vector** (force) scales respectively. Graphical approach becomes quite convenient when a beam carries a large number of transverse loads.

In case of uniformly distributed load of w N/m on the beam, the load is converted into a number of **small point loads** by dividing the loaded span into number of **equal sections**. Equivalent point load for 'w' uniformly distributed over a span 'dl_1', will be $P_1 = (w \cdot dl_1)$, $P_2 = w \cdot dl_2$, etc. These point loads are assumed to act at the center (or center of gravity) of wdl_1, wdl_2, ... For better **accuracy**, the udl is divided into **large number of point loads** considered over **large number** of **smaller spans** (Fig. 6.37).

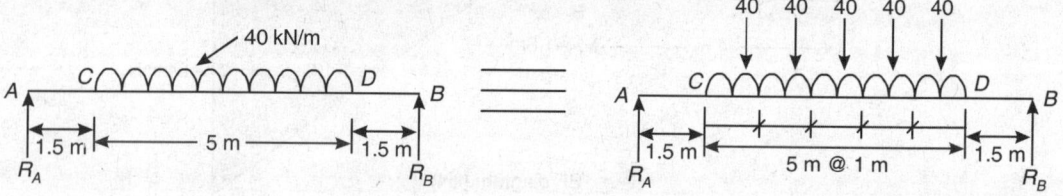

Fig. 6.37: Example equivalent loading

Consider a simply supported beam AB subjected to the point loads W_1, W_2, W_3 and W_4 at C, D, E and G respectively (Fig. 6.38). Draw the space diagram of beam assuming a linear scale (1 metre = m metre say). Mark the following forces acting on the beam AB in the direction of forces. Mark the spaces P, Q, R, S, T and U between the parallel forces W_1, W_2, W_3, W_4, R_B, and R_A starting from one end and moving in clockwise direction (as shown in Fig. 6.38).

Having marked forces W_1, W_2, W_3 and W_4 on the beam drawn to a **linear scale** of **1.0 metre = m metre**, mark the spaces P, Q, R. S. T and U as shown in Fig. 6.38a. Assume a **force sale** of 1 metre = n Newtons and draw force polygon pqrst (shown in Fig. 6.38b). Choose any **pole O** on the right of the forces pqrst, etc. Join the **pole O** with p, q, r, s and t respectively. Select any point (1) on the reaction R_A line. From the point (1) draw a line **1-2** in space diagram (Fig. 6.36a) parallel to the line **po** of force polygon (Fig. 6.38b).

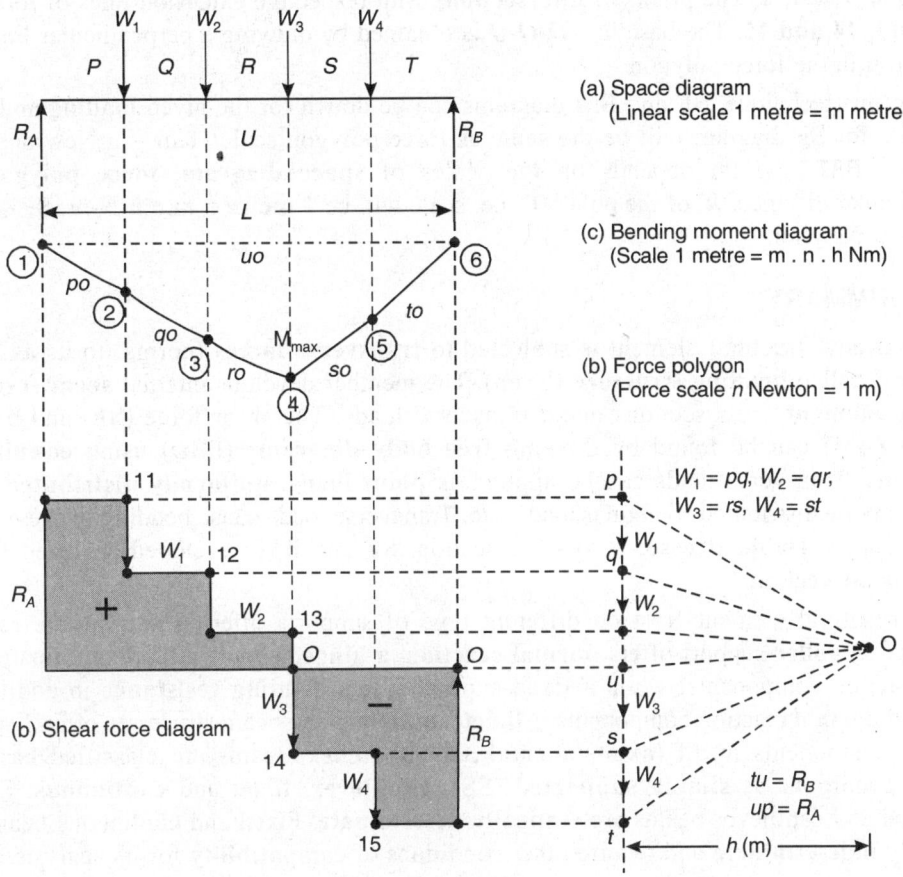

Fig. 6.38: Graphical Method for SF and BM

Line **1-2** intersects the line of action of the force W_1 in space diagram. Draw a line **2-3** parallel to *q-o* of the force polygon form the point of intersection **2**. This line **2-3** intersects the line of action of force W_2 at the point **3**. Continue this process of drawing these lines **3-4** parallel to *r-o*, **4-5** parallel to *s-o*, **5-6** parallel to *t-o* intersecting vertical reaction R_B line at the point **6**. Join the last point **6** on the reaction line R_B with the starting point **1** by a chain line **6-1** Draw a line *o-u* in the **force polygon** from the **pole O** parallel to **6-1**. This line **interests force line *p-q-r-s-t*** at the point *u*. *t-u* represents the reaction R_B and *u-p* represents the reaction R_A.

The diagram **1-2-3-4-5-6-1** represents the **bending moment** on certain scale. Let the **perpendicular** distance of the **pole** point **O** from the force line *p-q-r-s-t* be '*h*' metre. The **ordinates** from the **base line 1-6** represents **bending moment** to a scale 1 metre = *m.n.h* (N-m), if forces are in **N** and linear distances in '**m**'. It may be noted that the starting point **1** must be selected at the **hinge** support A or B so as to simplify the effect of inclined reactions, if any.

Shear force diagram can be drawn by extending the lines of actions of forces R_A, W_1, W_2, W_3, ... from the **space diagram** and projecting horizontal lines from the **force polygon** points **p, q, r, s, t, u**. The points of **intersections** with respective extension lines of forces are **11, 12, 13, 14 and 15**. The base line **O-O-O** is obtained by drawing a perpendicular line from the point **u** in the force polygon.

Using this technique, **SF** and **BM** diagrams can be drawn for the given **loading** and **span**. The scale for SF diagram will be the same as force polygon scale (1 m = n Newtons). The **scale for BM** diagram depends on the scales of **space diagram, force polygon** and perpendicular distance '**h**' of the pole '**O**'. i.e. scale will be **1 metre = m.n.h N-m**. SF scale: **1 metre = n Newton**.

6.14 SUMMARY

Whenever any structural element is subjected to **transverse forces** (normal to its axis), the member is called **bending structure** (beam). The member develops internal **shear force** and **bending moment** in its sections under transverse loads. The shear force (**SF**) and bending moment (**BM**) can be found by drawing **free body diagrams** (FBD) using **equilibrium conditions**. Transverse loads can be applied as **point loads, uniformly distributed loads** (udl), **variable** distributed triangular loads, etc. Transverse loads cause bending of the member axis causing **variable stresses** across the section. **SF** and **BM** developed are used for the **design of the section**.

The structural elements rest on different type of supports offering appropriate reactive resistance. A **roller** support offers **normal reaction**, a **hinge** support offers both **normal** and **axial** reaction components, while a **fixed** support offers **bending resistance** in addition to **axial** and **normal** reaction components. All **determinate** plane beam structures offer **3** support reaction components in all (**axial, normal** and **moment**). Beams are classified based on support conditions as **simply supported** (SS), **cantilever, fixed** and **continuous**. Simply supported and cantilever beams are **statically determinate**. Fixed and continuous beams are **statically indeterminate** and require other conditions of **compatibility** for its analysis.

Point or **concentrated** loads act on a very small area while **uniformly distributed** loads act on a large area and produce **less severe** SF and BM in beams. In case of **determinate** beams **equilibrium conditions** hold good for structure as a whole and also any part of the structure. **Equilibrium conditions** hold good both for external forces and internal resistances.

These conditions are: $\Sigma V = 0, \Sigma H = 0, \Sigma M = 0$, for plane determinate structures subjected to plane forces.

Internal **shear force** and **bending moment** at any section can easily be determined by drawing **free body diagrams** (FBD) and applying **equations of equilibrium**. Support reactions are determined by applying conditions of equilibrium considering external forces on the beam as a whole. The beams subjected to inclined loads also develop **Axial Force** (AF) resistances.

Shear force (SF) at any section of a beam is defined as **algebraic sum of all transverse forces** considered either on **left** or **right** of the section. Similarly, **Axial Force** (AF) at any section of a beam is defined as **algebraic sum of all axial forces** either on **left** or **right** of the section. **SF** will be considered **positive** if the **algebraic sum of transverse** forces is **upward on left** or **downward on right**.

Bending moment (BM) at any section of a beam is defined as **algebraic sum of moments of all transverse forces** either on **left** or **right about the section**. **Sagging moment** (i.e. **clockwise on left** or anticlockwise on right ⤻⤸) about the section will be considered as **positive**, while **hogging moment** (i.e. **anticlockwise on left** or clockwise on right ⤸⤻) about the section will be considered as **negative**. Sign convention is adopted for convenience and uniformity throughout the analysis. There is no hard and fast rule for the sign convention.

Tensile external axial forces shall be considered as **positive**, while **compressive external** axial forces shall be considered as **negative**. After

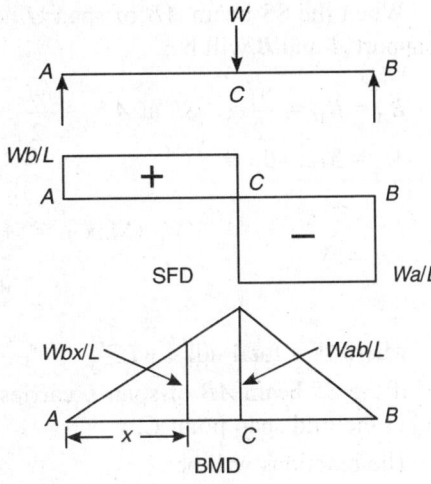

Fig. 6.39.

calculations of reactions, SF, AF and BM are calculated at different sections along the span of the beam. SF, AF and BM diagrams are drawn by assuming suitable linear scale (1 m = 10 mm, say) and force scale (1 kN = 1 mm, say) along the beam axis. SF, AF and BM values are drawn as ordinates normal to the beam axis.

A SS beam **AB** of span '**L**' carrying a point load **W** at C at a distance of '**a**' from A and '**b**' from the support B.

The reactions, *SF* and *BM* values will be:

$$R_A = \frac{W \cdot b}{L}, \quad R_B = \frac{W \cdot a}{L}, \quad M_C \text{ (Max.)} = \frac{W \cdot a \cdot b}{L}$$

$$\text{If } a = b = \frac{L}{2}, \quad R_A = R_B = \frac{W}{2}, \quad M_C \text{ (Max.)} = \frac{WL}{4}$$

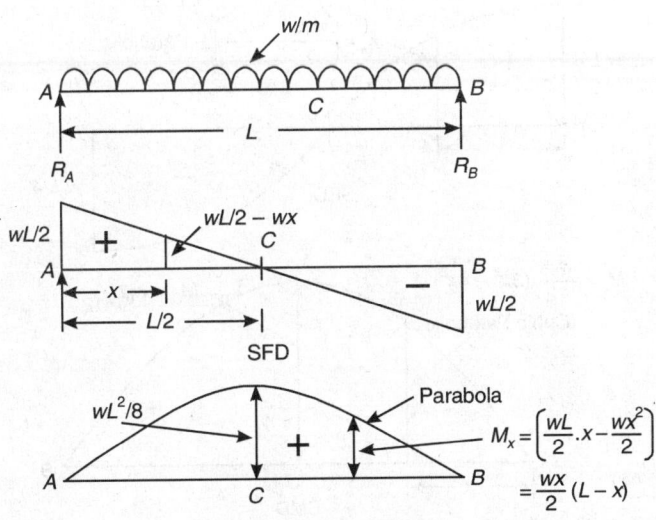

Fig. 6.40: BMD parabolic

When the SS beam AB of span 'L' carries udl of w/m over the entire span, the reactions at support A and B will be:

$$R_A = R_B = \frac{wL}{2}, \quad SF \text{ at } A = +\frac{wL}{2}, \quad SF \text{ at } B = -\frac{wL}{2}$$
$$M_A = M_B = 0,$$

$$M_C \text{ (Max.)} = +\frac{wL^2}{8}$$
$$= \frac{wL \cdot L}{8} = +\frac{WL}{8},$$

where W = total udl = wL.

If the SS beam AB of span L carries a variable triangular loading with maximum intensity w_0 at the mid span point C.

The reactions will be:

$$R_A = R_B = \frac{w_0 L}{4},$$

$$F_A = +\frac{w_0 L}{4}, \quad F_B = -\frac{w_0 L}{4}, \quad F_x = \left(\frac{w_0 L}{4} - \frac{w_0 x^2}{L}\right), \quad x = 0 \text{ to } \tfrac{1}{2} \quad \text{(Variation – parabolic)}$$

$$M_A = M_B = 0, \quad M_C \text{ (Max.)} = +\frac{w_0 L^2}{24}, \quad M_x = \frac{w_0 x}{12}(3L^2 - 4x^2), \quad x = 0 \text{ to } \tfrac{1}{2}$$

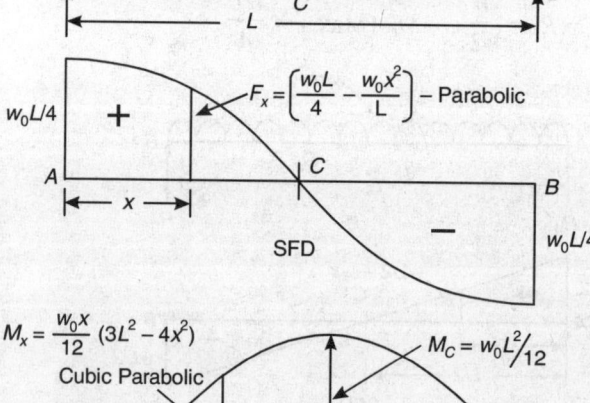

Fig. 6.41: BMD and SFD

In case of cantilever beam AB of span L carrying a point load W at the free end B, the reactions will be:

$R_A = W,\quad F_A = +W, 0 \quad$ and $\quad F_B = 0, +W,\quad F\,(B\text{ to }A) = +W\,(\text{constant})$

$M_A = -WL = M_{\max.},\quad M_B = 0,\quad Mx = -Wx\ (x \text{ from } B)$

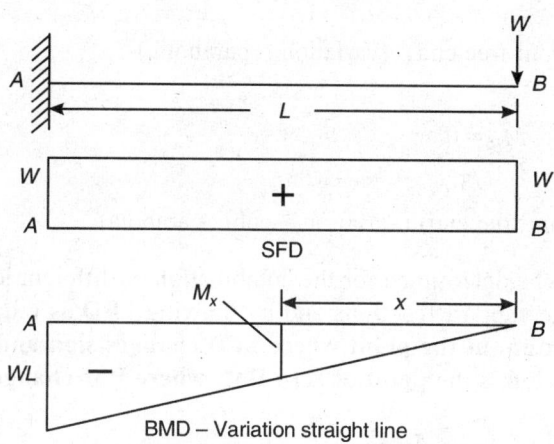

Fig. 6.42: SFD and BMD in cantilever

If cantilever beam **AB** of span '**L**' carries a udl w/m on the entire beam span L, the reactions will be:

$R_A\,(\text{fixed}) = wL,\quad F_A = +wL,\quad F_B = 0,\quad F_x = wx,\ \text{variation straight line.}$

$M_A\,(\text{Max.}) = -\dfrac{wL^2}{2},\quad M_B = 0,\ \text{variation parabolic},\ M_x = -\dfrac{wx^2}{2},\,(x \text{ from } B).$

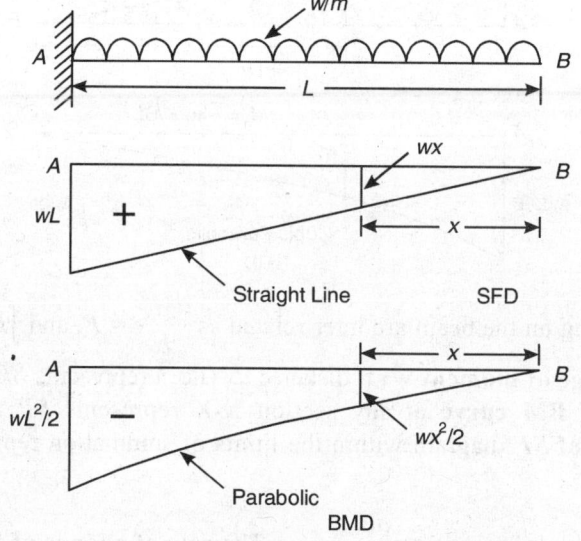

Fig. 6.43.

If the cantilever beam AB of span L carries a triangular load with maximum intensity w_0 at the fixed end A, the reactions will be:

$$R_A \text{ (fixed end)} = \frac{w_0 L}{2} \ (\uparrow), \quad F_A = +\frac{w_0 L}{2}, F_B = 0,$$

$$SF \ F_x = \frac{w_0 x^2}{2L}, \ (x \text{ from free end) (variation – parabolic)}$$

$$M_A \text{ (Max.)} = -\frac{w_0 L^2}{6}, \ M_B = 0,$$

$$M_x = -\frac{w_0 x^3}{6L}, \ (x \text{ from free end) (variation – cubic parabola)}$$

SF and BM can also be determined for the combination of different loads in similar manner by first determining the support reactions and by drawing FBD as required. It may be noted that **BM will be maximum at the point where SFD changes sign and SF = 0**. The **point of contraflexure** or **inflexion** is the **point of zero BM, where BM changes its sign**.

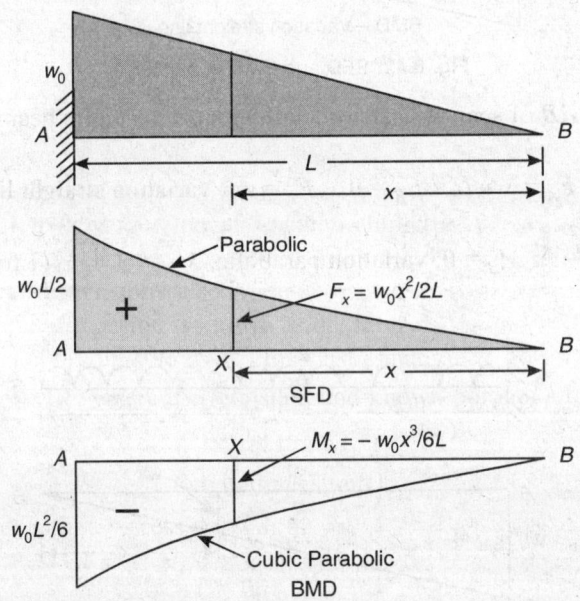

BM, SF and loading on the beam are inter-related as $\dfrac{dM_x}{dx} = F$, and $\int F \ dx = M$ or $\Sigma F \ \delta x = M$. The **rate of change of moment** w.r.t. distance dx (i.e.) represents SF 'F' at the section X-X (i.e. the **slope of BM curve** at any section X-X represents SF at the section). Also **summation of area of SF** diagram within the limits of summation represents the BM about the section.

Loading and SF are also related as $\dfrac{dF}{dx} = -w$. The **rate of change of SF** curve with respect to distance dx represents the **rate of loading** with opposite sign ($-w$) at the section X-X. In other words the **slope of the SF** diagram represents the **rate of distributed** load with opposite sign $(-w)$ at the section. Also $F = \int dF = \int -w\,dx = -\int w\,dx = \Sigma w\,\delta x$. The SF shall be equal to summation of area of loading diagram within the limits.

EXERCISE 6

Q.6.1. A simply supported beam AB of 5 m span is supported at A and B. The beam carries concentrated loads of 40 kN, 30 kN and 30 kN respectively at C ($AC = 2$ m), D ($AD = 3$ m) and E ($AE = 4$ m). The beam also caries a uniformly distributed load of 20 kN/m from A to D. Determine reactions, SF and BM values at various points, also determine the maximum bending moment and its location.

Ans. $R_A = 84$ kN, $R_B = 76$ kN, SF at $A = + 84$ kN and at $B = -76$ kN, SF is zero at $x = 2.20$ m from A, $M_{\max.} = + 128.4$ kN-m at the point X ($AX = 2.20$ m).

Q.6.2. A cantilever beam AB of 3 m span, carries a point load of 30 kN at a point 2 m from the fixed end. Another point load of 20 kN acts at the free end. A udl of 5 kN/m on 2 m span from the free end. Find SF, point of zero SF and maximum BM.

Ans. R_A (fixed end) = 60 kN, SF zero at free end and fixed end. $SF_A = -60$ kN, $SF_D = -25$ kN, -55 kN, $M_D = -22.5$ kN-m, $M_A = 140$ kN-m.

Q.6.3. A 4 m long cantilever beam carries a downward point load of 20 kN at the middle point of span, a udl of 5 kN/m over the span from mid point to the free end and a upward prop reaction of 10 kN at the free end. Calculate SF and BM at free end B, mid span C and fixed end A.

Ans. $R_A = 20$ kN, $SF_A = + 20$ kN, SF (mid span) = $+ 20$ kN, $SF_B = 0, -10$ kN, $M_A = -30$ kN-m, M_C ($M_{\max.}$) = $+ 10$ kN-m, $M_B = 0$.

Q.6.4. A 6 m long simply supported beam carries total udl of 60 kN over 2 m span on the left of the middle point. Determine the maximum BM and SF values with their locations.

Ans. $R_A = 40$ kN, $R_B = 20$ kN, $SF_A = 0, + 40$ kN, $SF_D = + 40$ kN, $SF_C = -20$ kN, $M_A = 0$, $M_B = 0$, $M_{\max.}$ (Ax = 2.333 m) = $+ 66.67$ kNm, $M_D = + 40$ kNm, $M_C = + 60$ kNm.

Q.6.5. A simply supported beam AB of 9 m span carries uniformly distributed load of 15 kN/m over the entire span. Beam also carries concentrated loads of 75 kN and 60 kN respectively at points 3 m (C) and 6 m (D) from the left hand support A. Determine reactions, shear forces and bending moment values. Also calculate maximum BM and its location.

Ans. $R_A = 137.5$ kN, $R_B = 132.5$ kN, $SF_A = + 137.5$ kN, $SF_C = + 92.5$ kN, $+ 17.5$ kN, $SF_D = -27.5$ kN, -87.5 kN, $SF_B = -132.5$ kN, 0, $M_A = 0$, $M_B = 0$, $M_C = + 345$ kN-m, $M_D = + 330$ kNm, $M_{\max.} = + 355.12$ kN-m at 4.166 m from A.

Q.6.6. An overhanging beam AF is supported at B and E on simple supports as shown in Fig. Q6.6. Determine SF, BM, values of maximum BM and locations of points of contraflexure.

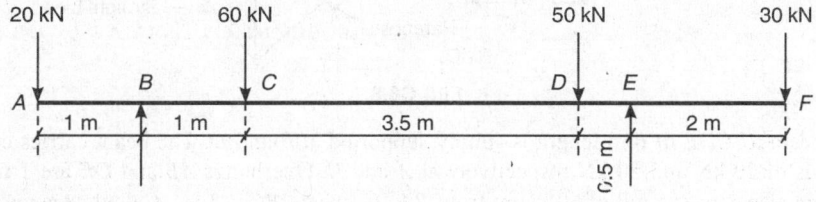

Fig. Q6.6.

Ans. $R_B = 65$ kN, $R_E = 95$ kN, $SF_A = 0, -20$ kN, $SF_B = -20$ kN, $+45$ kN, $SF_C = +45$ kN, -15 kN, $SF_D = -15$ kN, -65 kN, $SF_E = -65$ kN, $+30$ kN, $SF_F = 30$ kN, 0, $M_A = 0$, $M_B = -20$ kN-m, $M_C = +25$ kNm ($+M_{max.}$ at C), $M_D = -27.5$ kNm, $M_E = -60$ kNm, ($-M_{max.}$ at E). Points of inflexion from A, $x_1 = 1.444$ m, $x_2 = 3.666$ m.

Q.6.7. A simply supported beam ABC is shown in Fig. Q6.7. Draw SFD and BMD showing maximum and minimum values and their locations. Determine points of contraflexures.

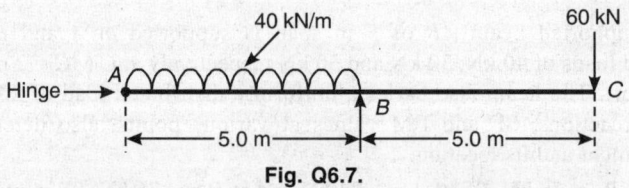

Fig. Q6.7.

Ans. $R_A = 40$ kN, $R_B = 220$ kN, $F_{max.} = +60$ kN, -160 kN (at B), $M_A = 0$, $M_{max.} = +20$ kNm (at 1 m from A), -300 kNm (at B). Point of contraflexure, $x = 2$ m from A.

Q.6.8. Determine reactions at B and C in case of beam $ABCD$ loaded as shown. Draw SF and BM diagrams showing maximum values along with locations. Determine the points of contraflexures.

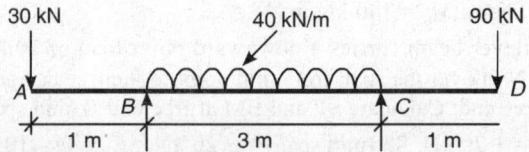

Ans. $R_B = 70$ kN, $R_C = 170$ kN, $F_{max.} = +90$ kN at D and -80 kN at C. $M_{max.} = -10.0$ kNm at 1 m from B and -90 kNm at C, no points of inflexion

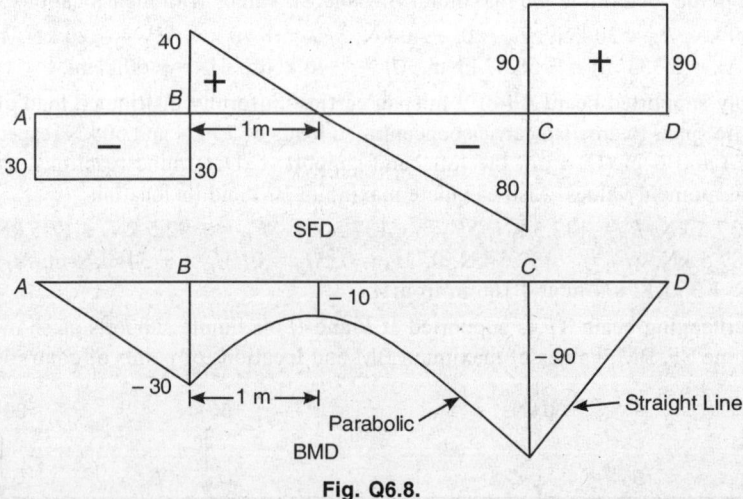

Fig. Q6.8.

Q.6.9. A beam $ABCDE$ of 6 m length is simply supported at B and D. The beam carries concentrated loads of 20 kN and 50 kN respectively at A and E. Overhangs AB and DE are 1 m each. The beam also carries a udl of 60 kN/m from B to C (where $BC = 3$ m). Calculate support reactions and draw SF and BM diagrams. Select absolute maximum values of SF and BM with their locations and indicate points of contraflexures.

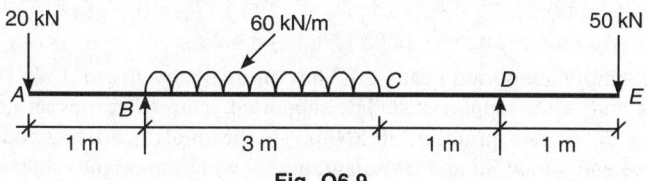

Fig. Q6.9.

Hints: $R_B = R_D = 125$ kN, $F_{max.} = +105$ kN at B, $M_{max.} = -50$ kNm (at D), $+71.875$ kNm at X ($BX = 1.75$ m). Points of contraflexures ($BX_1 = 0.2022$ m, $BX_2 = 3.33$ m)

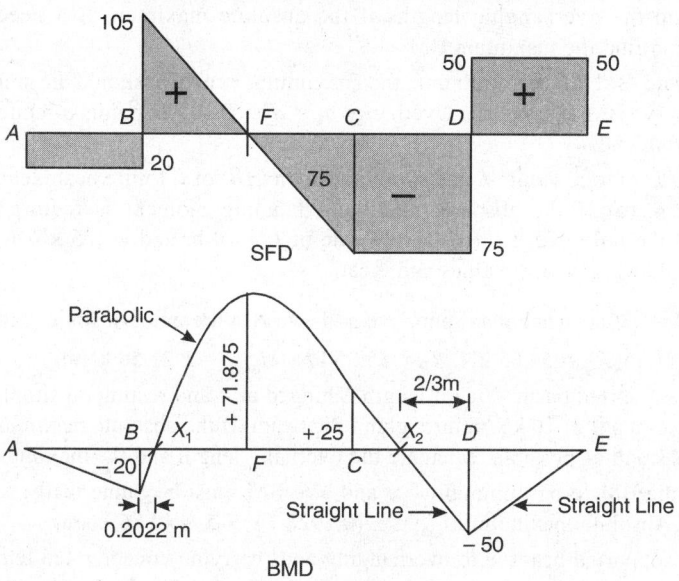

Fig. Q6.9.

Q.6.10. A loaded beam $ABCD$ is shown in Fig. Q6.10. Calculate reactions at B and C and draw BM and SF diagrams showing maximum values and locations wherever these occur. Also determine points of contraflexure.

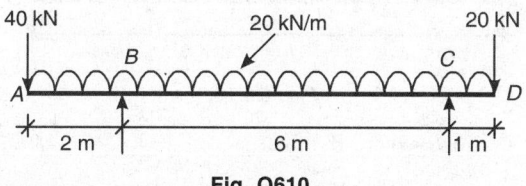

Fig. Q610.

Hints: $R_B = 155$ kN, $R_C = 85$ kN, $F_B = -80$ kN, $+75$ kN, $F = 0$ at $BX = 3.75$ m, BM at $B = -120$ kNm, $M_{3.75} = +20.625$ kNm, $M_C = -30$ kNm, points of contraflexure from A are at 4.314 m, and 7.186 m.

Q.6.11. A simply supported beam ABC of 7 m length carries an udl of 10 kN/m on the entire 7 m length. The beam overhangs by 2 m beyond the support B. Determine support reaction at A and B, shear force and bending moments at A, B, and C. Also determine the maximum values of SF and BM along with their locations. Locate points of contraflexure, if any.

Hints: $R_A = 21$ kN, $R_B = 49$ kN, $F_A = +21$ kN, $F_B = -29$ kN, $F_C = 0$, F_x $(AX - 2.1$ m$) = 0$, $M_A = 0$, M_B $= -20$ kNm, M_X $(AX = 2.1) = +22.05$ kNm, $x_i = 4.2$ m.

Q.6.12. A 8 m long simply supported beam ABC has an overhang BC of 3 m. The beam has hinged support at A and roller support at B. The supported span AB carries an udl of 20 kN/m while the overhang BC carries an udl of 30 kN/m. The beam also carries a concentrated load of 60 kN at the free end. Draw SF and BM diagrams showing important values and locations.

Hints: $R_A = -13$ kN, $R_B = +263$ kN, $F_A = -13$ kN, $F_B = -113$ kN, $+150$ kN, $F_C = +60$ kN, 0, $M_A = 0$, $M_B = -315$ kNm, $M_C = 0$, parabolic variation.

Q.6.13. A simply supported beam $ABCD$ of length L and equal overhangs on each side carries udl of w kN/m. Find the overhanging lengths if the absolute maximum BM needs to be minimum possible and find the maximum BM.

Hints: For absolute BM to be minimum, the maximum +ve BM should be numerically equal to maximum $-$ve BM at supports. Overhangs $a_1 = a_2 = 0.207\,L$, points of contraflexures $0.293\,L$, $0.707\,L$ from ends.

Q.6.14. A beam ABC of total length L and supported span (AB) of l. Entire beam length ABC carries udl of w kN/m. (a) If the absolute maximum bending moment is required to be minimum, determine the ratio of $L : l$. (b) If the value of $L = 10$ m and $w = 5$ kN/m, draw SF and BM diagrams showing critical values and locations.

Hints: a. $L : l = \sqrt{2} : 1$, when maximum $-$ve and +ve BM are numerically equal.

b. $L = 10$ m, $R_A = 14.65$ kN, $R_B = 35.35$ kN, $M_{max.} = \pm 21.50$ kNm.

Q.6.15. A simply supported beam ABC of length L hinged at A and resting on simple support at B. The beam carries a udl of 10 kN/m throughout the beam. If the absolute maximum bending moment is to be as small as possible, calculate the overhang length and the maximum BM if $L = 10$ m.

Hints: For absolute BM to be minimum $-$ve and +ve BM must be numerically equal, overhang $a = 0.2929\,L$, Absolute maximum $M_{max.} = \pm (0.2928\,L)^2 \times 5 = 42.86$ kNm.

Q.6.16. A simply supported beam with overhanging ends carrying concentrated loads each of equal to supported span total udl as shown in Fig. Q6.16. Determine the overhang on each side if the BM at the mid span E is zero. Draw SF and BM diagrams.

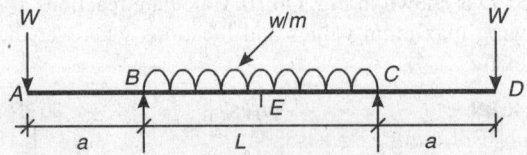

Fig. Q6.16.

Hints: $R_B = R_C = 1.5$ W $= 1.5\,wL$, where $W = wL$,

$$M_E = wL.\left(\frac{L}{2} + a\right) + \frac{w}{2} \cdot \frac{L}{2} \cdot \frac{L}{2} - 1.5\,wL \cdot \frac{L}{2} = 0,\ a = \frac{L}{8},\ F_B\,(\text{Left}) = -wL,$$

$$F_B\,(\text{Right}) = +0.5\,wL,\ M_B = \frac{-wL^2}{8} = M_C.$$

Q.6.17. A beam $ABCDE$ is loaded as shown in Fig. Q6.17. Draw SF and BM diagrams showing maximum values and their locations.

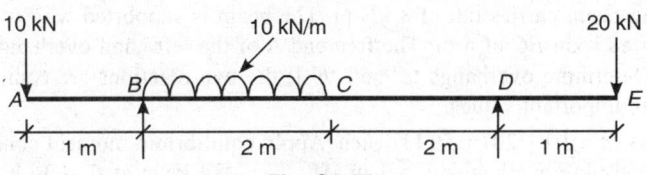

Fig. Q6.17.

Hints: R_B = 22.5 kN, R_D = 27.5 kN, F_B (Right) = + 12.5 kN, F_C = – 7.5 kN, F_D (Right) = + 20 kN, M_B = – 10 kNm, M_C = – 5 kNm, M_x (BX = 1.25 m) = + 2.19 kNm, M_D = – 20 kNm.

Q.6.18. A simply supported beam AB of 10 m span carries a triangular load with zero intensity at the support and maximum intensity of 30 kN/m at the mid span point. Calculate SF and BM at critical points and draw diagrams.

Hints: R_A = R_B = 75 kN, F_A (Right) = 75 kN, F_C = 0, M_A = 0, M_C = 250 kNm.

Q.6.19. A cantilever beam of 6 m span carries a trapezoidal load as shown in Fig. Q6.19. Calculate maximum BM and its location.

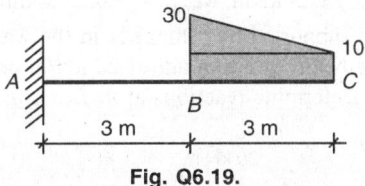

Fig. Q6.19.

Hints: F_C = 0, F_B = 60 kN, F_A = – 60 kN, M_A (Max.) = – 255 kNm.

Q.6.20. A hinged beam AB is subjected to a couple as shown in the Fig. Q6.20, find the reactions and draw SF and BM diagrams.

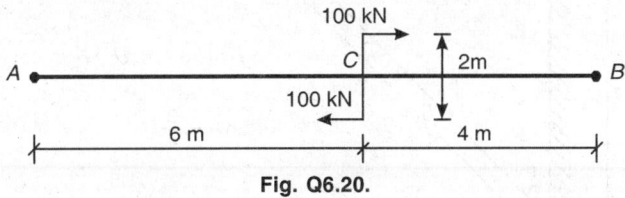

Fig. Q6.20.

Hints: $R_A = \dfrac{200}{10}$ = 20 kN ↓, R_B = 20 kN ↑, F_A = – 20 kN, F_C = – 20 kN, F_B = – 20 kN, 0, $M_A = M_B = 0$,

M_C = – 120 kNm, + 80 kNm, variation straight line.

Q.6.21. A 10 m long beam $ABCD$ has unequal overhangs on both simple supports B and C which are 6 m apart. Entire beam $ABCD$ carries udl of 40 kN/m. If the maximum bending moment has to be minimum possible, calculate the overhangs and draw BM diagram showing the maximum values.

Hints: Let the longer overhang be 'a' and smaller overhang will be $(4 – a)$. $R_B = \dfrac{200}{3}$ $(1 + a) = 215.33$

kN, $R_C = \dfrac{200}{3}$ $(5 – a) = 184.67$ kN, SF zero at x_1 (from free end A) $= \dfrac{5}{3}$ $(1 + a) = 5.3825$ m,

Max. –ve BM = – $20a^2$, Max. +ve BM at $x_1 = \dfrac{40}{2} \times \dfrac{25}{9} \times (1 + a)^2 - \dfrac{200}{3}$ $(1 + a)(x_1 – a)$, a

(larger) = 2.2295 m, b (smaller) = 1.7705 m, $M_{max.}$ = ± 99.46 kNm, at support B and at AX_1 = 5.3825 m.

Q.6.22. A 10 m long beam carries udl of 8 kN/m. The beam is supported with overhangs AB and DC with supported span BC of 6 m. The free end A of the left hand overhang carries a point load of 20 kN. Determine overhangs 'a' and 'b' if the two reactions are equal. Also draw SF and BM showing important values.

Hints: $R_B = R_C = \frac{1}{2}(8 \times 10 + 20) = 50$ kN each. Apply equilibrium moment conditions at supports B and C to determine overhangs, $a = 1$ m. $M_{max.} = -24$ kNm at B, -36 kNm at C, and $+6.25$ kNm at X at 2.75m from the support B.

Q.6.23. A 10 m long beam ABC is supported at A and B. The entire beam carries a udl of 20 kN/m and a point load of 50 kN at the free end C of overhang. Determine the supported span AB if the point of contraflexure E lies at the middle point (5 m from the support A). Also determine the maximum BM and their locations.

Hints: Let the supported span $AB = x$. Moment about A gives $R_B = \frac{1}{x}(50 \times 10 + 20 \times 10 \times 5) = \frac{1500}{x}$, $R_A = (250 - \frac{1500}{x})$. Moment about E (point of inflexion) $= 0 = 50 \times 5 + 20 \times 5 \times \frac{5}{2} - \frac{1500}{x}(x - 5)$, $x = 7.5$ m, $M_B = -187.5$ kNm, $M_{max.} = +62.5$ kNm at 2.5 m from A.

Q.6.24. A 6 m long beam ABC is supported by a hinge A in the wall and a 1m long prop BD at B as shown in Fig. Q6.24. The beam carries a udl of 20 kN/m on entire length and a point load of 50 kN at the free end C. Determine reactions at A, D and E. Also draw SF and BM diagrams for the beam ABC.

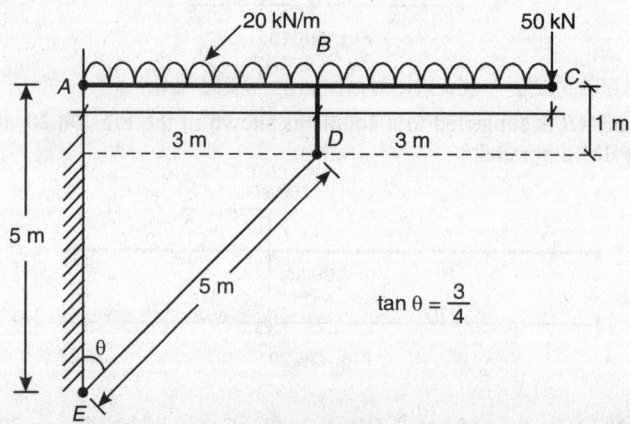

Fig. Q6.24.

Hints: Assume reaction R (axial) in strut DE. Vertical and horizontal forces at $D = \frac{4}{5}R$ (up), $\frac{3}{5}R$, left to right. FBD of ABC and BD, take moments of all forces about A : $R = 220$ kN, $V_B = 176$ kN, $H_B = 132$ kN. $M_{BC} = -240$ kNm, $M_{BA} = -108$ kNm (horizontal force at D causing moment couple), R_E (Vertical) $= 176$ kN, R_E (Horizontal) $= 132$ kN, $V_A = 6$ kN ($\downarrow$).

Q.6.25. A 8 m long bar $ABCD$ is bent and AB is 5 m and $ABC = 6$ m length as shown in Fig. Q6.25. The rod carries a load of 1 kN/m including self-weight. The bent rod is fixed at A. Draw SF and BM for the rod. The bends are rigid.

Ans. $F_D = 0$, $F_C = -2$ kN, $F_B = +3$ kN, $F_A = +8$ kN, $M_C = -2$ kNm, $M_B = +2$ kNm, $M_A = -25.5$ kNm.

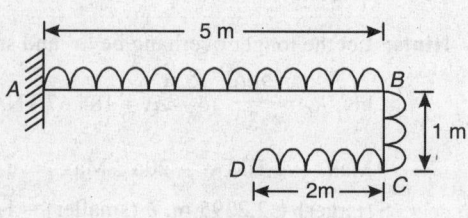

Fig. Q6.25.

Q.6.26. A 6 m long beam carries a brick wall transmitting wall loads at $45°$ with the horizontal from both end supports. The brick wall thickness is 400 mm. Brick masonry density is 21 kN/m^3.

Hints: Loading will be triangular having base 6 and apex 3 m high in the middle. $F_A = + 37.8$ kN, $F_B = - 37.8$ kN, M_C (Max.) $\dfrac{WL}{6} = + 75.6$ kNm (sagging).

Q.6.27. 5 m long simple beam AB carries a couple of anticlockwise moment 100 kNm at a point C, 2 m away from the support A. The beam also carries udl of 20 kN/m from C to B. Determine reactions, SF and BM in the beam ACB. Also determine point of inflexion and maximum BM.

Hints: $R_A = 38$ kN, $R_B = 22$ kN, $F_A = + 38$ kN, $F_C = + 38$ kN, $F_B = - 22$ kN, SF zero at 3.9 m from A. $M_{CA} = + 76$ kNm, $M_{CB} = - 24$ kNm, $M_{3.9}$ (Max.) $= + 12.1$ kNm, Inflexion 2.0 m and 2.80 m from A.

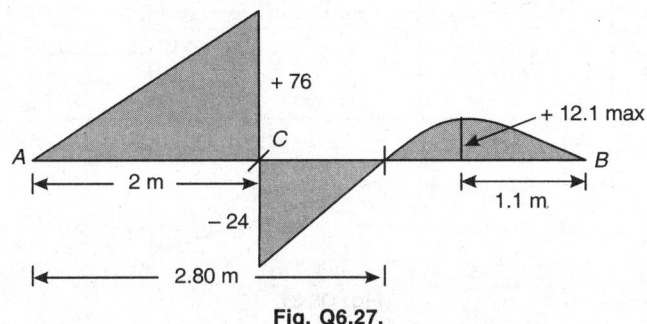

Fig. Q6.27.

Q.6.28. A lintel of 3 m span supports a brick wall of 300 mm thickness. The height of wall is 1 m at one end A and increases to 3 m at the other end B. The brick masonry weighs 20 kN/m^3. Draw SF and BM diagrams showing important values and locations.

Hints: Loading is considered in 2 parts i.e. udl + Triangular. $R_A = (9 + 6)$ kN $= F_A$, $R_B = (9 + 12) = 21$ kN, $w_x = (6 + \dfrac{12}{3} x) = (6 + 4x)$, $F_x = (R_A - \int w_x\, dx) = 15 - (6x + 2x^2)$, $F_x = 0$ at $x = $ **1.6225 m** from A, $M_x = 15x - 3x^2 - \dfrac{2}{3} x^3$, $M_{\text{max.}}$ $(x = 1.6225) = $ **13.592 kNm.**

Q.6.29. A beam AB of 20 m span is hinged at both the ends. Left half of the span AC carries a udl of 1 kN/m. It also carries a print load 16 kN at a point D 15 m from the support A. The beam also carries couples of 37.66 kNm in anticlockwise at the left support and 55.66 kNm in clockwise direction at the right hand support. Find the reactions and draw SF and BM diagrams showing important values.

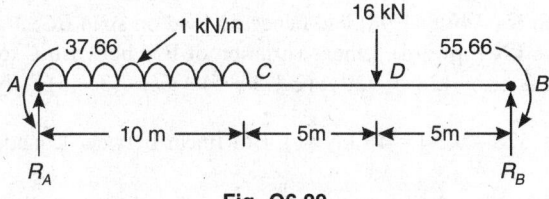

Fig. Q6.29.

Hints: $R_A = 10.6$ kN, $R_B = 15.4$ kN, $F_A = + 10.6$ kN, $F_C = + 0.6$ kN, $F_D = + 0.6$ kN, $- 15.4$ kN, $F_B = - 15.4$ kN, $M_A = - 37.66$ kNm, $M_C = + 18.33$, BM zero at $x = 4.51$ m, $M_D = + 21.34$ (Max.), M (A to C)-Parabolic, M (C to D)-Straight Line, M (D to B)-Straight Line, $M_B = - 55.66$ kNm.

Q.6.30. A ladder 5 m long weighs 400 N/m run, rests against a smooth vertical wall with bottom resting on rough ground. Vertical distance between the edges of the ladder is 3 m. A man weighing 800 N is standing on the ladder at 2 m (C) from the bottom and another person weighing 600 N standing at 4 m (D) from the bottom. Draw axial thrust, shear force and BM diagrams.

Hints: $\Sigma M_A = 0$, $R_{BH} = 2400$ N $= R_{AH}$, $\Sigma V = 0$, $R_{AV} = 3400$ N ($\uparrow$). Resolve forces along the ladder axis and transverse to it. $SF_A = -1280$ N, 0, $M_C = +1920$ N-m, $M_D = +1280$ N-m, $P_A = 3960$ N, $P_C = 3000$ N, $P_D = 2160$ N, $P_B = 1920$ N, $SFC = 0$, $SFD = +1120$ N, $SFB = 1440$ N.

Q.6.31. Fig. Q6.31 shows the shear force diagram (kN) for a beam which rests on supports, one at left end A. Determine the location of other support. Draw loading and BM diagrams showing important values and their locations.

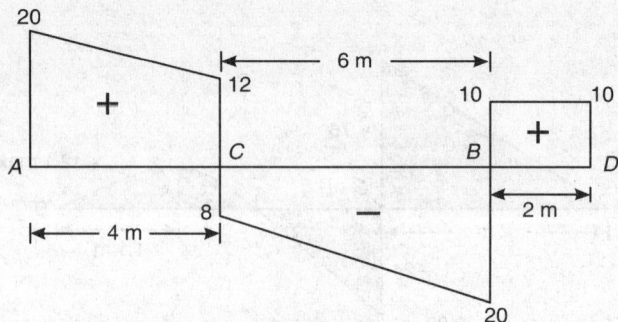

Fig. Q6.31.

Hints: $R_A = 20$ kN, SF from A to C sloping line, $w_1 = \left(\dfrac{20-12}{4}\right) = 2$ kN/m. Point load at $C = 12 - (-8)$

$= 20$ kN, C to B sloping straight line $w_2 = \left(\dfrac{20-8}{6}\right) = 2$ kN/m. Point load $(R_B) = -20 - 10$

$= -30$ kN (upward) i.e. support at B, constant SF (B to D) = 10 kN, point load at D (free end) $= (10 - 0) = 10$ kN (downward). Draw loading diagram and calculate BM from loading diagram or from SF diagram directly integrating between the points A to C, C to B and B to D. $M_{max.}$ (at C) = + 64 kNm, $M_B = -20$ kNm, point of inflexion $DX = 3.056$ m.

Q.6.32. The *BM* diagram for a beam *ABCDEF*, supported at B and E is shown in Fig. Q6.32. Draw the loading and shear force diagrams for the beam.

Hints: BM varies linearly between A to B indicating point load at $A = \dfrac{40}{2} = 20$ kN, *BM* also varies linearly from B to C (– 40 to + 98) and hence no load on span BC. $M_C = +98 = -20 \times (2+3) + R_B \times 3$, $R_B = 66$ kN (upward). Linear variation of BM between C to D indicates that there is point load W_C at C i.e. + 128 = – 20 × (2 + 3 + 5) + 66 × (3 + 5) – $W_C \times 5$,

$W_C = \dfrac{1}{5}[-200 + 528 - 128] = 40$ kN ($\downarrow$). BM linear between D and E, point load W_D at $D =$ 80 kN ($\downarrow$). BM varies parabolically between E and F, indicating udl on EF. $-20 = \dfrac{w(2)^2}{2}$, $w = 10$ kN/m. Reaction $R_E = 94$ kN ($\uparrow$). SF diagram can be drawn from loading diagram or directly by $SF = \dfrac{dM}{dx}$ equation.

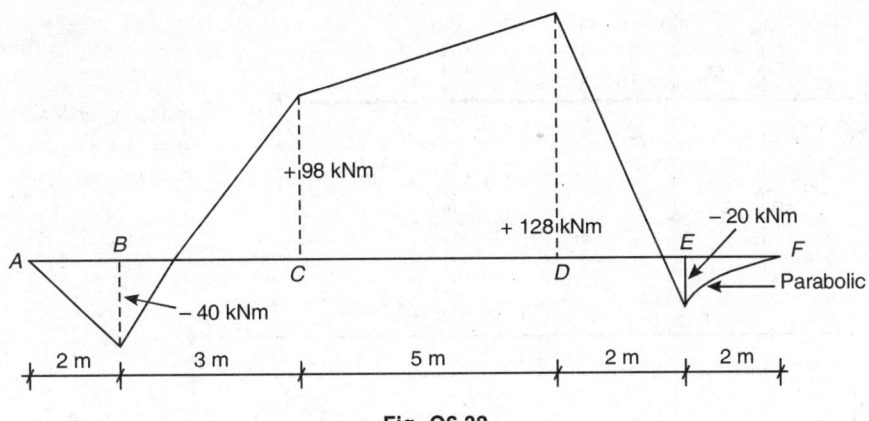

Fig. Q6.32.

Q.6.33. A simply supported horizontal beam *AB* is resting on roller support A and hinge support at *B*. The beam carries vertical loads as shown in Fig. Q6.33. Determine reactions and draw *SF* and *BM* diagrams graphically assuming suitable scales for loads and distances.

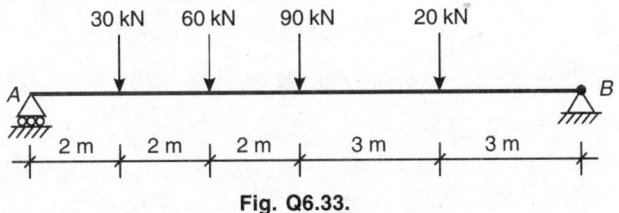

Fig. Q6.33.

Hints: Draw space diagram of beam assuming a linear scale (say 1 m = 10 mm). Mark spaces between forces as *P, Q, R, S, T* and *U*. Draw forces *pq, qr, rs, st*, etc. at certain scale (say 1 kN = 1 mm) parallel to applied loading in vector diagram. Take a pole '*O*' and join all the points *p, q, r,* .... etc. in the vector diagram. Reaction up R_A will be vertical. Extend force lines in space diagram. Draw link diagram (representing *BMD*) starting from the hinge point in space diagram, drawing lines parallel to *ot, os, or, oq, op,* .... etc. and marking points 1, 2, 3, ... etc. Points 1, 2, 3, ... etc. are obtained by intersecting force lines in space diagrams and lines drawn parallel *to, ot, os,* .... etc. Join starting point 1 with the last point on roller reaction line. From the pole point *O* draw a line parallel to the closing line of link diagram (1-2-3-..........) to get point u in the vector diagram. tu and up represents R_B and R_A respectively. Complete the process to draw *SF* and *BM* diagrams.

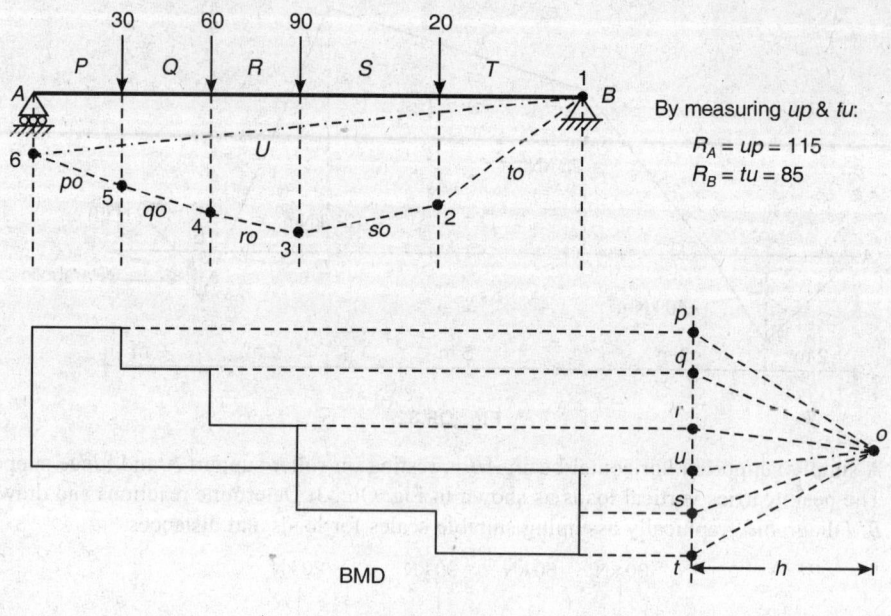

By measuring *up* & *tu*:

$R_A = up = 115$

$R_B = tu = 85$

BMD

Fig. Q6.33.

7

Centroid, Area and Moment of Inertia in Bending Structures

LEARNING OBJECTIVES

After studying this chapter, the learner **understands** Area, Centroid and Moment of Inertia of plane sections and will be able to:

7.1 **Explain** centre of **gravity** (centroid), axis of symmetry of a plane cross-sectional area.

7.2 **Calculate** position of **centroid** of a given plane area.

7.3 **Explain moment of inertia** (second moment of area) of a given cross-sectional area.

7.4 **Explain** parallel axes theorem.

7.5 **Calculate** moment of inertia of a given cross-sectional plane area.

7.6 **Calculate** moment of inertia of a given plane cross-sectional area about its normal NA.

7.7 **Explain** neutral axis (NA) of a plane cross-sectional area subjected to bending.

7.8 **Calculate** location of axis of symmetry and NA of a plane cross-sectional area.

7.1 INTRODUCTION

The bending elements offer resistance to bending depending on geometrical properties of cross-sectional area, location of centroid and plane of bending, etc. It is, therefore, most important to understand the geometrical properties before analysis of *bending stresses* in bending elements. To understand geometrical properties of cross-section, let us define *centre of gravity*. The *centre of gravity* of a body is a *point* through which the *resultant* of the *system of parallel* forces formed by the weights of all the particles of the body **passes**, for *all positions* of the body. Weights of various particles of the body act as parallel force system. The weights of particles of flat body with unknown thickness is directly proportional to areas covered by these particles. Thus in plane cross-sections only areas are considered for analysis of geometrical properties.

The plane areas, like *triangles, rectangles, circles*, etc. have only areas and no consideration for their masses. The *centre of gravity* or *centre of area* of such figures is known as *centroid*. In common teroninology, the cross-sectional area is taken to mean **plane surface**. **Area** is the **measure of size** of the plane surface. The determination of the **centroid** of the **plane cross-sectional** figure is the same as finding of the **centre of gravity** of a body. The term **centroid** and **centre of gravity** (cg) considering weights of particles are used interchangeably in general sense. The location of **centroid of a plane** cross-sectional **area** form an important **geometrical property** of the area and represents distribution pattern of area.

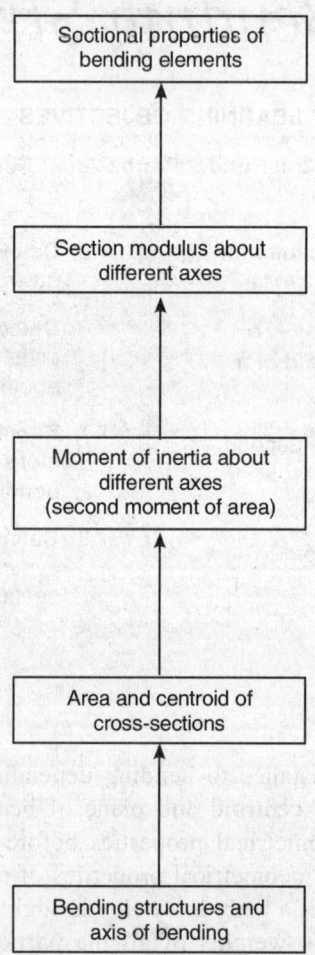

Fig. 7.0: Learning structure of geometrical properties of plane cross-sectional areas

7.2   AREAS AND CENTROIDS OF PLANE SECTIONS

Figure 7.1 shows a plane area (A), with its centroid $G(\bar{x}, \bar{y})$. Consider a small area δA with its centre (x, y).

The total area $\qquad A = \Sigma \delta A$

or $\qquad A = \int dA$ $\qquad\qquad$ ... 7.1

First moment of area about X and Y Axes are respectively:

$$M_x = \int y \, dA \quad \text{and} \quad M_y = \int x \, dA.$$

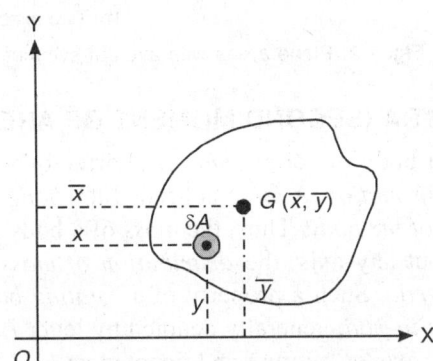

Fig. 7.1: Centroid of plane area

The coordinates $\bar{x}$ and $\bar{y}$ of the **centroid G** are:

$$\bar{x} = \frac{\int x \, dA}{\int dA} = \frac{\Sigma x \, dA}{\Sigma dA}, \quad \text{and} \quad \bar{y} = \frac{\int y \, dA}{\int dA} = \frac{\Sigma y \, dA}{\Sigma dA} \text{ (within limits)} \qquad ... 7.2$$

(I)   Centroid of an Area Symmetrical about an Axis

If a plane area has an **axis of symmetry**, the **centroid** of the area will **lie on that axis of symmetry**. It may be noted that the **first moment of area** about this axis of **symmetry will be zero**. Hence the centroid of this area will lie on this axis of symmetry. In channel section Fig. 7.2a, there is only **one axis of symmetry** and hence **centroid 'G'** lies on this axis of symmetry X-X. Distance of centroid G can be determined by equating **first moment of area** equal to **zero** about G. (Fig. 7.2a).

(II)   Centroid of an Area Symmetrical about Two Axes

If a plane area has two axes of symmetry, the centroid of such plane area will lie at the intersection of the two axes of symmetry. The centroid of such a plane sectional area can be obtained by fixing the axes of symmetry and their *intersection. First moment of area* about both the axes of symmetry *will be zero* (Fig. 7.2b).

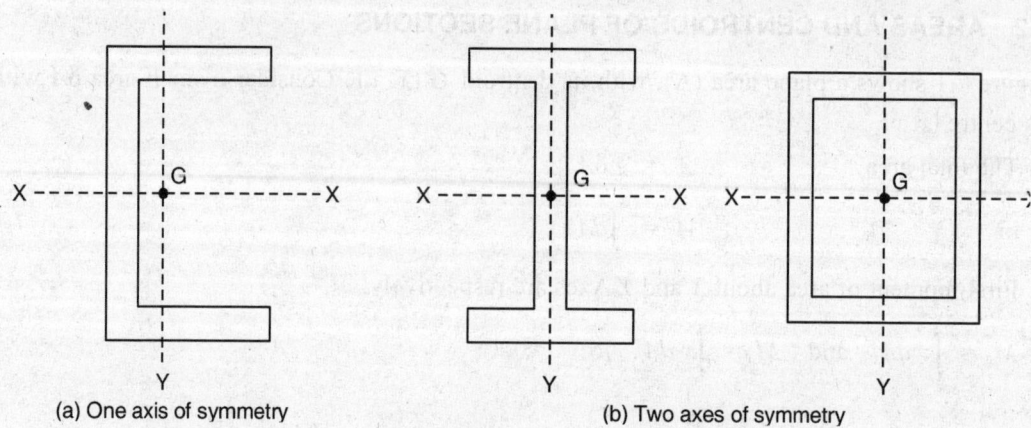

(a) One axis of symmetry   (b) Two axes of symmetry

Fig. 7.2: Plane areas with axes of symmetry

7.3 MOMENT OF INERTIA (SECOND MOMENT OF AREA)

Inertia is the property of a body of certain *mass* by virtue of which it tends to *maintain its own state*, i.e. state of *rest* or *motion*. A *force* is needed to *change the state*. The *magnitude* of force depends on the *mass of the body*. Thus, the *mass* of a body *measures the inertia*. In case of *rotating body mass* about any axis, the *distribution of mass* with respect to the axis of rotation also *affects the inertia*. Such a property of a *rotating body mass* is called *rotational inertia* or *simply moment of inertia*, generally denoted by letter *I*. Moment of inertia is always referred with respect to the *axis of rotation* and denoted as I_X, I_Y or I_Z. In case of *plane area the moment of inertia* is also known as *second moment of area*.

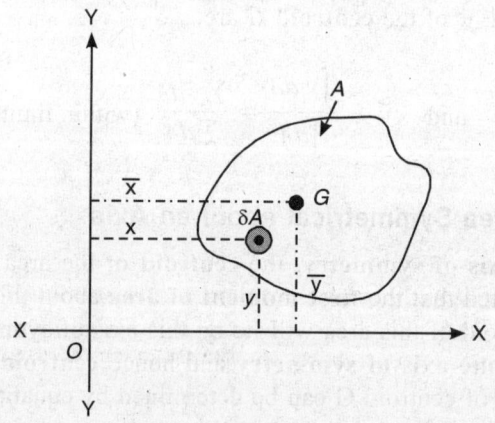

Fig. 7.3: Moment of inertia

Consider a plane area *A* and small element δA with its coordinates (*x*, *y*). *Second moment of area* δA about any axis X-X or **Y-Y** will be denoted as:

$$\delta I_X = \delta A y^2, \qquad\qquad \delta I_Y = \delta A x^2$$

Thus *second moment of the whole* area '*A*' of the lamina (plane area) will be *sum of the second moment* of all small elements forming the area *A*. i.e.

$$I_X = \Sigma \delta I_X = \Sigma y^2 \delta A = \int_{y_1}^{y_2} y^2 \, dA$$

and

$$I_Y = \Sigma \delta I_Y = \Sigma x^2 \delta A = \int_{x_1}^{x_2} x^2 \, dA$$

... 7.3

Small elemental area δA (dA) can be expressed in terms of element δx (dx) or δy (dy) as necessary for summation according to the *shape of the plane area*.

This second moment of area (or moment of inertia) is generally determined about the axes passing from the *centroid 'G'*. This second moment of area about an *axis passing through its centroid* is generally referred to as *moment of inertia* about that centroidal axis. These moment of inertias for a plane area X-Y are referred to *two normal axes X-X and Y-Y passing through the centroid* of the area. The axis passing through the centroid and perpendicular to the two normal axes X-X and Y-Y is known as *polar axis Z-Z* (Fig. 7.4).

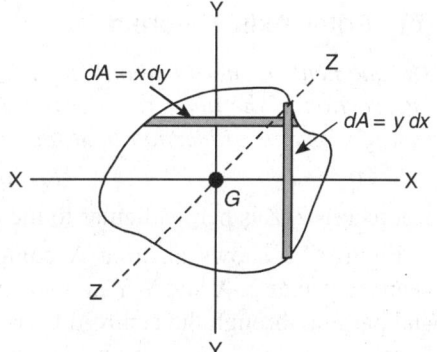

Polar moment of inertia can be expressed as:

$$I_{Z-Z} = \Sigma \delta A \cdot r^2 = \int r^2 \, dA \quad \text{... 7.4}$$

Where *r is the radial distance* of small area from the *polar axis Z-Z*. Polar moment of inertia of the whole area is *summation of product of square of radial distance 'r'* and the *elementary area δA* within the limits of the area.

Z-Z Normal to plane X-Y

Fig. 7.4: Polar axis Z-Z

7.4 THEOREMS OF MOMENT OF INERTIA

(a) Parallel Axis Theorem

The moment of inertia with respect to any axis parallel to the centroidal axis is equal to the moment of inertia with respect to the centroidal axis plus the product of the area of the figure and the square of the distance between the two parallel axes i.e.

$$I_{AB} = I_{XX} + Ah^2 \quad \text{... 7.5 (a)}$$

Also
$$I_{CD} = I_{YY} + Ak^2 \quad \text{... 7.5 (b)}$$

where I_{XX} and I_{YY} are moment of inertia about two normal *centroidal axes*.

AB is parallel axis to the centroidal axis **X-X** and situated at a distance of '*h*'.

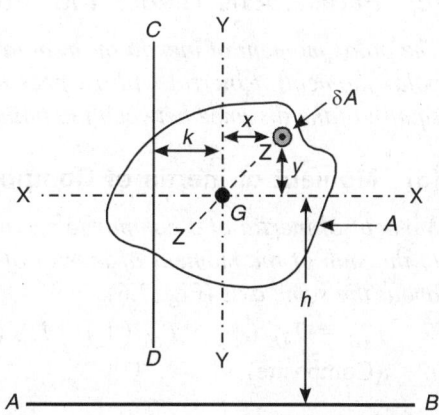

Fig. 7.5: Parallel axis theorem

CD is parallel axis to the centroidal axis **Y-Y** and situated at a distance of '**k**'.

This can be proved from the basic definition of moment of inertia.

Consider small elementary area δA with its centroid G_1 (x, y). The centroid of the whole area is G $(0, 0)$. The axes **X-X** and **Y-Y** pass through the centroid G.

$$I_{AB} = \Sigma\delta A\,(h+y)^2 = \Sigma y^2\,\delta A + \Sigma 2hy\,\delta A + \Sigma h^2\,\delta A$$
$$= I_{XX} + 2h\,\Sigma y\,\delta A + h^2\,\Sigma\delta A = I_{XX} + 0 + Ah^2$$

or
$$I_{AB} = I_{XX} + Ah^2 \qquad\qquad\qquad \text{... 7.5 (a)}$$

Similarly

$$I_{CD} = \Sigma\delta A\,(k+x)^2 = \Sigma x^2\,\delta A + \Sigma 2kx\,\delta A + \Sigma k^2\,\delta A = I_{YY} + 0 + Ak^2$$

or
$$I_{CD} = I_{YY} + Ak^2 \qquad\qquad\qquad \text{... 7.5 (b)}$$

First moment of area about centroidal axis is always zero (i.e. $\Sigma y\,\delta A = 0 = \Sigma x\,\delta A$).

(b) Polar Axis Theorem

The moment of inertia about an axis perpendicular to the plane and passing through the intersection of the other two normal axes X-X and Y-Y contained by the plane is equal to the sum of moments of inertia about the other two perpendicular axes X-X and Y-Y i.e.

$$I_Z \quad\text{or}\quad I_P = I_{XX} + I_{YY} \qquad\qquad\qquad \text{... 7.6}$$

where axis Z-Z is perpendicular to the plane of X-X and Y-Y.

Figure 7.5 shows an area A comprising of small elemental areas 'δA' and containing centroidal axes X-X and Y-Y in its plane. Let Z-Z axis be perpendicular to the plane of the area and passing through the centroid G (intersection of X-X and Y-Y axes).

The elementary area δA has its centroid at G_1 (x_1, y_1). By definition $I_{XX} = \Sigma y^2\,\delta A$, $I_{YY} = \Sigma x^2\,\delta A$, $I_{ZZ} = I_P = \Sigma r^2\,\delta A$. But radial distance r = diagonal. Thus, $r^2 = x^2 + y^2$.

$$I_Z\,(=I_P) = \Sigma r^2\,\delta A = \Sigma\,(x^2 + y^2)\,\delta A = \Sigma x^2\,\delta A + \Sigma y^2\,\delta A$$

or
$$I_Z = I_{YY} + I_{XX} \qquad\qquad\qquad \text{... 7.6}$$

It may be noted that *parallel axis theorem holds good for the polar axis theorem also.*

(c) Parallel Axis Theorem for Polar Moment of Inertia

The polar moment of inertia of an area with respect to any point O in its plane is equal to the polar moment of inertia with respect to the centroid G plus the product of the area and the square of the distance between the points 'O' and 'G'.

(d) Moment of Inertia of Composite Sections

Moment of inertia of a composite section about an axis is equal to the sum of the moment of inertia of its separate components about the same axis (Fig. 7.6).

$$I_{AB} = I_{AB}\,(A_1) + I_{AB}\,(A_2) + I_{AB}\,(A_3) + I_{AB}\,(A_4) \quad \text{... 7.7}$$
(Composite)

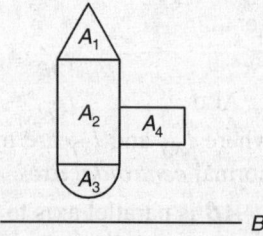

Fig. 7.6: Composite section

7.5 AREAS, CENTROIDS AND MOMENT OF INERTIA OF REGULAR SECTIONS

(a) Rectangle (Solid)

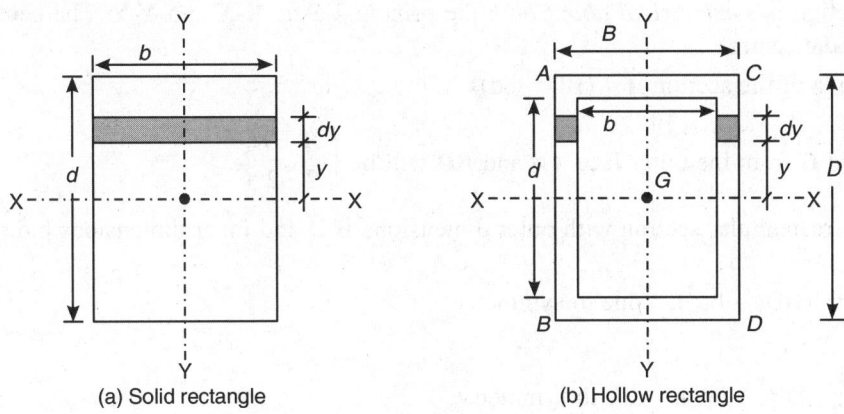

(a) Solid rectangle   (b) Hollow rectangle

Fig. 7.7: Rectangle

Consider a small strip of depth 'dy' and breadth 'b' located at a distance 'y' from the axis X-X (Fig. 7.7)

$$A = \int_{-d/2}^{+d/2} b \, dy = \left(b \cdot y\right)_{-d/2}^{+d/2} = b.d \qquad \qquad \text{... 7.8 (a)}$$

$\bar{x} = \dfrac{b}{2}$ from one longer side, $\bar{y} = \dfrac{d}{2}$ from one shorter side

Thus centroid with reference to sides will be located at $G\left(\dfrac{b}{2}, \dfrac{d}{2}\right)$. $\qquad$ ... 7.8 (b)

Second moment of area about X-X or moment of inertia about centroidal axis X-X will be:

$$I_{XX} = \int_{-d/2}^{+d/2} b \cdot y^2 \, dy = \left(\frac{b \cdot y^3}{3}\right)_{-d/2}^{+d/2} = \frac{bd^3}{12} \qquad \qquad \text{... 7.8 (c)}$$

Similarly

$$I_{YY} = \int_{-b/2}^{+b/2} d \cdot x^2 \, dx = \frac{d \cdot b^3}{12} \qquad \qquad \text{... 7.8 (d)}$$

(b) Hollow Rectangle

Consider a rectangle box-section with outer breadth '*B*' and depth '*D*'. Let the box has inner breadth '*b*' and depth '*d*'.

The section is *symmetrical about both* the principal axes **X-X** and **Y-Y**. The centroid *G* is located at the centre.

Solid area of the section $A = (BD - bd)$

Centroid *G* from the outer face *AB* and *BD* will be $\left(\dfrac{B}{2}, \dfrac{D}{2}\right)$.

Hollow rectangular section with outer dimensions B.D and inner dimensions b.d.

$$I_{XX} = \frac{1}{12}(BD^3 - bd^3), \quad \text{due to symmetry} \qquad \ldots 7.8\,(e)$$

$$I_{YY} = \frac{1}{12}(DB^3 - db^3), \quad \text{due to symmetry} \qquad \ldots 7.8\,(f)$$

(c) Triangular Section (Base *b* and Height *h*)

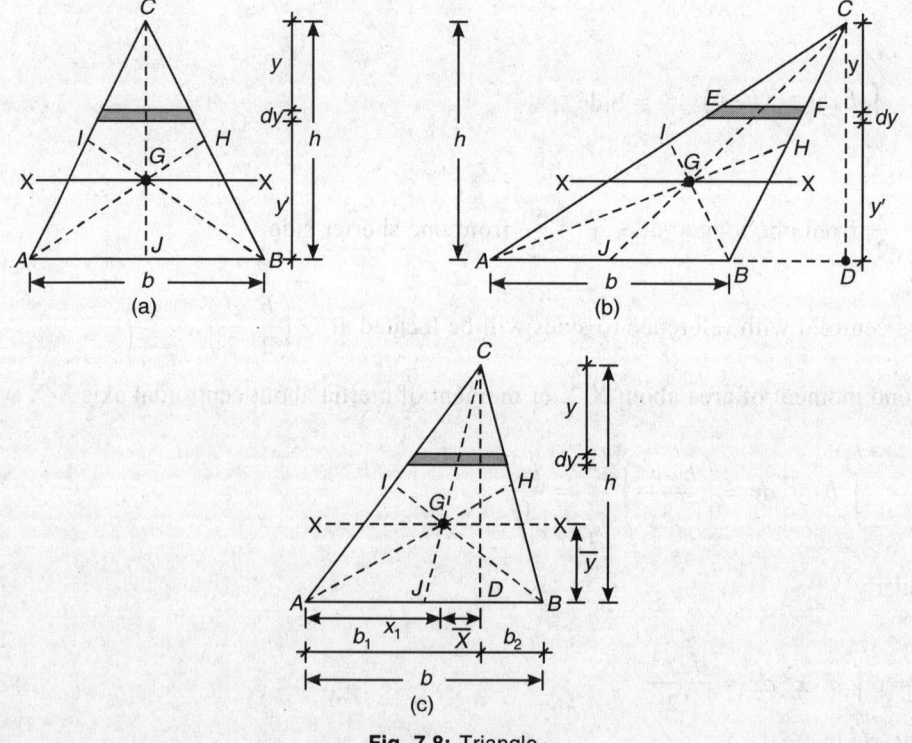

Fig. 7.8: Triangle

BI, CJ, AH are *bisectors*, *G* is intersection of bisectors.

Triangle *ABC* has base '*b*' and perpendicular *CD* of height '*h*' (Fig. 7.8c) in case of unsymmetrical triangle.

$$AD = b_1, \qquad BD = b_2, \qquad b_1 + b_2 = b$$

Centroid $G\ (\bar{x},\ \bar{y})$ is located from the point A at $\bar{x}_1$ and base AB at $\bar{y}$. G can also be specified from the end B as $\bar{x}_2, \bar{y}$.

$EF = \dfrac{b}{h} \cdot y$, where y is from C. Centroid G lies at the intersection of bisectors CJ, BI and AH (Fig. 7.8).

$$A\bar{y} = \int_0^h \left(\frac{b}{h} \cdot y\right)(h-y)\, dy = \frac{b}{h}\int_0^h (hy - y^2)\, dy$$

$$\frac{b \cdot h}{2}\, \bar{y} = \frac{b}{h}\left[\frac{hy^2}{2} - \frac{y^3}{3}\right]_0^h = \frac{b}{h} \cdot \frac{h^3}{6} = \frac{bh^2}{6}$$

$$\boxed{\bar{y} = \frac{h}{3}} \qquad \qquad \text{... 7.9 (a)}$$

Similarly by taking moment about A

$$\frac{b_1 h}{2}\left(\frac{2b_1}{3}\right) + \frac{b_2 h}{2}\left(b_1 + \frac{b_2}{3}\right) = \frac{b \cdot h}{2} \cdot \bar{x}$$

$$\bar{x} = \frac{b_1 + b}{3} = \frac{2b_1 + b_2}{3}, \qquad \bar{x}_2 \text{ (from B):} \quad \bar{x}_2 = \frac{b + b_2}{3} = \frac{b_1 + 2b_2}{3} \qquad \text{... 7.9 (b)}$$

Centroidal axis parallel to base is located at $\dfrac{h}{3}$ from the base of triangle.

If the triangle is unsymmetrical having 'b_1' and 'b_2' bases of two right angled triangles ADC and BDC. If resultant centroid is located at $\bar{x}$ from the perpendicular CD, then

$$A\bar{x} = A_1 \cdot \bar{x}_1 + A_2 \cdot \bar{x}_2, \quad \text{then} \quad \frac{1}{2} b \cdot h \cdot \bar{x} = \frac{1}{2} b_1 \cdot h \cdot \frac{b_1}{3} - \frac{1}{2} b_2 \cdot h \cdot \frac{b_2}{3}$$

$$\text{or} \quad \bar{x} = \frac{b_1^2}{3b} - \frac{b_2^2}{3b} = \frac{1}{3b}(b_1 + b_2)(b_1 - b_2) = \frac{b_1 - b_2}{3} = \frac{b - 2b_2}{3} \qquad \text{... 7.9 (c)}$$

Moment of inertia about X-X

$$I_X = \int_0^h \left(\frac{by}{h}\, dy\right)\left(\frac{2h}{3} - y\right)^2 = \frac{b}{h}\int_0^h \left(\frac{4h^2}{9} - \frac{4hy}{3} + y^2\right) y\, dy = \frac{b}{h}\left[\frac{4h^2}{9}\frac{y^2}{2} - \frac{4h}{3}\frac{y^3}{3} + \frac{y^4}{4}\right]_0^h$$

$$I_X = \frac{b}{h}\left[\frac{2h^4}{9} - \frac{4h^4}{9} + \frac{h^4}{4}\right] = \frac{bh^4}{36h}[8 - 16 + 9] = \frac{bh^3}{36}$$

$$I_X = \frac{bh^3}{36} \qquad \qquad \text{... 7.9 (d)}$$

$$I_{AB} \text{ (Base)} = \frac{bh^3}{36} + \frac{1}{2}bh\left(\frac{h}{3}\right)^2 = \frac{bh^3}{12} \qquad \qquad \text{... 7.9 (e)}$$

Similarly about Y-Y moment of inertia can be found by applying parallel axes theorem from centroids of individual triangles to the vertical CD.

$$I_{CD} = \left(I_{Y_1Y_1} + A_1 x_1^2\right) + \left(I_{Y_2Y_2} + A_2 x_2^2\right) = \frac{hb_1^3}{36} + \frac{b_1 h}{2}\left(\frac{b_1}{3}\right)^2 + \frac{hb_2^3}{36} + \frac{b_2 h}{2}\left(\frac{b_2}{3}\right)^2$$

$$I_{CD} = \frac{h}{12}\left(h_1^3 + b_2^3\right)$$

$$I_{YY} = I_{CD} - \frac{bh}{2}\frac{(b_1 - b_2)^2}{9} = \frac{h}{12}\left(b_1^3 + b_2^3\right) - \frac{bh}{18}(b_1 - b_2)^2$$

$$I_{YY} = \frac{h}{36}\left[3\left(b_1^3 + b_2^3\right) - 2b(b_1 - b_2)^2\right] = \frac{hb^3}{48}, \quad \text{when } b_1 = b_2 = \frac{b}{2}$$

(d) Circle (Radius R)

Area $= \pi r^2 = \dfrac{\pi}{4}(D^2)$, Centroid at centre,

Polar moment of inertia

$$I_P = \int_0^R 2\pi r \cdot dr \cdot r^2 = \int_0^R 2\pi r^3\, dr = \left[\frac{2\pi r^4}{4}\right]_0^R$$

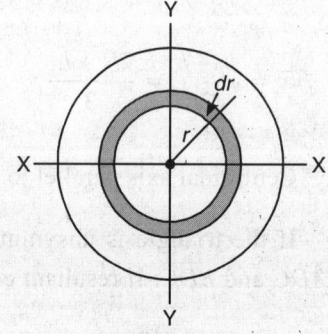

Fig. 7.9: Circle

$$I_P = \frac{\pi R^4}{2} = I_{XX} + I_{YY} = \frac{\pi D^4}{32} \qquad \qquad \text{... 7.10 (a)}$$

$$I_{XX} = I_{YY} = \frac{\pi R^4}{4} = \frac{\pi D^4}{64}$$

(e) Sector of a Circle

Figure 7.10 shows a sector of a circle subtending an angle of 2α at the centre O. OX is symmetrical axis X-X and OY is the perpendicular axis.

Centroid G is located on the axis of symmetry OX.

$\bar{y} = 0$, elementary area $\delta A = (r \cdot \delta\theta)\, r$.

δA is making an angle of θ with OX. δA has variable thickness zero to $rd\theta$ and subtends an angle of $d\theta$ at the centre. Cg of δA is situated at r from the centre O.

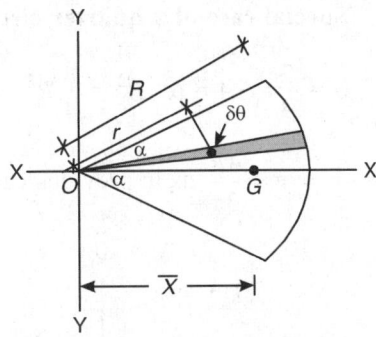

Fig. 7.10: Sector of a circle

$$\delta A = (r \cdot \delta\theta)\frac{1}{2}r, \qquad A = \int\limits_{-\alpha}^{+\alpha} dA = \int\limits_{-\alpha}^{+\alpha}\frac{1}{2}r^2\, d\theta = \frac{r^2}{2}(\theta)_{-\alpha}^{+\alpha} = \frac{r^2}{2}(\alpha + \alpha) = \alpha\, r^2$$

$$A = \alpha\, r^2 = \alpha\, R^2 \qquad\qquad \text{... 7.11 (a)}$$

Cg is $(r \cos\theta)$ away from the axis OY.

Moment about $OY = \delta M = \boldsymbol{\delta A \cdot r \cos\theta}$

Moment
$$M = \int\limits_{-\alpha}^{+\alpha} dM = \int\limits_{-\alpha}^{+\alpha}\delta A \cdot r \cos\theta = \int\limits_{-\alpha}^{+\alpha}\int\limits_{0}^{R} r^2\, d\theta\, dr \cos\theta$$

$$M = \int\limits_{-\alpha}^{+\alpha}\left(\frac{r^3 \cos\theta}{3}\, d\theta\right)_{0}^{R} = \left(\frac{R^3}{3}\sin\theta\right)_{-\alpha}^{+\alpha} = \frac{2}{3}\boldsymbol{R^3 \sin\alpha}$$

Area of the sector
$$A = \pi R^2 \cdot \frac{\alpha}{\pi} = \alpha \cdot R^2 \qquad\qquad \text{... 7.11 (b)}$$

$$\bar{x} = \frac{M}{A} = \frac{\dfrac{2}{3}R^3 \sin\alpha}{\alpha \cdot R^2} = \frac{2}{3}\frac{R \sin\alpha}{\alpha} \qquad\qquad \text{... 7.11 (c)}$$

Special case of a semicircle

$$\alpha = \frac{\pi}{2}, \quad A = \frac{\pi}{2}R^2, \qquad\qquad \text{... 7.11 (d)}$$

$$\bar{x} = \frac{2R}{3}\frac{\sin \pi/2}{\pi/2} = \frac{4R}{3\pi} \qquad\qquad \text{... 7.11 (e)}$$

$\bar{y} = 0$, if the semicircle is symmetrical about X-X.

Special case of a quarter circle

$$\alpha = \frac{\pi}{4}, \quad A = \alpha \cdot R^2 = \frac{\pi}{4} R^2, \qquad \qquad \ldots 7.11\ (f)$$

$$\bar{x} = \bar{y} = \frac{4R}{3\pi} \text{ as in previous case} \qquad \qquad \ldots 7.11\ (g)$$

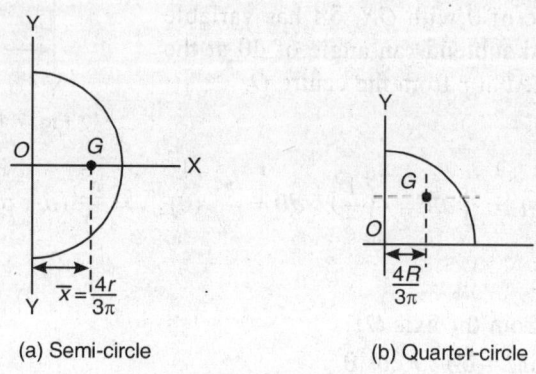

(a) Semi-circle (b) Quarter-circle

Fig. 7.11: Semi and quarter circle

(f) Symmetrical Hollow Circular Section

Consider a hollow circular section of external diameter '**D**' and internal diameter '**d**'.

By symmetry centroid G lies at the same centre of both the circles (external and internal).

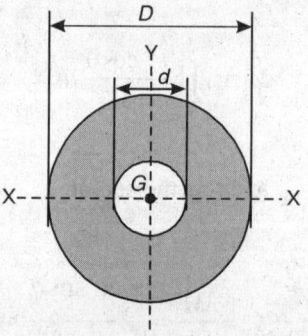

Area $A = \dfrac{\pi}{4}(D^2 - d^2)$ $\qquad \ldots 7.12\ (a)$

Moment of inertia:

$$I_{XX} = I_{YY} = \frac{\pi}{64}(D^4 - d^4) \qquad \ldots 7.12\ (b)$$

I_P (Polar) Moment of inertia $= \dfrac{\pi}{32}(D^4 - d^4) \qquad \ldots 7.12\ (c)$

Fig. 7.12: Hollow circular section

Section Modulus $\qquad Z_{XX} = \dfrac{I_{XX}}{y_{max.}} = \dfrac{\dfrac{\pi}{64}(D^4 - d^4)}{D\big/2}$

$$Z_{XX} = \frac{\pi}{32\,D}(D^4 - d^4) = Z_{YY} \qquad \qquad \ldots 7.12\ (d)$$

(g) Parabolic Semi Segment

Consider a parabola OAB as shown in Fig. 7.13 whose centroid lies at $G\ (\bar{x},\ \bar{y})$. The equation of parabolic variation is:-

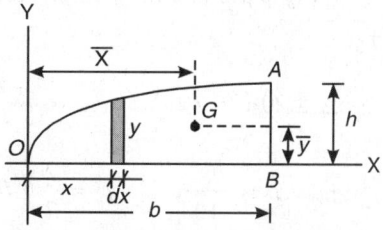

$$y^2 = \frac{h^2}{b}\cdot x, \qquad\qquad y = \frac{h}{\sqrt{b}}\cdot\sqrt{x}$$

Consider a small strip of height y and thickness dx at a distance of x from O.

Fig. 7.13: Parabolic semi segment

$$\text{Area }\delta A = y\cdot dx, \qquad\qquad A = \int_0^b y\,dx = \left(\frac{2h}{\sqrt{b}}\cdot\frac{x^{3/2}}{3}\right)_0^b = \frac{2}{3}bh \qquad\qquad\ldots\ 7.13\ (a)$$

Moment of elementary area about the axis OY:
$$x\,\delta A = x\cdot y\,dx$$

$$\bar{x} = \frac{\displaystyle\int_0^b x\cdot y\,dx}{\displaystyle\int_0^b y\,dx} = \frac{\displaystyle\int_0^b x\cdot\frac{h}{\sqrt{b}}\cdot\sqrt{x}\,dx}{2bh\big/3} = \frac{h}{\sqrt{b}\cdot\frac{2bh}{3}}\left(\frac{x^{5/2}}{5/2}\right)_0^b = \frac{h\cdot b^2\cdot\sqrt{b}}{\sqrt{b}\cdot\frac{2bh}{3}}\cdot\frac{2}{5} = \frac{3b}{5} \qquad\ldots\ 7.13\ (b)$$

Moment about OX:

$$\int_0^b y\,dx\cdot\frac{y}{2} = \frac{1}{2}\int_0^b y^2\,dx = \frac{1}{2}\int_0^b\frac{h^2}{b}\cdot x\,dx = \frac{h^2}{2b}\left(\frac{x^2}{2}\right)_0^b = \frac{h^2}{4b}(b^2)$$

$$\bar{y} = \frac{\text{Moment about }OX}{\text{Area}} = \frac{h^2 b^2}{4b\times\frac{2}{3}bh} = \frac{3h}{8} \qquad\qquad\ldots\ 7.13\ (c)$$

Second moment of area about OY:

$$I_Y = \int_0^b x^2\,y\,dx = \int_0^b x^2\cdot\frac{h}{\sqrt{b}}\cdot x^{1/2}\,dx = \left(\frac{h}{\sqrt{b}}\cdot\frac{2x^{7/2}}{7}\right)_0^b = \frac{2h}{7\sqrt{b}}b^3\sqrt{b} = \frac{2hb^3}{7} \qquad\ldots\ 7.13\ (d)$$

Second moment of area about OX:

$$I_X = \int_0^b y\,dx\left(\frac{y}{2}\right)^2 = \frac{1}{4}\int_0^b y^3\,dx = \frac{1}{4}\int_0^b\frac{h^3}{b\sqrt{b}}\cdot x^{3/2}\,dx = \frac{h^3}{4b\sqrt{b}}\left(\frac{x^{5/2}}{5/2}\right)_0^b = \frac{h^3}{10b\sqrt{b}}(b^2\sqrt{b})$$

$$I_X = \frac{bh^3}{10} \qquad\qquad\ldots\ 7.13\ (e)$$

Apply parallel axis theorem for finding I_{X_0} and I_{Y_0} about centroidal axes $X_0 - X_0$ and $Y_0 - Y_0$.

$$I_{x_0} + A\bar{y}_0^2 = I_X, \qquad I_{y_0} + A\bar{x}_0^2 = I_Y, \qquad \text{where } \bar{x}_0 = \frac{3}{5}b, \quad \bar{y}_0 = \frac{3}{8}h \qquad \text{... 7.13 (f)}$$

and $I_x = \dfrac{bh^3}{10}$, $\quad I_y = \dfrac{2}{7}b^3 h$

EXAMPLES OF COMBINED PLANE AREAS

EXAMPLE 7.1: A beam cross-section is shown in Fig. 7.14. Determine the area and centroid. Size of rectangle is 100 mm × 120 mm and triangular cut is 66 mm base × 45 mm perpendicular while semicircular cut has 33 mm radius as shown.

Solution:

Area of rectangle $A_1 = 100 \times 120$
$$= 12000 \text{ mm}^2$$

Area of triangular cut $A_2 = \dfrac{66}{2} \times 45$
$$= 1485 \text{ mm}^2$$

Area of circular cut $A_3 = \dfrac{\pi R^2}{2} = \dfrac{\pi (33)^2}{2}$
$$= 1711.3 \text{ mm}^2$$

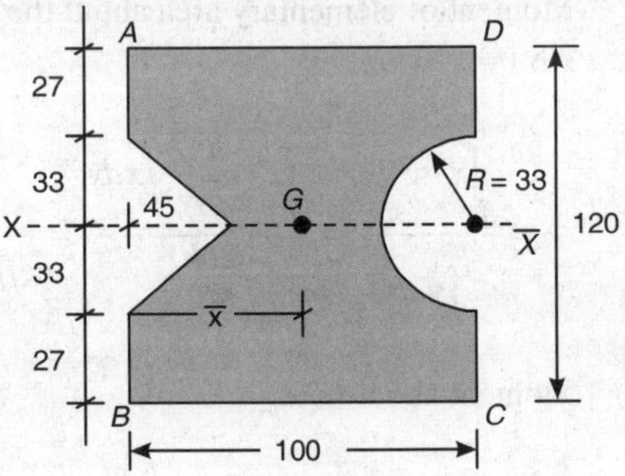

Fig. 7.14.

Net area $A_0 = A_1 - A_2 - A_3 = 12000 - 1485 - 1711.30 = \textbf{8803.70 mm}^2$

Section is symmetrical about axis X-X and let the centroid lies at the point G at $\bar{x}$ from the face AB.

Centroid of triangle from face $AB = \dfrac{45}{3} = 15$ mm

Centroid of semicircular from the face $CD = \dfrac{4R}{3\pi} = \dfrac{4 \times 33 \times 7}{3 \times 22} = 14$ mm

Centroid of semicircular from the face $AB = 100 - 14 = \textbf{86 mm}$

Centroid of rectangle $= \dfrac{b}{2} = \dfrac{100}{2} = 50$ mm

Centroid G at $\bar{x} = \dfrac{\Sigma A_1 x_1}{\Sigma A_1} = \dfrac{(12000 \times 50 - 1485 \times 15 - 1711.3 \times 86)}{12000 - 1485 - 1711.3}$

$\bar{x} = \dfrac{(600000 - 222715 - 147171.8)}{8803.7} = \dfrac{430553.2}{8803.7} = \textbf{48.906 mm}$

By symmetry $\bar{y} = \dfrac{120}{2} = 60$ mm above BC

Centroid G (48.906, 60) with reference to sides AB and BC.

Example 7.2: Determine centroid, area and moment of inertia about two principal centroidal axes X-X and Y-Y for a T-section 120 mm $\times$ 150 mm $\times$ 10 mm as shown in Fig. 7.15.

Solution:

Let the centroid of the total T-section be at G (intersection of Y-Y and X-X. T-section comprises of two parts $ABCD$ and $EFHJ$ having their individual centroids at G_1 and G_2 as shown in Fig. 7.15.

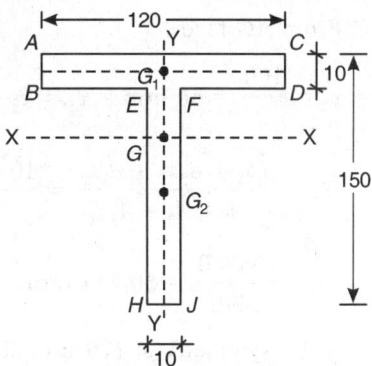

Fig. 7.15: T-section

Area A_1 $(ABCD) = 120 \times 10 = 1200$ mm^2,

Centroid G_1 at **5 mm** below AC

Area A_2 $(EFHJ) = (150 - 10) \times 10 = 1400$ mm^2,

Centroid G_2 from $AC = 10 + \dfrac{140}{2} = 80$ mm

Total area $A = A_1 + A_2 = 1200 + 1400 = \mathbf{2600\ mm^2}$

Centroid $\boldsymbol{G}$ ($\bar{y}$) from $\boldsymbol{AC} = \dfrac{A_1 y_1 + A_2 y_2}{A_1 + A_2} = \dfrac{1200 \times 5 + 1400 \times 80}{2600} = 45.385$ mm

$\bar{y} = \mathbf{45.385\ mm}$ below AC (**104.62** mm above HJ)

Moment of Inertia

I_1 (about G_1) $= \dfrac{1}{12} \times 120 \times 10^3 = \mathbf{1 \times 10^4\ mm^4}$,

I_2 (about G_2) $= \dfrac{10}{12} \times 140^3 = 10^4\ \mathbf{(228.67)\ mm^4}$

I_{XX} (About principal axis through centroid G) $= I_1 + A_1\ (y_1^2) + I_2 + A_2\ (y_2^2)$

$I_{XX} = 1 \times 10^4 + 1200\ (45.385 - 5)^2 + 228.67 \times 10^4 + 1400\ (80 - 45.385)^2$

$I_{XX} = 10^4\ [1 + 195.714 + 228.67 + 167.748] = \mathbf{593.13 \times 10^4\ mm^4}$

$I_{YY} = I_{Y_1 Y_1} + I_{Y_1 Y_1} = \dfrac{10 \times 120^3}{12} + \dfrac{140 \times 10^3}{12} = 10^4\ (144 + 1.167) = \mathbf{145.167 \times 10^4\ mm^4}$

(G, G_1 and G_2 all lie along Y-Y).

EXAMPLE 7.3: Find the centroid, area and moment of inertia about two principal axes passing through the centroid for channel section shown in Fig. 7.16.

Solution: Let the principal axes passing through the centroid G be X-X and Y-Y. $G(\bar{x}, \bar{y})$ with reference to face AC and CD.

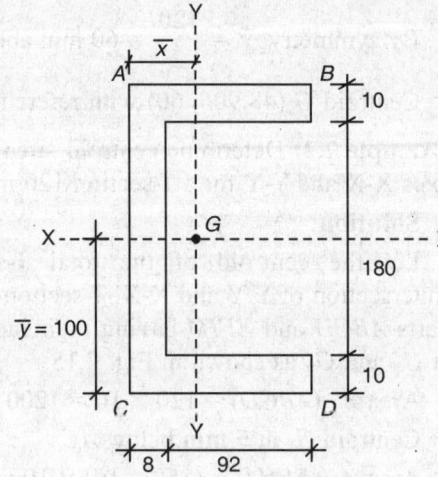

Area: $A_1 = 100 \times 10 = 1000 \text{ mm}^2 = A_3$,

$A_2 = (200 - 20) \, 8 = 1440 \text{ mm}^2$

Total area: $A = 1000 + 1440 + 1000 = \mathbf{3440 \text{ mm}^2}$

From AC face

$$x_1 = \frac{100}{2} = 50 \text{ mm} = x_3, \quad x_2 = \frac{8}{2} = 4 \text{ mm}$$

$$\bar{x} = \frac{A_1 x_1 + A_2 x_2 + A_3 x_3}{A_1 + A_2 + A_3} = \frac{10^3 \times 50 \times 2 + 1440 \times 4}{3440}$$

$$= \frac{105760}{3440} = \mathbf{30.744 \text{ mm}}$$

Fig. 7.16: Channel section

$\bar{y}$ by symmetry at **100 mm** above CD.

Moment of inertia about X-X by symmetry

$$I_X = \frac{1}{12} [B.D^3 - b.d^3] = \frac{1}{12} [100 \times 200^3 - 92 \times 180^3] = 2195.47 \times 10^4 \text{ mm}^4$$

Alternatively (by individual areas)

$$I_X = \left(I_{X_1} + A_1 y_1^2\right) + I_{X_2} + \left(I_{X_3} + A_3 y_3^2\right) = 2\left[\frac{100 \times 10^3}{12} + 1000 \, (95)^2\right] + \frac{8 \times 180^3}{12}$$

$$I_X = 10^4 \left[\frac{20}{12} + 20(9.5)^2\right] + 388.8 = (1806.67 + 388.8) = \mathbf{2195.47 \times 10^4 \text{ mm}^4}$$

$$I_Y = I_{Y_1} + A_1 \, (x_1 - \bar{x})^2 + I_{Y_2} + A_2 \, (\bar{x} - x_2)^2 + I_{Y_3} + A_3 \, (x_3 - \bar{x})^2$$

$$I_Y = \left\{\frac{10 \times 100^3}{12} + 1000 \, (50 - 30.744)^2\right\} 2 + \frac{180}{12} (8)^3 + 1440 \, (30.744 - 4)^2$$

$$I_Y = 10^4 \left[\frac{2000}{12} + 0.2 \, (19.256)^2\right] + 0.7680 \times 10^4 + 102.9948 \times 10^4$$

$$I_Y = 10^4 \, [166.667 + 74.1587 + 103.7628] = \mathbf{344.5882 \times 10^4 \text{ mm}^4}$$

Alternatively (from face AC)

$$I_Y = \left\{\frac{10 \times 100^3}{3} \times 2 + \frac{180 \times 8^3}{3}\right\} - 3440 \, (30.744)^2 = 10^4 \left[\frac{2000}{3} + 3.072 - 325.1466\right]$$

$$I_Y = \mathbf{344.59 \times 10^4 \text{ mm}^4}$$

EXAMPLE 7.4: An unsymmetrical I-section is shown in Fig. 7.17. Determine the area, location of centroid from bottom face and moment of inertia about two principal axes **X-X** and **Y-Y** passing from the centroid. Also calculate **section modulus** (max. and min.)

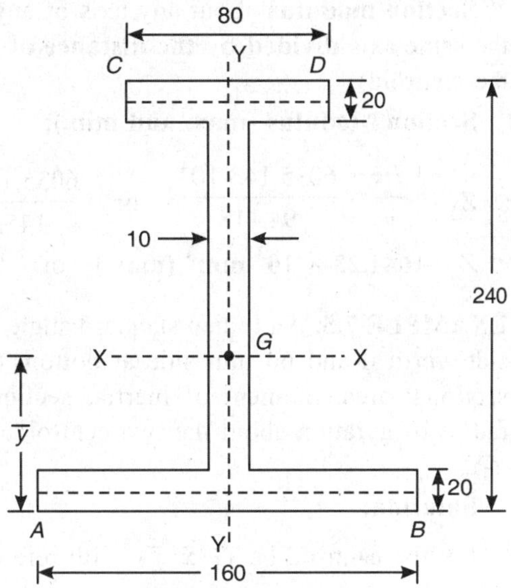

Solution:

Area: $A_1 = 160 \times 20 = 3200$ mm^2,

$y_1 = 10$ mm

$A_2 = (240 - 40)\ 10 = 2000$ mm^2,

$y_2 = 20 + 100 = 120$ mm

$A_3 = (80 \times 20) = 1600$ mm^2,

$y_3 = 200 + 20 + 10 = 230$ mm

Total Area:

$A = A_1 + A_2 + A_3$

$= 3200 + 2000 + 1600$

$= 6800$ mm^2

Fig. 7.17: I-section
(All dimensions in mm)

$$\bar{y}\ (\text{from } AB) = \frac{A_1 y_1 + A_2 y_2 + A_3 y_3}{A_1 + A_2 + A_3}$$

$$= \frac{3200 \times 10 + 2000 \times 120 + 1600 \times 230}{6800}$$

$$\bar{y} = \frac{640000}{6800} = \mathbf{94.118\ mm}\ (\text{from bottom})$$

$$\bar{y}\ (\text{from top}) = (240 - 94.118) = \mathbf{145.882\ mm}$$

$\bar{x}$ is in the middle due to symmetry = 80 mm from the edge A.

Moment of Inertia:

$$I_X = \left\{ \frac{160 \times 20^3}{12} + 3200\ (94.118 - 10)^2 \right\} + \frac{10 \times 200^3}{12} + 2000\ (120 - 94.118)^2 + \frac{80 \times 20^3}{12}$$

$$+ 1600\ (230 - 94.118)^2$$

$$I_X = 10^4\ [10.667 + 2264.268 + 666.667 + 133.976 + 5.333 + 2954.227]$$

$$= \mathbf{6035.14 \times 10^4\ mm^4}$$

$$I_Y = \frac{20 \times 80^3}{12} + \frac{200 \times 10^3}{12} + \frac{20 \times 160^3}{12} = 10^4\ [85.333 + 1.667 + 682.667]$$

$$= \mathbf{769.67 \times 10^4\ mm^4}$$

Section modulus about any axis of any bending element is its **moment of inertia** about the same axis **divided by the distance** of extreme fibre from the same axis (passing through the centroid).

Section Modulus (max. and min.):

$$Z_X = \frac{I_x}{y} = \frac{6035.14 \times 10^4}{94.118} \quad \text{or} \quad \frac{6035.14 \times 10^4}{145.882}$$

$Z_X = 641.23 \times 10^3 \text{ mm}^3 \text{ (max.)} \quad \text{or} \quad 413.7 \times 10^3 \text{ mm}^3 \text{ (min.)}$

EXAMPLE 7.5: An Indian standard angle of 90 mm × 60 mm × 10 mm is placed with 90 mm side vertical and 60 mm side at bottom (Fig. 7.18). Calculate location of centroid, cross-sectional area, moment of inertia, section modulus and radius of gyration about the two centroidal axes X-X and Y-Y.

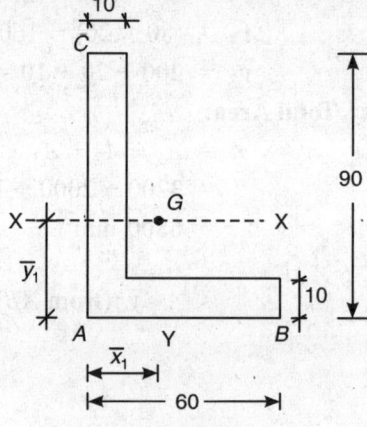

Fig. 7.18
(All dimensions in mm)

Solution:

Let the centroid be $G(\bar{x}, \bar{y})$ with reference to AC and AB.

$A_1 = (60 - 10) \, 10 = 500 \text{ mm}^2$,

$x_1 (AC) = \dfrac{50}{2} + 10 = 35 \text{ mm}$,

$y_1 (AB) = \dfrac{10}{2} = 5 \text{ mm}$

$A_2 = 90 \times 10 = 900 \text{ mm}^2$,

$x_2 (AC) = \dfrac{10}{2} = 5 \text{ mm}, \qquad y_2 (AB) = \dfrac{90}{2} = 45 \text{ mm}$,

$A = 500 + 900 = 1400 \text{ mm}^2$

$\bar{x} (AC) = \dfrac{A_1 x_1 + A_2 x_2}{A_1 + A_2} = \dfrac{500 \times 35 + 900 \times 5}{1400} = 15.71 \text{ mm}$,

$\bar{x}_2 \text{ (from } B) = 60 - 15.71 = 44.29 \text{ mm}$

$\bar{y} (AB) = \dfrac{500 \times 5 + 900 \times 45}{1400} = \mathbf{30.71 \text{ mm}}$

$\bar{y}_2 \text{ (from } C) = 90 - 30.71 = \mathbf{59.29 \text{ mm}}$

Moment of Inertia:

$$I_{AB} = \frac{1}{3}(50 \times 10^3) + \frac{10 \times 90^3}{3} = 10^4 \left(\frac{5}{3} + 243\right) = \mathbf{244.67 \times 10^4 \text{ mm}^4}$$

$I_{XX} + A\bar{y}_1^2 = I_{AB}$

$I_{XX} = 244.67 \times 10^4 - 1400 \, (30.71)^2 = (244.67 - 132.03) \, 10^4 = \mathbf{112.64 \times 10^4 \text{ mm}^4}$

$$I_{AC} = \frac{1}{3}[10 \times 60^3 + 80 \times 10^3] = 10^4 \left(\frac{224}{3}\right) = \textbf{74.67} \times \textbf{10}^4 \textbf{ mm}^4$$

$$I_{YY} = I_{AC} - A\bar{x}_1^2 = 74.67 \times 10^4 - 1400\,(15.71)^2 = 74.67 - 34.55)\,10^4 = \textbf{40.12} \times \textbf{10}^4 \textbf{ mm}^4$$

Section Modulus:

$$Z_X = \frac{I_x}{y} = \frac{112.64 \times 10^4}{30.71} = \textbf{36.68} \times \textbf{10}^3 \textbf{ mm}^3 \text{ (max.)},$$

$$Z_X \text{(min.)} = \frac{112.64 \times 10^4}{(90 - 30.71)} = \textbf{19.0} \times \textbf{10}^3 \textbf{ mm}^3$$

$$Z_Y = \frac{I_Y}{x} = \frac{40.12 \times 10^4}{15.71} = \textbf{25.54} \times \textbf{10}^3 \textbf{ mm}^3 \text{ (max.)},$$

$$Z_Y \text{(min.)} = \frac{40.12 \times 10^4}{(60 - 15.71)} = \textbf{9.06} \times \textbf{10}^3 \textbf{ mm}^3$$

Radius of Gyration:

$$AK_{XX}^2 = I_{XX}, \qquad\qquad\qquad A K_{YY}^2 = I_{YY}$$

$$K_x = \sqrt{\frac{I_x}{A}} = \sqrt{\frac{112.64 \times 10^4}{1400}} = \textbf{28.360 mm}, \quad K_y = \sqrt{\frac{I_y}{A}} = \sqrt{\frac{40.12 \times 10^4}{1400}} = \textbf{16.93 mm}$$

EXAMPLE 7.6: A hollow equal hexagonal section with outer side of 60 mm and uniform thickness of 4 mm is placed with 60 mm side horizontal as shown in Fig. 7.19. Determine moment of inertia, centroid, and section modulus about horizontal axis passing through the centroid '*G*'.

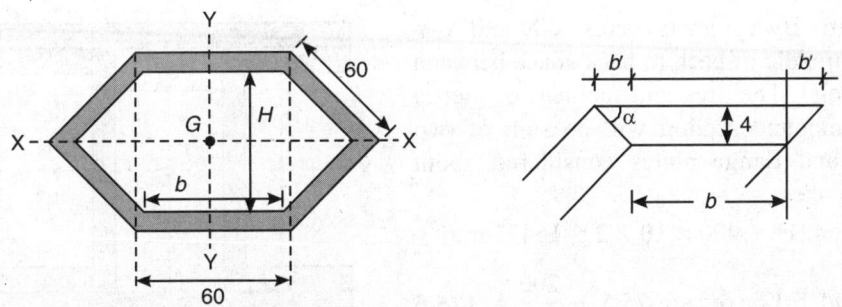

Fig. 7.19: Hexagonal hollow section

Solution:

Outer side = 60 mm, Thickness = 4 mm, In hexagon each angle = $\dfrac{720}{6} = 120°$

By symmetry $\theta = 60°$, $\alpha = 180 - 90 - 60 = 30°$, $b' = 4 \tan 30° = 4 \times \dfrac{1}{\sqrt{3}}$,

Inside width $b = B - 2b' = 60 - 2 \times 4 \times \dfrac{1}{\sqrt{3}} = \textbf{55.38 mm}$

$H = 60 \sin 60 = 30\sqrt{3} = \textbf{51.96 mm}\ (= y)$, $\quad h = H - t = 51.96 - 4.0 = \textbf{47.96 mm}$

Moment of Inertia of outer portion about X-X:

$$I_{X_0} = \frac{B}{12}(2H)^3 + \frac{B\cos 60° \times H^3}{12} \times 4 = \frac{2}{3}BH^3 + BH^3 \cdot \frac{4}{2\times 12} = \left(\frac{2}{3}+\frac{1}{6}\right)BH^3 = \frac{5}{6}BH^3$$

Similarly,

$$I_{X_1} = \frac{5}{6}bh^3$$

$$I_X = I_{X_0} - I_{X_1} = \frac{5}{6}(BH^3 - bh^3) = \{60 \times (51.96)^3 - 55.38 \times (47.96)3\}\frac{5}{6}$$

$$I_X = \frac{5}{6}[841.703 - 610.929] \times 10^4\ \text{mm}^4 = \frac{230.774 \times 5}{6} \times 10^4 = \textbf{192.31} \times \textbf{10}^4\ \textbf{mm}^4$$

Section Modulus:

$$Z_X = \frac{I_X}{y} = \frac{192.31 \times 10^4}{51.96} = \textbf{37009.24 mm}^3$$

EXAMPLE 7.7: A compound section is formed by joining two IS channels of 300 mm × 100 mm with the help of 400 mm × 10 mm plates (one at top and one at bottom) as shown in Fig. 7.20. Channel webs are separated 200 mm back to back. Calculate I_X and I_Y of the compound (built-up) section. Also calculate the section modulus about both the principal axes X-X and Y-Y. Each channel has: Area = 4211 mm^2, Centroid $\bar{x}_1 = 25.5$ mm, from back face, $I_X = 6048 \times 10^4$ mm^4, $I_Y = 346 \times 10^4$ mm^4.

Solution: By symmetry axes X-X and Y-Y lies at G (middle of back to back space between the channels). The area and moment of inertia of the compound section will be sum of two channels and flange plates considered about respective axes.

$$A = 2 \times 4211 + 400 \times 10 \times 2 = 16422\ \text{mm}^2$$

Y_1Y_1 and Y_2Y_2 axes are $25.5 + \dfrac{200}{2} = \textbf{125.5}$

mm from the Y-Y axis of the compound section passing from the centroid G.

Axis X-X of the compound section lies at the same point and axis of the individual channels.

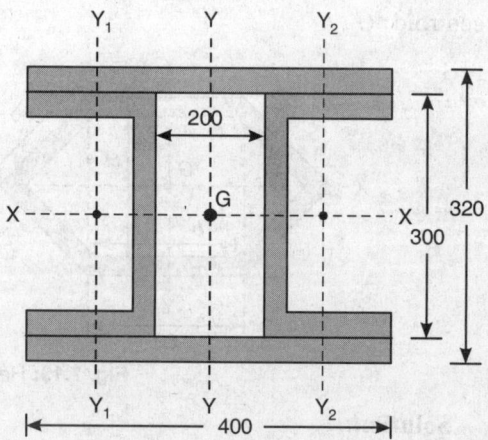

Fig. 7.20: Compound section

$$I_X = 2 \times 6048 \times 10^4 + \frac{1}{12} \times 400 \times (320^3 - 300^3) = 12096 \times 10^4 + \frac{10^5}{3}(32768 - 27000)$$

$$I_X = \{12096 + 19226.67\} \times 10^4 = 31322.67 \times 10^4\ \text{mm}^4$$

$$I_Y = 2[346 \times 10^4 + 4211 \times (125.5)^2] + \frac{1}{12}(320 \times 400^3 - 300 \times 400^3)$$

$$I_Y = 2 \times 10^4 (346 + 6632.43) + \frac{1}{12} \times 400^3 (20) = 13956.861 \times 10^4 + \frac{128}{12} \times 10^7$$

$$I_Y = 10^4 (13956.861 + 10666.67) = \mathbf{24623.53 \times 10^4 \ mm^4}$$

$$Z_X = \frac{I_X}{y_1} = \frac{32322.67 \times 10^4}{160} = \mathbf{1957.6667 \times 10^3 \ mm^3}$$

$$Z_Y = \frac{I_Y}{x_1} = \frac{24623.53 \times 10^4}{200} = \mathbf{1231.18 \times 10^3 \ mm^3}$$

EXAMPLE 7.8: Indian standard I-section of ISLB 300 mm × 150 mm has another IS joist of ISLB 200 mm × 100 mm attached with the top flange and bottom flange as shown in Fig. 7.21. Determine the values of M.I. and section modulus compound section about X-X and Y-Y axes.

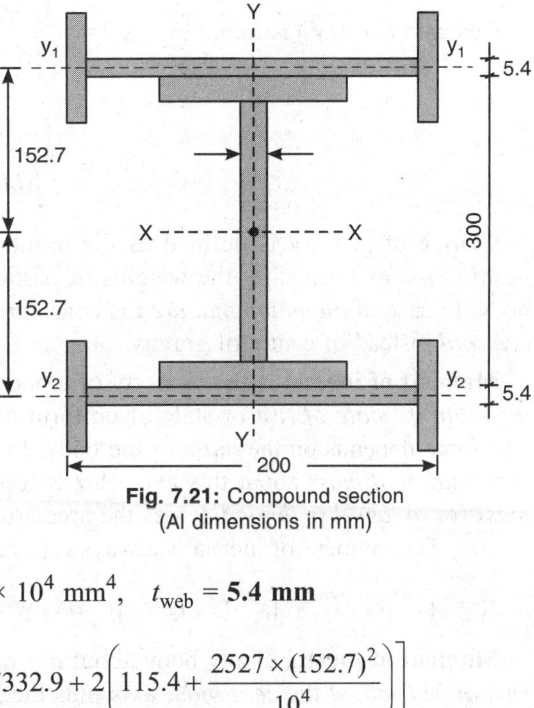

Solution:

For ISLB$_0$ 300 × 150:

$A_0 = 4808 \ mm^2$,

Individual $I_{X_0} = 7332.9 \times 10^4 \ mm^4$,

$I_{Y_0} = 376.2 \times 10^4 \ mm^4$, $t_{web} = 6.7$ mm

For each ISLB$_1$:

$A_1 = 2527 \ mm^2$,

$I_{X_1} = 1696.6 \times 10^4 \ mm^4$, $\quad I_{Y_1} = 115.4 \times 10^4 \ mm^4$, $\quad t_{web} = \mathbf{5.4 \ mm}$

Fig. 7.21: Compound section (Al dimensions in mm)

$$I_X \text{ (Compound)} = I_{X_0} + \left[2I_{Y_1} + A_1 y_1^2\right]10^4\left[7332.9 + 2\left(115.4 + \frac{2527 \times (152.7)^2}{10^4}\right)\right]$$

$$I_X \text{ (Compound)} = 10^4[7332.9 + 230.8 + 5892.3 \times 2] = \mathbf{19348.3 \times 10^4 \ mm^4}$$

$$I_Y \text{ (Compound)} = \left[I_{Y_0} + 2I_{X_1}\right] = 376.2 \times 10^4 + 2(1696.6) \times 10^4 = \mathbf{3769.4 \times 10^4 \ mm^4}$$

$$Z_X = \frac{I_X}{y_{max.}} = \frac{19348.3 \times 10^4}{\left(\frac{305.4}{2} + \frac{100}{2}\right)} = \frac{2 \times 19348.3 \times 10^4}{405.4} = \mathbf{954.53 \times 10^3 \ mm^3}$$

$$Z_Y = \frac{I_Y}{x} = \frac{3769.4 \times 10^4}{\dfrac{200}{2}} = \mathbf{376.94 \times 10^3 \ mm^3}$$

7.6 SUMMARY

Geometrical properties of cross-sectional areas of bounding elements play an important role in analysis of bending stress caused by bending moment. Geometrical properties comprise of **shape** of cross-sectional area, its **area**, **centroid**, and **moment of inertia** (also known as second moment of area).

Area of any figure $\qquad A = \int dA$, within limits.

Centroid $G(\bar{x}, \bar{y})$ is found by: $\quad \bar{x} = \dfrac{\int x \, dA}{\int dA}$, within limits.

$$\bar{y} = \frac{\int y \, dA}{\int dA}, \text{ within limits.}$$

Centre of gravity is defined as the point through which the *resultant* of the system of *parallel forces* formed by the weights of particles of the body *passes* for *all positions* of the body. In case of *plane lamina, area* is considered instead of weights and the point is known as *centroid* instead of centre of gravity.

Moment of inertia is the property of a body of certain *mass* by virtue of which it *tends to maintain* its *state of rest* or state of uniform motion. *A force is required* to change this state. The force depends on the *mass* of the body. In case of *rotating body mass* about any axis, the *distribution of mass* about this axis *also affects* the force. The geometrical property based on *mass* (or *areas*) *distribution* across the area around the *axis of rotation* is called its *moment of inertia*. The moment of inertia is always referred about certain axis of rotation, i.e.

$$I_X = \int y^2 \, dA, \ I_Y = \int x^2 \, dA, \ I_Z = \int r^2 \, dA, \text{ within limits of summation or integration.} \quad \dots 7.3$$

Moment of inertia of any body about *any axis* other than centroidal axis is obtained as the *sum of M.I.* about the *centroidal* axis plus the *product of the area* and *square of the distance* between the parallel axes. i.e. $I_{AB} = I_0 + Ah^2$, where h is distance between *parallel axes AB* and OO through centroid.

Moment of inertia about a *polar axis* Z-Z passing through the intersection of the two normal axes X-X and Y-Y will be *equal to the sum of the two normal axes*, i.e.

$$I_P = \int r^2 \, dA = \Sigma A r^2 = A \, \Sigma(x^2 + y^2) = A \, \Sigma x^2 + A \, \Sigma y^2 = I_X + I_Y \qquad \dots 7.4$$

Different sections have geometrical properties as given in the Table: 7.1.

Table 7.1: Properties of plane areas

S. No.	Shape of Plane Areas	Geometrical Properties
1	**Rectangle:** $(b \times d)$	Area $A = b \cdot d;\ \overline{x} = \dfrac{b}{2};\ \overline{y} = \dfrac{d}{2}$ $I_X = \dfrac{1}{12}bd^3;\ I_Y = \dfrac{1}{12}db^3;\ I_{XY} = 0$ $I_{BD} = \dfrac{1}{3}b \cdot d^3;\ I_{AB} = \dfrac{1}{3}d \cdot b^3;\ I_{X'Y'} = \dfrac{1}{4}b^2d^2$
2	**Triangle:** $(b \times h)$ $(b = b_1 + b_2)$	$A = \dfrac{1}{2}b \cdot h;\ \overline{x}\ (\text{from } C) = \left(\dfrac{b + b_1}{3}\right);\ \overline{y} = \dfrac{h}{3}$ $\overline{x}'\ (\text{from } B) = \left(\dfrac{b + b_2}{3}\right);\ I_X = \dfrac{1}{36}bh^3;$ $I_Y = \dfrac{bh}{36}(b^2 - bb_1 + b_1^2) = \dfrac{bh}{36}(b^2 - bb_2 + b_2^2)$ $I_{XY} = \dfrac{bh^2}{72}(b - 2b_1);\ I'_X\ (BC) = \dfrac{1}{12}bh^3$ $I_{X'-Y'} = \dfrac{bh^2}{24}(3b - 2b_1)$
3	**Trapezoid:** $\{(a, b) \times h\}$	$A = \left(\dfrac{a + b}{2}\right)h;\ \overline{y} = \dfrac{h}{3}\left(\dfrac{2a + b}{a + b}\right)$ from CD $I_X = \dfrac{h^3(a^2 + 4ab + b^2)}{36(a + b)};\ I'_X(CD) = \dfrac{h^3}{12}(3a + b)$
4	**Circle: Diameter** d	$A = \dfrac{\pi}{4}d^2;\ \overline{y} = \dfrac{d}{2};\ \overline{x} = 0$ $I_X = \dfrac{\pi}{64}d^4 = I_Y;\ I_Z = \dfrac{\pi}{32}d^4$ $I'_X = \dfrac{5\pi d^4}{64};\ I_{XY} = 0$

5	**Half Circle: Radius r**	$A = \dfrac{\pi}{8}d^2 = \dfrac{\pi}{2}r^2 \; ; \; \bar{y} = \dfrac{4r}{3\pi}$
		$I_X = \dfrac{(9\pi^2 - 64)r^4}{72\pi} = 0.10976\, r^4 \; ; \; I_Y = \dfrac{\pi r^4}{8}$
		$I'_X = \dfrac{\pi r^4}{8} \; ; \; I_{XY} = 0$
6	**Quarter Circle: Radius r**	$A = \dfrac{\pi}{4}r^2 \; ; \; \bar{x} = \bar{y} = \dfrac{4r}{3\pi}$
		$I_X = I_Y = \dfrac{(9\pi^2 - 64)r^4}{144\,\pi} = 0.05488\, r^4$
		$I'_X = I'_Y = \dfrac{\pi r^4}{16}$
7	**Parabolic Semi-Segment:**	$y = f(x) = h\left(1 - \dfrac{x^2}{b^2}\right) \; ; \; A = \dfrac{2}{3}bd$
		$\bar{x} = \dfrac{3b}{8} \; ; \; \bar{y} = \dfrac{2d}{5}$
		$I_X = \dfrac{16bd^3}{105} \; ; \; I_Y = \dfrac{2db^3}{15} \; ; \; I_{XY} = \dfrac{b^2 d^2}{12}$
8	**Semi-Segment: n^{th} degree**	$y = f(x) = h\left(1 - \dfrac{x^n}{b^n}\right), \; n > 0$
		$A = \left(\dfrac{n}{n+1}\right)bd; \; \bar{x} = \dfrac{b\,(n+1)}{2\,(n+2)} \; ; \; \bar{y} = \dfrac{n \cdot d}{(2n+1)}$
		$I_X = \dfrac{2bd^3 n^3}{(n+1)\,(2n+1)\,(3n+1)} \; ; \; I_Y = \dfrac{d \cdot b^3 \cdot n}{3\,(n+3)}$
9	**Parabolic Spandrel:**	$y = f(x) = d\left(\dfrac{x^2}{b^2}\right) \; ; \; A = \dfrac{bd}{3}$
		$\bar{x} = \dfrac{3b}{4} \; ; \; \bar{y} = \dfrac{3d}{10}$
		$I_X = \dfrac{bd^3}{21} \; ; \; I_Y = \dfrac{db^3}{5} \; ; \; I_{XY} = \dfrac{b^2 d^2}{12}$
10	**Spandrel: n^{th} degree:**	$y = f(x) = d\left(\dfrac{x^n}{b^n}\right), \; n > 0; \; A = \dfrac{bd}{(n+1)}$
		$\bar{x} = \dfrac{b\,(n+1)}{(n+2)} \; ; \; \bar{y} = \dfrac{d(n+1)}{2(2n+1)}$
		$I_X = \dfrac{bd^3}{3\,(3n+1)} \; ; \; I_Y = \dfrac{db^3}{(n+3)} \; ; \; I_{XY} = \dfrac{b^2 d^2}{4\,(n+1)}$

EXERCISE 7

Q.7.1. Determine area, centroid and second moment of area of a L-section shown in Fig. Q7.1.

Ans. $A = 1200 + 2000 = 3200 \text{ mm}^2$

$$\bar{y} = \frac{1200 \times 10 + 2000 \times 70}{3200} = 47.5 \text{ mm, from } AB$$

$$\bar{x} = \frac{1200 \times 30 + 2000 \times 70}{3200} = 17.5 \text{ mm, from } AC$$

$$I_{AB} = \frac{20 \times 120^3}{3} + \frac{40 \times 20^3}{3} = \mathbf{11626.67 \times 10^3 \ mm^4}$$

$$I_{XX} = I_{AB} - A\bar{y}^2$$

$$I_{XX} = 11626.67 \times 10^3 - 3200 \, (47.5)^2 = 440.667 \times 10^4 \text{ mm}^4$$

$$I_{YY} = \frac{20 \times 60^3}{12} + 1200(30 - 17.5)^2 + \frac{100 \times 20^3}{12} + 2000(17.5 - 10)^2$$

$$I_{YY} = 72.667 \times 10^4 \text{ mm}^4$$

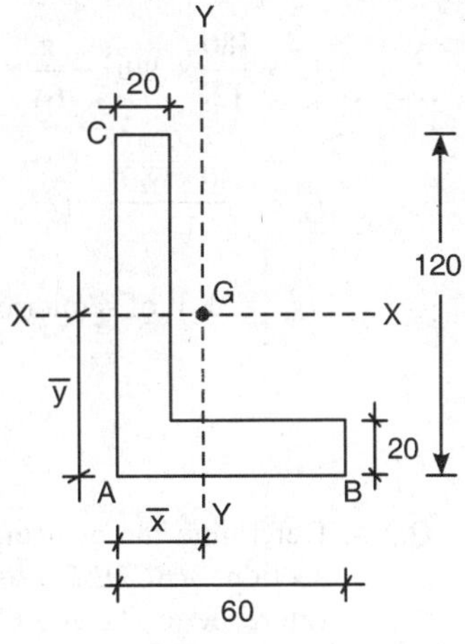

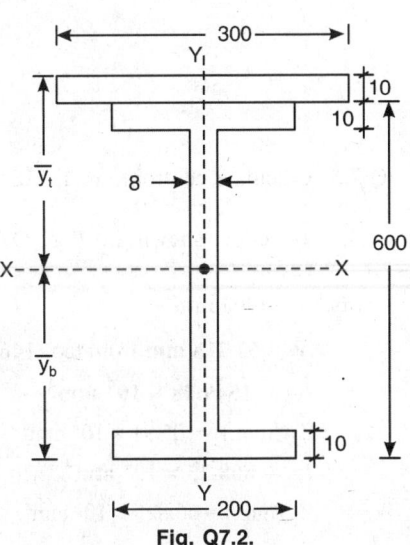

Fig. Q7.1.
(All dimensions in mm)

Q.7.2. Determine area, centroid and moment of inertia about two principal axes for a gantry girder section shown in Fig. Q7.2.

Ans. $A = 300 \times 10 + 200 \times 10 \times 2 + 580 \times 8 = \mathbf{11640 \ mm^2}$

$$\bar{y}_t = \frac{3000 \times 5 + 2000 \times 15 + 580 \times 8 \times 310 + 2000 \times 605}{11640}$$

$$\bar{y}_t = \frac{2693.4}{11.64} = \mathbf{231.39 \ mm}, \quad (\bar{y}_b = \mathbf{378.61 \ mm})$$

$$I_X = \frac{200 \times 10^3}{12} + 2000 \, (373.61)2 + \frac{8 \times 580^3}{12}$$

$$+ 4640 \times (78.61)^2 + \frac{200 \times 10^3}{12} + 2000 \, (595 - 378.61)^2$$

$$+ \frac{300 \times 10^3}{12} + 3000 \, (231.39 - 5.0)^2$$

$$I_X = \mathbf{68538.145 \times 10^4 \ mm^4}$$

$$I_Y = \frac{10 \times 300^3}{12} + \frac{10 \times 200^3}{12} \times 2 + \frac{580 \times 8^3}{12}$$

$$= \mathbf{3585.71 \times 10^4 \ mm^4}$$

Fig. Q7.2.
(All dimensions in mm)

Q.7.3. A beam section with circular cuts is shown in Fig. Q7.3. Determine net area and moment of inertia about two principal axes.

Ans. $A = 180 \times 240 - \pi(80)^2 = 23086 \text{ mm}^2$

$$I_X = \frac{180}{12} \times 240^3 - \frac{\pi}{64} \times 160^4 = \mathbf{17518.0 \times 10^4 \text{ mm}^4}$$

$$I_Y = \frac{240 \times 180^3}{12} - [2I_{Y_0}] = \mathbf{4456.02 \times 10^4 \text{ mm}^4}$$

I_{Y_0} = M. I. of semicircle about Y-Y axis

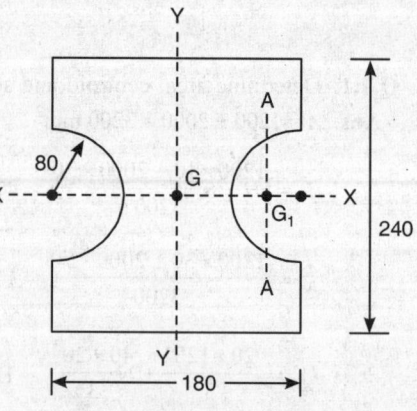

Fig. Q7.3.

Q.7.4. Determine the centroid and area of a concrete dam-section *ABCDEFG* shown in Fig. Q 7.4 with reference to *BC* and *CD*.

Ans. $A = \mathbf{43.5 \text{ m}^2}$

$\bar{y} = \mathbf{4.6207 \text{ m}}$ from the bottom horizontal face *CD*

$\bar{x} = \mathbf{2.1724 \text{ m}}$ from the vertical face *BC*.

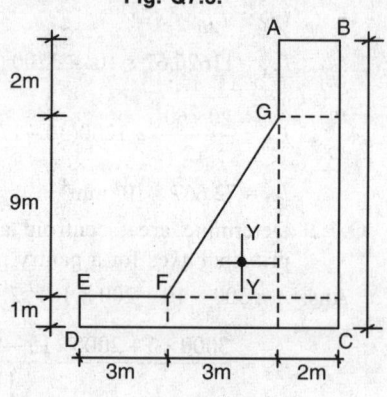

Fig. Q7.4.

Q.7.5. Calculate centroid, area, M.I. and section modulus of a
T-section shown in Fig. Q7.5 about two principal axes.

Ans. $A = \mathbf{3900 \text{ mm}^2}$

$\bar{y} = \mathbf{53.718 \text{ mm}}$ from top (146.282 mm from bottom).

$I_{XX} = \mathbf{1549.28 \times 10^4 \text{ mm}^4}$,

$Z_X \text{(min.)} = 105.91 \times 10^3 \text{ mm}^3$

$I_{YY} = \mathbf{668.25 \times 10^4 \text{ mm}^4}$

$Z_Y \text{(min.)} = 66.825 \times 10^3 \text{ mm}^3$.

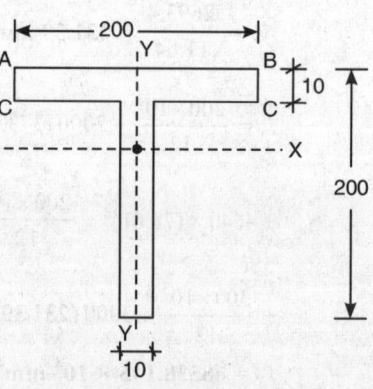

Fig. Q7.5.

Q.7.6. Determine area, centroid, M.I. and section modulus w.r.t. principal axes X-X and Y-Y for a section shown in Fig. Q7.6.

Ans. $A = 11500$ mm²

$\bar{y}$ (from bottom flange)

= **123.696 mm**, (176.304 mm from top)
$I_{XX} = 13982.63 \times 10^4$ mm⁴,
Z_X (max.) = **1130.4 × 10³ mm³**
Z_X (min.) = **793.1 × 10³ mm**
$I_{YY} = 1027.083 \times 10^4$ mm⁴
Z_Y (min.) = **136.94 × 10³ mm³**

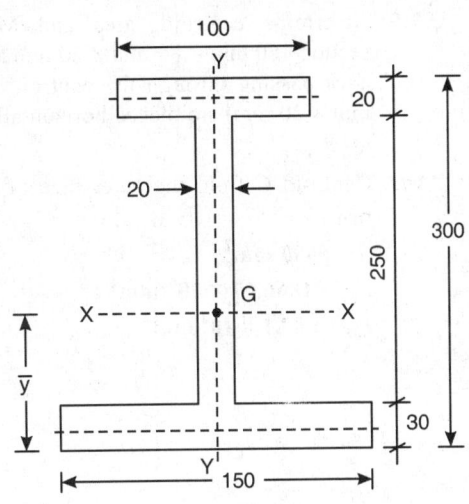

Fig. Q7.6.

Q.7.7. Determine centroid, area and moment of inertia about X-X and Y-Y axes passing from the centroid of the section shown in Fig. Q7.7.

Ans. $\bar{x}$ = **24.2262 mm**,

$\bar{y}$ = **39.2262 mm**

$A = 2100$ mm²,
$I_{XX} = 330.75 \times 10^4$ mm⁴,
$I_{YY} = 166.37 \times 10^4$ mm⁴.

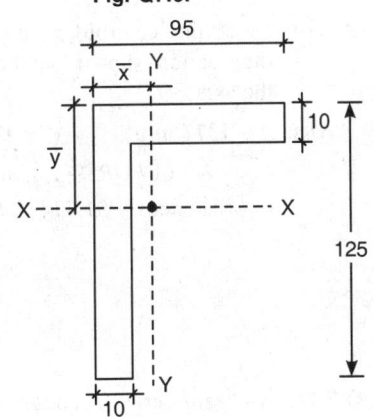

Fig. Q7.7.

Q.7.8. Determine centroid, area, and M.I. about X-X axis passing from the centroid for the section shown in Fig. Q7.8.

Ans. $A = 8400$ mm²

$\bar{y}$ = (bottom) = **81.67 mm**
$I_{XX} = 4017.71 \times 10^4$ mm⁴.

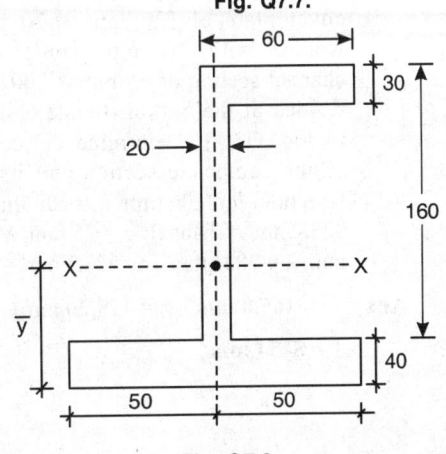

Fig. Q7.8.

Q.7.9. Determine centroid, area and M.I. of a channel section 160 mm × 80 mm × 20 mm about X-X and Y-Y axes passing through the centroid if the flanges (80 mm × 20 mm) are placed horizontally as shown in Fig. Q7.9.

Ans. Centroid G from the outer face of web, $\overline{x}$ = 27.143 mm

$A = 5600\ \text{mm}^2$

$I_{XX} = 1866.67 \times 10^4\ \text{mm}^4$,

$I_{YY} = 302.1 \times 10^4\ \text{mm}^4$.

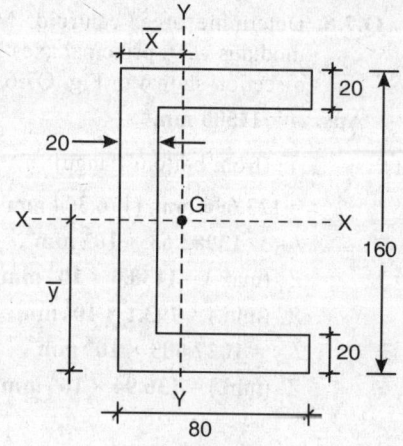

Fig. Q7.9.

Q.7.10. Determine centroid, area and moment of inertia of the section shown hatched in Fig. Q7.10 about the axis PQ.

Ans. $A = 1372\ \text{mm}^2$, $\overline{x}$ = 17.444 mm

$I_{PQ} = I_{PQ}$ of $PQRS - I_{PQ}$ of semicircle

I_{PQ} (hatched) = $51.714 \times 10^4\ \text{mm}^4$.

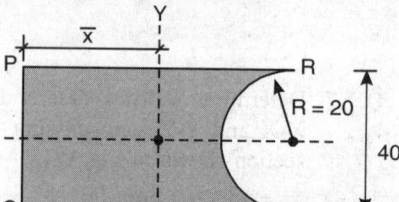

Fig. Q7.10.
(All dimensions in mm)

Q.7.11. A beam cross-section is fabricated with a I-section 100 mm × 200 mm. A cover plate of 120 mm × 20 mm is welded with the top flange and a channel section of 75 mm × 200 mm is welded at the bottom flange as shown in Fig. Q7.11. Determine the centroid of the composite section and its area. I-section: $b = 100$ mm, $h = 200$ mm, $A = 3233\ \text{mm}^2$. Channel: $b = 75$ mm, $h = 200$ mm, $A = 2821\ \text{mm}^2$, $c_y = 21.7$ mm.

Ans. $\overline{y}$ = 165.8 mm, (y_t = 129.20 mm)

$A = 8454\ \text{mm}^2$

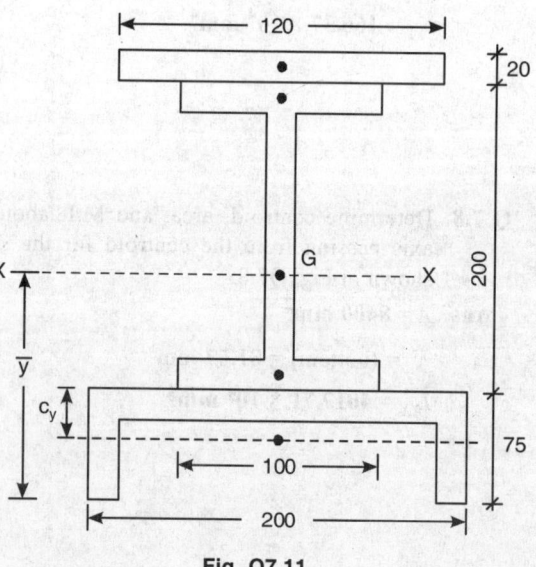

Fig. Q7.11.

Q.7.12. A compound section comprises of two rolled steel joists 250×125 @ 27.9 kg/m placed side by side with joist webs separated by 175 mm and 300 mm $\times$ 20 mm cover plates welded to each top and bottom flange. Calculate the area and moment of inertia about two principal axes X-X and Y-Y. For each $250 \times 125 \times 27.9$ kg joist area = 3553 mm^2, I_{XX} = 3718 $\times$ 10^4 mm^4, I_{YY} = 193.4 $\times$ 10^4 mm^4.

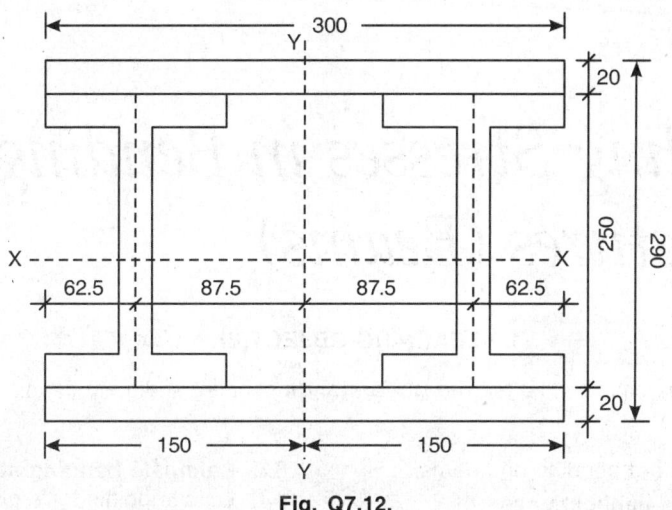

Fig. Q7.12.

Hints: Total area = 2 (3553 + 300 $\times$ 20) = **19106 mm^2**

M.I. is to be calculated by using parallel axes theorem and individual M.I., Area and distances between parallel axes.

$I_X = 2 \times 3718 \times 10^4 + 2 \times 10955 \times 104 = 29346 \times 10^4$ mm^4

I_{YY} (2 joists) = 2 [193.4 $\times$ 10^4 + 3553 (87.5)^2] = 5827 $\times$ 10^4 mm^4

$$I_Y \text{ (Total)} = \frac{40}{12} \times 300^3 + 5827 \times 10^4 \text{ mm}^4 = \textbf{14827} \times \textbf{10}^4 \textbf{ mm}^4$$

8

Bending Stresses in Bending Structures (Beams)

LEARNING OBJECTIVES

After studying this chapter, the learner **understands** simple bending in beams and will be able to:

8.1 **State effect** of bending on beams.

8.2 **Explain** the **implications** of the **assumptions** made in theory of simple bending.

8.3 **Explain** the **neutral axis** in relation to theory of simple bending.

8.4 **Explain** the **bending** and **curvature effect** in fibres of flexural members.

8.5 **Evaluate** and **explain longitudinal stress distribution** caused by bending moments across the section of a beam (flexural member).

8.6 **Evaluate** and **explain** the **moment of resistance** of sections of flexural members.

8.7 **Explain** the **meaning** and **units** of terms in bending equation.

8.8 **Calculate bending stresses** at various points for a given section subjected to a given bending moment.

8.9 **Calculate moment of resistance** of a given section with given permissible bending stresses in the material of the member.

8.10 **Calculate maximum permissible load** on a beam of given section, span and allowable bending stresses in the material of the beam.

8.11 **Choose** from the given simple sections, the **safe** and **economical** section required to carry a given load on a beam of given span and bending stresses in the material.

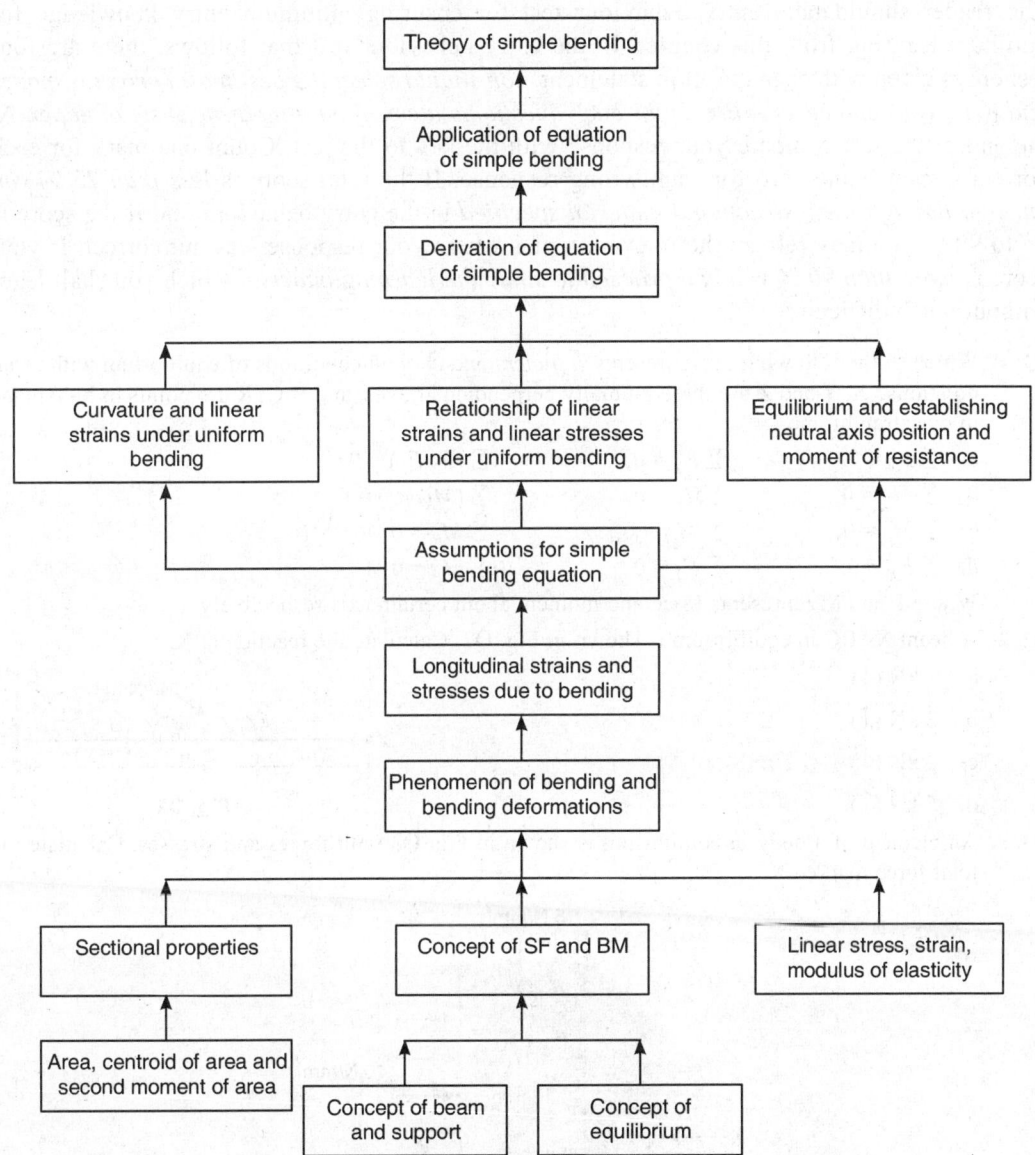

Fig. 8.0: Learning structure of theory of simple bending in beams

8.0 ENTRY BEHAVIOUR TEST

The reader should take entry behaviour test for ensuring minimum entry knowledge for optimum learning from this chapter. In the entry behaviour test that follows, there are four responses given with each question/statement. *You should select the best/most correct response and write your choice as A/B/C/D for each question/statement on a separate sheet of paper.* At the end of the test compare your response with the key to the test. Count one mark for each correct response and zero for each wrong response. If the total score is *less than 75 % you must go through the instructional material specified* in the entry behaviour and if the score is 75 to 90 % you may refresh the relevant topics where your response was not correct. If your score is *more than 90 % you can proceed to study the learning material* which you shall learn without much difficulty.

Q. 1. Which of the following set represents *sufficient* and *enough* conditions of equilibrium with usual notations? X, Y and Z are three mutually perpendicular axes, and P, Q, R are points in X-Y plane in one straight line.

a. $\Sigma F_X = 0$, $\quad\quad$ $\Sigma F_Y = 0$, $\quad\quad$ $\Sigma (F_X.F_Y) = 0$

b. $\Sigma M_X = 0$, $\quad\quad$ $\Sigma M_Y = 0$, $\quad\quad$ $\Sigma (M_X.M_Y) = 0$

c. $\Sigma M_P = 0$, $\quad\quad$ $\Sigma M_Q = 0$, $\quad\quad$ $\Sigma M_R = 0$

d. $\Sigma F_X = 0$, $\quad\quad$ $\Sigma F_Y = 0$ $\quad\quad$ $\Sigma M_P = 0$

Where F and M represents force and moment about certain axis respectively.

Q. 2. A beam XABC in equilibrium is shown in Fig. Q2. Calculate the reaction at X.

a. 1 kN ($\downarrow$)

b. 2 kN ($\uparrow$)

c. 2 kN ($\downarrow$)

d. 5 kN ($\uparrow$)

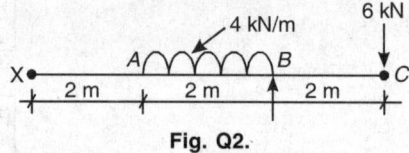

Fig. Q2.

Q. 3. An element of a body in equilibrium is shown in Fig. Q3 with forces and stresses. Calculate the total force marked X.

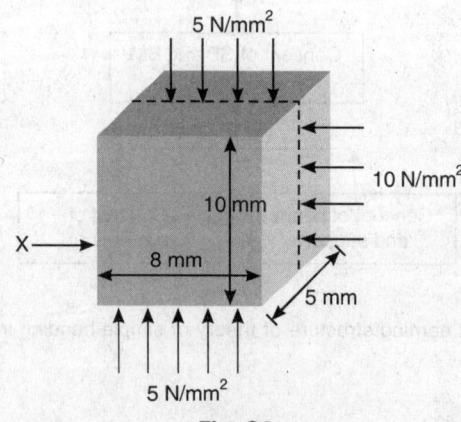

Fig. Q3.

a. 200 N $\quad\quad\quad\quad\quad\quad$ b. 400 N

c. 500 N $\quad\quad\quad\quad\quad\quad$ d. 800 N

Q. 4. Which of the following represents the definition of unit stress?

   a. External force acting per unit length.

   b. Change in force acting per unit length.

   c. Internal resistance developed per unit area.

   d. Internal resistance developed per unit volume.

Q. 5. Which of the following represents the definition of longitudinal strain?

   a. Internal resistance per unit length along the force.

   b. Change in dimension per unit dimension in the direction of force.

   c. Change in dimension per unit lateral dimension.

   d. Change in length per unit force along the length.

Q. 6. Which of the following statements defines Hooke's Law of elasticity?

   a. Within elastic limits the linear stress is proportional to the lateral stress.

   b. Within elastic limits the longitudinal stress is proportional to the corresponding strain.

   c. In case of elastic materials the stress strain are linearly proportional for all values of stress.

   d. In case of elastic materials the longitudinal strains are linearly proportional to corresponding lateral strains.

Q. 7. Which of the following statements defines elastic limit?

   a. The limiting ratio of linear stress to corresponding strain.

   b. The limiting linear strain upto which the material behaves as elastic.

   c. The maximum stress upto which the material regains its shape and size completely on unloading.

   d. The maximum stress achieved in an elastic material before it fails.

Q. 8. Which of the following give mathematical meaning of shear force at a section of beam?

   a. Algebraic sum of all forces on one side of the section of beam.

   b. Arithmetic sum of all transverse forces on either side of the section of beam.

   c. Resultant of all forces on left or right side of the section of beam.

   d. Resultant of all transverse forces on either side of the section of beam.

Q. 9. Calculate the shear force at 'X' for a beam shown in Fig. Q9.

   a. 16 kN ($\downarrow\uparrow$)

   b. 14 kN ($\downarrow\uparrow$)

   c. 11 kN ($\uparrow\downarrow$)

   d. 4 kN ($\uparrow\downarrow$)

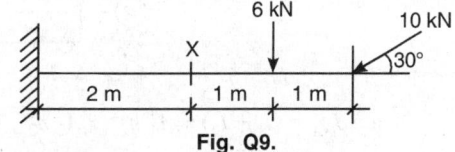

Fig. Q9.

Q. 10. Give mathematical definition of bending moment at section of a beam.

   a. Algebraic sum of moments about the section of transverse forces considered on one side of the section.

   b. Arithmetic sum of moments of transverse forces considered on both sides of the section about the section.

   c. Arithmetic sum of moments of transverse forces considered on either side of the section about the section.

   d. Algebraic sum of moments about the support of transverse forces considered on one side of the section.

Q. 11. Calculate bending moment at section X in a beam shown in Fig. Q11.

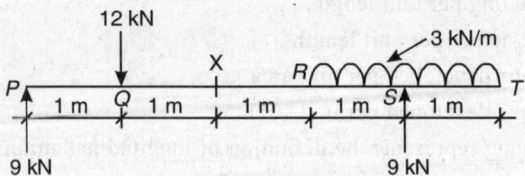

Fig. Q11.

a. 3 kNm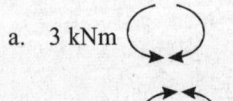

b. 6 kNm

c. 12 kNm

d. 36 kNm

Q. 12. Calculate numerical values of bending moment at Y in the beam shown in Fig. Q12.

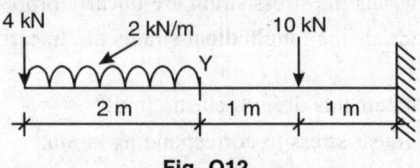

Fig. Q12.

a. 4 kNm

b. 10 kNm

c. 12 kNm

d. 38 kNm

Q. 13. Which of the following represents the correct application of parallel axes theorem for finding the second moment of area I_{XX} of the total area A_1 and A_2 about the common axis X-X if I_1 and I_2 are moment of inertia of area A_1 and A_2 respectively about their own centroids?

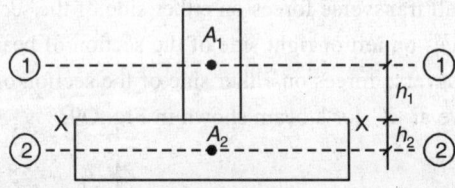

Fig. Q13.

a. $I_{XX} = (I_1 + I_2) + A_1 h_1^2 + A_2 h_2^2$,

b. $I_{XX} = (I_1 + I_2) + (A_1 + A_2)(h_1 + h_2)^2$

c. $I_{XX} = (I_1 + I_2) + (A_1 + A_2)(h_1^2 + h_2^2)$,

d. $I_{XX} = (I_1 + I_2) + \dfrac{(A_1 + A_2)}{2} \dfrac{(h_1 + h_2)^2}{2}$

Q. 14. Find section modulus (Z) of the hollow circular section having inner diameter d_1 and outer diameter d_2.

a. $Z = \dfrac{\pi}{64}(d_2^3 - d_1^3)$

b. $Z = \dfrac{\pi}{64 d_2}(d_2^4 - d_1^4)$

c. $Z = \dfrac{\pi}{32}(d_2^3 - d_1^3)$

d. $Z = \dfrac{\pi}{32 d_2}(d_2^4 - d_1^4)$

Q. 15. Calculate the moment of inertia about X-X axis for a steel joist shown in Fig. Q15.

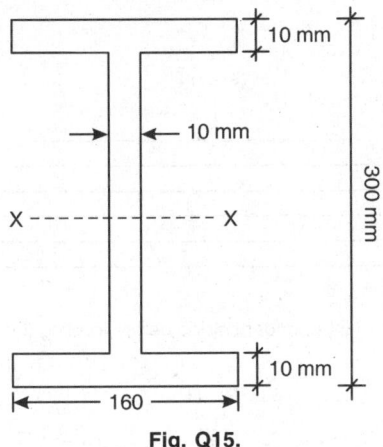

Fig. Q15.

a. $I_{XX} = \dfrac{1}{3} (16 \times 30^3 - 15 \times 28^3) \text{ cm}^4,$ b. $I_{XX} = \dfrac{1}{12} (16 \times 30^3 - 15 \times 28^3) \text{ cm}^4$

c. $I_{XX} = \dfrac{1}{12} (2 \times 16^3 + 28 \times 1^3) \text{ cm}^4,$ d. $I_{XX} = \dfrac{1}{12} (30 \times 16^3 - 28 \times 15^3) \text{ cm}^4$

KEY TO ENTRY BEHAVIOUR TEST

Q. No.	1	2	3	4	5	6	7	8	9	10	11	12	13	14	15
Correct Response	(d)	(a)	(c)	(c)	(b)	(b)	(c)	(d)	(c)	(a)	(b)	(c)	(a)	(d)	(b)

8.1   INTRODUCTION

This chapter on theory of simple bending in beams and bending stress forms the *foundation* of all advance courses in *theory of structures, machine design, structural design*, etc. Most of *engineering students do not understand* theory of simple bending and bending stresses. This chapter is specially, written with modular approach for *mastery learning* to clarify all concepts and principles. There are inbuilt *practice tasks* to facilitate learning and evaluate learning.

All *structural elements* are required to transfer different types of loads safely *without excessive deformations and internal stresses*. The structural elements subjected to any transverse loading undergo bending deformations. As a result of this bending deformation, the longitudinal fibres which are straight becomes curved to develop stable condition of the member (Figs 8.1a to c).

Consider a simply supported beam element consisting of number of independent parallel fibres and subjected to a *transverse load 'P'* (Fig. 8.1a). Under the action of this transverse load the *fibres*, if independent, shall continue to deform as shown in Fig. 8.1b. But in reality, these fibres are *not completely independent* and hence a *resistance* across the surfaces of these *fibres in longitudinal direction is set up* as shown in Fig. 8.1c. It is development of these *internal resistances* which brings about the *stability of the beam* elements subjected to *external transverse forces.*

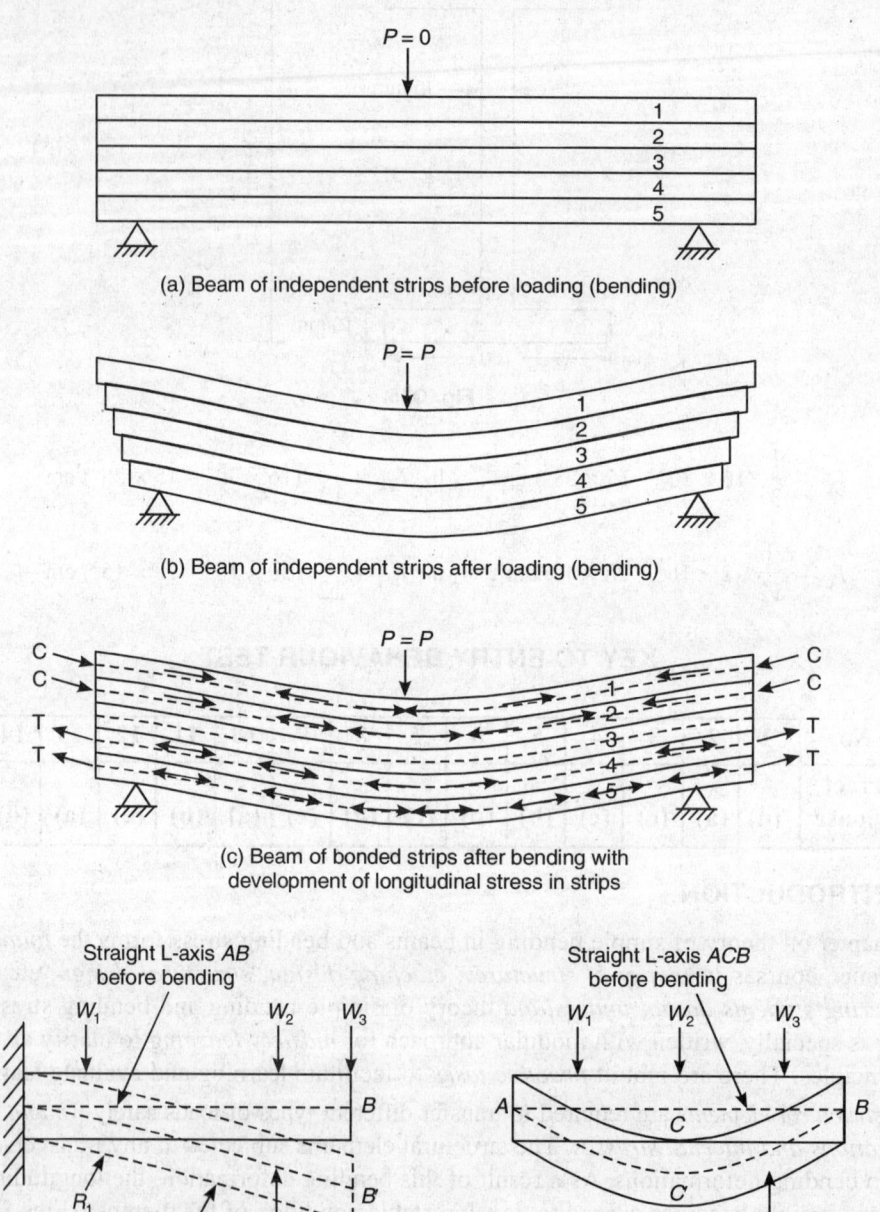

(a) Beam of independent strips before loading (bending)

(b) Beam of independent strips after loading (bending)

(c) Beam of bonded strips after bending with development of longitudinal stress in strips

(d) Cantilever beam

(e) Simply supported beam

Fig. 8.1.

Different beam elements under transverse loading undergo bending deformation till an *equilibrium* is established between the *external loads* and the corresponding *internal resistances* developed by *bending deformation* (Figs 8.1d and e). When the *bending is caused by a couple*, it is called pure bending, while bending caused by transverse loads not forming a couple is called *simple bending*. The external bending moment caused by transverse loads is *balanced* by development of internal moment of resistance through internal stresses caused as a result of bending deformation (curvature). These stresses developed in the longitudinal fibres due to bending shall be evaluated in subsequent paras. Since the external bending moment varies from section to section along the longitudinal axis, the curvature also varies along the longitudinal axis.

After undergoing bending deformation, the beam attains a stable condition and hence is in static equilibrium. Since beam as a whole is in stable equilibrium condition, all the elements comprising the beam shall be considered in equilibrium under external and internal forces. Thus by this simple concept of equilibrium of element, under external and internal forces, shall be used in evaluating the internal stresses. The diagram of small element in equilibrium is known indicating all internal and external forces (free body diagram) and using the condition of equilibrium the internal stresses are evaluated as indicated in subsequent para.

To simplify the evaluation of internal stresses caused by bending, certain assumptions are necessary. These assumptions, although not realistic, leads to evaluation of stresses of sufficient accuracy for most of practical purposes. These assumptions are dealt in next para.

8.2   ASSUMPTIONS IN THEORY OF SIMPLE BENDING

Various assumptions in theory of simple bending are assumed to simplify the evaluation of bending stresses. The assumptions along with its implications are listed below:

1. The longitudinal *axis* of the beam is assumed to be *initially straight* so that the total curvature caused is considered due to bending moment alone. This simplifies the calculations of *strains* caused *due to bending* of beam.

2. The *length* of the flexural member (beam) is assumed to be *sufficiently longer as compared to the cross-sectional dimensions*. This assumption implies that the bending is mainly occurring in the longitudinal axis of the flexural member and hence *no effect* need to be considered in evaluation of longitudinal strains due to *lateral bending* or other deformations.

3. The *cross-section* of the flexural member is *uniform* throughout the length. This assumption eliminates the consideration of the effect of uneven bending curvature along a small length under the same bending moment.

4. The *cross-section is symmetrical* about the *longitudinal axis* along which the loads are also symmetrically placed. This assumption *avoids* the consideration of any *torsional moment* together with bending moment along the longitudinal axis.

5. The *material* of the beam is assumed to be *homogeneous and isotropic* (similar properties in all directions). This assumption implies that the behaviour of material in *compression* as well as in *tension is the same* throughout. Hence the *modulus of elasticity* both in *tension* and *compression* is the same throughout the beam.

6. The beam is assumed to consist of *parallel fibres* which are capable of bending independently from each other. This *assumption although highly unrealistic, neglects* the effect of the *horizontal shear stresses* within the various parallel fibres while computing the bending stress. This introduces a very small error in the analysis of bending stresses and is well accepted for all practical purposes (Fig. 8.2).

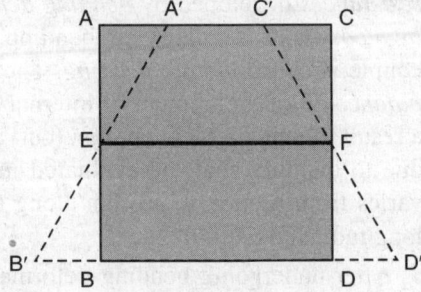

Fig. 8.2: Longitudinal element (AC = EF = BD = dx)

7. The beam is subjected to such a loading that the *stresses* developed remain *within elastic limit*. This assumption indicates that the *Hooke's Law* is applicable to stress and strain relationship throughout the beam.

8. The *cross-sections which are plane before bending remain plane after bending*. This important assumption implies that the *strain in any parallel fibre due to bending is linearly proportional to its distance from the neutral fibre* (EF). The *neutral fibre* is one which *does not undergo any change in length* (i.e. deformation) due to bending (Fig. 8.2).

9. The beam is assumed to be in *static equilibrium* and therefore, any element of the beam will also be in *equilibrium* for the analysis of stresses.

SYMBOLS

Unless otherwise mentioned the following symbols shall be used in the book:

A	Area of cross-section (mm^2).
B	Width of cross-section (mm).
L	Length of member (i.e. longitudinal axis) (mm).
X–X	*Longitudinal axis* through centroid of section.
Y–Y	*Vertical axis* of symmetry through centroid of the section (*minor* principal axis of the section).
Z–Z	*Horizontal axis* through centroid of the section (*major* principal axis about which bending occurs).
Z	Section modulus (mm^3).
x, y, z	Distances along respective directions (or axes) (mm).
$\delta x, \delta y, \delta z$	*Small increment* distances of elements along respective directions.
$\bar{x}, \bar{y}, \bar{z}$	Distances of centroid along respective axes from the origin of axes (mm).
I	*Second moment of area* (also known as *moment of inertia* of section) (mm^4).
I_X, I_Y, I_Z	*Second moment of areas* (i.e. moment of inertia of section *about respective axes*) (mm^4).
M_r	*Moment of resistance* and/or bending moment at a section (N-mm, kN-m).

F	*Shear force* at a section (kN, N).
A, B, C,	Points on beam.
D, E, F, ...	
E	*Modulus of elasticity* of material (N/mm^2, kN/mm^2).
R	Radius of curvature of bent neutral fibre of a beam at a section (mm, m).
y, y_1, y_2, y_t, y_c	Distance from the N.A. of bending (mm).
f, f_1, f_2, f_t, f_c	Fibre bending stresses (N/mm^2, kN/mm^2).
$d\theta$	Small elemental angle in radians.
da	Small elemental area.
Σ	Summation of the quantity.
$\int$	Integration, i.e. summation of a quantity.

8.3   THEORY OF SIMPLE BENDING

Beams subjected to transverse loading or a couple undergo bending. Transverse forces cause bending moments and shear forces along the longitudinal axis of the beam. The bending moment may be constant if a couple is acting and variable if general transverse loading is acting along the longitudinal axis. The beam as a whole and any element of the beam undergoes deformation (bending) when subjected to transverse loads as shown in Figs 8.3 and 8.4.

Consider a straight longitudinal *element MNPQ* of length dx (Fig. 8.5b) and subjected to pure bending moment M. The cross-section is also shown in Fig. 8.5a. The element after bending is shown in Fig. 8.5e. The vertical transverse and parallel sections at MN and PQ which are *plane before bending remain plane even after bending* but rotate through an angle θ on the application of the moment. M as shown in Fig. 8.5e. The beam as a whole and hence the element also becomes curved in the plane of bending. The curvature produced in the beam is small and also the bending moment be assumed constant and hence the element of the beam bends to form *circular arc*. The centre of circle, the circumference of which contains this element is called the *centre of curvature* at this point. The radius of this circular arc at the *neutral fibre* (*EF*) is called the *radius of curvature* '*R*'. The *inner fibres* such as *AB, CD, NP*, etc. *shorten* while the *outer fibres* such as *GH, NQ*, etc. *elongates* due to bending as shown in Fig. 8.5e. In between these fibres there will be some fibre which neither increases nor decreases in length. Such a fibre which does not undergo any change in length is called *neutral fibre* and the surface of the beam containing such fibres is known as *neutral surface*. The intersection of this neutral surface with the cross-section of the beam is termed as *neutral axis*. All the fibres on the inner side of this *neutral axis* are in *compression* while all those fibres on *outer side* of this axis are in *tension*.

Consider any *fibre CD*, located at a distance of *y from the neutral axis* and which becomes *C'D'* after bending. The *change in length* of CD = *(C'D' − CD)*.

Original length of CD = dx = *length of EF* = *E'F'* after bending (being unchanged as neutral) = $R.\theta$, if radius upto *E'F'* is R and *rotation* of transverse plane is θ.

Changed length of CD = C'D' = *(R − y)* θ

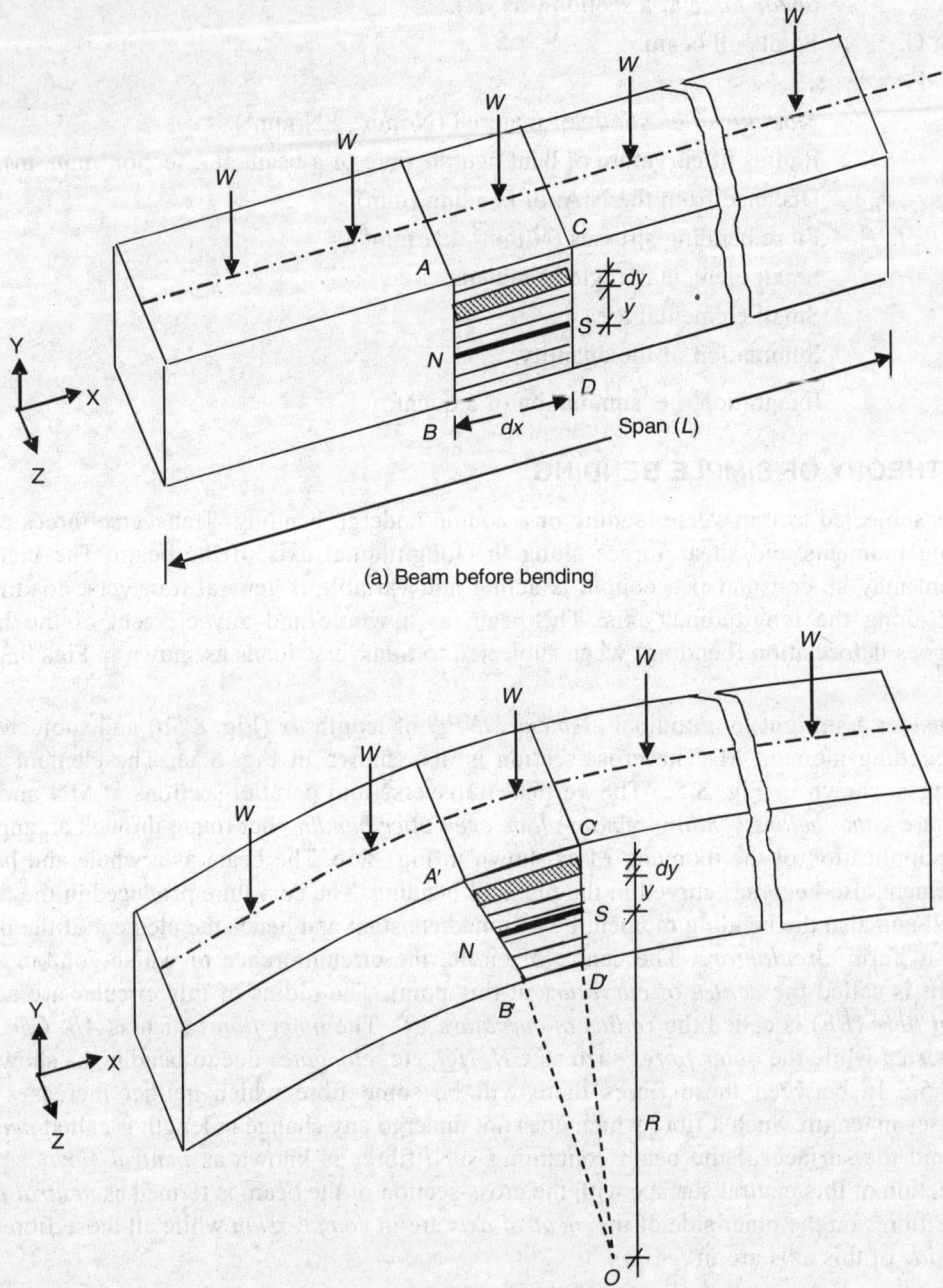

(a) Beam before bending

(a) Beam after bending

Fig. 8.3: Cantilever beam (longitudinal view)

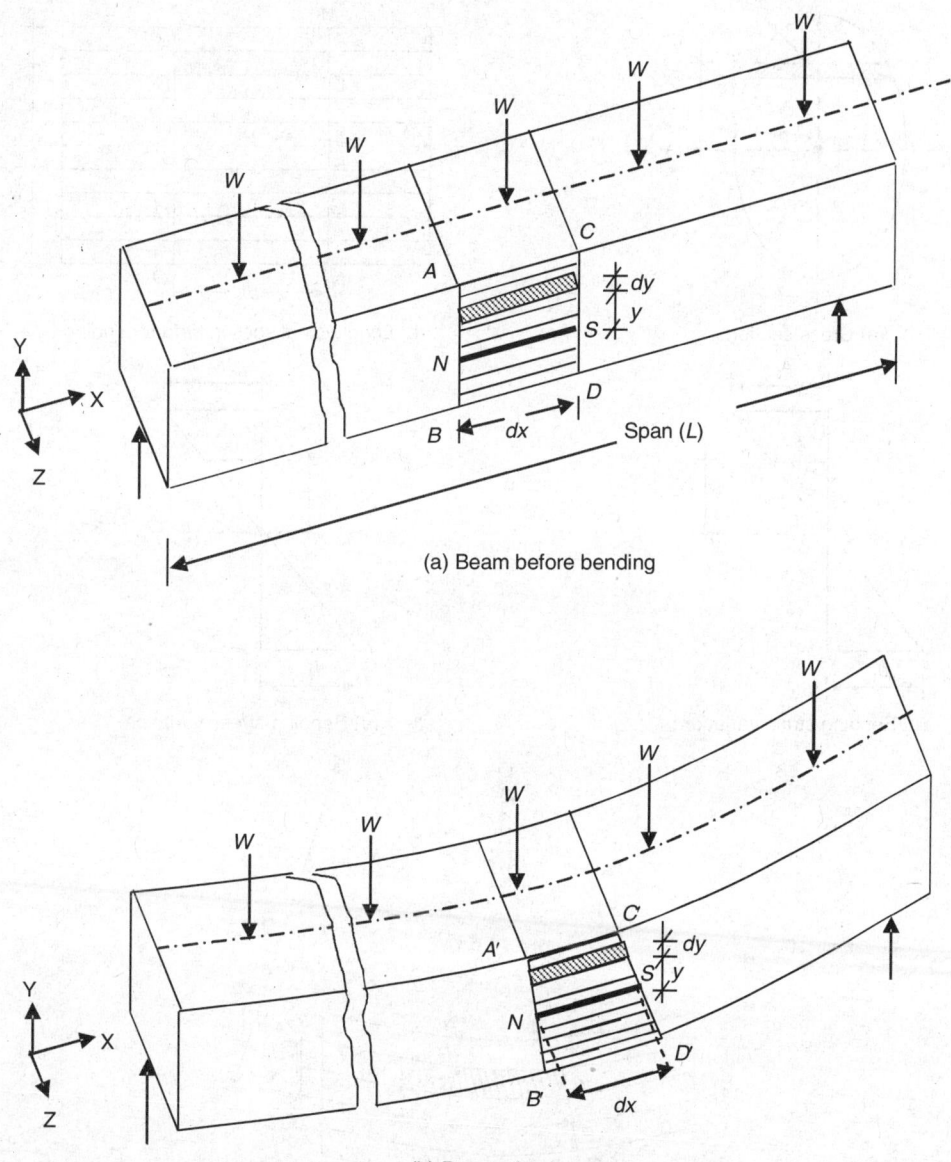

(a) Beam before bending

(b) Beam after bending

`Fig. 8.4:` Simply supported beam (longitudinal view)

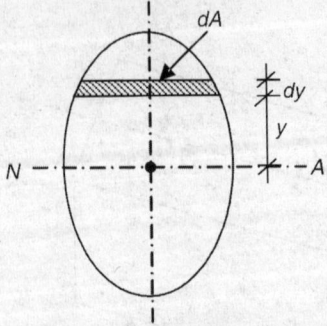

(a) Cross-section

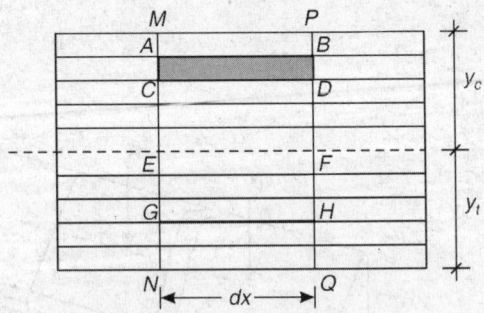

(b) Longitudinal section before bending

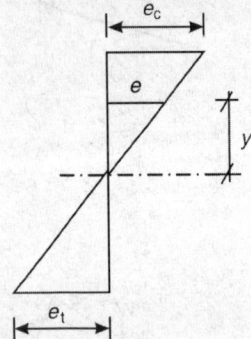

(c) Bending strain variation

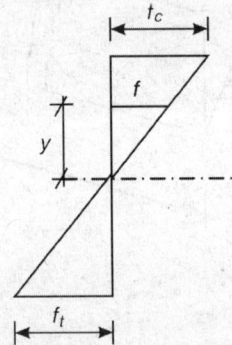

(d) Bending stress variation

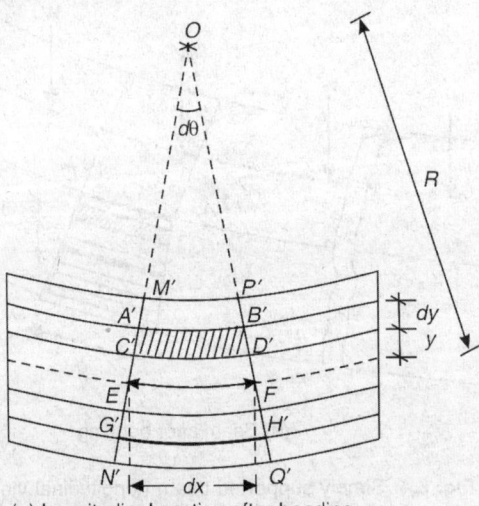

(e) Longitudinal section after bending

Fig. 8.5: Bending and stress variation

$\therefore$ *Change in length of CD = C′D′ − CD = (R − y) θ − R . θ*

$= − y . \theta$ (− sign indicates *reduction*)

Strain 'e' $= \dfrac{\text{Change in length of } CD}{\text{Original length of } CD} = \dfrac{y \cdot \theta}{R \cdot \theta} = \dfrac{y}{R}$ ... (i)

If the material is within elastic limit and obeys *Hooke's Law, longitudinal strain 'e'* due to *bending,*

$$e = \frac{\text{Longitudinal Stress 'f' due to Bending}}{\text{Modulus of Elasticity 'E'}} = \frac{f}{E} \qquad \qquad \text{... (ii)}$$

From (i) and (ii)
$$e = \frac{f}{E} = \frac{y}{R}$$

or
$$\boxed{\frac{f}{y} = \frac{E}{R}} \qquad \qquad \text{... (8.1)}$$

8.4 POSITION OF NEUTRAL AXIS

Since the beam is in equilibrium its every section will also be in equilibrium under the effect of forces. Consider free body diagram of cross-section of a beam together with longitudinal stress distribution across the section. Since there is *no resultant external force* in the longitudinal direction on the section, the *resultant of internal longitudinal forces* must also be *zero*. In other words the total tensile force developed on the section must be equal to the total compressive force developed.

By definition of neutral axis, it is the line in the cross-section along which no longitudinal strain (hence no longitudinal stress) develops. Thus the position of this axis will be such that the total internal tensile force developed on one side of this axis and the total internal compressive force developed on the other side of this axis are equal. In other words the total internal longitudinal force (normal to cross-section) developed on the cross section will be zero if there is no external longitudinal force acting on the beam section.

Consider small elementary area δa in the cross-section (Fig. 8.5a) located at a distance y from the NA. The longitudinal stress in a fibre at a distance of y from the NA

$$= f = \frac{E}{R} \cdot y \text{ (refer equation 8.1).}$$

Therefore the internal force developed in the elementary area $\delta a = f \cdot \delta a = \frac{E}{R} \cdot y \cdot \delta a.$

The total internal longitudinal force on the whole cross-section = summation of all such parallel forces on elementary strip $= \Sigma \dfrac{E}{R} \cdot \delta a \cdot y.$

Since there is no *resultant* longitudinal force on the cross-section, this summation

$\Sigma \dfrac{E}{R} \cdot \delta a \cdot y = 0$. For a given material and given beam section the term $\dfrac{E}{R}$ is constant and hence:

$\Sigma y \delta a = 0,$ or $A \bar{y} = 0,$ since $A \ne 0,$ and hence $\bar{y} = 0$ ... (8.2)

$\bar{y}$ **lies at the centroid of the section**.

This represents the sum of moments of all the small areas about the neutral axis and is zero. This is only possible about the centroid of the area. *Thus the neutral axis passes through the centroid of the cross-section.*

8.5 EVALUATION AND EXPLANATION OF MOMENT OF RESISTANCE OF FLEXURAL SECTION

We have seen that the internal longitudinal tensile force and internal longitudinal compressive force developed on the section above and below the neutral axis forms an internal couple. Since the beam is in *equilibrium* after bending, the *moment of internal couple* developed must be *equal and opposite to the moment caused* by the *external couple* (i.e. bending moment). To calculate the moment of this *internal couple*, consider the same cross-section and elementary *strip* as shown in Fig. 8.5a.

The internal force developed in the strip = $f\,dA$.

Moment of this force about the **NA** = $f\,.\,dA\,.\,y$.

∴   Total *internal moment* developed $M = \Sigma f\,.\,y\,.\,dA$,

But bending stress f at any distance y from NA = $\dfrac{E}{R}\,.\,y$

Substituting in *moment equation*, we have

$$M = \Sigma \left(\frac{E}{R}\cdot y\right)\cdot y \cdot dA = \frac{E}{R}\,\Sigma\, y^2.\,dA \qquad \left(\text{since } \frac{E}{R} = \text{constant at a particular section}\right)$$

or $M = \dfrac{E}{R}\,\Sigma\, y^2.\,dA$ ... (8.3)

Quantity $\Sigma\, y^2.\,dA$ is *second moment of area* of the *cross-section about the neutral axis* of bending and is generally denoted by the letter I. In the present section considered the NA about which bending takes place is Z-Z axis and hence the value of $I = I_{ZZ}$, thus, the above equation reduces to

$$M = \frac{E}{R}\,.\,I \qquad \text{or} \qquad \frac{M}{E\,I} = \frac{1}{R} \qquad \text{or} \qquad \frac{M}{I} = \frac{E}{R} \qquad \text{... (8.4)}$$

In equation (8.1), we have already seen that $\dfrac{E}{R} = \dfrac{f}{y}$, hence the general equation of bending

becomes $\dfrac{M}{I} = \dfrac{f}{y} = \dfrac{E}{R}$ ... (8.5)

Equation (8.4) gives relationship of moment (M), curvature $\left(\dfrac{1}{R}\right)$ and *flexural rigidity* (*stiffness factor EI*). Since *EI* is constant for a *uniform section*, it is evident that the curvature $\left(\dfrac{1}{R}\right)$ is *directly proportional* to the bending moment (M) and *inversely proportional* to the *flexural rigidity* (EI) of the *beam section*.

From the general bending equation (8.5), it is obvious that for a given beam section of moment of *inertia I* and subjected to a certain bending moment *M*, the *bending stress 'f'* developed in *any fibre* across the section is **directly proportional to its distance 'y' from the neutral fibres**. From this equation it can also be seen that in a given fibre at a given distance

'y' in a certain section, the bending stress 'f' is *directly proportional to the applied bending moment 'M'.*

It must be understood here that it is the bending moment which causes *bending* (i.e. *curvature*) and results in *longitudinal strains* in various fibres. These longitudinal **strains are responsible for setting up of longitudinal stresses** which in turn give rise to internal moment of resistance. Thus, for beams in equilibrium, the *external bending moment at a section* and the *moment of resistance developed at that section* are *numerically the same*. If the beam section and the *maximum permissible stress* in the *material are known*, the **maximum permissible moment of resistance** of the section can be found by the equation

$$M = \frac{I}{y_{max}} \cdot f_{max}. \qquad \qquad ... (8.6)$$

In equation (8.6), the term $\dfrac{I}{y_{max.}}$ depends on the *shape* and *size* of the cross-section and is termed as *section modulus* (*Z*).

8.6 UNITS OF QUANTITIES IN BENDING EQUATION

Various quantities in bending equation have basic units involving **force** and **linear dimensions**. While substituting values of various quantities in the bending equation **consistency in these units of force** and **linear dimensions** must be maintained. In this book generally the *S.I. units* have been used along with *M.K.S. system of units*.

In *S.I. units* the basic unit of force has been taken as '*Newton*' (*N*). *Newton is defined as a force which when applied on a **mass of 1 kg produces an acceleration of 1 m/sec^2.** Thus weight of 1 kg mass* (due to *gravitational force*) will be *equal to 9.80 Newton* (considering *acceleration* due to gravity 'g' = 9.80 ms^{-2}). In *M.K.S. system of units* the *weight of 1kg mass shall be taken as 1 kg.f* (equivalent to *9.80 N, approximately equal to 10 N*).

Following table indicate certain set of *consistent units* for various quantities in *bending equation.*

Quantities	Force	N	kN	kgf	kgf
	Linear Dimensions	mm	m	cm	m
M (Bending moment or moment of resistance)		N-mm	kN-m	kgf-cm	kgf-m
I (Moment of inertia or second moment of area)		mm^4	m^4	cm^4	m^4
f (Bending Stress)		N/mm^2	kN/m^2	kgf/cm^2	kgf/m^2
y (Distance from NA)		mm	m	cm	m
E (Modulus of Elasticity)		N/mm^2	kN/m^2	kgf/cm^2	kgf/m^2
R (Radius of Curvature)		mm	m	cm	m

Other combination of force and linear units can also be derived keeping in mind the principle of similarity of *dimensions* (*dimensional homogeneity*).

PRACTICE TASK I

Select the correct response/answer from the four choices provided to each of the following questions and write choice number on the response sheet.

1. A long vertical member subjected to a vertical load applied longitudinally at the centre of the cross-section. What will be the nature of major stress in the member across the section?

   a. Uniform axial stress.   b. Uniform shear stress.

   c. Bending stress.   d. Transverse stress.

2. A long vertical steel member is fixed rigidly in ground. A object flying horizontally from North to South strikes the member at its top free end. What type of stresses will be set up in the steel member fibres in its north face?

   a. Tangential stress.   b. Longitudinal compressive stress.

   c. Longitudinal tensile stress.   d. No stress.

3. Which assumption leads to linear variation of bending strain with respect to the distance from the neutral axis?

   a. Longitudinal axis is assumed to be initially straight.

   b. Cross-section of the beam is assumed to be uniform throughout.

   c. Cross-sections which are plane before bending remain plane after bending.

   d. The material of the beam is assumed to be homogeneous and isotropic.

4. For simple bending theory which assumption is responsible for eliminating the effect of torsion T.

   a. The longitudinal axis of the beam is initially straight.

   b. The length of beam is much larger than the cross-sectional dimensions.

   c. The cross-section is symmetrical about the longitudinal axis along which the loads are symmetrically placed.

   d. The cross-sections which are plane before bending remains plane after bending.

5. Which one is **not** an assumption in simple theory of bending?

   a. Loading is such that the material is within elastic limit.

   b. The cross-section of the flexural member is such that it is symmetrical above and below the neutral surface.

   c. Flexural member consists of parallel fibres capable of bending independently.

   d. All elements of the flexural members are in static equilibrium.

6. Where does the elastic neutral axis lie in an uneven cross-section of a flexural member?

   a. Along a line dividing the cross-section into equal areas.

   b. Along a line passing through the centroid of the cross-section.

   c. Along a line passing through zero shear.

   d. Along a line at right angles to the plane of the section and passing through the centroid.

7. On what factor the magnitude of the bending stress in any fibre under a given moment at a given section of elastic homogeneous material depends?

   a. Distance of the fibre from the neutral axis.

   b. Width of the fibre.

   c. Modulus of elasticity.

   d. Depth of the section.

8. In the bending equation the term moment of inertia (second moment of area) refers to which axis?
   a. Centroidal axis having the least moment of inertia.
   b. Centroidal axis having the maximum moment of inertia.
   c. Centroidal axis lying in the loading plane.
   d. Centroidal axis perpendicular to the loading plane.
9. A simply supported beam of uniform circular section of homogeneous material is subjected to UDL What is the type of bending deformation along the longitudinal axis of the beam?
   a. Uniform circular radius of curvature throughout the length.
   b. Variable circular radius of curvature in relation to shear force.
   c. Maximum **radius of curvature** at the supports and minimum radius of curvature at the mid span.
   d. Minimum radius of curvature at supports and maximum radius of curvature at the mid span.
10. Why assumptions are necessary in derivation of simple theory of bending?
   a. To make derivation simple without introducing error of any practical importance.
   b. To make derivation simple without introducing any error.
   c. To introduce error in equation for rough use.
   d. To make the equation applicable for simply supported beams.

KEY TO PRACTICE TASK I

Q. No.	1	2	3	4	5	6	7	8	9	10
Correct Response	(a)	(c)	(c)	(c)	(b)	(b)	(a)	(d)	(c)	(a)

8.7   APPLICATION OF BENDING EQUATION IN SOLUTION OF PROBLEMS OF BENDING STRESS ANALYSIS AND SIMPLE DESIGN

Calculate Bending Stresses Caused by Bending Moment

EXAMPLE 8.1: A simply supported beam consists of a uniform rectangular section 60 mm × 150 mm deep. At certain section along the length, the beam is subjected to a sagging bending moment of 90 kN-m. Calculate maximum longitudinal stress in the beam section and show bending stress variation across the section. If the modulus of elasticity for the material is 2 × 10^5 N/mm², find the radius of curvature of the bent beam.

Solution:

The beam cross-section is rectangular and its centroid is symmetrically placed at 75 mm from top and bottom fibres.

Second moment of area (about axis of bending)

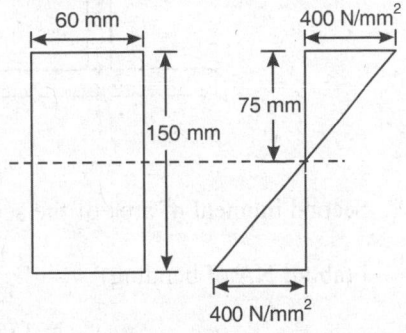

$$= \frac{1}{12} \times 60 \times 150^3$$

$$= 16875 \times 10^3 \text{ mm}^4$$

Fig. 8.6.

Maximum BM at section $= 90$ kN-m

$= 90 \times 10^6$ N-mm

Since beam section is in equilibrium the moment of resistance = Max. BM $= 90 \times 10^6$ N-mm

Using the bending equation: $\dfrac{M}{I} = \dfrac{f}{y} = \dfrac{E}{R}$

or $\dfrac{90 \times 10^6}{16875 \times 10^3} = \dfrac{f}{75} = \dfrac{2 \times 10^5}{R}$, or $f = \dfrac{90 \times 10^6 \times 75}{16875 \times 10^3} = \textbf{400 N/mm}^2$

Since the section is symmetrical, the stress varies from 400 N/mm^2 tensile at bottom to 400 N/mm^2 (compressive) at top.

$$R = \frac{2 \times 10^5 \times 16875 \times 10^3}{90 \times 10^6} = 20 \times 1875 = 37500 \text{ mm} = \textbf{37.5 m}$$

EXAMPLE 8.2: A cantilever beam is subjected to maximum hogging BM of 27.0 kN-m. If the beam section consists of a rolled steel joist, shown in sketch. Calculate maximum and minimum longitudinal stresses setup in the section. Also show the stress variation across the section.

Solution:

Moment of resistance = max. BM $= 27.0$ kN-m $= 27 \times 10^6$ N-mm.

The joist is symmetrical about the centroid, where the NA lies.

Thus, the NA lies at $= 110$ mm

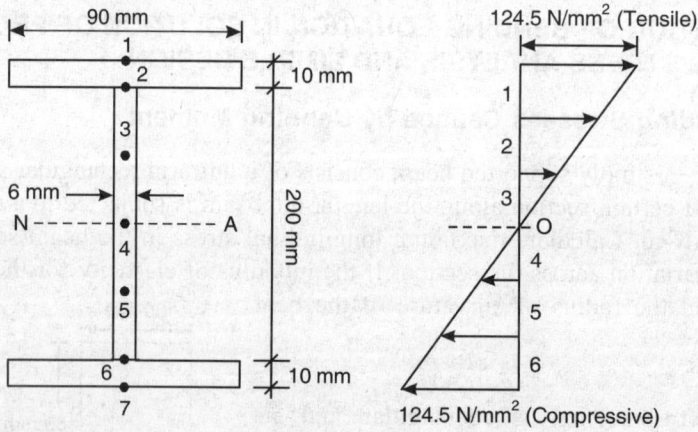

Fig. 8.7.

Second moment of area of the section about the NA of bending is given as under:

$$I \text{ (about NA of bending)} = [\frac{1}{12} \times 90 \times 10^3 + (90 \times 10)(110 - 5)^2] \times 2 + \frac{6 \times 200^3}{12}$$

$$= [7500 + 9922500] \times 2 + 4 \times 10^6 = 19860000 + 4 \times 10^6$$

$$= 2386 \times 10^4 \text{ mm}^4$$

Using the equation of bending: $\dfrac{M}{I} = \dfrac{f}{y}$

$$f_{max.} = \frac{M}{I} \cdot y_{max.} = \frac{27 \times 10^6}{2386 \times 10^4} \times 110 = 124.5 \text{ N/mm}^2$$

At top the maximum bending stress is 124.5 N/mm^2 (tensile), while maximum bending stress is 124.5 N/mm^2 (compressive) in bottom fibres.

$$f_2 \text{ (at any fibre 100 mm away from NA)} = \frac{27 \times 10^6}{2386 \times 10^4} \times 100 = 113.2 \text{ N/mm}^2$$

(tensile in top and compressive in bottom)

$$f_3 \text{ (at any fibre 50 mm away from NA)} = \frac{27 \times 10^6}{2386 \times 10^4} \times 50 = 56.6 \text{ N/mm}^2$$

EXAMPLE 8.3: An equal angle section 150 mm × 150 mm × 10 mm is subjected to a maximum BM of 20 kN-m (hogging) at certain section. If the angle is placed in such a manner that one of the 150 mm leg is kept horizontally in top face, calculate the maximum and minimum bending stresses across the section. Also find the stresses in the inner face of horizontal leg. Calculate the stresses on a separate sheet and compare the results.

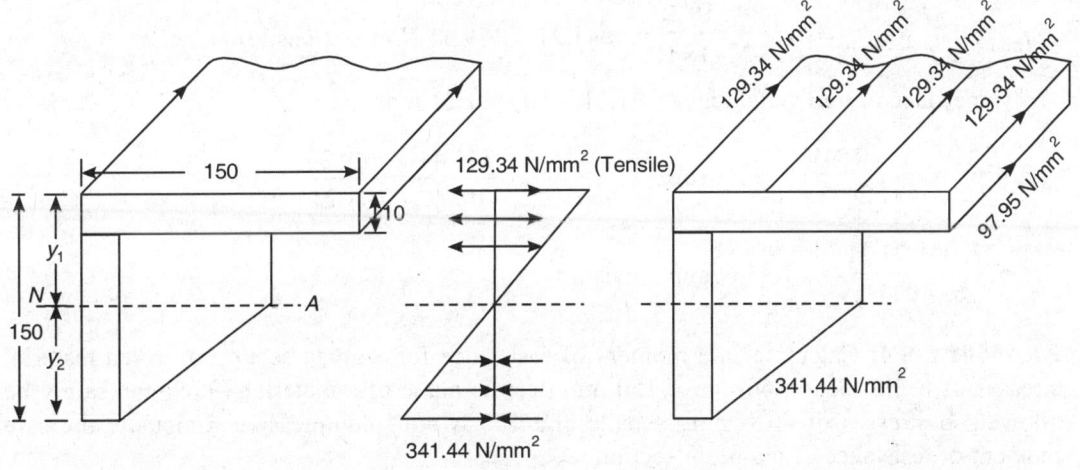

Fig. 8.8.

Solution:

Here the section is unsymmetrical and hence first establish the neutral axis of bending. **Recall that the NA passes through the centroid of the section.**

Thus if the NA is y_1, from top horizontal leg than:

$$y_1 = \frac{(150-10)\times10\times\dfrac{10}{2}+10\times150\times\dfrac{150}{2}}{(150-10)\times10+10\times150} = \frac{7000+112500}{1400+1500} = \frac{119500}{2900} = 41.21 \text{ mm}$$

$y_2 = 150 - 41.21 = \mathbf{108.79 \text{ mm}}$

Second moment of area of the cross-section about the N.A. of bending:

$$I = \frac{1}{12}\times140\times10^3 + 1400\times(41.21-5)^2 + \frac{10\times150^3}{12} + 1500\times(75-41.21)^2$$

$$= 11667 + 1835630 + 2812500 + 1712646 = 6372443 \text{ mm}^4 = \mathbf{637.2443\times10^4 \text{ mm}^4}$$

BM is hogging and hence the top fibre shall be subjected to tensile stress while the bottom fibres shall be subjected to compressive stress.

Maximum stress shall be in the fibres located at maximum distance from the neutral axis.

$y_{\text{max.}} = y_2 = 108.79 \text{ mm}$, $\qquad y_{\text{min.}} = y_1 = 41.21 \text{ mm}$

$M = $ BM at the section $= 20 \text{ kN-m} = 20\times10^6 \text{ N-mm}$

Using the simple equation of bending: $\dfrac{M}{I} = \dfrac{f}{y}$

$$f_{\text{max.}} = \frac{M}{I}\cdot y_{\text{max.}} = \frac{20\times10^6}{637.2443\times10^4}\times108.79 = \mathbf{341.44 \text{ N/mm}^2} \text{ (compressive)}$$

$$f_{\text{min.}} = \frac{M}{I}\cdot y_{\text{min.}} = \frac{20\times10^6}{637.2443\times10^4}\times41.21 = 129.34 \text{ N/mm}^2 \text{ (tensile)}$$

y (inner face of horizontal legs) $= 41.21 - 10 = 31.21 \text{ mm}$

$$\frac{f}{31.21} = \frac{20\times10^6}{637.2443\times10^4} \qquad \text{also} \qquad \frac{f}{31.21} = \frac{341.44}{108.9} = \frac{129.34}{41.21}$$

$$f = \frac{2000\times31.21}{637.2443} = 97.95 \text{ N/mm}^2 \text{ (tensile)}$$

EXAMPLE 8.4: Calculate safe moment of resistance for a given section of given material stresses. A beam section 60 mm × 150 mm deep is made of a material which can safely be allowed to stresses of 40 N/mm^2 tensile and 100 N/mm^2 compressive. Calculate the safe moment of resistance of the beam section.

Solution:

Recall the bending equation: $\dfrac{M}{I} = \dfrac{f}{y}$, which gives, $M = \dfrac{I}{y}\cdot f$

For the given rectangular section $b = 60 \text{ mm}$ and $d = 150 \text{ mm}$

$$I = \frac{1}{12}\times60\times150^3 = 16875\times10^3 \text{ mm}^4$$

$$M \text{ (tension)} = \frac{16875 \times 10^3}{75} \times 40 = 9 \times 10^6 \text{ N-mm} = \textbf{9 kN-m}$$

$$M \text{ (compression)} = \frac{16875 \times 10^3}{75} \times 100 = \textbf{225} \times \textbf{10}^5 \text{ N-mm} = \textbf{22.5 kN-m}$$

Thus safe moment of resistance = **9 kN-m**

EXAMPLE 8.5: A cantilever beam consists of rolled steel angle section 150mm × 150mm × 10mm with one 150 mm leg horizontal in top face. If the material can be stressed to a maximum compressive stress of 200 N/mm^2 and maximum tensile stress of 100 N/mm^2, calculate the safe moment of resistance of the section. For the given cantilever the tension zone is above NA and compression zone is below NA. Refer example 3 where, values of y_1 = 41.21 mm (tension zone), y_2 = 108.79 mm (compression zone), I = 637.2443 × 10^4 mm^4.

Solution:

Using bending equation: $M = \dfrac{I}{y} \cdot f$

For tension face:

$$M \text{ (tension)} = \frac{I}{y_t} \cdot f_t = \frac{637.2443 \times 10^4}{41.21} \times 100 = 154.63 \times 10^5 \text{ N-mm}$$

For compressive face:

$$M \text{ (compression)} = M = \frac{I}{y_c} \cdot f_c = \frac{637.2443 \times 10^4}{108.79} \times 200 = 117.15 \times 10^5 \text{ N-mm}$$

Thus the safe moment of resistance will be **117.5 × 10^5 N-mm** (i.e. 11.715 kN-m).

Evaluation of Safe Load on a Given Beam of Given Section and Material

EXAMPLE 8.6: A simply supported beam of 4 m span consists of steel joist shown in Fig. 8.9 carries uniformly distributed load. If the material can safely be stressed of a tensile stress of 200N/mm^2 and a compressive stress of 150 N/mm^2, find the maximum load on the beam which can be safely carried without exceeding any stress. Also find the actual stresses for this loading.

Solution:

The section is symmetrical and its sectional properties are:

$I = [90 \times 220^3 - 84 \times 200^3] = 2386 \times 10^4 \text{ mm}^4$

$$y_1 = y_2 = \frac{220}{2} = 110 \text{ mm}$$

Moment of resistance

$$M_1 \text{ (compression zone)} = \frac{2386 \times 10^4}{110} \times 150$$

$$= 32.536 \times 10^6 \text{ N-mm } (32.536 \text{ kN-m})$$

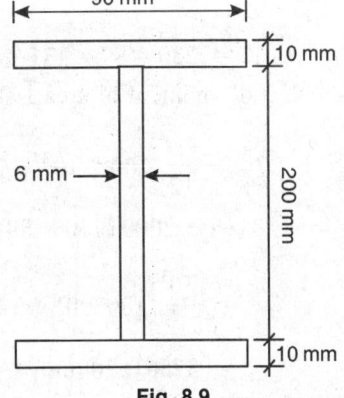

Fig. 8.9.

$$M_2 \text{ (tensile zone)} = \frac{2386 \times 10^4}{110} \times 200 = 43.3682 \times 10^6 \text{ N-mm (43.3682 kN-m)}$$

Thus safe moment of resistance = **32.536 kN-m.**

In case of simply supported beam the maximum BM occurs at mid span and is equal to

$\frac{w}{8} \cdot L^2$, where w = load in kN/m.

The maximum BM should not exceed the safe moment of resistance and hence the maximum BM = safe moment of resistance i.e.

$$\frac{w}{8} \cdot L^2 = 32.536 \text{ kN-m}, \quad \text{i.e.} \quad w = \frac{32.536 \times 8}{4 \times 4} = \mathbf{16.268 \text{ kN/m}}$$

The section is symmetrical and hence both the extreme stresses shall be equal i.e. actual stresses developed corresponding to the given moment of resistance shall be **150 N/mm²**.

EXAMPLE 8.7: A cantilever beam of 2 m span carries a concentrated load of W placed at its free end. The permissible bending stresses in the material are 220 N/mm² (tensile) and 100 N/mm² (compressive), calculate the safe value of W in kN, if the section of the cantilever beam is as shown in the sketch.

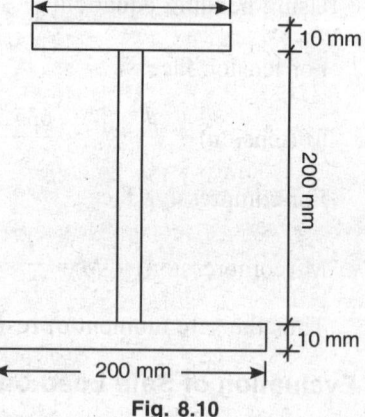

Fig. 8.10

Solution.

For the load W the maximum BM

$$= WL$$
$$= W \times 2 = 2W \text{ kN-m.}$$

From the sketch, the position of NA (centroid) from the bottom (compression) flange

$$y_c = \frac{200 \times 10 \times 5 + 200 \times 10 \times (100+10) + 100 \times 10 \times (210+5)}{2000 + 2000 + 1000} = \mathbf{89 \text{ mm}} \text{ (8.9 cm)}$$

$\therefore \ y_t = 220 - 89 = \mathbf{131 \text{ mm}}$ (13.1 cm)

Second moment of area I about the neutral axis of bending

$$I_z = \frac{1}{12} [200 \times 10^3 + 10 \times 200^3 + 100 \times 10^3] + 2000 (89-5)^2$$
$$+ 2000 (110-89)^2 + 1000 (215-89)^2$$
$$= \frac{10^3}{12} [80300] + 14112000 + 882000 + 1587600 = 6691670 + 30870000$$
$$= 37561670 \text{ mm}^4 = \mathbf{3756.167 \text{ cm}^4}$$

$$MR_t = \frac{I}{y_t} \cdot f_t = \frac{37561670 \times 220}{131} = 61401280 \text{ N-mm} = \mathbf{61.40128 \text{ kN-m}}$$

$$MR_c = \frac{I}{y_c} \cdot f_c = \frac{37561670 \times 100}{89} = 41080500 \text{ N-mm} = \mathbf{41.0805 \text{ kN-m}}$$

Thus safe MR is lesser of the two values and will be equal to 41.0805 kN-m.

Maximum BM $\le$ Safe MR

∴ $W \le \mathbf{20.5402 \text{ kN}}$ (20.5 kN say)

Evaluation of Safe and Economical Section for a given Flexural Member

EXAMPLE 8.8: A simple beam supported over an effective span of 6 m carries a concentrated load of 40 kN at its mid span. The maximum permissible bending stress (both in tension and compression) in the material is 150 N/mm². Neglecting the effect of self weight of beam, suggest with reasons the most suitable section of the beam from the following available sections.

a. Solid circular section 160 mm diameter having cross-section of $A = 200.96$ cm², and section modulus $Z = 401.92$ cm³.

b. Solid circular section of 140 mm diameter having cross-section of $A = 153.86$ cm², and section modulus $Z = 269.3$ cm³.

c. Solid rectangular section of 100 mm width × 155 mm depth having cross-section $A = 155$ cm², and section modulus $Z = 400.21$ cm³.

d. Solid rectangular section of 155 mm width × 100 mm depth having cross-section $A = 155$ cm², and section modulus $Z = 258.2$ cm³.

Solution:

For the given beam and loading maximum BM (M) occurs at centre

$$= \frac{WL}{4} = \frac{40 \times 6}{4} = 60 \text{ kN-m} = 60 \times 10^6 \text{ N-mm}$$

For the given material the section modulus $Z = \dfrac{I}{y_{\text{max.}}}$

Required $Z = \dfrac{M}{f} = \dfrac{60 \times 10^6}{150} = 4 \times 10^5 \text{ mm}^3 = \mathbf{400 \text{ cm}^3}$

From the four available sections only (a) and (c) satisfy the safety requirement of $Z \ge \mathbf{400}$ cm³.

Between (a) and (c), the area of cross-section 'A' is less for (c) and hence its mass/metre shall be less and therefore (c) will be economical. Thus solid rectangular section 100 mm wide × 155 mm deep is recommended for the given beam.

EXAMPLE 8.9: A cantilever beam projecting 3 m from the face of the wall carries uniformly distributed load of 16 kN/m (including self weight). The material can safely be stressed to a maximum bending stress of 180 N/mm². Suggest the most suitable beam section with reasons for selection from the following available sections.

a. Solid rectangular section 100 mm × 155 mm deep having cross-section $A = 155$ cm^2 and section modulus $Z = 400.21$ cm^3.

b. Solid square section 134 mm × 134 mm having cross-section $A = 179.56$ cm^2 and section modulus $Z = 401.2$ cm^3.

c. **Hollow rectangular** section 120 mm wide × 240 mm deep outer dimension with 10 mm uniform wall thickness all around having cross-section $A = 68$ cm^2, $Z = 412.5$ cm^3.

d. Hollow circular section of 240 mm outer diameter with 10 mm uniform wall thickness all around having cross-section $A = 72.22$ cm^2 and section modulus $Z = 400.0$ cm^3.

Solution:

For the given cantilever the maximum BM $= \dfrac{wL^2}{2} = \dfrac{16 \times 3 \times 3}{2} = $ **72 kN-m.**

Thus required moment of resistance $M = 72$ kN-m $= 72 \times 10^6$ N-mm.

Section modulus required $Z = \dfrac{M}{f} = \dfrac{72 \times 10^6}{180} = 400000$ mm$^3 = 400$ cm^3

All the available sections are having section modulus greater than 400 cm^3 and hence can be used. **Section (c)** having lowest cross-sectional area ($A = 68$ cm^2) is the lightest and shall be adopted from *economy point of view*. Recommended section is 120 mm wide × 240 mm deep (outer) with 10 mm uniform wall thickness.

EXAMPLE 8.10: For a certain loading and span, the beam requires a section modulus $Z = 415$ cm^3. Suggest with reasons which of the following rolled steel beam sections will be most suitable.

a. ISLC 350 at the rate of 38.8 kg/m, $Z_{XX} = 532.1$ cm^3.

b. ISLB 300 at the rate of 37.7 kg/m, $Z_{XX} = 488.9$ cm^3.

c. ISWB 250 at the rate of 40.9 kg/m, $Z_{XX} = 475.4$ cm^3.

d. ISMC 300 at the rate of 35.8 kg/m, $Z_{XX} = 424.2$ cm^3.

e. ISLB 325 at the rate of 43.1 kg/m, $Z_{XX} = 607.7$ cm^3.

Solution: Here all sections except ISMC 300 at the rate of 35.8 kg/m has more than the required section modulus, and hence can be safely adopted. But from the economy point of view ISLB 300 at the rate of 37.7 kg/m is recommended as it has the least weight and also satisfies the safety requirement.

PRACTICE TASK P II

Select the correct response to the following multiple choice questions and compare your response with the key to practice task P II. By comparing your response, determine your unlearned instructional material and again study the relevant portion of instructional resource material.

1. A beam of rectangular section 60 mm × 150 mm is subjected to a maximum BM of 18 × 10^6 N-mm. The modulus of elasticity of the beam material is 2 × 10^5 N/mm^2. What will be the radius of curvature due to bending?

    a. 360000 mm.   b. 187500 mm.

    c. 2500 mm.   d. 90 mm.

2. A rectangular section 300 mm × 500 mm is placed with 500 mm side vertical and subjected to certain loading which causes maximum fibre bending stress of 300 N/mm². What will be the maximum fibre bending stress under the same loading if the beam is placed with 300 mm side vertical?

    a. 180 N/mm²   b. 300 N/mm²

    c. 500 N/mm²   d. 833.3 N/mm²

3. A T-section shown in Fig. 8.11 is subjected to a maximum sagging BM of 100000 N-mm. Determine the nature and magnitude of the maximum fibre bending stress developed due to the bending moment. Area = 2900 mm², I_{NA} = 6372443 mm⁴ (about neutral axis). NA = 41.21 mm from top and 108.70 mm from bottom.

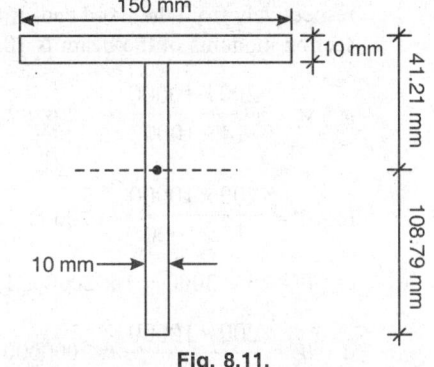

Fig. 8.11.

    a. $\dfrac{41.21 \times 100000}{6372443} = 0.65$ N/mm² (compression)

    b. $\dfrac{108.70 \times 100000}{6372443} = 1.70$ N/mm² (tension)

    c. $\dfrac{100000}{2900} = 34.48$ N/mm² (tension)

    d. $\dfrac{6372443}{2900} = 2197.39$ N/mm² (compression)

4. A section in Fig. 8.12 has safe permissible stresses of 2000 kgf/cm² in tension and 1200 kgf/cm² in compression. Calculate the safe moment of resistance of the section used for a horizontal cantilever beam with one flange horizontal in top face as shown. Area = 29 cm², I_Z = 637.0 cm⁴.

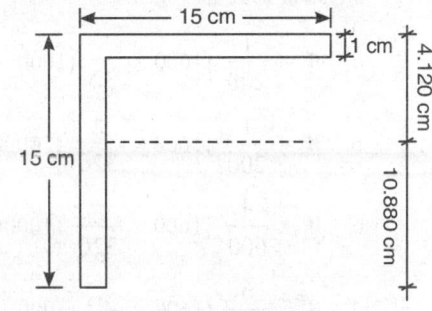

Fig. 8.12.

    a. $\dfrac{637.0}{4.12} \times 2000 = 309223$ kgf-cm.

    b. $\dfrac{637.0}{4.12} \times 1200 = 195534$ kgf-cm.

    c. $\dfrac{637.0}{10.88} \times 2000 = 117096$ kgf-cm.

    d. $\dfrac{637.0}{10.88} \times 1200 = 702570$ kgf-cm.

5. A simply supported beam of 5 m span carries a concentrated load of 2000 kgf at 2 m from the left hand support. If the material is permitted to be stressed to a maximum of 1000 kgf/cm² tensile stress and 800 kgf/cm² compression, find the safe section modulus required for a symmetrical section. Neglect the self weight of the beam.

    a. $\dfrac{2000 \times 2 \times 3}{5} \times \dfrac{100}{800} = 300$ cm³.   b. $\dfrac{2000 \times 5}{4} \times \dfrac{100}{1000} = 250$ cm³.

    c. $\dfrac{2000 \times 2 \times 3}{5} \times \dfrac{100}{1000} = 240$ cm³.   d. $\dfrac{2000 \times 5}{4} \times \dfrac{100}{1000} = 125$ cm³.

6. A cantilever beam of 4 m span carries a **UDL** of 2000 N/m over the entire span. Find the section modulus required for the beam section at 1 m from the fixed end if the maximum permissible bending stress is 1000 N/mm^2.

   a. $\dfrac{2000 \times 1 \times 3 \times 1000}{4 \times 1000} = 1500 \text{ mm}^3.$ b. $\dfrac{2000 \times 1000}{1000} = 2000 \text{ mm}^3.$

   c. $\dfrac{2000 \times 3 \times 3 \times 1000}{2 \times 1000} = 9000 \text{ mm}^3.$ d. $\dfrac{2000 \times 4 \times 4 \times 1000}{2 \times 1000} = 16000 \text{ mm}^3.$

7. A 3 m long cantilever beam carries a concentrated loads of 3W, 2W and W at 1 m, 2 m and 3 m respectively from the fixed end. If the maximum permissible bending stress is 200 N/mm^2 and the section modulus of the beam is 10, 000, find the safe value of W.

   a. $W = \dfrac{200 \times 10000}{14 \times 1000} = 1428.57 \text{ N}.$

   b. $W = \dfrac{200 \times 10000}{10 \times 1000} = 200 \text{ N}.$

   c. $W = (3 \times 3000 + 2 \times 2000 + 1 \times 1000) = 14000 \text{ N}.$

   d. $W = \dfrac{200 \times 10000}{10} = 2000000 \text{ N}.$

8. A simply supported beam of 6 m span carries two concentrated loads of W each at one third span points. If the beam section consists of a hollow circular section having outer diameter 10 cm and wall thickness of 1 cm, calculate the safe value of W when the maximum permissible bending stress is 1600 kgf/cm^2.

   a. $W = \dfrac{1}{200} [1600 \times \dfrac{\pi}{32} (1000 - 512)] = 383.43 \text{ kgf}.$

   b. $W = \dfrac{1}{200} [1600 \times \dfrac{\pi}{320} (10000 - 4096)] = 463.99 \text{ kgf}.$

   c. $W = \dfrac{4}{600} [1600 \times \dfrac{\pi}{320} (10000 - 4096)] = 618.52 \text{ kgf}.$

   d. $W = \dfrac{6}{600} [1600 \times \dfrac{\pi}{32} (1000 - 512)] = 766.86 \text{ kgf}.$

9. A uniform section of 20 cm side is used for a simply supported beam of 4 m span and can safely carry a concentrated load of 3000 kgf at its mid span without exceeding permissible bending stress. If the effect of self weight of beam is neglected, calculate the single concentrated load which may be carried safely at the quarter span from one support.

   a. $W = \dfrac{3000 \times 4}{4} \times \dfrac{2}{1} = 6000 \text{ kgf}.$ b. $W = 3000 \times \dfrac{6}{4} = 4500 \text{ kgf}.$

   c. $W = \dfrac{3000 \times 4 \times 4}{4 \times 3} = 4000 \text{ kgf}.$ d. $W = \dfrac{3000 \times 4}{4} \times \dfrac{1}{1} = 3000 \text{ kgf}.$

10. Different beam sections are available for a simply supported beam of 8 m span and carrying UDL of 200 N/m over the entire span (including self weight). If the permissible bending stress is 160 N/mm^2, suggest a most suitable section for this from the following 4 sections.

   a. Solid circular section of 110 mm diameter having section modulus = 130660 mm^3 and area = 9500 mm^2.

   b. Solid rectangular section 60 mm × 100 mm having section modulus = 100000 mm^3 and area = 6000 mm^2.

   c. T-section 200 mm × 200 mm × 10 mm thickness having section modulus = 105780 mm^3 and area = 3900 mm^2.

   d. I-section of 80 mm × 150 mm deep × 10 mm thick flanges and web having section modulus = 12100 mm^3 and area 2900 mm^2.

KEY TO PRACTICE TASK II

Q. No.	1	2	3	4	5	6	7	8	9	10
Correct Response	(b)	(c)	(b)	(d)	(a)	(c)	(b)	(b)	(c)	(d)

8.8 BEAMS OF COMPOSITE SECTIONS (FLITCHED BEAM SECTIONS)

When the beam section of a single material of lower strength becomes too large, the beam *section can be strengthened* by using stronger material jointly. Such beam sections comprising of *more than one material* are called *composite* beam sections such as: (i) *Flitched* beams, (ii) *Sandwitch* beams, (iii) *Bimetallic* beams, and (iv) *Reinforced* beams.

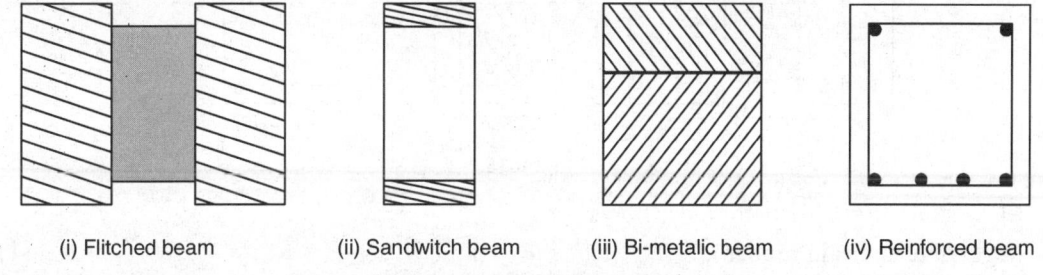

| (i) Flitched beam | (ii) Sandwitch beam | (iii) Bi-metalic beam | (iv) Reinforced beam |

Fig. 8.13: Composite beam sections

These beams of composite sections can be *analyzed* by using the same principles of *bending theory* since the assumption that the cross-sections *plane before bending* remain *plane after bending* holds good *regardless of the material*. The *strains* developed in different fibres will be *proportionate* to its *distance* from the *neutral fibre*. Using this assumption the strains in two materials can be determined according to their location with respect to the neutral fibre. The strains at a given distance will be the same in the two materials. The strain variation along the depth will be linear.

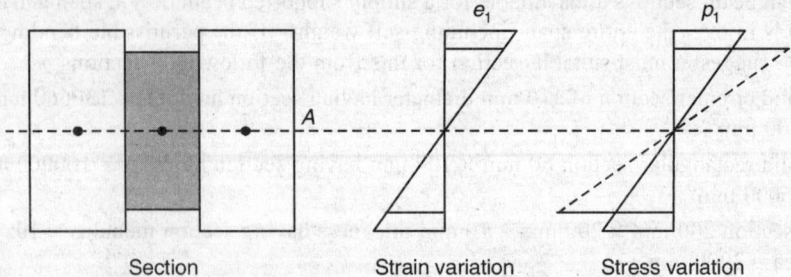

Section   Strain variation   Stress variation

Fig. 8.14: Strain distribution along depth

If the two material elements are rigidly joined together, they will behave like one element and the bending will take place about the common centroidal axis. The sum of the *moment of resistance* offered by the *two elements* will be *equal* to the *bending moment* acting at the section.

Consider a *symmetrical section* comprising of two rectangular timber elements joined rigidly by a steel strip. Let each of the timber elements be $\dfrac{b}{2} \times D$ and steel strip joining the two timber elements symmetrically be $(t \times d)$ as shown in Fig. 8.15.

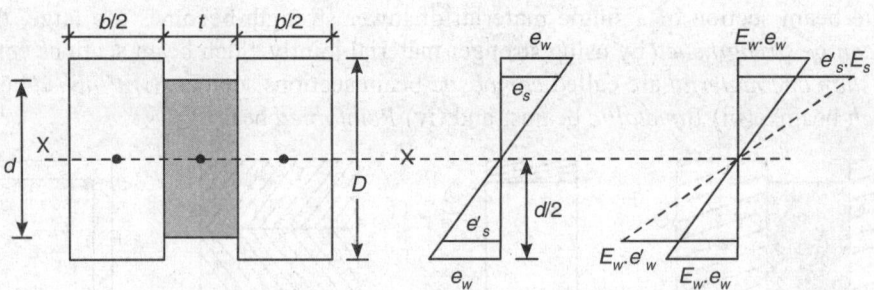

Fig. 8.15.

Max. strain in timber outer most fibres = e_w

Max. fibre stress in timber $f_w = E_w \cdot e_w$ ... (1)

Max. strain in steel outer most fibres = $e_s = \dfrac{e_w}{D/2} \cdot \dfrac{d}{2} = e_w \cdot \dfrac{d}{D}$ ... (2)

Max. stress in steel $f_s = E_s \cdot e_s = E_s \cdot \left(e_w \cdot \dfrac{d}{D}\right) = E_s \cdot e_w \cdot \dfrac{d}{D}$ ... (3)

At the same point strain e_w = strain e_s

or $\dfrac{f_w}{E_w} = \dfrac{f_s}{E_s}$ or $f_s = \dfrac{E_s}{E_w} \cdot f_w$

Let $\dfrac{E_s}{E_w}$ = modular ratio of steel to wood = m (say), $\therefore$ $f_s = m \cdot f_w$ ... (4)

Moment of inertia about X-X (considering steel '*m*' times stronger to wood)

$$I_w = \frac{1}{12} \cdot b \cdot D^3, \qquad I_s = \frac{t_s \cdot d_s^3}{12}$$

Equivalent wooden section $I_X = \dfrac{b \cdot D^3}{12} + m\left(\dfrac{t_s \cdot d_s^3}{12}\right)$

Total moment of resistance M will be sum of the moment of resistances of two materials, i.e.

$$M = M_1 + M_2 = M_w + M_s \qquad \text{... (5)}$$

or

$$M = f_w \cdot \frac{I_w}{y} + \frac{(m \cdot f_w)\, I_s}{y} = \frac{f_w}{y}\,(I_w + m \cdot I_s) \qquad \text{... (6)}$$

Total moment of resistance can be calculated using above equations if geometry of sections and maximum permissible stress in timber are given.

Stresses can be calculated for the given moment and geometry of the sections by using these equations and considering radii of curvature of two materials equal i.e. $R_1 = R_2$

$$\frac{E_1}{R_1} = \frac{M_1}{I_1} \qquad \text{and} \qquad \frac{E_2}{R_2} = \frac{M_2}{I_2}$$

$$\text{sinc} \quad R_1 = R_2 \quad \text{or} \quad \frac{E_1 I_1}{M_1} = \frac{E_2 I_2}{M_2} \quad \text{or} \quad \boxed{\frac{M_1}{M_2} = \frac{E_1 I_1}{E_2 I_2}} \qquad \text{... (7)}$$

From equations (5) and (7), M_1 and M_2 can be calculated from which *fw* and *fs* can also be calculated.

Consider an overall unsymmetrical section ($b \times D$) with ($b \times d$) comprising of weaker material (stress f_1) and stronger material section ($b \times t$) at the bottom (stress f_2).

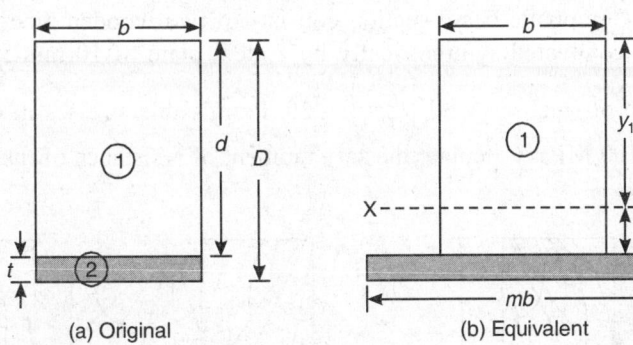

(a) Original   (b) Equivalent

For equivalent section on the width '*b*' of the stronger material is considered equal to *mb* of weaker material, where $m = \dfrac{E_2}{E_1}$.

Maximum allowable stress $f_1 = \dfrac{f_2}{m}$ ... (8)

Also maximum allowable stress $f_2 = m \cdot f_1$

For finding allowable *moments* or *stresses* above equations can be used. Geometrical properties of the *equivalent section* can be determined by considering width (*b*) equal to *mb*. Centroid and section modulus (Fig. 8.16) for the equivalent section can be calculated in usual manner.

Consider a sandwitch beam section $b \times D$ with $b \times t$ stronger material each at top and bottom faces as shown in Fig. 8.17. These beam sections can be analysed by drawing *equivalent* section as shown in Fig. 8.17b. It can also be analysed by considering section modulus of each individual material and its moment of resistances.

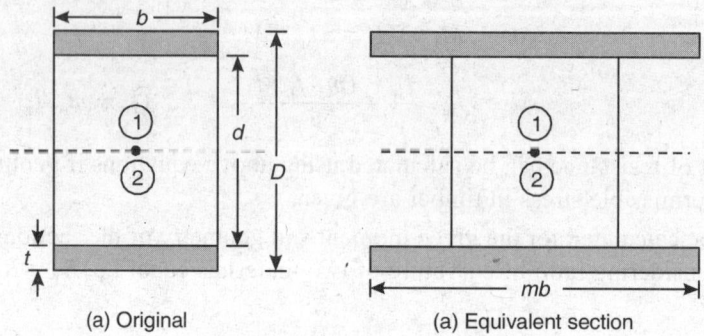

(a) Original   (a) Equivalent section

Fig. 8.17: Sandwitch beam section

For calculating moment of resistance ($M = f.Z$), consider stress in outermost extreme fibres.

The stress in extreme fibre must be taken as lesser of the two values (mf_1 or f_2) for finding safe design with respect to both the materials.

For equivalent section approach, the stress in extreme fibre must be that of equivalent material (i.e. f_1 or whichever is less) for finding *safe design*. These concepts of flitched or composite beam sections will be clear by studying following numerical examples.

EXAMPLE 8.11: A Composite beam section consists of two wooden pieces of section 100 mm × 200 mm each connected symmetrically by a steel plate of 10 mm × 160 mm in the middle. Modular ratio of steel to wood $\left(\dfrac{E_s}{E_w}\right) = 20$. Permissible stresses in timber (f_w) = 7.5 Mpa and steel (f_s) = 100 MPa. Calculate the safe moment of resistance of the section.

Solution:

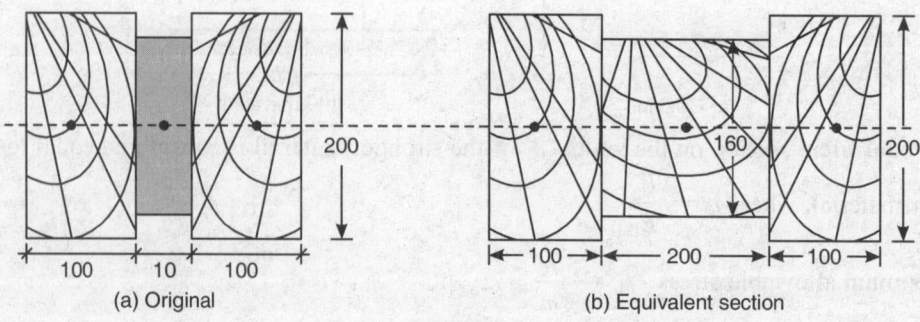

(a) Original   (b) Equivalent section

Fig. 8.18.

Wooden: $2 \times 100 \text{ mm} \times 200 \text{ mm}$

Steel: $10 \text{ mm} \times 160 \text{ mm}$

$\dfrac{E_s}{E_w} = m = 20,\qquad f_w = 7.5 \text{ MPa},\qquad f_s = 100 \text{ MPa},$

$mf_w = 20 \times 7.5 = 150 \text{ MPa (unsafe as more than 100 MPa)}$

$f'_w \text{ (safe)} = \dfrac{100}{16/2} \times \dfrac{20/2}{m} = \dfrac{100 \times 5}{4 \times 20} = \dfrac{25}{4} = 6.25 \text{ Mpa}$

$I_X \text{ (equivalent)} = \dfrac{1}{12}\,(2 \times 100 \times 200^3 + 20 \times 10 \times 160^3) = \dfrac{200}{12}\,(8000 + 4096)10^3 \text{ mm}^4$

$Z_X \text{ (equivalent)} = \dfrac{I_X}{200/2} = \dfrac{12096 \times 2 \times 10^5}{12 \times 100} \text{ mm}^3 = \dfrac{12096 \times 10^3}{6} \text{ mm}^3$

$M_X = f \times Z_X = 6.25 \times \dfrac{12096 \times 10^3}{6} \text{ N-mm} = \dfrac{6.25 \times 12096}{6 \times 10^3} \text{ kN-m} = \mathbf{12.6 \ kN\text{-}m}$

Alternatively: safe 'f_w' $= 6.25$ Mpa instead of 7.5 Mpa, f_s (safe) $= 100$ Mpa.

$M = M_w + M_s = f_w\,(Z_w) + f_s\,(Z_s) = \left[6.25\left(\dfrac{2 \times 100 \times 200^3}{12 \times 100} \right) + 100\left(\dfrac{10 \times 160^3}{12 \times 80} \right) \right] \dfrac{1}{10^6}$

$= \left[\dfrac{6.25 \times 1600}{1200} + \dfrac{16 \times 256}{12 \times 80} \right] \text{ kN-m} = 8.3333 + 4.2667 \approx \mathbf{12.6 \ kN\text{-}m}$

EXAMPLE 8.12: A wooden section of 200 mm width $\times$ 240 mm depth is sandwitched between two 200 mm wide $\times$ 5 mm thick steel plates at top and bottom. The composite beam section is used over a simply supported beam of 6 m span and carrying UDL of 10 kN/m over the entire span. Find the maximum stresses in steel and wood if modular ratio of steel to wood is 20.

Solution:

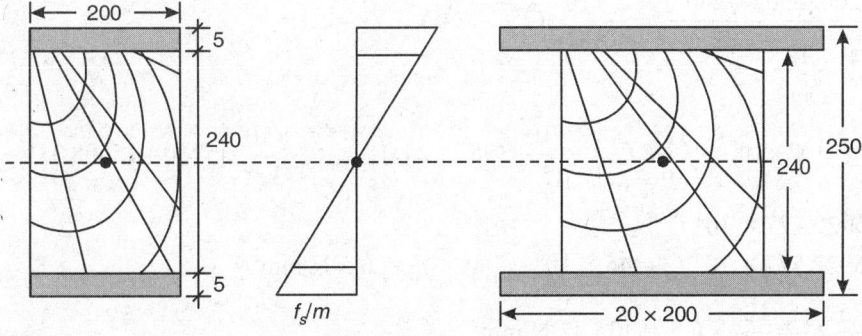

Fig. 8.19: Sandwitched section

$m = 20$, UDL $= 10$ kN/m, $L = 6$ m

$$\text{BM (max.)} = \frac{w \cdot L^2}{8}$$

$$= \frac{10 \times 6^2}{8} = \textbf{45 kN-m}$$

Z_X (equivalent for wood) $= \dfrac{2}{250 \times 12} \left[200 \times 240^3 + 200 \times 20 \times (250^3 - 240^3)\right]$

$$Z_X = \frac{200}{1500} \times 10^3 \left[24^3 + (25^3 - 24^3)\,20\right] = \frac{2}{15} \times 10^3 \left[25^3 \times 20 - 19 \times 24^3\right]$$

$$= \frac{2 \times 10^3}{15} \left[312500 - 262656\right] = 6645.8667 \times 10^3 \text{ mm}^3$$

Let extreme stress in equivalent wooden fibres $= f'_w$

$$M = f'_w \times Z = 45 \times 10^6 \text{ N-mm}, \qquad f'_w = \frac{45 \times 10^6}{6645.8667 \times 10^3} = \textbf{6.771 N/mm}^2$$

(Stress in actual extreme wooden fibres $= f_w$)

$$f_w = \frac{f'_w}{125} \times 120 = \frac{6.77 \times 120}{125} = \textbf{6.5 N/mm}^2, \qquad f_s = mf'_w = 20 \times 6.77 = \textbf{135.4 N/mm}^2$$

Alternative Approach:

Total moment of resistance = Sum of moment of resistances of wood and steel i.e.

$M = M_w + M_s = f_w \times Z_w + f_s \times Z_s$

At the level of steel $\quad \dfrac{f_s}{m\,(125)} = \dfrac{f_w}{120} \quad$ or $\quad f_s = \dfrac{125}{120} \times 20 \quad f_w = \dfrac{125}{6}$

$$Z_w = \frac{1}{6} \times 200 \times 240^2 \text{ mm}^3 = 8 \times 24 \times 10^4 \text{ mm}^3 = 192 \times 10^4 \text{ mm}^3$$

$$Z_s = \frac{200\,(250^3 - 240^3)}{12 \times \dfrac{250}{2}} = \frac{2 \times 10^3}{15}\,(25^3 - 24^3)$$

$$M = f_w (192 \times 10^4) + \frac{125}{6} f_w \left\{ \frac{2 \times 10^3}{15}\,(25^3 - 24^3) \right\} = \frac{f_w \times 10^3}{3}\,[5760 + 15008.33]$$

$$= 6922.777 \times 10^3 f_w$$

Thus $6922.777 \times 10^3 f_w = 45 \times 10^6 \qquad \therefore \quad f_w = 6.5$ N/mm^2

$$f_s = \frac{125}{6} f_w = \frac{125}{6} \times 6.5 = 135.42 \text{ N/mm}^2$$

EXAMPLE 8.13: A steel plate 50 mm wide × 20 mm thick is embedded in a wooden section of 120 mm width by 250 mm depth as shown in Fig. 8.20. The steel plate is fixed symmetrically in wooden section at 20 mm above the bottom face of wooden section. Modular ratio $\dfrac{E_s}{E_w} = 16$, permissible stresses in steel and wood are respectively 100 Mpa and 8 Mpa. Determine safe U.D.L. on a simply supported beam.

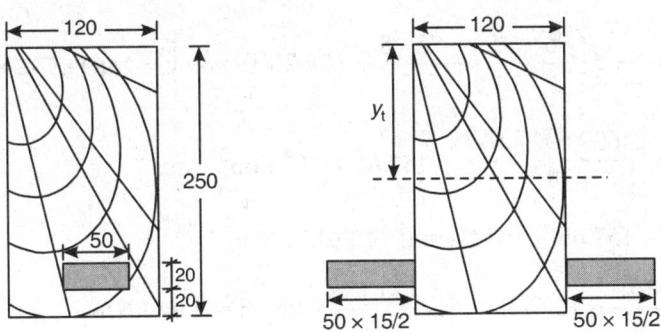

Fig. 8.20: Composite section

Solution:

First make a equivalent section of the reinforced wooden beam section (Fig. 8.20b). The composite section is not symmetrical about X-X axis and hence first determine location of $\bar{y}_b$.

$$\bar{y}_b = \frac{120 \times 250 \times \dfrac{250}{2} + (16-1)50 \times 20 \times 30}{(120 \times 250 + 15 \times 50 \times 20)} = \frac{30000 \times 125 + 15000 \times 30}{(30000 + 15000)} = \frac{280}{3}$$

$$= 93.333 \text{ mm}, \qquad \left(y_t = 250 - \frac{280}{3} = \frac{470}{3} \text{ mm} \right)$$

If safe wooden stress is taken as $f_w = 8$ Mpa,

$$f_s \text{ (max.)} = m \times f_w = 16 \left\{ \frac{8}{470/3} \times \left(\frac{280}{3} - 20 \right) \right\}$$

$$= 16 \left\{ \frac{3 \times 8}{470} \times \frac{220}{3} \right\} = 16 \times \frac{1760}{470} = 59.91 \text{ Mpa (OK as less than 100 Mpa)}$$

If $f_s = 100$ Mpa, $\qquad f_w = \dfrac{100}{\left(\dfrac{280}{3} - 20 \right)} \times \dfrac{470}{3} \times \dfrac{1}{16} = \dfrac{100}{220} \times \dfrac{470}{16}$

$$= 13.35 \text{ Mpa} > 8 \text{ Mpa (More than permissible } f_w = 8 \text{ Mpa)}$$

This can not be permitted. Thus safe f_w (max.) = 8 Mpa for safe loading.

MI about X-X

$$\text{Equivalent } I_X = \frac{120}{12} \times 250^3 + 120 \times 250 \left(\frac{250}{2} - \frac{280}{3}\right)^2 + \frac{15 \times 50 \times 20^3}{12}$$

$$+ 15 \times 50 \times 20 \left(\frac{280}{3} - 30\right)$$

$$= 10^4 \left[15625 + 3\left(\frac{95}{3} \times \frac{95}{3}\right) + \frac{75 \times 8}{12} + 75 \times 2 \times 19 \times 19\right] = \textbf{24699.962} \times \textbf{10}^4 \textbf{ mm}^4$$

$$Z_{XX} = \frac{I}{y_{\text{max.}}} = \frac{24699.692 \times 10^4}{470\!\!\Big/\!\!3} = \textbf{157.66} \times \textbf{10}^4 \textbf{ mm}^3$$

$$M = f_w \times Z = 8 \times 157.66 \times 10^4 \text{ N-mm} = 12.613 \text{ kN-m}$$

$$\text{BM (UDL)} = \frac{w(5)^2}{8} \leq 12.613, \quad w \leq \frac{12.613 \times 8}{25} \text{ kN/m} = \textbf{4.04 kN/m}$$

Safe UDL = **4.0 kN/m**

8.9 SUMMARY

To understand theory of simple bending, it is essential to know **Conditions of equilibrium, internal resistance, freebody diagram, support conditions, stresses,** shear force, **bending moment,** transverse loading, bending of structural elements, **internal and external equilibriums,** etc.

Equation of bending stresses and moment of resistance is derived by making certain **assumptions** such as: **Plane sections** before bending remains **plane** after bending, beam **material** is **homogeneos and isotropic,** stresses are **within elastic range** following Hooke's law, longitudinal axis is considered **initially straight,** the cross-section is **Symmetrical and uniform** throughout the length and beam is in **static equilibrium,** etc.

Beams subjected to **transverse loading** undergo bending of longitudinal axis which develops **longitudinal stresses and strains** and internal **moment of resistance** equal to external BM. The **neutral axis of bending** (zero strain and zero stress) passes through the **centroid of the Cross-section.** Bending stresses are **tensile on one side** of the neutral axis (NA) and **compressive on the other side** of the NA.

The **bending equation** with usual notations is given as: $\dfrac{M}{I} = \dfrac{f}{y} = \dfrac{E}{R}$

According to Hoke's law longituidinal strain 'e' is given as:

$$e = \frac{\text{Longitudinal bending stress } 'f'}{\text{Modulus of Elasticity } 'E'}, \text{ within elastic limits.}$$

$\dfrac{I}{y} = Z$ (known as **Section Modulus**)

The **moment of resistance (M)** is equal to **f.Z**, with relation to **bending stress** (f) and Section Modulus (z).

The units of various quantities in the bending equation must be **consistent** *i.e.* all force related quantities are taken in say 'N' and all linear quantities are taken in say 'mm'.

Rectangular section has moment of intertia (MI) about the major axis $I_{xx} = \dfrac{bd^3}{12}$, minor axis $I_{yy} = \dfrac{b^3 d}{12}$, while Section Modulus $Z_{xx} = \dfrac{bd^2}{6}$, and $Z_{yy} = \dfrac{b^2 d}{6}$, about respective axes.

Circular Section has $I_{xx} = I_{yy} = \dfrac{\pi}{64}(d)^4$, $Z_{xx} = Z_{yy} = \dfrac{\pi(d)^3}{32}$

Hollow Circular Section has $I_{xx} = I_{yy} = \dfrac{\pi}{64}(D^4 - d^4)$, $Z = \dfrac{\pi}{32D}(D^4 - d^4)$,

Composite beam sections are made by combining stronger material (viz. steel) and comparatively weaker and cheaper material to **enhance** structural **resistance** to carry greater transverse loads and keep the **section size moderate**. Analysis of composite beam sections is done by converting composite section into **equivalent** one material section by using modular ratio

$$m = \frac{E_1}{E_2} \quad \text{and} \quad f_1 = mf_2.$$

Total moment of resistance of the Composite Section will be sum of the **moment of resistance of the two materials**, i.e.

$$M = M_1 + M_2$$

Strains in the two materials at the **same point** will be equal

i.e. $$e = \frac{f_1}{E_1} = \frac{f_2}{E_2}$$

Flitched, Sandwitch beam section etc. are the examples of Composite Sections.

EXERCISE 8

Answer the questions and solve the problems given below on separate sheets of paper.

Q.8.1. Explain the effect of transverse loading on a structural member. Explain the development of stresses and strains.

Ans. Refer § 8.1 on introduction.

Q.8.2. List five important assumptions made in simple theory of bending and explain implications of each.

Ans. Refer § 8.2 on assumptions.

Q.8.3. Define neutral axis. Derive the position of neutral axis in a beam section under bending.

Ans. Refer § 8.3 Theory of simple bending.

Q.8.4. Derive the equation of bending stress at any distance 'y' from the neutral axis.

Ans. Refer § 8.3 Theory of simple bending.

Q.8.5. Derive the equation of moment of resistance of beam section in terms of radius of curvature.

   Ans. Refer § 8.3 Theory of simple bending.

Q.8.6. Write bending equation and explain each term including their units.

   Ans. Refer § 8.5 and § 8.6 for units and equation 8.5.

Q.8.7. A 6 m simply supported beam consists of 120 mm × 200 mm rectangular section. The beam carries UDL of 20 kN/m run over the entire span. Find the maximum bending stress induced in the beam section and show its variation across the section.

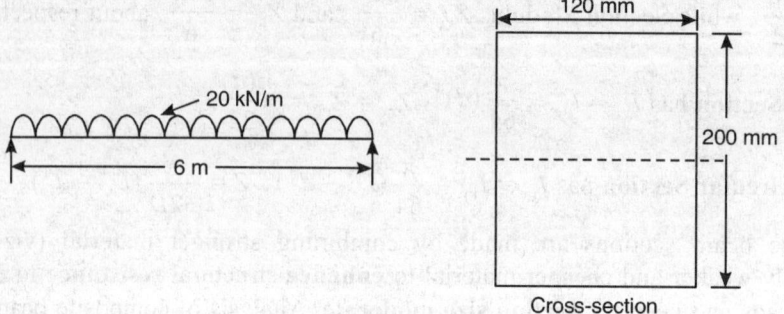

Fig. Solution Q8.7.

   Ans. The beam with loading is shown in fig. Q8.7.

   Find the maximum B.M. in the beam

$$M_{max.} = \frac{w \cdot L^2}{8} = \frac{20 \times 6 \times 6}{8} = 90 \text{ kN-m} = \mathbf{90 \times 10^6 \text{ N-mm}}$$

   Cross-section properties:

   Moment of inertia (i.e. second moment of area) about the neutral axis of bending

$$I_Z = \frac{1}{12} \times 120 \times 200^3 = 8 \times 10^7 \text{ mm}^4, \qquad y_{max.} = \frac{200}{2} = 100 \text{ mm}$$

   Bending stress using bending equation: $\dfrac{M}{I} = \dfrac{f}{y}$

$$f_{max.} = \frac{M}{I_Z} \cdot y_{max.} = \frac{90 \times 10^6 \times 100}{8 \times 10^7} = 112.5 \text{ N/mm}^2$$

   Section is symmetrical fmax. = 112.5 N/mm^2 on both faces.

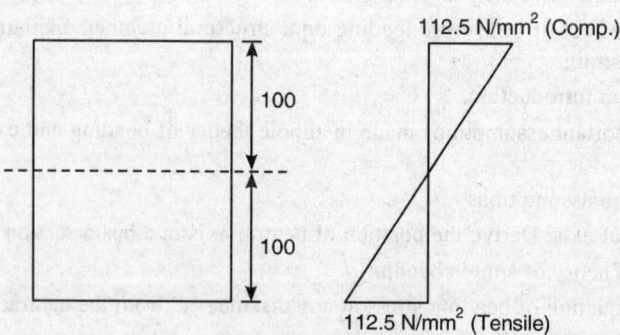

Fig. Solution Q8.7.

Q.8.8. A cantilever beam of 3 m span carries a concentrated load of 3 kN at 1 m from the free end and 4 kN at the free end. Calculate the nature and magnitude of maximum bending stress and distribution across the section as shown in Fig. Q8.8.

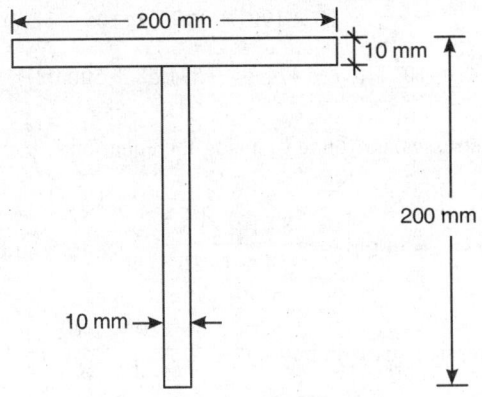

Fig. Solution Q8.8.

Hints: The beam with its stress and cross-section are shown in fig. solution Q8.8.

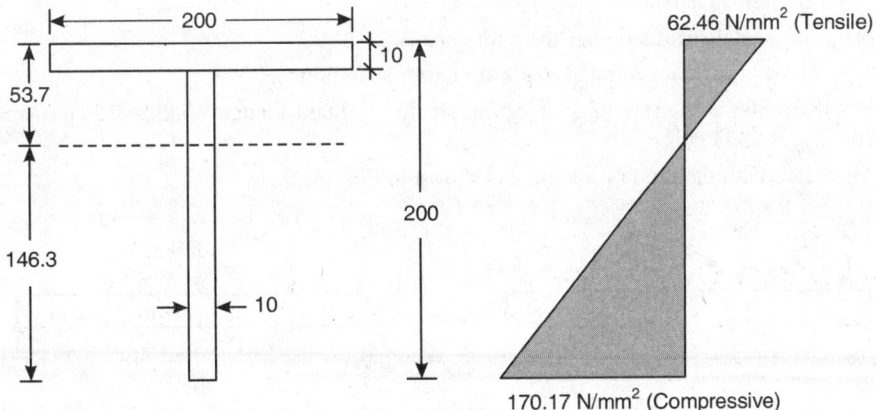

Fig. Solution Q8.8.

Maximum BM will occur at the fixed end A.

$M_{max.} = 4 \times 3 + 3 (3 - 1) = 18$ kN-m $= 18 \times 10^6$ N-mm.

For the cross-section shown calculate centroid and second moment of area about the NA of bending.

$$y_1 \text{ (from top flange)} = \frac{200 \times 10 \times 5 + 190 \times 10 \times 105}{200 \times 10 + 190 \times 10} = \frac{2095}{39} = \mathbf{53.7 \ mm}$$

y_2 (from bottom face of web) = 200 − 53.7 = **146.3 mm**

$$I_Z \text{ (N.A. of bending)} = \frac{1}{12} \times 200 \times 10^3 + 2000 \times (53.7 - 5)^2$$

$$+ \frac{1}{12} \times 10 \times 190^3 + 1900 \times (105 - 53.7)^2$$

$$= 10^4 [1.67 + 474.34 + 571.58 + 500.02] = \mathbf{1547.61 \times 10^4 \ mm^4}$$

Maximum bending stresses are found by using the equation: $\dfrac{M}{I} = \dfrac{f}{y}$

f_t (maximum tensile stress in top) $= \dfrac{18 \times 10^6 \times 53.7}{1547.61 \times 10^4} = \mathbf{62.46 \ N/mm^2}$

f_c (maximum compressive stress in bottom) $= \dfrac{18 \times 10^6 \times 146.3}{1547.61 \times 10^4} = 170.17 \ N/mm^2$

Q.8.9. An I-section ISLB 300 at the rate of 37.7 kgf/m is used as a beam simply supported over a span of 8 m. Calculate the total safe load which can be carried on the beam without exceeding the bending stress of 2000 kgf/cm² intension and 1500 kgf/cm² in compression any where in the beam if the load acts as:

(a) Uniform distributed over the entire span.

(b) Three equal concentrated loads at quarter span points.

For ISLB 300 at the rate of 37.7 kgf/m, Depth = 30 cm, Flange Width = 15 cm, Area = 48.08 cm², I_{ZZ} = 7333 cm⁴.

Ans. The beam with the type of loadings is shown in Fig. Q8.9.

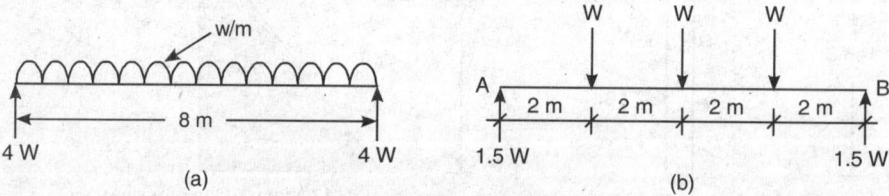

Fig. Solution Q8.9.

(a) Maximum BM $= \dfrac{w \cdot L^2}{8} = \dfrac{w \times 8 \times 8}{8}$ = 8w kgf-m = 800w kg-cm

(b) Maximum BM = 1.5W × 4 − W × 2 = 4W kgf-m = 400W kgf-cm

Given I-section is symmetrical hence its section modulus

$$Z = \frac{I}{y_{max.}} = \frac{7333}{30/2} = \frac{7333}{15} = 488.87 \ cm^3$$

Maximum permissible stress in compression is 1500 kgf/cm² while maximum permissible stress in tension is 2000 kgf/cm². For the section to be safe both in tension and compression the maximum permissible stress in bending shall be taken as 1500 kgf/cm².

Equating maximum BM and MR, we have

(a) $800w = f \cdot Z = 1500 \times 488.87 = 733300$

$$w = \frac{733300}{800} = 916.63 \text{ kgf/m}$$

Total UDL = 7333 kgf

(b) $400W = f \cdot Z = 1500 \times 488.87$

$W = 1833.30$ kgf

Total of all the 3 loads = 5500 kgf

Q.8.10. Compare weights of beam sections to develop same strengths in bending in following cases.

(a)   Solid circular section v/s hollow circular section with inner diameter $3/4^{th}$ of outer diameter.

(b)   Square section v/s rectangular section with breadth half of depth.

(c)   I-section with flange width of half the depth and thicknesses of flange and web equal to $1/20^{th}$ depth.

Ans. Since the bending strengths are same, their section modulus must be the same. Weights shall be directly in the ratio of areas of cross-section the span will be the same in each case.

(a) Z_s for solid circular section $= \dfrac{733300}{800}$

Z_h for hollow circular section $= \dfrac{\pi}{32 d_o}(d_o^4 - d_i^4)$

$$\left(d_i = \frac{3}{4}d_o\right) = \frac{\pi}{32 d_o}\left(d_o^4 - \frac{81}{256}d_o^4\right) = \frac{175}{256} \times \frac{\pi}{32}d_o^3$$

since $Z_s = Z_h$

$$\therefore \qquad \frac{\pi}{32}D^3 = \frac{175}{256} \times \frac{\pi}{32}d_o^3$$

or $\dfrac{D^3}{d_o^3} = \dfrac{175}{256}$ or $\dfrac{D^2}{d_o^2} = \left(\dfrac{175}{256}\right)^{2/3}$ ... (i)

Ratio of weights of solid to hollow substituting $(D/d_o)^2 = \dfrac{\dfrac{\pi}{4}D^2}{\dfrac{\pi}{4}\left(d_o^2 - \dfrac{9}{16}d_o^2\right)}$

$$= \frac{16 D^2}{7 d_o^2} \qquad \text{... (ii)}$$

$$\frac{W_s}{W_h} = \frac{16}{7} \times \left(\frac{175}{256}\right)^{2/3} = \mathbf{1.774}$$

(b) Z_s for square $= \dfrac{1}{6}d_s^3$

$$Z_r \text{ for rectangular} = \frac{1}{6} \cdot bd_r^2 = \frac{1}{6} \cdot \frac{d_r}{2} \cdot d_r^2 = \frac{1}{12} \cdot d_r^3$$

$$\because \quad Zs = Zr, \quad \therefore \quad \frac{1}{6} d_s^3 = \frac{1}{12} d_r^3$$

or $\quad \left(\dfrac{d_s}{d_r}\right)^3 = \dfrac{1}{2} \quad$ or $\quad \left(\dfrac{d_s}{d_r}\right)^2 = \left(\dfrac{1}{2}\right)^{2/3} \qquad \qquad \dots \text{(iii)}$

Weights ratio $\qquad \dfrac{W_s}{W_r} = \dfrac{d_s^2}{\frac{1}{2} d_r^2} = 2\left(\dfrac{d_s}{d_r}\right)^2 \quad \dots \text{(iv)}$

Substituting $\left(\dfrac{d_s}{d_r}\right)$,

$$\frac{W_s}{W_r} = 2\left[\left(\frac{1}{2}\right)^{\frac{2}{3}}\right] = 1.2599 = 1.26$$

(c) $\quad Z_i = \dfrac{I}{d/2} = \dfrac{2I}{d} = \dfrac{2}{d}\left[\dfrac{I}{12}\left\{\dfrac{d}{2} \cdot d^3 - \dfrac{9d}{20}\left(\dfrac{18}{20}d\right)^3\right\}\right] = \dfrac{d^3}{6}\left[\dfrac{1}{2} - \dfrac{52488}{160000}\right]$

$$= \frac{d^3}{6} \cdot \frac{27512}{160000}$$

$$Z_r = \frac{1}{6} \cdot \frac{d_r}{2} \cdot d_r^2 = \frac{d_r^3}{12}$$

$$\because \qquad \qquad Z_i = Z_r$$

$$\therefore \qquad \qquad \frac{d^3}{6} \cdot \frac{27512}{160000} = \frac{d_r^3}{12}$$

or $\qquad \qquad \left(\dfrac{d}{d_r}\right)^3 = \dfrac{80000}{27512}$

Weight of I-section $w_i = \left[\dfrac{d}{2} \cdot \dfrac{d}{20} \times 2 + \dfrac{18}{20}d \times \dfrac{d}{20}\right] \times \text{constant}$

$$= \frac{19\,d^2}{200} \times \text{constant} \qquad \qquad \dots \text{(v)}$$

Weight of rectangular section $w_r = \left[\dfrac{d_r}{2} \cdot d_r\right] \times \text{constant}$

$$= \frac{d_r^2}{2} \times \text{constant} \qquad \qquad \dots \text{(vii)}$$

$$\frac{w_i}{w_r} = \frac{19d^2}{200} \Big/ \frac{d_r^2}{2} = \frac{19 \times 2}{200} \left(\frac{d}{dr}\right)^2$$

Substituting $\left(\dfrac{d}{d_r}\right)$,

$$\frac{w_i}{w_r} = \frac{19}{100} \left(\frac{80000}{27512}\right)^{\frac{2}{3}} = 0.3871$$

Q.8.11. Find the dimensions of rectangular beam of maximum strength which can be cut from a circular log section of 30 cm diameter.

Ans. The circular section with possible rectangular cut-section is shown. Let the rectangular section which gives maximum bending strength has width to b and depth d.

From the figure diagonal of rectangle = diameter of log

$\therefore \quad b^2 + d^2 = 30^2$ ... (i)

Bending strength depends on section modulus Z.

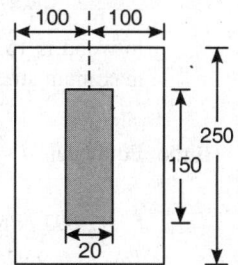

Fig. Solution Q8.11.

$$Z = \frac{1}{6} \cdot bd^2 = \frac{b}{6}(30^2 - b^2) \qquad \ldots \text{(ii)}$$

For Z to be maximum w.r.t. b, we have $\dfrac{dZ}{db} = 0$

$\therefore \quad \dfrac{d}{db}(30^2 b - b^3) = 0$

or $900 - 3b^2 = 0$

or $b^2 = 300, \qquad b = \mathbf{17.32\ cm}$

$\therefore \quad d^2 = 30^2 - 300 = 600, \qquad d = \mathbf{24.50}$

Thus the rectangular beam of 17.32 cm × 24.50 cm giving maximum bending strength can be cut from a circular section of 30 cm.

Q.8.12. Two wooden beam elements each of 100 mm width × 250 mm depth are joined by a steel plate of 20 mm × 150 mm depth in the middle to form a composite beam of 200 mm × 250 mm external dimensions. Safe permissible stress in wood = 8 N/mm², Modular ratio of steel to wood = $\dfrac{50}{3}$. Determine moment of resistance of the composite section and maximum stress in element.

Hints: Beam section is shown in the Fig. Q8.12.

$$m = \frac{E_s}{E_w} = \frac{50}{3}$$

$$f_s = m \cdot f_w = \frac{50}{3}\left(\frac{f_w}{125} \times 75\right) = \mathbf{80\ N/mm^2}$$

$Z_{\text{equiv.}} = 2788.33 \times 10^3\ \text{mm}^3$

$M = f_w \cdot (Z_{\text{equiv.}}) = \mathbf{22.3\ kN\text{-}m}$

Fig. Q8.12.

Q.8.13. (a) A composite sandwitch beam section is formed by fixing steel plates of 5 mm × 300 mm each on **vertical faces** of a wooden beam section of 300 mm × 300 mm. Modular ratio of steel to wood is 20. The composite beam is simply supported over a span of 6 m and carries a U.D.L. of 20 kN/m on the entire span. Calculate maximum stresses in wood and steel.

(b) If the steel plates are attached to top and bottom **horizontal faces** of the wooden beam in stead of **vertical faces**, determine these stresses.

Hints:

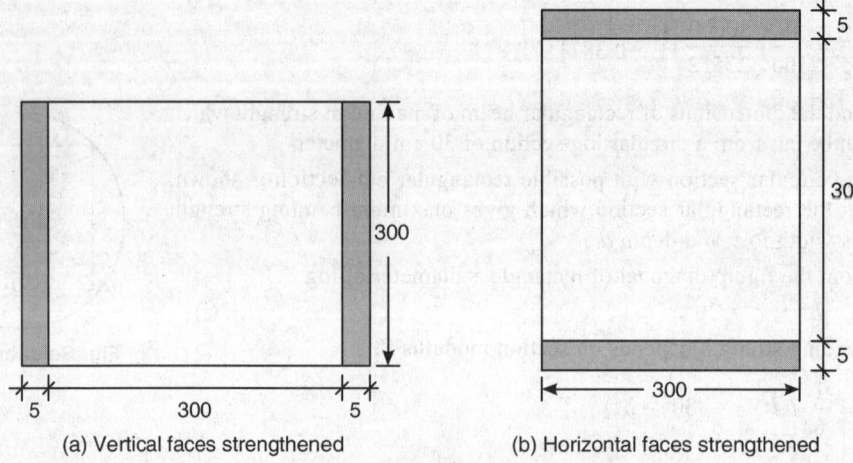

(a) Vertical faces strengthened   (b) Horizontal faces strengthened

Fig. Q8.13.

Two sections shown in Fig. Q8.13. Find equivalent sections and find Z_{XX}.

$MR = f_w \cdot Z_{XX}$

$f_s = m \cdot f_w$ at the same place

(a) $f_w = 12$ N/mm^2, $f_s = 240$ N/mm^2

     $M_w = 54$ kN-m, $M_s = 36$ kN-m

(b) $f_w = 6.52$ N/mm^2, $f_s = 134.82$ N/mm^2

     $M_w = 29.35$ kN-m, $M_s = 60.65$ kN-m

Q.8.14. A wooden beam section of 50 mm × 150 mm depth is attached with a steel plate of 50 mm × 6 mm thick at the bottom face. The steel plate is attached by 6 mm diameter screws spaced at 80 mm pitch along the span. The composite beam is simply supported over a span of 2.5 m and it carries a concentrated load of 1000 N at the middle point. Modular ratio of steel to wood is 15. Neglecting the effect of self weight and weakness due to screws, calculate the maximum stresses developed in wood and steel.

Hints: Equivalent $I = 3118.7 \times 10^4$ mm^4, $\bar{y} = 51.75$ mm $\left(q_{screw} = \dfrac{q \times 80 \times 50}{\pi(3)^2} \right)$

$f_w = 2.092$ N/mm^2 (compressive), $f_s = 15.55$ N/mm^2 (tensile)

$\{q_{screw} = 9.975$ N/mm^2 (shear)$\}$ – optional.

Shear Stresses in Bending Structures (Beams)

9.1 INTRODUCTION

In simple theory of bending, we made certain assumptions for the determination of *longitudinal stresses* caused by *bending moments*. The most important assumption to simplify the bending stress analysis is *"the plane sections before bending* remains *plane sections after bending"*. This assumption is based on the fact that the effect of shear stress is not considered in *simple theory of bending*. This assumption of plane section *before and after* bending does not account for *relative movement of different layers* and warping due to shear stresses. The effect of this assumption of *neglecting shear stress* on bending stresses is practically of no significance. The bending stresses analyzed by simple theory of bending are quite satisfactory for design purposes. However, the *shear stresses caused by shear forces* on beam sections cannot be ignored altogether due to its *own independent importance*.

We shall analyze shear stress distribution 'q' across the *cross-section parallel to the applied shear force*. Complimentary shear developed shall also be considered.

9.2 HORIZONTAL AND VERTICAL SHEAR STRESSES

Consider a rectangular beam section of *width B* and *depth D* subjected to vertical *shear force F* at any section X–X. Consider an element of small length *dx* (between X–X and X′-X′) and height *dy* between the two planes parallel to neutral surface. Since the beam is in equilibrium, the small element *dx* × *dy* × *B* will also be in *equilibrium*.

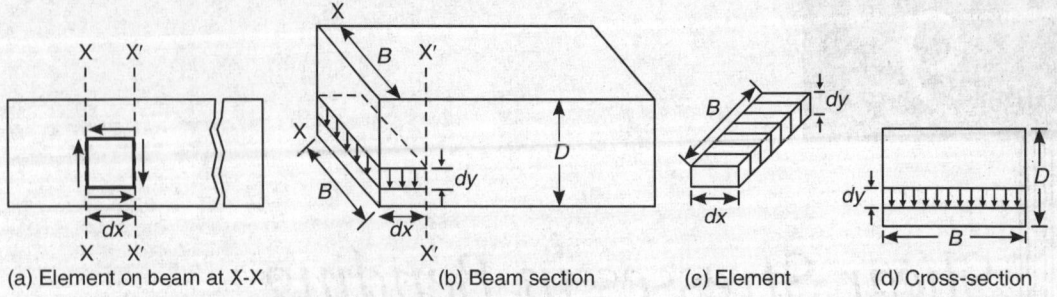

(a) Element on beam at X-X (b) Beam section (c) Element (d) Cross-section

Fig. 9.1: Beam section and complimentary shear

The element $B.dx.dy$ is subjected to a transverse shear force F which causes shear stress 'q' on vertical faces of the element and act parallel to the applied shear force 'F'. It may be assumed that the *shear stress 'q' is uniform across the width B of the cross-section*. Since the element is in *equilibrium, the shear stress q* in vertical faces is accompanied by a *perpendicular shear stress q' on horizontal faces* of the element (shear stress in horizontal layers). Consider the *equilibrium of the element* and take *moment* of the forces acting on the element, we have

$$B.dy.q.dx - B.dx.q'.dy = 0, \qquad \text{or} \qquad B.dy.q.dx = B.dx.q'.dy$$

i.e. $q' = q$, *Complimentary horizontal shear stress q' = Vertical shear stress q*

The *horizontal shear stresses* on the *outermost surfaces are zero*, the vertical shear stresses will also be zero on these surfaces. It is important to note that the *vertical shear stress q* becomes zero at the outermost top and bottom surfaces. The *shear stresses vary* across the depth i.e. $q = 0$ at $y = \pm\dfrac{D}{2}$ from the neutral surface. We shall now determine the shear stress variation across the depth with respect to neutral surface.

9.3 SHEAR STRESS VARIATION ACROSS DEPTH

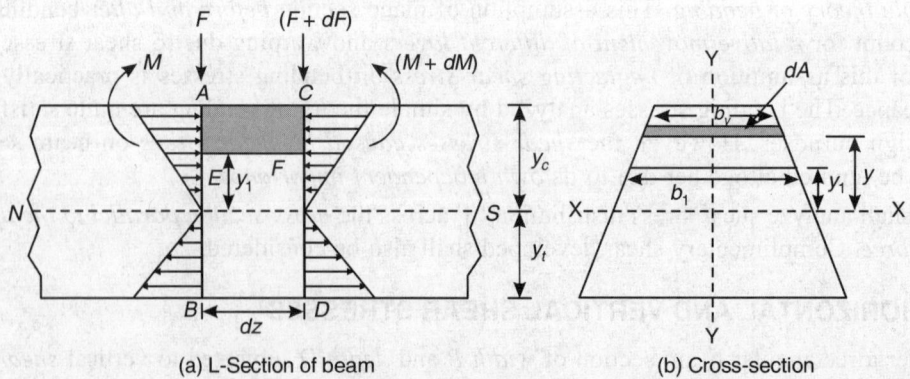

(a) L-Section of beam (b) Cross-section

Fig. 9.2: Shear stress

Consider longitudinal sections $ABCD$ of small length dz along the axis. The beam is subjected to bending moment of M at AB and $(M + dM)$ at CD (at a small distance of dz along the axis).

Let the SF at AB be F and that at CD be $(F + dF)$. Let Neutral surface of the cross–section be at X-X. Consider a small element $AEFC$ with the surface EF at a distance y_1 above neutral axis X-X. In cross-section consider a elementary strip of width by located at a distance y above NA (X-X). Consider *equilibrium* of $AEFC$, forces on the element are:

  i. **Horizontal compressive force** C due to bending moment M on face AE;

  ii. **Horizontal compressive force** $(C + dC)$ due to bending moment $(M + dM)$ on face CF;

  iii. **Horizontal shear force** on layer EF due to horizontal **shear stress** $q_1 = d_z{\cdot}b_1{\cdot}q_1$;

  iv. **Vertical force** on face AE resisting vertical shear force F;

  v. **Vertical force** on face CF resisting vertical shear force $(F + dF)$;

  vi. **Vertical force** on the **portion** AC, if any.

Consider an elementary area dA at a distance of y from the neutral axis X-X (Fig. 9.2b – cross-section).

Bending stress on elementary strip of dA area at y from NA due to bending moment will be

$$f = \frac{M}{I_x} \cdot y.$$

Force on small strip area on face $AB = f.dA = \dfrac{M}{I_x} \cdot y.\ dA$

Similarly force on small strip area on face CD due to BM $(M + dM)$ will be $\dfrac{(M + dM)}{I_x} \cdot y\ dA.$

Unbalanced force on the small area dA will be $\dfrac{(M + dM)}{I_x} \cdot y\ dA - \dfrac{M}{I_x} \cdot y\ dA = \dfrac{dM}{I_x} \cdot y\ dA$

Total unbalanced force caused due to BM on the element $AEFC$ will be $= \displaystyle\int_{y_1}^{y_c} \dfrac{dM}{I_x} \cdot y\ dA$

But the element $AEFC$ is in *stable equilibrium* and hence $\sum H = 0$, i.e. *Horizontal shear force on EF* will be $(q_1 {\cdot} b_1 {\cdot} dz) = \displaystyle\int_{y_1}^{y_c} \dfrac{dM}{I_x} \cdot y\ dA$

or $\qquad q_1 = \dfrac{dM}{I_x} \cdot \dfrac{1}{b_1 . dz} \displaystyle\int_{y_1}^{y_c} y\ dA = \dfrac{1}{b_1 . I_x} \cdot \dfrac{dM}{dz} \displaystyle\int_{y_1}^{y_c} y\ dA$

$\dfrac{dM}{dz} = F$ (shear force at the section) and $\displaystyle\int_{y_1}^{y_c} y\ dA$ is the moment of shaded area above EF

(away from NA) about NA   X-X $= A \cdot \bar{y}$

i.e. $$q_1 = \frac{F.A.\overline{y}}{b_1.I_x} \qquad \qquad ... (9.1)$$

Thus shear stress at any layer $\quad q = \dfrac{F.A.\overline{y}}{b_1.I_x}$

In equation (9.1):

q	is shear stress in any *layer at y above* NA;
F	is SF at the section under consideration;
b	is *width* of the cross-section at the *layer considered* for q;
I_x	is *moment of inertia* about axis of bending X-X;
$A\overline{y}$	is *moment of shaded* cross-sectional area *away from the* NA;
A	is shaded area away from the layer considered;
$\overline{y}$	is *distance of centroid* of shaded area from the NA of the cross-section.

Assumptions: The above analysis is based on certain assumptions stated below:

i. 'q' is assumed *uniform across full width of the cross-section* for all values of y, irrespective of the shape. This assumption is *not strictly true* because the tangential value must be zero at the boundaries of the section. Hence q may be taken as *average value* across the section.

ii. The value $F = \dfrac{dM}{dx}$, is based on the assumption that the *bending stress* 'f' varies linearly across the depth of the section and is zero at the centroid. The bending stress 'f' is *not perfectly straight* line but can be approximately assumed straight line and passing through the centroid of the cross-section.

iii. The material is homogeneous and isotropic. The values of modulii of elasticity E in tension and compression will be the same.

9.4 SHEAR STRESS DISTRIBUTION ACROSS DIFFERENT SHAPE OF SECTIONS

(A) Solid Rectangular Sections

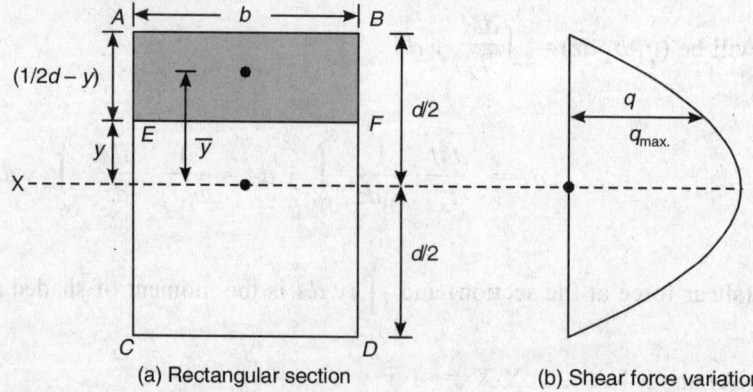

(a) Rectangular section   (b) Shear force variation

Fig. 9.3: Shear stress distribution

Consider a rectangular section *ABCD* subjected to SF of *F*. Let us determine shear stress '*q*' at any surface *EF* located at a distance '*y*' above the NA passing through centroid of the section width of *EF* = *b* as the section is uniform (rectangular). The shaded area $A = b \cdot (\frac{d}{2} - y)$ and centroid of shaded area will be at

$$\bar{y} = y + \quad -\frac{y}{2} = \left(\frac{d}{4} + \frac{y}{2}\right) = \frac{1}{2}\left(\frac{d}{2} + y\right)$$

$$A\,\bar{y} = b \cdot (\frac{d}{2} - y) \cdot \frac{1}{2}\left(\frac{d}{2} + y\right) = \frac{b}{2}\left(\frac{d^2}{4} - y^2\right)$$

$$q = \frac{F.A\bar{y}}{I_x.b} = \frac{F\frac{b}{2}\left(\frac{d^2}{4} - y^2\right)}{\frac{b.d^3}{12}.b} = \frac{6F\left(\frac{d^2}{4} - y^2\right)}{b.d^3}$$

or
$$q = \frac{6F}{b.d^3}\left(\frac{d^2}{4} - y^2\right) = \frac{F}{2I_x}\left(\frac{d^2}{4} - y^2\right) \qquad \text{... (9.2)}$$

q varies parabolically and is inversely proportional to square of distance '*y*' from the NA i.e. *q* will be *minimum (zero)* at the fardest surface ($y \neq \frac{d}{2}$) and will be *maximum when y = 0*.

$$q_{max.}\ (y = 0, \text{ at NA}) = \frac{F.d^2}{8I_x} = \frac{F.d^2}{8.\frac{b.d^3}{12}} = \frac{3F}{2b.d} = 1.5\frac{F}{b.d}$$

or
$$q_{max.} = 1.5\frac{F}{b.d} = 1.5\ q_{average} \qquad \text{... (9.3)}$$

$q = 0$, when $y = \frac{d}{2}$ i.e. at the top or bottom surface.

(B) Solid Circular Section (Diameter d)

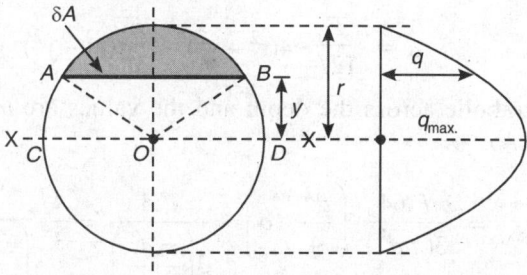

Fig. 9.4: Shear stress across solid circular section

Let the solid circular beam section has diameter 'd' and radius r. Consider a layer AB at a distance 'y' from the neutral axis X-X.

Width AB of the chord $= b = 2\left(\sqrt{r^2 - y^2}\right) = 2\,(r^2 - y^2)^{\frac{1}{2}}$

Consider a small strip of dy thickness above AB. Area δA of strip
$$\delta A = b.\delta y = 2\,(r^2 - y^2)^{\frac{1}{2}}.\delta y$$
Moment of this small area about X-X $= \delta A \cdot y = 2\delta y\,(r^2 - y^2)^{\frac{1}{2}} \cdot y$

Total moment of shaded area from $y = y$ to $y = r$ will be $\int y\ da = \int\limits_{y}^{r} 2\,(r^2 - y^2)^{\frac{1}{2}}\,y\ dy$

$$A\,\overline{y} = \int\limits_{y=y}^{y=r} b.y\ dy$$

Where
$$b = 2\,(r^2 - y^2)^{\frac{1}{2}}$$

or $\qquad\qquad b^2 = 4\,(r^2 - y^2),\qquad$ by differentiating $2b\dfrac{db}{dy} = 4\,(0 - 2y)$

or $\qquad\qquad b\dfrac{db}{dy} = -y.dy,\qquad\qquad$ when $y = r,\ b = 0,$

$$y = y,\ b = b$$

$$A\,\overline{y} = -\int\limits_{y=y}^{y=r} b.\frac{b}{4}\ db = \frac{1}{4}\int\limits_{y=y}^{y=r} b^2\ db = \frac{1}{4}\left(\frac{b^3}{3}\right)_{y=r}^{y=y}$$

or $\qquad\qquad A\,\overline{y} = \dfrac{1}{4}\left(\dfrac{b^3}{3} - 0\right) = \dfrac{b^3}{12}\qquad$ since $b = 0$, when $y = r$ and $b = b$ when $y = y$

Shear stress $\qquad\qquad q = \dfrac{F.A\overline{y}}{I.b} = \dfrac{F}{I.b}\cdot\dfrac{b^3}{12} = \dfrac{F}{I}\cdot\dfrac{b^2}{12}$

or $\qquad\qquad q = \dfrac{F}{12I}\cdot 4(r^2 - y^2) = \dfrac{F}{3I}(r^2 - y^2)\qquad\qquad$ ... (9.4)

Variation of q is parabolic across the depth and the values are *maximum* with $y = 0$ and *minimum (zero)* with $y = r$.

$$q_{max.} = \frac{F.64}{3(\pi.d^4)}\cdot\left(\frac{d^2}{4} - o\right) = \frac{F.4}{3\left(\dfrac{\pi.d^2}{4}\right)} = \frac{4}{3}\cdot\frac{F}{\left(\dfrac{\pi.d^2}{4}\right)} = \frac{4}{3}\cdot q_{average},$$

Cross-section area of circular section $= \dfrac{\pi}{4}d^2$

i.e. in circular section $q_{max.}$ at NA $= \dfrac{4}{3} \cdot q_{average}$ ... (9.5)

(C) Triangluar Solid Section with Base 'B' and Height 'H'

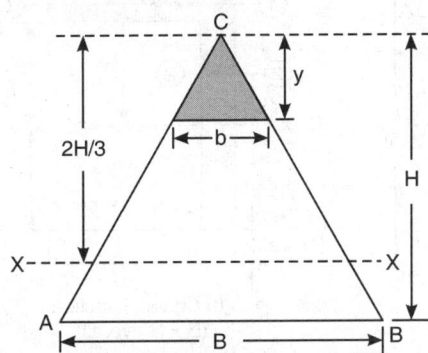

Fig. 9.5: Triangular solid section

Consider a strip of width 'b' at a distance of 'y' form the vertex.

$$\dfrac{b}{y} = \dfrac{B}{H}, b = \dfrac{B}{H} \cdot y$$

Shear stress q in any strip at $y = \dfrac{FA\bar{y}}{I \cdot b}, I_x = \dfrac{BH^3}{36}$

$A\bar{y}$ (shaded) $= \dfrac{1}{2} \cdot b \cdot y \left(\dfrac{2H}{3} - \dfrac{2}{3}y \right) = \dfrac{b.y(H-y)}{3}$

$q = \dfrac{F(A\bar{y})}{I_x.b} = \dfrac{F}{I.b} \cdot \dfrac{by(H-y)}{3} = \dfrac{F}{31} y(H-y)$ parabolic variation

For maximum value of 'q', $\dfrac{dq}{dy} = 0$,

$\dfrac{d}{dy} \dfrac{F}{3I}(Hy - y^2) = 0$, i.e. $(H - 2y) = 0, y = \dfrac{H}{2}$, q is maximum at $\dfrac{H}{2}$

$q_{max} \left(\text{at } y = \dfrac{H}{2} \right) = \dfrac{36F \dfrac{H}{2} \cdot \dfrac{H}{2}}{3(BH^3)} = \dfrac{3F}{BH} = \mathbf{1.5} \; q_{\text{avg.}}$

$$q_{N.A.}\left(\text{at } y = \frac{2H}{3}\right) = \frac{12F \cdot \frac{2H}{3} \cdot \frac{H}{3}}{(BH^3)} = \frac{8F}{3BH} = \frac{4}{3} q_{avg}.$$

(D) Rolled Steel symmetrical I and Channel Sections (Outer B × D and Web t × d)

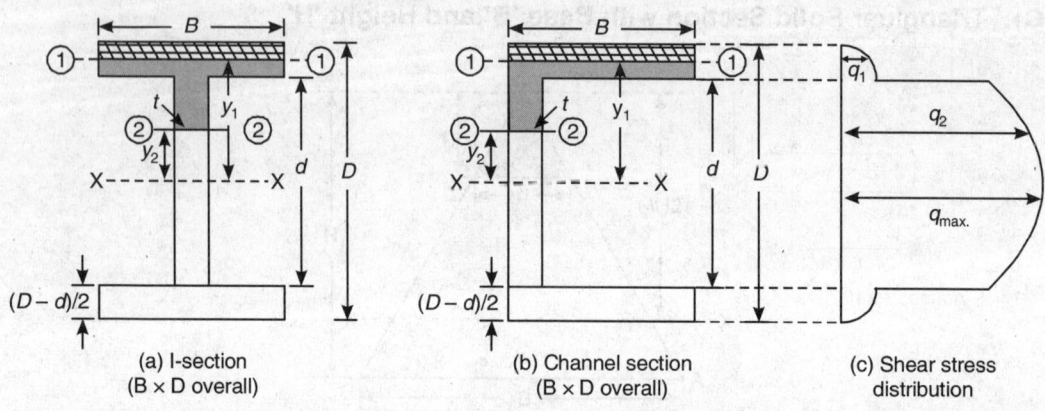

(a) I-section
(B × D overall)

(b) Channel section
(B × D overall)

(c) Shear stress
distribution

Fig. 9.6: Rolled steel joists (I and channel sections)

Consider rolled steel sections symmetrical about X–X (*I* and channel sections of $B \times D$ overall $t \times d$ webs) subjected to shear force F.

(i) Shear Stress in Flange

Consider section (layer) 1-1 *in the flange* at a *distance* of y_1 from X-X.

Moment of shaded area $\quad A\bar{y} = B\left(\frac{D}{2} - y_1\right)\left\{y_1 + \frac{1}{2}\left(\frac{D}{2} - y_1\right)\right\}$

$$= \frac{B}{2}\left(\frac{D}{2} - y_1\right)\left(\frac{D}{2} + y_1\right) = \frac{B}{2}\left(\frac{D^2}{4} - y_1^2\right)$$

Shear stress q_1 **(in flange)** $= \dfrac{F.A\bar{y}}{I.B} = \dfrac{F}{21}\left(\dfrac{D^2}{4} - y_1^2\right)$ ... (9.6)

(Parabolic $q_1 = 0$ at $y_1 = \dfrac{D}{2}$)

q'_{flange} (flange and web joint) $= \dfrac{F}{8I}(D^2 - d^2), \qquad y_1 = \dfrac{D}{2}$

ii. Shear Stress in Web-Flange Function

$$q_w \text{ (flange and web joint)} = \frac{F}{8I}(D^2 - d^2) \cdot \frac{B}{t}$$

Thus the shear stress q at the joint of flange and web suddenly increases from

$$q_f = \frac{F}{8I}(D^2 - d^2) \text{ in flange to } q_w = \frac{F}{8I}(D^2 - d^2) \cdot \frac{B}{t} \text{ in web of thickness '}t\text{' and } y_1 = \frac{d}{2}.$$

iii. Shear Stress in Web at y_2 above NA

Let the point 2 lie in the web at y_2 above the NA

Shaded area moment about the NA $= A\,\bar{y} = A_1\bar{y}_1 + A_2\bar{y}_2$

i.e.
$$A\,\bar{y} = B\left(\frac{D-d}{2}\right)\left(\frac{D+d}{2}\right)\frac{1}{2} + t\cdot\left(\frac{d}{2} - y_2\right)\left\{\frac{1}{2}\left(\frac{d}{2} - y_2\right) + y_2\right\}$$

$$= \frac{B}{8}(D^2 - d^2) + \frac{t}{2}\left(\frac{d}{2} - y_2\right)\left(\frac{d}{2} + y_2\right)$$

$$= \frac{B}{8}(D^2 - d^2) + \frac{t}{2}\left(\frac{d}{4} - y_2^2\right)$$

Shear stress in web at $y_2 = \dfrac{F \cdot A \cdot \bar{y}}{t \cdot I_x} = \dfrac{F}{t \cdot I_x}\left\{\dfrac{B}{8}(D^2 - d^2) + \dfrac{t}{2}\left(\dfrac{d^2}{4} - y_2^2\right)\right\}$

i.e. $\quad q_2 = \dfrac{F}{Ix}\left\{\dfrac{B}{8t}(D^2 - d^2) + \dfrac{1}{8}\left(d^2 - 4y_2^2\right)\right\}$ (Parabolic variation) ... (9.7)

when $y_2 = 0$,

$$q_{max.} = \frac{F}{Ix}\left\{\frac{B}{8t}(D^2 - d^2) + \frac{d^2}{8}\right\} = \frac{F}{8I_x \cdot t}[B(D^2 - d^2) + t \cdot d^2]$$

$$q_{max.} = \frac{F}{8I_x} \cdot \frac{B}{t}(D^2 - d^2) + \frac{F}{8I_x}d^2 \qquad\qquad ... (9.8)$$

From above analysis of shear stress it is quite evident that most of the shear resistance is offered by the web only. The *flanges of the joist* offer most of the *bending resistance*. i.e. the flanges resist most of the bending stresses, while the web resists most of the shear stresses.

Thus in design of rolled steel joists, the *flanges are designed for bending moment* while the *web is designed for shear force*.

(E) Shear Stress on other Sections

If we use the shear stress equation $q = \dfrac{F \cdot A \cdot \overline{y}}{I_x \cdot b}$, we shall get different shear stress variation as

shown in Fig. 9.7. These variations can be understood by actual examples.

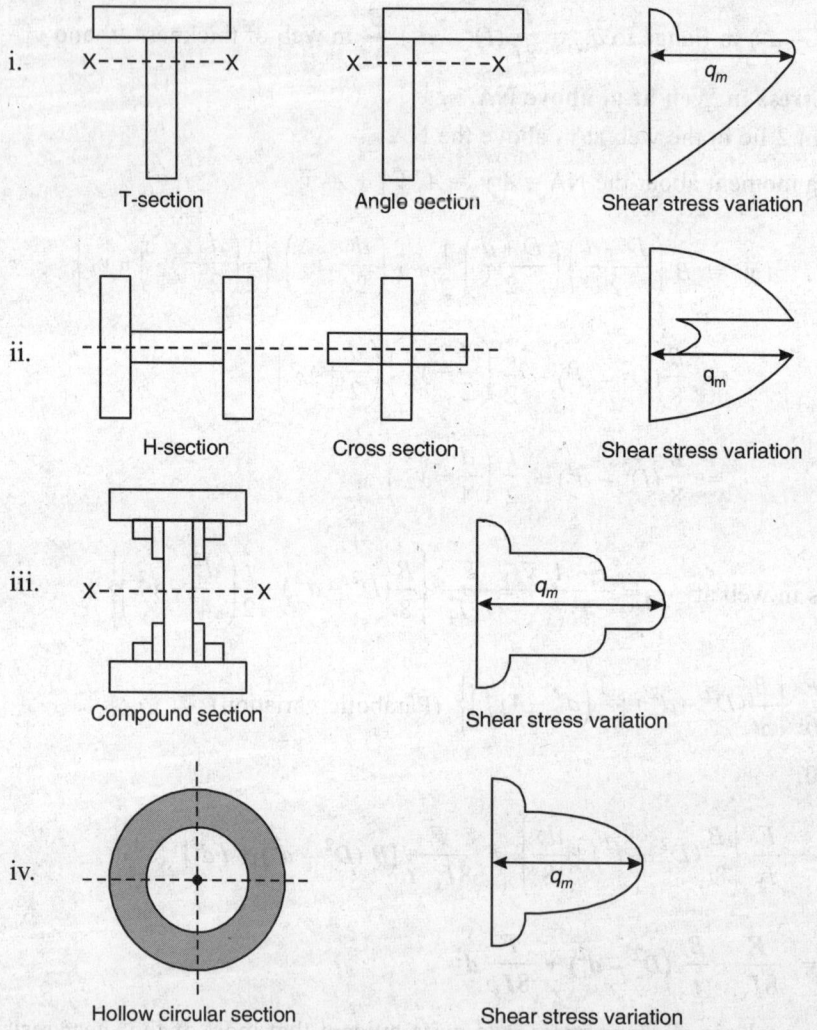

Fig. 9.7: Shear stress variation across different sections

EXAMPLE 9.1: A rolled steel section of 300 mm × 150 mm with flanges 150 mm × 15 mm and web of 270 mm × 10 mm. If a beam of above section carries a shear force of 100 kN at certain section. Determine the shear stress variation across the section showing the values of shear stress at (i) Joint of flange and web, (ii) 80 mm above the NA and (iii) at the NA.

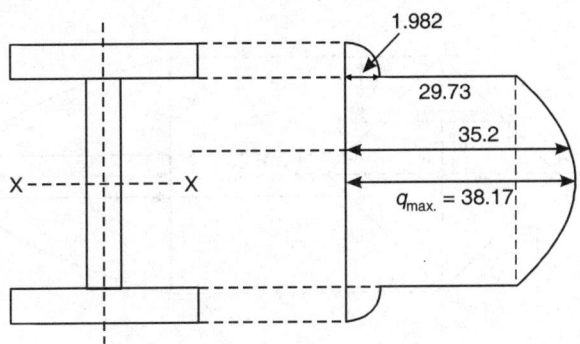

Fig. 9.8: Shear stress diagram

Solution:

$$D = 300 \text{ mm}, \qquad d = (300 - 2 \times 15) = 270 \text{ mm}, \qquad B = 150 \text{ mm},$$
$$b = (150 - 10) = 140 \text{ mm}$$

$$I_x = \left(\frac{BD^3}{12} - \frac{bd^3}{12} \right) = \left(\frac{150 \times (300)^3}{12} - \frac{140 \times (270)^3}{12} \right)$$

$$= 10^4 (33750 - 22963.5) = 10786.5 \times 10^4 \text{ mm}^4$$

$$q_f \text{ (Joint)} = \frac{FA\bar{y}}{I_x \cdot b_f} = \frac{100 \times 10^3 \times 150 \times 15 \times 142.5}{107865 \times 10^3 \times 150} = \mathbf{1.982 \text{ N/mm}^2}$$

$$q_w \text{ (Joint)} = \frac{FA\bar{y}}{I_x \cdot b_w} = \frac{100 \times 10^3 \times 150 \times 15 \times 142.5}{107865 \times 10^3 \times 10} = \mathbf{29.73 \text{ N/mm}^2}$$

$$q \text{ (80 mm)} = \frac{100 \times 10^3 \times \left\{ 150 \times 15 \times 142.5 + 10 \times (135 - 80) \times \left(80 + \dfrac{135 - 80}{2} \right) \right\}}{107865 \times 10^3 \times 10}$$

$$= \frac{10\{320625 + 59125\}}{1078625} = \frac{379750 \times 10}{107865} = \mathbf{35.206 \text{ N/mm}^2}$$

$$q_{max} = \frac{100 \times 10^3}{107865 \times 10^3 \times 10} = \left\{ 150 \times 15 \times 142.5 + 10 \times 135 \times \frac{135}{2} \right\}$$

$$= \frac{10^5 \times 411750}{107865 \times 10^4} = \mathbf{38.173 \text{ N/mm}^2}$$

EXAMPLE 9.2: A 150 mm × 150 mm square timber beam section $ABCD$ is placed with its one of the diagonal AC horizontal and vertex B at the top as shown in Fig. 9.8a. The beam carries a SF of **400 kN** at the mid span of 8 m long beam. Show the shear stress distribution and calculate maximum values and at the NA at AC.

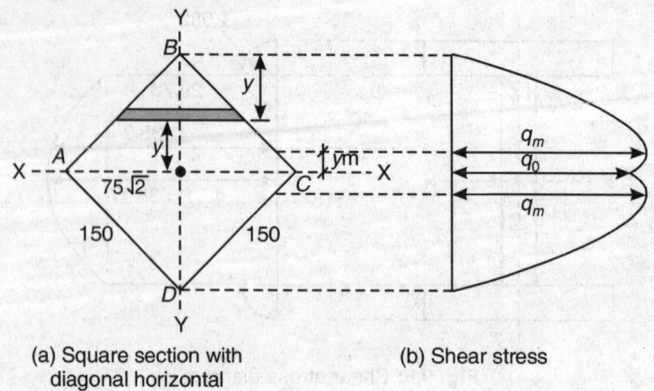

(a) Square section with   (b) Shear stress
diagonal horizontal

Fig. 9.9: Shear stress variation

Solution:

Width of diagonal $b = 150\sqrt{2}$ mm,

$$I_x = \left(\frac{1}{12}\cdot b\cdot h^3\right) \times 2 = \frac{150\sqrt{2} \times \left(75\sqrt{2}\right)^3}{12} \times 2 = \frac{150 \times 75 \times 75 \times 75 \times 4 \times 2}{12}$$

$$= 4218.75 \times 10^4 \text{ mm}^4$$

Consider a small strip of dy thickness and breadth $b = 2y$

Shaded area $A\,\bar{y} = \frac{b \times y}{2} \times \left(\frac{y}{3} + 75\sqrt{2} - y\right) = \frac{2y\cdot y}{2}\left(75\sqrt{2} - \frac{2}{3}y\right)$

or $\quad A\,\bar{y} = y^2\left(75\sqrt{2} - \frac{2}{3}y\right)$

$$\therefore \quad q = \frac{F\cdot A\bar{y}}{I\cdot b} = \frac{400 \times 1000}{I_x\cdot b} \times y^2\left(75\sqrt{2} - \frac{2}{3}y\right) = \frac{4 \times 10^5\left(75\sqrt{2}y^2 - \frac{2}{3}y^3\right)}{2y \times 4218\cdot75 \times 10^4}$$

$$q = \frac{4 \times 10 \times \left(75\sqrt{2}y - \frac{2}{3}y^2\right)}{2 \times 4218.75}$$

For q to be maximum: $\dfrac{dq}{dy} = 0 = 75\sqrt{2} - \dfrac{2}{3}\cdot 2y,$

or $\quad y = \dfrac{75\sqrt{2} \times 3}{4} = \dfrac{225\sqrt{2}}{4}\quad \left(\dfrac{3}{4}\text{of height }75\sqrt{2}\right)$

$$\therefore \quad q_{max.}\left(\text{for } y = \frac{225\sqrt{2}}{4}\right) = \frac{20}{4218.75}\left(75\sqrt{2}\times\frac{225\sqrt{2}}{4} - \frac{2}{3}\times\frac{225\times225\times2}{4\times4}\right)$$

$$= \frac{20}{4218.75}\left(\frac{75\times225}{2} - \frac{75\times225}{4}\right) = \frac{20}{4218.75}\times\frac{75\times225}{4} = \textbf{20 N/mm}^2$$

$$q_0\left(\text{at } y = 75\sqrt{2}\right) = \frac{20}{4218.75}\left(75\sqrt{2}\times75\sqrt{2} - \frac{2}{3}(75\sqrt{2})^2\right) = \frac{20\times\dfrac{1}{3}\times75\times75\times2}{4218.75}$$

$$= \textbf{17.775 N/mm}^2$$

$$q \text{ (at } y = 0) = 0$$

EXAMPLE 9.3: A 160 mm × 120 mm × 10 mm uniform T-section is used for 2.5 m cantilever carrying a udl of 2 kN/m over the entire span. Determine maximum shear and bending stresses in the section. Also calculate the maximum shear and bending stresses at a section 2 m from the free end.

Solution:

$$L = 2.5 \text{ m} = \textbf{2500 mm}, \ w = \frac{2000}{1000} \text{ N/mm} = 2 \text{ N/mm}.$$

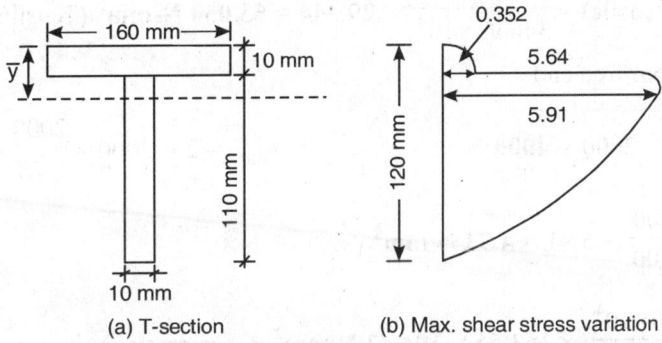

(a) T-section   (b) Max. shear stress variation

Fig. 9.10.

$$\bar{y} \text{ (from top)} = \frac{(160-10)10\times5 + 120\times10\times\dfrac{120}{2}}{150\times10 + 120\times10}$$

$$\bar{y}_t = \frac{79500}{2700} = \textbf{29.44 mm}$$

$$I_{xx} = I_{AB} - A{y_t}^2 = \frac{1}{3}\times150\times10^3 + \frac{1}{3}\times10\times120^3 - 2700\,(29.444)^2$$

$$= 5\times10^4 + 576\times10^4 - 234.08\times10^4 = 346.92\times10^4 \text{ mm}^4$$

i.   Maximum SF = $wl = 2\times2500 = \textbf{5000 N}$

$$\text{Maximum BM} = \frac{wl^2}{2} = 2 \times 2500 \times \frac{2500}{2} = \textbf{625} \times \textbf{10}^4 \textbf{ N-mm}$$

$$\text{Maximum shear stress } q_{max.} \text{ at NA} = \frac{F_{max.} \cdot A\overline{y}}{I \cdot b}$$

$$q_{max.} = \frac{5000}{346.92 \times 10^4 \times 10} \left\{ 1600 \times 24.44 + 19.44 \times 10 \times \frac{19.44}{2} \right\}$$

$$q_{max.} = \frac{5}{34692} (39110.4 + 1890.983) = \frac{5 \times 41000.79}{34692} = \textbf{5.910 N/mm}^2$$

$$\text{Maximum} \qquad M - 625 \times 10^4 \text{ N-m}$$

$$f_{max.} \text{ (bending, bottom fibre – compressive)} = \frac{M}{I} \cdot y_{max.}$$

$$= \frac{2650000}{346.92 \times 10^4} \times 90.5555$$

$$= \textbf{163.15 N/mm}^2 \text{ (Compressive)}$$

$$f_{max.} \text{ (top tensile)} = \frac{2650000}{346.92 \times 10^4} \times 29.444 = \textbf{53.050 N-mm}^2 \text{ (Tensile)}$$

ii.   At 2 m from free end

$$F_{max.} = 2 \times 2000 = \textbf{4000 N}, \qquad\qquad M_{max.} = 2 \times 2000 \times \frac{2000}{2} = \textbf{4} \times \textbf{10}^6 \textbf{ N-mm}$$

$$q_{max.} = \frac{4000}{5000} \times 5.91 = \textbf{4.73 N/mm}^2$$

$$f_{max.} = \frac{400 \times 10^4}{625 \times 10^4} \times 163.55 = \textbf{104.42 N/mm}^2 \text{ (Compressive)}$$

$$f_{max.} \text{ (top)} = \frac{400}{625} \times 53.05 = \textbf{33.95 N/mm}^2 \text{ (Tensile)}$$

EXAMPLE 9.4: A wooden beam section 80 mm × 240 mm (deep) is strengthened by a steel plate 80 mm wide × 10 mm thick throughout the span at the bottom face with steel screws of 10 mm diameter placed at a pitch of 120 mm. The beam is simply supported at a span of 3 m and carries a concentrated load of 4 kN at the centre of the span. If self weight of the beam is neglected and modulus ratio of steel to wood = 15, determine the *maximum stresses* in steel plate, screws and wood.

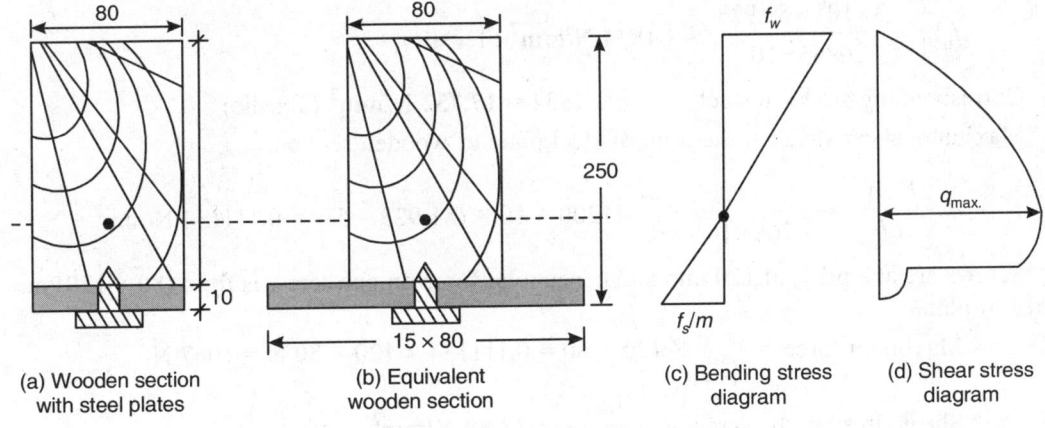

(a) Wooden section with steel plates

(b) Equivalent wooden section

(c) Bending stress diagram

(d) Shear stress diagram

Fig. 9.11.

Solution:

Equivalent section width bottom plate $= m.b = 15 \times 80 = 1200$ mm

Total depth $= 240 + 10 = 250$ mm

$$\overline{y}_b \ (C_g \text{ from bottom}) = \frac{1200 \times 10 \times 5 + 240 \times 80 \times (120 + 10)}{1200 \times 10 + 240 \times 80} = \frac{60000 + 2496000}{12000 + 19200}$$

$$y_b = \frac{256000}{31200} = \textbf{81.923 mm}$$

$y_t \ (\text{top}) = 250 - 81.923 = \textbf{168.077 mm}$

I_{xx} of equivalent section (wooden) $= I_{AB} - Ay_b^2$

$$= \frac{80 \times 250^3}{3} + \frac{1120 \times 10^3}{3} - 31200(81.923)^2$$

$$= \textbf{20765} \times \textbf{10}^4 \textbf{ mm}^4$$

Maximum BM at the centre $= \dfrac{wl}{4} = \dfrac{4000 \times 3}{4}$ N-m $= \textbf{3} \times \textbf{10}^6$ **N-mm**

Maximum SF at ends $= \dfrac{w}{2} = \textbf{2000 N}$

Bending stress equation $\dfrac{M}{I} = \dfrac{f_t}{y_t} = \dfrac{f_b}{y_b}$ in equilibrium wooden section.

$$f_{\text{top}} = \frac{M}{I} \cdot y_{\text{top}} = \frac{3 \times 10^6 \times 168.077}{20765 \times 10^4} = \textbf{2.428 N/mm}^2 \text{ (Compressive)}$$

$$f_{\text{bott.}} = \frac{3 \times 10^6 \times 81.923}{20765 \times 10^4} = \textbf{1.1834 N/mm}^2 \text{ (Tensile)}$$

Corresponding stress in steel = 15 × 1.1834 = **17.752 N/mm^2** (Tenslie)

Maximum shear stress at the joint of steel plate of wooden section

$$q = \frac{F \cdot A\overline{y}}{I.b} = \frac{2000}{20765 \times 10^4 \times 80} \{1200 \times 10 \times (81.923 - 5)\} = 0.111134 \text{ N/mm}^2$$

Screws are at a pitch of 120 mm and hence total force on one screw is from (80 × 120) mm^2 area of plate.

Maximum force = $q_{\text{max.}}$ × 120 × 80 = 0.111134 × 120 × 80 N = 1067 N

Shear stress in the screw = $\dfrac{1067}{\dfrac{\pi}{4}(10)^2}$ = **13.59 N/mm^2**

EXAMPLE 9.5: A wooden beam of rectangular section 150 mm × 250 mm is simply supported over a span of 5 m. The beam carries a point load W at a point 2 m from the left support. If the permissible bending stress is 10 N/mm^2 and permissible shear stress is 1 N/mm^2, find the safe value of the load W.

Solution:

$B = 150$ mm, $\qquad D = 250$ mm, $\qquad L = 5$ m = 5000 mm, $\qquad a = 2000$ mm

$$\text{Maximum BM} = \frac{W \cdot a \cdot b}{L} = W \cdot \frac{2000 \times 3000}{5000} = \textbf{1200 W N-mm.}$$

$$\text{Maximum SF} = \frac{3W}{5} \text{ N,}$$

$$I_{\text{XX}} = \frac{150}{2} \times (250)^3 = 19531.25 \times 10^4 \text{ mm}^4$$

$$f = \frac{M}{I} \cdot y \qquad\qquad \therefore 10 = \frac{1200 \times 125 W}{19531.25 \times 10^4} = 7.68 \times 10^{-4} \, W$$

$$\therefore W_1 = \frac{10 \times 10^4}{7.68} = \textbf{13020 N}$$

$$\text{Maximum shear stress} = 1 \text{ N/mm}^2 = \frac{F \cdot A\overline{y}}{I \cdot b} = \frac{\dfrac{3W_2}{5} \cdot \dfrac{B \cdot D}{2} \cdot \dfrac{D}{4}}{\dfrac{1}{12} B \cdot D^3 \cdot B} = \frac{3W_2 \cdot 12}{5 \times 8 \times BD}$$

or $1 = \dfrac{36W_2}{40 \times 150 \times 250}$, $W_2 = \dfrac{40 \times 150 \times 250}{36} = \dfrac{125000}{3}$

$$= \textbf{41667 N}$$

Thus, load $W = \textbf{13020 N}$ will be safe both in shear and bending.

EXAMPLE 9.6: A laminated wooden beam 120 mm wide and 180 mm deep (made of three 120 mm × 60 mm deep planks glued together to resist longitudinal shear). The beam is simply supported over a span of 2 m. If the allowable shear stress in the glued joint is 1.0 N/mm², find the safe u.d.l. on the beam.

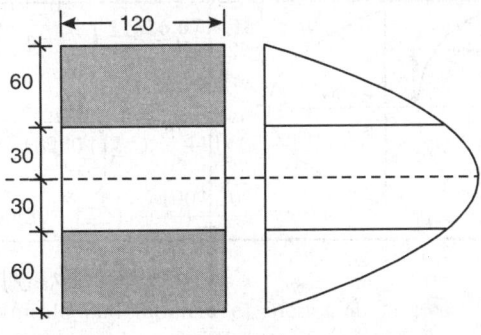

Fig. 9.12.

Solution:

$$I_{XX} = \frac{1}{12} \times 120 \times (180)^3 = 10^4 \times 5832 \text{ mm}^4$$

Max. SF $= F = \dfrac{wL}{2} = \dfrac{w}{2}(2000) = w \times 10^{+3}$ N-mm

Let w be udl N/mm

Shear stress at the glued portion

$$q = \frac{F \cdot A\overline{y}}{I \cdot b} = \frac{1000w}{5832 \times 10^4 \times 120} \times 120 \times 60 \times \left(30 + \frac{60}{2}\right)$$

$$1.0 = \frac{w}{58320} \times 60 \times 60 = \frac{10w}{162}$$

$\therefore$ $w = \dfrac{162}{10} = \textbf{16.2 N/mm} = \textbf{16200 N/m}$

9.5   SUMMARY – SHEAR STRESS IN BENDING STRUCTURES

The effect of shear forces is neglected in simple theory of bending by assuming cross-sectional planes before bending to remain planes after bending without practically affecting the *bending equation*. The *shear-forces* develop direct *shear stresses* in bending elements which cannot be

Table 9.1: Maximum shear stresses in various sections

S. No.	SHAPE OF PLANE AREAS	GEOMETRICAL PROPERTIES
1.		$q = \dfrac{F}{2I_x}\left(\dfrac{d^2}{4} - y^2\right)$ $q_m(NA) = \dfrac{3F}{2b \cdot d} = 1.5 q_{avg.}$
2.		$q = \dfrac{F}{3I}\left(r^2 - y^2\right)$ $q_m = \dfrac{4F}{3\left(\dfrac{\pi}{4}d^2\right)} = \dfrac{4}{3}\,q_{avg.}$
3.		$q = \dfrac{12F}{bh^3}\left(h - y\right)y$, where y from vertex $q_{max}\left(y = \dfrac{1}{2}h\right) = \dfrac{3F}{2\left(\dfrac{bh}{2}\right)} = \dfrac{3}{2}\,q_{avg.}$ $q_{NA}\left(y = \dfrac{2h}{3}\right) = \dfrac{24F}{9bh} = \dfrac{4F}{3\left(\dfrac{bh}{2}\right)} = \dfrac{4}{3}\,q_{avg.}$
4.		$q_f = \dfrac{F}{2I}\left(\dfrac{D^2}{4} - y^2\right)$ $q'_f\,(Junction) = \dfrac{F}{8I}\left(D^2 - d^2\right)$ $q'_w\,(Junction) = \dfrac{F\left(D^2 - d^2\right)}{8I} \cdot \dfrac{B}{t}$ $q_m\,(NA) = \dfrac{F}{2I_x} \cdot \dfrac{B}{t}\left(D^2 - d^2\right) + \dfrac{F \cdot d^2}{2I_x}$
5.		$q = \dfrac{F \cdot A\bar{y}}{I \cdot b}$ $q_m = \dfrac{F}{I} \cdot \dfrac{A\bar{y}}{t_w} = \dfrac{F}{2I}\left(D - \bar{y}\right)^2$

neglected. In addition to longitudinal bending stresses developed due to bending, shear stresses are also developed in bending elements due to shear forces.

Horizontal (longitudinal) and vertical (transverse) shear stresses exist simultaneously in bending elements which can be determined by considering free body diagram of small element at any section. A *shear stress 'q'* on any fibre of *width 'b'* at a distance of *'y' from the neutral axis* in any section is given as:

$$q = \frac{F \cdot A\overline{y}}{I \cdot b}, \text{ with usual notations} \qquad \text{... (9.1)}$$

In case of composite beam sections, the connecting media (screws) are subjected to a total shear force depending on the pitch or surface area covered by each screw. *Total shear force acting on the area between the two consecutive screws can be calculated on the basis of maximum shear stress at the surface. This shear force is borne by each screw and accordingly shear stress setup in the screw can be calculated.*

From equation (9.1), it can easily be found that the *maximum shear stress exists at the neutral surface* in different sections as given in the Table 9.1.

EXERCISE 9

Q.9.1. Prove that shear stress 'q' in any layer of width 'b' in cross-section at a distance 'y' above the neutral surface subjected to shear force 'F' will be 'q' = $\dfrac{F \cdot A\overline{y}}{I \cdot b}$, where $A\overline{y}$ is the moment of cross-sectional area away from the layer of width 'b' and I is the second moment of area about the neutral axis.

Q.9.2. Prove that the *maximum shear stress* in a solid rectangular beam section b × d subjected to a transverse shear force of 'F' will be 1.5 *times the average shear stress* across the cross-section.

Q.9.3. Prove that the *maximum shear stress* in a solid circular beam section of diameter 'd' subjected to shear force 'F' will be $\dfrac{4}{3}$ *times the average shear stress* across the section.

Q.9.4. Determine the maximum shear stress in a solid triangular beam section of base 'b' and height 'h' placed with base horizontal in bottom face and subjected to transverse shear Force F.

Ans. $q_{max} = \dfrac{3F}{bh} = 1.5q_{avg.}$, $q_{NA} = \dfrac{4}{3}q_{avg.}$

Q.9.5. A T-section 100 mm × 160 mm × 10 mm uniform thickness is placed with 100 mm flange horizontal. The section is used for a cantilever beam of 3 m span subjected to a udl of 2 kN/m. Determine the maximum shear stress.

Ans. $\overline{y}$ = 53 mm, $\qquad\qquad$ $I = 666.083 \times 10^4$ mm^4, $\qquad\qquad$ q_m = **5.16 N/nm^2**

Q.9.6. A symmetrical RSJ comprises of two flanges each of 200 mm × 10 mm connected by a web plate of 400 mm × 8 mm. The section is used for a simply supported beam of 4 m span carrying a point load of 100 kN at the middle point. Determine the maximum shear stress and show shear stress variation across the cross-section. Calculate the values of the shear stress at the junction of the flange and web. Also calculate the maximum bending stress. Neglect effect of self weight.

Ans. $I = 2108 \times 10^5$ mm^4, $\qquad\qquad$ $F_{max.}$ = 50 kN, $\qquad\qquad$ $q_{max.}$ = 16.9 N/mm^2,

q_w (junction) = 12.156 N/mm^2, $\qquad\qquad\qquad\qquad\qquad$ q_f (junction) = 0.486 N/mm^2,

Maximum bending stress f = **99.62 N/mm^2**.

Q.9.7. An angle section of 160 mm × 120 mm × 10 mm placed with 160 mm flange horizontal and in the bottom as shown in Fig. Q9.7. Draw shear stress variation across the depth and calculate the maximum value if the section is used for a cantilever of 2 m span and carrying a udl of 5 kN/m. Calculate the maximum bending stress and check if it is safe in bending and shear if permissible stresses are bending tension = 120 N/mm^2, bending compression = 100 N/mm^2 and shear stress = 10.0 N/mm^2.

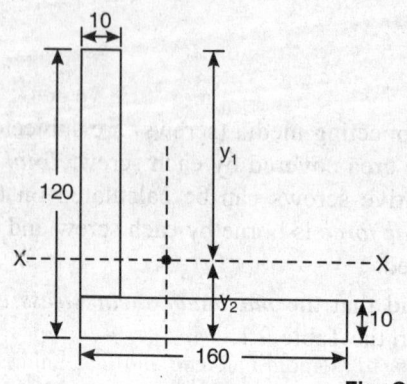
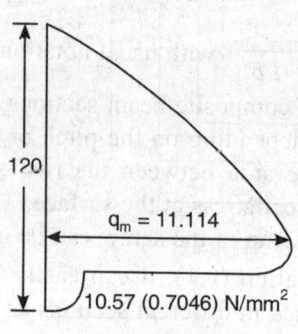

Fig. Q9.7.

Ans. $F_{max.} = 10^4$ N, $M_{max.} = 10^7$ N-mm, $y_1 = 90.556$ mm,
$y_2 = 29.444$ mm, $I_X = 346.92 \times 10^4$ mm^4,
$q_{max.} = \mathbf{11.114}$ **N/mm^2** > 10 N/mm^2 **(unsafe)**, q_w (junction) = 10.57 N/mm^2, **(unsafe)**
q_f *(junction)* = 0.7046 N/mm^2, $f_t = \mathbf{261.03} > 120$ N/mm^2 **(unsafe)**,
$f_c = 84.87 < 100$ N/mm^2 **(safe)**.

Q.9.8. A wooden section of 120 mm × 180 mm is used for a simply supported beam of 3 m span. Find the safe load W which can be placed at 1 m from the left support if the maximum permissible stresses are: f (bending) = 10 N/mm^2, q (shear) = 1.0 N/mm^2.

Ans. $F_{max} = \dfrac{2W}{3} N,$ $M_{max} = \dfrac{2000W}{3}$ N–mm, $I = 5832 \times 10^4$ mm^4

Safe: $W_1 = 9.72$ kN (bending), $W_2 = 21.6$ kN (shear)
Safe load both in bending and shear $W = 9.72$ kN.

Q.9.9. A beam simply supported over a span of 5 m is loaded with a central point load of 40 kN. The reactangular section is 200 mm wide × 400 mm deep. Calculate maximum bending and shear stresses at a cross-section 2 m from the support. Also calculate bending and shear stresses at points 200 mm, 150 mm, 100 mm, 50 mm and 0 mm from the NA at the same cross-section.

Ans. $F_2 = 20 \times 10^3$ N, $M_2 = 40 \times 10^6$ N-mm, $I = 10^8\left(\dfrac{32}{3}\right)$ mm^4,
$q_{max.} = 0.375$ N/mm^2, $f_{max.} = 7.5$ N/mm^2,
$q_{200} = 0$, $f_{200} = 7.5$ N/mm^2, $q_{150} = 0.164$ N/mm^2,
$f_{150} = 5.625$ N/mm^2, $q_{100} = 0.28125$ N/mm^2, $f_{100} = 3.75$ N/mm^2,
$q_0 = 0.375$ N/mm^2, $f_0 = 0$.

Q.9.10. A 200 mm × 400 mm (overall) RSJ has flanges 20 mm thick and web 16 mm thick. The beam section at certain point is subjected to a bending moment of 200 kN-m and a shear force of 400 kN. Find the bending stresses and shear stresses at the NA, junction of the web and flange and extreme fibres of the flange.

Ans. $I = 351.275 \times 10^6$ mm^4

   i. $f_{NA} = 0$, q_{NA} (max.) = 72.54 N/mm^2.

   ii. $f_{junction}$ (flange-web) = 102.48 N/mm^2, q_w (junction) = 54.09 N/mm^2

       q_f (junction) = 4.33 N/mm^2.

   iii. f_{flange} (max.) = 113.87 N/mm^2, $q_f = 0$

       f_{100} = **56.94 N/mm^2**, q_{100} = **66.843 N/mm^2**

Q.9.11. A wooden beam section 40 mm × 120 mm deep is strengthened by a steel plate 40 mm × 10 mm thick throughout the span at the bottom face with the steel screws of 10 mm diameter and placed at a pitch of 100 mm along the span. The cantilever beam has 2 m span and supports a concentrated load of 3 kN at the free end. Determine the maximum bending and shear stresses in the steel plate, wooden section and screws if the self weight is neglected and modular ratio of steel to wood = 15.

Ans. y_t (top) = 96.111 mm, y_c (bottom) = 33.889 mm, $I_X = 17077.7 \times 10^3$ mm^4,

   $M_{max.} = 6 \times 10^6$ N-mm, $F_{max.}$ = 3000 N,

   f_c (max.) in steel = 178.6 N/mm^2, f_c (junction) in steel = 125.9 N/mm^2,

   f_c (junction) in wood = 8.394 N/mm^2, f_t (top) in wood = 33.77 N/mm^2,

   q (junction) in steel = 0.7613 N/mm^2, q (screw) in shear = 38.8 N/mm^2.

Unit IV

Bending Structures: Deformations

10

Slope and Deflections

LEARNING OBJECTIVES

After studying this chapter, the learner **understands Slope** and **Deflections** produced in beams due to transverse loading and will be able to:

10.1 **Explain** development of deformations (**Slope and Deflections**) due to **transverse loading** on beams.

10.2 **Derive** basic relationship of **deflection 'y'** and **bending moment 'M'** by double integration approach.

10.3 **Derive** equations of **slope and deflection** for **specific loading** on a given beam by double integration method.

10.4 **Calculate maximum** values of **slope and deflection** for a given loading on a given beam by double integration method.

10.5 **Calculate slopes and deflections** at given points for a given type of loading on a given beam by double integration and **Macaulay's Method**.

10.6 **Explain the area moment method** for **slope and deflection** in beams carrying different transverse loading.

10.7 **Explain the conjugate beam method** for slope and deflection in beams carrying different transverse loading.

10.8 **Explain the Strain Energy method** for slope and deflection in beams carrying different transverse loading.

10.9 **Explain the importance of slope and deflection** in design of beams and stiffness calculations.

10.10 **Explain the importance of deflections** caused by shear stress in the design of beams.

10.1 INTRODUCTION

In earlier chapters, we have studied longitudinal and transverse stresses caused in beams due to bending moments and shear forces. We designed the beam sections in such a way that the stresses caused by bending moment or shear force *remain below the safe stress* values of the material. Apart from the *limits of stresses*, the *deformations* in beams must also *remain within certain limits* of material. The deformations in beams depend on the *stiffness of the beam sections*, which in turn depends on the *material* and *sectional* properties. It is observed that the deformations in beams (slope and deflections) are inversely proportional to the stiffnesses of

336

the beams. Thus beams are designed to withstand safely bending moments and shear forces but must also have *adequate stiffness* to keep *slope and deflections* within certain permissible limits.

We observed that the longitudinal axis of the beam member under transverse loading deflects. The *deflected shape* of the longitudinal axis under transverse loads within certain limits is called *elastic curve* (Fig. 10.1).

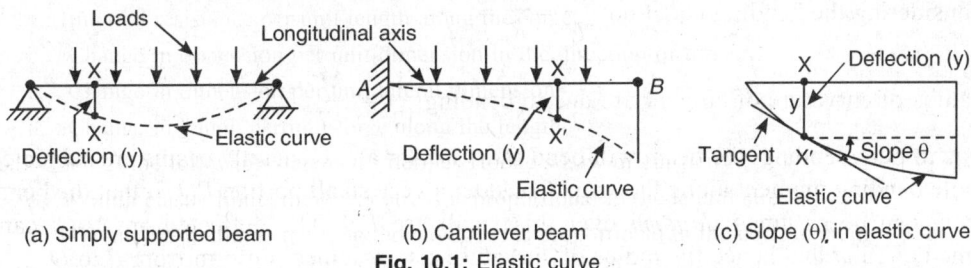

| (a) Simply supported beam | (b) Cantilever beam | (c) Slope (θ) in elastic curve |

Fig. 10.1: Elastic curve

Various points on the longitudinal axis under the transverse loads move in the direction of the loads. The *curve* joining these *deflected points* forms the *elastic curve*. The transverse *displacement* (XX') in the direction of the load with reference to the *original point* (X) on the *longitudinal axis*, is called *deflection* ($y = XX'$). The inclination θ_x of the *tangent to the elastic curve* at any point X' with reference to the *original axis* before loading is called *slope* ($dy/dx = \theta_x$). The *slope* and *deflection* in any beam element should *not exceed* certain limit to keep good appearance of the structural element. Large *deflections and slopes* in beams cause cracks in ceiling plasters and give rise to unsightly appearance. Such unsightly conditions and cracks in plasters give rise to the feeling of *insecurity* to the occupants affecting their *health, productivity* and *efficiency*.

In case of *statically indeterminate* structures, the *slope and deflection* equations are used for finding the indeterminate forces (or stresses) by forming *additional equations of compatibility* (geometric *consistency* and *continuity*) at any point in additions to the basic *equations of static equilibrium*.

10.2 THE EQUATION OF DEFLECTION CURVE

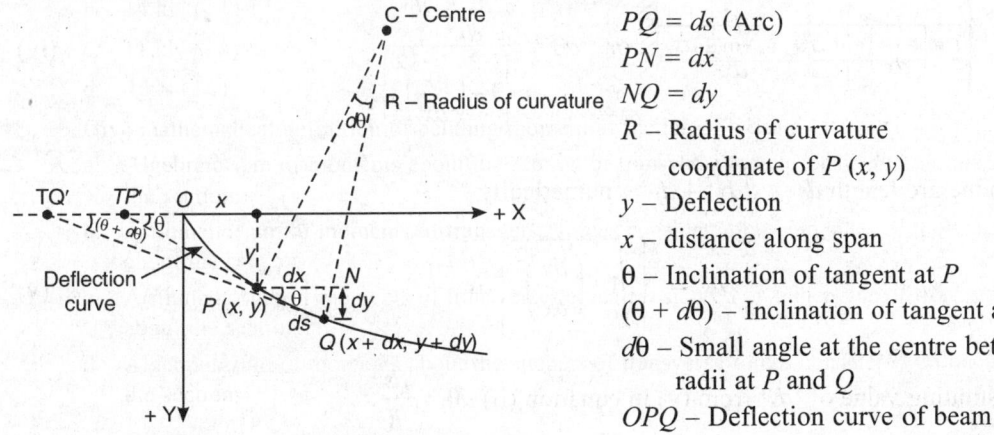

$PQ = ds$ (Arc)
$PN = dx$
$NQ = dy$
R – Radius of curvature
  coordinate of P (x, y)
y – Deflection
x – distance along span
θ – Inclination of tangent at P
($\theta + d\theta$) – Inclination of tangent at Q
$d\theta$ – Small angle at the centre between
  radii at P and Q
OPQ – Deflection curve of beam

Fig. 10.2: Elastic deflection curve of beam

Figure 10.2 shows a beam axis *OX* with support, say at *O*. The beam is subjected to a *pure bending* moment (couple), say '*M*'. The beam axis under pure bending moment *M* bends in a *circular arc*. Consider a small *arc PQ* (length *ds*). Let *P* (*x*, *y*) and *Q* (*x* + *dx*, *y* + *dy*) be two points of a small circular arc *PQ*. *PN* (*dx*), is small distance along the beam axis *OX*. '*dy*' be the *deflection difference* between the two points *P* and *Q* of the small arc.

Considering the bending equation: $\dfrac{M}{I} = \dfrac{E}{R} = \dfrac{f}{d_1}$

Radius of curvature of circular arc due to bending: $\dfrac{1}{R} = \dfrac{M}{EI}$ ... (i)

Due to pure bending, the beam will bend into circular arc. Generally beams are subjected to variable bending moment along the span. Consider a very small portion *PQ* so that the bending *moment* can be assumed *uniform* over this small arc *PQ*. The deflected arc '*PQ*' can be assumed circular and hence the radius of curvature also assumed uniform from *P* to *Q*.

The length of arc *PQ* (*ds*) = *R*. *d*θ, where '*d*θ' is the *angle* subtended at the center of the circular arc between two radii drawn at *P* and *Q*. Since the *arc* is very small the angle *d*θ will be also very small.

$$d\theta = \frac{ds}{R}$$ ... (ii)

Consider triangle formed by *P*, *Q* and *N*. The tangent at *P* makes an angle θ with the original horizontal axis *OX*. From the triangle *PNQ*:

$$\tan\theta = \frac{dy}{dx}$$ ... (iii)

Differentiating (iii):

$$\sec^2\theta \cdot d\theta = \frac{d^2y}{dx^2} \cdot dx, \quad \text{or} \quad \left(1 + \tan^2\theta\right) \cdot d\theta = \frac{d^2y}{dx^2} \cdot dx$$

or $\left\{1 + \left(\dfrac{dy}{dx}\right)^2\right\} \cdot d\theta = \dfrac{d^2y}{dx^2} \cdot dx$, or $d\theta = \dfrac{\dfrac{d^2y}{dx^2} \cdot dx}{\left\{1 + \left(\dfrac{dy}{dx}\right)^2\right\}}$ ... (iv)

Also the arc length $ds = \sqrt{dx^2 + dy^2}$, numerically

or $$ds = \left\{\sqrt{1 + \left(\frac{dy}{dx}\right)^2}\right\} dx$$ ... (v)

Substituting value of '*ds*' from (v) in equation (ii) $d\theta = \dfrac{ds}{R}$,

$$\left\{\pm\sqrt{1+\left(\frac{dy}{dx}\right)^2}\right\}\frac{dx}{R} = d\theta \qquad \qquad \text{... (vi)}$$

from (iv),
$$d\theta = \frac{\dfrac{d^2y}{dx^2}}{\left\{1+\left(\dfrac{dy^2}{dx}\right)^2\right\}}\cdot dx$$

from (vi) and (iv) we have,
$$\left\{\pm\sqrt{1+\left(\frac{dy}{dx}\right)^2}\right\}\frac{dx}{R} = \frac{\dfrac{d^2y}{dx^2}}{\left\{1+\left(\dfrac{dy^2}{dx}\right)^2\right\}}\cdot dx$$

or
$$\frac{1}{R} = \pm\frac{\dfrac{d^2y}{dx^2}}{\left\{1+\left(\dfrac{dy}{dx}\right)^2\right\}\left\{\sqrt{1+\left(\dfrac{dy}{dx}\right)^2}\right\}} = \pm\frac{\dfrac{d^2y}{dx^2}}{\left\{1+\left(\dfrac{dy}{dx}\right)^2\right\}^{\frac{3}{2}}} \qquad \text{... (vii)}$$

Since dy/dx is small, $(dy/dx)^2$ will be very small, and can be ignored for simplicity, therefore equation (vii) reduces to

$$\frac{1}{R} = \pm\frac{\dfrac{d^2y}{dx^2}}{(1)^{\frac{3}{2}}} \text{ or } \frac{1}{R} = \pm\frac{d^2y}{dx^2}$$

From equation (i) $\dfrac{1}{R} = \dfrac{M}{EI} = \pm\dfrac{d^2y}{dx^2}$

or
$$EI\cdot\frac{d^2y}{dx^2} = \pm M \qquad \qquad \text{... (10.1a)}$$

This equation is applied over a *small length*, so that the *arc remains circular* and the equation (10.1) holds good for the deflected curve portion PQ. i.e. $EI\cdot\dfrac{d^2y}{dx^2} = \pm M$ ... (10.1a)

Generalized differential equation of deflection curve $EI\cdot\dfrac{d^2y}{dx^2} = \pm M_x$, depends on the sign conventions for M_x, deflection y and slope $\dfrac{dy}{dx}$.

By *integrating* differential equation (10.1a) once, we get the *slope of deflection* curve by substituting the values of '*x*' and *constants* of integration which are found by *boundary/support* conditions. By *integrating* the differential equation (10.1) *twice*, we get *deflection* by substituting the values of '*x*' and *constants* of integration which are found by *boundary/support* conditions. Equation (10.1a) is *general equation* and can be converted into *appropriate equation* by considering proper *sign conventions*.

Deflection '*y*' shall be considered *positive* in *downward* direction and *negative* in *upward* direction *above* the original axis. The values of *bending moment* causing deflection is considered *positive* when it is *sagging* and *negative* when it is *hogging*. The slope is considered *positive* when *clockwise* and *negative* when *anticlockwise*. Considering these sign conventions the *differential equation* of *deflection curve* becomes:

$$EI\frac{d^2y}{dx^2} = -M_x$$

... (10.1)

Figure 10.3 shows deflection curve with *sign conventions* of various quantities.

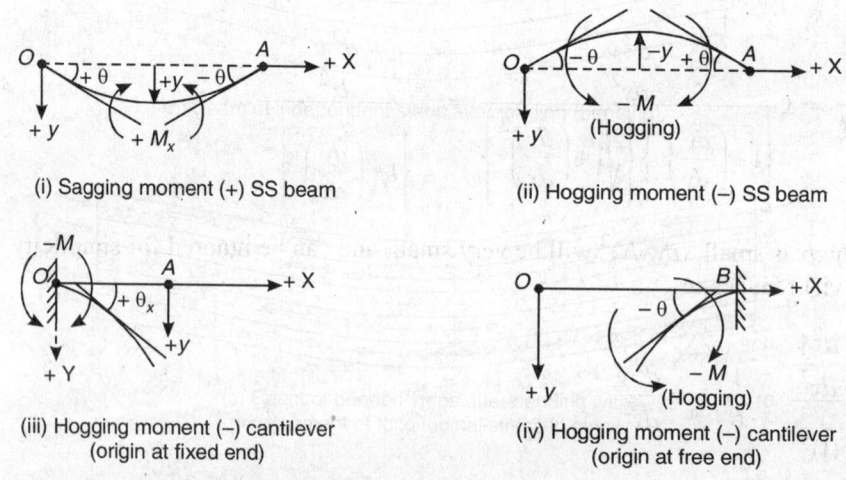

(i) Sagging moment (+) SS beam

(ii) Hogging moment (–) SS beam

(iii) Hogging moment (–) cantilever
(origin at fixed end)

(iv) Hogging moment (–) cantilever
(origin at free end)

Fig. 10.3: Sign convention for deflection curve

Using equation (10.1) with proper sign conventions, various quantities (θ, y, M, x, etc.) are shown in Fig. 10.3. By considering sagging moment as positive, we get slope (θ), $dy/dx = +$ ve at $x = 0$, (Fig 10.3i), similarly $dy/dx = -$ ve at $x = L$ in SS beam.

Similarly, different cases are shown in Fig. 10.3 (i – iv), for proper understanding of *deflection curve* and *signs* of various quantities (θ, y, M, x, etc.).

Slope and deflections can be found by *double integration* of differential equation $EI\frac{d^2y}{dx^2} = -M_x$, by finding *constants of integration* from known boundary *support conditions*.

Different engineers devised different methods of finding slope-deflection by application of the

basic equation $EI\dfrac{d^2y}{dx^2} = -M_x$, with proper signs.

These methods are based on simple *double integration, Macaulay's* method of integration, *Mohr's theorems* based on graphical approaches such as *moment area method*, and *conjugate beam* method, etc. *Strain energy* method can also be suitably used for finding slope and deflections.

We shall first use basic *Double integration* method for finding *slope and deflections* in cases of beams with different type of loadings. Double integration of differential equation

$$EI\dfrac{d^2y}{dx^2} = -M_x.$$

First Integration of the equation gives: $EI\dfrac{dy}{dx} = -M_x \cdot \dfrac{x}{n} + C_1$, n depends on total powers of x in the expression M_x.

Slope $\dfrac{dy}{dx} = \theta = \dfrac{1}{EI}\left[-M_x \cdot \dfrac{x}{n} + C_1\right]$ ... (i)

Second integration, $EI \cdot y = \left[(-M_x)\dfrac{x^2}{(n+1)} + C_1x + C_2\right]$

Deflection $y = \dfrac{1}{EI}\left[-M_x \cdot \dfrac{x^2}{n+1} + C_1x + C_2\right]$ ... (ii)

10.3 DOUBLE INTEGRATION METHOD FOR SLOPE AND DEFLECTION IN BEAMS

10.3.1 Cantilever Beams

Case 1a: Concentrated load W on free end A.

Cantilever AB of span L,

$M_x = -W(x)$ at any point X at a distance of x from the free end A.

Free end A and Fixed end B.

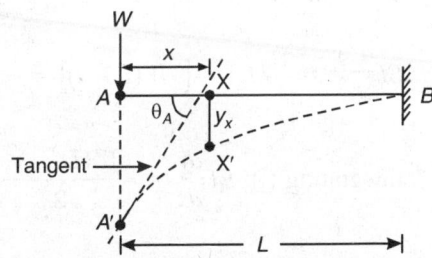

Fig. 10.4: Cantilever AB with point load W at A (free end)

Diferential equation: $EI\dfrac{d^2y}{dx^2} = -M_x = -(-Wx)$

$$= +Wx$$

Integrating: $EI\dfrac{dy}{dx} = W \cdot \dfrac{x^2}{2} + C_1$ ... (i)

Slope $\dfrac{dy}{dx} = 0$, at fixed end i.e. $x = L$

$$0 = \frac{W}{2}(L^2) + C_1, \qquad C_1 = -\frac{WL^2}{2}$$

$$EI\frac{dy}{dx} = \frac{Wx^2}{2} - \frac{WL^2}{2} \qquad \qquad \dots \text{(ii)}$$

Integrating again: $EI \cdot y = \frac{W}{2} \cdot \frac{x^3}{3} - \frac{WL^2}{2} \cdot x + C_2$

At fixed end, $x = L$, $y = 0$, i.e. $0 = \frac{W}{6}L^3 - \frac{WL^2}{2} \cdot L + C_2 \quad \therefore C_2 = +\frac{WL^3}{3}$

i.e. $EIy = \frac{W}{6}x^3 - \frac{WL^2}{2} \cdot x + \frac{WL^3}{3} \qquad \qquad \dots \text{(iii)}$

Free end $x = 0$:

Free end slope $\underset{\text{max}}{\theta_A}(x = 0) = -\frac{WL^2}{2EI}$ radian $\qquad \qquad \dots \text{(10.2)}$

Free end deflection $\underset{\text{max}}{y_A}(x = 0) = +\frac{WL^3}{3EI}$ units $\qquad \qquad \dots \text{(10.3)}$

Case 1b: Consider cantilever AB with A as fixed and B as free and carrying point load W at the free end.

Span $AB = L$, A – Fixed, B – Free end.

x from fixed end A

$M_x = -W(L-x)$

$EI\frac{d^2y}{dx^2} = -M_x = -[-W(L-x)] = +W(L-x) \qquad \dots \text{(i)}$

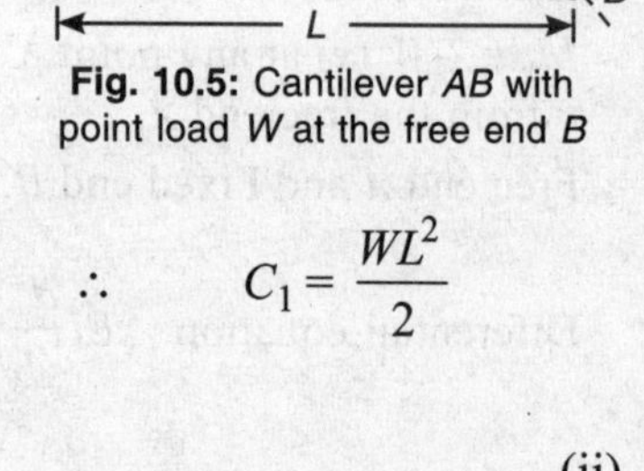

Fig. 10.5: Cantilever AB with point load W at the free end B

Integrating (i): $EI\frac{dy}{dx} = -\frac{W(L-x)^2}{2} + C_1$

At $x = 0$, $\frac{dy}{dx} = 0$, $\therefore \quad 0 = -\frac{W(L^2)}{2} + C_1, \qquad \therefore \quad C_1 = \frac{WL^2}{2}$

$\therefore \qquad EI\frac{dy}{dx} = -\frac{W(L-x)^2}{2} + \frac{WL^2}{2} \qquad \qquad \dots \text{(ii)}$

Integrating again: $EI \cdot y = +\frac{W(L-x)^3}{6} + \frac{WL^2}{2} \cdot x + C_2$

$$At\ x = 0,\ y = 0, \quad \text{i.e.} \quad 0 = +\frac{WL^3}{6} + 0 + C_2, \qquad\qquad \therefore \qquad C_2 = -\frac{WL^3}{6}$$

$$\therefore \qquad EIy = +\frac{W(L-x)^3}{6} + \frac{WL^2}{2} \cdot x - \frac{WL^3}{6} \qquad\qquad \text{... (iii)}$$

Maximum values are at the *free end* ($x = L$):

Free end $\quad \underset{\text{max}}{\theta_B} = +\dfrac{WL^2}{2EI}$ radian $\qquad\qquad\qquad\qquad$ ... (10.4)

Free end $\quad \underset{\text{max}}{y_B} = \dfrac{\dfrac{WL^3}{2} - \dfrac{WL^3}{6}}{EI} = \dfrac{WL^3}{3EI}$ $\qquad\qquad\qquad$ ... (10.5)

From equation (ii) and (iii) the values of slope θ and deflection y can be found at any point x from the fixed end A.

Case 2: Cantilever beam AB with concentrated load 'W' at a distance of 'l_1' from the fixed end.

$\qquad\qquad AB$ – Span L, $A_1 C_1$ – Straight

$\qquad\qquad BC_1$ – Curved, BC – l_1 from the fixed end

$\qquad\qquad$ Load W at C, $AA_2 = y_c$, $AA_1 = y_c + A_1A_2$

Consider a cantilever AB fixed at B and carrying a point load W at a distance of l_1 from the fixed end 'B'.

From earlier derivations:

Slope at C: $\theta_c = -\dfrac{Wl_1^2}{2EI}$, $\qquad$ (Anticlockwise)

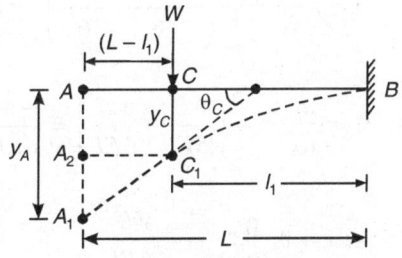

Fig. 10.6: Cantilever with point load W at l_1 from the fixed end

$$\underset{\text{max}}{\theta_A} = \theta_c = -\frac{Wl_1^2}{2EI}, \qquad\qquad\qquad\qquad \text{... (10.6)}$$

Deflection $y_c = \dfrac{Wl_1^3}{3EI}$

Since there is no bending between free end A and the load point C, the tangent to deflection curve at C remains straight, i.e. slope from C to A remains constant as θ_C.

Deflection at A = Deflection y_C at C + Slope $\theta_C \times (L - l_1)$

or $\underset{\text{(max)}}{y_A} = AA_2 + A_2A_1 = \dfrac{Wl_1^3}{3EI} + \dfrac{-Wl_1^2}{2EI}\,(-CA) = \dfrac{Wl_1^3}{3EI} + \dfrac{Wl_1^2(L-l_1)}{2EI}$ $\qquad$ ... (10.7)

Case 3a: Cantilever AB (span L) carrying uniformly distributed load 'w' on the entire span.

$\qquad\qquad AB$ – Span L, $\quad AX - x$ from the free end

$\qquad\qquad$ udl – w/m, $\quad$ Deflection at $X = y_x$

Deflection at $A = y_A$ (max.)

Consider free end A as the origin. Bending moment M_x at any section $X\text{-}X'$ at a distance of x from the free end A:

$$M_x = -\frac{wx^2}{2}$$

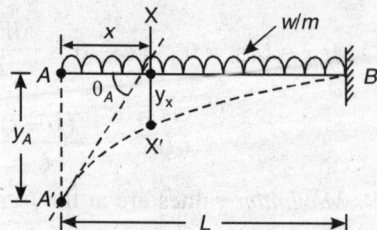

Fig. 10.7: Cantilever carrying udl w/m on entire span

We have, $EI\dfrac{d^2 y}{dx^2} = -M_x = -\left[-\dfrac{wx^2}{2}\right] = \dfrac{wx^2}{2}$ ... (i)

Integrating (i): $EI\dfrac{dy}{dx} = +\dfrac{w \cdot x^3}{2 \times 3} + C_1 = \dfrac{wx^3}{6} + C_1$

Slope $\dfrac{dy}{dx} = 0$, at B (i.e. $x = L$), $0 = \dfrac{wL^3}{6} + C_1$, $\therefore C_1 = -\dfrac{wL^3}{6}$

$\therefore EI\dfrac{dy}{dx} = -\left(+\dfrac{wL^3}{6} - \dfrac{wx^3}{6}\right),$

$\theta_x = \dfrac{dy}{dx} = -\left(\dfrac{wL^3}{6EI} - \dfrac{wx^3}{6EI}\right) = \dfrac{-w}{6EI}(L^3 - x^3)$ ... (ii)

$At\ x = 0,\ \underset{(max)}{\theta_A} = -\dfrac{wL^3}{6EI}$ ... (10.8)

Integrating (ii): $EIy = +\dfrac{wx^4}{24} - \dfrac{wL^3 \cdot x}{6} + C_2$

$At\ x = L$, (fixed support B), $y = 0 = +\dfrac{wL^4}{24} - \dfrac{wL^4}{6} + C_2$, $\therefore C_2 = +\dfrac{wL^4}{8}$

$\therefore EIy = \dfrac{w \cdot x^4}{24} - \dfrac{wL^3}{6} \cdot x + \dfrac{wL^4}{8}$

$y_x = \dfrac{wx^4}{24EI} - \dfrac{wL^3 \cdot x}{6EI} + \dfrac{wL^4}{8EI}$ ... (iii)

$\underset{(max)}{y_A}\ (x = 0) = \dfrac{wL^4}{8EI}$ ... (10.9)

Proper units may be used in the formula, so that there is no discrepancy of units and all units are consistent.

Case 3b: Cantilever AB of span L, carrying udl over partial span L_1 from the fixed end. Span $AB = L$, $AC = (L - L_1)$, $BC = L_1$ (loaded span),

Slope in BC – Variable, Deflection in BC – Variable
Deflection in AC – Variable

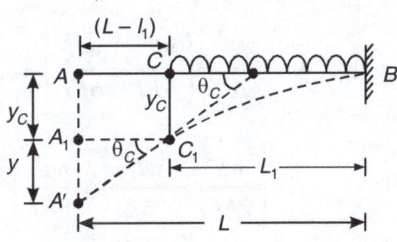

$$\frac{dy}{dx} = \theta_x = \frac{wx^3}{6EI} - \frac{wL_1^3}{6EI}, \text{ where } x \text{ is from } C$$

$$\theta_c = -\frac{wL_1^3}{6EI} \text{ (maximum at } C\text{)}$$

Fig. 10.8: Cantilever with partial udl
(on span L_1 from the fixed support)

From C to A, $\underset{\text{(max)}}{\theta_A} = -\frac{wL_1^3}{6EI} = \theta_c$ ... (10.10)
(straight line)

Deflection from C to B (x from C): $y_x = \dfrac{1}{EI}\dfrac{wx^4}{24} - \dfrac{wL_1^3 x}{6} + \dfrac{wL_1^4}{8}$

$$\underset{(x=0,\,atC)}{y_c} \text{ (At point } C\text{)} = \frac{wL_1^4}{8EI} \qquad \qquad ... (10.11)$$

From A to C: $y = y_c \quad _c(L \quad x \quad L_1)$, where x from A.

$$y = \left\{\frac{wL_1^4}{8EI}\right\} + \left\{\frac{wL_1^3}{6EI}\right\} (L \quad x \quad L_1), x \text{ from the end } A.$$

$$y \text{ (max.)} = \underset{(x=0)}{y_A} = \left\{\frac{wL_1^4}{8EI}\right\} + \left\{\frac{wL_1^3}{6EI}\right\}\{L - L_1\} \qquad ... (10.12)$$

Use appropriate units for loading (total or per metre) so that it matches with other linear and force units.

Case 3c: Cantilever AB (span L) and carrying udl on partial span $(L - L_1)$ from the free end.

    AB – Span L, CB – Empty Span L_1
    udl $+ w$ on AC, $- ve$ udl $(-w)$ on BC
    Maximum Deflection at $A - y_{max}$.

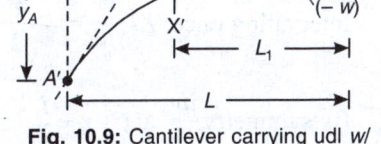

Consider the span AC loaded with udl 'w'. This can be considered *equivalent* to whole span AB loaded with 'w' while portion BC loaded with negative loading $(- w)$.

Fig. 10.9: Cantilever carrying udl w/
m on the span AC

(a)   Consider X-X' within AC i.e. $x \le (L - L_1)$
(b)   Consider X-X' within CB i.e. $x \ge (L - L_1)$

(a) $\theta_x = \dfrac{dy}{dx} = \left(\dfrac{wx^3}{6EI} - \dfrac{wL^3}{6EI}\right) - \left\{\dfrac{wL_1^3}{6EI}\right\} = \dfrac{w}{6EI} \quad x^3 \quad L_1^3 \quad L^3$ ... (10.13)

$$y = \left[\frac{wx^4}{24EI} - \frac{wL^3 \cdot x}{6EI} + \frac{wL^4}{8EI}\right] - \left[\frac{wL_1^4}{8EI} + \frac{wL_1^3}{6EI}\cdot(L - L_1 - x)\right] \qquad ... (10.14)$$

(b) $\quad \theta_x = \dfrac{dy}{dx} = \left(\dfrac{wx^3}{6EI} - \dfrac{wL^3}{6EI} \right) - \left[\dfrac{w}{6EI}(x - L + L_1)^3 - \dfrac{wL_1^3}{6EI} \right]$ $\qquad$ ... (10.15)

$$y = \left[\dfrac{wx^4}{24EI} - \dfrac{wL^3 \cdot x}{6EI} + \dfrac{wL^4}{8EI} \right] - \left[\dfrac{w(x - L + L_1)^4}{24EI} - \dfrac{wL_1^3}{6EI}(x - L + L_1) + \dfrac{wL_1^4}{8EI} \right] \quad \text{... (10.16)}$$

10.3.2 Simply Supported Beams

Case 1: Concentrated load W at the mid span point C.

Simply Supported Span $AB = L$

Load W at the Centre C

Reaction: $R_A = R_B = \dfrac{W}{2}$

BM (Sagging) $M_x = + \dfrac{W}{2} \cdot x$

(from A to C)

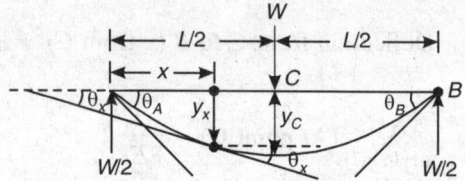

Fig. 10.10: Simply supported beam AB with point load W at centre

Consider a SS beam AB carrying a point load W at the mid span point C. By symmetry

$R_A = R_B = \dfrac{W}{2}$.

$$\underset{(A\,to\,C)}{M_x} = + \dfrac{W}{2} \cdot x$$

$$EI \dfrac{d^2y}{dx^2} = - \left(\dfrac{W}{2} \cdot x \right) \qquad \text{... (i)}$$

Integrating once: $EI \dfrac{dy}{dx} = - \dfrac{W}{2} \cdot \dfrac{x^2}{2} + C_1 = - \dfrac{W}{4} \cdot x^2 + C_1,$ $\qquad$ ... (ii)

By symmetry $\dfrac{dy}{dx}$ at $C\left(x = \dfrac{L}{2} \right) = 0 = - \dfrac{W}{4}\left(\dfrac{L^2}{4} \right) + C_1, \quad \therefore \quad C_1 = \dfrac{WL^2}{16}$ $\qquad$ ...(iii)

$$EI \dfrac{dy}{dx} = - \dfrac{W}{4} \cdot x^2 + \dfrac{WL^2}{16} \qquad \text{... (10.17)}$$

$$\underset{(x=0)}{\theta_A} = \dfrac{WL^2}{16EI} = - \theta_B \qquad \text{... (10.18)}$$

Integrating once again: $EIy = - \dfrac{W}{4} \cdot \dfrac{x^3}{3} + \dfrac{WL^2}{16} \cdot x + C_2$ $\qquad$ ... (iv)

At $x = 0 \cdot y = 0$ gives, $0 = 0 + 0 + C_2$, $\therefore C_2 = 0$ ... (v)

$$EIy = -\frac{W}{12} \cdot x^3 + \frac{WL^2}{16} \cdot x \qquad \qquad \text{... (10.19)}$$

y_m = maximum deflection at the centre $\left(x = \dfrac{L}{2} \right)$

$$y_{max} = \frac{1}{EI} \left[-\frac{WL^3}{96} + \frac{WL^3}{32} \right] = \boldsymbol{\frac{WL^3}{48EI}} \qquad \qquad \text{... (10.20)}$$

Case 2: Uniformly distributed load w/m over the entire span 'L' of SS beam AB.

SS Span AB (L), udl = w/m on full span AB

C – Centre of span ($x = \dfrac{L}{2}$)

θ_x – Slope at any point X

y_x – Deflection at any point X

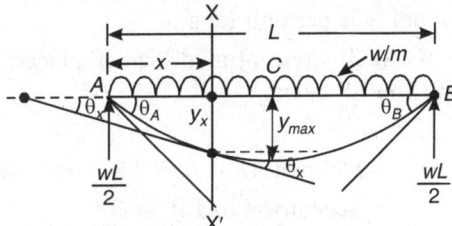

Fig. 10.11: SS beam AB carrying udl w/m on the entire span

By symmetry BM: $M_x = +\dfrac{wL}{2} x - \dfrac{wx^2}{2}$ (Sagging)

Differential equation of deflection curve $EI\dfrac{d^2y}{dx^2} = -M_x = -\dfrac{wL}{2} \cdot x + \dfrac{wx^2}{2}$

$$EI\frac{d^2y}{dx^2} = -\frac{wL}{2} \cdot x + \frac{wx^2}{2} \qquad \qquad \text{... (i)}$$

Integrating once: $EI\dfrac{dy}{dx} = -\dfrac{wL}{2} \cdot \dfrac{x^2}{2} + \dfrac{w}{2} \cdot \dfrac{x^3}{3} + C_1$ ... (ii)

At the centre $\left(x = \dfrac{L}{2} \right)$, $\dfrac{dy}{dx} = 0 = -\dfrac{wL}{4}\left(\dfrac{L^2}{4} \right) + \dfrac{w}{6}\left(\dfrac{L^3}{8} \right) + C_1$, $\therefore C_1 = +\dfrac{wL^3}{24}$ ... (iii)

$$\therefore EI\frac{dy}{dx} = -\frac{wL}{4} \cdot x^2 + \frac{w}{6} \cdot x^3 + \frac{wL^3}{24} \qquad \qquad \text{... (10.21)}$$

$$\begin{array}{l} \theta_A \\ {\scriptstyle (x=0)} \end{array} = \frac{wL^3}{24EI} = -\theta_B \qquad \qquad \text{... (10.22)}$$

Integrating once more: $EIy = -\dfrac{wl}{4} \cdot \dfrac{x^3}{3} + \dfrac{w}{6} \cdot \dfrac{x^4}{4} + \dfrac{wL^3}{24} \cdot x + C_2$ ... (iv)

$y = 0$ at $x = 0$, hence $0 = -0 + 0 + 0 + C_2$, $\therefore$ $C_2 = 0$ ... (v)

$$EIy = \frac{wL}{12} \cdot x^3 + \frac{w}{24} \cdot x^4 + \frac{wL^3}{24} \cdot x + 0$$ ... (10.23)

y_{max} at the centre $\left(x = \frac{L}{2} \right)$, $\therefore$ $y_{max} = \frac{1}{EI} \left[-\frac{wL^4}{96} + \frac{w}{24} \cdot \frac{L^4}{16} + \frac{wL^3}{24} \cdot \frac{L}{2} \right]$

$$y_{max} \left(x = \frac{L}{2} \right) = \frac{wL^4}{EI} \left[\frac{-4+1+8}{96 \times 4} \right] = \frac{5}{384} \frac{wL^4}{EI}$$ ... (10.24)

udl w is per unit length.

Case 3: Concentrated load W placed eccentrically at C on SS beam.

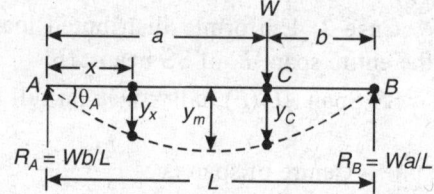

SS Span $AB = L$, $AC = a$, $CB = b$, $a > b$
Eccentric Load W at C.

Fig. 10.12: Eccentric load W on SS beam AB

Figure 10.12 shows a beam AB of SS span L and loaded with a point load W at a point C, distance 'a' from the end A and 'b' from the end B.

Reactions: $R_A = \dfrac{Wb}{L}$, $R_B = \dfrac{Wa}{L}$

Span AC $(x < a)$: $EI \dfrac{d^2y}{dx^2} = -M_x = -\dfrac{Wb}{L} \cdot x$ ... (i)

Integrating once: $EI \dfrac{dy}{dx} = -\dfrac{Wb}{L} \cdot \dfrac{x^2}{2} + C_1$ ... (ii)

Integrating again: $EIy = -\dfrac{Wb}{L} \cdot \dfrac{x^3}{6} + C_1 x + C_2$ ... (iii)

Span CB $(x > a)$: $EI \dfrac{d^2y}{dx^2} = -\dfrac{Wb}{L} \cdot x + W(x-a)$ ... (iv)

Integrating once: $EI \dfrac{dy}{dx} = -\dfrac{Wb}{L} \cdot \dfrac{x^2}{2} + \dfrac{W}{2}(x-a)^2 + C_3$ ... (v)

Integrating again: $EIy = -\dfrac{Wb}{L} \cdot \dfrac{x^3}{6} + \dfrac{W}{6}(x-a)^3 + C_3 x + C_4$ ... (vi)

Equation (ii) and (iii) give values of *slope* and *deflections* for the *portion AC* while equation (v) and (vi) give values of *slope* and *deflections* for the portion **CB**.

The constants of integration C_1, C_2, C_3 and C_4 can be found from various boundary conditions for the portions *AC* and *CB*. When $x = a$, $\dfrac{dy}{dx} = \theta_c$. This must be same from both the equations (ii) and (v) for *AC* or *CB*.

$$-\frac{Wb}{L} \cdot \frac{x^2}{2} + C_1 = -\frac{Wb}{L} \cdot \frac{x^2}{2} + \frac{W}{2} \cdot (0)^2 + C_3, \therefore \; \boldsymbol{C_1 = C_3} \qquad \text{...(vii)}$$

Also deflection at *C* shall be equal from equation (iii) and (vi) when $x = a$.

$$EIy = -\frac{Wba^3}{6L} + C_1 \cdot a + C_2 = -\frac{Wba^3}{6L} + 0 + C_3 \, (a) + C_4$$

or $C_2 = C_4$, Since $C_1 = C_3$ [equation (vii)].

Substituting $C_3 = C_1$, and $C_2 = C_4$ in equation (vi) and (iii) we have

$$EIy = -\frac{Wb}{6L} \cdot x^3 + C_1 x + C_2, \text{ for } \boldsymbol{AC}$$

$$EIy = -\frac{Wb}{6L} \cdot x^3 + \frac{W}{6} \, (x - a)^3 + C_1 x + C_2, \text{ for } \boldsymbol{CB}$$

At support A, $x = 0$, $y = 0 = 0 + 0 + C_2$, $C_2 = 0 = C_4$ $\qquad \text{... (viii)}$

At support B, $x = L$, $y = 0 = -\dfrac{Wb}{6L} \cdot L^3 + \dfrac{W}{6} \, (L - a)^3 + C_1 L + 0$

or $C_1 = \left(\dfrac{WbL^2}{6} - \dfrac{W}{6} \cdot b^3 \right) \dfrac{1}{L} = \dfrac{Wb}{6L} \, (L^2 - b^2)$

Thus $C_1 = C_3 = \dfrac{Wb}{6L} \, (L^2 - b^2) = \dfrac{Wba(a + 2b)}{6L}$ $\qquad \text{... (ix)}$

Putting these values in the slope-deflection equations we have

AC $(x < a)$

$$EI\frac{dy}{dx} = -\frac{Wbx^2}{2L} + \frac{Wb}{6L} \, (L^2 - b^2) = \frac{Wb}{6L} \, [L^2 - b^2 - 3x^2] \qquad \text{... (10.25)}$$

$$EIy = -\frac{Wbx^3}{6L} + \frac{Wbx}{6L} \, (L^2 - b^2) + 0 = \frac{Wbx}{6L} \, [L^2 - b^2 - x^2] \qquad \text{... (10.26)}$$

CB $(x > a)$

$$EI\frac{dy}{dx} = -\frac{Wbx^2}{2L} + \frac{W}{2} \, (x - a)^2 + \frac{Wb}{6L} \, (L^2 - b^2)$$

$$EI\frac{dy}{dx} = \frac{Wb}{6L}[L^2 - b^2 - 3x^2] + \frac{W}{2}(x - a)^2 \qquad \qquad \text{... (10.27)}$$

$$EIy = -\frac{Wbx^3}{6L} + \frac{Wb}{6L}(L^2 - b^2)x + \frac{W}{6}(x - a)^3$$

$$EIy = -\frac{Wb \cdot x}{6L}[L^2 - b^2 - x^2] + \frac{W}{6}(x - a)^3 \qquad \qquad \text{... (10.28)}$$

From these equations (10.25), (10.26), (10.27) and (10.28), slope and deflections can be found anywhere on the beam A to C and C to B.

Slopes at $x = 0$ and $x = L$

$$\underset{(x=0)}{\theta_A} = \frac{Wb}{6EIL}[L^2 - b^2 - 0] = \frac{Wb \cdot a(L + b)}{6EIL} \qquad \qquad \text{... (10.29(a))}$$

$$\underset{(x=L)}{\theta_B} = \frac{Wb}{6EIL}[L^2 - b^2 - 3L^2] + \frac{W}{2EI}(L - a)^2 = + \frac{Wb}{6EIL}[L^2 - b^2 - 3L^2 + 3bL]$$

$$= \frac{Wb}{6EIL}[-b^2 - 2L^2 + 3Lb] = -\frac{Wb}{6EIL}[-Lb + b^2 - 2Lb + 2L^2]$$

or $\quad \theta_B = -\frac{Wb}{6EIL}[-b(L - b) + 2L(L - b)] = -\frac{Wb}{6EIL}[(L - b)(2L - b)] = -\frac{Wab}{6EIL}(L + a)$

or $\quad \theta_B = -\frac{Wab}{6EIL}(L + a) \qquad \qquad \text{... (10.29(b))}$

Deflection at C ($x = a$): $y_c = \frac{Wab}{6EIL}[L^2 - b^2 - a^2] = \frac{Wa^2b^2}{3EIL}$

Thus deflection under the load: $y_c = \dfrac{Wa^2b^2}{3EIL} \qquad \qquad \text{... (10.30)}$

Deflection at the centre $\left(x = \dfrac{L}{2}\right)$: $y_o = \dfrac{Wb \cdot \dfrac{L}{2}}{6EIL}\left[L^2 - b^2 - \dfrac{L^2}{4}\right] = \dfrac{Wb}{12EI}\left[\dfrac{3L^2}{4} - b^2\right]$

$$\underset{\left(x=\frac{L}{2}\right)}{y_o} = \frac{Wb}{48EI}(3L^2 - 4b^2) \qquad \qquad \text{... (10.31)}$$

Maximum Deflection

Since $a > b$, span AC is more than half and hence maximum deflection will lie in span AC. For y to be maximum, $\dfrac{dy}{dx} = 0$,

Thus $EI\dfrac{dy}{dx} = 0 = \dfrac{Wb}{6L}(L^2 - b^2 - 3x^2)$

i.e. $L^2 - b^2 - 3x^2 = 0$ or $x^2 = \left(\dfrac{L^2 - b^2}{3}\right)$, $x = \sqrt{\left(\dfrac{L^2 - b^2}{3}\right)}$

$$y_{max.}\left[x = \sqrt{\left(\dfrac{L^2 - b^2}{3}\right)}\right] = \dfrac{Wb}{6EIL}\cdot\left(\dfrac{L^2 - b^2}{3}\right)^{\!\!1/2}\left[L^2 - b^2 - \dfrac{L^2 - b^2}{3}\right]$$

$$y_{max.} = \dfrac{Wb}{6EIL}\cdot\dfrac{2}{3}\dfrac{(L^2 - b^2)^{3/2}}{\sqrt{3}} = \dfrac{Wb}{9\sqrt{3}EIL}(L^2 - b^2)^{3/2} \qquad\qquad \text{... (10.32)}$$

It may be noted that the maximum deflection is very near the mid span point and this can

be found by assuming the portion CB ($= b$) very small. In this case, $x_{max} = \dfrac{L}{\sqrt{3}} = \mathbf{0.577\ L}$ i.e.

the point of maximum deflection exists only $0.077\ L$ away from the centre. For approximate practical designs the maximum deflection may be worked out at the centre for shortcut.

$$y_{max}\left(\text{when } a = b = \dfrac{L}{2}\right) = \dfrac{W\dfrac{L}{2}}{9\sqrt{3}EIL}\left(L^2 - \dfrac{L^2}{4}\right)^{\!\!3/2} = \dfrac{W[L^3]\cdot 3\sqrt{3}}{18\sqrt{3}EI\cdot 4\sqrt{4}} = \dfrac{WL^3}{48EI}$$

For $a = b = \dfrac{L}{2}$, $y_{max} = \dfrac{WL^3}{48EI}$ (same as for central load) $\qquad\qquad$ ... (10.33)

Case 4: Simply Supported Beam AB (L) carrying udl w/m over the entire span

Span $AB = l$, $\quad AC = CB = l/2$

Symmetry $\theta_A = -\theta_B$, $\quad$ udl $= w/m$

Total $W = w.L$

Consider any section X-X at a distance x from A.

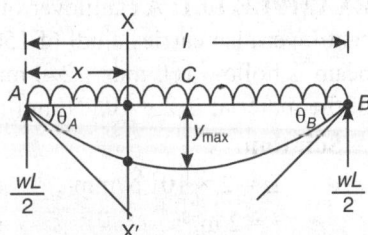

Fig. 10.13: SS beam with udl

$$M_x = \left(\dfrac{wl}{2}\cdot x - \dfrac{w\cdot x^2}{2}\right)$$

$$EI\dfrac{d^2y}{dx^2} = -M_x = -\dfrac{wl}{2}\cdot x + \dfrac{wx^2}{2} \qquad\qquad \text{... (i)}$$

Integrating once: $EI\dfrac{dy}{dx} = -\dfrac{wl}{2}\cdot\dfrac{x^2}{2} + \dfrac{w}{2}\cdot\dfrac{x^3}{3} + C_1 \qquad\qquad$ ... (ii)

By symmetry $\dfrac{dy}{dx} = 0$ at $x = \dfrac{l}{2}$, and hence

$$0 = -\frac{wl}{4}\left(\frac{l}{2}\right)^2 + \frac{w}{6}\left(\frac{l}{2}\right)^3 + C_1, C_1 = wl^3\left(\frac{1}{16} - \frac{1}{48}\right) = \frac{wl^3}{24}$$

$$EI\frac{dy}{dx} = -\frac{wl}{4}x^2 + \frac{w}{6}\cdot x^3 + \frac{wl^3}{24} \qquad \text{... (10.34)}$$

Integrating: $EIy = -\frac{wl}{4}\cdot\frac{x^3}{3} + \frac{w}{6}\cdot\frac{x^4}{4} + \frac{wl^3}{24}\cdot x + C_2$ ... (iii)

at $x = 0$, $y = 0$, gives $0 = -0 + 0 + 0 + C_2$, $C_2 = 0$

$$EIy_x = -\frac{wl}{12}\cdot x^3 + \frac{w}{24}\cdot x^4 + \frac{wl^3}{24}\cdot x + 0 \qquad \text{... (10.35)}$$

Equations (10.34) and (10.35) gives slope and deflection at any point x.

$$\theta_A(x=0) = +\frac{wl^3}{24EI}, \quad \theta_B(x=L) = \frac{wl^3}{EI}\left(\frac{1}{24} + \frac{1}{6} - \frac{1}{4}\right) = -\frac{wl^3}{24EI} \qquad \text{... (10.36)}$$

$$y_{max}_{\left(x=\frac{l}{2}\right)} = \left(-\frac{wl^4}{96} + \frac{wl^4}{384} + \frac{wl^4}{48}\right)\frac{1}{EI} = \frac{wl^4}{384EI}(-4 + 1 + 8) = \frac{5wl^4}{384EI}$$

or $\quad y_{max}_{\left(x=\frac{l}{2}\right)} = \frac{5wl^4}{384EI} = \frac{5Wl^3}{384EI}, \quad W = wl$ ... (10.37)

EXAMPLE 10.1: A cantilever of 3 m span carries a point load of 2000 N at the free end. The cantilever also carries a udl of 1500 N/m for 2 m length from the fixed end. The section of the beam is hollow 160 mm × 240 mm outer and 120 mm × 200 mm inside. Modulus of Elasticity of the material is 2×10^5 N/mm². Find the maximum slope and deflection in the beam.

Solution:

E = 2×10^5 N/mm², $\qquad$ L = 3 m,

l = 2 m, $\qquad\qquad\qquad$ w = 1500 N/m,

W = 2000 N $\qquad\qquad\quad$ = 1.5 N/mm

Size: $\qquad\qquad\qquad\quad$ 160 mm × 240 mm

Hollow: $\qquad\qquad\qquad$ 120 mm × 200 mm

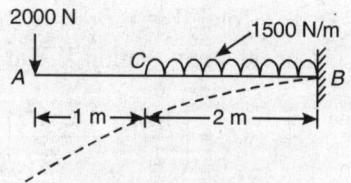

Fig. 10.14: Cantilever with udl and point load

$$I_x = \frac{1}{12}(160 \times 240^3 - 120 \times 200^3)$$

$$= 10^4(32 \times 576 - 8 \times 10^3)$$

$$= 8 \times 10^4(2304 - 1000) = \mathbf{10432 \times 10^4 \ mm^4}$$

Maximum deflection at the free end will be sum of deflections due to point load and udl i.e.

$$y_{max} = \frac{WL^3}{3EI} + \left[\frac{w \cdot l_1^4}{8EI} + \frac{w \cdot l_1^3}{6EI} \cdot (l - l_1)\right] mm$$

$$y = \frac{1}{EI}\left[\frac{2000 \times 3000^3}{3} + \frac{1500}{1000} \times \frac{2000^4}{8} + \frac{1500}{1000} \times \frac{2000^3}{6}(1000)\right]$$

$$y = \frac{10^{12}}{2 \times 10^5 \times 10432 \times 10^4}\left[18 + \frac{1.5 \times 16}{8} + \frac{1.5 \times 8}{6}\right] = \frac{23 \times 10^{12}}{2 \times 10432 \times 10^9} = \mathbf{1.1024\ mm}$$

$$\text{Maximum slope} = \frac{WL^2}{2EI} + \frac{wL_1^3}{6EI}$$

$$= \frac{2000 \times 3000^2}{2 \times 2 \times 10^5 \times 10432 \times 10^4} + \frac{1500}{1000} \times \frac{2000^3}{6 \times 2 \times 10^5 \times 10432 \times 10^4}$$

$$\theta_A\,(max) = \frac{10^9}{10^9}\left[\frac{18}{4 \times 10432} + \frac{12}{12 \times 10432}\right] = \frac{5.5}{10432} = \mathbf{0.000527\ radian}$$

$$= \frac{5.5 \times 180}{10432\pi}\ \text{degrees} = \mathbf{0.030223\ degrees}$$

EXAMPLE 10.2: A wooden beam of 1.5 m length is supported on 1.0 span centrally with equal overhangs of 250 mm on either side. The beam carries concentrated load of 500 N on either overhang end. E for wood = 1×10^4 N/mm^2. Determine deflection at the centre of span and slopes at the supports if the section is 60 mm × 100 mm.

Solution:

$AB = 1000$ mm, $CD = 1500$ mm

$W_C = W_D = 500$ N $E = 1 \times 10^4$ N/mm^2

$I = \dfrac{60}{12} \times 100^3 = 5 \times 10^6\text{mm}^4$

By Symmetry $R_A = R_B = 500$ N

$BM : C$ to $A : 500x$

$M_A = 125000$ N-mm

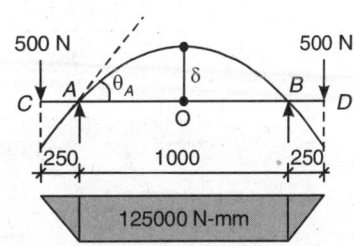

Fig. 10.15: Overhanging beam

M_A to $M_B = 125000$ N-mm constant. Deflected shape between A to B will be circular as the BM is constant.

If radius of circle = R, then $\dfrac{L}{2} \cdot \dfrac{L}{2} = (2R - \delta)\delta$

or $\dfrac{L^2}{4} = 2R \cdot \delta - \delta^2$, neglecting square of small quantities' or $\quad \dfrac{L^2}{4} = 2R \cdot \delta, \ \delta = \dfrac{L^2}{8R} = \dfrac{L^2}{8\dfrac{EI}{M}} = \dfrac{ML^2}{8EI}$

$$\delta = \dfrac{ML^2}{8EI} = \dfrac{125000(1000)^2}{8 \times 1 \times 10^4 \times 5 \times 10^6} = \dfrac{12.5}{40} \text{ mm} = \mathbf{0.3125mm}$$

$$\text{Maximum slope } \theta_A = \dfrac{L}{2R} = \dfrac{L}{2\dfrac{EI}{M}} = \dfrac{ML}{2EI} = \dfrac{125000 \times 1000}{2 \times 1 \times 10^4 \times 5 \times 10^6}$$

$$\theta_A = -\theta_B = \dfrac{125}{10^5} = \mathbf{0.00125 \ radians \ (0.07166 \ degrees)}$$

EXAMPLE 10.3: A simply supported beam of 6 m span has uniform cross-section with uniform depth of **400 mm**. (i) The beam carries a udl which creates maximum bending stress of **90 N/mm²**. Find the maximum slope and deflection. (ii) If the beam flanges are so made that the bending stress developed is **110 N/mm²** uniform through out, find the maximum slope and deflections in the beam. $E = 2.1 \times 10^5$ N/mm².

Solution:

$\quad L = 6 \text{ m} = 6000 \text{ mm}, \qquad d = 400 \text{ mm}, w = ?, E = 2.1 \times 10^5 \text{ N/mm}^2$

$\quad$ (i) $f = 90$ N/mm², $\qquad$ (ii) $f = 110$ N/mm²

(i) $\quad$ Maximum BM $= \dfrac{wL^2}{8} = \dfrac{w(6000)^2}{8}$ N-mm $= \mathbf{4500000 \ w}$ **N-mm**

$$f_{max} = \dfrac{M}{I/y} = \dfrac{45w \times 10^5}{I/200} = \dfrac{9000w \times 10^5}{I} \text{ N/mm}^2$$

$$90 = \dfrac{9000w \times 10^5}{I} \text{ N/mm}^2, \dfrac{w}{I} = \dfrac{90}{9000 \times 10^5} = \dfrac{1}{10^7}$$

$$\text{Maximum } \delta = \dfrac{5}{384} \cdot \dfrac{w(L)^4}{EI} = \dfrac{5 \times (6000)^4}{384 \times 2.1 \times 10^5} \cdot \dfrac{w}{I} = \dfrac{5 \times 36 \times 36 \times 10^{12}}{384 \times 2.1 \times 10^5} \times \dfrac{1}{10^7}$$

$$\delta = \dfrac{5 \times 3 \times 900}{8 \times 210} = \dfrac{450}{56} \text{ mm} = 8.0357 \approx \mathbf{8.04 \ mm}$$

$$\theta_A = -\theta_B = \dfrac{wL^3}{24EI} = \dfrac{1}{10^7} \times \dfrac{(6000)^3}{24 \times 21 \times 10^4} = \dfrac{3}{700}$$

$$= \mathbf{0.004268 \ radian} \ (0.246 \ degrees)$$

(ii)   Flanges are so designed that the bending stress is uniform $\left(\text{i.e. } \dfrac{f}{y} = \dfrac{E}{R} = \text{Constant} \right)$

and the radius of curvature 'R' is also uniform, i.e. circular arc.

$$\text{Maximum } \delta = \frac{L^2}{8R} = \frac{L^2}{8E} \cdot \frac{f}{y} = \frac{6000 \times 6000 \times 110}{8 \times 21 \times 10^4 \times 200} = \frac{36 \times 110}{8 \times 21 \times 2} = \frac{990}{84} \text{ mm}$$

$\delta_{\max} = \mathbf{11.786 \ mm}$

$$\theta_A = -\theta_B = \frac{L}{2R} = \frac{600}{2} \cdot \frac{110}{21 \times 10^4 \times 200} = \frac{11}{1400} = \mathbf{0.00786 \ redian} \ (0.45 \text{ degrees})$$

EXAMPLE 10.4: A cantilever beam of 4 m span has uniform depth of 400 mm. The beam carries a udl of 5000 N/m on the entire span. The beam cross-section has moment of inertia I = 2 × 10⁸ mm⁴ and modulus of elasticity E = 2 × 10⁵ N/mm². (i) Determine maximum deflection and maximum bending stress. (ii) Determine how much additional point load 'W' can be applied at 3m from the fixed end so that the maximum deflection remains within 10 mm and the maximum bending stress remains within 100 N/mm².

Solution:

Span L = 4 m (4000 mm),   Depth D = 400 mm,
I = 2 × 10⁸ mm⁴, E = 2 × 10⁵ N/mm²,

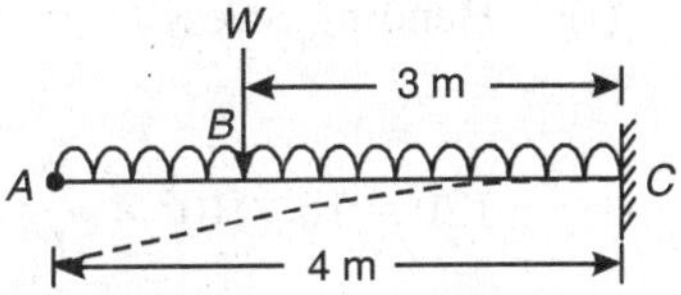

$$w = \frac{5000}{1000} = \mathbf{5 \ N/mm}$$

Fig. 10.16: Cantilever

(i) $y_{\max} = \dfrac{wL^4}{8EI} = \dfrac{5(4000)^4}{8 \times 2 \times 10^5 \times 2 \times 10^8} = \dfrac{5 \times 256 \times 10^{12}}{8 \times 4 \times 4 \times 10^{13}} = \mathbf{4.0 \ mm}$

$$\text{Maximum } BM = \frac{wL^2}{2} = \frac{5(4000)^2}{2} = \mathbf{40 \times 10^6 \ N\text{-}mm}$$

$$\text{Mamimum Bending Stress} = \frac{40 \times 10^6}{2 \times 10^8} \times \frac{400}{2} = \frac{160}{4} = \mathbf{40 \ N/mm^2}$$

(ii) Maximum deflection due to additional point load $W = \dfrac{W(3000)^3}{3EI} + \dfrac{W(3000)^2}{2EI} \times 1000$

$\delta_2 = \dfrac{W(27 \times 10^9)}{3 \times 2 \times 10^8 \times 2 \times 10^5} + \dfrac{W(9 \times 10^9)}{2 \times 2 \times 10^8 \times 2 \times 10^5} = W\left[\dfrac{9}{40000} + \dfrac{9}{80000} \right] = \dfrac{9 \times 3W}{80000}$

Total deflection 10 mm = 4.0 mm + δ_2

or   6 mm = $\dfrac{9 \times 3W}{80000}$, $W_1 = \dfrac{160000}{9} \text{N} = \dfrac{160}{9} \text{kN} = \mathbf{17.778 \ kN}$

Additional bending stress + $(100 - 40) = 60$ N/mm^2

$$f = \frac{M}{I} \cdot y = \frac{W(3000)}{2 \times 10^8} \cdot \frac{400}{2} = \frac{12 \times 10^5 W}{4 \times 10^8} = \frac{3W}{10^3} = 60,$$

$$W_2 = \frac{60 \times 10^3}{3} = 20000 \text{ N} = 20 \text{ kN}$$

Safe in both conditions lesser of W_1 or W_2 i.e. 160000/9 N = 17778 N (17.778 kN)
Thus *additional safe point* load at 3 m will be 17778 N.

EXAMPLE 10.5: A simply supported beam of 8 m span and 400 mm uniform depth of section. (i) If the maximum bending stress developed under udl is 120 N/mm^2, find the maximum deflection. (ii) If this maximum bending stress is developed by a concentrated load W acting at the mid span, find the maximum deflection. $E = 2 \times 10^5$ N/mm^2.

Solution:

$L = 8$ m (8000 mm), $d = 400$ mm, $f = 120$ N/mm^2, $E = 2 \times 10^5$ N/mm^2

(i)   Bending Stress $f = \dfrac{M}{I} \cdot y = \dfrac{wl^2}{8I} (200) = \dfrac{w(8000)^2}{I} \cdot \dfrac{200}{8} = \dfrac{w}{I} \times 10^8 \times 16$

$120 = 16 \times 10^8 \times \dfrac{w}{I}$, or $\dfrac{w}{I} = \dfrac{120}{16 \times 10^8} = 0.75 \times 10^{-7}$ ... (i)

Maximum $\delta = \dfrac{5}{384} \cdot \dfrac{wL^4}{EI} = \dfrac{5(8000)^4}{384 \times 2 \times 10^5} \cdot \dfrac{w}{I} = \dfrac{5 \times 10^{12} \times 64 \times 64}{768 \times 10^5} (0.75 \times 10^{-7})$

$\delta = \dfrac{5 \times 64 \times 64 \times 0.75}{768} = \dfrac{5 \times 12}{3} = \textbf{20 mm}$

(ii)   In case of concentrated load W at the mid span point

Bending Stress $= 120 = \dfrac{M}{I} \cdot y = \dfrac{W(8000)}{4I} \cdot 200 = 4 \times 10^5 \times \dfrac{W}{I}$

$\dfrac{W}{I} = \dfrac{120}{4 \times 10^5} = \textbf{3} \times \textbf{10}^{-4}$

Maximum $\delta = \dfrac{WL^3}{48EI} = \dfrac{W}{I} \left[\dfrac{(8000)^3}{48 \times 2 \times 10^5} \right] = \dfrac{3 \times 10^{-4} \times 8 \times 64 \times 10^9}{48 \times 2 \times 10^5} = \textbf{16 mm}$

EXAMPLE 10.6: A simply supported overhanging beam of total length 'L' with overhang on each side 'a'. The beam carries a point load W at the centre causing a downward deflection 'δ' at the centre and also upward deflection 'δ' at both the free ends of overhangs. Determine the ratio of overhangs 'a' to the total length 'L'

Solution:

Central Point Load $= W$,

Supported Span $= (L - 2a)$,

Slope at the Support $A = \theta_A = -\theta_B$,

Deflection at Free Ends $\delta_D = \theta_A(a) = \theta_B(a)$

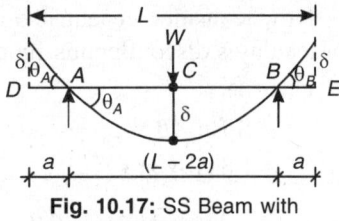

Fig. 10.17: SS Beam with overhangs

Deflection at the mid point C

$$\delta_c = \frac{W(L-2a)^3}{48EI}, \qquad \text{Slope } \theta_A = -\theta_B = \frac{W(L-2a)^2}{16EI} \qquad \text{... (i)}$$

$$\therefore \quad \delta_D = \frac{W(L-2a)^3}{16EI}(a) \qquad \text{... (ii)}$$

$$\delta_D = \delta_C \text{ i.e. } \frac{W(L-2a)^3}{48EI} = \frac{W(L-2a)^2 \cdot (a)}{16EI}$$

or $\quad \dfrac{(L-2a)}{3} = \dfrac{a}{1}$ or $(L-2a) = 3a$

or $\quad L = 5a$ or $\dfrac{a}{L} = \dfrac{1}{5}$ i.e. Ratio of overhang to total length $= 1 : 5$

10.4   MACAULAY'S METHOD (DISCONTINUITY FUNCTION)

In the case of a simply supported beam with an *eccentric point load* or more than one load, the simple *double integration* approach becomes quite *cumbersome* because *separate bending equations* are necessary for *each segment*. Constants of integrations are also required separately for each segment making the whole process quite lengthy and involved. It involves number of boundary conditions for solving for the constant of integration. Double integration approach also becomes cumbersome in case of *combination of loads* and *partially loaded* span with uniformly distributed load. However *Macaulay* has developed a function which uses *discontinuity functions* for different segments. *Macaulay's* approach has become more popular due to its *general application* for variety of loads. The *unique feature* of Macaulay's *discontinuous function* is a *single expression*. In conventional method a discontinuous function is described by a series of expressions, one for each segment in which the function is distinctly applicable. The use of *special brackets* for discontinuity function was introduced by the English Mathematician *W. H.* Macaulay and hence this approach is known as Macaulay's approach.

These brackets are with vertical lines and marked as $\vdots$, { } or < >.

The vertical line is marked to show separation of functions. The Macaulay's approach essentially comprises of successive integration for the expression for the bending moment. Although the bending moment expression *varies from section to section* between the loads but the *constants of integration* are valid for all the sections.

Now consider eccentric load W on a simply supported beam (case 3) to demonstrate Macaulay's discontinuous function approach.

Case 1:

Span $AB = L$, $\qquad AC = a,\ CB = b$

Load W at C:

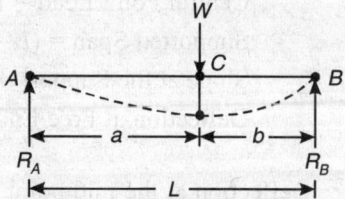

$$R_A = \frac{Wb}{L},\ R_B = \frac{Wa}{L}$$

Fig.10.18: SS beam with point load W at any section C ($AC = a$)

$$M_x = R_A \cdot x \Big| -W(x-a)$$

$$EI\frac{d^2y}{dx^2} = -\frac{Wb}{L}\cdot x + \Big| W(x-a) \qquad \text{... (i)}$$

Integrating once: $EI\dfrac{dy}{dx} = -\dfrac{Wb}{L}\cdot \dfrac{x^2}{2}\Big| + \dfrac{W(x-a)^2}{2} + C_1$

$$EI\frac{dy}{dx} = -\frac{Wb\cdot x^2}{2L} + C_1 \Big| + \frac{W(x-a)^2}{2} \qquad \text{... (ii)}$$

Integrating: $EIy = -\dfrac{Wbx^3}{6L} + C_1x + C_2 \Big| + \dfrac{W(x-a)^3}{6} \qquad \text{... (iii)}$

$y = 0$, at $x = 0$, $0 = -0 + C_1 \cdot 0 + C_2 = 0$

$$EIy = -\frac{Wbx^3}{6L} + C_1x \Big| + \frac{W(x-a)^3}{6} \qquad \text{... (iv)}$$

$y = 0$, $x = L$, $0 = -\dfrac{Wb}{6L}\cdot L^3 + C_1 \cdot L + \dfrac{W}{6}(L-a)^3$

$$C_1L = \frac{WbL^3}{6L} - \frac{Wb^3}{6},\ C_1 = \frac{Wb}{6L}(L^2 - b^2) \qquad \text{... (v)}$$

Substituting values of $C_1 = \dfrac{Wb}{6L}(L^2 - b^2)$ in equation (ii) and (iv)

$$EI\frac{dy}{dx} = -\frac{Wb\cdot x^2}{2L} + \frac{Wb(L^2 - b^2)}{6L} \Big| + \frac{W}{2}(x-a)^2$$

For AC: $\dfrac{dy}{dx} = \dfrac{Wb(L^2 - b^2 - 3x^3)}{6EIL} \qquad \text{... same as (10.26)}$

$$EIy = -\frac{Wbx^3}{6L} + \frac{Wb(L^2 - b^2)}{6L}\cdot x \Big| + \frac{W(x-a)^3}{6}$$

For AC: $y = -\dfrac{Wbx(L^2 - b^2 - x^2)}{6EIL}$ ... same as (10.27)

At $x = 0$, $\theta_A = +\dfrac{Wb(L^2 - b^2)}{6EIL} = +\dfrac{Wab}{6EIL}(L + b)$

At $x = L$, $\theta_B = \left[-\dfrac{WbL}{2} + \dfrac{Wb}{6L}(L^2 - b^2) + \dfrac{Wb^2}{2}\right]\dfrac{1}{EI} = -\dfrac{Wb}{6EIL}[3L^2 - a(L + b) - 3bL]$

$= -\dfrac{Wb}{6EIL}[3L(L - b) - aL - ab] = -\dfrac{Wb}{6EIL}[3La - La - ab]$

$= -\dfrac{Wb}{6EIL}[(2L - b)a] = -\dfrac{Wba}{6EIL}(L + a)$

i.e. $\theta_A = +\dfrac{Wab(L + b)}{6EIL}$, $\theta_B = \dfrac{Wab(L + a)}{6EIL}$ ... same as (10.28)

$EIy_c\ (x = a) = -\dfrac{Wb \cdot a^3}{6L} + \dfrac{Wb(L^2 - b^2)a}{6L} + 0 = \dfrac{Wba}{6L}(L^2 - a^2 - b^2)$

$y_c = -\dfrac{Wab}{6EIL}(L^2 - a^2 - b^2) = \dfrac{Wa^2b^2}{3EIL}$ ... same as derived earlier in (10.30)

Case 2: Partially loaded SS beam with udl

Span $AB = L$, udl $= w/m$

Loaded Span $CB = l$,

Unloaded portion $(L - l) = AC$

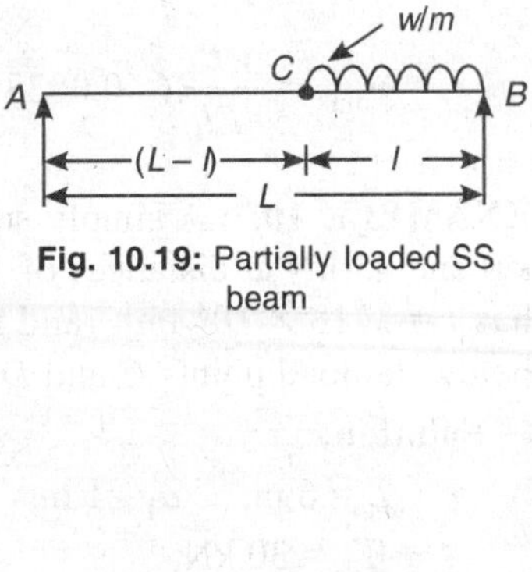

Fig. 10.19: Partially loaded SS beam

$R_A = \dfrac{w \cdot l^2}{2L}$,

$M_x = \dfrac{wl^2}{2L} \cdot x \left|-\dfrac{w}{2}\{x - (L - l)\}^2\right.$

$EI\dfrac{d^2y}{dx^2} = -M_x = -\dfrac{wl^2}{2L} \cdot x \left|+\dfrac{w}{2}\{x - (L - l)\}^2\right.$

Integrating: $EI\dfrac{dy}{dx} = -\dfrac{wl^2}{2L} \cdot \dfrac{x^2}{2} + C_1 \left|+\dfrac{w}{6}\{x - (L - l)\}^3\right.$ ... (i)

Integrating again: $EIy = -\dfrac{wl^2 \cdot x^3}{12L} + C_1 \cdot x + C_2 \left|+\dfrac{w}{24}\{x - (L - l)\}^4\right.$ ... (ii)

At $x = 0$, $y = 0$, $C_2 = 0$

At $x = L$, $y = 0$, $\qquad 0 = -\dfrac{wl^2 \cdot L^3}{12L} + C_1 L + 0 \left| + \dfrac{w}{24}\{L-(L-l)\}^4 \right.$

$$0 = -\frac{wl^2 \cdot L^2}{12} + C_1 L \left| + \frac{W}{24} \cdot l^4 \right., \quad C_1 = \frac{wl^2 \cdot L^2}{12L} - \frac{wl^4}{24L} = \frac{wl^2}{24L}(2L^2 - l^2)$$

$$\therefore \quad EI\frac{dy}{dx} = -\frac{wl^2 \cdot x^2}{4L} + \frac{wl^2(2L^2 - l^2)}{24L} \left| + \frac{w}{6}\{x-(L-l)\}^3 \right. \qquad \text{... (10.41)}$$

$$EIy = -\frac{wl^2 \cdot x^3}{12L} + \frac{wl^2 x(2L^2 - l^2)}{24L} \left| = \frac{w}{24}\{x-(L-l)\}^4 \right. \qquad \text{... (10.42)}$$

If maximum deflection is given to be at the point C, find the maximum deflection. For y to be maximum, dy/dx at the point is zero.

$x = (L - l)$, $0 = -\dfrac{wl^2(L-l)^2}{4L} + \dfrac{wl^2(2L^2 - l^2)}{24L}$, or $0 = -6l^2(L-l)^2 + l^2(2L^2 - l^2)$

or $\quad l^2 [-6(L^2 - 2L \cdot l + l^2) + (2L^2 - l^2)] = 0$

or $\quad 7l^2 - 12L \cdot l + 4L^2 = 0$, or $l = \dfrac{12 \pm \sqrt{144 - 4 \times 7 \times 4}}{2 \times 7} \cdot L = \left(\dfrac{12 \pm 5.656}{14} \right) \cdot L = \mathbf{0.453L}$

Substituting $l = 0.453L$, $x = (L - l) = 0.547\,L$

$$y_{max} = \frac{wL^4}{EI}(-0.0028 + 0.0084) = 0.0056\frac{wL^4}{EI} = \frac{wL^4}{\mathbf{178.85EI}}$$

EXAMPLE 10.7: A simply supported beam AB of 6 m span carries concentrated loads of 30 kN and 45 kN at distances of 1 m and 4 m from the support A respectively. The beam section has $I = 3718 \times 10^4$ mm^4 and material $E = 2 \times 10^5$ N/mm^2. Determine slope and deflections below the load points C and D. Also find the maximum deflection and its ratio with the span.

Solution:

$L = 6$ m, $\quad a_1 = 1$ m, $\quad a_2 = 4$ m

$W_1 = 30$ kN, $\qquad W_2 = 45$ kN

$E = 200$ kN/mm^2

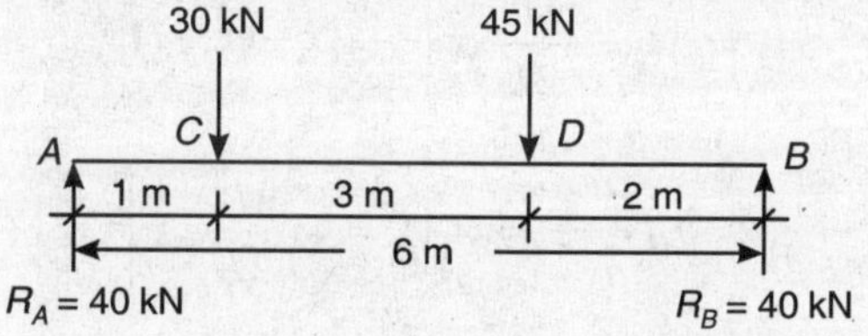

$R_A = \dfrac{1}{6}[45 \times 2 + 30 \times 5] = 40$ kN

$R_B = \dfrac{1}{6}[30 \times 1 + 45 \times 4] = 35$ kN

Fig. 10.20.

$M_x = R_A \cdot x \left| -30(x-1) \right. \left| -45(x-4) \right. = 40x \left| -30(x-1) \right. \left| -45(x-4) \right.$

$$\therefore \quad EI\frac{d^2y}{dx^2} = -40x \Big| +30(x-1)+45(x-4) \qquad \text{... (i)}$$

Integrating: $EI\dfrac{dy}{dx} = -\dfrac{40x^2}{2} + C_1 \Big| +\dfrac{30}{2}(x-1)^2 + \dfrac{45}{2}(x-4)^2 \qquad \text{... (ii)}$

Integrating: $EIy = -\dfrac{40x^3}{6} + C_1x + C_2 \Big| +\dfrac{30}{6}(x-1)^3 + \dfrac{45}{6}(x-4)^3 \qquad \text{... (iii)}$

At $x = 0$, $y = 0$: $0 = C_2$, $EIy = -\dfrac{40}{6}x^3 + C_1x \Big| +5(x-1)^3 + \dfrac{15}{2}(x-4)^3 \qquad \text{... (iv)}$

At $x = L = 6$ m, $y = 0$: $0 = -\dfrac{40}{6}(6)^3 + C_1(6) \Big| +5(6-1)^3 + \dfrac{15}{2}(6-4)^3$

or $\quad C_1 = \dfrac{1}{6}\left[40\times36 - 5\times125 - \dfrac{15}{2}\times8\right] = \left[40\times6 - \dfrac{685}{6}\right] = \textbf{125.84 m units}$

$$EI\frac{dy}{dx} = -20x^2 + 125.84\Big| +15(x-1)^2 + \dfrac{45}{2}(x-4)^2, \qquad \text{... (v)}$$

$$EIy = -\frac{20x^3}{3} + 125.84x\Big| +5(x-1)^3 + \dfrac{15}{2}(x-4)^3 \qquad \text{... (vi)}$$

In these equations the distances are taken in 'm' while E and I must also be taken in similar units. Loads in kN, E also in kN/m^2.

$$EI \text{ in kN-m}^2 = 200 \times 10^6 \times \frac{3718\times10^4}{(1000)^4} = \frac{7436\times10^{12}}{10^{12}} = \textbf{7436 kN-m}^2 \textbf{ units}$$

Equations (v) and (vi) are general equations and can be used to find slope and deflection at C and D.

$x = 1$ m: $\theta_c = \dfrac{1}{7436}[-20(1)^2 + 125.84] = \dfrac{105.84}{7436} = \textbf{+ 0.01423 radian}$

$x = 4$ m: $\theta_D = \dfrac{1}{7436}[-20(4)^2 + 125.84 + 15\,(4-1)^2 + 0] = -\dfrac{59.16}{7436} = \textbf{- 0.007956 radian}$

$y_c\,(x = 1\text{m}) = \dfrac{1}{7436}\left[-\dfrac{20}{3}(1)^3 + 125.84(1)\right] = \dfrac{119.17}{7436} = \textbf{0.0160265 m = 16.0265 mm}$

$y_D\,(x = 4\text{m}) = \dfrac{1}{7436}\left[-\dfrac{20}{3}(4)^3 + 125.84(4)\Big| + 5(4-1)^3\right] = \dfrac{211.7}{7436} = \textbf{0.02849 m = 28.49 mm}$

Since slope $\theta_C = +\,0.01423$ radian changes to $-\,0.007956$ radian at D, the value will be zero between C and D. i.e.

$$-20x^2 + 125.84 + 15\,(x-1)^2 = 0, \quad \text{or } 5x^2 + 30x - 140.84 = 0$$

$$x = \frac{30 \quad \sqrt{900 \quad 20 \quad 140.84}}{10} = \textbf{3.097 m}$$

$$y_{max}\,(x = 3.094 \text{ m}) = \frac{1}{7436} \quad \frac{20}{3}(3.097)^3 \quad 125.84(3.097) \quad 5(2.097)^3$$

$$= \frac{1}{7436}\,[-198.00 + 389.7 + 46.1] = \frac{237.80}{7436}\,\text{m} = \textbf{0.03198 m}$$

$$y_{max} = \textbf{31.98 mm}$$

$$\text{Ratio with span} = \frac{31.98}{6000} = \frac{1}{187.6}$$

EXAMPLE 10.8: A simply supported beam ACB (span $L = 10$ m) carries udl of w/mm ($w = 10$ kN/m) on part of span from C to B ($AC = a = 5$ m). (i) Determine the general equations of slope and deflection for the whole beam. (ii) Determine the maximum deflection in the beam and the point where it occurs. The beam cross-section has $I = 6000 \times 10^4$ mm^4 and beam material has $E = 2 \times 10^5$ N/mm^2.

Solution:

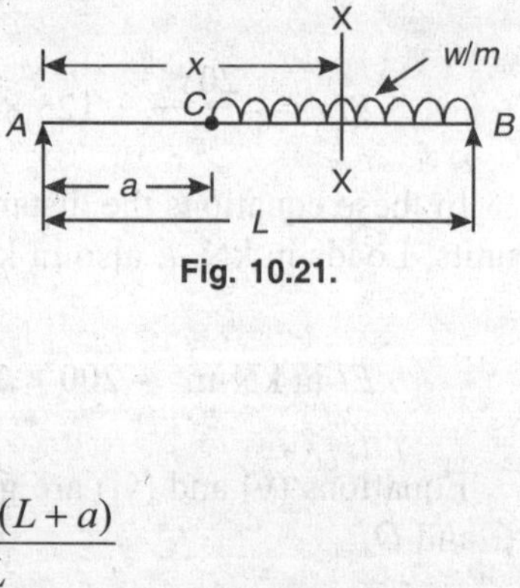

Span $AB = L = 10$ m udl $= $ w/m on CB $(L - a)$

$I = 6 \times 10^7$ mm^4,

$E = 2 \times 10^5$ N/mm^2

Fig. 10.21.

$$EI = \frac{6 \times 2 \times 10^{12}}{10^9}\,\text{kN-m}^2 = 12000 \text{ kN-m}^2$$

$$R_A = \frac{w(L-a)^2}{2L}, \quad R_B = \frac{w(L-a)\left(a + \dfrac{L-a}{a}\right)}{L} = \frac{w(L-a)(L+a)}{2L}$$

(i) $\quad EI\dfrac{d^2y}{dx^2} = -\,\dfrac{w(L-a)^2 \cdot x}{2L}\Bigg|+\dfrac{w}{2}(x-a)^2$ \hfill ... (i)

Integrating: $EI\dfrac{dy}{dx} = -\,\dfrac{w(L-a)^2 \cdot x^2}{4L} + C_1\Bigg|+\dfrac{w}{2 \times 3}(x-a)^3$ \hfill ... (ii)

Integrating again: $EIy = -\,\dfrac{w(L-a)^2 \cdot x^3}{12L} + C_1 x + C_2 + \Bigg|\dfrac{w(x-a)^4}{24}$ \hfill ... (iii)

At $x = 0$, $y = 0$, gives $C_2 = 0$, $EIy = -\dfrac{w(L-a)^2 \cdot x^3}{12L} + C_1 x \left|+\dfrac{w(x-a)^4}{24}\right.$ ... (iv)

At $x = L$, $y = 0$, $0 = -\dfrac{w(L-a)^2}{12L} \cdot L^3 + C_1 \cdot L \left|+\dfrac{w(L-a)^4}{24}\right.$

or $C_1 = -\dfrac{w(L-a)^4}{24} + \dfrac{w(L-a)^2 \cdot L^3}{12L^2} = \dfrac{2w(L-a)^2 \cdot L^2 - w(L-a)^4}{24L}$

or $C_1 = \dfrac{w}{24L}(L-a)^2 \left\{2L^2 - (L-a)^2\right\} = \dfrac{w(L-a)^2}{24L}\left[2L^2 - (L^2 - 2aL + a^2)\right]$

or $C_1 = \dfrac{w(L-a)^2}{24L}(L^2 - a^2 + 2aL)$ ... (v)

Thus $EI\dfrac{dy}{dx} = -\dfrac{w(L-a)^2 \cdot x^2}{4L} + \dfrac{w(L-a)^2(L^2 - a^2 + 2aL)}{24L} \left|+\dfrac{w}{6}(x-a)^3\right.$ ... (vi)

$EIy = \dfrac{w(L-a)^2 \cdot x^3}{12L} + \dfrac{w(L-a)^2(L^2 - a^2 + 2aL)}{24L} \cdot x \left|+\dfrac{w}{24}(x-a)^4\right.$ ... (vii)

Equations (vi) and (vii) represent general equations of slope and deflections anywhere on the beam.

(ii) $L = 10$ m, $a = 5$ m, $w = 10$ kN/m, $EI = \dfrac{12 \times 10^{12}}{10^9} =$ **12000 kN-m²**

Putting values in (vii)

$12000y = -\dfrac{10(10-5)^2}{12 \times 10} \cdot x^3 + \dfrac{10(10-5)^2}{240}(10^2 - 5^2 + 2 \times 5 \times 10)x + \left|\dfrac{10}{24}(x-5)^4\right.$

$= -\dfrac{250x^3}{120} + \dfrac{250(175)}{240} \cdot x + \left|\dfrac{10}{24}(x-5)^4\right. = \dfrac{250x^3}{120} + \dfrac{250 \times 175}{240} \cdot x \left|\dfrac{10}{24}(x-5)^4\right.$

$EI\dfrac{dy}{dx} = -\dfrac{25 \times 10 \cdot x^2}{40} + \dfrac{250}{240}(175) \left|+\dfrac{10}{6}(x-5)^3\right. = -6.25x^2 + \dfrac{25.0 \times 175}{24.0} + \dfrac{10}{6}(x-5)^3$

At $x = 5$m, $\dfrac{dy}{dx} = \dfrac{-\dfrac{6.25 \times 25}{1} + \dfrac{175 \times 25}{24.0} + 0}{EI} = \dfrac{-156.25 + 182.3}{EI} = \dfrac{+26.05}{EI}$

$\theta_5 = \dfrac{(-156.25 + 182.3)}{12000} = +\dfrac{26.05}{12000} =$ **+ 0.0021708 radian**

At $x = 6$m, $EI\dfrac{dy}{dx} = -\dfrac{250 \times 36}{40} + \dfrac{250 \times 175}{240} \Bigg| + \dfrac{10}{6}(6-5)^3$

$$\theta_6 = \dfrac{-225 + 182.3 + 1.667}{12000} = \dfrac{-41.042}{12000} = -\textbf{0.00342 radian}$$

dy/dx is $+ve$ upto $x = 5$m, while it becomes $-ve$ at $x = 6$m. Therefore $dy/dx = 0$ between $x = 5$m to 6m.

$$\dfrac{dy}{dx} = 0 = -\dfrac{25}{4}x^2 + \dfrac{175 \times 25}{24} + \dfrac{10}{6}(x-5)^3$$

or $\quad \dfrac{25}{4}x^2 - \dfrac{10}{6}(x-5)^3 - \dfrac{175 \times 25}{24} = 0$ or $150x^2 - 40(x-5)^3 - 175 \times 25 = 0$

By trial consider $x = 5.4023$ m, $150(5.4023)^2 - 40(0.4023)^3 - 175 \times 25 = 0$
$$4377.73 - 2.6044 - 4375 \approx 0$$

$dy/dx = 0$ at $x \approx 5.4023$, hence y_{max} at $x = \textbf{5.4023 m}$

$$EIy = -\dfrac{250}{120}x^3 + \dfrac{250 \times 175x}{240} \Bigg| + \dfrac{10}{24}(x-5)^4$$

$12000y = \dfrac{1}{24}\left[50(5.4023)^3 \quad 25 \quad 175(5.4023) \quad 10(0.4023)^4 \right]$

$$= \dfrac{-7883.26 + 23635.06 + 0.262}{24} = \dfrac{+15752.06}{24} = \textbf{656.336}$$

$$y_{max} = \dfrac{656.336}{12000} = \textbf{0.0547 m} = \textbf{54.7 mm} \text{ at } 5.4023 \text{ m from } A.$$

EXAMPLE 10.9: A simply supported beam $ACDB$ (span L) carries udl of w/m partly on portion CD. (i) Determine general equations of slope and deflection by Macaulay's method. (ii) Determine deflections at C, D and mid span point if span $AB = 10$ m, $AC = 2$ m, $CD = 5$ m, udl $w = 2$ kN/m, material $E = 200$ kN/mm^2, MI of cross-section $I = 49 \times 10^6$ mm^4.

Solution:

$AB = L = 10$ m, $\quad AC = a_1 = 2$ m,

$AD = a_2 = 7$ m, $\quad w = 2$ kN/m,

$E = 200$ kN/mm$^2 = 2 \times 10^8$ kN/m^2

$I = 49 \times 10^6$ mm$^4 = 49 \times 10^{-6}$ m^4

$EI = 2 \times 10^8 \times 49 \times 10^{-6} = \textbf{9800 kN-m}^2$

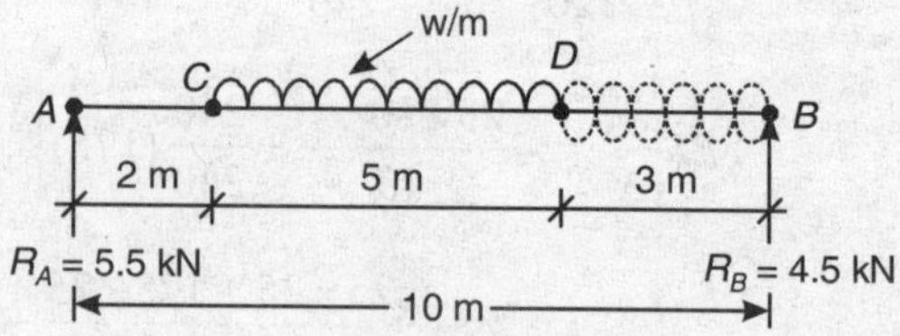

Imaginary load w upward D to B.

$$R_A = \dfrac{1}{10}[2 \times 5(5.5)] = 5.5 \text{ kN}, \ R_B = 4.5 \text{ kN}$$

Fig. 10.22.

$$M_x = + R_A \cdot x \left| - \frac{w(x-2)^2}{2} \right| + \frac{w(x-7)^2}{2}$$

Deflection equation:

$$EI\frac{d^2y}{dx^2} = - M_x = - R_A \cdot x \left| + \frac{w}{2}(x-2)^2 \right| - \frac{w}{2}(x-7)^2$$

$$EI\frac{d^2y}{dx^2} = - 5.5x \left| \frac{2}{2}(x \quad 2)^2 \right| \frac{2}{2}(x \quad 7)^2 \qquad \dots \text{(i)}$$

Integrating: $EI\dfrac{dy}{dx} = - \dfrac{5.5x^2}{2} + C_1 \left| + \dfrac{(x-2)^3}{3} \right| - \dfrac{(x-7)^3}{3}$

or $\quad 9800\dfrac{dy}{dx} = - \dfrac{5.5x^2}{2} + C_1 \left| + \dfrac{(x-2)^3}{3} \right| - \dfrac{(x-7)^3}{3} \qquad \dots \text{(ii)}$

Integrating again: $9800y = - \dfrac{5.5x^3}{6} + C_1 x + C_2 \left| + \dfrac{(x-2)^4}{12} \right| - \dfrac{(x-7)^4}{12} \qquad \dots \text{(iii)}$

$y = 0$, at $x = 0$ and at $x = L = 10$ m. Substituting in (iii)

$$C_2 = 0, \text{ and } 0 = - \frac{5.5}{6}(10)^3 + C_1(10) \left| + \frac{8^4}{12} - \frac{3^4}{12} \right., C_1 = \frac{5500}{60} - \frac{64 \times 64}{120} + \frac{81}{120} = 58.208$$

$$9800y = - \frac{5.5x^3}{6} + 58.208x \left| + \frac{(x-2)^4}{12} \right| - \frac{(x-7)^4}{12}$$

$$y = - \frac{5.5x^3}{6 \times 9800} + \frac{58.208}{9800}x \left| + \frac{(x-2)^4}{12 \times 9800} \right| - \frac{(x-7)^4}{12 \times 9800} \qquad \dots \text{(iv)}$$

Deflections at C $(x = 2$ m$)$, D $(x = 7$ m$)$ and mid span $(x = 5$ m$)$

$$y = - \frac{11x^3}{12 \times 9800} + \frac{58.208}{9800}x \left| + \frac{(x-2)^4}{12 \times 9800} \right| - \frac{(x-7)^4}{12 \times 9800}$$

$$y_C (x = 2) = - \frac{11 \times 8}{12 \times 9800} + \frac{0.594(2)}{100} + 0$$

$$= - 0.0007483 + 0.01188 = 0.01113 \text{ m} = \mathbf{11.13 \text{ mm}}$$

$$y_D (x = 7) = - \frac{11 \times 7^3}{12 \times 9800} + \frac{0.594(7)}{100} \left| + \frac{5^4}{12 \times 9800} \right.$$

$$= - 0.03208 + 0.04158 + 0.005315 = 0.0148146 \text{ m} = \mathbf{14.815 \text{ mm}}$$

$$y_{mid} \, (x = 5) = \frac{11 \times 5^3}{12 \times 9800} + \frac{0.594(5)}{100} \bigg| + \frac{3^4}{12 \times 9800}$$

$$= -0.011692 + 0.0297 + 0.0006888 = 0.0186967 \text{ m} = \textbf{18.697 mm}$$

EXAMPLE 10.10: A simply supported beam AB of span 4 m carries an anticlockwise moment couple of 80 kN-m at the support B as shown in Fig. 10.23. Derive general equations of slope and deflection and also determine the maximum values. $E = 200$ kN/mm^2, $I = 3717.8 \times 10^4$ mm^4.

Solution:

$L = 4$ m, $\qquad\qquad\qquad M = 80$ kNm

$E = 200 \times 10^6$ kN/m^2, $\quad I = 3717.8 \times 10^{-8}$ m^4

$EI = 200 \times 10^6 \times 3717.8 \times 10^{-8}$ kN-m^2

$EI = 7435.6$ kN-m^2

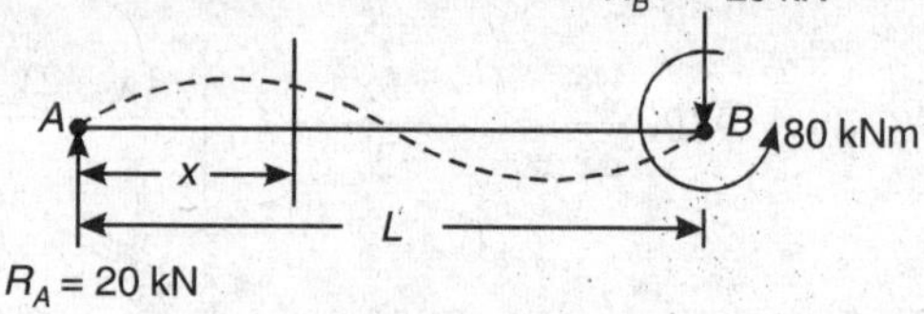

$R_A = \dfrac{80}{4} = 20$ kN (Upward)

$R_B = \dfrac{80}{4} = -20$ kN (Downward)

$M_x = 20x$

Fig. 10.23.

$$EI\frac{d^2y}{dx^2} = -M_x = -20x \qquad\qquad \text{... (i)}$$

Integrating: $EI\dfrac{dy}{dx} = -\dfrac{20x^2}{2} + C_1 = -10x^2 + C_1 \qquad\qquad$ ... (ii)

Integrating again: $EIy = -\dfrac{10x^3}{3} + C_1 x + C_1 \qquad\qquad$ ... (iii)

$y = 0$, at $x = 0$, gives $C_2 = 0$.

$y = 0$, at $x = L = 4$ m: $0 = -\dfrac{10}{3}(4)^3 + C_1(4) + 0$, $C_1 = \dfrac{160}{3}$

$$\therefore \qquad EI\frac{dy}{dx} = -10x^2 + \frac{160}{3} \qquad\qquad \text{... (iv)}$$

$$EIy = -\frac{10x^3}{3} + \frac{160}{3}x \qquad\qquad \text{... (v)}$$

Maximum slope at A and B ($x = 0$ and $x = 4$ m),

$$\theta_A = \frac{160}{3EI} = \frac{160}{3 \times 7435.6} = +0.0071724 \text{ radian}$$

$$\theta_B = -\frac{160}{EI} + \frac{160}{3EI} = -\frac{320}{3EI} = -0.0143454 \text{ radian}$$

For maximum y, $\dfrac{dy}{dx} = 0 = -10(x^2) + \dfrac{160}{3}$, $x = \dfrac{4}{\sqrt{3}}$ m

$$y_{max}\left(x = \dfrac{4}{\sqrt{3}} \text{m}\right) = \dfrac{1}{EI}\left(-\dfrac{10}{3}\cdot\dfrac{64}{3\sqrt{3}} + \dfrac{160}{3}\cdot\dfrac{4}{\sqrt{3}}\right)$$

$$= \dfrac{1}{7435.6\times3}\dfrac{1}{\sqrt{3}}(-213.3+640) \ = +\ \textbf{0.0110432 m} = \textbf{11.043 mm}$$

EXAMPLE 10.11: A simply supported beam AB of length L is subjected to a couple of moment 'M' anticlockwise at C at a distance of 'a' from the support 'A' ($AC = a$). Derive general equations of slope and deflections and determine these values at the point C.

Solution:

Refer Fig. 10.24

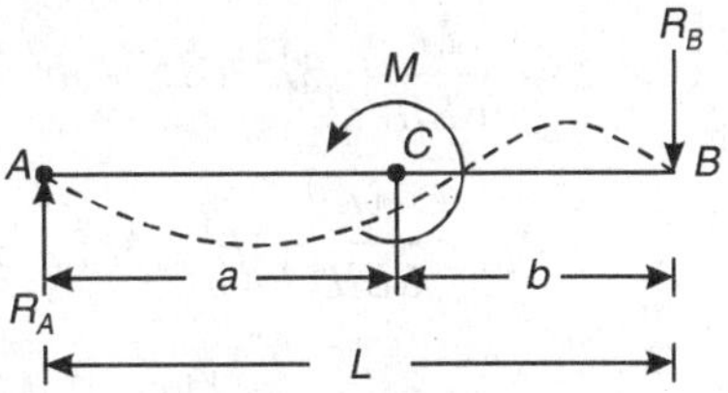

$R_A = \dfrac{M}{L}$, $R_B = \dfrac{-M}{L}$

M_x (x from A) $= R_A \cdot x \left|-M(x-a)^0\right.$

M_x (AC) $= R_A \cdot x$

M_x (CB) $= R_A \cdot x \left|-M(x-a)^0\right.$

Differential equation of deflection curve

$$EI\dfrac{d^2y}{dx^2} = -M_x = R_A \cdot x \left|+M(x-a)^0\right.$$

$$EI\dfrac{d^2y}{dx^2} = -\dfrac{M}{L}\cdot x \left|+M(x-a)^0\right. \qquad \text{... (i)}$$

Integrating: $EI\dfrac{dy}{dx} = -\dfrac{M}{L}\cdot\dfrac{x^2}{2} + C_1 \left|+\dfrac{M(x-a)^1}{1}\right. \qquad \text{... (ii)}$

Integrating again: $EIy = -\dfrac{M}{L}\cdot\dfrac{x^3}{6} + C_1 x + C_2 \left|+\dfrac{M(x-a)^2}{2}\right. \qquad \text{... (iii)}$

At $x = 0$, $y = 0$, gives, $C_2 = 0$

At $x = L$, $y = 0$, gives, $C_1 = \dfrac{ML^2}{6L} - \dfrac{M(L-a)^2}{2L} = \dfrac{ML^2}{6L} - \dfrac{Mb^2}{2L} = \dfrac{M}{6L}(L^2 - 3b^2)$

Therefore $EI\dfrac{dy}{dx} = -\dfrac{M}{2L}\cdot x^2 + \dfrac{M}{6L}(L^2 - 3b^2) \left|+M(x-a)\right. \qquad \text{... (iv)}$

Fig. 10.24.

$$EIy = -\frac{M}{6L} \cdot x^3 + \frac{M}{6L}(L^2 - 3b^2) \cdot x \bigg| + \frac{M}{2}(x-a)^2 \qquad \ldots \text{(v)}$$

$$\theta_C (x = a) = \frac{1}{EI}\left[-\frac{Ma^2}{2L} + \frac{M}{6L}(L^2 - 3b^2)\right] = \frac{M}{6EIL}(L^2 - 3b^2 - 3a^2) \qquad \ldots \text{(10.40)}$$

$$\theta_A (x = 0) = \frac{1}{EI}\left[\frac{M}{6L}(L^2 - 3b^2)\right] = \frac{M}{6EIL}(L^2 - 3b^2) \qquad \ldots \text{(10.41)}$$

$$\theta_B (x = L) = \frac{1}{EI}\left[-\frac{ML^2}{2L} + \frac{M}{6L}(L^2 - 3b^2) + M(L - a)\right]$$

$$= \frac{M}{6EIL}(-3L^2 + L^2 - 3b^2 + 6L^2 - 6La)$$

$$\theta_B = \frac{M}{6EIL}(4L^2 - 3b^2 - 6La) \qquad \ldots \text{(10.42)}$$

$$y_C (x = a) = -\frac{Ma^3}{6EIL} + \frac{M}{6EIL}(L^2 - 3b^2) \cdot a = \frac{Ma}{6EIL}[(L + a)b - 3b^2]$$

$$y_C = \frac{Ma \cdot b}{6EIL}(L + a - 3b) = \frac{Mab}{3EIL}(a - b) \qquad \ldots \text{(10.43)}$$

EXAMPLE 10.12: A cantilever beam of AB of span L and fixed at B. Free end A is subjected to a couple of moment M. Determine the equation of deflection curve if the value of EI is constant.

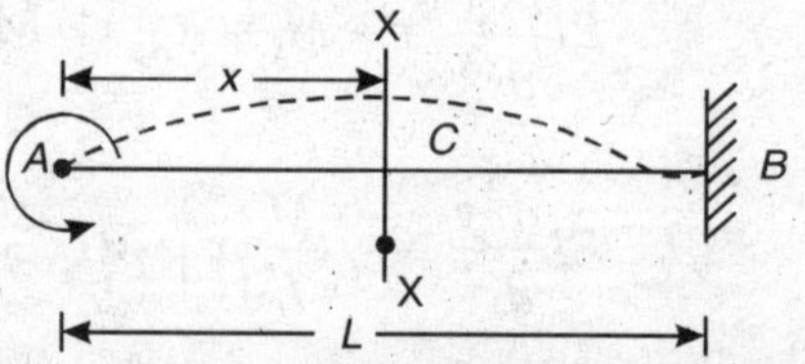

Solution:

$M = -M$ (anticlockwise on left side)

$$EI\frac{d^2y}{dx^2} = -M_x = -(-M)$$

Fig. 10.25.

Integrating: $EI\dfrac{dy}{dx} = M \cdot x + C_1 \qquad \ldots \text{(1)}$

$dy/dx = 0$, at $x = L$ (i.e. fixed point).

$$0 = M.L + C_1, \qquad C_1 = -ML$$

$$EI\frac{dy}{dx} = M \cdot x - ML = -M(L - x) \qquad \ldots \text{(2)}$$

Integrating: $EIy = -\dfrac{M(L - x)^2}{-2} + C_2 = \dfrac{M}{2}(L - x)^2 + C_2$

$y = 0$, at $x = L$, i.e. $0 = \dfrac{M}{2}(L - L)^2 + C_2 \quad \therefore C_2 = 0$

$$\therefore \quad EIy = \frac{M}{2}(L-x)^2 \qquad \qquad \text{... (3)}$$

$$\theta_{max} \text{ at } x = 0, \; \theta_A = -\frac{M(L-0)}{EI} = -\frac{ML}{EI} \qquad \qquad \text{... (10.44)}$$

$$y_{max} \text{ at } x = 0, \; y_A = -\frac{M(L-0)^2}{2EI} = +\frac{ML^2}{2EI} \qquad \qquad \text{... (10.45)}$$

EXAMPLE 10.13: A simply supported beam AB of 8 m span is loaded as shown in Fig. 10.26. (a) Determine deflection at the centre point. (b) Find the maximum deflection and (c) Slope at the support A and B. $E = 2 \times 10^5$ N/mm^2, $I = 2 \times 10^7$ mm^4.

Solution:

$$AB = L = 8 \text{ m}, \qquad w = 2 \text{ kN/m},$$
$$EI = 2 \times 10^8 \times 2 \times 10^{-5} = 4000 \text{ kN-m}^2$$

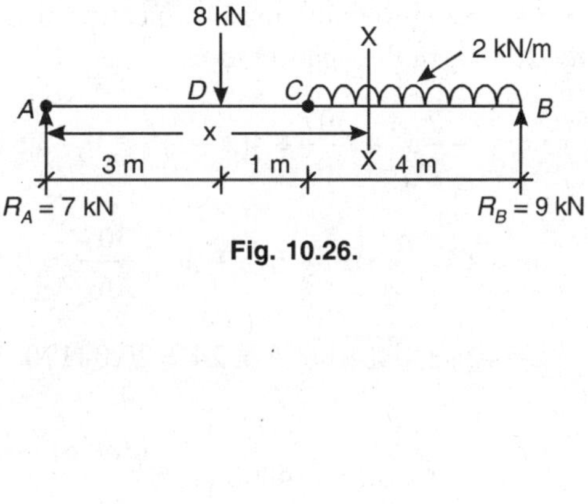

$$R_A = \frac{1}{8}[2 \times 4 \times 2 + 8 \times 5] = 7 \text{ kN}$$

Fig. 10.26.

$$R_B = \frac{1}{8}[2 \times 4 \times 6 + 8 \times 3] = 9 \text{ kN}$$

$$M_x = 7x \Big| \; - 8(x-3) \Big| - \frac{2(x-4)^2}{2}$$

$$EI\frac{d^2y}{dx^2} = -M_x = -7x \Big| + 8(x-3) \Big| + (x-4)^2$$

Integrating: $EI\dfrac{dy}{dx} = -\dfrac{7x^2}{2} + C_1 \Bigg| + \dfrac{8(x-3)^2}{2} \Bigg| + \dfrac{(x-4)^3}{3}$ ... (i)

Integrating again: $EIy = -\dfrac{7x^3}{6} + C_1x + C_2 \Bigg| + \dfrac{8(x-3)^3}{6} \Bigg| + \dfrac{(x-4)^4}{12}$ ... (ii)

$x = 0, y = 0,$ gives: $C_2 = 0$

$$x = L, y = 0, \text{ gives: } C_1 = \frac{1}{L}\left[\frac{7L^3}{6} - \frac{8(L-3)^3}{6} - \frac{(L-4)^4}{12}\right]$$

$$= \frac{7L^2}{6} - \frac{8(L-3)^3}{6L} - \frac{(L-4)^4}{12L}$$

Putting $L = 8$ m: $C_1 = \dfrac{7}{6}(8)^2 - \dfrac{8}{6 \times 8}(5)^3 - \dfrac{1}{12 \times 8}(4)^4 = \dfrac{224}{3} - \dfrac{125}{6} - \dfrac{8}{3} = \dfrac{448 - 125 - 16}{6}$

$$C_1 = \frac{307}{6} = \mathbf{51.667} \qquad\qquad \text{... (iii)}$$

Thus $EI\dfrac{dy}{dx} = -\dfrac{7x^2}{2} + \dfrac{307}{6} \Bigg| + \dfrac{8(x-3)^2}{2} \Bigg| + \dfrac{(x-4)^3}{3}$ ... (iv)

$$EIy = -\frac{7x^3}{6} + \frac{307}{6}x \Bigg| + \frac{8(x-3)^3}{6} \Bigg| + \frac{(x-4)^4}{12} \qquad\qquad \text{... (v)}$$

Equations (iv) and (v) provides slope and deflections at any point.

(a) Deflection at $x = 4$ m, $y_4 = \dfrac{1}{4000}\left[-\dfrac{7}{6}(4)^3 + \dfrac{307}{6}(4) \Bigg| + \dfrac{8}{6}(4-3)^3 \right]$

$$= \mathbf{0.03033 \ m \ (30.33 \ mm)}$$

Assuming the maximum deflection near centre, we have $dy/dx = 0$ for the span D to C. i.e. $dy/dx = 0$, in the span D to C.

i.e. $-\dfrac{7}{2}x^2 + \dfrac{307}{6} + 4(x-3)^2 = 0, -3.5x^2 + \dfrac{307}{6} + 4(x^2 - 6x + 9) = 0$

or $(4 - 3.5)x^2 - 24x + 36 + \dfrac{307}{6} = 0, \ x^2 - 48x + 174.33 = 0,$

$$x = +24 \pm \sqrt{401.66} = +24 \pm 20.04170 = 3.9584 \approx \mathbf{3.958 \ m}$$

(b) $\underset{(x=3.958\,m)}{y_{max}} = \dfrac{1}{4000}\left[-\dfrac{7}{6}(3.958)^3 + \dfrac{307}{6}(3.958) + \dfrac{8}{6}(0.958)^3 \right] = \dfrac{1}{4000}[122.485]$

$$= \frac{122.845}{4000} = \mathbf{0.03062 \ m = 30.62 \ mm}$$

This (30.62 mm) is almost same as the deflection at the centre (30.33 mm) and therefore, deflection can be considered maximum at the centre approximately for the purpose of design.

(c)   Slope at A $(x = 0)$ from equation (iv):

$\underset{(x=0)}{EI\theta_A} = -\dfrac{7}{2}(0)^2 + \dfrac{307}{6} = \dfrac{307}{6}, \ \therefore \ \theta_A = \dfrac{307}{6 \times 4000} = +\ 0.0127916$ radian

$\underset{(x=8\,m)}{EI\theta_B} = -\dfrac{7}{2}(8)^2 + \dfrac{307}{6} + \dfrac{8}{2}(8-3)^2 + \dfrac{(8-4)^3}{3}, \ \theta_B = \dfrac{1}{4000}[\ 224 \quad 51.167 \quad 100 \quad 21.333]$

$$= -\ \mathbf{0.012875 \ radian}$$

EXAMPLE 10.14: A cantilever AB of span L is loaded with uniformly varying load of intensity zero at the free end A and w/m length at the fixed end B. Derive the expression for the deflection and slope at any point X from the free end. Also find the slope and deflection at the free end A.

Solution:

Rate of loading at any point X from free end A

$$w_x = \frac{w}{L} \cdot x, \qquad \text{Span } AB = L, \text{ free end } A.$$

Differential equation for rate of loading

$$EI\frac{d^4y}{dx^4} = wx = \frac{w}{L} \cdot x \qquad \text{... (i)}$$

Integrating: $EI\dfrac{d^3y}{dx^3} = -F_x = \dfrac{w}{L} \cdot \dfrac{x^2}{2} + C_1 \qquad \text{... (ii)}$

Integrating again $EI\dfrac{d^2y}{dx^2} = -M_x = \dfrac{w}{6L}x^3 + C_2$

At $x = 0$, $M_A = 0$, hence $C_2 = 0$

$$EI\frac{d^2y}{dx^2} = \frac{w}{6L} \cdot x^3 \qquad \text{... (iii)}$$

This can also be derived directly by BM equation.

$$M_x = -\frac{w}{L} \cdot x \cdot \frac{x}{2} \cdot \frac{x}{3} = -\frac{w}{6L} \cdot x^3$$

$$EI\frac{d^2y}{dx^2} = -M_x = -\left(-\frac{w}{6L} \cdot x^3\right) = +\frac{wx^3}{6L} \qquad \text{... same as (iii)}$$

Integrating again $EI\dfrac{dy}{dx} = \dfrac{w}{6L} \cdot \dfrac{x^4}{4} + C_3,$

At $x = L$, $dy/dx = 0$,

$$\therefore \quad 0 = \frac{w}{6L} \cdot \frac{L^4}{4} + C_3, \; C_3 = -\frac{wL^4}{24L} = -\frac{wL^3}{24}$$

$$\therefore \quad EI\frac{dy}{dx} = \frac{w}{24L} \cdot x^4 - \frac{wL^3}{24} \qquad \text{... (iv)}$$

Integrating again $EIy = \dfrac{w}{24L} \cdot \dfrac{x^5}{5} - \dfrac{wL^3}{24}x + C_4$

At $x = L$, $y = 0$,

$$0 = \frac{wL^4}{120} - \frac{wL^4}{24} + C_4, \; C_4 = \frac{wL^4}{30}$$

Thus, $EIy = \dfrac{w}{120L}\cdot x^5 - \dfrac{wL^3}{24}\cdot x + \dfrac{wL^4}{30}$ ... (v)

θ_A (free end) $= -\dfrac{wL^3}{24EI}$, θ_B (fixed end) $= 0$ ... (10.46)

$y_A\,(x = 0) = \dfrac{wL^4}{30EI}$ ... (10.47)

Case 3: Slope-deflection in propped cantilever

Some times the cantilevers are additionally supported at free ends by using *props* which may be *rigid* or *elastic* (Fig. 10.28). The *rigid props* do not yield and the net deflection of the free end remains zero. The elastic props *yield* to some extent and put up a reaction to reduce yielding. The reaction of the prop can be found by *equating deflection* at the free end by loading and negative deflection due to the prop reaction in case of rigid props.

i.e. $y_{\text{loading}} = -y_{\text{prop reaction}}$ for rigid props.

or $y_{\text{loading}} - y_{\text{prop reaction}} = 0$ ... (10.48)

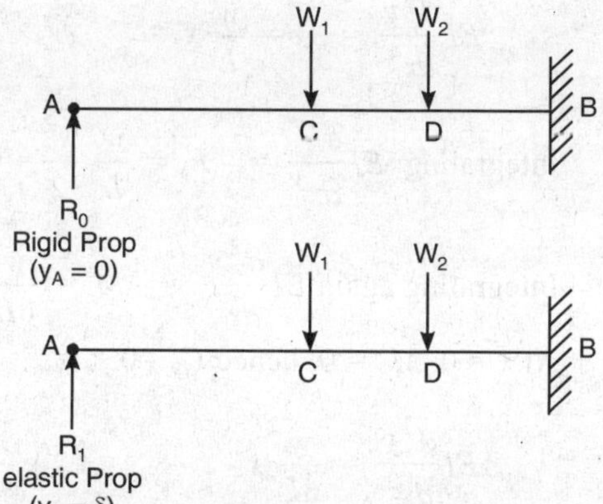

The net deflection at free support of rigid prop will be zero. In case of elastic prop the yielding (δ) is equated to the difference of deflections caused by loading and the prop reaction (R).

i.e. $y_{\text{loading}} - y_{\text{reaction}} = \delta_{\text{yielding}}$ for elastic props ... (10.45)

These equations are used to determine the prop reaction. The prop reaction and the loading are used to formulate the *equation of deflection curve*. The constants of integration are determined as usual by applying the boundary conditions (i.e. fixed end, $y = 0$, $dy/dx = 0$). This is explained by some examples.

EXAMPLE 10.15: A cantilever beam of 4 m span carries a load of 20 kN at a point 3 m from the fixed end B. (a) Determine deflection at the free end A when there is no prop. (b) If the free end is supported by a rigid prop at the same level as the fixed end, determine the prop reaction. (c) If the elastic prop at the free end yields by $\dfrac{110}{EI}$ m, determine the elastic prop reactions and draw bending moment diagram.

Solution:

$L = 4$ m, $AC = a = 1$ m, $CB = 3$ m

$W = 20$ kN, $EI - $ kN-m^2, $\delta = \dfrac{110}{EI}$ m

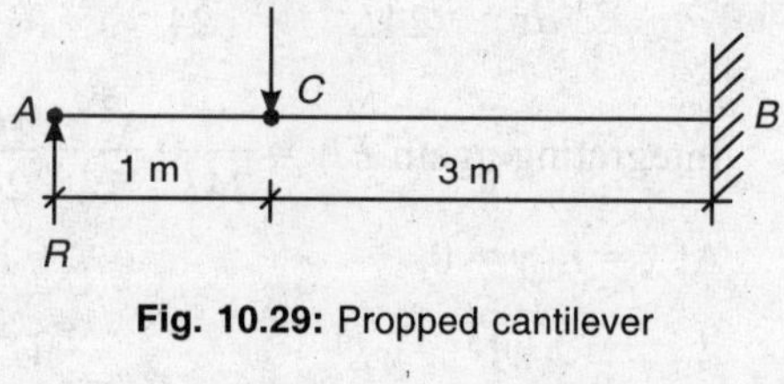

Fig. 10.28.

Fig. 10.29: Propped cantilever

(a) $\delta_C = \dfrac{WL_1^3}{3EI}$

$\delta_A = \dfrac{WL_1^3}{3EI} + \dfrac{WL_1^2}{2EI}(L - L_1) = \dfrac{20(3)^3}{3EI} + \dfrac{20(3)^2 \cdot 1}{2EI} = \dfrac{270}{EI}$ m

(b) Let the rigid prop reaction = R

Upward deflection at A due to prop reaction $R = \dfrac{RL^3}{3EI} = \dfrac{R(4)^3}{3EI}$

Downward deflection at A due to loading $= \dfrac{270}{EI}$

Since the prop is rigid, net deflection = zero

i.e. $\dfrac{R(64)}{3EI} - \dfrac{270}{EI} = 0$, or $R = \dfrac{270 \times 3}{64} = \dfrac{810}{64} = 12.656$ kN

$M_x = \dfrac{810}{64} \cdot x \Bigg| - 20(x-1), \ EI \dfrac{d^2y}{dx^2} = - \dfrac{810}{64} x \Bigg| + 20(x-1)$ ... (i)

Integrating: $EI\dfrac{dy}{dx} = - \dfrac{810x^2}{128} + C_1 \Bigg| + 10(x-1)^2$ ... (ii)

Integrating: $EIy = - \dfrac{810x^3}{128 \times 3} + C_1 \cdot x + C_2 \Bigg| + \dfrac{10}{3}(x-1)^3$ ... (iii)

equation (ii) $\dfrac{dy}{dx} = 0$, at $x = 4$ m, $0 = - \dfrac{810}{128}(4)^2 + C_1 \Bigg| + 10(4-1)^2$

$\therefore \quad C_1 = \dfrac{810}{8} - 90 = \dfrac{90}{8}$

equation (iii) $EIy = - \dfrac{810x^3}{128 \times 3} + \dfrac{90}{8}x + C_2 \Bigg| + \dfrac{10}{3}(x-1)^3$ ... (iv)

$y = 0$, at $x = 4$ m (fixed end), $\therefore \ 0 = \dfrac{-810}{128}\left(\dfrac{64}{3}\right) + \dfrac{90(4)}{8} + C_2 + \dfrac{10}{3}(3)^3$

$\therefore \quad C_2 = \dfrac{810}{6} - \dfrac{90}{2} - 90 = \dfrac{810 - 270 - 540}{6} = 0$

Hence equation (iv) $EIy = - \dfrac{810}{128}\dfrac{x^3}{3} + \dfrac{90}{8} \Bigg| + \dfrac{10}{3}(x-1)^3$ ... (v)

equation (ii) $EI\dfrac{dy}{dx} = - \dfrac{810x^2}{128} + \dfrac{90}{8}x \Bigg| + 10(x-1)^2$ ... (vi)

Equations (v) and (vi) are sufficient to determine slope and deflection anywhere along the beam.

(c)   The prop yields by $\dfrac{110}{EI}$ m.

$\therefore \quad y_{\text{loading}} - y_{\text{reation}} = \delta = \dfrac{110}{EI}$

or $\dfrac{270}{EI} - \dfrac{RL^3}{3EI} = \dfrac{110}{EI}$, or $\dfrac{270}{EI} - \dfrac{110}{EI} = \dfrac{R(4)^3}{3EI}$

or $R = \dfrac{160 \times 3}{64} = 7.5$ kN

The bending moment diagram will be prepared with $R = 7.5$ kN.

A to C: $M_x = 7.5\, x$ kN-m,
$\qquad\qquad M_C = 7.5$ kN-m

C to B: $M_x = 7.5\, x - 20\,(x - 1)$,
$\qquad\qquad M_B = 7.5 \times 4 - 20\,(3)$
$\qquad\qquad\quad = -30$ kN-m

Diagram shown in Fig. 10.30.

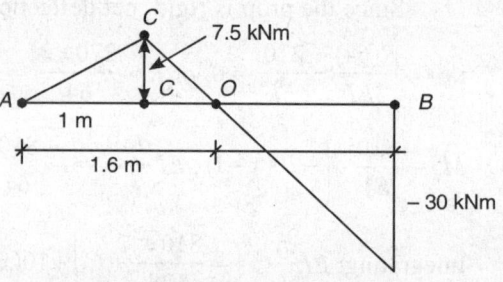

Fig. 10.30: BM diagram (kN-m)

EXAMPLE 10.16: A cantilever AB of span 4 m carries udl 10 kN/m on the entire span. (a) Determine the deflection at the free end. (b) If the free end is supported by a rigid prop at the level of fixed end, determine the prop reaction. Also determine the deflection curve equation and draw BM diagram. $E = 2 \times 10^5$ N/mm^2, $I = 3718 \times 10^4$ mm^4.

Solution:

$L = 4$ m, $w = 10$ kN/m

$EI = 2 \times \dfrac{10^5}{10^3} \times 10^6 \times \dfrac{3718 \times 10^4}{10^{12}}$ kNm^2

$\quad = 3718 \times 2$ kN-m$^2 = 7436$ kN-m^2

(a) δ_A (no prop) $= \dfrac{wL^4}{8EI} = \dfrac{10 \times 4^4}{8 \times 7436}$

$\qquad\qquad = \dfrac{320}{7436} = 0.043034$ m

$\qquad\qquad = (43.034 \text{ mm})$

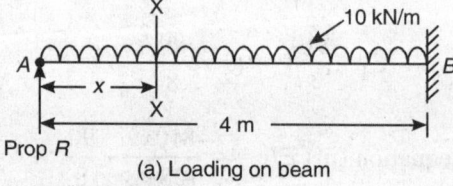

(a) Loading on beam

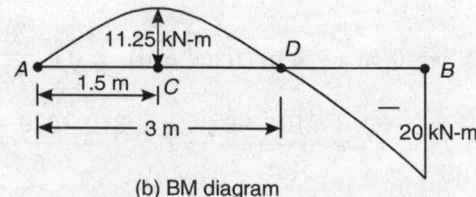

(b) BM diagram

Fig 10.31: Propped cantilever

(b)   Rigid prop reaction $= R$.

'δ' due to loading is balanced by prop reaction to make deflection zero.

i.e. $\dfrac{RL^3}{3EI} - 0.043034 = 0$ or $R = \dfrac{320 \times 3EI}{7436 \times 64} = \dfrac{320 \times 3}{64} = 15$ kN

BM $M_x = 15x - \dfrac{10}{2}(x)^2 = (15x - 5x^2)$, $BM = 0$ at $x = $ **3 m,**

BM maximum at $\dfrac{d}{dx}(M_x) = 0$ or $15 - 10x = 0$, $x = $ **1.5 m**

M_{max} $(x = 1.5$ m$) = 15(1.5) - 5(1.5)^2 = 22.5 - 11.25 = +$ **11.25 kN m**

Point of contraflexure at $x = 3$m, Point of Maximum BM = 1.50 m

M_B $(x = 4$ m$) = 15 \times 4 - 5(4)^2 = 60 - 80 = -$ **20 kN-m**

EXAMPLE 10.17: A simply supported beam of 10 m span is supported at ends A and B, and at the centre C at the *same level*. If the beam carries udl of 10 kN/m over the entire 10 m span. Determine the reactions at A, B and C and draw BM diagram. Also determine deflection curve equation and maximum deflection.

$E = 2 \times 10^5$ N/mm^2, $I = 7333 \times 10^4$ mm^4.

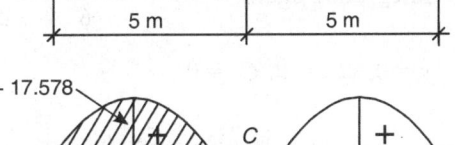

Solution:

$EI = 2 \times 10^5 \times 10^{-3} \times 10^6$

$\times 7333 \times 10^{-8}$ kN-m$^2 = 14666$ kN-m^2

$w = 10$ kN/m, $\qquad L = 10$ m

Reactions at the central prop support $= R$

End reactions are equal due to symmetry $R_A = R_B = \dfrac{wl - P}{2}$

Notional deflection at the centre $y_C = \dfrac{5wL^4}{384EI} = \dfrac{5}{384} \dfrac{wL^4}{EI}$

Notional upward deflection due to prop reaction $R = \dfrac{RL^3}{48EI}$

Since net deflection at $C = 0$, $\dfrac{R \cdot L^3}{48EI} - \dfrac{5}{384} \cdot \dfrac{wL^4}{EI} = 0$

or $\quad R = wl \cdot \dfrac{5}{384} \times 48 = 10 \times 10 \times \dfrac{5}{8} = $ **62.5 kN**

$R_A = R_B = \dfrac{10 \times 10 - 62.5}{2} = $ **18.75 kN** each, $F_x = 0$ at $\dfrac{18.75}{10} = $ **1.875 m**

$BM\ M_x = 18.75x - \dfrac{10}{2}x^2 \Big| + 62.5\ (x - 5)$ $\qquad\qquad$ ... (i)

SF $= 0$ at $x = 1.875$ m and hence M_x will be maximum.

M_{max} $(x = 1.875$ m$) = 18.75\ (1.875) - 5\ (1.875)^2 = $ **17.578 kN-m**

M_C $(x = 5$ m$) = 18.75\ (5) - 5\ (5)^2 = 93.75 - 125 = -$ **31.25 kN-m**

$$M_x \text{ (beyond } C) = 18.75x - 5x^2 \Big| + 62.5\,(x - 5)$$

$$M_B\,(x = 10 \text{ m}) = 187.5 - 500 + 312.5 = 0$$

General equation of deflection

$$EI\frac{d^2y}{dx^2} = -18.75x + 5x^2 \Big| - 62.5\,(x - 5) \qquad \text{... (ii)}$$

Integrating: $EI\dfrac{dy}{dx} = -\dfrac{18.75x^2}{2} + \dfrac{5x^3}{3} + C_1 \Big| - 62.5\dfrac{(x-5)^2}{2}$ \qquad ... (iii)

Integrating again: $EIy = -\dfrac{18.75x^3}{6} + \dfrac{5x^4}{12} + C_1 x + C_2 \Big| - 62.5\dfrac{(x-5)^3}{6}$ \qquad ... (iv)

$y = 0$, at $x = 0$, $C_2 = 0$

$$EIy = -\frac{18.75x^3}{6} + \frac{5x^4}{12} + C_1 x \Big| - 62.5\frac{(x-5)^3}{6} \qquad \text{... (v)}$$

$y = 0$, at $x = 5$ gives: $0 = -\dfrac{18.75(5)^3}{6} + \dfrac{5(5)^4}{12} + C_1(5)$

$$C_1 = \frac{18.75(25)}{6} - \frac{5(125)}{12} = 78.125 - 52.083 = 26.042$$

Deflection curve equation is:

$$EIy = -\frac{18.75x^3}{6} + \frac{5x^4}{12} + 26.042x \Big| - 62.5\frac{(x-5)^3}{6} \qquad \text{... (vi)}$$

y_{max} in zone $x = 0$ to 5 m, $dy/dx = 0$

$$0 = -\frac{18.75x^2}{2} + \frac{5x^3}{3} + 26.042, \text{ or } 56.25x^2 - 10x^3 - 156.252 = 0$$

By trial $x = \mathbf{2.1078}$ m

$$y_m\,(x = 2.1078 \text{ m}) = \frac{1}{EI}\left[-\frac{18.75}{6}(2.1078)^3 + \frac{5}{12}(2.1078)^4 + 26.042(2.1078) \right]$$

$$= \frac{-29.264 + 8.22244 + 54.8913}{EI} = \frac{33.85140}{14666} = .002308 \text{ m (2.308 mm)}$$

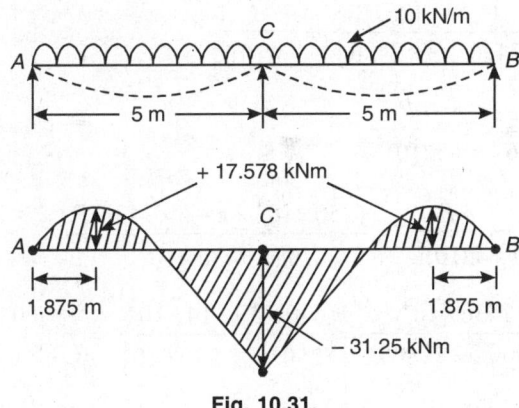

Fig. 10.31.

EXAMPLE 10.18: Two cantilever beams of 5 m span and the same cross-section are fixed one above the other vertically in the same plane as shown in Fig. 10.32. Free ends of both the beams are connected by a vertical steel rod of 20 mm diameter and 2 m length. Upper cantilever carries an udl of 15 kN/m for 4 m length from the fixed end. Determine the rod reaction and deflection of the free end A of upper cantilever if 'E' of beams and the rod is $= 2 \times 10^5$ N/mm², I of beams $= 3718 \times 10^4$ mm⁴.

Solution:

$L = 5$ m, $\qquad L_1 = 4$ m,

$w = 15$ kN/m, $\quad EI = 7436$ kN-m²

Rod diameter = 20 mm

Rod length = 2 m

Rod reaction = P kN

Change in rod length due

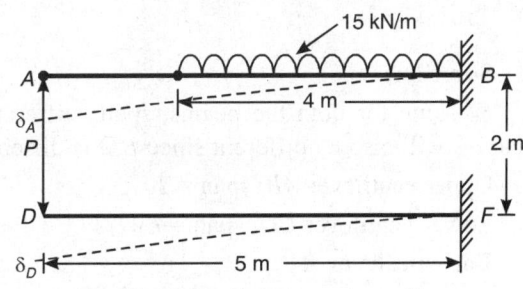

Fig. 10.32: Elastic rod prop in cantilever

$$\text{compression in prop} = \frac{P(L)}{AE} = \frac{P(2)}{\dfrac{\pi}{4}(0.02)^2 \times 2 \times 10^8} = \frac{P}{\pi(10)^4}$$

Deflection at the free end (δ_A):

$$\delta_A = \left\{ \frac{wL_1^4}{8EI} + \frac{wL_1^3(L - L_1)}{6EI} \right\} - \frac{PL^3}{3EI} = \left[\frac{15(4)^4}{8} + \frac{15(4)^3(5-4)}{6} - \frac{P(5)^3}{3} \right] \frac{1}{EI}$$

or $\quad \delta_A = \dfrac{640}{EI} - \dfrac{125P}{3EI} = \left(640 - \dfrac{125P}{3} \right) \dfrac{1}{7436}$ $\qquad$ ... (ii)

δ_D (Lower beam) due to prop reaction $P = \dfrac{P(L)^3}{3EI} = \dfrac{P(125)}{3 \times 7436}$ $\qquad$ ... (iii)

$(\delta_A - \delta_D) =$ comp. deformation in the rod.

i.e. $\left\{\left(640-\dfrac{125P}{3}\right)\dfrac{1}{7436}-\dfrac{125P}{3\times7436}\right\} = \dfrac{P}{\pi(10)^4}$

or $\left\{\dfrac{640}{7436}-\dfrac{2\times125P}{3\times7436}\right\} = \dfrac{P}{\pi(10)^4}$

or $\dfrac{640}{7436} = \dfrac{250P}{3\times7436}+\dfrac{P}{\pi(10)^4} = \dfrac{(250\times10^4\times\pi+3\times7436)P}{3\pi\times7436\times10^4}$

$\therefore P = \dfrac{640}{7436}\times\dfrac{3\pi\times7436\times10^4}{(2500000\pi+22308)} = \dfrac{640\times3\times3{\cdot}14\times10^4}{(250\pi+2.2308)10^4} = \dfrac{6028.8}{(785+2.2308)} = \textbf{7.660 kN}$

$\therefore \delta_A = \left(640-\dfrac{125\times7.66}{3}\right)\dfrac{1}{7436} = \textbf{0.043157 m (43.157 mm)}$

EXAMPLE 10.19: A cantilever of span L carries a point load W at the free end. This cantilever rests on another cantilever of the same cross-section but span L_1 from the fixed end as shown in Fig. 10.33. Prove that the lower cantilever gets a pressure of $P = \dfrac{3W}{4}\left(\dfrac{L}{L_1}-\dfrac{1}{3}\right)$.

[AMIE India]

Solution:

EI same for both the beams, spans different and hence BM will be different. Deflection curves will also be different since BM is different.

Upper cantilever AB, span $= L$

Lower cantilever CD, span $= L_1$

For cantilever AB

$M_x = -Wx$ (when no lower beam and x is from the free end)

$EI\dfrac{d^2y}{dx^2} = -(-W\cdot x) = +W\cdot x$ ... (i)

Integrating: $EI\dfrac{dy}{dx} = +\dfrac{W\cdot x^2}{2}+C_1$ ... (ii)

Integrating again: $EIy = +\dfrac{W\cdot x^3}{6}+C_1x+C_2$ ... (iii)

$\dfrac{dy}{dx} = 0$, at $x = L$ (fixed end),

From equation (ii) $= -\dfrac{WL^2}{2}$

From equation (iii)

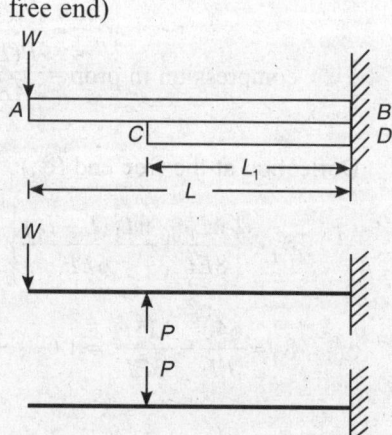

Fig. 10.33: Cantilever resting on cantilever

$y = 0$, at $x = L$ (fixed end), $C_2 = + \dfrac{WL^3}{3}$

$\therefore \quad EIy = + \dfrac{W \cdot x^3}{6} - \dfrac{WL^2 \cdot x}{2} + \dfrac{WL^3}{3}$ ... (iv)

$y_{C_1}(x = L - L_1) = \dfrac{W}{6EI}\left[(L - L_1)^3 - 3L^2(L - L_1) + 2L^3\right] = \dfrac{W}{6EI}\left[3L \cdot L_1^2 - L_1^3\right]$

Lower cantilever reaction P at C causes upward deflection $y_{C_2} = - \dfrac{PL_1^3}{3EI}$,

Deflection y_{C_3} due to reaction P on lower cantilecer $= + \dfrac{PL_1^3}{3EI}$

Net deflection of point $C = \dfrac{PL_1^3}{3EI} = \dfrac{W}{6EI}(3LL_1^2 - L_1^3) - \dfrac{PL_1^3}{3EI}$

or $\dfrac{2PL_1^3}{3EI} = \dfrac{WL_1^2}{6EI}(3L - L_1)$, $\therefore P = \dfrac{W}{4L_1}(3L - L_1) = \dfrac{3W}{4}\left(\dfrac{L}{L_1} - \dfrac{1}{3}\right)$

EXAMPLE 10.20: An overhanging beam $ABCD$ of length 'L' is supported at a hinge A and simple support 'C'. The portion 'CD' of length '$L/3$' carries a concentrated load 'W' at the free end 'D'. Half of the supported span 'BC' carries a total udl of W. The portion $AB = BC = CD = L/3$. Determine the deflection at the free end D if the beam comprises of uniform cross-section throughout (Fig. 10.34).

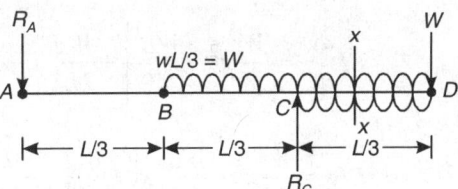

Fig. 10.34: Overhang beam $ABCD$

Solution:

Reactions at A and C are calculated by taking moment about C and A respectively.

$R_C = \dfrac{3}{2L}\left[W \cdot L + W\left(\dfrac{L}{3} + \dfrac{L}{6}\right)\right] = \dfrac{9}{4}W = 2.25\ W$

$R_A = \left(-W \cdot \dfrac{L}{3} + W \cdot \dfrac{L}{6}\right)\dfrac{3}{2L} = -0.25\ W \text{ (Downward)}$

$w = \dfrac{3W}{L}$

Bending equation for overhanging portion CD:

$M_x = -0.25W \cdot x \left|-\dfrac{w}{2}\left(x - \dfrac{L}{3}\right)^2\right| + 2.25W\left(x - \dfrac{2L}{3}\right) + \dfrac{w}{2}\left(x - \dfrac{2L}{3}\right)^2$

$$EI\frac{d^2y}{dx^2} = +0.25W \cdot x \bigg| + \frac{w}{2}\left(x - \frac{L}{3}\right)^2 \bigg| -2.25W\left(x - \frac{2L}{3}\right) \bigg| -\frac{w}{2}\left(x - \frac{2L}{3}\right)^2 \qquad \dots \text{(i)}$$

Integrating: $EI\dfrac{dy}{dx} = 0.25W \cdot \dfrac{x^2}{2} + C_1 \bigg| + \dfrac{w}{6}\left(x - \dfrac{L}{3}\right)^3 \bigg| -\dfrac{2.25W}{2}\left(x - \dfrac{2L}{3}\right)^2 -\dfrac{w}{6}\left(x - \dfrac{2L}{3}\right)^3 \quad \dots \text{(ii)}$

Integrating again:

$$EIy = 0.25W \cdot \frac{x^3}{6} + C_1 x + C_2 \bigg| + \frac{w}{24}\left(x - \frac{L}{3}\right)^4 \bigg| -\frac{w}{24}\left(x - \frac{2L}{3}\right)^4 -\frac{2.25W}{6}\left(x - \frac{2L}{3}\right)^3 \qquad \dots \text{(iii)}$$

At $x = 0$, $y = 0$, gives $C_2 = 0$.

At $x = \dfrac{2}{3}L$, $y = 0$, gives: $0 = \dfrac{0.25W}{6}\left(\dfrac{2L}{3}\right)^3 + C_1\left(\dfrac{2L}{3}\right) + \dfrac{w}{24}\left(\dfrac{L}{3}\right)^4$

$$C_1\left(\frac{2L}{3}\right) = -\frac{W}{24}\left(\frac{8L^3}{27}\right) - \frac{3W}{L \cdot 24}\left(\frac{L^4}{81}\right) = -\frac{WL^3}{81} - \frac{WL^3}{648} = -\frac{WL^3}{648}(9) = -\frac{WL^3}{72}$$

$$C_1 = -\frac{WL^3}{72} \times \frac{3}{2L} = -\frac{WL^2}{48}$$

Putting values in slope and deflection equation (ii) and (iii)

$$EI\frac{dy}{dx} = \frac{W}{8}x^2 - \frac{WL^2}{48} \bigg| + \frac{W}{2L}\left(x - \frac{L}{3}\right)^3 \bigg| -\frac{W}{2L}\left(x - \frac{2L}{3}\right)^3 - \frac{9W}{8}\left(x - \frac{2L}{3}\right)^2 \qquad \dots \text{(iv)}$$

$$EIy = \frac{W}{24}x^3 - \frac{WL^2}{48}x \bigg| + \frac{W}{8L}\left(x - \frac{L}{3}\right)^4 \bigg| -\frac{W}{8L}\left(x - \frac{2L}{3}\right)^4 - \frac{9W}{24}\left(x - \frac{2L}{3}\right)^3 \qquad \dots \text{(v)}$$

Deflection at free end of overhang at D $(x = L)$

$$EIy_D = \frac{WL^3}{24} - \frac{WL^3}{48} + \frac{W}{8L}\frac{16L^4}{81} - \frac{W}{8L}\frac{L^4}{81} - \frac{9W}{24}\frac{L^3}{27} \cdot$$

$$y_D = \frac{WL^3}{24EI}\left[1 - \frac{1}{2} + \frac{16}{27} - \frac{1}{27} - \frac{1}{3}\right] = \frac{WL^3}{24EI}\left(\frac{54 - 27 + 32 - 2 - 18}{2 \times 27}\right) = \frac{39WL^3}{2 \times 24 \times 27 EI}$$

$$y_D = \frac{13WL^3}{432EI}$$

EXAMPLE 10.21: A beam ABC is supported at A and B ($AB = L$) and overhangs from B to C ($BC = a$). It carries a point load W at the free end C. Determine the maximum deflection between the supports, deflection at C and slope at A.

Solution:

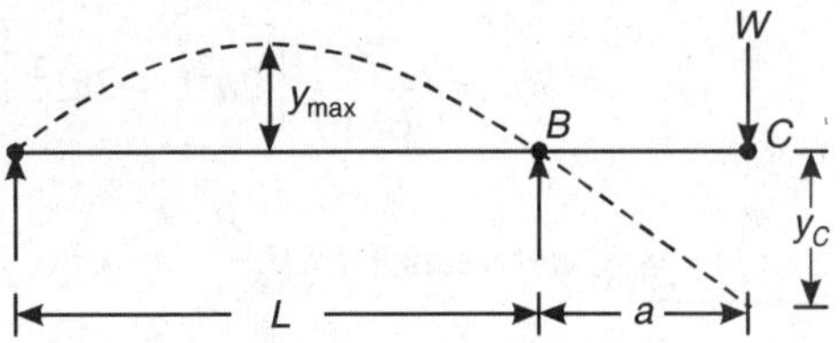

$$R_A = -\frac{1}{L}(Wa) \text{ downward}$$

$$R_B = \frac{W(L\ a)}{L} \text{ upward}$$

Fig. 10.35: Overhanging beam

$$M_x = -\frac{Wa}{L}\cdot x \ \Big| \ \frac{W(L\ a)}{L}(x\ L)$$

$$EI\frac{d^2y}{dx^2} = \frac{Wa}{L}\cdot x \ \Big| \ \frac{W(L\ a)(x\ L)}{L} \qquad \dots \text{(i)}$$

Integrating: $EI\dfrac{dy}{dx} = \dfrac{Wa\cdot x^2}{2L} + C_1 \ \Big| -\dfrac{W(L+a)}{L}\dfrac{(x-L)^2}{2} \qquad \dots \text{(ii)}$

Integrating again: $EIy = \dfrac{Wa\cdot x^2}{6L} + C_1 x + C_2 \ \Big| \dfrac{W(L+a)(x-L)^3}{6L} \qquad \dots \text{(iii)}$

Boundary conditions: $x = 0$, $y = 0$, gives $C_2 = 0$

$$x = L,\ y = 0:\ 0 = \frac{WaL^2}{6} + C_1(L),\ \mathbf{C_1 = -\frac{WaL}{6}}$$

Hence $EI\dfrac{dy}{dx} = \dfrac{Wa\cdot x^2}{2L} - \dfrac{WaL}{6} \ \Big| -\dfrac{W(L+a)(x-L)^2}{2L} \qquad \dots \text{(iv)}$

$$EIy = \frac{Wa\cdot x^3}{6L} - \frac{WaL}{6}x \ \Big| \ \frac{W(L+a)(x-L)^3}{6L} \qquad \dots \text{(v)}$$

From (iv) slope at A $(x = 0)$: $EI\theta_A = -\dfrac{WaL}{6}$, $\boldsymbol{\theta_A = -\dfrac{WaL}{6EI}}$

Slope at B $(x = L)$: $\theta_B = \dfrac{1}{EI}\left[\dfrac{WaL}{2} - \dfrac{WaL}{6}\right] = +\dfrac{WaL}{3EI}$

$$\begin{aligned} y_C \atop {\scriptstyle(x=L+a)} &= \frac{1}{EI}\left[\frac{Wa}{6L}(L+a)^3 - \frac{WaL}{6}(L+a) - \frac{W(L+a)}{6L}(a)^3\right] \\ &= \frac{Wa}{6LEI}\left[(L^3 + a^3 + 3a^2L + 3aL^2) - L^2(L+a) - (L+a)a^2\right] \\ &= \frac{Wa}{6LEI}\left[3a^2L + 3aL^2 - aL^2 - a^2L\right] \end{aligned}$$

$$y_C = \frac{Wa}{6LEI}\left[2a^2L + 2aL^2\right] = \frac{Wa^2L}{3EIL}(a+L) = \frac{Wa^2(a+L)}{3EI}$$

y_{max} between A to B, $\dfrac{dy}{dx} = 0$ i.e. $0 = \dfrac{Wax^2}{2L} - \dfrac{WaL}{6}$ or $x = \dfrac{L}{\sqrt{3}}$

$$y_{max}\left(x = \frac{L}{\sqrt{3}}\right) = \frac{1}{EI}\left[\frac{Wa}{6L}\left(\frac{L}{\sqrt{3}}\right)^3 - \frac{WaL}{6}\left(\frac{L}{\sqrt{3}}\right)\right] = \frac{WaL^2}{6EI}\left(\frac{1}{3\sqrt{3}} - \frac{1}{\sqrt{3}}\right)$$

$$y_{max} = \frac{WaL^2}{6EI}\frac{(-2)}{3\sqrt{3}} = -\frac{WaL^2}{9\sqrt{3}EI} \quad \text{(Negative sign indicates upward deflection)}$$

EXAMPLE 10.22: A cantilever AB of 4 m span is propped at 1 m from the free end. The cantilever carries a concentrated load of 10 kN at the free end. If the prop is rigid and remains at the level of fixed support. Determine the prop reaction, slope and deflection at the free end. Also draw the BM diagram. $EI = 4000$ kN-m^2.

Solution:

Span $AB = 4$ m, $\qquad AC = 1$ m,

Prop reaction $= P$

$EI = 4000$ kN-m^2

$M_x = -10\,x$, when there is no prop

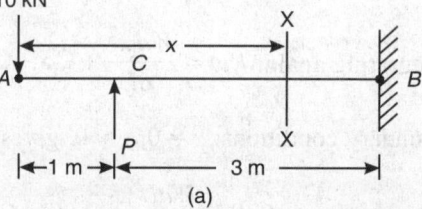

(a)

$$EI\frac{d^2y}{dx^2} = -(-10x) = 10x$$

Integrating: $EI\dfrac{dy}{dx} = \dfrac{10x^2}{2} + C_1$

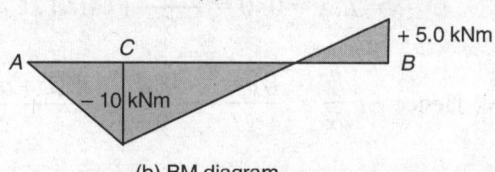

(b) BM diagram

Fig. 10.36: Propped cantilever

Integrating again: $EIy = \dfrac{10x^3}{6} + C_1x + C_2$

$x = L$, $dy/dx = 0$, $5(L)^2 + C_1 = 0$,

$C_1 = -5(4)^2 = -80$ kN-m units

$x = L$, $y = 0$, $\dfrac{10}{6}(L)^3 - 80(L) + C_2 = 0$,

$C_2 = 80(4) - \dfrac{10}{6}(64) = \dfrac{320}{3} \times 2 = \dfrac{640}{3}$ kN-m units

$$EI\frac{dy}{dx} = 5x^2 - 80 \qquad\qquad\qquad \text{... (i)}$$

$$EIy = \frac{5x^3}{3} - 80x + \frac{640}{3}$$ ... (ii)

$$y_C \ (x = 1 \text{ m}) = \left(\frac{5}{3} - 80 + \frac{640}{3}\right)\frac{1}{EI} = +\frac{405}{3EI} = \frac{135}{EI}, \text{ when no prop.}$$

Upward balancing deflection due to prop 'P' $= \dfrac{PL_1^3}{3EI}$

i.e. $\dfrac{P(3)^3}{3EI} = \dfrac{135}{EI}$, $P = \dfrac{135}{9} = \textbf{15 kN}$

BM Diagram A to C: $M_x = -10 \ (x)$ ... straight line,
        $M_C \ (x = 1 \text{ m}) = -\textbf{10 kNm}$
    C to B: $M_x = -10 \ (x) + P \ (x - 1) = -10x + 15 \ (x - 1) = (5x - 15)$
        $M_B \ (x = 4 \text{ m}) = 20 - 15 = +\textbf{5 kNm}$

BM diagram shown in Fig. 10.36b.
General equation of BM

$$M_x = -10x \Big| + 15(x - 1)$$

$$EI\frac{d^2y}{dx^2} = +10x \Big| - 15(x - 1)$$ ... (iii)

Integrating: $EI\dfrac{dy}{dx} = +\dfrac{10}{2}x^2 + C_3 \Big| -\dfrac{15}{2}(x - 1)^2$ ... (iv)

Integrating again: $EIy = +\dfrac{10}{6}x^3 + C_3 x + C_4 \Big| -\dfrac{15}{6}(x - 1)^3$ ... (v)

$\dfrac{dy}{dx} = 0$ at $x = 4$, i.e. $0 = 5(4)^2 + C_3 - \dfrac{15}{2}(3)^2$, $C_3 = -80 + \dfrac{135}{2} = -\dfrac{25}{2}$ ... (vi)

Also $y = 0$ at $x = 1$, i.e. $0 = \dfrac{10}{6}(1)^3 - \dfrac{25}{2}(1) + C_4$, $C_4 = \dfrac{25}{2} - \dfrac{10}{6} = +\dfrac{65}{6}$ ... (vii)

$\therefore$ $EI\dfrac{d^2y}{dx^2} = +5x^2 - \dfrac{25}{2} \Big| -\dfrac{15}{2}(x - 1)^2$ ... (viii)

$$EIy = \frac{5}{3}x^3 - \frac{25}{2}x + \frac{65}{6} \Big| -\frac{15}{6}(x - 1)^3$$ ... (ix)

$$\theta_A \ (x = 0) = -\frac{25}{2EI} = \frac{25}{2 \times 4000} = -\textbf{0.003125 radians}$$

$$y_A \,(x = 0) = \frac{65}{6EI} = \frac{65}{6 \times 4000} = \mathbf{0.002708 \text{ m}} \text{ (2.708 mm)}$$

Case 4: Triangular Loading

EXAMPLE 10.23: Simply supported beam having variable loading with rate of loading zero at supports and maximum rate w at the centre of the span as shown in Fig. 10.37. Derive the equations of slope and deflection.

The SS beam AB carries variable distributed load. The reactions are symmetrical and each reaction will be half of the total load.

i.e. $R_A = R_B = \dfrac{wL}{4}$

General equation of bending moment will be (x from A):

$$M_x = \frac{wL}{4} \cdot x - \frac{2w}{L} \cdot x \cdot \frac{x}{2} \cdot \frac{x}{3} \Bigg| + \frac{2w}{L/2} \cdot \frac{\left(x - \dfrac{L}{2}\right)^3}{6}$$

$$= \frac{wLx}{4} - \frac{wx^3}{3L} \Bigg| + \frac{2w}{3L}\left(x - \frac{L}{2}\right)^3$$

$$EI\frac{d^2 y}{dx^2} = -M_x = -\frac{wL}{4}x + \frac{w}{3L}x^3 \Bigg| - \frac{2w}{3L}\left(x - \frac{L}{2}\right)^3 \qquad \qquad \dots \text{(i)}$$

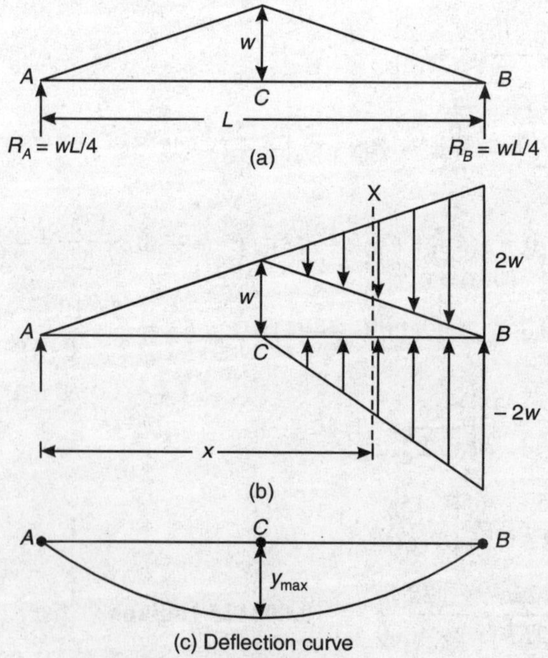

(a)

(b)

(c) Deflection curve

Fig. 10.37: Variable loading on SS beam

Integrating: $EI\dfrac{dy}{dx} = -\dfrac{wL}{8}x^2 + \dfrac{w}{3L}\dfrac{x^4}{4} + C_1 \left| -\dfrac{2w}{3L}\dfrac{\left(x-\dfrac{L}{2}\right)^4}{4} \right.$ .. (ii)

Integrating: $EIy = -\dfrac{wL}{8}\dfrac{x^3}{3} + \dfrac{w}{3L}\dfrac{x^5}{20} + C_1x + C_2 \left| -\dfrac{2w}{12L}\dfrac{\left(x-\dfrac{L}{2}\right)^5}{5} \right.$ ... (iii)

$y = 0$ at $x = 0$, gives $C_2 = 0$

$y = 0$ at $x = L$ gives, $0 = -\dfrac{wL^4}{24} + \dfrac{wL^4}{60} + C_1L - \dfrac{w}{30}\left(\dfrac{L^4}{32}\right)$

or $C_1 = \dfrac{wL^4}{12L}\left(\dfrac{1}{2} - \dfrac{1}{5} + \dfrac{1}{80}\right) = \dfrac{wL^4}{12L}\left(\dfrac{40-16+1}{80}\right) = \dfrac{25wL^4}{12\times 80L} = \dfrac{5wL^3}{192}$

$\therefore$ General equation

$$EI\dfrac{dy}{dx} = -\dfrac{wL}{8}x^2 + \dfrac{w}{12L}x^4 + \dfrac{5wL^3}{192} \left| -\dfrac{w}{6L}\left(x-\dfrac{L}{2}\right)^4 \right.$$ ... (iv)

$$EIy = -\dfrac{wL}{24}x^3 + \dfrac{w}{60L}x^5 + \dfrac{5wL^3}{192}x \left| -\dfrac{w}{30L}\left(x-\dfrac{L}{2}\right)^5 \right.$$ ... (v)

$\theta_A\,(x=0) = \dfrac{1}{EI}\dfrac{5wL^3}{192} = \dfrac{5}{192}\dfrac{wL^3}{EI}$ ... (10.50)

$\theta_B\,(x=L) = \dfrac{1}{EI}\left[-\dfrac{wL^3}{8} + \dfrac{wL^3}{12} + \dfrac{5wL^3}{192} - \dfrac{wL^3}{96}\right] = \dfrac{wL^3}{192EI}[-24+16+5-2] = -\dfrac{5wL^3}{192EI}$

... (10.51)

$y_C\left(x=\dfrac{L}{2}\right) = \dfrac{1}{EI}\left[-\dfrac{wL^4}{192} + \dfrac{wL^4}{1920} + \dfrac{5wL^4}{192\times 2} \right| -0\left.\right] = \dfrac{wL^4}{1920EI}[-10+1+25] = \dfrac{16\,wL^4}{1920EI}$

$$y_C = \dfrac{wL^4}{120EI}$$ ... (10.52)

EXAMPLE 10.24: A simply supported beam AB of *span* 8 m carries a load varying uniformly from *zero* at the support A to 120 kN/m at the right support B. Derive general equations for slope and deflections, and find the slopes at the supports A and B, and the maximum deflection in the beam in terms of EI which is constant.

Solution:

Span $AB = 8$ m, $\qquad w = 120$ kN/m

$$w_x = \frac{120}{8} \cdot x = 15x \text{ kN/m}$$

Reaction $R_A = \dfrac{1}{8}\left[\dfrac{120 \times 8}{2} \times \dfrac{8}{3}\right] = \textbf{160 kN}$

$$R_B = \frac{1}{8}\left[\frac{120 \times 8}{2} \times \frac{2 \times 8}{3}\right] = \textbf{320 kN}$$

General equation of BM

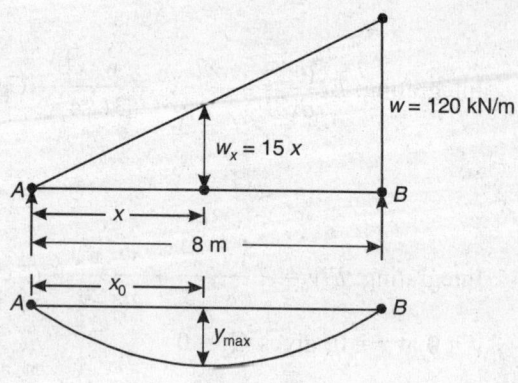

Fig. 10.38.

$$M_x = 160x - 15x \times \frac{x}{2} \cdot \frac{x}{3} = 160x - \frac{5}{2}x^3$$

$\therefore \qquad\qquad EI\dfrac{d^2 y}{dx^2} = -\textbf{160}x + \dfrac{\textbf{5}}{\textbf{2}}x^3$ $\qquad\qquad$ ... (i)

Integrating: $\qquad EI\dfrac{dy}{dx} = -80x^2 + \dfrac{5}{8}x^4 + C_1$ $\qquad\qquad$ ... (ii)

Integrating: $\qquad EIy = -\dfrac{80}{3}x^3 + \dfrac{1}{8}x^5 + C_1 x + C_2$ $\qquad\qquad$ ... (iii)

$y = 0$, at $x = 0$, gives $C_2 = 0$

$y = 0$, at $x = L$, gives $C_1 L = \dfrac{80}{3}L^3 - \dfrac{1}{8}L^5$, $C_1 = \left(\dfrac{80}{3} - \dfrac{L^2}{8}\right)L^2 = \left(\dfrac{80}{3} - \dfrac{1 \times 8^2}{8}\right)8^2$

or $\qquad\qquad C_1 = \dfrac{56 \times 64}{3} = 1194.667$ kN-m units

$$EI\frac{dy}{dx} = -80x^2 + \frac{5}{8}x^4 + 1194.667 \qquad\qquad \text{... (iv)}$$

$$EIy = -\frac{80}{3}x^3 + \frac{1}{8}x^5 + 1194.667x \qquad\qquad \text{... (v)}$$

Slope $\qquad \underset{(x=0)}{\theta_A} = \dfrac{1}{EI}(1194.667) = \dfrac{\textbf{1194.667}}{\textbf{\textit{EI}}}$

$$\underset{(x=8)}{\theta_B} = \frac{1}{EI}\left[-80 \times 64 + \frac{5}{8}(8)^4 + 1194.667\right]$$

$$= \frac{1}{EI}\left[-5120 + 2560 + 1194.667\right] = -\frac{\textbf{1365.33}}{\textbf{\textit{EI}}}$$

For y to be maximum $\dfrac{dy}{dx} = 0 = -80x^2 + \dfrac{5}{8}x^4 + 1194.667$, $x^2 = \dfrac{21.5763}{1.25} = 17.261$

$\therefore$ $x = 4.1546 \text{ m} \approx \textbf{4.155 m}$

$$y_{max} = \dfrac{1}{EI}\left[-\dfrac{80}{3}(4.155)^3 + \dfrac{1}{8}(4.155)^5 + 1194.667(4.155) \right] = \dfrac{3205.84}{EI}\ \textbf{m}$$

EXAMPLE 10.25: A cantilever of span 3 m, carries a uniformly varying load from zero at the free end to 90 kN/m at the fixed end. Calculate slope and deflection at the free end, if value of $EI = 12000$ kN-m².

Solution:

$$w_x = \dfrac{90}{3}x = 30x$$

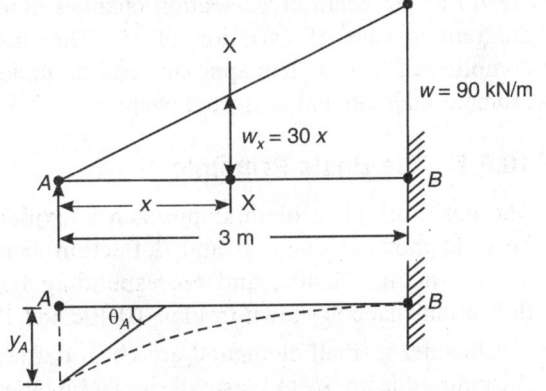

$w = 90$ kN/m

$w_x = 30\,x$

$$M_x = -\dfrac{30x \cdot x}{2} \cdot \dfrac{x}{3} = -5x^3$$

3 m

$$EI\dfrac{d^2 y}{dx^2} = +5x^3 \qquad \ldots \text{(i)}$$

Integrating once:

Fig. 10.39: Cantilever with variable load

$$EI\dfrac{dy}{dx} = \dfrac{5x^4}{4} + C_1 \qquad \ldots \text{(ii)}$$

Integrating again: $\qquad EIy = \dfrac{5x^5}{4 \times 5} + C_1 x + C_2 \qquad\qquad \ldots \text{(iii)}$

$\dfrac{dy}{dx} = 0$ at $x = L = 3$ m, gives $C_1 = -\dfrac{5}{4}L^4 = -101.25$

$y = 0$, at $x = L = 3$ m, gives $C_2 = \dfrac{5}{4}L^5 - \dfrac{L^5}{4} = L^5 = 243$

Slope equation: $\qquad EI\dfrac{dy}{dx} = \dfrac{5x^4}{4} - 101.25 \qquad\qquad \ldots \text{(iv)}$

Deflection equation: $\qquad EIy = \dfrac{x^5}{4} - 101.25x + 243 \qquad\qquad \ldots \text{(v)}$

Slope at free end A $(x = 0) = -\dfrac{101.25}{EI} = \dfrac{101.25}{12000} = -0.0084375$ radoam

Deflection at free end $(x = 0)$: $y_A = \dfrac{243}{EI} = \dfrac{243}{12000} = 0.02025$ m (20.25 mm)

The deflected shape is shown in Fig. 10.39.

10.5 THE AREA-MOMENT METHOD

The slopes and deflections in determinate beams can easily be determined by using area of bending moment diagram of a loaded beam. If we consider the basic equation of bending-

$$\frac{M}{I} = \frac{E}{R} \quad \text{or} \quad \frac{M}{EI} = \frac{1}{R} = \text{Curvature of deflection curve}$$

The *area* of bending moment diagram and *moment of area* of bending moment diagram relate to the slope and deflections at a particular point. It gives slope and deflection with reference to certain reference point. This method becomes useful specially when *moment of inertia* of the beam cross-section changes along the span. The area is considered for (M/EI) diagram in case of variation of 'I'. The method is quite simple and can be applied for complicated beams. The approach can be understood by considering a loaded beam, bending moment diagram and deflected shape.

10.5.1 The Basic Principle

The basic principle of this approach is explained with the help of beam, loading diagram, bending moment diagram and deflection shape shown in Fig. 10.40. Arbitrary loading is shown in Fig. 10.40a, and corresponding bending moment is shown in Fig. 10.40b and deflection shape is shown in Figs 10.40c and 10.40d.

Consider a small elemental arc ds in deflection curve at a distance x from the support A. Assuming the arc ds to be small, it can be taken as circular. The angle subtended at the centre between the two radii at point (1) and (2) will be $d\theta$.

Arc Length $ds = R\,d\theta$, or $d\theta = \dfrac{ds}{R}$

Bending equation $\qquad \dfrac{M}{I} = \dfrac{E}{R}$, gives $\dfrac{1}{R} = \dfrac{M}{EI}$

Therefore, $\qquad\qquad d\theta = \dfrac{M}{EI} \cdot ds$

Since deflections are very small as compared to span, we can take

$ds \approx dx$ i.e. $\qquad d\theta = \dfrac{M}{EI} \cdot dx$ ... (i)

or $\qquad\qquad\qquad \dfrac{d\theta}{dx} = \dfrac{M}{EI}$ ... (ii)

Also $\theta = \dfrac{dy}{dx}$, $\therefore$ $\qquad \dfrac{d\theta}{dx} = \dfrac{d^2y}{dx^2}$ ... (iii)

(ii) and (iii) gives $\qquad \dfrac{d\theta}{dx} = \dfrac{d^2y}{dx^2} = \dfrac{M}{EI}$ (Deflection equation) ... (iv)

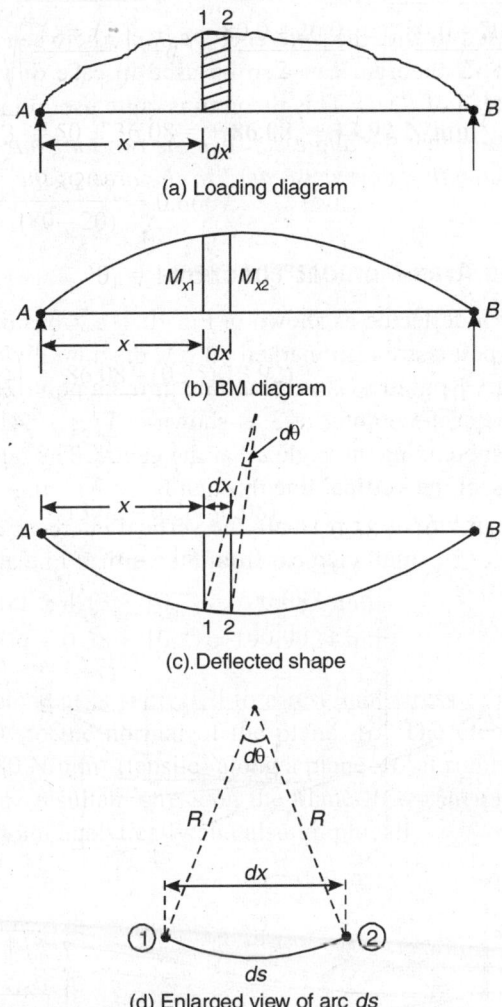

(a) Loading diagram

(b) BM diagram

(c). Deflected shape

(d) Enlarged view of arc *ds*

Fig. 10.40: Area moment method

From (i)
$$d\theta = \frac{M}{EI} \cdot dx$$

or
$$\int_{\theta_1}^{\theta_2} d\theta = \int_{1}^{2} \frac{M}{EI} \cdot dx = \frac{\text{Area of BM Diagram between (1) and (2)}}{EI}$$

or $(\theta_1 - \theta_2) = \dfrac{\text{Area of BM Diagram}}{EI}$ ... (10.53)

10.5.2 First-Moment Area – Mohr's Theorem

This theorem (*I*) can be used to find the *difference of slopes* between the two points along the span in a loaded beam. This theorem can also be used in case of variable '*EI*' by drawing revised *M/EI* diagram instead of *BMD*. This theorem is quite useful in finding slope indirectly.

Mohr's first theorem states that the *angle between the two tangents drawn at any two points to the elastic curve (or deflection curve) is equal to the area of the (M/EI) diagram between these two points.*

10.5.3 Mohr's Second Area-Moment Theorem

Consider a loaded beam *AB* deflected as shown in Fig.10.41a. Consider two points 1 and 2 on the beam. Corresponding points are also marked on *BM* diagram divided by *EI*. Draw tangent at point 1 to the elastic curve and draw a vertical line through point 2. Let the tangent at point 1 intersect the vertical line at the point 2 at 3 as shown in Figs 10.41a and b. Consider small elemental arc length '*ds*' subtending an angle *dθ* at the centre. The tangents at the two ends of the elemental length intersect the vertical line through point 2 giving a vertical intercept *dv*.

Since the elemental length '*ds*' is very small, the vertical intercept *dV* can be approximately $= x_1 . d\theta$, where x_1 is distance of small element from the vertical line at the point 2.

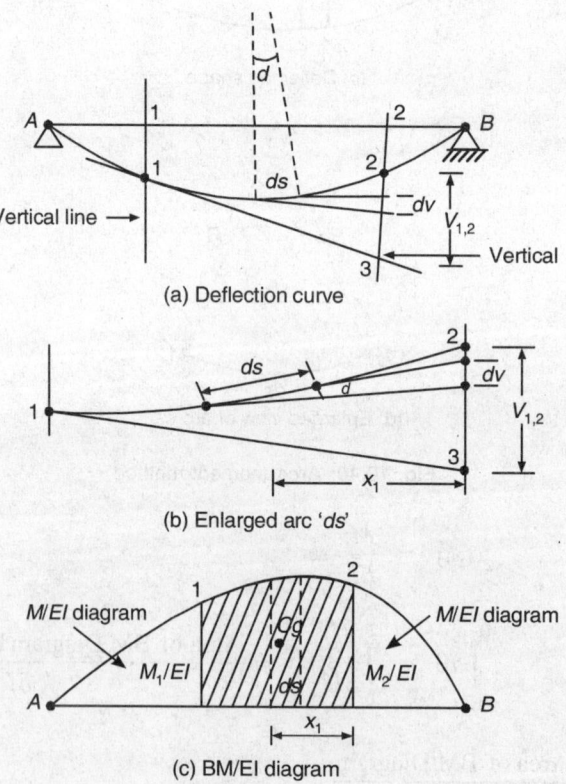

(a) Deflection curve

(b) Enlarged arc '*ds*'

(c) BM/EI diagram

Fig. 10.41: Area-moment theorem

i.e.
$$dv = x_1 \cdot d\theta = x_1 \cdot \frac{M}{EI} dx$$

$\dfrac{M}{EI} \cdot dx$ is the *area* of M/EI diagram over the length dx.

$\dfrac{M}{EI} \cdot dx \cdot x_1$ is the *moment of the area* of M/EI diagram about the point 2 as shown in Fig. 10.41c.

Integration of $dV = \dfrac{M}{EI} \cdot dx \cdot x_1$ over the length 1 to 2, we get

$$\int_1^2 dV = \int_1^2 \frac{M}{EI} \cdot dx \cdot x_1$$

LH expression $\int_1^2 dV$ over points 1 to 2 gives intercept 2-3 = $V_{1,2}$.

RH expression $\int_1^2 \dfrac{M}{EI} \cdot dx \cdot x_1$ between the points 1 and 2 represents the *moment of area* of M/EI diagram between 1 and 2 about 2. [Refer Figs 10.41b and 10.41c].

$$V_{1\text{-}2} = \int_1^2 \frac{M}{EI} \cdot x_1 dx \qquad \qquad \dots (10.54)$$

Thus the *vertical intercept at a point* (2) *cut by the two tangents drawn at 1 and 2 will be equal to the moment of area of M/EI diagram* between the points 1 and 2 *about the point 2.* Thus a generalized statement of theorem II can be stated.

Mohr's second theorem states that the *moment of the area of the M/EI diagram* between two points of a beam *about one of these* points is equal to the *vertical intercept made by the tangent* drawn at one point *on a vertical line through the second point* (about which the moment is taken). It may be noted that the intercept will be about the same point about which the moment of M/EI area is taken.

The *tangent at 1* cuts on intercept at the vertical line *through 2* as $V_{1\text{-}2}$ and tangent at 2 cuts an intercept at the vertical line through 1 as $V_{2\text{-}1}$ (as shown in Fig. 10.42). It may be noted that the moment is taken about the point where vertical intercept is taken. It may be noted that $V_{1\text{-}2}$ may not be equal to $V_{2\text{-}1}$ except in symmetrical cases.

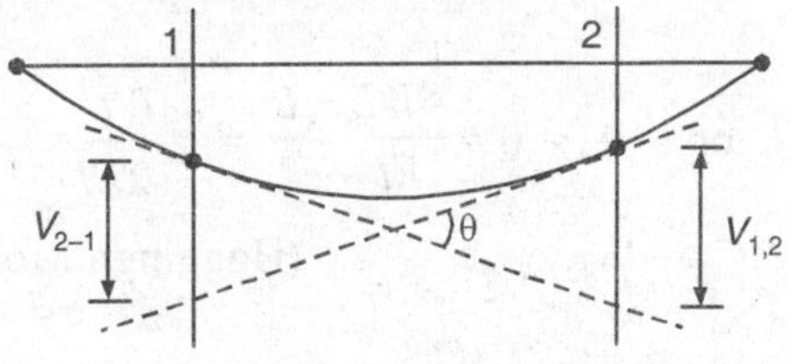

Fig. 10.42: Second theorem
moment area method

Two theorems help in finding slope and deflections.

Sign Conventions

We have already considered sign convention of bending moment as **positive for sagging** and **negative for hogging bending moments**. Similarly **downward deflections** were considered *positive* while ***upward deflections*** were considered as **negative**. The slopes were *positive if the tangent rotates clockwise* while ***negative if it rotates anticlockwise***. Considering similar sign conventions the ***vertical intercept above the tangent*** *may be taken as **positive*** (moment of *M/EI* area also positive) while ***intercept below the tangent*** *may be taken as **negative*** (moments of area of M/EI diagram also negative.

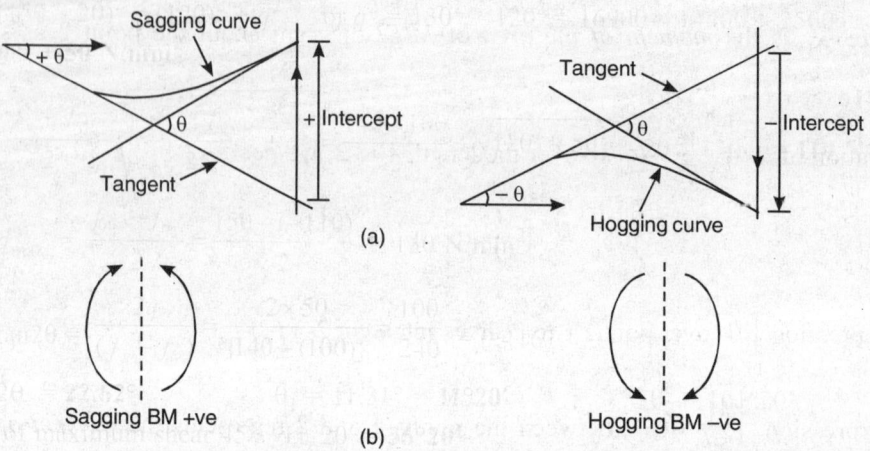

Fig. 10.43: Sign conventions of vertical intercept in moment area method

We shall solve some problems using *M/EI area* method.

EXAMPLE 10.26: A cantilever *AB* of span *L* carries a load *W* at the free end *A*. Calculate slope and deflection at the free end in terms of constant *EI* by using moment area method.

Solution:

Span *AB* = *L*, Point load *W* at *A*. *BM* is shown in Fig. 10.44.

Draw a tangent at *B* (fixed end), it will be horizontal. Draw a tangent at *A*, the slope of the tangent $A = \theta_A$, θ_A is also the angle between the two tangents.

According to the first theorem

$$(\theta_A - \theta_B) = \text{Area of } \frac{M}{EI} \text{ between } A \text{ and } B$$

or $\quad \theta_A - 0 = \dfrac{-WL}{EI} \times \dfrac{L}{2} = -\dfrac{WL^2}{2EI} \qquad \dots \text{(i)}$

(Hoggingn Moment – *ve*)

i.e. $\quad \theta_A = -\dfrac{WL^2}{2EI} = \text{Slope at the free end}$

Moment of $\dfrac{M}{EI}$ area about $A = \left(-\dfrac{WL}{EI} \cdot \dfrac{L}{2} \cdot \dfrac{2L}{3} \right)$

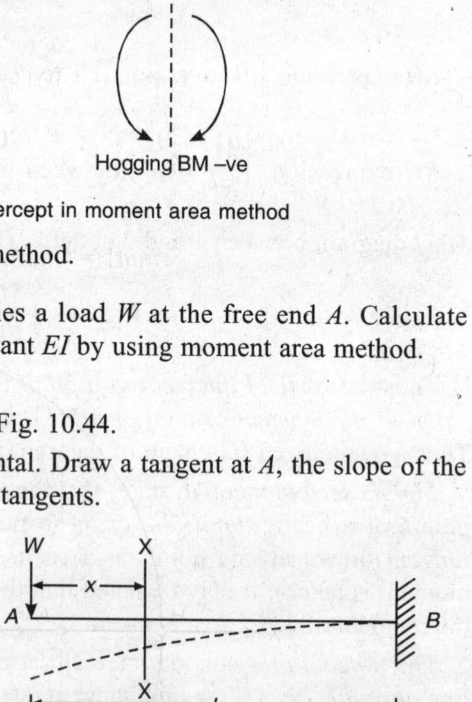

Fig. 10.44: Cantilever

= Vertical intercept at A by tangent at B and A (intercept below the tangent – ve)

Vertical intercept at $A = V_{B-A} = y_A = -\left(-\dfrac{WL^3}{3EI}\right) = +\dfrac{WL^3}{3EI}$ ... (ii)

Thus deflection at A (free end) $= +\dfrac{WL^3}{3EI}$ ... same as found by double integration.

EXAMPLE 10.27: Calculate the slope and deflection for the cantilever AB of span L and carrying a point load W at a point C at a distance of 'a' from the fixed end B (with $a < L$) by using area moment method. EI is constant.

Solution:

    BM at $B = -W.a$

Tangent at B cuts vertical line at C. Tangent at C subtends angle θ_C with the tangent at drawn B.

By Mohr's first theorem

$(\theta_C - \theta_B)$ = Area of BMD/EI, between C and B

$\theta_B = 0$

$\theta_C = -\dfrac{1}{2}\dfrac{Wa \cdot a}{EI} = \dfrac{Wa^2}{2EI}$ ... (i)

i.e. anticlockwise rotation

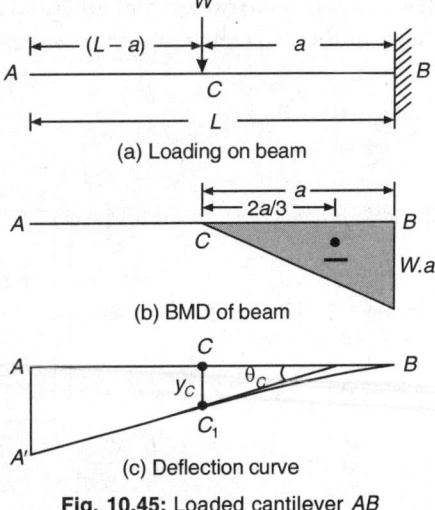

W

|← (L – a) →|← a →|

A   C   B

|← L →|

(a) Loading on beam

|← a →|
|← 2a/3 →|

A   C   B

$W.a$

(b) BMD of beam

C

A y_C θ_C B

C_1

A'

(c) Deflection curve

Fig. 10.45: Loaded cantilever AB

Vertical intercept at C due to tangent at B (below the tangent and hence – ve)

   V_{B-C} = Moment of area of M/EI between B and C about C

or $y_C = -\left(-\dfrac{Wa}{EI}\cdot\dfrac{a}{2}\cdot\dfrac{2a}{3}\right) = +\dfrac{Wa^3}{3EI}$

    $y_C = +\dfrac{Wa^2}{2EI}-\left(\dfrac{2}{3}a\right) = +\dfrac{Wa^3}{3EI}$ ... (ii)

Slope and deflection at A

There is no change in elastic curve from C to A and the slope remains constant (i.e. $\theta_c = \theta_A$).

   $\theta_c = \theta_A = -\dfrac{Wa^2}{2EI}$ (Slope in anticlockwise)   ... (iii)

Vertical intercept at A cut by the tangent drawn from B

$V_{B-A} = -\dfrac{Wa^2}{2EI}\cdot\left(L-a+\dfrac{2}{3}a\right) = -\dfrac{Wa^2}{6EI}(3L-3a+2a) = -\dfrac{Wa^2}{6EI}(3L-a)$

$$(y_A - y_B) = y_A = -V_{B-A} = -\left(-\frac{Wa^2(3L-a)}{6EI}\right) = +\frac{Wa^2(3L-a)}{6EI}$$

$(-ve$ intercept below the tangent $BA = +$ deflection)

EXAMPLE 10.28: A cantilever AB of span L carries a udl w/m over the entire span. EI is constant through out the beam. Calculate slope and deflection at the free end A and at any point x from the free end by using M/EI diagram area and moment method.

Solution:

Span $AB = L$, udl $= w/m$

EI = constant

M/EI is similar to BM diagram

Tangent at A meets the tangent at B in A as the tangent BA is horizontal $(\theta_B = 0)$.

Difference of slope at A and B

$$(\theta_A - \theta_B) = \theta_A$$

Difference of slope between the two points A and B will be equal to area of M/EI diagram between the two points.

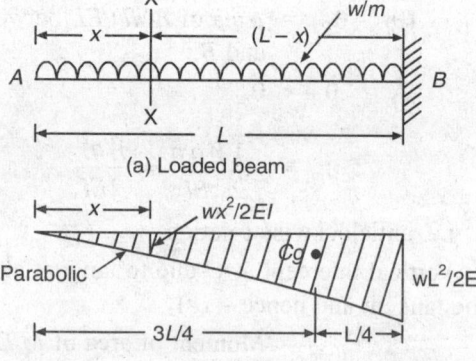

(a) Loaded beam

i.e. $\quad \theta_A - \theta_B = -\dfrac{1}{3} \cdot \dfrac{wL^2}{2EI} \cdot L$

or $\quad \theta_A = -\dfrac{wL^3}{6EI}$ (antilockwise) ... (i)

(b) BMD – M/EI

Intercept V_{B-A} at A will be equal to the moment of M/EI diagram between A and B about A.

Intercept $V_{B-A} = y_A$ at A.

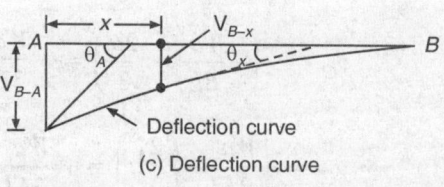

$$y_A = -V_{B-A} = +\frac{L}{3} \cdot \frac{wL^2}{2EI} \cdot \frac{3L}{4}$$

(c) Deflection curve

Fig.: 10.46.

$$y_A = +\frac{wL^4}{8EI} \qquad \text{... (ii)}$$

Intercept at A is negative as the intercept is below the tangent from B but represents $+ve$ deflection in the direction of loading. Slope and deflection at the point X from the free end can be found similarly.

$\theta_x - \theta_B = $ Area of M/EI between B and X.

$\theta_B = 0, \qquad\qquad\qquad \theta_x - \theta_B = \theta_x$

$$\therefore \quad \theta_x = -\left(\frac{wL^2}{2EI} \cdot \frac{L}{3} - \frac{wx^2}{2EI} \cdot \frac{x}{3}\right) = -\frac{w}{6EI}(L^3 - x^3) \qquad \text{... (iii)}$$

$$Cg \text{ of } \frac{M}{EI} \text{ from } A = \frac{\left(\dfrac{wL^4}{8EI} - \dfrac{wx^4}{8EI}\right)}{\left(\dfrac{wl^3}{6EI} - \dfrac{wx^3}{6EI}\right)} = \frac{3\,(L^4 - x^4)}{4\,(L^3 - x^3)}$$

$$y_x = -V_{B-x} = +\frac{w(L^3 - x^3)}{6EI}\left\{\frac{3\,(L^4 - x^4)}{4\,(L^3 - x^3)} - x\right\} = \frac{w}{8EI}(L^4 - x^4) - \frac{w}{6EI}(L^3 - x^3) \cdot x$$

$$y_x = \left\{\frac{w(L^4 - x^4)}{8EI} - \frac{w(L^3 - x^3)x}{6EI}\right\} = \frac{w}{48\,EI}\left\{6(L^4 - x^4) - 8(L^3 - x^3)x\right\} \qquad \text{... (iv)}$$

EXAMPLE 10.29: A simply supported beam of span L carries udl of w/m on the entire span. Determine the slope at ends and deflection at the centre of span by using M/EI diagram principles (Mohr's theorems). EI is constant.

Solution:

BMD is parabolic with

BM at the centre $C = wL^2/8$,

M/EI at $C = wL^2/8EI$

Slope of tangent at C will be zero due to symmetry. By Mohr's theorem

$(\theta_A - \theta_C)$ = Area of M/EI diagram between A and C

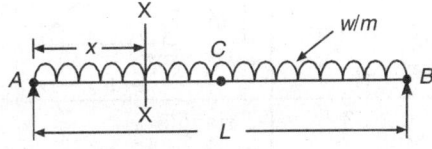

or $\quad \theta_A - 0 = \dfrac{2}{3} \cdot \dfrac{L}{2} \cdot \dfrac{wL^2}{8EI} = \dfrac{wL^3}{24EI} \qquad$ ... (i)

Similarly,

$\theta_C - \theta_B$ = Area of M/EI between C and B.

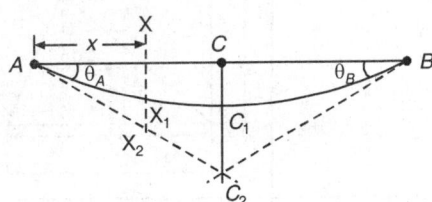

i.e. $\quad -\theta_B = \dfrac{wL^3}{24EI}$,

Fig.: 10.47.

or $\quad \theta_B = -\dfrac{wL^3}{24EI}$ (Anticlockwise) $\qquad$ ... (ii)

Vertical intercept at C ($C_1 - C_2$) will be equal to moment of M/EI diagram area between A and C about C.

i.e. $\quad V_{A-C} = \dfrac{2}{3} \cdot \dfrac{L}{2} \cdot \dfrac{wL^2}{8EI}\left(\dfrac{3}{8} \cdot \dfrac{L}{2}\right) = \dfrac{wL^4}{128EI} = C_1 C_2 \qquad$ ... (iii)

$$CC_2 = \theta_A \cdot \frac{L}{2} = \frac{wL^3}{24EI} \cdot \frac{L}{2} = \frac{wL^4}{48EI} \qquad \text{... (iv)}$$

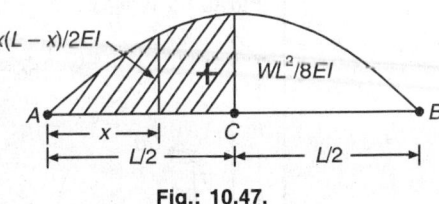

Table 10.1: Areas and centroids of different shapes

S. NO.	DIAGRAM SHAPE	AREA	CENTROID
1.	Rectangular	$b.h$	$\bar{x} = \dfrac{b}{2}, \bar{y} = \dfrac{h}{2}$
2.	Triangle	$\frac{1}{2}\ b.h$	$\bar{x} = \dfrac{b}{3}, \bar{y} = \dfrac{h}{3}$
3.	Triangle	$\dfrac{(a+b)h}{2}$	$\bar{x}_a = \dfrac{2a+b}{3}, \bar{x}_b = \dfrac{a+2b}{3}$ $\bar{y} = \dfrac{h}{3}$
4.	Parabola $y = kx^2$	$\dfrac{2}{3}bh$	$\bar{x}_1 = \dfrac{3b}{8}, \bar{y}_1 = \dfrac{2h}{5}$
5.	Parabola $y = kx^n$	$\dfrac{n.b.h}{(n+1)}$	$\bar{x}_1 = \dfrac{(n+1)b}{2(n+2)}, \bar{y}_1 = \dfrac{n.h}{(2n+1)}$

S. NO.	DIAGRAM SHAPE	AREA	CENTROID
6. (a)	Parabola $y = kx^2$ 	$\dfrac{1}{3}b.h$	$\overline{x}_1 = \dfrac{b}{4}, \overline{y}_1 = \dfrac{3h}{10}$
6. (b)	Parabola $y = kx^2$ 	$\dfrac{1}{3}b.h$	$\overline{x}_1 = \dfrac{3b}{4}, \overline{y}_1 = \dfrac{3h}{10}$ $\overline{x}_2 = \dfrac{b}{4}$
7.	Parabola $y = kx^3$ 	$\dfrac{1}{4}b.h$	$x_1 = \dfrac{4b}{5}, x_2 = \dfrac{b}{5}$ $y_1 = \dfrac{2h}{7}, y_2 = \dfrac{5h}{7}$
8.	Parabola $y = kx^n$ 	$\dfrac{b.h}{(n+1)}$	$\overline{x}_1 = \dfrac{(n+1)b}{(n+2)}, \overline{x}_2 = \dfrac{b}{(n+2)}$ $\overline{y}_1 = \dfrac{(n+1)h}{2(2n+1)}$

Deflection at C $(y_C) = CC_1 = CC_2 - C_1C_2 = \dfrac{wL^4}{48EI} - \dfrac{wL^4}{128EI} = \dfrac{(8-3)wL^4}{384EI}$

$$y_C = \dfrac{5wL^4}{384EI} \qquad \qquad \text{... (v)}$$

Same as derived in double integration method. $\dfrac{M_x}{EI} = \dfrac{wx(L-x)}{2EI}$

Deflection at any point $x = (\theta_A{\cdot}x - V_{A-x}) = \dfrac{wL^3{\cdot}x}{24EI} - \dfrac{wx^3(L-x)}{8EI} = \dfrac{wx}{24EI}[L^3 - x^2{\cdot}3(L-x)]$

$$y_x = \dfrac{wx}{24EI}[L^3 - 3Lx^2 + 3x^3] \qquad \qquad \text{... (vi)}$$

EXAMPLE 10.30: An overhanging beam ABE of 12 m total length overhangs at B. The supported span AB carries point loads of 20 kN and 40 kN at C and D respectively. Length AC = 2 m, CD = 5 m and DB = 3 m. The overhang BE = 2 m and carries a point load of 20 kN at E. If EI is constant throughout and equal to 20000 kN-m² units. Find the deflections at the load points C, D and E.

Solution:

Reactions at A and B.

$$R_A = \dfrac{20 \times 8 + 40 \times 3 - 20 \times 2}{10} = \mathbf{24\ kN}$$

$R_B = \mathbf{56\ kN}$

$M_C = 48$ kN-m, $\qquad M_D = 68$ kN-m

$M_B = -40$ kN-m, $\qquad M_F = 0$,

or $DF = 1.89$ m.

Areas of M/EI diagram

$$ACC' = \dfrac{48 \times 2}{2EI} = \dfrac{48}{EI},\ Cg\dfrac{4}{3}\text{ m from } A$$

$$CC'DD' = \dfrac{290}{EI},\ Cg \text{ at } 2.6437 \text{ m from } C$$

$$DD'F = \dfrac{68 \times 1.89}{2EI} = \dfrac{64.22}{EI}\text{ kNm,}$$

Cg at 0.63 m from D

$$BB'F = \dfrac{40 \times 1.11}{2} = 22.2\text{ kNm,}$$

Cg at 0.37 m from B

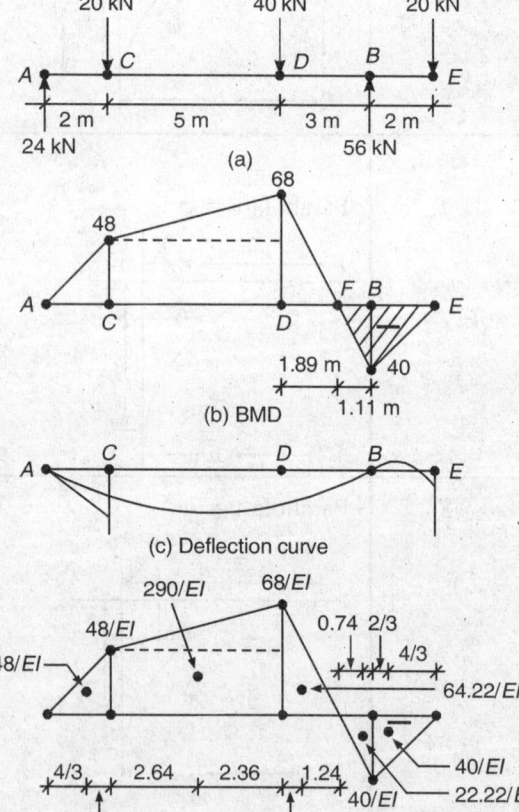

(a)

(b) BMD

(c) Deflection curve

(d) M/EI diagram

Fig. 10.48.

$$BB'E = \frac{40}{EI} \text{ kNm}, \ Cg \text{ at } \frac{2}{3} \text{ m from } B$$

Deflection at the support $B = 0$.

$$\therefore \qquad \theta_A = \frac{V_{AB}}{10}$$

$$= \frac{1}{10EI} \left[48 \times \left(8 + \frac{2}{3} \right) + 290(3 + 2.3563) + 64.22 \times 2.37 - 22.22 \times 0.37 \right]$$

$$= \frac{211.35}{EI} = \frac{211.35}{20000} = 0.01057 \text{ radians}$$

Similarly

$$\theta_B = \frac{V_{BA}}{10}$$

$$= \frac{1}{10EI} \left[48 \times \frac{4}{3} + 290 \times 4.6437 + 64.22 \times 7.63 - 22.22 \times 9.63 - \right]$$

$$= \frac{169}{2000} = \textbf{0.00845 radians}$$

Deflection at C, $y_C = \theta_A \cdot 2 - V_{A-C} = 0.01057 \times 2 - \dfrac{48}{EI} \times \dfrac{2}{2} \times \dfrac{2}{3}$

$$y_C = 0.01057 \times 2 - \frac{32}{20000} = \textbf{0.01954 m} \ (19.54 \text{ mm})$$

Deflection at D, $y_D = \theta_A \cdot 7 - V_{A-D} = 0.01057 \times 7 - \dfrac{1}{EI} \left(48 \times \dfrac{17}{2} + 290 \times 2.3563 \right)$

$$y_D = \frac{169}{20000} \times 2 - \frac{40}{EI} \times \frac{4}{3} = \left(\frac{338 - 53.33}{20000} \right) = \frac{284.67}{20000} = \textbf{0.01423 m} \ (14.23 \text{ mm})$$

EXAMPLE 10.31: A simple supported beam AB of 5 m span is subjected to a couple of 150 kN-m (anticlockwise) at C, 3 m from the left hand support. Calculate the slopes at ends A and B and deflections at the point 'C'. EI is constant throughout the span. Value of $EI = 7500$ kN-m^2.

Solution:

$$R_A = R_B = \frac{150}{5} = 30 \text{ kN}$$

BM at $C = R_A \times 3 = 30 \times 3 = 90$ kN-m

$M_C = + 90$ kN-m,

M'_C (right of C) $= + 90 - 150 = - \textbf{60 kN-m}$

BM diagram is shown in Fig. 10.49. Deflected shape is also shown in Fig. 10.49.

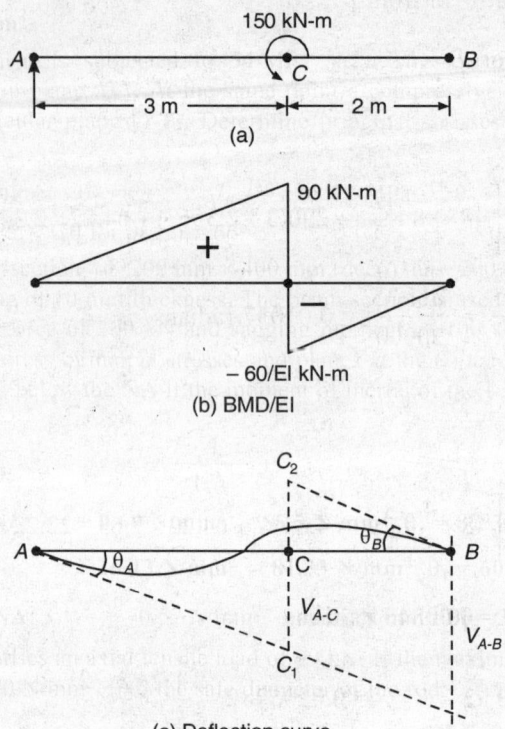

(a)

(b) BMD/EI

(c) Deflection curve

Fig. 10.49.

$$\text{Slope} \quad \theta_A = \frac{V_{A-B}}{L} = \left[-\frac{60 \times 2}{2EI} \cdot \frac{2 \times 2}{3} + \frac{90}{EI} \times \frac{3}{2}\left(2 + \frac{3}{3}\right)\right]\frac{1}{5}$$

$$\theta_A = +\frac{65}{EI} \text{ (Clockwise)}$$

$$\theta_B = \frac{V_{A-B}}{L} = \frac{1}{EIL}\left[+\frac{90 \times 3}{2} \times \frac{2}{3} \times 3 - \frac{60 \times 2}{2}\left(3 + \frac{2}{3}\right)\right]$$

$$= \frac{1}{5EI}[+270 - 220]$$

$$\theta_B = +\frac{50}{5EI} = +\frac{10}{EI}, \text{ (Clockwise)}$$

$$EI = 7500 \text{ kN-m}^2$$

$$\theta_A = +\frac{65}{7500} = 0.00867 \text{ radians}$$

$$\theta_B = +\frac{10}{7500} = 0.00133 \text{ radians}$$

Deflection at C

$$y_C = V_{AC} - CC_1 = \frac{90}{EI} \times \frac{3}{2} \times \frac{3}{3} - \frac{65}{EI} \times 3$$

$$= \frac{1}{EI}[135 - 195] = -\frac{60}{EI}$$

$$y_C = -\frac{60}{7500} = -0.008 \text{ m (8 mm above)}$$

EXAMPLE 10.32: A simply supported beam AB of 6 m span carries a uniformly varying load from zero at A to 60 kN/m at B. Determine the *slopes* at the supports and deflection at the middle of the span. EI is constant and is equal to 16000 kN-m^2.

Solution:

$$R_A = \frac{1}{6}\left[\frac{60 \times 6}{2} \times \frac{6}{3}\right] = 60 \text{ kN,}$$

$$R_B = 120 \text{ kN}$$

BM at any point X from A will be:

$$M_x = R_A \cdot x - 10x \cdot \frac{x}{2} \cdot \frac{x}{3} = \left(60x - \frac{5x^3}{3}\right)$$

M_x/EI diagram comprises of two different terms and the area and moment of such area can be found by integration only. But if two terms are drawn separately, their areas and moment of areas can be found by standard formulae. For area-moment method it is better to draws BM diagram separately for each term. Thus

$M_x = 60x - \dfrac{5x^3}{3}$ comprises of two diagrams $M_{x_1} = +60x$ (straight line), $M_{x_2} = -\dfrac{5x^3}{3}$ (cubic

parabola). The areas and centroids can be found by standard formulae for standard geometrical shapes (refer Table 10.1).

Slope $\theta_A = \dfrac{V_{A-B}}{L}$, $\theta_B = \dfrac{V_{B-A}}{L}$

$$\theta_A = \frac{1}{6EI}\left[\frac{360 \times 6}{2} \times \frac{6}{3} - \frac{360 \times 6}{4} \times \frac{6}{5}\right]$$

$$\theta_A = \frac{(2160 - 648)}{6EI} = \frac{1512}{6EI}$$

$$= \frac{252}{EI} = \frac{252}{16000} = 0.01575 \text{ radian}$$

$$\theta_B = \frac{1}{6EI}\left[\frac{360\times6}{2}\times4 - \frac{360\times6}{4}\times\frac{24}{5}\right]$$

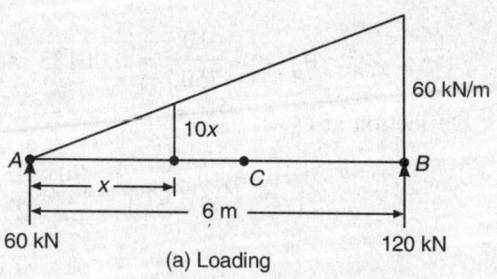

(a) Loading

$$= \frac{(4320-2592)}{6\times16000} = 0.018 \text{ radians}$$

Deflection at C (mid point)

$$y_C = (\theta_A\cdot3 \text{ m} - C_1C_2)$$
$$= (0.01575 \times 3 - V_{A-C})$$

or $y_C = \left[0.04725 - \left(\frac{360}{6}\times3\times\frac{3}{4}\times\frac{3}{5}\right)\frac{1}{EI}\right]$

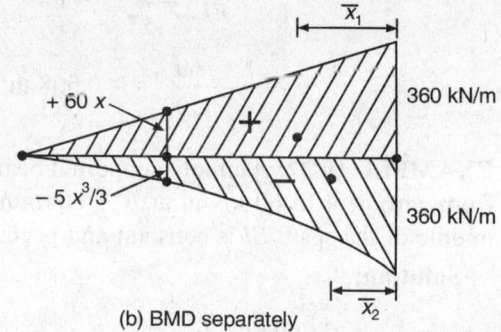

(b) BMD separately

$$= \left[0.04725 - \left(\frac{270-20.25}{EI}\right)\right]$$

or $y_C = \left(0.04725 - \frac{249.75}{16000}\right)$

$$= (0.04725 - 0.015609)$$
$$= \mathbf{0.03164 \text{ m } (31.64 \text{ mm})}$$

For y_{max} from A at x say.

$$EIy_x = 252x - \left[\frac{60x^3}{6} - \frac{5x^3}{3}\cdot\frac{x}{4}\cdot\frac{x}{5}\right]$$

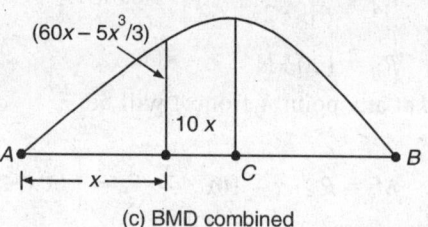

(c) BMD combined

$$= 252x - 10x^3 + \frac{x^5}{12}$$

For y_{max}, $\frac{dy}{dx} = 0$, $0 = 252 - 30x^2 + \frac{5x^4}{12}$,

or $5(x^2)^2 - 360(x^2) + 3024 = 0$

$x^2 = 9.7093$, $x = \mathbf{3.116 \text{ m}}$

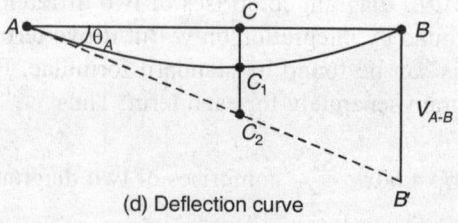

(d) Deflection curve

Fig. 10.50.

$$y_{max} (x = 3.116 \text{ m}) = \frac{1}{EI}\left[252\times3.116 - \left\{10\times(3.116)^3 - \frac{(3.116)^5}{12}\right\}\right]$$

$$= \frac{1}{16000}[785.23 - 278.07] = 0.0317 \text{ m}$$

$$y_{max} = \mathbf{0.0317 \text{ m } (31.70 \text{ mm})} \approx y_C.$$

EXAMPLE 10.33: A simply supported beam AB of 10 m span carries udl of 20 kN/m over left hand half span and a point load of 40 kN at D (centre of right hand half span). $E = 200$ kN/mm^2, $I = 28000$ cm^4. Determine the slopes at ends, and the deflection at the mid span and under the point load of 40 kN at D.

Solution:

$EI = 200 \times 10^6 \times 28000 \times 10^{-8}$ kN-m^2 = 56000 kN-m^2

$L = 10$ m,

udl = 20 kN/m over 5 m (A to C),

Point load $W = 40$ kN at 7.5 m

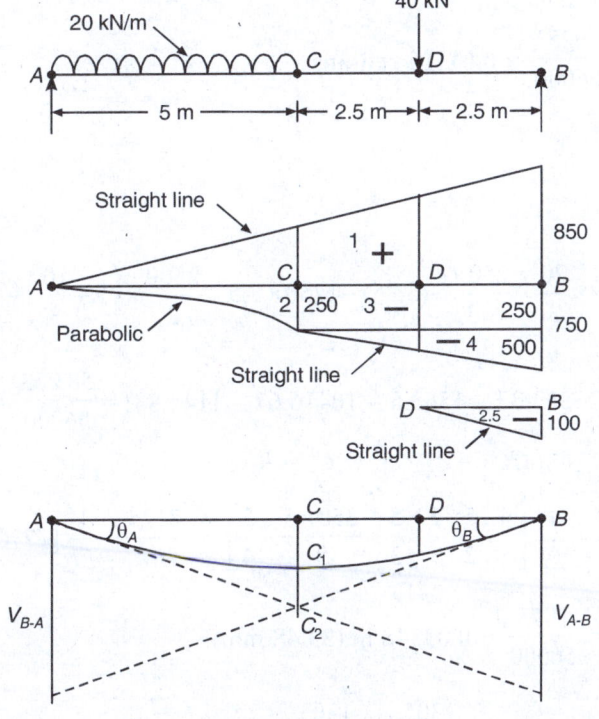

Fig. 10.51.

Reactions: $R_A = \dfrac{1}{10}[40{\times}2.5+100{\times}7.5] = +\,85$ kN, $R_B = +\,55$ kN (upward)

A to C: $M_x = +\,85x - \dfrac{20.x^2}{2}$ ... (i)

C to D: $M_x = +\,85x - 20 \times 5\,(x - 2.5)$ ... (ii)

D to B: $M_x = +\,85x - 100\,(x - 2.5) - 40\,(x - 7.5)$ ... (iii)

M_x comprises of *one parabolic* and *one straight line* variation from A to C.

M_x comprises of *two straight lines* from C to D.

M_x comprises of *three straight lines* from D to B.

These variations are shown in Fig. 10.51.

$$\theta_A = \frac{V_{A-B}}{L}$$

$$= \frac{1}{10EI}\left[\frac{850\times10}{2}\times\frac{10}{3} - \frac{250\times5}{3}\times\left(5+\frac{5}{4}\right) - 250\times5\times2.5 - \frac{500\times5}{2}\times\frac{5}{3} - \frac{250}{2}\times\frac{2.5}{3}\right]$$

or $\quad \theta_A = \frac{1}{10EI}[14166.67 - 2604.13 - 3125 - 2083.33 - 104.17]$

or $\quad \theta_A - \frac{625}{EI} = \frac{625}{56000} = \mathbf{0.01116\ radians}$

Similarly

$$\theta_B = \frac{V_{B-A}}{10}$$

$$= \frac{1}{10EI}\left[\frac{850\times10}{2}\times\frac{20}{3} - \frac{250\times5}{3}(3.75) - 1250\times7.5 - \frac{2500}{2}\times\left(5+\frac{10}{3}\right) - \frac{250}{2}\times\left(7.5+\frac{5}{3}\right)\right]$$

or $\quad \theta_B = \frac{1}{10EI}[28333.33 - 1562.5 - 10416.67 - 1145.83] = \frac{583.33}{56000} = 0.01042$ radians

$$y_C = CC_2 - C_1C_2, \quad CC_2 = \theta_A \times 5, \quad C_1C_2 = V_{A-C}$$

$$y_C = 5\times\frac{625}{EI} - \frac{1}{EI}\left[\frac{4.25\times5}{2}\times\frac{5}{3} - \frac{250\times5}{3}\times\frac{5}{4}\right] = \frac{3125}{EI} - \frac{1}{EI}[1770.83 - 520.83]$$

$$y_C = \frac{1875}{EI} = \frac{1875}{56000} = 0.03348\ \text{m (33.48 mm)}$$

Similarly $y_D = 7.5\theta_A - V_{A-D} = \frac{1305}{EI} = \frac{1305}{56000} = \mathbf{0.02330\ m\ (23.30\ mm)}$

10.6 CONJUGATE BEAM METHOD (MOHR'S THEOREM III AND IV)

Mohr's first two theorems are based on *area and moment of area of M/EI diagram*. These areas and moments about certain point are used to calculate *difference of slopes* and *vertical intercepts by tangents* to indirectly calculate *deflections*. Similarly Mohr's *theorem III and IV* are used to determine *slopes and deflections* of real beam from shear force and bending moments of equivalent beam carrying a **load equal to M/EI diagram** of the real beam. **The equivalent beam carrying M/EI diagram as loading** is known as *conjugate beam*.

(a) SS Beams

The fundamental relationship of $\dfrac{1}{R} = \dfrac{d^2y}{dx^2} = \dfrac{M}{EI}$, was already established in double integration approach. Thus equation $EI\dfrac{d^2y}{dx^2} = M$, was mainly used for finding slope and deflection by double integration.

In earlier units of *shear force-bending moment*, we have already established the relationship

$$\frac{dM}{dx} = F \quad \text{and} \quad \frac{dM}{dx} = w. \qquad \text{... (i)}$$

i.e. $\dfrac{d}{dx}\left(EI\dfrac{d^2y}{dx^2}\right) = F$ or $EI\dfrac{d^3y}{dx^3} = F = \int w\,dx$, (shear force) $\qquad$ ... (ii)

and $\dfrac{d}{dx}\left(EI\dfrac{d^3y}{dx^3}\right) = w$

i.e. $EI\dfrac{d^4y}{dx^4} = w$ (rate of loading) $\qquad$ ... (iii)

$$EI\frac{d^3y}{dx^3} = \int w\,dx, \text{ F (shear force)} \qquad \text{... (ii)}$$

$$EI\frac{d^2y}{dx^2} = \int F\,dx = \text{M (bending moment)} \qquad \text{... (iv)}$$

$\dfrac{d^2y}{dx^2} = \dfrac{M}{EI}$, if M/EI is loading on the conjugate beam then $\int$ loading will be SF i.e.

$$\frac{dy}{dx} = \int \left(\frac{M}{EI}\right)dx = \text{slope of real beam} = \text{SF of conjugate beam} \qquad \text{... (10.55)}$$

Also

$$y = \iint \left(\frac{M}{EI}\right)dx = \int F\,dx = \text{bending moment of equivalent beam with} \frac{M}{EI} \text{ as loading}$$

$$\text{... (10.56)}$$

These relations are quite similar to relationships between rate of loading w, *SF* and *BM* in case of beams to the relations of *BM*, slope and deflection. *Slope-deflection* equations can be derived from the load intensity function as in SS beams carrying udl w per unit length.

i.e. $EI\dfrac{d^4y}{dx^4} = w$

Integrating these functions four times provide *deflection* by finding constants of integration by applying suitable boundary conditions. A *simplified approach of conjugate beam* can be used by suitably *forming conjugate beam* using appropriate boundary or support conditions

To understand the conjugate beam, consider a simply supported beam AB of span L and carrying udl w/m (Fig. 10.52).

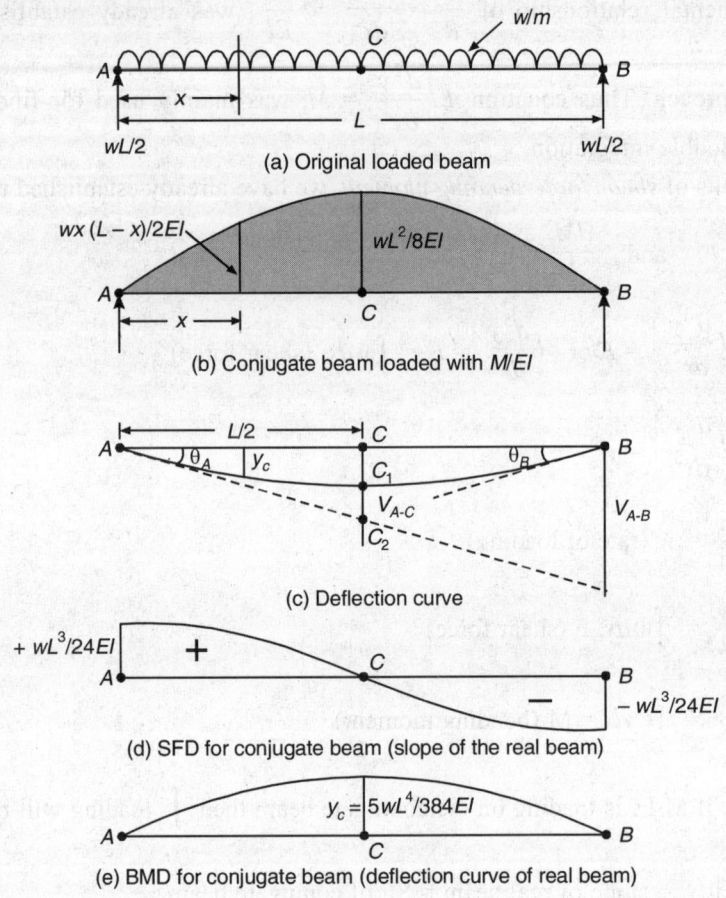

(a) Original loaded beam

(b) Conjugate beam loaded with M/EI

(c) Deflection curve

(d) SFD for conjugate beam (slope of the real beam)

(e) BMD for conjugate beam (deflection curve of real beam)

Fig. 10.52.

Total load on the conjugate beam $= \dfrac{2}{3} \cdot \dfrac{wL^2}{8EI} \cdot L = \dfrac{wL^3}{12EI}$ ... (i)

The reactions of conjugate beam by symmetry:

$$R'_A = \frac{1}{2} \cdot \frac{wL^3}{12EI} = R'_B$$

SF at $A = F'_A = +R'_A = +\dfrac{wL^3}{24EI} = \theta_A$ ...(ii)

SF at $B = F'_B = \left(R'_A - \dfrac{wL^3}{12EI}\right) = \dfrac{wL^3}{24EI} - \dfrac{wL^3}{12EI} = -\dfrac{wL^3}{24EI} = \theta_B$ ...(iii)

Deflection at C by conjugate beam theorem = BM of conjugate beam

i.e. $y_C = M'_A = R'_A \cdot \dfrac{L}{2} - \dfrac{2}{3} \times \dfrac{wL^2}{8EI} \cdot \dfrac{L}{2} \times \dfrac{3}{8} \cdot \dfrac{L}{2}$ (Moment of BM/EI area about C)

or $y_C = \dfrac{wL^3}{24EI} \cdot \dfrac{L}{2} - \dfrac{wL^4}{128EI} = \dfrac{wL^4(8-3)}{384EI} = \dfrac{\mathbf{5wL^4}}{\mathbf{384EI}}$ ... (iv)

Deflection is same as derived by double integration or moment area method.

Figure 10.52d shows **SF diagram of the conjugate beam** and represents *slope diagram of the real beam*. Figure 10.52e shows the **BM diagram of the conjugate beam** and represents the **deflection curve of the real beam**.

Thus two theorems related to conjugate beams can be stated as:

If the BMD/EI of real beam represents the loading diagram of an equivalent conjugate beam then the shear force of the conjugate beam represents the slope of the real beam and the bending moment of the conjugate beam represents the deflection of the real beam.

b. Conjugate Beams for Cantilevers and Overhanging Beams

In case of cantilevers, the BM diagram is negative for the downward loading, hence the loading for the conjugate beam will be negative (i.e. upward). Since the shear force at any point in the conjugate beam *represents the slope* at the *corresponding point of the real beam*, the *free end of the conjugate beam* will be corresponding to the *fixed end of the real beam*, so as to give SF and BM of free end of conjugate beam equals to zero (i.e. SF$' = 0$, and BM$' = 0$). To satisfy these conditions, the **equivalent conjugate beam will have fixed end at the free end of the real beam and the conjugate beam will have free end at the fixed end of the real beam**.

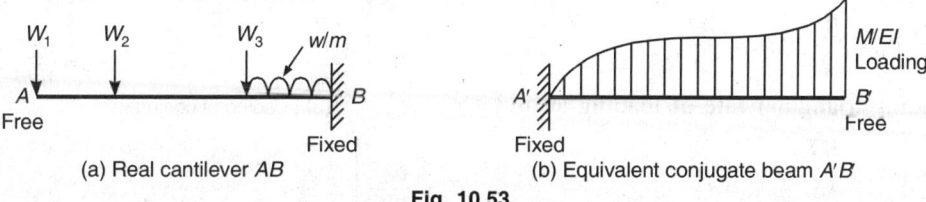

(a) Real cantilever AB · (b) Equivalent conjugate beam $A'B$

Fig. 10.53.

Overhanging beams are also modified to suit the requirements of slope-deflection and shear force-bending moments.

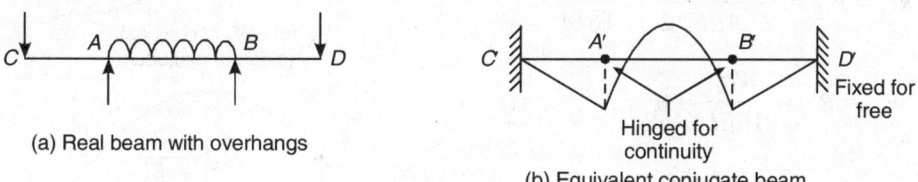

(a) Real beam with overhangs · (b) Equivalent conjugate beam

Fig. 10.54.

Similarly a fixed beam AB with fixed ends at A and B of real beam will have hinge supports at A' and B' of the equivalent conjugate beam.

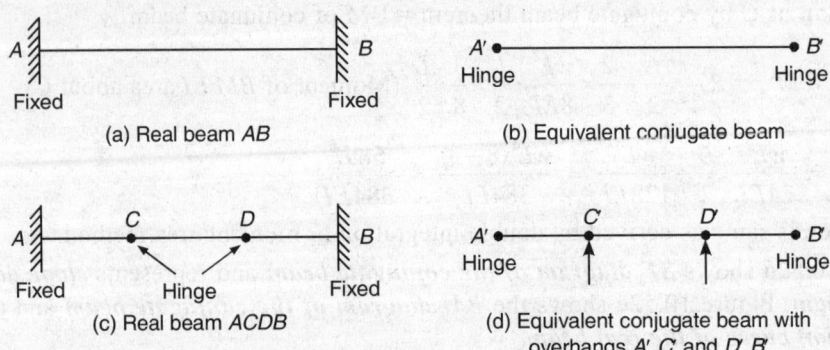

(a) Real beam *AB*   (b) Equivalent conjugate beam

(c) Real beam *ACDB*   (d) Equivalent conjugate beam with overhangs *A' C* and *D' B*

Fig. 10.55: Equivalent conjugate beams

Thus in conjugate beam approach, the *free end of the real beam* is replaced by *fixed support* in equivalent *conjugate beam*. *Fixed support of real beam* becomes *free end of the equivalent conjugate beam*. Similarly if any point of *real beam has zero deflection*, the *equivalent conjugate beam* will have *either free end or hinge* point.

These equivalent conjugate beams will be explained with examples.

EXAMPLE 10.34: A simply supported beam *AB*, span *L* (*L* = 8 m) carries a point load *W* (*W* = 20 kN) at the centre *C*, derive slope equations for supports and deflection at the centre *C*, if *EI* is constant (*EI* = 5000 kN-m^2).

Solution:

$$R_A = R_B = \frac{W}{2}, \quad M_C = \frac{WL}{4}$$

Loading triangular rate of loading at mid span point $= \dfrac{WL}{4EI}$.

Reactions of conjugate beam

$$R'_A = R'_B = \frac{1}{2} \cdot \frac{WL}{4EI} \cdot \frac{L}{2} = \frac{WL^2}{16EI}$$

$$SF \quad F'_A = R'_A = \frac{WL^2}{16EI} = \theta_A \qquad \text{... (i)}$$

$$SF \quad F'_B = R'_A - \left(\frac{WL}{4EI} \cdot \frac{L}{2}\right)$$

$$= \frac{WL^2}{16EI} - \frac{WL^2}{8EI} = -\frac{WL^2}{16EI} = \theta_B$$

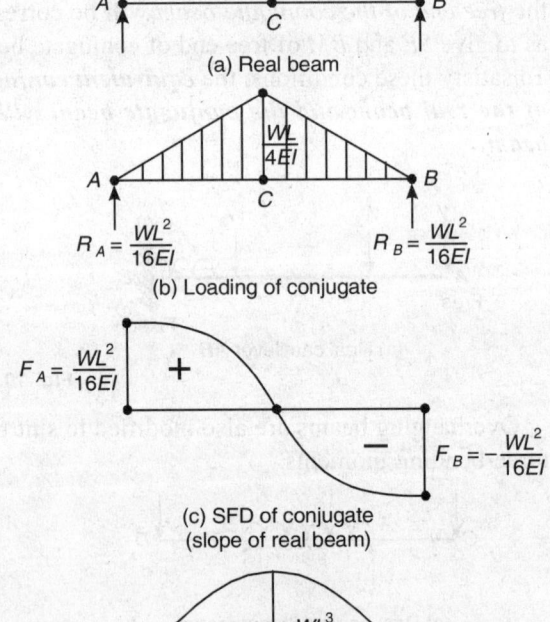

(a) Real beam

$$R_A = \frac{WL^2}{16EI} \qquad R_B = \frac{WL^2}{16EI}$$

(b) Loading of conjugate

$$F_A = \frac{WL^2}{16EI} \qquad F_B = -\frac{WL^2}{16EI}$$

(c) SFD of conjugate (slope of real beam)

$$M_A = 0 \qquad M_C = \frac{WL^3}{48EI} \qquad M_B = 0$$

(d) BMD of conjugate (deflection curve of real beam)

Fig. 10.56.

$$F'_C = \frac{WL^2}{16EI} - \frac{WL^2}{4EI}\cdot\frac{1}{2}\cdot\frac{L}{2} = 0 = \theta_C \qquad \text{... (iii)}$$

$BM \quad M'_A = 0 = y_A, \quad M'_B = 0 = y_B \qquad \text{... (iv)}$

$$BM \quad M'_C = y_C = R'_A\cdot\frac{1}{2} - \frac{1}{2}\cdot\frac{L}{2}\cdot\frac{WL}{4EI}\left(\frac{1}{3}\cdot\frac{L}{2}\right) = \frac{WL^2}{16EI}\cdot\frac{L}{2} - \frac{WL^3}{96EI} = \frac{WL^3(3-1)}{96EI}$$

or
$$y_C = \frac{\boldsymbol{WL^3}}{\boldsymbol{48EI}} \qquad \text{... (v)}$$

$$\theta_A = -\theta_B = \frac{WL^2}{16EI}$$

$$= \frac{20\times8\times8}{16\times5000} = 0.016 \text{ radian}$$

$$y_C = \frac{WL^3}{48EI} = \frac{20\times8^3}{48\times5000} = 0.04267 \text{ (42.67 mm)}$$

EXAMPLE 10.35: A wooden beam of 1.5 m length is symmetrically supported over 1.0 m span with equal overhangs on each side. The overhangs carry a load of 10 kN on each free end. E for wood is 1×10^7 kN/m², the cross-section is rectangular, 60 mm wide × 100 mm deep. Determine the slopes at the two supports and deflections at the mid span point by conjugate beam method.

Solution:

$$I = \frac{0.06}{12}\times0.10^3 = 10^{-5}\times\frac{1}{2} = 5\times10^{-6}\text{m}^4$$

$E = 1\times 10^7$ kN/m², $\qquad EI = 50$ kN-m²

Span $AB = 1.0$ m, $\qquad$ Overhang = 0.25 m

Real Beam

$M_A = 10\times 0.25 = -2.5$ kN-m $= M_B$

M_A to M_B remains constant at (hogging) -2.5 kN-m.

Equivalent conjugate beam A' amd B' will be hinged as deflections are zero free ends D' and E' are fixed and supported ends A' and B' are hinged.

Loading on Conjugate beam for:

D' to A': $\quad -\dfrac{2.5}{50}$ kNm, Triangular, zero at D'

A' to B': $\quad -\dfrac{2.5}{50}$ kNm, udl (upward)

B' to E': $\quad -\dfrac{2.5}{50}$ kNm, Triangular, zero at E'

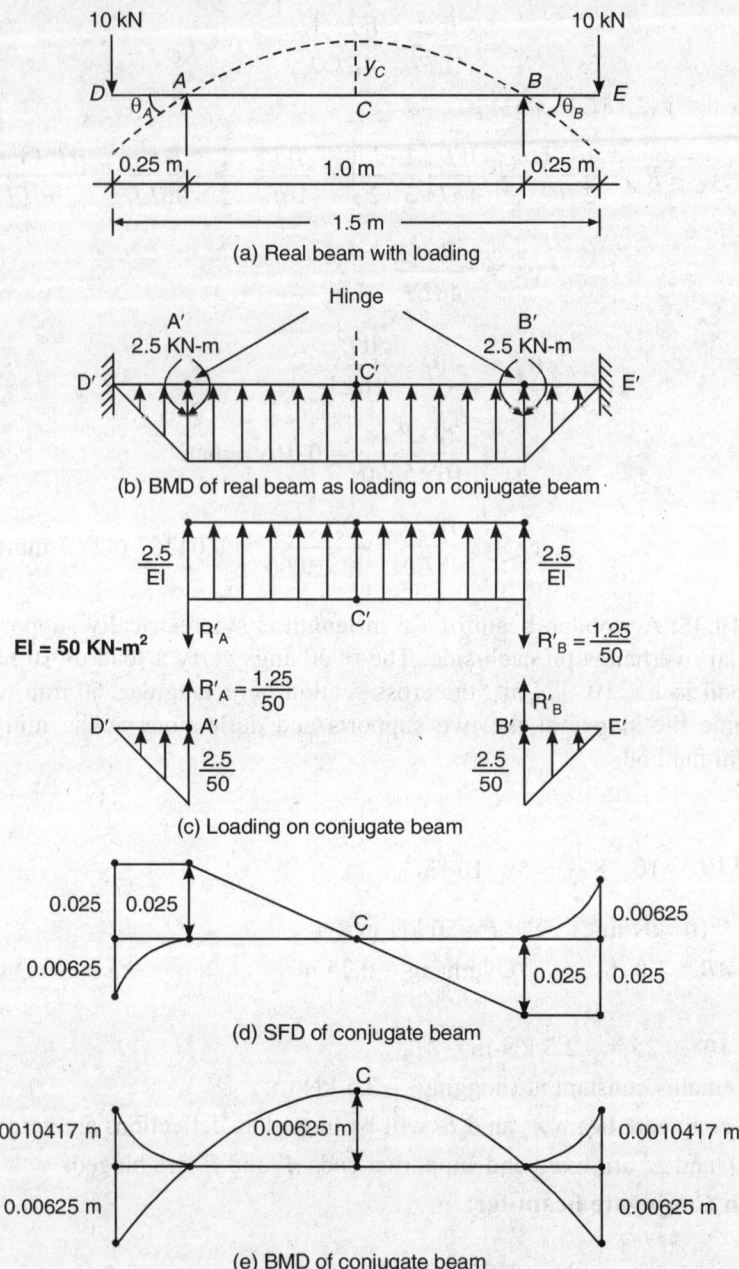

(a) Real beam with loading

(b) BMD of real beam as loading on conjugate beam

(c) Loading on conjugate beam

(d) SFD of conjugate beam

(e) BMD of conjugate beam

Fig. 10.57: Conjugate beam method

Reactions ($A'\ B'$)

$$R'_A = R'_B = -\frac{2.5 \times 1.0}{2EI} = \frac{1.25}{50}, \text{ downward}$$

This loading becomes zero at the fixed ends D' and E'.

SF in conjugate beam

$$SFA' = -\frac{2.50}{50} \times \frac{1.0}{2} = -\frac{2.50}{100} = -0.025 \text{ radian (anticlockwise)}$$

$$SFD' = -\frac{2.50}{50} - \frac{2.50}{50} \times \frac{0.25}{2} = -0.03125 \text{ radian (anticlockwise)}$$

$$SF \text{ at } C' = -\frac{2.5}{100} + \frac{2.5}{50} \times 0.5 = 0.0, \text{ Tangent is horizontal}$$

$$SF \text{ at } B' = -\frac{2.5}{100} + \frac{2.50}{50} \times 1 = +\frac{2.50}{100} = +0.025 \text{ radian (clockwise)}$$

$$SF \text{ at } E' = +\frac{2.5}{100} + \frac{2.50}{50} \times \frac{0.25}{2} = +0.03125 \text{ radian (clockwise)}$$

Bending moment of conjugate beam

$$y_A = M_A, \text{ Hinge, } y_A \text{ (deflection)} = 0$$

$$y_D = M_D = \frac{1.25}{50} \times 0.25 + \frac{2.50}{50} \times \frac{0.25}{2} \times \frac{0.25 \times 2}{3}$$

$$= 0.00625 + 0.0010417 = 0.00729 \text{ m (7.292 mm)}$$

$$y_C = M_C = -\frac{1.25}{50} \times 0.50 + \frac{2.5}{50} \times 0.5 \times \frac{0.5}{2} = -0.00625 \text{ m (6.25 mm above beam axis)}$$

EXAMPLE 10.36: A simply supported beam AB of 8 m span carries udl of 10 kN/m on the entire span and a concentrated load of 40 kN at the mid span point C. Derive the equations of slope and deflection at any point X at a distance of x from the LH support A. $AX \le AC$. EI is constant throughout and is equal to 46854 kN-m^2. Also determine the slopes at A and B and deflection at C by conjugate beam method.

Solution:

$$AB = L = 8 \text{ m}, \qquad AC = 4 \text{ m}, \qquad W_C = 40 \text{ kN}, \qquad w = 10 \text{ kN/m}$$
$$EI = 46854 \text{ kN-m}^2$$

Real beam

$$R_A = R_B = \frac{(40 + 10 \times 8)}{2} = 60 \text{ kN}$$

$$M_x = 60x - \frac{10x^2}{2} \Bigg| - 40(x - 4)$$

$$M_x = 60x - 5x^2 \Big| - 40(x - 4)$$

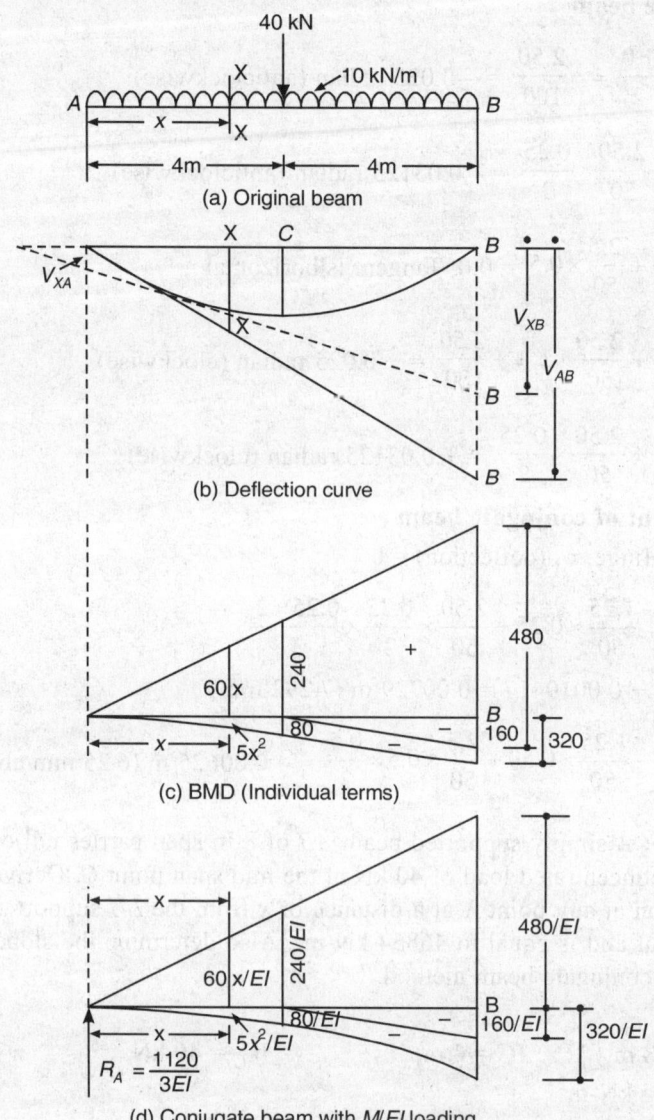

(a) Original beam

(b) Deflection curve

(c) BMD (Individual terms)

(d) Conjugate beam with M/EI loading

Fig. 10.58: Conjugate beam

Conjugate beam (Fig. 10.58d) and loading

$$R'_A = \frac{1}{8EI}\left[+\frac{480\times 8}{2}\times\frac{8}{3}-\frac{320\times 8}{3}\times\frac{8}{4}-\frac{160\times 4}{2}\times\frac{4}{3}\right]$$

$$= \frac{1}{EI}\left[640-\frac{640}{3}-\frac{160}{3}\right]=\frac{\mathbf{1120}}{\mathbf{3EI}}$$

$$R'_B = \frac{1}{8EI}\left[\frac{480\times8}{2}\times\frac{2}{3}\times8 - \frac{320\times8}{3}\times\frac{3}{4}\times8 - \frac{160\times4}{2}\left(4+\frac{8}{3}\right)\right]$$

$$= \frac{1}{EI}\left[1280 - 640 - 40\times\frac{20}{3}\right] = \frac{1120}{3EI}$$

By symmetry $R'_A = R'_B = \dfrac{1120}{3EI}$

Shear force of conjugate beam

$$F_A = R_A = \frac{1120}{3EI} = \theta_A \text{ by definition}$$

$$\theta_A = \frac{1120}{3\times46854} = +\,\textbf{0.007968 radian}$$

$$\theta_C = F'_C = +\frac{1120}{3EI} - \frac{240\times4}{2EI} + \frac{80\times4}{3EI} = \frac{1120 - 480\times3 + 320}{3EI} = 0, \text{ Also by symmetry } \theta_C = 0$$

$$\theta_B = F'_B = \left[\frac{1120}{3EI} - \frac{480\times8}{2EI} + \frac{320\times8}{3EI} + \frac{160\times4}{2EI}\right] = -\frac{1120}{3EI} = -\,0.007968 \text{ radian}$$

Bending Moment Diagram of Conjugate Beam

$$y_x = M_x = R_A \cdot x - \frac{60x}{EI}\cdot\frac{x}{2}\cdot\frac{x}{3} + \frac{5x^2}{EI}\cdot\frac{x}{3}\cdot\frac{x}{4} = \left[\frac{1120\cdot x}{3EI} - \frac{10x^3}{EI} + \frac{5x^4}{3EI\times4}\right]$$

$$\begin{array}{c} y_c \\ {\scriptstyle(x=4m)} \end{array} = M'_C = \frac{1120\times4 - 10\times64\times3 + 5\times64}{3EI} = \frac{2880}{3EI} = \frac{960}{EI}$$

or $y_C = \dfrac{960}{46854} = 0.0204892 \approx 0.02049 \text{ m (20.49 mm)}$

EXAMPLE 10.37: A simply supported beam ACB of span L carries a point load W at the middle point C. The cross-section of beam has moment of inertia from A to $C = 2I$ and from C to $B = I$. Determine deflection at the middle point C and slope at ends A and B by using conjugate beam method.

Solution:

$R_A = R_B = W/2$ by symmetry

Bending moment at the centre $C = M_C = WL/4$

EI from A to $C = 2\,EI$

EI from C to $B = EI$

Figure 10.59 shows *real beam* with *loading, bending moment diagram*, and *BM/EI diagram* (Fig. 10.59c) *loading on conjugate beam.*

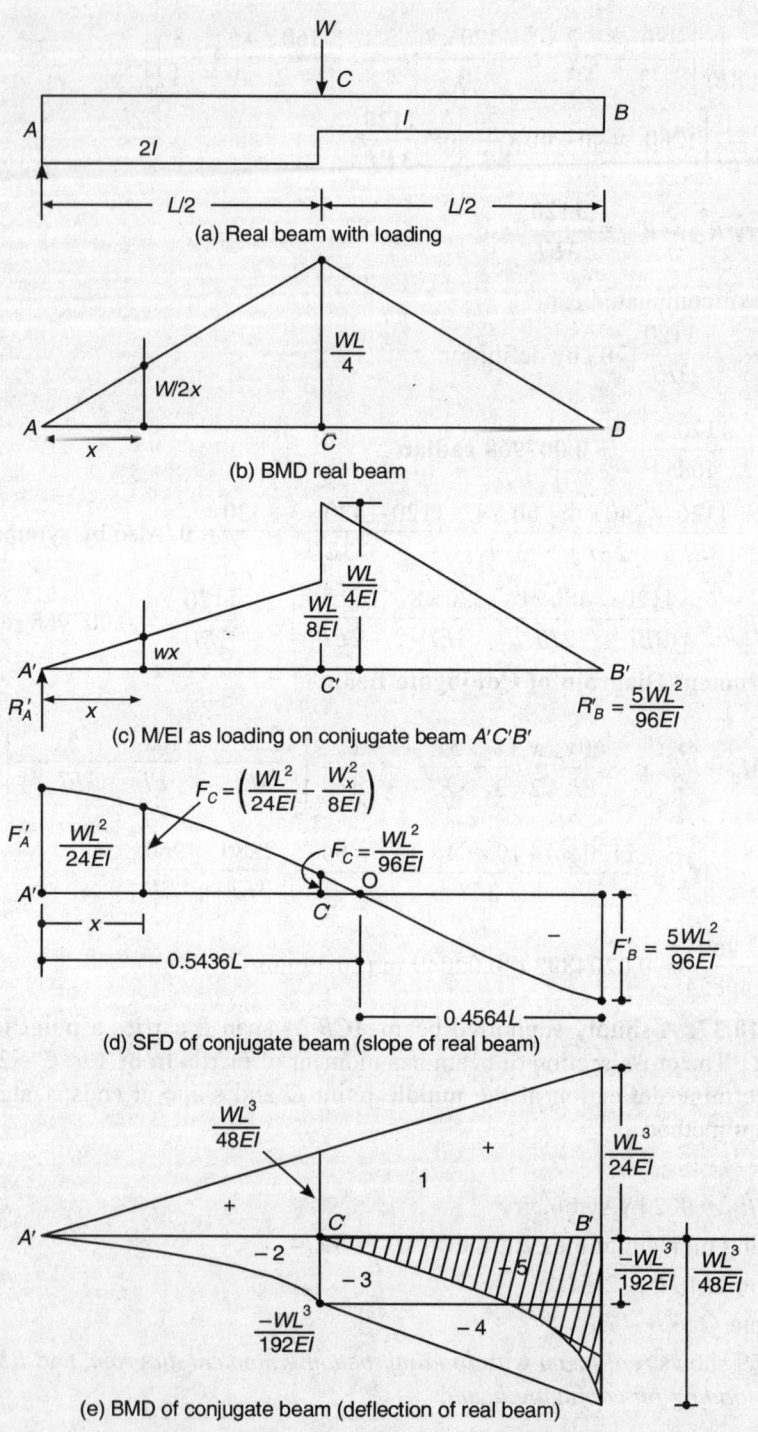

(a) Real beam with loading

(b) BMD real beam

(c) M/EI as loading on conjugate beam A'C'B'

(d) SFD of conjugate beam (slope of real beam)

(e) BMD of conjugate beam (deflection of real beam)

Fig. 10.59: Conjugae beam SFD and BMD

Reactions of conjugate beam

$$R'_A = \frac{1}{L}\left[\frac{WL}{8EI}\cdot\frac{L}{4}\left(\frac{L}{2}+\frac{L}{6}\right)+\frac{WL}{4EI}\cdot\frac{L}{4}\left(\frac{2}{3}\cdot\frac{L}{2}\right)\right] = \frac{WL^2}{48EI}[1+1] = \frac{WL^2}{24EI}$$

$$R'_B = \frac{1}{L}\left[\frac{WL^2}{32EI}\left(\frac{L}{3}\right)+\frac{WL^2}{16EI}\left(\frac{L}{2}+\frac{L}{6}\right)\right] = \frac{WL^2}{16EI}\left[\frac{1}{6}+\frac{4}{6}\right] = \frac{5WL^2}{96EI}$$

Shear Force Diagram Conjugate Beam (Figs 10.59c and d)

$$F'_A = R'_A = \frac{WL^2}{24EI}$$

$$\theta_A = F'_A = \frac{WL^2}{24EI} \qquad \ldots (1)$$

$$F'_x = R'_A = -\frac{w_x\cdot x}{2} = \frac{WL^2}{24EI} - \frac{WL}{8EI}\cdot\frac{2}{L}\cdot\frac{x^2}{2}$$

or $\displaystyle F'_x \atop (atoc)$ $\displaystyle = \frac{WL}{8EI}\left(\frac{L}{3} - \frac{x^2}{L}\right) = \frac{WL^2}{24EI} - \frac{Wx^2}{8EI}$

$$\frac{F'_C}{(x=L/2)} = \frac{WL^2}{24EI} - \frac{W}{8EI}\cdot\frac{L^2}{4} = +\frac{WL^2}{96EI}$$

$$F'_B = -\frac{5}{96}\cdot\frac{WL^2}{EI} = \theta_B \qquad \ldots (2)$$

F'_x will be zero beyond C. Let $F'_{xb} = 0$, at a distance x_b from the end B'.

$$0 = -\frac{5}{96}\frac{WL^2}{EI} + \frac{WL}{4EI}\cdot\frac{2x_b}{L}\cdot\frac{x_b}{2} = -\frac{W}{4EI}\left[\frac{5L^2}{24} - x_b^2\right] \text{ or } \frac{5L^2}{24} - x_b^2 = 0$$

$$x_b = \frac{L}{2}\sqrt{\frac{5}{6}} = 0.4564L, \ x_a = 0.5436L$$

∴ *BM will be maximum at the point O.*

BM at $A = 0$, *BM* at $B = 0$, *BM* at $C = \left(R'_A\cdot\frac{L}{2} - \frac{WL}{8EI}\cdot\frac{1}{2}\cdot\frac{L}{2}\cdot\frac{L}{6}\right)$

$$y_C = M'_C = \left(\frac{WL^2}{24EI}\cdot\frac{L}{2} - \frac{WL^3}{192EI}\right) = +\frac{WL^3}{64EI} \qquad \ldots (3)$$

$$y_{max} (xb = 0.4564L) = M_{max} = \frac{5WL^2}{96EI}(0.4564L) - \frac{WL}{4EI} \cdot \frac{2}{L}(0.4564L) \times \frac{(0.4564L)^2}{2 \times 3}$$

$$= \frac{WL^3}{EI}\left[\frac{5 \times 0.4564}{96} - \frac{(0.4564)^3}{12}\right] = \frac{WL^3}{63.1EI}, \text{ almost same as at the centre } C.$$

EXAMPLE 10.38: A simply supported beam AB of span L carries a point load of W at the centre. The beam cross-section is such that the middle half span has moment of inertia equal to I while the two end spans of $L/4$ has moment of inertia equal to $I/2$. Determine the slopes at end supports and deflection at the middle point. E of the material is constant for the entire span.

Solution:

Real Beam BMD:

$$M_C = \frac{WL}{4}, M_D = \frac{WL}{8}$$

Loading on Conjugate Beam

Variable triangular loading at $A' = 0$,

$$W'_D = \frac{WL}{8EI}, \frac{WL}{4EI} \text{ and } W'_E = \frac{WL}{4EI} \text{ and } \frac{WL}{8EI}$$

$$W'_C = \frac{WL}{4EI}$$

Reactions Conjugate Beam

$$R'_A = \frac{1}{L}\left[\frac{WL}{4EI} \cdot \frac{L}{8} \cdot \frac{2}{3} \cdot \frac{L}{4} + \left(\frac{WL}{4EI} + \frac{WL}{8EI}\right)\frac{1}{2} \cdot \frac{L}{2} \cdot \frac{L}{2} + \frac{WL}{4EI} \cdot \frac{L}{8}\left(\frac{1}{3} \cdot \frac{L}{4} + \frac{3L}{4}\right)\right]$$

$$= \frac{WL^2}{32EI}\left[\frac{1}{6} + \frac{3}{2} + \frac{10}{12}\right] = \frac{30WL^2}{32 \times 12EI} = \frac{5WL^2}{64EI}$$

$$R'_B = R'_A = \frac{5WL^2}{64EI} \qquad \qquad \dots (1)$$

Shear Force-Diagram conjugate Beam

$$F'_A = R'_A = +\frac{5WL^2}{64EI}$$

$$\theta_A = +\frac{5WL^2}{64EI} \qquad \qquad \dots (2)$$

$$F'_B = R'_B = -\frac{5WL^2}{64EI} = \theta_B \qquad \qquad \dots (3)$$

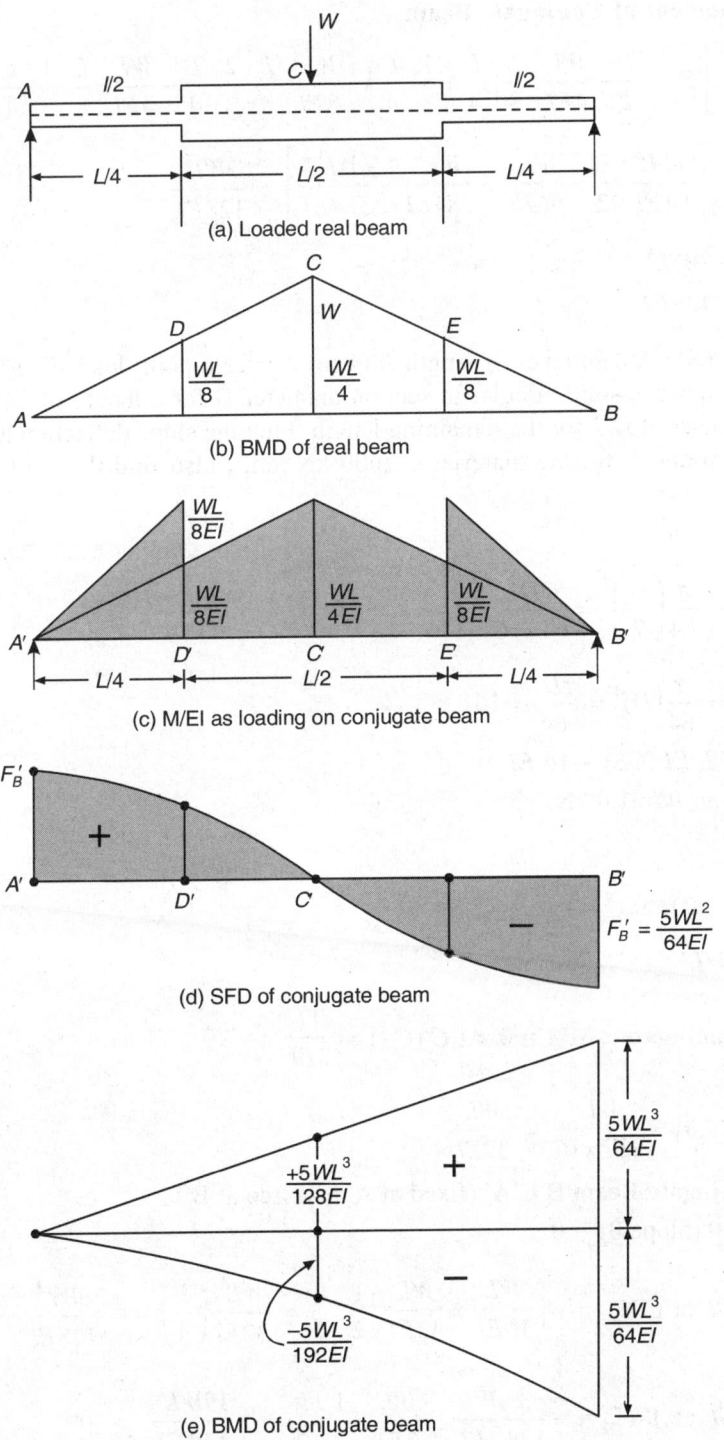

(a) Loaded real beam

(b) BMD of real beam

(c) M/EI as loading on conjugate beam

(d) SFD of conjugate beam

(e) BMD of conjugate beam

Fig. 10.60: Conjugate beam method

Bending moment of Conjugate Beam

$$M'_C = \left[R'_A \cdot \frac{L}{2} - \frac{WL}{4EI} \cdot \frac{L}{8} \left(\frac{L}{4} + \frac{1}{3} \cdot \frac{L}{4} \right) - \frac{WL}{8EI} \cdot \frac{L}{8} \cdot \frac{2}{3} \cdot \frac{L}{4} - \frac{WL}{4EI} \cdot \frac{L}{8} \cdot \frac{1}{3} \cdot \frac{L}{4} \right]_1$$

$$= \left[\frac{5WL^2}{64EI} \cdot \frac{L}{2} - \frac{WL^3}{96EI} - \frac{WL^3}{384EI} - \frac{Wl^3}{384EI} \right] = \frac{3WL^3}{128EI}$$

$$y_C = M'_C = \frac{3WL^3}{128EI} \qquad\qquad\qquad \dots (4)$$

EXAMPLE 10.39: A cantilever of length L (4 m) carries a point load W (1.0 kN) at its free end A. The member is solid circular in section diameter D for a length of $L/2$ from the fixed end and a diameter of $D/2$ for the remaining length. Find the slope deflection at point C and A. If D = 100 mm and E for the material = 2000 kN/mm^2, also find the values of slopes and deflections.

Solution:

$$I_1 \ (A \text{ to } C) = \frac{\pi}{64} \left(\frac{D}{2} \right)^4 = \frac{\pi D^4}{64 \times 16} = I$$

$$I_2 \ (C \text{ to } B) = \frac{\pi}{64} (D)^4 = \frac{\pi D^4}{64} = 161$$

$EI \ (AC) = EI, \ EI \ (CB) = 16 \ EI$

Span $L = 4$ m, $W = 1.0$ kN

Real Beam

BM at $A = 0$, BM at $C = \dfrac{WL}{2}$

BM at $B = WL$

Loading (conjugate): At $A = 0$, At $C \ (CA) = \dfrac{WL}{2EI}$

$$\text{At } C \ (CB) = \frac{WL}{2EI \times 16} = \frac{WL}{32EI}$$

SFD – Conjugate Beam B′C′A′ (fixed at A' and free at B')

SF $\quad F'_B = 0$, Slope $\theta_B = 0$ $\qquad\qquad\qquad \dots (1)$

$$\theta_C = SF \text{ at } C' \ F'_C = - \left(\frac{WL}{16EI} + \frac{WL}{32EI} \right) \frac{1}{2} \cdot \frac{L}{2} = \frac{WL^3}{32EI} \left(\frac{3}{4} \right) = - \frac{3WL^2}{128EI} \qquad \dots (2)$$

$$\theta_A = SF \text{ at } A' \ F'_A = - \frac{3}{128} \frac{WL^2}{EI} - \frac{WL}{2EI} \times \frac{1}{2} \cdot \frac{L}{2} = - \frac{19WL^2}{128EI} \qquad \dots (3)$$

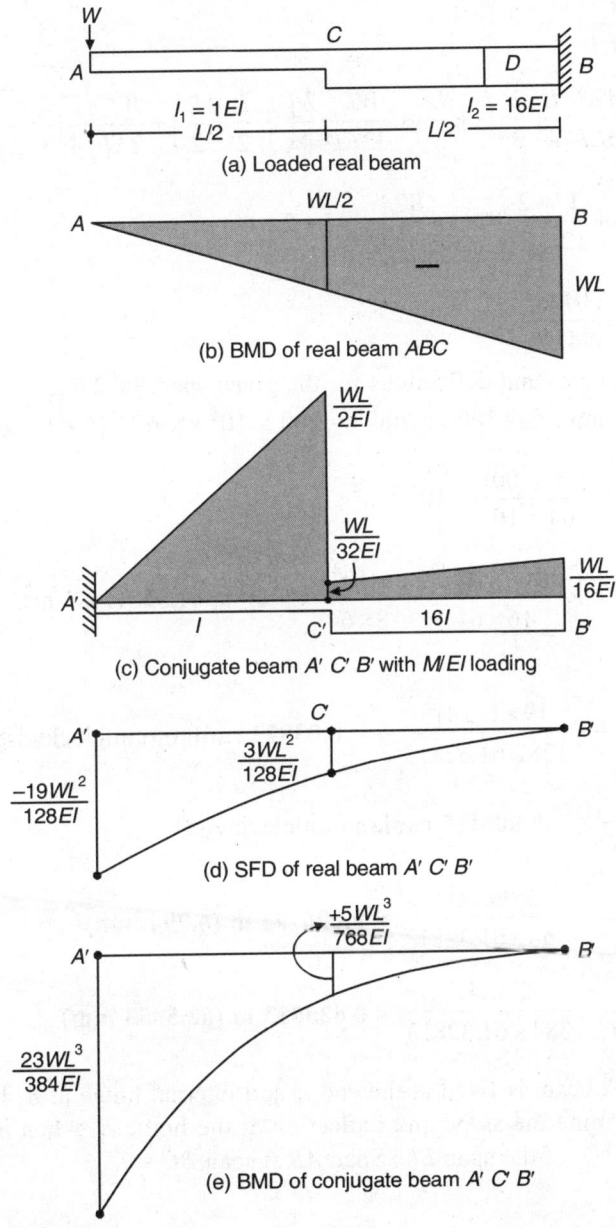

(a) Loaded real beam

(b) BMD of real beam *ABC*

(c) Conjugate beam *A' C' B'* with *M/EI* loading

(d) SFD of real beam *A' C' B'*

(e) BMD of conjugate beam *A' C' B'*

Fig. 10.61.

BMD–Conjugate Beam

$$M'_B = 0, \; y_B = M'_B = 0 \qquad \qquad \dots (4)$$

$$M'_C = y_C = \frac{WL}{16EI} \cdot \frac{L}{4} \cdot \frac{L}{3} + \frac{WL}{32EI} \cdot \frac{L}{4} \cdot \frac{L}{6} = \frac{WL^3}{192EI} + \frac{WL^3}{768EI} = + \frac{5WL^3}{768EI}$$

Thus $y_C = +\dfrac{5WL^3}{768EI}$ ... (5)

$$M'_A = +\left[\frac{WL}{16EI}\cdot\frac{L}{4}\left(\frac{2}{3}\cdot\frac{L}{2}+\frac{L}{2}\right)+\frac{WL}{32EI}\cdot\frac{L}{4}\left(\frac{1}{3}\cdot\frac{L}{2}+\frac{L}{2}\right)+\frac{WL}{2EI}\cdot\frac{L}{4}\cdot\frac{2}{3}\cdot\frac{L}{2}\right]$$

$$=\frac{WL^3}{16EI}\left[\frac{5}{24}+\frac{1}{12}+\frac{2}{3}\right]=+\frac{WL^3}{16\times24EI}(5+2+16)$$

$$y_A = M'_A = \frac{23WL^3}{384EI}$$ ... (6)

Actual values of slopes and deflections for the given span, load and EI.

$L = 4$ m, $D = 100$ mm, $E = 200$ kN/mm^2 $= 200 \times 10^6$ kN/m^2, $W = 1.0$ kN

$$I = \frac{\pi}{64}\left(\frac{D}{2}\right)^4 = \frac{\pi}{64}\frac{(100)^4}{16}\times10^{-12}\,\text{m}^4$$

$$EI = 200\times10^6 \times\frac{\pi10^8\times10^{-12}}{16\times64} = \frac{\pi(10^4)}{8\times64}\,\text{kN-m}^2 = \mathbf{61.32812\,kN\text{-}m^2}$$

$\theta_B = 0$

$$\theta_A = \frac{19WL^2}{128EI} = -\frac{19\times1\times(4)^2}{128\times61.32813} = -\,\mathbf{0.03873\ radian}\ \text{(anticlockwise)}$$

$$\theta_C = -\frac{3WL^2}{128EI} = -\,\mathbf{0.006115\ radian}\ \text{(anticlockwise)}$$

$$y_C = +\frac{5WL^3}{768EI} = \frac{5.0\times(4)^3}{768\times61.32813} = \mathbf{0.006794\ m\ (6.794\ mm)}$$

$$y_A = +\frac{23WL^3}{384EI} = \frac{230(4)^3}{384\times61.32813} = \mathbf{0.625053\ m\ (62.5053\ mm)}$$

EXAMPLE 10.40: A beam is fixed at the end A and internal hinge at B. Its end C rests on a roller support. Determine the slope and deflection at the hinge B, when loaded with a point load W at the centre 'D' of the span BC. Span AB = span $BC = L$.

Solution:

Span AB = Span $BC = L$, $BD = DC = \dfrac{L}{2}$

Since there are 3 reactions, and there are 3 supports with given additional condition of $M_B = 0$ (internal hinge). The beam will be determinate.

Three conditions of equilibrium: $\Sigma V = 0$, $\Sigma M_C = 0$, $\Sigma M_B = 0$

Reactions at $C = R_C$, taking moment about B, $R_C \cdot L - \dfrac{WL}{2} = 0$, $\therefore$ $R_C = \dfrac{W}{2}$

$\Sigma V = 0$, $R_A + R_C - W = 0$, $\therefore$ $R_A = \dfrac{W}{2}$

Thus, BM for the real beam:

$$M_D = \frac{W}{2} \cdot \frac{L}{2} = \frac{WL}{4}, \quad M_A = \frac{W}{2} \cdot 2L - W \frac{3L}{2} = -\frac{WL}{2}$$

Draw M/EI diagram as loading on conjugate beam.

The conjugate beam with M/EI as loading is shown in Fig. 10.62c.

the fixed end A will be free end A' in conjugate beam and B' as interior support.

Conjugate beam Fig. 10.62c

Moment about C

$$R'_B \cdot L + \frac{WL}{4EI} \cdot \frac{L}{2} \cdot \frac{L}{2} - \left(\frac{WL}{2EI} \cdot \frac{L}{2} \cdot \frac{5L}{3} \right) = 0$$

$$R'_B = \left(\frac{5}{12} - \frac{1}{16} \right) \frac{WL^2}{EI} = \frac{17WL^2}{48EI} \downarrow$$

$$R'_C = -\frac{WL}{2EI} \cdot \frac{L}{2} + \frac{17WL^2}{48EI} + \frac{WL}{4EI} \cdot \frac{L}{2} = \frac{WL^2(-12 + 17 + 6)}{48EI} = \frac{11WL^2}{48EI} \uparrow$$

SFD of Conjugate beam (Fig. 10.62d)

$$F'_C = \theta_C = \frac{11WL^2}{48EI} \text{ (anticlockwise)}$$

$$F'_B \text{ (Right)} = (\theta_B)_R = -\frac{5WL^2}{48EI}$$

$$F'_B \text{ (Left)} = (\theta_B)_L = \left(-\frac{5}{48} + \frac{17}{48} \right) \frac{WL^2}{EI} = +\frac{WL^2}{4EI} \text{ (clockwise)}$$

BMD of conjugate beam (deflection of real beam)

$$M'_B = 0 = y_c, \quad M'_A = 0 = y_A, \quad M'_B = y_B = \frac{WL}{2EI} \cdot \frac{L}{2} \cdot \frac{2L}{3} = \frac{WL^3}{6EI}$$

The conjugate beam with M/EI as loading is shown in Fig. 10.62c

The fixed end A of real beam will be free end A' in *conjugate beam* and B' as interior support.

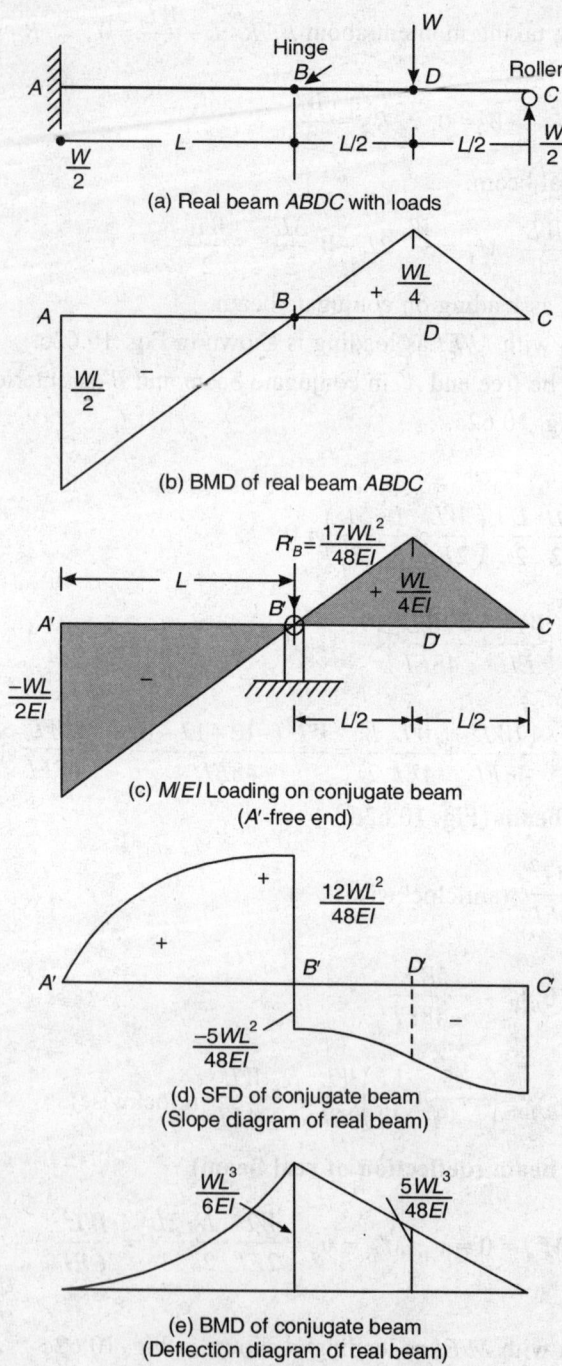

(a) Real beam *ABDC* with loads

(b) BMD of real beam *ABDC*

(c) *M/EI* Loading on conjugate beam
(*A'*-free end)

(d) SFD of conjugate beam
(Slope diagram of real beam)

(e) BMD of conjugate beam
(Deflection diagram of real beam)

Fig. 10.62: Conjugate beam method

Conjugate beam Fig. 10.62c

Moment about C

$$R'_B \cdot L + \frac{WL}{4EI} \cdot \frac{L}{2} \cdot \frac{L}{2} - \left(\frac{WL}{2EI} \cdot \frac{L}{2} \cdot \frac{5L}{3} \right) = 0$$

$$R'_B = \left(\frac{5}{12} - \frac{1}{16} \right) \frac{WL^2}{EI} = \frac{17WL^2}{48EI} \downarrow$$

$$R'_C = -\frac{WL}{2EI} \cdot \frac{L}{2} + \frac{17WL^2}{48EI} + \frac{WL}{4EI} \cdot \frac{L}{2} = \frac{WL^2(-12+17+6)}{48EI} = \frac{11WL^2}{48EI} \uparrow$$

SFD of Conjugate beam (Fig. 10.62d)

$$F'_C = \theta_C = -\frac{11WL^2}{48EI} \quad \text{(anticlockwise)}$$

$$F'_B \text{ (Right)} = (\theta_B)_R = -\frac{5WL^2}{48EI} \quad \text{(anticlockwise)}$$

$$F'_B \text{ (Left)} = (\theta_B)_L = \left(-\frac{5}{48} + \frac{17}{48} \right) \frac{WL^2}{EI} = +\frac{WL^2}{4EI} \quad \text{(clockwise)}$$

BMD of comjugate beam (deflection of real beam)

$$M'_B = 0 = y_C, \ M'_A = 0 = y_A, \ M_B = y_B = \frac{WL}{2EI} \cdot \frac{L}{2} \cdot \frac{2L}{3} = \frac{WL^3}{6EI}$$

$$y_D = +\frac{11WL^2}{48EI} \cdot \frac{L}{2} - \frac{WL}{4EI} \cdot \frac{L}{4} \cdot \frac{L}{6} = +\frac{11WL^3}{96EI} - \frac{WL^3}{96EI} = -\frac{10WL^3}{96EI}$$

Thus,

$$\theta_C = F'_C = -\frac{11WL^2}{48EI}, \ \theta_B \text{ right} = \frac{5WL^2}{48EI}, \ \theta_B \text{ left} = +\frac{WL^2}{4EI}, \ \theta_A = 0$$

$$y_C = 0, \ y_B = \frac{WL^3}{6EI}, \ y_D = \frac{10WL^3}{96EI} = \frac{5WL^3}{48EI}$$

10.7 DEFLECTIONS BY STRAIN ENERGY METHOD

10.7.1 Introduction

Whenever an element is loaded, it undergoes deformation and work is done in this process by external load. Applying the principle of **conservation of energy**, it is found that this *external work* is stored in the element as **strain energy**. In case of axial load in the element, the *strain energy* stored $U = \frac{f^2}{2E} \times$ volume where f is axial stress and E is the modulus of elasticity.

Let us now apply the concept of *strain energy* to *bending moment* in beams. Consider a simply supported beam of span L, subjected to a uniform moment M. The beam is subjected to uniform couple of moment 'M' and hence the beam deflection curve will be a circular arc having uniform curvature. Radius of curvature

$$\frac{1}{R} = \frac{M}{EI} \qquad \text{... (i)}$$

If θ is the angle subtended at the centre of circular arc then,

$$\theta = \frac{L}{R} \qquad \text{... (ii)}$$

$$\frac{1}{R} = \frac{\theta}{L} = \frac{M}{EI} \quad \text{or} \quad \theta = \frac{ML}{EI} \qquad \text{... (iii)}$$

i.e.
$$\theta = \frac{ML}{EI} \qquad \text{... (10.57)}$$

Thus θ is linearly related to bending moment M (Fig. 10.63). If the moment 'M' is applied gradually starting with zero, having zero rotation θ. The work done by 'M' in rotating beam by θ shall be

$$W = \frac{1}{2} \cdot M \cdot \theta \qquad \text{... (iv)}$$

This work is stored as strain energy U, and *by conservation of energy, we have*

i.e.
$$U = \frac{1}{2} \cdot M \cdot \theta = \frac{1}{2} M \cdot \frac{ML}{EI} = \frac{M^2 L}{2EI} \qquad \text{... (v)}$$

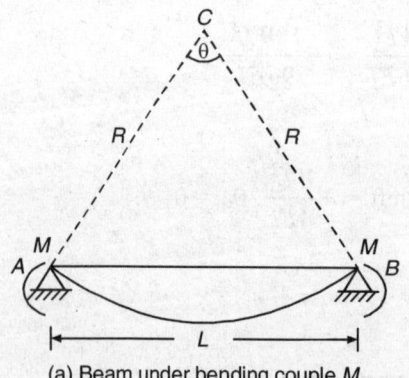

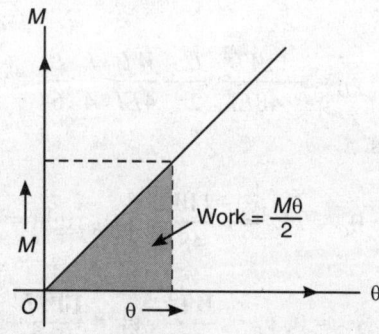

(a) Beam under bending couple M (b) Relation of rotation θ & moment M

Fig. 10.63: Relation of BM 'M' and 'θ'

or
$$U = \frac{M^2 L}{2EI} \qquad \text{... (10.58a)}$$

Also
$$U = \frac{1}{2}\theta \cdot \frac{EI\theta}{L} = \frac{EI\theta^2}{2L} \qquad \text{... (10.58b)}$$

Table 10.2: Maximum slope and deflection in beams

S. No.	Type of beam and diagram	Maximum slope	Maximum deflection
1.	Cantilever with point load W at free end A. W A —————— B L	$\theta_A = \dfrac{WL^2}{2EI}$ $\theta_B = 0$	$y_A = \dfrac{WL^3}{3EI}$ $y_B = 0$
2.	Cantilever with point load W at any point l_1 from B. W C A —————— B $\vert\!\leftarrow L_1 \rightarrow\!\vert$ $\vert\!\leftarrow\quad L \quad\rightarrow\!\vert$	$\theta_A = \theta_C = \dfrac{Wl_1^2}{2EI}$ $\theta_B = 0$	$y_A = \dfrac{Wl_1^3}{3EI} + \dfrac{Wl_1^2}{2EI}(L - l_1)$
3.	Cantilever with udl $W' = wL$. w/m A ~~~~~~~~~ B L_1 $\vert\!\leftarrow\quad L \quad\rightarrow\!\vert$	$\theta_A = -\dfrac{wL^3}{6EI} = -\dfrac{W'L^2}{6EI}$	$y_A = \dfrac{wL^4}{8EI} = \dfrac{W'L^3}{8EI}$, $W' = wL$
4.	Cantilever with partial udl w/m on L_1 span. w/m C ~~~~~ A —————— B $\vert\!\leftarrow L_1 \rightarrow\!\vert$ $\vert\!\leftarrow\quad L \quad\rightarrow\!\vert$	$\theta_A = \theta_C = \dfrac{wL_1^3}{6EI}$	$y_A_{(y\max)} = \dfrac{wL_1^4}{8EI} + \dfrac{wL_1^3}{6EI}(L - L_1)$
5.	Cantilever with partial udl on free end side. A ~~~~~~ C $\quad\quad\vert\!\leftarrow L_1 \rightarrow\!\vert B$ $\vert\!\leftarrow\quad L \quad\rightarrow\!\vert$	$\theta_A = -\left(\dfrac{wL^3}{6EI} - \dfrac{wL_1^3}{6EI} \right)$	$y_A_{(y\max)} = \dfrac{wL^4}{8EI} - \left[\dfrac{wL_1^4}{8EI} + \dfrac{wL_1^3}{6EI}(L - L_1) \right]$
6.	Cantilever with variable udl w/m $w_x = \dfrac{w.x}{L}$ A —————— B $\vert\!\leftarrow x \rightarrow\!\vert$ $\vert\!\leftarrow\quad L \quad\rightarrow\!\vert$	$\theta_A = -\dfrac{wL^3}{24EI}$ $\theta_x = \dfrac{w}{24EIL}(x^4 - L^4)$	$y_A = \dfrac{wL^4}{30EI}$ $y_x = \dfrac{w}{120EIL}(x^5 - 5L^4x + 4L^5)$

S. No.	Type of beam and diagram	Maximum slope	Maximum deflection
7.	Cantilever with moment couple 'M' at the free end.	$\theta_A = -\dfrac{ML}{EI}$ $\theta_x = -\dfrac{M(L-x)}{EI}$	$y_A = \dfrac{ML^2}{2EI}$ $y_x = \dfrac{M(L-x)^2}{2EI}$
8.	Simply supported beam with central load W.	$\theta_A = -\theta_B = \dfrac{WL^2}{16EI}$	$y_C = \dfrac{WL^3}{48EI}$
9.	Simply supported beam with point load W at any point X, $AX = a$, $BX = b$, $a \geq b$.	$\theta_A = \dfrac{Wab(L+b)}{6EIL}$ $\theta_B = -\dfrac{Wab(L+a)}{6EIL}$	$y_{max} = \dfrac{Wb(L^2 - b^2)^{3/2}}{9\sqrt{3}EIL}$ $a > b$ $y_C = \dfrac{Wa^2b^2}{3EIL}$
10.	SS beam AB with udl w/m on entires pan $W = wL$	$\theta_A = -\theta_B = \dfrac{wL^3}{24EI} = \dfrac{W'L^2}{24EI}$	y_C (max) $= \dfrac{5wL^4}{384EI} = \dfrac{5W'L^3}{384EI}$
11.	SS beam with variable udl	$\theta_A = -\theta_B = \dfrac{5wL^3}{192EI}$	$y_C = \dfrac{wL^4}{120EI} = y_{max}$ $y_x = \dfrac{wx}{960EIL}[25L^4 - 40L^2 \cdot x^2 + 16x^4]$ $\left(x \leq \dfrac{L}{2} \right)$
12.	SS beam with variable loading zero at one end and w/m at the other end	$\theta_A = \dfrac{7wL^3}{360EI}$ $\theta_B = \dfrac{-wL^3}{45EI}$ $\theta_x = 0$, at $x = 0.5193L$	$y_{max} = \dfrac{2.5wL^4}{384EI} = \dfrac{0.00652wL^4}{EI}$ ($x = 0.5193L$ from A)

S. No.	Type of beam and diagram	Maximum slope	Maximum deflection
13.	SS beam with moment couple 'M' at B	$\theta_A = \dfrac{ML}{6EI}, \theta_B = \dfrac{-ML}{3EI}$ $\theta_x = \dfrac{M(L^2 - 3x^2)}{6EIL}$ $\theta = 0, \text{at } x = \dfrac{L}{\sqrt{3}}$	$y_{max} = \dfrac{ML^2}{9\sqrt{3}EI}, x = \dfrac{L}{\sqrt{3}}$ $y_x = \dfrac{Mx(L^2 - x^2)}{6EIL}$
14.	SS beam with moment couple 'M' at any point C	$\theta_A = \dfrac{M}{6EIL}(L^2 - 3b^2)$ $\theta_x = \dfrac{M}{6EIL}(L^2 - 3b^2 - x^2)$ $x \le a$	$y_{max} = \dfrac{M}{9EIL}(L^2 - 3b^2)$ $y_x = \dfrac{M \cdot x(L^2 - 3b^2 - x^2)}{6EIL}$ $(x \le a)$

10.7.2 Deflection by strain energy method

Consider a normal beam carrying transverse loads, the beam undergoes *bending* with variable 'M' according to location. At any point X, bending moment is M_x as function of distance x taken from the support. Bending moment at a point $(x + dx)$ distance will be $(M_x + dM_x)$. Consider the element of very small length **dx**, for this small length dx the arc may be **assumed as circular** and the small rotation of tangents be '**dθ**'.

$$\frac{d\theta}{dx} = \frac{d}{dx} \cdot \frac{dy}{dx} = \frac{d^2 y}{dx^2}$$

or $\quad d\theta = \dfrac{d^2 y}{dx^2} \cdot dx = \dfrac{M_x}{EI} \cdot dx \qquad\qquad$ ... (vi)

Strain energy stored in the element dx shall be $dU = \dfrac{1}{2} M_x \cdot d\theta = \dfrac{M_x^2}{2EI} \cdot dx$

$$dU = \frac{M_x^2}{2EI} \cdot dx = \frac{EI}{2}\left(\frac{d^2 y}{dx^2}\right)^2 \cdot dx \qquad\qquad \text{... (vii)}$$

Total strain energy in the whole beam

$$U = \int_0^L dU = \int_0^L \frac{M_x^2}{2EI} \cdot dx = \int_0^L \left(\frac{d^2 y}{dx^2}\right)^2 \cdot dx \qquad\qquad \text{... (10.59)}$$

Equation 10.59 is used when *bending moment equation* or the *deflection curve* equation is known.

Deflections can be found under the point load by equating total *strain energy 'U'* equal to the **work done by the point load** in deflecting the beam by **'y'**.

i.e. $\quad \dfrac{1}{2} \cdot W \cdot y = U = \int_0^L \dfrac{M_x^2}{2EI} \cdot dx \;$ or $\; y = \dfrac{1}{W} \int_0^L \dfrac{M_x^2}{EI} dx$ ... (10.60)

In this method the *deflections* can be found *only under the load*.

Simitarly, when a couple 'μ' acts at any point, the rotation 'θ' at the point can only be found.

$$\dfrac{1}{2} \cdot \mu \cdot \theta = U = \int_0^L \dfrac{M_x^2}{2EI} \cdot dx \;\; \text{or} \;\; \theta = \dfrac{1}{\mu} \int_0^L \dfrac{M_x^2}{EI} \cdot dx \qquad \text{... (10.61)}$$

This direct method of strain encrgy is useful only to calculate **slope and deflections** at the point of application of **moment couple** and **load** respectively. For determining the *slope and deflection* at any other point other methods of **unit load** or **fictitious** load shall be used. The method is illustrated by examples.

EXAMPLE 10.41: A cantilever beam *AB* of span *L* carries a concentrated load *W* at the free end. Determine the deflection at the free end by strain energy method if *EI* is constant throughout.

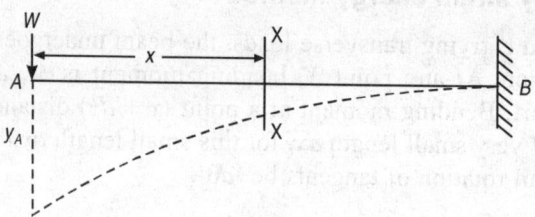

Fig. 10.64.

Solution:

Span $AB = L$, Point Load $A = W$, $M_x = -W \cdot x$ at a distance x from A

Strain Energy $U = \int_0^L \dfrac{M_x^2}{2EI} \cdot dx \; = \text{Workdone} = \dfrac{1}{2} \cdot W \cdot y$

or $\quad \dfrac{1}{2} \cdot W \cdot y_A = \int_0^L \dfrac{(-W \cdot x)^2}{2EI} dx \; = \int_0^L \dfrac{W^2 \cdot x^2}{2EI} \cdot dx$

or $\quad y_A = \dfrac{2}{W} \left[\dfrac{W^2}{2EI} \cdot \dfrac{x^3}{3} + C \right]_0^L = \dfrac{2}{W} \cdot \dfrac{W^2}{6EI} (L^3 - 0) = \dfrac{WL^3}{3EI}$

i.e. $\quad y_A = \dfrac{WL^3}{3EI}$, same as found by other methods.

EXAMPLE 10.42: A cantilever beam AB of span L is subjected to a moment couple 'M' at the free end A. Calculate the rotation of tangent at the free end A by strain energy method.

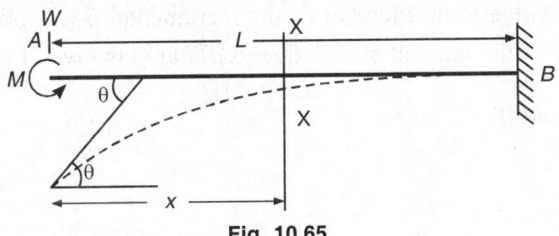

Fig. 10.65.

Solution:

Span $AB = L$, Couple Moment $= M$, $M_x = -M$ (Hogging)

$$du = \frac{M_x}{2} d\theta = \frac{M_x}{2} \cdot \frac{M_x \cdot dx}{EI} = \frac{M_x^2 \, dx}{2EI}$$

$$U = \frac{M \cdot \theta_A}{2} = \int_0^L \frac{M_x^2 \cdot dx}{2EI} = \int_0^L \frac{(-M)^2}{2EI} \cdot dx , \qquad \left\{ d\theta = \frac{M_x \cdot dx}{EI} \text{ for small element } dx \right\}$$

or $U = \dfrac{M \cdot \theta_A}{2} = \dfrac{(M^2 \cdot x)_0^L}{2EI} = \dfrac{M^2 L}{2EI}$, or $\theta_A = \dfrac{M^2 L}{2EI} \cdot \dfrac{2}{M} = \dfrac{ML}{EI}$ (anticlockwise)

i.e. $\theta_A = \dfrac{ML}{EI}$, same as found earlier by other methods.

10.7.3 Deflection by Unit Load Method

We have seen that we can find the deflection and slope in beam only at the point of application of load or moment respectively by direct method of strain energy. We shall find the deflection of beam at any other point (C) by applying a *unit load* at that point. Let the deflection is required at any point (C) distant x_C from the LHS due to external loading. Consider any section X-X at a distance x from the LHS A and another section at a small distance 'dx' from X-X. Let Mx be the BM at the section. Apply an imaginary point load dW at the point C where

the deflection is required. **Work** supplied by this load dW will be $\dfrac{1}{2} \cdot dW \cdot y_C$ ... (1)

Where y_C is required deflection at C.

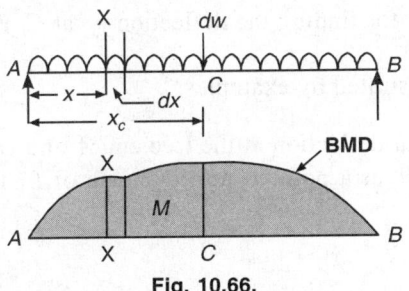

Fig. 10.66.

The *BM* at *X* due to the additional load of *dW* will be increased to $(M + dM)$, where *dM* is the increment due to additional load *dW*.

If *m* is the *BM* at *X-X* due to unit load at *C*, the incremental *BM* $dM = m.dW$.

The change in slope of the tangent at *X-X* due to *dW* at *C* is given by:

$$EI\frac{d\theta}{dx} = dM = mdW$$

or $$d\theta = \frac{mdW}{EI}\cdot dx \qquad\qquad \text{... (2)}$$

Work stored in the elementary strip *dx* by substituting $d\theta = \dfrac{mdW}{EI}\cdot dx$

$$= \frac{1}{2}(M + dM)\cdot d0 = \frac{1}{2}(M + mdW)\frac{mdW}{EI}dx.$$

Work stored in the whole beam due to load *dW* at *C* will be

$$\frac{1}{2EI}\left[\int_0^L M\cdot m\cdot dW\cdot dx + \int_0^L m^2\cdot d_w^2\cdot dx\right] = \frac{1}{2EI}\int_0^L M\cdot m\cdot dW\cdot dx, \qquad\qquad \text{... (3)}$$

neglecting small quatities of higher order.

Equating the work done by load *dW* to the energy stored in the beam due to the load *dW*, we get

$$\frac{1}{2}dW\cdot y_C = \frac{1}{2EI}\int_0^L M\cdot m\cdot dW\cdot dx, \text{ from equation (1) and (3)}$$

or $$y_C = \frac{1}{EI}\int_0^L M\cdot m\cdot dx = \int_0^L \frac{M\cdot m}{EI}\cdot dx \qquad\qquad \text{... (10.62)}$$

Where *M* is *BM* at any point *X-X* as function of *x*, and *m* is *BM* at *X-X* due to unit load at *C* where the deflection is required to be found.

These steps can be summarized for finding deflection at any point *C* by applying a *unit load at C* as follows:

(i) Find *BM* M_x at any section *X-X* at a distance *x* from the LHS *A* in terms of *x*.

(ii) Apply a *unit load* at the *point C* under consideration and find the *BM* 'm_x' at the section *X-X* in terms of *x* due to unit load at *C*.

(iii) Integrate $\int_0^L \dfrac{M_x\cdot m_x\cdot d_x}{EI}$ for finding the deflection y_C at *C*. $y_C = \int_0^L \dfrac{M_x\cdot m_x}{EI}\cdot dx$

The procedure will be illustrated by examples.

EXAMPLE 10.43: Prove that deflection at the free end *A* of a cantilever *AB* of span *L* due to application of a point load *W* at a point *C* at a distance of L_1 from the fixed end *B* will be

$$y_A = \frac{WL_1^3}{3EI} + \frac{WL_1^2}{2EI}(L - L_1).$$

Solution:

Load at $C = W$, Imaginary Load at $A = 1$.

Span $AB = L$, $BC = L_1$, $AC = (L - L_1)$

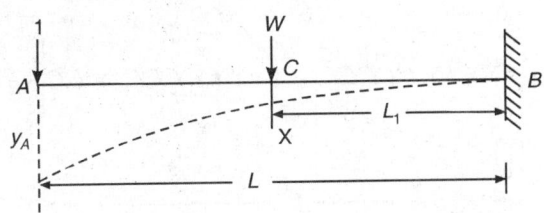

Fig. 10.67.

$AC:$ $M = 0,$ $m = -1 . x = -x$

$CB:$ $M = -W \{x - (L - L_1)\} = -W . x + W(L - L_1)$

         $m = -1(x)$

For AC: $\displaystyle\int_0^{(L-L_1)} \frac{M \cdot m \cdot dx}{EI} = \int_0^{(L-L_1)} \frac{0 \cdot (-x) \cdot dx}{EI} = 0$

For CB: $\displaystyle\int_{(L-L_1)}^{L} \frac{M \cdot m \cdot dx}{EI} = \int_{(L-L_1)}^{L} -W \frac{\{x - (L - L_1)\}\{-x\}}{EI} dx = \int_{(L-L_1)}^{L} +W \frac{[x^2 - x(L - L_1)]}{EI} dx$

$$y_A = \frac{W}{EI}\left[\frac{x^3}{3} - \frac{x^2}{2}(L - L_1)\right]_{(L-L_1)}^{L}$$

$$= \frac{W}{EI}\left[\left\{\frac{L^3}{3} - \frac{L^2}{2}(L - L_1)\right\} - \left\{\frac{(L-L_1)^3}{3} - \frac{(L-L_1)^2}{2}(L - L_1)\right\}\right]$$

or $\displaystyle y_A = \left\{\frac{WL^3}{3EI} - \frac{WL^2(L - L_1)}{2EI}\right\} - \frac{W}{6EI}\{2(L - L_1)^3 - 3(L - L_1)^3\}$

$$= \frac{WL^3}{3EI} - \frac{WL^2(L - L_1)}{2EI} - \frac{W}{6EI}\{-(L - L_1)^3\}$$

or $\displaystyle y_A = \frac{WL^2}{6EI}[2L - 3(L - L_1)] + \frac{W}{6EI}(L - L_1)^3$

$$= \frac{W}{6EI}[2L^3 - 3L^3 + +3L^2 L_1 + L^3 - 3L^2 L_1 + 3LL_1^2 - L_1^3]$$

or $\displaystyle y_A = \frac{W}{6EI}[3L \cdot L_1^2 - 3L_1^3 + 2L_1^3] = \frac{W}{6EI}[3L_1^2(L - L_1) + 2L_1^3] = \frac{WL_1^2(L - L_1)}{2EI} + \frac{WL_1^3}{3EI}$

Same as derived by other methods. *Hence Proved*

EXAMPLE 10.44: A cantilever AB of span L and carries an udl of w/m over the entire span. Calculate deflections at the free end A and a point C at a distance of L_1, from the free end. EI is constant throughout.

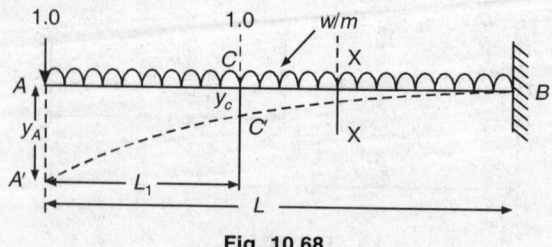

Fig. 10.68.

Solution:

Span $AB = L$, A free end, $AC = L_1$

$AA' = y_A$, $CC' = y_C$, EI constant

(i)   Deflection at the free end A :

Unit Load at A : $M_x = -\dfrac{wx^2}{2}$, $m_x = -1 \cdot x = -x$

$$y_A = \int_0^L \frac{M_x \cdot m}{EI} \cdot dx = \int_0^L \frac{wx^3}{2EI} \cdot dx = \frac{w}{2EI}\left[\frac{x^4}{4}\right]_0^L$$

$$y_A = \left(\frac{w \cdot x^4}{8EI}\right)_0^L = +\frac{wL^4}{8EI}, \text{ (same as found from other approaches)}$$

(ii)   Deflection at any point C at a distance of L_1 from the free end.
       Apply an imaginary unit load at C.

AC (0 to L_1): $M_x = -\dfrac{wx^2}{2}$, $m = 0$

C to B (L_1 to L): $M_x = -\dfrac{wx^2}{2}$, $m = -1\ (x - L_1)$

$$y_C = \int_0^{L_1} \frac{M_x \cdot m_x \cdot dx}{EI} + \int_{L_1}^L \frac{M_x \cdot m_x \cdot dx}{EI} = \int_0^{L_1} -\frac{wx^2}{2EI}(0) \cdot dx + \int_{L_1}^L +\frac{wx^2}{2EI}(x - L_1) \cdot dx$$

or $$y_C = 0 + \int_{L_1}^L \left\{+\frac{wx^3}{2EI} - \frac{wx^2 L_1}{2EI}\right\} \cdot dx = \left[\frac{w \cdot x^4}{8EI} - \frac{wx^3 \cdot L_1}{6EI}\right]_{L_1}^L$$

$$= \frac{w}{48EI}[(6L^4 - 8L_1 L^3) - (6L_1^4 - 8L_1^4)]$$

or $y_C = \dfrac{w}{48EI}[6L^4 - 8L_1L^3 - 6L_1^4 + 8L_1^4] = \dfrac{w}{48EI}[6L^4 - 8L_1L^3 + 2L_1^4]$

$$= \dfrac{w}{24EI}[3L^4 - 4L_1L^3 + L_1^4]$$

Thus y_C (deflection at C) = $\dfrac{w}{24EI}[3L^4 - 4L_1L^3 + L_1^4]$

EXAMPLE 10.45: Find the central deflection by unit load method in a simply supported beam AB of 10 m span and loaded with a point load of 20 kN at a point 6 m from the *LH* support A. EI is constant = 30000 kN-m^2.

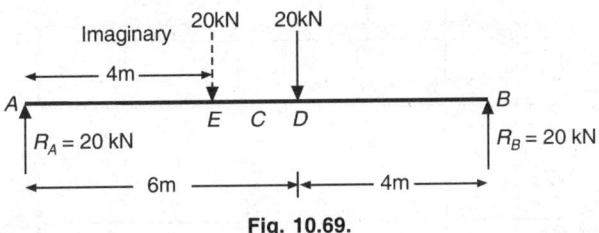

Fig. 10.69.

Solution:

$AB = 10$ m, $AD = 6$ m, $DB = 4$ m, $EI = 30000$ kN-m^2

For simplicity, a load of 20 kN is applied at 6 m from the RHS (at E) and this makes the central deflection double due to symmetrical placement of loads. Apply an unit load at the centre point C.

A to E (0 to 4 m): M_x (due to external loads) = + **20x**, m (unit load at C) = + $\dfrac{1}{2} \cdot x$

E to C (4 to 5 m): $M_x = 20x - 20(x - 4)$, $m = + \dfrac{1}{2} \cdot x$

Deflection: y_C'

$$y_C' = \int_0^L \dfrac{M_x \cdot m_x \cdot dx}{EI} = \dfrac{2}{EI} \int_0^{L/2} M_x \cdot m_x dx \text{ due to symmetry}$$

or $y_C' = \dfrac{2}{EI}\left[\int_0^4 M_x \cdot m_x dx + \int_4^5 M_x \cdot m_x dx\right]$

$x = 0$ to 4m: $\int_0^4 M_x \cdot m_x dx = \int_0^4 20 \cdot x \cdot \dfrac{x}{2} dx = \left[\dfrac{10x^3}{3}\right]_0^4 = \dfrac{10}{3}(4^3) = \dfrac{640}{3}$ ... (1)

$x = 4$ to 5m: $\int_4^5 M_x \cdot m_x dx = \int_4^5 \{20x - 20(x - 4)\}\dfrac{x}{2} dx = \int_4^5 \dfrac{80x}{2} dx = \left[\dfrac{40x^2}{2}\right]_4^5$

$$\therefore \quad \underset{(2loads)}{y'_C} = \frac{2}{EI}\left[\frac{640}{3}+180\right] = \frac{2\times1180}{3\times30000} = 0.02622\text{m (26.22mm)}$$

$$\underset{(\text{single load})}{y'_C} = \frac{y'_C}{2} = \frac{0.02622}{2} = \textbf{0.01311 m (13.11 mm)}$$

EXAMPLE 10.46: A simply supported beam AB of span L carries udl of w/m on the entire length. Find the slope at the end A and deflection at the centre C, if EI is constant throughout by using unit load (moment) method.

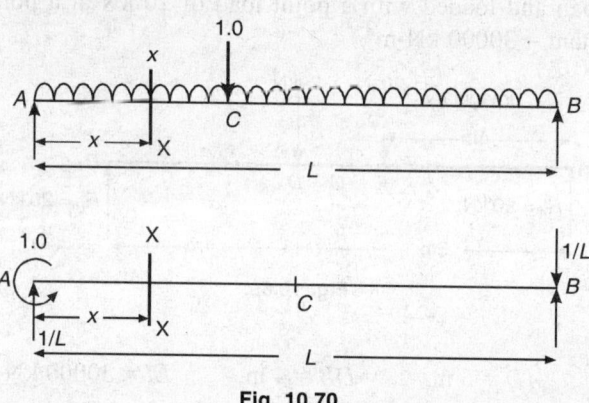

Fig. 10.70.

Solution:

$$AB = L, AC = \frac{L}{2}, EI = \text{constant}$$

$$y_C = \frac{1}{EI}\int_0^L M_x \cdot m_x \cdot dx = \frac{2}{EI}\int_0^{L/2} M_x m_x dx, \text{ due to symmetry}$$

A to C: $M_x = \left(\frac{wL}{2}\cdot x - \frac{wx^2}{2}\right)$, $m = \frac{x}{2}$

$$\int_0^{L/2} M_x \cdot m_x dx = \int_0^{L/2}\left(\frac{wL}{2}\cdot x - \frac{w}{2}\cdot x^2\right)\cdot\frac{x}{2}\cdot dx$$

or $$\int_0^{L/2} M_x \cdot m_x dx = \frac{wL}{4}\cdot\left(\frac{x^3}{3}\right)_0^{L/2} - \frac{w}{4}\left(\frac{x^4}{4}\right)_0^{L/2} = \frac{(8-3)wL^4}{3\times16\times16} = \frac{5}{768}wL^4$$

$$\therefore \quad y_C = \frac{2}{EI}\int_0^{L/2} M_x \cdot m_x \cdot dx = \frac{2}{EI}\times\frac{5}{768}wL^4$$

$$= \frac{5wL^4}{384EI} \text{ (same as found earlier by double integration)}$$

Apply unit moment couple at A and reactions at $A = \dfrac{1}{L}\uparrow$.

At any point X: $M_x = \left(\dfrac{wL}{2}\cdot x - \dfrac{wx^2}{2}\right)$, $m_x = \left(1 - \dfrac{x}{L}\right)$

Slope at $A = \dfrac{1}{EI}\int_0^L Mx\cdot mx\cdot dx = \dfrac{1}{EI}\int_0^L\left(\dfrac{wLx}{2} - \dfrac{wx^2}{2}\right)\left(1 - \dfrac{x}{L}\right)dx = \dfrac{w}{2EI}\int_0^L\left[(Lx - x^2)\left(\dfrac{L-x}{L}\right)\right]dx$

$= \dfrac{w}{2EI}\int_0^L\dfrac{x}{L}(L^2 - 2Lx + x^2)dx = \dfrac{w}{2EIL}\left[\dfrac{L^2\cdot x^2}{2} - 2\dfrac{Lx^3}{3} + \dfrac{x^4}{4}\right]_0^L$

$= \dfrac{w}{2EIL}\left[\dfrac{L^4}{2} - \dfrac{2L^4}{3} + \dfrac{L^4}{4}\right] = \dfrac{wL^3}{24EI}[6 - 8 + 3] = \dfrac{wL^3}{24EI}$, same as found earlier.

10.8   DEFLECTION BY FICTITIOUS LOAD METHOD (CASTIGLIANO'S FIRST THEOREM)

Deflections at various points under a load in a structural element can be studied using the concept of *elastic strain energy*. In previous examples the *external work done* by a load was equated directly to the strain energy in the structural element. *Castigliano* has simplified the calculation of *deflection* by introducing the concept of *partial derivative of the strain energy with respect to that of force* where deflection is required. The general statement of *Castigliano's First Theorem* is as under:

"If any structural elastic system is in equilibrium under the action of a set of forces W_1, W_2, W_3, ..., W_n, and undergo the corresponding displacements (deflections and slopes) δ_1, δ_2, δ_3, ..., δ_n and a set of moments M_1, M_2, M_3, ..., M_n, and undergo the corresponding rotations θ_1, θ_2, θ_3, ..., θ_n, then the partial derivative of the total strain energy 'U' with respect to any one of the forces or moments, taken individually would provide its corresponding displacement in its direction of action."

Mathematically : $\dfrac{\partial U}{\partial W_1} = \delta_1$ and $\dfrac{\partial U}{\partial M_1} = \theta_1$ ... (10.63)

This theorem can be used for finding the deflection and slope at any point of the beam. Consider a beam subjected to loads W_1, W_2, ..., W_n, etc and it is required to find deflection under the load W_n.

The strain energy of bending is given by:

$$U = \int_0^L \dfrac{M_x^2 dx}{2EI} \qquad\qquad ... (10.64)$$

(where M is bending moment expressed in terms of W_n)

By definition (*Castigliano's Theorem*):

$$\delta_1 = \frac{\partial U}{\partial W_1} = \int_0^L M \left(\frac{\partial U}{\partial W_1} \right) \frac{dx}{EI} \qquad \qquad \dots (10.65)$$

In this expression, $\dfrac{\partial U}{\partial W_1}$ is evaluated by *differenting inside the integral sign, before integrating.* If there is no load acting at a point where deflection is required, a *fictitious load* W may be applied at the point, in the required direction of the deflection. After finding partial differential coefficient $\dfrac{\partial M}{\partial W}$, the value of the fictitious load is *set to zero* before undertaking integration.

i.e. $\delta_1 = \int_0^L M \dfrac{\partial M}{\partial W_1} \dfrac{dx}{EI}$

Here fictitious load W_1 shall be applied at the points where deflection is required. This method is also known as *fictitious load method*. If rotation of the beam axis is required, the equation (10.65) may be modified as:

$$\theta_1 = \frac{\partial U}{\partial M_1} \int_0^L M \left(\frac{\partial M}{\partial M_1} \right) \frac{dx}{EI} \qquad \qquad \dots (10.66)$$

Here the BM M_x is function of fictitious moment M_1 applied at the point where slope or rotation is required to be found. The *fictitious load* method will be illustrated through examples.

EXAMPLE 10.47: Determine slope at ends and deflection at the middle point of a simply supported beam AB of a span L (6 m) carrying udl w/m (10 kN/m) over the entire span by using Castigliano's first theorem. EI (4500 kN-m^2) is constant throughout.

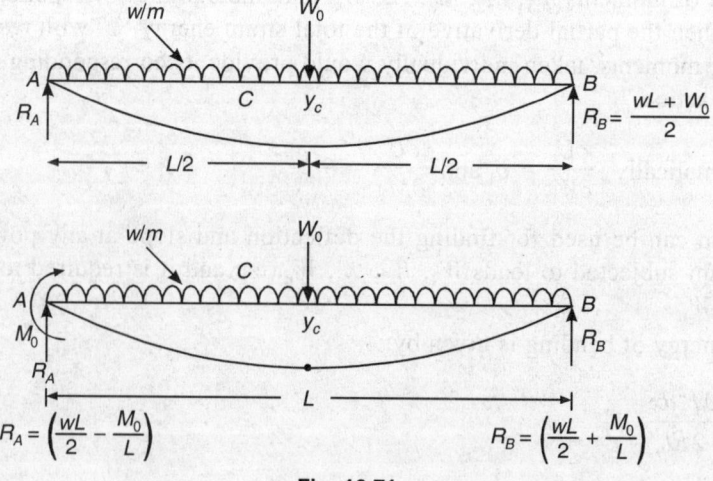

Fig. 10.71.

Solution:

$AB = L = 6$ m, $\qquad$ $AC = CB = 3$ m, $\qquad$ $EI = 4500$ kN-m^2

$w = 10$ kN/m, $\qquad$ Fictitious load W_0 at C, $\qquad$ Fictitious moment M_0 at A and B

(i) **Deflection at C**

$$R_A = \left(\frac{wL}{2} + \frac{W_0}{2} \right), \quad M_x = \left(\frac{wL}{2} + \frac{W_0}{2} \right) x - \frac{w \cdot x^2}{2}, \frac{\partial M}{\partial W_0} = \frac{x}{2}$$

$$y_C = \frac{1}{EI} \int_0^L M_x \cdot \frac{\partial M_x}{\partial W_0} \cdot dx = \frac{2}{EI} \int_0^{L/2} M_x \cdot \frac{\partial M_x}{\partial W_0} \cdot dx , \text{ by symmetry}$$

or $\quad$ $$y_C = \frac{2}{EI} \int_0^{L/2} \left\{ \left(\frac{wL}{2} + \frac{W_0}{2} \right) x - \frac{wx^2}{2} \right\} \frac{x}{2} \cdot dx$$

Putting $W_0 = 0$, before integration

$$y_C = \frac{2}{EI} \int_0^{L/2} \left\{ \frac{wL}{2} \cdot x - \frac{wx^2}{2} \right\} \frac{x}{2} \cdot dx = \frac{2}{EI} \left[\frac{wL}{4} \cdot \frac{x^3}{3} - \frac{w}{4} \cdot \frac{x^4}{4} \right]_0^{L/2}$$

$$= \frac{2}{EI} \cdot \frac{w}{4} \left[\frac{L}{3} \cdot \frac{L^3}{8} - \frac{1}{4} \cdot \frac{L^4}{16} \right] = \frac{2wL^4}{4EI} \left[\frac{8-3}{3 \times 64} \right] = \frac{5wL^4}{384EI} \text{ (same as found earlier)}$$

$\therefore \quad$ $$y_C = \frac{5}{384} \cdot \frac{10(6)^4}{4500} = \mathbf{0.0375} \text{ m } \mathbf{(37.5 \text{ mm})}$$

(ii) **Slope at A**

Apply a moment M_0 at the end A to find slope at A. Reactions at A and B will

be $\left(\dfrac{wL}{2} - \dfrac{M_0}{L} \right)$ and $\left(\dfrac{wL}{2} + \dfrac{M_0}{L} \right)$ respectively.

$$M_x \text{ (x from A)} = \left(\frac{wL}{2} - \frac{M_0}{L} \right) x - \frac{wx^2}{2} + M_0,$$

$$M_x \text{ (x from B)} = \left(\frac{wL}{2} + \frac{M_0}{L} \right) x - \frac{wx^2}{2} \qquad \qquad \text{... (i)}$$

$$U = \int_0^L \frac{M^2 \cdot dx}{2EI}, \quad \theta_A = \frac{\partial U}{\partial M_0} = \int_0^L M \frac{\partial M}{\partial M_0} \frac{dx}{EI}, \quad \frac{\partial M_x}{\partial M_0} = \left(-\frac{x}{L} + 1 \right)$$

$$\theta_A = \int_0^L \left[\left(\frac{wL}{2} - \frac{M_0}{L} \right) x - \frac{wx^2}{2} + M_0 \right] \left(1 - \frac{x}{L} \right) \frac{dx}{EI}$$

or $\quad \theta_A = \int_0^L \left(\frac{wL}{2} \cdot x - \frac{wx^2}{2} \right) \left(1 - \frac{x}{L} \right) \frac{dx}{EI} = \frac{w}{2EI} \int_0^L \left(Lx - x^2 - x^2 + \frac{x^3}{L} \right) dx$

or $\quad \theta_A = \frac{w}{2EI} \left[L\frac{x^2}{2} - \frac{2x^3}{3} + \frac{x^4}{4L} \right]_0^L = \frac{wL^3}{2EI} \left[\frac{1}{2} - \frac{2}{3} + \frac{1}{4} \right] = \frac{wL^3}{24EI} (6 - 8 + 3) = \frac{wL^3}{24EI}$

Similary $\quad \theta_B = -\frac{wL^3}{24EI}, \quad \theta_A = \frac{216(10)}{24 \times 4500} = 0.020$ redians

EXAMPLE 10.48: Determine slope and deflection at the free end A of a cantilever beam AB ($L = 3$ m) carrying a point load of 10 kN at 2 m from the fixed end B. EI is constant (4000 kN-m²).

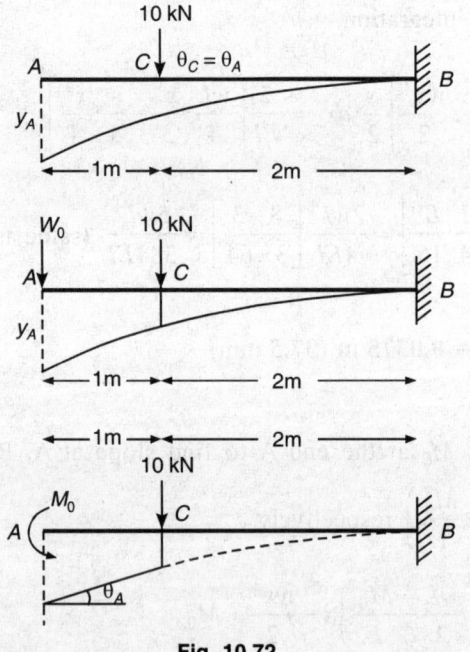

Fig. 10.72.

Solution:

$AB = L = 3$ m, $\qquad CB = L_1 = 2$ m, $\qquad\qquad W = 10$ kN

$EI = 4000$ kN-m² $\qquad$ Fictitious moment M_0 at A

i. **Deflection at A (Fictitious load W_0 at A)**

A to C : M_x (x from A) $= - (0 + W_0 \cdot x)$

C to B : M_x (x from A) $= - 10 (x - 1) - W_0 \cdot x$

$$\frac{\partial M}{\partial W_0} = -x$$

$$y_A = \frac{1}{EI} \int_0^L M_x \cdot \frac{\partial M_x}{\partial W_0} \cdot dx = \frac{1}{EI} \left[\int_0^1 M_x \cdot \frac{\partial M}{\partial W_0} \cdot dx + \int_1^3 M_x \cdot \frac{\partial M}{\partial W_0} \cdot dx \right]$$

$$= \frac{1}{EI} \int_0^1 -(W_0 \cdot x)(-x)dx + \frac{1}{EI} \int_1^3 \{-10(x-1) - W_0 \cdot x\}(-x)dx$$

Putting $W_0 = 0$

$$y_A = \frac{1}{EI} \int_0^1 0.dx + \frac{1}{EI} \int_1^3 10(x-1)xdx = 0 + \frac{1}{EI} \left[\frac{10x^3}{3} - \frac{10x^2}{2} \right]_1^3$$

$$= \frac{1}{EI} \left[\left\{ \frac{10}{3}(3^3 - 1^3) \right\} - \frac{10}{2}(3^2 - 1^2) \right] = \frac{1}{EI} \left[\frac{260}{3} - \frac{80}{2} \right]$$

$$= \frac{280}{6EI} = \frac{280}{6 \times 4000} = \textbf{0.01167 m (11.67 mm)}$$

ii. **Slope at A (fictitious moment M_0 at A)**

A to C : $M_x = -(M_0 + 0), \dfrac{\partial M_x}{\partial M_0} = (-1)$

C to B : $M_x = -[M_0 + 10 (x-1), \dfrac{\partial M_x}{\partial M_0} = (-1)$

$$\theta_A = \int_0^3 M \frac{\partial M}{\partial M_0} \cdot \frac{dx}{EI} = \int_0^1 M \frac{\partial M}{\partial M_0} \cdot \frac{dx}{EI} + \int_1^3 M \frac{\partial M}{\partial M_0} \cdot \frac{dx}{EI}$$

$$= \int_0^1 -(M_0)(-1) \frac{dx}{EI} + \int_1^3 -[M_0 + 10(x-1)(-1)] \frac{dx}{EI}$$

Putting $M_0 = 0$

$$\theta_A = \int_0^1 0(-1) \frac{dx}{EI} + \int_1^3 10(x-1)(-1) \frac{dx}{EI} = \int_1^3 (-10x + 10) \frac{dx}{EI}$$

$$= \frac{1}{EI} \left[-\frac{10x^2}{2} + 10x \right]_1^3 = \frac{1}{EI} [-5(3^2 - 1^2) + 10(3 - 1)]$$

$$= \left[\frac{-40 + 20}{EI} \right] = -\frac{20}{EI} = -\frac{20}{4000} = \textbf{- 0.005 radian} \text{ (anticlockwise } -ve)$$

EXAMPLE 10.49: An overhanging beam ABC of 8 m length is supported at A and B ($AB = 6$ m). The beam carries a moment M (20 kN–m) at the overhanging end C. Find the slope and deflection of the overhanging end C by using Castigliano's theorem. EI is constant. $E = 200$ kN/mm^2, $I = 4000$ cm^4.

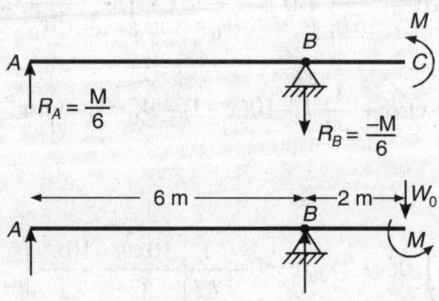

Fig. 10.73.

Solution:

$$AB = 6 \text{ m}, \qquad BC = 2 \text{ m}, \qquad ABC = 8 \text{ m}, \qquad M = 20 \text{ kN-m}$$
$$EI = 200 \times 10^6 \times 4000 \times 10^{-8} = \mathbf{8000 \text{ kN-m}^2}$$

(a)   Rotation at C

The reactions at A and B : $R_A = \dfrac{M}{6}, R_B = -\dfrac{M}{6}(\downarrow)$

$$\theta_C = \frac{\partial U}{\partial M} = \frac{1}{EI}\int_A^B M_x \cdot \frac{\partial M_x}{\partial M}dx + \frac{1}{EI}\int_B^C M_x \cdot \frac{\partial M_x}{\partial M}dx \qquad \text{... (i)}$$

(A to B) x from A

$$M_x = \frac{M}{6} \cdot x, \quad \frac{\partial M_x}{\partial M} = \frac{x}{6} \qquad \text{... (ii)}$$

(B to C) x from A

$$M_x = \frac{M}{6} \cdot x - \frac{M}{6}(x - 6) = \frac{M}{6} \cdot x - \frac{M}{6} \cdot x + M = M, \frac{\partial M_x}{\partial M} = 1$$

$$\therefore \quad \theta_C = \frac{1}{EI}\int_0^6 \frac{M}{6} \cdot x \frac{\partial M_x}{\partial M}dx + \frac{1}{EI}\int_6^8 M_x \frac{\partial M_x}{\partial M}dx = \frac{1}{EI}\left[\int_0^6 \frac{M}{6} \cdot \frac{x^2 \cdot dx}{6} + \frac{1}{EI}\int_6^8 M(1)dx\right]$$

or $\quad \theta_C = \dfrac{1}{EI}\left[\left(\dfrac{M}{36} \cdot \dfrac{x^3}{3}\right)_0^6 + (M \cdot x)_6^8\right] = \dfrac{1}{EI}\left[\dfrac{M}{36} \cdot \dfrac{216}{3} + M(8-6)\right] = \dfrac{1}{EI}[2M + 2M]$

or $\quad \theta_C = \dfrac{4M}{EI} = \dfrac{4M}{8000} = \dfrac{M}{2000} \text{ radian} = \dfrac{20}{2000} = \dfrac{1}{100} = \mathbf{0.010 \text{ radian}}$

(b) Deflection at C

Apply a fictitious load W_0 at C in upward direction.

$$R_A = \left(\frac{M + W_0 \cdot 2}{6}\right)\uparrow, \, R_B = \left(\frac{M + W_0 \cdot 8}{6}\right)\downarrow$$

$$y_C = \frac{\partial U}{\partial W_0} = \frac{1}{EI}\int_A^B M_x \cdot \frac{\partial M_x}{\partial W_0}dx + \frac{1}{EI}\int_B^C M_x \cdot \frac{\partial M_x}{\partial W_0}dx$$

For AB (x from A) : $M_x = \left(\frac{M + 2W_0}{6}\right)x, \, \frac{\partial M_x}{\partial W_0} = \frac{x}{3}$

For CB (x from B) : $M_x = M + W_0 \cdot (2 - x), \, \dfrac{\partial M_x}{\partial W_0} = (2 - x)$, x from B for BC for

simplicity (0 to 2)

$$y_C \atop (w_0=0) = \frac{1}{EI}\left[\int_0^6 \left(\frac{M + 2W_0}{6}\right)\cdot x\frac{x}{3}dx + \int_0^2 \{M + W_0(2 - x)\}(2 - x)dx\right]$$

$$= \frac{1}{EI}\left[\left(\frac{M}{18}\times\frac{x^3}{3}\right)_0^6 + \left\{\frac{M(2 - x)^2}{-2}\right\}_0^2\right]$$

$$y_C = \frac{1}{EI}[4M - (-2M)] = \frac{6M}{EI} = \frac{6M}{8000} = 0.00075 \times 20 = \mathbf{0.0150m \, (15.0mm)}$$

EXAMPLE 10.50: A basket ball rigid frame AB of height 'L' and horizontal rod BC span $L/2$ carries a basket board of vertical weight W. Neglecting axial deformations, find the horizontal and vertical displacements of the board point C as shown in Fig. 10.74. EI is constant throughout.

Solution:

(i) Vertical Deflection

$$y_C \text{ (Vertical)} = \frac{\partial U}{\partial W} = \frac{1}{EI}\int M_x \cdot \frac{\partial M_x}{\partial W} \cdot dx$$

For CB (0 to $L/2$) : x from C $M_x = -W \cdot x, \dfrac{\partial M_x}{\partial W} = -x$ ... (i)

$AB = L$, $BC = L/2$, Load at $C = W$, EI = Constant

For BA (0 to L) : x from B $M_x = -\dfrac{WL}{2}$ (constant), $\dfrac{\partial M_x}{\partial W} = -\dfrac{L}{2}$

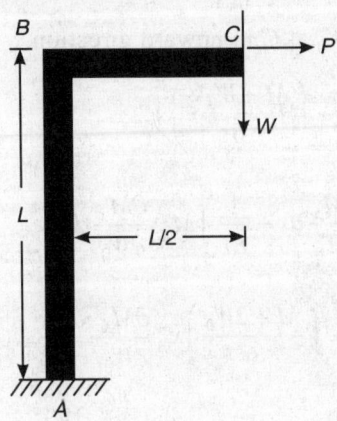

Fig. 10.74.

$$\therefore \quad y_{CV} = \frac{1}{EI}\left[\int_0^{L/2} + W \cdot x^2 \cdot dx + \int_0^L \frac{WL}{2} \cdot \frac{L}{2} \cdot dx\right] \qquad \text{... (ii)}$$

$$y_{CV} = \frac{1}{EI}\left[\left(\frac{W \cdot x^3}{3}\right)_0^{L/2} + \left(\frac{WL^2}{4} \cdot x\right)_0^L\right] = \frac{1}{EI}\left[\frac{W \cdot L^3}{24} + \frac{WL^3}{4}\right] = \frac{7WL^3}{24EI} \text{ (downward)}$$

(ii) Hrizontal displacement of C : Apply a fictitious horizontal load P at C. Horizontal displacement $y_{CH} = \dfrac{\partial U}{\partial P} = \dfrac{1}{EI}\int M_x \cdot \dfrac{\partial M_x}{\partial P} \cdot dx$ for the whole frame ABC.

For CB (0 to $L/2$) :　　　x from C, $M_x = -W \cdot x$,　　　$\dfrac{\partial M_x}{\partial P} = 0$

For BA (0 to L) :　　　x from B, $M_x = -\dfrac{WL}{2} - P \cdot x$,　　　$\dfrac{\partial M_x}{\partial P} = -x$

Thus $y_{CH} = \dfrac{1}{EI}\left[\int_0^{L/2}(-W \cdot x)(0)\cdot dx + \int_0^L\left(-\dfrac{WL}{2} - P \cdot x\right)(-x)\cdot dx\right]$

or $\quad y_{CH} = \dfrac{1}{EI}\left[0 + \int_0^L\left(\dfrac{WL}{2} + 0\right)(x)dx\right] = \dfrac{1}{EI}\left[\dfrac{WL}{2} \cdot \dfrac{x^2}{2}\right]_0^L = \dfrac{WL^3}{4EI} \text{ (outward} \rightarrow)$

EXAMPLE 10.51: A cantilever AB of span L carries uniformly varying load with maximum intensity w/m at the fixed end B and zero at the free end A. By using Castigliano's theorem determine the slope and deflection at the free end A. EI is constant throughout.

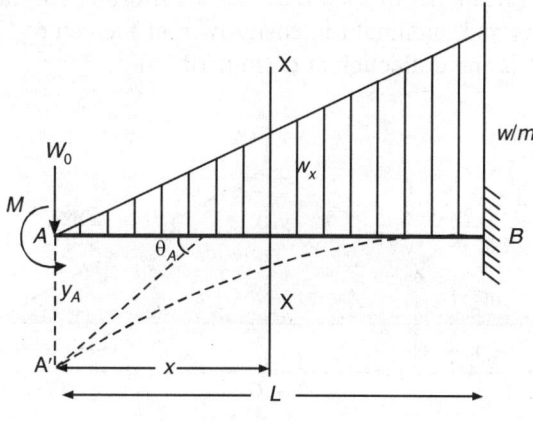

Fig. 10.75.

Solution:

$$AB = L, \qquad AX = x, \qquad w_x = \frac{w}{L} \cdot x, \qquad EI = \text{constant}$$

(i) Deflection at A:

Apply a fictitious load W_0 at A.

$$M_x = -\left[W_0 \cdot x + \frac{w}{L} \cdot x \cdot \frac{x}{2} \cdot \frac{x}{3} \right] = -\left[W_0 \cdot x + \frac{w}{6L} \cdot x^3 \right], \quad \frac{\partial M_x}{\partial W_0} = -x$$

$$y_A = \int_0^L M_x \cdot \frac{\partial M_x}{\partial W_0} \cdot \frac{dx}{EI} = \int_0^L -\left[W_0 \cdot x + \frac{w}{6} \cdot \frac{x^3}{L} \right](-x) \frac{dx}{EI}$$

Put $W_0 = 0$, before integation :

$$y_A = \int_0^L +\left(\frac{wx^4}{6L} \right) \frac{dx}{EI} = \frac{w}{6EIL} \left(\frac{x^5}{5} \right)_0^L = \frac{wL^4}{30EI} \qquad\qquad \text{... (10.67)}$$

(ii) Slope at A:

Apply a fictitious moment M_0 (anticlockwise) at the free end A.

$$M_x = -\left[M_0 + \frac{wx^3}{6L} \right], \frac{\partial M_x}{\partial M_0} = (-1)$$

$$\theta_A = \int_0^L M_x \cdot \frac{\partial M_x}{\partial M_0} \cdot \frac{dx}{EI} = \int_0^L -\left[M_0 + \frac{wx^3}{6L} \right](-1) \frac{dx}{EI}$$

Put $M_0 = 0$ before integration :

$$\theta_A = \int_0^L +\frac{wx^3}{6L} \cdot \frac{dx}{EI} = \left[\frac{wx^4}{24EIL} \right]_0^L = \mathbf{\frac{wL^3}{24EI}} \qquad\qquad \text{... (10.68)}$$

EXAMPLE 10.52: A SS beam AB of span L carries a uniformly varying (triangular) load with zero intensity at supports and maximum intensity w/m at the centre C. If EI is constant, find the slopes at ends A and B and deflection at C (Fig. 10.76).

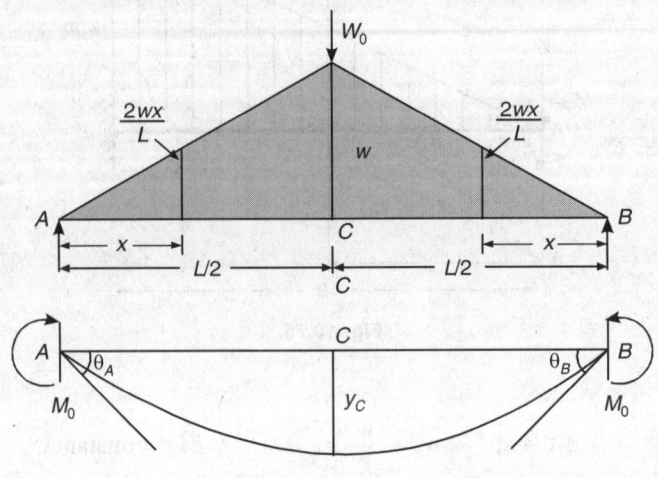

Fig. 10.76.

Solution:

$$AB = L, \qquad AC = L/2 = CB, \qquad w_C = w, \qquad w_x = \frac{2w}{L}\cdot x, \qquad x \text{ from ends}$$

$$EI = \text{constant}$$

(i) Slope at A:

Apply fictitious moment M_0 (clockwise)

Reactions: $R_A = \left(\dfrac{wL}{4} - \dfrac{M_0}{L}\right),\ R_B = \left(\dfrac{wL}{4} + \dfrac{M_0}{L}\right)$

A to C (x from A):

$$M_x = \left(\frac{wL}{4} - \frac{M_0}{L}\right)x + M_0 - \frac{w_x \cdot x^2}{6} = \left\{\frac{wL}{4}x + M_0\left(1 - \frac{x}{L}\right) - \frac{wx^3}{3L}\right\}, \frac{\partial M_x}{\partial M_0} = \left(1 - \frac{x}{L}\right)$$

B to C (x from B):

$$M_x = \left\{\frac{wL}{4}x + \frac{M_0}{L} - \frac{wx^3}{3L}\right\}, \frac{\partial M_x}{\partial M_0} = \frac{x}{L}$$

$$\theta_A = \frac{1}{EI}\int_A^C M_x \cdot \frac{\partial M_x}{\partial M_0}\cdot dx + \frac{1}{EI}\int_B^C M_x \cdot \frac{\partial M_x}{\partial M_0}\cdot dx = \int_0^{L/2} M_x \cdot \frac{\partial M_x}{\partial M_0}\cdot \frac{dx}{EI} + \int_0^{L/2} M_x \cdot \frac{\partial M_x}{\partial M_0}\cdot \frac{dx}{EI}$$

$$= \frac{1}{EI}\left[\int_0^{L/2}\left\{\frac{wL}{4}x + M_0\left(1-\frac{x}{L}\right) - \frac{wx^3}{3L}\right\}\left(1-\frac{x}{L}\right)dx + \int_0^{L/2}\left\{\frac{wL}{4}x + M_0\frac{x}{L} - \frac{wx^3}{3L}\right\}\left(\frac{x}{L}\right)dx\right]$$

Putting $M_0 = 0$ before integration

$$\theta_A = \frac{1}{EI}\left[\int_0^{L/2}\left\{\frac{wL}{4}x + 0 - \frac{wx^3}{3L}\right\}\left(1-\frac{x}{L}\right)dx + \int_0^{L/2}\left\{\frac{wL}{4}x + 0 - \frac{wx^3}{3L}\right\}\left(\frac{x}{L}\right)dx\right]$$

$$= \frac{1}{EI}\left[\left\{\frac{wL}{4}\cdot\frac{x^2}{2} - \frac{w}{3L}\cdot\frac{x^4}{4} - \frac{w}{4}\cdot\frac{x^3}{3} + \frac{w}{3L^2}\cdot\frac{x^5}{5}\right\}_0^{L/2} + \left\{\frac{wL}{4L}\cdot\frac{x^3}{3} - \frac{w}{3L^2}\cdot\frac{x^5}{5}\right\}_0^{L/2}\right]$$

$$= \frac{wL^3}{EI}\left[\frac{1}{32} - \frac{1}{12\times16} - \frac{1}{12\times8} + \frac{1}{15\times32}\right] + \frac{wL^3}{EI}\left[\frac{1}{12\times8} - \frac{1}{15\times32}\right]$$

or $\quad \theta_A = \dfrac{wL^3}{EI}\left[\dfrac{6-1}{192}\right] = \dfrac{5wL^3}{192EI}$, Similarly $\theta_B = -\dfrac{5wL^3}{192EI}$ $\qquad$ ... (10.69)

(ii) **Deflection at C:[7]**

Apply fictitious load W_0 at C downward

$$R_A = \left(\frac{wL}{4} + \frac{W_0}{2}\right), \qquad R_B = \left(\frac{wL}{4} + \frac{W_0}{2}\right), \qquad w_x = \frac{2w}{L}\cdot x$$

$$M_x \atop {(x \text{ from } A)} = \left(\frac{wL}{4} + \frac{W_0}{2}\right)x - \frac{w_x\cdot x^2}{6} = \left(\frac{wL}{4}\cdot x + \frac{W_0\cdot x}{2} - \frac{wx^3}{3L}\right), \ \frac{\partial M}{\partial W_0} = \frac{x}{2}, \text{ Loading Symmetrical}$$

$$y_C = \frac{1}{EI}\int_0^L M_x\cdot\frac{\partial M_x}{\partial W_0}\cdot dx = \frac{2}{EI}\int_0^{L/2} M_x\cdot\frac{\partial M_x}{\partial W_0}\cdot dx, \text{ due to symmetry}$$

$$y_C = \frac{2}{EI}\int_0^{L/2}\left\{\frac{wL}{4}\cdot x + \frac{W_0}{2}\cdot x - \frac{w}{3L}\cdot x^3\right\}\left(\frac{x}{2}\right)dx = \frac{2}{EI}\left[\frac{wL}{8}\cdot\frac{x^3}{3} - \frac{w}{6L}\cdot\frac{x^5}{5}\right]_0^{L/2}$$

Putting $W_0 = 0$, before integration

or $\quad y_C = \dfrac{2}{EI}wL^4\left[\dfrac{1}{24}\cdot\dfrac{1}{8} - \dfrac{1}{30}\cdot\dfrac{1}{32}\right] = \dfrac{2wL^4}{EI}\left(\dfrac{5-1}{960}\right) = \dfrac{wL^4}{120EI}$

Maximum deflection at $C = y_{max} = \dfrac{wL^4}{120EI}$ $\qquad$ ... (10.70)

EXAMPLE 10.53: A SS beam AB of span L carries a uniformly varying *udl* with *zero* intensity at A and w/m intensity at the end B. If EI is constant, find the slopes at the supports and deflection at C (midspan point).

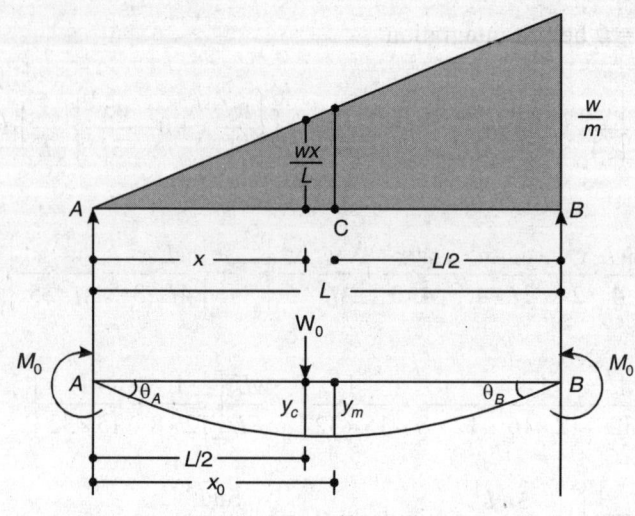

Fig. 10.77.

Solution:

$$AB = L, \ AC = \frac{L}{2}, \ x \text{ from } A, \ EI = \text{constant.}$$

$$w_A = 0, \ w_B = w, \ w_x = \frac{w}{L} \cdot x$$

(i)   Slope at A:

Apply a fictitious moment M_0 clockwise at A.

$$R_A = \frac{1}{L}\left[\frac{wL}{2} \cdot \frac{L}{3} - \frac{M_0}{L}\right], \ R_B = \frac{1}{L}\left[\frac{wL}{2} \cdot \frac{2L}{3} + \frac{M_0}{L}\right]$$

$$R_A = \left(\frac{wL}{6} - \frac{M_0}{L}\right), \left(\frac{wL}{3} + \frac{M_0}{L}\right)$$

$$M_x \ (x \text{ from } A) = \left(\frac{wL}{6} - \frac{M_0}{L}\right)x + M_0 - \frac{w}{L} \cdot x \cdot \frac{x}{2} \cdot \frac{x}{3} = \left(\frac{wL}{6} \cdot x + M_0 - M_0 \frac{x}{L} - \frac{w}{6} \frac{x^3}{L}\right)$$

$$\frac{\partial M_x}{\partial M_0} = \left(1 - \frac{x}{L}\right)$$

$$\theta_A = \frac{1}{EI} \int_0^L \left\{\frac{wL}{6} x + M_0\left(1 - \frac{x}{L}\right) - \frac{w}{6L}x^3\right\}\left(1 - \frac{x}{L}\right)dx$$

Put $M_0 = 0$, Prior to integration.

$$\theta_A = \frac{1}{EI}\int_0^L\left\{\frac{wL}{6}x - \frac{w\cdot x^3}{6L} - \frac{wL}{6L}x^2 + \frac{wx^4}{6L^2}\right\}dx = \frac{1}{EI}\left[\frac{wL}{6}\cdot\frac{x^2}{2} - \frac{w}{6L}\cdot 1\cdot\frac{x^4}{4} - \frac{w}{6}\frac{x^3}{3} + \frac{w}{6L^2}\cdot\frac{x^5}{5}\right]_0^L$$

or $\quad \theta_A = \dfrac{wL^3}{EI}\left[\dfrac{1}{12} - \dfrac{1}{24} - \dfrac{1}{18} + \dfrac{1}{30}\right] = \dfrac{wL^3}{360EI}(30-15-20+12)$

or $\quad \boldsymbol{\theta_A = \dfrac{7wL^3}{360EI}}$ (clockwise) $\hspace{4cm}$ ... (10.71)

(ii) Slope at B:

Apply a fictitious moment M_0 (anti clockwise at B.

$$R_A = \left(\frac{wL}{6} + \frac{M_0}{L}\right), R_B = \left(\frac{wL}{3} - \frac{M_0}{L}\right)$$

$$M_x \text{ (}x \text{ from } A\text{)} = \left\{\left(\frac{wL}{6}\cdot x + \frac{M_0}{L}\cdot x\right) - \frac{w}{L}\frac{x^3}{6}\right\}, \qquad \frac{\partial M_x}{\partial M_0} = +\left(\frac{x}{L}\right)$$

$$\theta_B = \frac{1}{EI}\int_0^L M_x\cdot\frac{\partial M_x}{\partial M_0}\cdot dx = \frac{1}{EI}\int_0^L\left\{\frac{wL}{6} + M_0\cdot\frac{x}{L} - \frac{w}{6L}\cdot x^3\right\}\left(\frac{x}{L}\right)dx,,$$

Put $M_0 = 0$, before integration

$$\theta_B = \frac{1}{EI}\left[\frac{w}{6}\cdot\frac{x^3}{3} - \frac{w}{6L^2}\cdot\frac{x^5}{5}\right]_0^L = \frac{wL^3}{EI}\left(\frac{5-3}{90}\right)$$

$$\boldsymbol{\theta_B = \dfrac{-wL^3}{45EI}}\text{ (anticlockwise)} \hspace{4cm} ... (10.72)$$

(iii) Maximum Deflection at x_0 from A:

Apply fictitious moment M_0 at x_0 from A.

For maximum deflection, the rotation or slope will be zero at $x_0.$

$$R_A = \left(\frac{wL}{6} - \frac{M_0}{L}\right), R_B = \left(\frac{wL}{3} + \frac{M_0}{L}\right)$$

$$\underset{M_x(x_0 \text{ from } A)}{(A \text{ to } x)} = \left\{\left(\frac{wL}{6} - \frac{M_0}{L}\right)x - \frac{w}{L}\frac{x^3}{6}\right\}, \qquad \frac{\partial M_x}{\partial M_0} = \left(-\frac{x}{L}\right)$$

$$\theta_x = \frac{1}{EI}\int_0^{x_0} M_x\frac{\partial M_x}{\partial M_0}\cdot dx + \frac{1}{EI}\int_{x_0}^L M_x\cdot\frac{\partial M_x}{\partial M_0}\cdot = 0\text{, for maximum deflection }\left(\theta_{x_0} = 0\right).$$

$$\frac{1}{EI}\left[\int_0^{x_0}\left\{\frac{wL}{6}\cdot x-\frac{M_0}{L}\cdot x-\frac{w}{6L}\cdot x^3\right\}\left(-\frac{x}{L}\right)dx+\right.$$

$$\left.\int_{x_0}^L\left\{\frac{wL}{6}x-\frac{M_0}{L}\cdot x-\frac{w}{6L}\cdot x^3+M_0\right\}\left(1-\frac{x}{L}\right)dx\right]=0$$

Putting $M_0 = 0$, before integration

or $\quad\int_0^{x_0}\left\{-\frac{w}{6}\cdot x^2+\frac{w}{6L^2}\cdot x^4\right\}dx+\int_{x_0}^L\left\{\frac{wL}{6}x-\frac{wx^3}{6L}-\frac{w}{6}\cdot x^2+\frac{w}{6L^2}\cdot x^4\right\}dx=0$

or $\quad\dfrac{w}{6}\left\{-\dfrac{x^3}{3}+\dfrac{x^5}{5L^2}\right\}_0^{x_0}+\dfrac{w}{6}\left\{L\cdot\dfrac{x^2}{2}-\dfrac{x^4}{4L}-\dfrac{x^3}{3}+\dfrac{x^5}{5L^2}\right\}_{x_0}^L=0$

or $\quad\left\{-\dfrac{x_0^3}{3}+\dfrac{x_0^5}{5L^2}\right\}+\left\{\dfrac{L}{2}(L^2-x_0^2)-\dfrac{1}{4L}(L^4-x_0^4)-\dfrac{1}{3}(L^3-x_0^3)+\dfrac{1}{5L^2}(L^5-x_0^5)\right\}=0$

or $\quad L^3\left(\dfrac{1}{2}-\dfrac{1}{4}-\dfrac{1}{3}+\dfrac{1}{5}\right)-\dfrac{Lx_0^2}{2}+\dfrac{x_0^4}{4L}=7L^3-30Lx_0^2+\dfrac{15}{L}x_0^4=0$

or $\quad 7L^4-30L^2x_0^2+15x_0^4=0$,

By trial or quadratic equation $x_0^2=0.2697L^2$, $x_0 = \mathbf{0.51933L}$

For Maximum deflection apply fictitious load W_0 at $x_0 = 0.5193L$ from A.

$$R_A=\left(\frac{wL}{6}+\frac{0.481LW_0}{L}\right),\quad R_B=\left(\frac{wL}{3}+\frac{0.5193LW_0}{L}\right)$$

$$y_{\max}=\int_0^{0.5193L}M_x\cdot\frac{\partial M_x}{\partial W_0}\cdot\frac{dx}{EI}+\int_{0.5193L}^L M_x\frac{\partial M_x}{\partial W_0}\cdot\frac{dx}{EI}$$

$$y_m=\frac{1}{EI}\left[\int_0^{0.5193L}\left\{\frac{wL\cdot x}{6}+0.481W_0x-\frac{W}{6L}\cdot x^3\right\}(0.481x)dx+\right.$$

$$\left.\int_{0.5193L}^L\left\{\frac{wL}{6}x+0.481W_0x-\frac{w}{6L}x^3-W_0(x-0.5193L)\right\}(0.5193L(L-x)dx\right]$$

Putting $W_0 = 0$, before integrating

$$y_m=\frac{1}{EI}\left[\int_0^{0.5193L}\left(0.481\frac{wL}{6}x^2-\frac{0.481w}{6L}x^4\right)dx\right.$$

$$\left.+\int_{0.5193L}^L\left\{0.5193\frac{wL^2}{6}x-\frac{0.5193x^3}{6}-\frac{0.5193wL}{6}x^2+\frac{0.5193w}{6L}x^4\right\}dx\right]$$

$$= \frac{1}{EI}\left[\frac{0.481wL}{6}\cdot\frac{x^3}{3} - \frac{0.481w}{6L}\cdot\frac{x^5}{5}\right]_0^{0.5193L}$$

$$+ \frac{1}{EI}\left[\frac{0.5193wL^2}{6}\cdot\frac{x^2}{2} - \frac{0.5193w}{6}\frac{x^4}{4} - \frac{0.5193wL}{6}\frac{x^3}{3} + \frac{0.5193wx^5}{30L}\right]_{0.5193L}^{L}$$

$$= \frac{w}{EI}\left[\frac{0.481L}{18}(0.5193L)^3 - \frac{0.481}{30L}(0.5193L)^5\right]$$

$$+ \frac{w}{EI}\left[\frac{0.5193L^2}{12}(L^2 - 0.5193^2 L^2) - \frac{0.5193}{24}(L^4 - 0.5193^4 L^4)\right.$$

$$\left. - \frac{0.5193L}{18}(L^3 - 0.5193^3 L^3) + \frac{0.5193}{30L}(L^5 - 0.5193^5 \cdot L^5)\right]$$

$$= \frac{wL^4}{EI}\left[\frac{0.481 \times 0.5193^3}{18} - \frac{0.481(0.5193)^5}{30} + \frac{0.5193}{12}(1 - 0.1593^2)\right.$$

$$\left. - \frac{0.5193}{24}(1 - 0.5193^4) - \frac{0.5193}{18}(1 - 0.5193^3) + \frac{0.5193}{30}(1 - 0.5193^5)\right]$$

$$= \frac{wL^4}{360\,EI}[1.34486 - 0.217352 + 11.37605 - 7.22016 - 8.928893 + 5.993477]$$

$$= \frac{2.348wL^4}{360EI}$$

$$y_{max} = \frac{2.348wL^4}{360EI}wL^4 \approx \frac{2.5045}{384EI}wL^4 \qquad\qquad \dots (10.73)$$

(iv) Deflection at the centre C:

Apply a fictitious load W_0 at C.

$$R_A = \left(\frac{wL}{6} + \frac{W_0}{2}\right),\ R_B = \left(\frac{wL}{3} + \frac{W_0}{2}\right)$$

$$A \text{ to } C: M_x = \left\{\frac{wL}{6}\cdot x + \frac{W_0}{2}\cdot x - \frac{w}{L}\cdot\frac{x^3}{6}\right\},\ \frac{\partial M_x}{\partial W_0} = +\frac{x}{2}$$

$$C \text{ to } B: M_x = \left\{\frac{wL}{6}\cdot x + \frac{W_0}{2}x - \frac{w}{6L}\cdot x^3\right\} - W_0\left(x - \frac{L}{2}\right),\ \frac{\partial M_x}{\partial W_0} = \frac{x}{2} - \left(x - \frac{L}{2}\right) = \left(\frac{L-x}{2}\right)$$

$$y_C = \frac{1}{EI}\left[\int_0^{\frac{L}{2}} M_x \cdot \frac{\partial M_x}{\partial W_0}\cdot dx + \int_{\frac{L}{2}}^{L} M_x \cdot \frac{\partial M_x}{\partial W_0}\cdot dx\right]$$

Putting $W_0 = 0$, before integration

$$y_C = \frac{1}{EI}\left[\int_0^{\frac{L}{2}} \left\{ \frac{wL}{6} \cdot x - \frac{w}{6L} \cdot x^3 \right\}\left(\frac{x}{2}\right) dx + \int_{\frac{L}{2}}^{L} \left\{ \frac{wL}{6} \cdot x - \frac{w}{6L} \cdot x^3 \right\}\left(\frac{L-x}{2}\right) \cdot dx \right]$$

or $$y_C = \frac{w}{6EI}\left[\int_0^{\frac{L}{2}} \left\{ \frac{Lx^2}{2} - \frac{x^4}{2L} \right\} dx + \int_{\frac{L}{2}}^{L} \left\{ \frac{L^2 x}{2} - \frac{x^3}{2} - \frac{Lx^2}{2} + \frac{x^4}{2L} \right\} dx \right]$$

$$= \frac{w}{12EI}\left[\left\{ L \cdot \frac{x^3}{3} - \frac{x^5}{5L} \right\}_0^{\frac{L}{2}} + \left\{ L^2 \cdot \frac{x^2}{2} - \frac{x^4}{4} - \frac{L \cdot x^3}{3} + \frac{x^5}{5L} \right\}_{\frac{L}{2}}^{L} \right]$$

$$= \frac{w}{12EI}\left[\left\{ \frac{L^4}{24} - \frac{L^4}{160} \right\} + \left\{ \left(\frac{L^4}{2} - \frac{L^4}{8} \right) - \left(\frac{L^4}{4} - \frac{L^4}{64} \right) - \frac{L^4}{3}\left(1 - \frac{1}{8} \right) + \frac{L^4}{5}\left(1 - \frac{1}{32} \right) \right\} \right]$$

$$= \frac{wL^4}{12EI}\left[\frac{1}{24} - \frac{1}{160} + \frac{1}{2} - \frac{1}{8} - \frac{1}{4} + \frac{1}{64} - \frac{1}{3} + \frac{1}{24} + \frac{1}{5} - \frac{1}{160} \right]$$

$$= \frac{wL^4}{12EI} \times \frac{1}{960}[40 - 6 + 480 - 120 - 240 + 15 - 320 + 40 + 192 - 6]$$

$$y_C = \frac{5wL^4}{768EI} = \frac{2.5}{384EI} \quad \text{(Almost same as maximum deflection)}.$$

10.9 DEFLECTION DUE TO SHEAR

"Any bending structural member (beam) also develops shear force and shear stresses in addition to bending moment and bending stresses. Strain energy is developed due to shear stresses caused by shear forces in beams by external loading. Consider an elementary strip dx at a distance 'x' from the support where the section is subjected to shear force 'F' and shear stress 'q'. *Shear* stress q at a layer dy at a distance 'y' from the neutral axis of the section is given as $q = \dfrac{F(A\bar{y})}{Ib}$

Where : I = moment of inertia of the whole section,

b = width of the section at the layer 'dy'

$A\bar{y}$ = moment of the area *of section away* from the layer (*i.e. above or below the strip* 'dy') about the NA

"The *volume* of the elementary strip 'dx', thickness 'dy' and width 'b' will be $v = dx \cdot b \cdot dy$

Strain energy stored in the elementary strip $= du = \dfrac{q^2}{2G} \times Vol.$

i.e. $du = \dfrac{q^2}{2G} \times b \cdot dx \cdot dy$

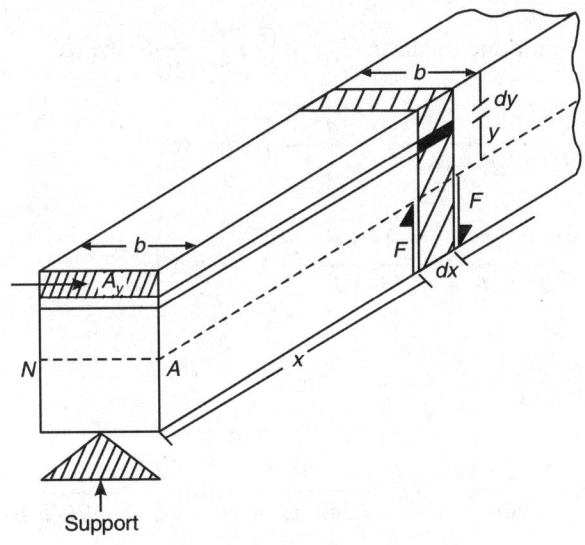

Fig. 10.78: Beam under shear stresses

Total energy in the elementary section $dx = \displaystyle\int_{yt}^{ye} \dfrac{q^2}{2G} \cdot b \cdot dx \cdot dy$, where '$q$' is the shear stress at layer 'dy' and 'G' is the shear modulus (Modulus of rigidity)

Total strain energy stored in the whole beam $= U_{\text{shear}} = \displaystyle\int_{0}^{L} \int_{yt}^{ye} \dfrac{q^2}{2G} \cdot b \cdot dx \cdot dy$ ... (10.74)

This will be explained by simple examples.

EXAMPLE 10.54: A simply supported beam AB of span L (10m) carries a point load W (20kN) at the centre. The section comprises of a rectangular section of width b (90 mm) and depth d (200 mm). The shear modulus of the beam material is G (80 kN/mm^2). Calculate the strain energy developed due to shear stress in the whole beam.

Solution: SF at x from the support $F_x = \dfrac{W}{2}$ (10 kN) upto midspan point.

$I = \dfrac{1}{12}bd^3$ (6×10^7 mm^4), $G = 80$ kN/mm^2 (8×10^4 N/mm^2)

q (Shear stress at any layer dy) $= \dfrac{F}{2I}\left(\dfrac{d^2}{4} - y^2\right) = \dfrac{W12}{2 \times 2bd^3}\left(\dfrac{d^2}{4} - y^2\right)$

or $q = \dfrac{3W}{bd^3}\left(\dfrac{d^2}{4} - y^2\right)$, $q^2 = \dfrac{9W^2}{b^2d^6}\left(\dfrac{d^4}{16} + y^4 - \dfrac{d^2y^2}{2}\right)$

Strain energy in the strip dx, $\quad du = \int_{yc}^{yt} \dfrac{q^2}{2G} \cdot b \cdot dy \cdot dx$

Total energy in the beam due to shear $U_s = 2 \int_0^{\frac{L}{2}} \int_{-d/2}^{+d/2} \dfrac{q^2}{2G} b \cdot dy \cdot dx$

$$U_s = \dfrac{2}{2G} \int_0^{\frac{L}{2}} \int_{-d/2}^{+d/2} \dfrac{9W^2}{b^2 d^6} \left(\dfrac{d^4}{16} + y^4 - \dfrac{d^2 y^2}{2} \right) b \cdot dy \cdot dx$$

$$= \dfrac{1}{G} \int_0^{L/2} \left[\dfrac{9W^2}{b^2 d^6} \left\{ \dfrac{d^4}{16} \cdot y + \dfrac{y^5}{5} - \dfrac{d^2}{2} \cdot \dfrac{y^3}{3} \right\}_{-d/2}^{+d/2} b dx \right]$$

$$= \dfrac{9W^2}{Gb \cdot d^6} \int_0^{\frac{L}{2}} \left\{ \dfrac{d^5}{16} + \dfrac{d^5}{5 \times 16} - \dfrac{d^5}{6 \times 4} \right\} dx$$

$$= \dfrac{9W^2 \cdot d^5}{Gb \cdot d^6} \left[\dfrac{15 + 3 - 10}{240} \right] (x)_0^{\frac{L}{2}} = \dfrac{9W^2}{240 \cdot G \cdot b \cdot d} (8) \left(\dfrac{L}{2} \right) = \dfrac{3 \cdot W^2 \cdot L}{20G \cdot b \cdot d} \qquad \text{... (i)}$$

Substituting the values of $W = 20{,}000$N, $L = 10{,}000$ mm, $G = 8 \times 10^4$ N/mm^2

$$U_s = \dfrac{3(2 \times 10^4)^2 (10^4)}{20(8 \times 10^4)(90)(200)} = \dfrac{3 \times 4 \times 10^{12}}{2 \times 8 \times 9 \times 2 \times 10^8} = \dfrac{10^4}{24} \text{ N-mm}$$

$$= \textbf{416.67 N-mm.}$$

10.9.1 Deflection due to Shear Stresses

Shear stresses cause deformations resulting in additional dflections. The deflections caused by shear stresses are, however, small in comparison to deflections caused by bending. Shear stresses vary across the section along the depth of the section (Fig. 10.79c).

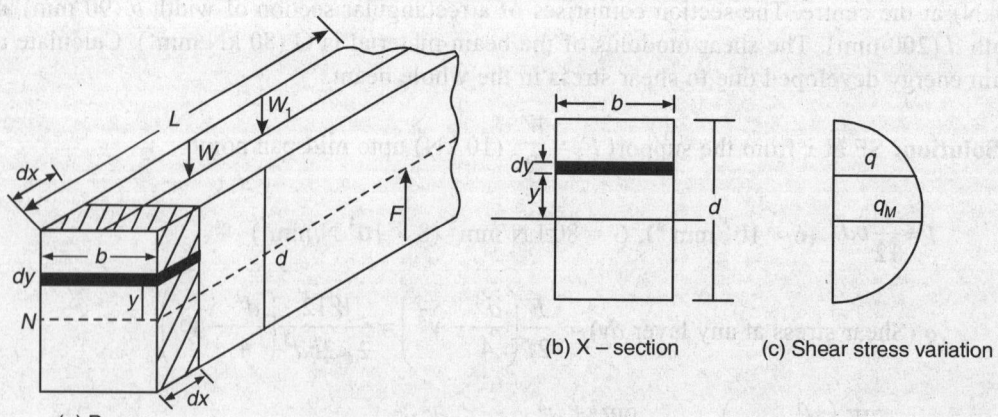

(a) Beam

(b) X – section

(c) Shear stress variation

Fig. 10.79: Beam and shear stresses

Due to variation in shear stress along the depth, the cross-sections become curved surfaces. The original axis say '*ab*' at the NA deform to line '*ac*' due to **shear strain**. The angle between the line *ac* and *ab r* represents shear strain. Shear strain '*r*' will be maximum at the NA as the shear stress is maximum at the NA The value of *r* becomes *zero* at the *outermost* fibres. Thus the outermost sides will remain unchanged. The **plane sections assumed to remain planes** *after bending infact does not remain planes* **due to shear**. The plane sections get warped due to shear strains. The rectangular elements become rhombus. The deflection curve due to shear depends on the shear strain at the NA The *slope* of the deflection curve is equal to the shear strain at the NA The appoximate deflection 'y_s' due to shear is found by equating *work done by loading* causing shear and *the shear* strain energy stored in the beam.

Strain energy stored in the elementary strip $dx = du = \dfrac{q^2}{2G} \times$ Volume $= \dfrac{q^2}{2G} b \cdot dx \cdot dy$

Total energy stored in the section of length $dx = \displaystyle\int_{y_t}^{y_c} \dfrac{q^2}{2G} b \cdot dx \cdot dy$

Thus strain energy stored in the *whole beam* due to shear

$$U_s = \int_0^L \int_{y_t}^{y_c} \frac{q^2}{2G} b \cdot dx \cdot dy \qquad \qquad \dots \text{(i)}$$

work done by the load $W = \dfrac{1}{2} W y_s$ where y_s is the deflection due to shear force, Thus equating

$$\frac{1}{2} W y_s = \int_0^L \int_{y_t}^{y_c} \frac{q^2}{2G} b \cdot dx \cdot dy , \quad \text{or} \quad y_s = \frac{2}{W} \int_0^L \int_{y+}^{y_c} \frac{q^2}{2G} \cdot b \cdot dx \, dy \qquad \dots \text{(10.75)}$$

Thus from the equation 10.75, we can find the deflection y_s due to shear by integrating. This process will be explained by examples.

EXAMPLE 10.55: A simply supported beam AB of span $L = 10$m and carrying a point load W of 20 kN at the centre. The cross-section of beam is rectangular of 90 mm width $\times$ 200 mm depth. The shear modulus of the material $G = 80$ kN/mm^2. The modulus of Elasticity E = 200 kN/mm^2. Calculate the total deflection at the centre due to bending and shear.

Solution:

$L = 10$m, $W = 20$kN, $I = \dfrac{1}{12} \times 90 \times 200^3 \times 10^{-12}$ m$^4 = 6 \times 10^{-5}$ m^4.

$G = 80 \times 10^6$ kN/m^2, $E = 200 \times 10^6$ kN/m^2,

$EI = 200 \times 10^6 \times 6 \times 10^{-5} = 12 \times 10^{+3}$ kN-m^2

Work done by the load W due to shear deflection $y_s = \dfrac{1}{2} W \cdot y_s = 10 y_s$ kN-m

Energy stored due to shear strain $U_s = \displaystyle\int_0^L \int_{y_t}^{y_c} \frac{q^2}{2G} b \cdot dx \cdot dy$

or $\quad U_s = \dfrac{3}{20}\dfrac{W^2 \cdot L}{G \cdot b \cdot d}$ (*As found in example* 10.54), Equating the work done to the shear strain energy

$$\frac{1}{2}W \cdot y_s = \frac{3}{20}\frac{W^2 \cdot L}{G \cdot b \cdot d}$$

or $\quad y_s = \dfrac{6}{20}\dfrac{W \cdot L}{G \cdot b \cdot d} = \dfrac{1\times 16 \times 20 \times 10}{20 \times 80 \times 10^6 \times \dfrac{90}{1000} \times \dfrac{200}{1000}} = 0.000041666$ m.

Deflection due to bending $y_b = \dfrac{WL^3}{48EI} = \dfrac{20 \times 10^3}{48 \times 12 \times 10^3} = 0.03472222$ m.

Total deflection = 0.0347222 + 0.00041667 = 0.0347639 m (34.7639 mm)

It may be noted that the deflection caused by shear is only

$$\frac{0.000041667}{0.0347639} \times 100 = 0.12 \text{ \% of the total deflection.}$$

The deflection caused by *shear is so small* that it is generally neglected in *design problems for deflection check.*

10.10 DEFLECTION DUE TO IMPACT LOADING

We have determined the deflections in beam due to static loading gradually applied. In this case the work done by static load was equated to the strain energy stored in the beam. If the load is applied suddenly or dropped from the height, the *instantaneous deflectin* caused will be much higher than the *static deflection* (δ_s). In case of sudden or impact loading, the elastic beam vibrates for some time before coming to *rest position. The instantaneous deflection* (δ_i), caused by impact load or suddenly applied load can be calculated by making certain assumption. Let there be static gradual load 'P' which caused *deflection* same as *instantaneous deflection* caused by a *dynamic impact load.*

Work done by a falling load 'W' through height 'h' = $W(h + \delta)$

Work done by equivalent gradual static load 'P' = $\dfrac{1}{2}P \cdot \delta \cdot$

Equating the work done to equivalent work done

i.e. $\quad W(h + \delta) = \dfrac{1}{2}P \cdot \delta \cdot$ ... (i)

Also static gradually applied loading 'P' in the beam

$$\delta = \frac{PL^3}{48EI} \text{ or } P = \frac{48EI\delta}{L^3} \qquad\qquad \text{... (ii)}$$

Substituting value of P in equation (i), we have

$$W(h + \delta) = \frac{1}{2} \times \frac{48EI \cdot \delta^2}{L^3}$$

or $\dfrac{24EI}{L^3} \cdot \delta^2 - W \cdot \delta - Wh = 0$ ... (iii)

Quadratic equation solution gives :

$$\delta = \frac{W \pm \sqrt{W^2 + 4Wh \cdot \dfrac{24EI}{L^3}}}{\dfrac{48EI}{L^3}}$$

or $\delta = \dfrac{WL^3}{48EI} + \sqrt{\left(\dfrac{WL^3}{48EI}\right)^2 + 2h\dfrac{WL^3}{48EI}}$,

$$\delta_s = \frac{WL^3}{48EI}$$ ... (10.76)

Putting static deflection = δ_s, we have dynamic deflection

$$\delta = \delta_s + \sqrt{(\delta_s)^2 + 2h(\delta_s)}$$ ... (10.77)

If $h = 0$, and load is suddenly applied, then

$$\delta = \delta_s + \sqrt{(\delta_s)^2 + 0} = 2\delta_s$$ ... (10.78)

If h is very large compared to δ_s, then $(\delta_s)^2$ can be neglected and terms with h can only be considered for approximate determination of dynamic deflection.

i.e. $\delta_d = \sqrt{2h\delta_s}$ ... (10.79)

For calculating deflection due to dynamic load, we generally make certain assumptions to simplify the calculations, which are quite valid for general design purposes as the dynamic deflection equations provide alomost the maximum values.

Assumptions

i. The falling weight remains in contact with the beam and moves with it.

ii. No energy loss takes place due to impact.

iii. The beam is linearly elastic.

iv. The potential energy of the beam due to change in position is neglected.

v. The deflected shape of the beam *remains* the same under dynamic load as under the static load.

These assumptions are made to simplify the determination of deflection due to impact loading. It is quite clear that the deflection caused by *impact loading* is *much higher* than the deflection caused by *gradual loading*.

EXAMPLE 10.56: A Simply supported beam AB is of 10m span. A load of 20 kN is dropped on to the beam at the midspan point from a height of 30 mm.

The cross-section of the beam comprises of 90mm × 200mm rectangular section. Determine the instantaneous deflection and the stress developed in the beam. E (material) = 200 kN/mm².

Solution:

$L = 10$m, $h = 30$mm (0.030m), $E = 200 \times 10^6$ kN / m²

$I = \dfrac{1}{12} \cdot 90 \times 200^3 \times 10^{-12}\text{m}^4 = \mathbf{60 \times 10^{-6}\text{m}^4}$, $EI = 2 \times 10^8 \times 6 \times 10^{-5} = \mathbf{12 \times 10^3}$ **kN-m²**

$$\text{W} (0.030 + \delta) = P \cdot \frac{\delta}{2} \text{ or } P \cdot \delta = 2 \times 20 (0.03 + \delta) \qquad \dots \text{(i)}$$

Also $\delta = \dfrac{PL^3}{48EI} = \dfrac{P(10)^3}{48 \times 12 \times 10^3}$, P is equivalent static load

or $P = (48 \times 12)\delta = 576\delta$ $\qquad \dots$ (ii)

Substituting in (i), we get $(48 \times 12)\delta^2 = 40 (0.03 + \delta)$

or $\mathbf{576\delta^2 - 40\delta - 1.2 = 0}$ or $\mathbf{P^2 - 40P - 691.2 = 0}$ $\qquad \dots$ (iii)

Solving $\delta = \dfrac{+40 + \sqrt{1600 + 4 \times 576 \times 1.2}}{2 \times 576} = \dfrac{40 + \sqrt{4364.8}}{1152} = \mathbf{0.0920717}$ **m**

$\delta = \mathbf{92.0717}$ **mm**, $P = (48 \times 12) \times 0.0920717 = \mathbf{53.033}$ **kN**

$\delta_{\text{static}} = \dfrac{20 \times (10)^3}{48 \times 12 \times 10^3} = \mathbf{0.03472}$ **m (34.72 mm)**

Max. $BM = \dfrac{PL}{4} = \dfrac{53.033 \times 10}{4}$ kN-m = 132.5825 kN-m

Bending stress $= \dfrac{M}{I} \cdot y = \dfrac{132.5825}{6 \times 10^{-5}} \times \dfrac{100}{1000} = 220970.823$ kN/m² (220.9708 N/mm²)

EXAMPLE 10.57: A cantilever of length L carries a concentrated load of W kN at the free end. The cantilever has *M.I.* = I for the half length from the free end while *M.I.* becomes $16I$ for the remaining half length upto the fixed end. The material has modulus of Elasticity = E.

Find the deflection at the free end.

Solution:

$AB = BC = \dfrac{L}{2}$.

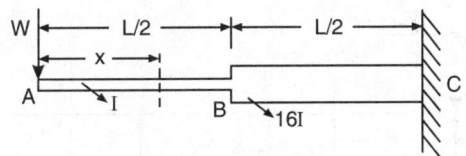

Fig. 10.58.

$I_{AB} = I$, $I_{BC} = 16I$, Load at $A = W$

Since load is directly acting at A, direct method can be applied.

Strain Energy $U = \int_0^L \dfrac{M_x^2}{2EI_x} \cdot dx$,   ($x$ measured from free end A).

$$U = \int_0^{\frac{L}{2}} \frac{M_x^2 \cdot dx}{2EI_{ab}} + \int_{\frac{L}{2}}^L \frac{M_x^2}{2EI_{CB}} dx$$

$$U_{AB} = \int_0^{\frac{L}{2}} \frac{(W \cdot x)^2}{2EI_{AB}} dx = \left(\frac{W^2 \cdot x^3}{6EI}\right)_0^{\frac{L}{2}} = \frac{W^2 L^3}{48EI} \qquad \ldots \text{(i)}$$

$$U_{BC} = \int_{L/2}^L \frac{(W \cdot x)^2}{2EI_{BC}} dx = \left\{\frac{W^2 \cdot x^3}{6E(16I)}\right\}_{L/2}^L = \frac{W^2 L^3}{96EI}\left[1 - \frac{1}{8}\right] = \frac{7W^2 L^3}{96 \times 8EI} \qquad \ldots \text{(ii)}$$

$$U = U_{AB} + U_{BC} = \frac{W^2 L^3}{48EI} + \frac{W^2 L^3 \times 7}{96 \times 8EI} = \frac{W^2 L^3}{96 \times 8EI}(16 + 7) = \frac{23W^2 L^3}{768EI} \qquad \ldots \text{(iii)}$$

Work done by the load W in undergoing deflection $y_A = \dfrac{1}{2}W \cdot y_A$

Thus $\dfrac{1}{2}W \cdot y_A = \dfrac{23W^2 L^3}{768EI}$, or $y_A = \dfrac{\mathbf{23WL^3}}{\mathbf{384EI}}$

EXAMPLE 10.58: A simply supported beam ABC is hinged at A and overhangs over B. Supported length $AB = L$ and overhang $BC = L_1$. A point load of W acts at the overhanging end C. Calculate the deflection under the load W at C. EI is constant.

(i) Use Castigliano's theorem (ii) Unit load method.

Solution:

$$R_A = -\frac{W}{L}L_1, \ R_B = \frac{W}{L}(L + L_1)$$

(i)   Unit load method.

AB (x from A) : $M_x = -\dfrac{W \cdot L_1}{L} \cdot x$, $m_x = -\dfrac{L_1}{L} \cdot x$

BC (x_1 from C) : $M_x = W \cdot x_1$, $m_{x_1} = -x_1$ (for simplification x_1 from C)

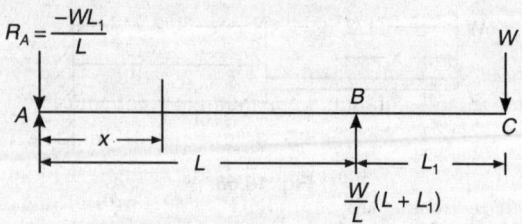

Fig. 10.59.

$$y_C = \int_0^{L+L_1} \frac{M_x m_x dx}{EI} = \int_0^L \frac{M_x \cdot m_x dx}{EI} + \int_0^{L_1} \frac{M_{x_1} \cdot m_{x_1} dx}{EI}$$

or $\quad y_C = \dfrac{1}{EI} \int_0^L -\dfrac{WL_1}{L} \cdot x \left(-\dfrac{L_1}{L} x \right) dx + \dfrac{1}{EI} \int_0^{L_1} -Wx_1(-x_1)dx$

or $\quad y_C = \dfrac{WL_1^2}{EIL^2} \left(\dfrac{x^3}{3} \right)_0^L + \dfrac{W}{EI} \left(\dfrac{x_1^3}{3} \right)_0^{L_1} = \dfrac{WL_1^2 \cdot L^3}{3EIL^2} + \dfrac{W}{3EI} \cdot L_1^3 = \dfrac{WL_1^2}{3EI}(L+L_1)$

$$y_C = \frac{WL_1^2(L+L_1)}{3EI}$$

(ii) **Castigliano's First Theorem**

$$y_C = \frac{\partial U}{\partial W} = \frac{1}{EI} \int M_x \cdot \frac{\partial M_x}{\partial W} \cdot dx = \frac{1}{EI} \left[\int_0^L M_x \cdot \frac{\partial M_x}{\partial W} \cdot + \int_0^{L_1} M_{x_1} \cdot \frac{\partial M_{x_1}}{\partial W} \cdot dx_1 \right]$$

Portion AB ($x = 0$ to L)

$$M_x = -\frac{WL_1}{L} \cdot x, \quad \frac{\partial M_x}{\partial W} = -\frac{L_1 \cdot x}{L}.$$

$$\frac{1}{EI} \int_0^L -\frac{WL_1}{L} \cdot x \cdot \left(-\frac{L_1}{L} x \right) dx = \frac{WL_1^2}{EIL^2} \left(\frac{x^3}{3} \right)_0^L = \frac{WL_1^2 \cdot L}{3EI} \qquad \ldots \text{(i)}$$

Portion BC (x_1 from C)

$$M_{x_1} = -W \cdot x_1, \quad \frac{\partial M_{x_1}}{\partial W} = -x_1$$

$$\frac{1}{EI} \int_0^{L_1} -W \cdot x_1(-x_1)dx = \frac{W}{EI} \left(\frac{x_1^3}{3} \right)_0^{L_1} = \frac{WL_1^3}{3EI} \qquad \ldots \text{(ii)}$$

Total $y_C = \dfrac{1}{EI} \int_0^L \dfrac{WL_1^2}{L^2} \cdot x^2 dx + \dfrac{1}{EI} \int_0^{L_1} W \cdot (x_1^2)dx = \dfrac{WL_1^2 L}{3EI} + \dfrac{WL_1^3}{3EI} = \dfrac{WL_1^2}{3EI}(L+L_1) \qquad \ldots \text{(iii)}$

EXAMPLE 10.59: A simply supported beam AB of span L and uniformly variable X – section. The MI at the supports is $= I$ and at the centre C, it is $= 2I$. Determine the deflection at the centre and compare with the central deflection having unform $M.I.$

Solution:

A to C (x from A)

$$I_x = I + \frac{I}{L} \cdot 2x = \left(I + \frac{2I}{L}x\right),$$

By strain energy method

$$y_C = \frac{\partial U}{\partial W} = \frac{1}{E}\int_0^L \frac{M_x}{I_x} \cdot \frac{\partial M_x}{\partial W} \cdot dx = \frac{2}{E}\int_0^{\frac{L}{2}} \frac{M_x}{I_x}, \frac{\partial M_x}{\partial W} \cdot dx$$

Due to symmetrical section and loading.

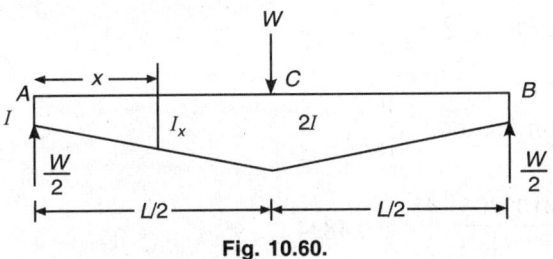

Fig. 10.60.

$$R_A = R_B = \frac{W}{2}, \ M_x = \frac{W}{2} \cdot x, \ \frac{\partial M_x}{\partial W} = \frac{x}{2}$$

$$y_C = \frac{2}{E}\int_0^{L/2} \frac{M_x}{I_x} \frac{\partial M_x}{\partial W} \cdot dx = \frac{2}{E}\int_0^{\frac{L}{2}} \frac{\dfrac{W}{2} \cdot x \left(\dfrac{x}{2}\right)}{\left(I + 2I \cdot \dfrac{x}{L}\right)} dx = \frac{2W}{E \cdot 4}\int_0^{\frac{L}{2}} \frac{x^2}{I\left(1 + \dfrac{2x}{L}\right)} \cdot dx$$

$$= \frac{WL}{2EI}\int_0^{\frac{L}{2}} \frac{x^2}{(L + 2x)} dx, \ \text{Let } (L + 2x) = t, \text{ for simplication, limits} : t = L, \text{ when } x = 0$$

$$t = 2L, \text{ when } x = \frac{L}{2}, \text{ Also } x = \left(\frac{t - L}{2}\right), \ dt = 2dx, \text{ or } dx = \frac{dt}{2}$$

$$y_C = \frac{WL}{2EI}\int_L^{2L} \frac{\left(\dfrac{t - L}{2}\right)^2}{t} \cdot \frac{dt}{2} = \frac{WL}{2EI}\int_L^{2L} \left(\frac{t^2 - 2tL + L^2}{4t}\right)\frac{dt}{2}$$

or $\quad y_C = \dfrac{WL}{16EI} \displaystyle\int_L^{2L}\left(t - 2L + \dfrac{L^2}{t}\right)dt = \dfrac{WL}{16EI}\left\{\dfrac{t^2}{2} - 2Lt + L^2\log_e t\right\}_L^{2L}$

$\quad\quad = \dfrac{WL}{16EI}\left\{\dfrac{1}{2}(4L^2 - L^2) - 2L(2L - L) + L^2(\log_e^{2L} - \log_e^{L})\right\}$

$\quad\quad = \dfrac{WL}{16EI}\left\{\dfrac{3}{2}L^2 - 2L^2 + L^2(\log_e^{2L} - \log_e^{L})\right\}, \quad \left(\log_e^{2L} - \log_e^{L} = \log_e\dfrac{2L}{L}\right)$

$\quad\quad = \dfrac{WL^3}{16EI}\left\{\dfrac{3}{2} - 2 + 1 \times 2.196240\log_{10}^2\right\} = \dfrac{WL^3}{16EI}[-0.5 + 0.66112938]$

$\quad\quad = \dfrac{WL^3}{EI}\ (0.0100706)$

y'_C with uniform section $= \dfrac{WL^3}{48EI}$

Ratio $\dfrac{y_c}{y'_c} = \dfrac{0.0100706 \times 48}{1} = \mathbf{0.4834}$

EXAMPLE 10.60: A cantilever $ABCD$ bend of uniform x – section carries a point load W at D as shown in Fig. 10.61a. Calculate vertical displacement of the load point D.

Solution:

The vertical displacement of the load point D canbe found by strain energy method.

$y_d = \dfrac{\partial U}{\partial W}$, Where U is strain energy of the whole bend due to BM

or $\quad y_d = \dfrac{1}{E} \cdot \displaystyle\int_D^A \dfrac{M_x}{I_x}\dfrac{\partial M_x}{\partial W} \cdot dx$

$I_x = I$, uniform throughout.

$y_d = \dfrac{1}{EI}\left[\displaystyle\int_D^C M_x \cdot \dfrac{\partial M_x}{\partial W} + \int_C^B M_x \dfrac{\partial M_x}{\partial W} dx + \int_B^A M_x \dfrac{\partial M_x}{\partial W} \cdot dx\right]$ ... (1)

BM causing convexity in side is $-ve$.

DC (x from D)

$M_x = - W \cdot x \quad \dfrac{\partial M_x}{\partial W} = - x$, limits 0 at D to 'a' at C.

CB (x from C and limits 0, at C to $2a$ at B)

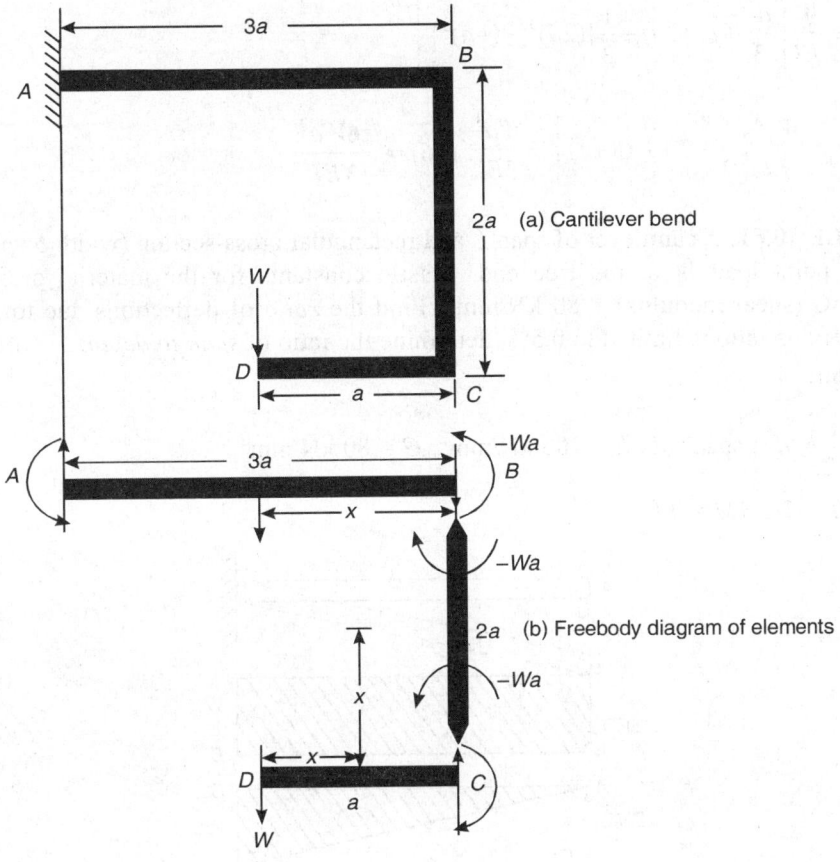

Fig. 10.61.

$$M_x = -Wa; \frac{\partial M_x}{\partial W} = -a$$

BA (x from B and limits 0, at B to $3a$ at A)

At B : Moment $-Wa$, and load W downward.

$$M_x = -Wa + W \cdot x; \frac{\partial M_x}{\partial W} = (-a + x)$$

Thus $y_d = \dfrac{1}{EI} \left[\displaystyle\int_0^a -W \cdot x(-x)dx + \int_0^{2a} -Wa(-a)dx + \int_0^{3a} (-Wa + Wx)(-a + x)dx \right]$

or $y_d = \dfrac{1}{EI} \left[\left(\dfrac{W \cdot x^3}{3} \right)_0^a + (Wa^2 \cdot x)_0^{2a} + W \left\{ \dfrac{(x-a)^3}{3} \right\}_0^{3a} \right]$

$$= \frac{W}{EI}\left[\frac{a^3}{3} + a^2(2a) + \frac{1}{3}\{(2a)^3 - (-a)^3\}\right]$$

or $y_d = \dfrac{Wa^3}{EI}\left[\dfrac{1}{3} + 2 + \dfrac{1}{3}(8+1)\right] = \dfrac{Wa^3}{3EI}(16) = \dfrac{16Wa^3}{3EI}$

EXAMPLE 10.61: A cantilever of span L and rectangular cross-section (width b and depth d) carries a point load W at the free end. Elastic constants for the material are: $E = 200$ kN/mm^2, G (shear modulus) $= 80$ kN/mm^2. Find the *ratio* of deflections due to *shear and bending*. If this ratio is limited to 0.5%, determine the ratio of *span to depth*.

Solution:

$\dfrac{1}{12}b \cdot d^3$, span $= L$, $E = 200$ kN/mm^2, $G = 80$ kN/mm^2

$SF = W$, $BM = WL$,

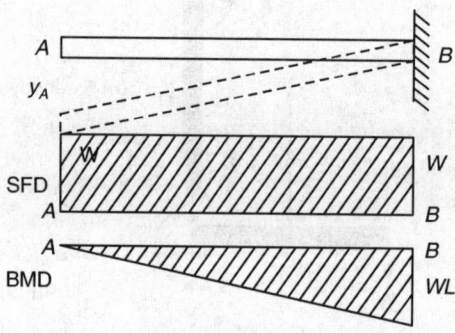

Fig. 10.62.

Shear stress at any layer 'y' from the NA

$$q = \frac{F}{2I}\left(\frac{d^2}{4} - y^2\right), \quad q = \frac{6W}{bd^3}\left(\frac{d^2}{4} - y^2\right)$$

Total shear strain energy $= U$.

$$U = \int_0^L \int_{-\frac{d}{2}}^{+\frac{d}{2}} \frac{b}{2G} \cdot q^2 \cdot dy \cdot dx = \frac{b}{2G}\int_0^L \int_{-\frac{d}{2}}^{+\frac{d}{2}} \left\{\frac{6W}{bd^3}\left(\frac{d^2}{4} - y^2\right)\right\}^2 \cdot dy \cdot dx$$

or $U = \dfrac{b}{2G}\int_0^L \int_{-\frac{d}{2}}^{+\frac{d}{2}} \left\{\dfrac{36W^2}{b^2 \cdot d^6}\left(\dfrac{d^4}{16} - \dfrac{d^2 y^2}{2} + y^4\right)\right\} dydx$

$$= \frac{18W^2 b}{b^2 \cdot d^6 \cdot G} \int_0^L \left(\frac{d^4}{16} \cdot y - \frac{d^2}{2} \cdot \frac{y^3}{3} + \frac{y^5}{5} \right)_{-\frac{d}{2}}^{+\frac{d}{2}} \cdot dx = \frac{18W^2}{bd^6 G} \int_0^L \left\{ \frac{d^5}{16} - \frac{d^2}{6} \left(\frac{d^3}{4} \right) + \frac{d^5}{5} \cdot \frac{1}{16} \right\} dx$$

$$= \frac{18W^2}{G \cdot bd^6} \int_0^L d^5 \left(\frac{15 - 10 + 3}{240} \right) dx = \frac{18 \times 8W^2 \cdot d^5}{Gbd^6 \times 240} (x)_0^L = \frac{3W^2 \cdot L}{5Gb \cdot d}$$

Work done by load $= \frac{1}{2} W \cdot y_s =$ strain energy $= \dfrac{3W^2 \cdot L}{5Gbd}$

$\therefore \quad y_s = \dfrac{6WL}{5Gbd}$ ... (i)

Deflection due to bending $y_b = \dfrac{WL^3}{3EI}$ at the free end.

$$y_b = \frac{WL^3}{3E \left(\frac{1}{12} bd^3 \right)} = \frac{4WL^3}{E \cdot bd^3} \qquad \text{... (ii)}$$

Ratio $\dfrac{y_s}{y_b} = \dfrac{6WL}{5Gbd} \times \dfrac{Ebd^3}{4WL^3} = \dfrac{3}{10} \dfrac{d^2}{L^2} \cdot \dfrac{E}{G} = \dfrac{3E}{10G} \left(\dfrac{d}{L} \right)^2$ ... (iii)

Substituting the values $\dfrac{y_s}{y_b} = \dfrac{3 \times 200}{10(80)} \left(\dfrac{d}{L} \right)^2 = \mathbf{0.75} \left(\dfrac{d}{L} \right)^2$ ... (iv)

For $y_s = \dfrac{0.5}{100} (y_s + y_b)$, Where $200y_s = y_s + y_b$ or $\mathbf{199 y_s = y_b}$ ... (v)

from (iv) and (v)

$$\frac{y_s}{y_b} = \frac{1}{199} = 0.75 \left(\frac{d}{L} \right)^2$$

or $\left(\dfrac{d}{L} \right)^2 = \dfrac{1}{199 \times .75} = \dfrac{1}{149.25}$

$\therefore \quad \left(\dfrac{L}{d} \right)$ ratio $= \sqrt{149.25} = \mathbf{12.2168}$

10.11 SUMMARY – SLOPE AND DEFLECTIONS

In any structural analysis we determine the type and amount of *stresses* caused by *external loading* so as to keep the structural member within *safe permissible limits*. For *safe design* deformations are also required to remain within safe limit for the given loading. For bending members (beams) apart from *bending stresses*, it is necessary to determine *slope and deflections* for safe design of members. The *slope and deflections* are necessary to ensure *adequate stiffness* of the structural member.

"The **slope** is the *angle* that the **tangent to the elastic curve** (deflection curve) makes with the *initial unbent axis* of the beam. The **deflection** is the **vertical displacement** of a point of the beam due to *bending under loading*.

"The bending equation $\dfrac{M}{I} = \dfrac{E}{R}$ or $\dfrac{1}{R} = \dfrac{M}{EI}$ is used to derive the fundamental *differential equation* for the *slope* and the *deflection*. Since beam deflections are very small compared to its length, the curvature $\dfrac{1}{R}$ canbe equated to $\dfrac{d^2y}{dx^2}$, i.e. $\dfrac{1}{R} = \dfrac{M}{EI} = -\dfrac{d^2y}{dx^2}$ or $\boxed{EI\dfrac{d^2y}{dx^2} = -M}$

BM is taken according to sign conventions of bending moment (sagging moment as +ve and hogging moment as negative). Thus the **differential equation** is used to calculate the slope and the deflection by **double integration**.

$$EI\frac{d^2y}{dx^2} = -M_x \qquad \qquad \text{... (i)}$$

$$\text{Slope}: \ EI\frac{dy}{dx} = -M_x \cdot x + c_1 \qquad \qquad \text{... (ii)}$$

$$\text{Deflection}: \ EI\,y = -M_x \cdot \frac{x^2}{2} + c_1 x + c_2 \qquad \qquad \text{... (iii)}$$

Constants of integration are found from known boundary (support) conditions.

The differential equation : $EI\dfrac{d^2y}{dx^2} = -M_x$, was further modified for *general application* in *Macaulay's* approach to solve problems of beams carrying number of loads. In *Macaulay's* method the *constants of integration* are found for the beam *as a whole* while *terms* are considered for the *respective portion* of the beam span. Macaulay has introduced discontinuity lime : for showing the specific term aplicable to specific portion only.

$$EI\frac{d^2y}{dx^2} = -M_x = [M_{x_1} \vdots + M_{x_2} \vdots + Mx_3 \vdots + \cdots]$$

$$\text{Slope}: \ EI\frac{dy}{dx} = [-M_{x_1} \cdot x + c_1 \vdots + M_{x_2} \cdot x \vdots + M_{x_3} \cdot x \cdots]$$

Deflection : $EIy = -M_{x_1} \cdot \dfrac{x^2}{2} + c_1 x + c_2 \vdots + M_{x_2} \cdot \dfrac{x^2}{2} \vdots + M_{x_3} \cdot \dfrac{x^2}{2} \cdots$

$$EI \frac{d^3 y}{dx^3} = -\frac{dM_x}{dx} = F_x \text{ (S.F.)}$$

$$EI \frac{d^4 y}{dx^4} = \frac{dF_x}{dx} = w \text{ (rate of loading).}$$

The *area-moment method* is also based on the same basic equation $\dfrac{1}{R} = \dfrac{M}{EI}$. Mohr has developed two theorems for indirectly determining the slope and deflections of beams by calculating *areas* and *moments* of areas of $\dfrac{M}{EI}$ *diagram* between certain points.

Mohr's *second theorem* states that the *moment of the area* of the $\dfrac{M}{EI}$ *diagram between any two points* is equal to the **vertical intercept** *at the point* about which *the moment is taken* between the *elastic curve and the tangent drawn at the other point.*

"These two theorems help *to calculate slope and deflection indirecly* by applying *suitable known boundary* conditions. When there are number of loads, the areas and moments can be calculated by drawing *BMD in parts separately for each term* of the *bending equation. Areas and centroids* for various shapes are provided in Table 10.1 for reference.

Mohr has further devised another approach by drawing a *equivalent conjugate beam* with $\dfrac{M}{EI}$ *diagram* as its loading using the same principle of area and moment of area of BMD of original beam. According to this method, the **shear force and bending moment** of the *conjugate* beam with $\dfrac{M}{EI}$ of original beam as loading at any point represents **slope and deflection** *respectively.* The conjugate beam is specially formed with equivalent support conditions as under.

Other important conditions for conjugate beam

- A stable and **statically determinate** *real beam* will have a stable and **statically determinate conjugate beam**.
- An *unstable* real beam will have *statically indeterminate conjugate beam*. No analysis of such an equaivalent beam is necessary.
- A *stativally indeterminate real beam* will have *unstable conjugate beam*.

Real beam support condition **Equivalent conjugate beam condition**

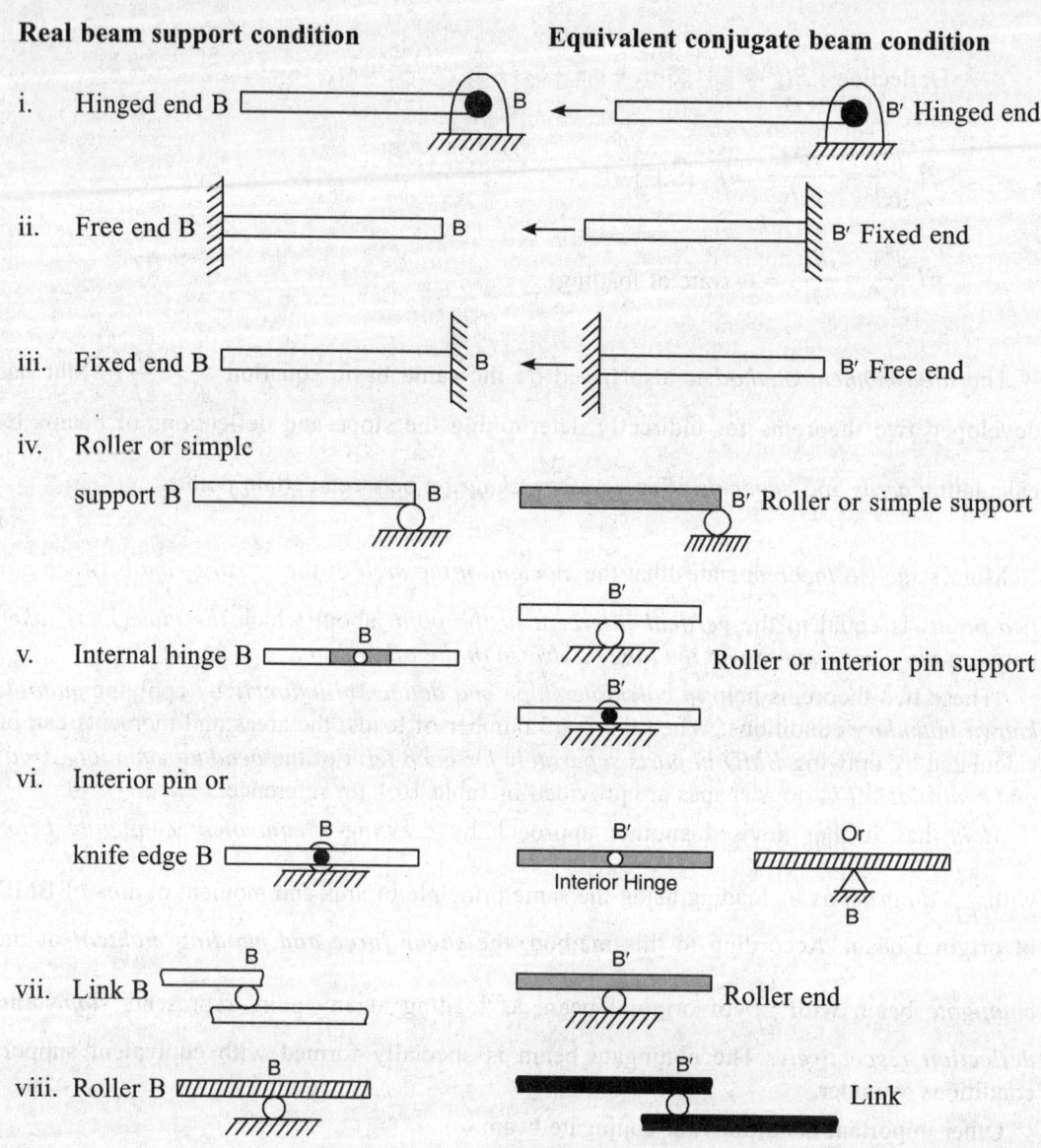

i.   Hinged end B   B′ Hinged end

ii. Free end B   B′ Fixed end

iii. Fixed end B   B′ Free end

iv. Roller or simple
     support B   B′ Roller or simple support

v.   Internal hinge B   Roller or interior pin support

vi. Interior pin or
     knife edge B   Interior Hinge   Or   B

vii. Link B   Roller end

viii. Roller B   Link

Bending elements also undergo deformations (deflection and rotations) and work is stored in this process as **strain energy**. If a beam is subjected to gradually applied moment couple 'M', the beam deform and the strain energy stored will be

$$U = \frac{1}{2} M \cdot \theta = \frac{M^2 \cdot L}{2EI} = \frac{EI\theta^2}{2L}$$

In case of variable moment Mx, consider a small strip dx, the rotation in this small strip due to moment M_x will be $d\theta$. The strain energy **du** in the dx strip will be

$$du = \frac{M_x^2 \cdot dx}{2EI} = \frac{EI}{2}\left(\frac{d^2y}{dx^2}\right)^2 \cdot dx$$

Total strain energy $U = \int_0^L \frac{M_x^2 \cdot dx}{2EI}$ = Load × average deformation.

Applying this equation, deflection 'y' or rotation θ can be found by applying the principle of conservation of energy (i.e. work done = strain energy stored)

A cantilever of span 'L' carrying a point load W at the free end will have

$$\frac{1}{2}W \cdot y = U = \int_0^L \frac{(-W \cdot x)^2}{2EI} \cdot dx, \text{ This gives : } y = \frac{WL^3}{3EI}, \text{ and}$$

$$\frac{1}{2}M \cdot \theta = U = \int_0^L \frac{(-M_x)^2 dx}{2EI}, \text{ This gives : } \theta = \frac{ML}{EI}.$$

Strain energy approach can also be used for finding rotation and deflection at any point by applying a *unit load* at the point i.e.

$$y_C = \frac{1}{EI}\int_0^L M_x \cdot m_x \cdot dx, \text{ where } M_x \text{ is BM at any point } x - x \text{ due to given loading as function}$$

of x, while m_x is the moment at point $x - x$ due to *unit load* applied at the point where deflection is desired.

Consider a cantilever of span L and carrying a udl w/m on the entire span.

$$y_C = \frac{2}{EI}\int_0^L \left(-\frac{wx^2}{2}\right)(-x)dx = \int_0^L \frac{wx^3}{2EI} \cdot dx = \frac{wL^4}{8EI}$$

Consider a simply supported beam AB of span L carrying uniformly distributed load w/m on the entire span, the slope at ends and deflection at the centre shall be :

$$\theta_A = -\theta_B = \frac{wL^3}{24EI}, \quad y_C = \frac{5wL^4}{384EI}.$$

$$\theta_A = \frac{1}{EI}\int_0^L M_x \cdot m_x \cdot d_x = \frac{wL^3}{24EI}, \text{ where } m_x = \text{moment at any section } x - x \text{ due to unit}$$

moment couple m_0 applied at A.

$$y_C = \frac{2}{EI}\int_0^{\frac{L}{2}} M_x \cdot m_x \cdot d_x = \frac{5wL^4}{384EI}$$

Consider a simply supported beam AB of span 'L' carrying a concentrated load W at the centre of the span, the deflection at the centre and rotation at the ends will be:

$$y_C = \frac{2}{EI}\int_0^{\frac{L}{2}} M_x \cdot m_x \cdot dx = \frac{WL^3}{48EI}, \quad \theta_A = \frac{WL^2}{16EI} = -\theta_B, \text{ (due to Symmetry)}$$

Elastic strain energy concept can be used by applyin a *fictitious load* W_0 or *moment* at the points where deformations (deflection or rotations) are desired.

i.e. deflection $y = \dfrac{\partial U}{\partial W_0}$ and $\theta = \dfrac{\partial U}{\partial M_0}$, where U is the strain energy due to moment M_x.

$$U = \int_0^L \frac{M_x^2 \cdot dx}{2EI}, \; y = \delta = \frac{\partial U}{\partial W_0} = \int_0^L M_x \cdot \left(\frac{\partial M_x}{\partial W_0} \right) \frac{dx}{EI},$$

Fictitious load W_0 or moment M_0 is applied where ever deformations are desired.

$$\theta = \frac{\partial U}{\partial M_0} = \int_0^L M_x \left(\frac{\partial M_x}{\partial M_0} \right) \frac{dx}{EI},$$

Fictitious loads are put equal to zero after partial derivation and before integration.

Applying this approach of fictitious load for a cantilever of span L carrying a uniformly variable distributed load **w/m** at the fixed end and zero at the free end, the slope and deflections will be

$$\theta = \int_0^L M_x \cdot \frac{\partial M_x}{\partial M_0} \cdot \frac{dx}{EI} = \frac{wL^3}{24EI}, \; y = \int_0^L M_x \cdot \frac{\partial M_x}{\partial W_0} \cdot \frac{dx}{EI} = \frac{wL^4}{30EI}$$

Consider a simply supported beam **AB of span L** carrying uniformly variable load zero at supports and w/m at the midspan point, the slope at ends and deflection at the centre will be :

$$\theta_A = \frac{2}{EI} \int_0^{L/2} M_x \cdot \frac{\partial M_x}{M_0} \cdot dx = \frac{5wL^3}{192EI}, \; \theta_B = \frac{-5wL^3}{192EI}, \text{ (symmetry of loading)}$$

$$y_C = \frac{2}{EI} \int_0^{L/2} M_x \cdot \frac{\partial M_x}{\partial W_0} \cdot dx = \frac{wL^4}{120EI},$$

Strain energy concept can also be applied for finding *deflection due to shear*.

Strain energy in a small strip due to shear $dU = \dfrac{q^2}{2G} \times$ **Volume** $= \dfrac{q^2}{2G} \cdot b \cdot dx \cdot dy$, for a small strip. $(b \cdot dx \cdot dy)$.

$$U = \int_0^L \int_{y_t}^{y_c} \frac{q^2}{2G} \cdot b \cdot dx \cdot dy \text{ or } \frac{1}{2} W \cdot y_s = U_s = \int_0^L \int_{y_t}^{y_c} \frac{q^2}{2G} \cdot b \cdot dx \cdot dy, \text{ where } G \text{ is shear Modulus of}$$

Elasticity and 'q' is shear stress at any layer 'dy' in a small strip of length 'dx' at any section $x - x$.

Dynamic loads causing impact on structural members develop much higher stresses and instantaneous deformations as compared to gradually applied static loads.

A dynamic load **W** falling through a height '**h**' will cause higher deflection which can be found by

$$\delta = \frac{WL^3}{48EI} + \sqrt{\left(\frac{WL^3}{48EI} \right) + 2h \cdot \frac{WL^3}{48EI}}, \text{ where static deflection } \delta_s = \frac{WL^3}{48EI},$$

$\delta = \delta_s + \sqrt{\delta_s^2 + 2h \cdot \delta_s} = 2\delta_s$, if h is zero.

Similarly if h is very high compared to δ_s, then

$\delta_s = \sqrt{2h \cdot \delta_s}$, by neglecting δ_s^2 being small compared to the height h.

PRACTICE EXERCISE 10

Q.10.1. Derive the relationship of deflection and bending moment in beams due to transverse loading giving various assumptions and proper sign conventions.

Ans. $EI \dfrac{d^2 y}{dx^2} = -M_x$, where y is deflection and M_x is BM at any section $x - x$.

Q.10.2. Derive slope and deflection at the free end of a cantiliver of span L and carrying udl of **w/m** over the entire span.

Ans. $\theta = -\dfrac{wL^3}{6EI}, \dfrac{wL^4}{8EI}$

Q.10.3. Determine maximum deflection in case of a cantilever of span L and carrying a concetrated load W at a distance of L_1 from the fixed end. EI is constant throughout.

Ans. y_{max} (free end) $= \dfrac{WL_1^3}{3EI} + \dfrac{WL_1^2}{2EI}(L - L_1)$

Q.10.4. A simply supported beam AB of span L carries a udl of w/m over the entire span. Determine the maximum deflection and ratio of maximum deflection and maximum bending moment. EI is constant throughout.

Ans. $y_{max} = \dfrac{5wL^4}{384EI}$, **Ratio** $\dfrac{y}{M} = \dfrac{5L^2}{48EI}$

Q.10.5. A simply supported beam AB of span 'L' carries a point load 'W' at a distance of 'a' from the left support and 'b' from the RH support. Find the slopes at the supports and deflection under the load. Also find the maximum deflection if $a > b$. EI is constant throughout.

Ans. $\theta_A = \dfrac{Wa \cdot b(L + b)}{6EIL}$, $\theta_B = -\dfrac{Wa \cdot b(L + a)}{6EIL}$, $y_C = \dfrac{Wa \cdot b(L^2 - b^2 - a^2)}{6EIL} = \dfrac{Wa^2 b^2}{3EIL}$,

$y_{max} = \dfrac{Wa \cdot b(L^2 - b^2)^{\frac{3}{2}}}{9\sqrt{3}EIL}$.

Q.10.6. A cantilever of 3m span carries a point load of 4kN at the free end in addition to udl of 3kN/m for 2m length from the fixed end. The beam section is hollow rectangular with constant $EI = 10432$ kN-m^2. Determine the maximum slope and deflection in the beam.

Ans. $\theta = 0.002108$ **radian** (0.1209 degree), $y_{max} = $ **4.41 mm**

Q.10.7. A beam of 3m length is supported cantrally over 2.0m with equal overhangs of 0.5m on either side. The beam supports concentrated load of 1kN on each overhang end. EI for the beam is constant and is equal to 50kN-m^2. Find the deflection of the mid point and slopes at the supports.

Ans. $y_C = 0.005$m, $\theta_A = -\theta_B = 0.010$ *radiands* (0.5733°)

Q.10.8. A simply supported beam, of 6m *span* has uniform cross-section with uniform depth of 300 mm. The beam carries a central point load which creates maximum bending stress of 120 N/mm^2. Find the maximum slope and deflection. If the beam flanges are so designed that the bending stress developed is 150 N/mm^2 uniform throughout, find the maximum slope and deflection in the beam. $E = 2 \times 10^5$ N/mm^2.

Ans. (i) $\delta = 12.0$ mm, $\theta_A = -\theta_B = 0.006$ radian, (ii) $\delta = 22.5$ mm, $\theta_A = -\theta_B = 0.015$ radian.

Q.10.9. A cantilever beam AB of span 3m has uniform depth of 300mm. The beam carries an udl of 4kN/m on the entire length. Moment of inertia of the beam section $I = 2 \times 10^8$ mm^4 and material $E = 2 \times 10^5$ N/mm^2, (i) Determine maximum deflection and maximum bending stress. (ii) Determine maximum additional point load W applied at 2m from the fixed end so that the maximum deflection does not exceed 2.0mm and the maximum bending stress does notexceed 120.0 N/mm^2.

Ans. (i) $\delta_{max} = 1.013$ mm, $f_{max} = 90.0$ N/mm^2 (ii) $W = 8.464$ kN to satisfy both the requirements.

Q.10.10. A simply supported beam of 6m span and 300mm uniform depth of the section develops a maximum bending stress of 120 N/mm^2 under an *udl* on the entire span. Find the maximum deflection if the same total *udl* acts as concentrated load at the mid span point. $E = 2 \times 10^5$ N/mm^2. Also find the maximum bending stress in that case.

Ans. δ_m (conc. load) = **24mm**, f_{max} in case of concentrated load = **240 N/mm^2**

Q.10.11. A simply supported overhanging beam has equal overhags 'a' on each side. The beam carries a point load W at the midspan point. If the beam section is uniform and the upward deflections at the free ends of overhangs are numerically equal to the mid – span deflection, find the ratio of the overhang length to the supported span.

Ans. Ratio of span (L) to the over hang (a) = 3:1.

Q.10.12. A simply supported beam AB of 6m span carries two concentrated loads of 40 kN and 60 kN at 1m and 4m distance from the LH support A respectively. I of beam section = 49573 × 10^3 mm^4, $E = 2 \times 10^5$ N/mm^2. Determine the slope and deflections below the load points at C and D. Also find the maximum deflection to span ratio.

Ans. $\theta_C = 0.01423$ radian, $\theta_D = -0.007956$ radians,

$\delta_C = $ **16.026 mm**, $\delta_D = $ **28.47 mm**, δ_{max} ($x = 3.097$ m) = **31.98 mm**.

Q.10.13. A SS beam ACB of 10m span carries an *udl* of 8 kN/m on RHS 5m span (CB). LHS span AC (5m) does not carry any load. Determine the maximum deflection in the beam and the point where it occurs. EI is constant and equal to 12 × 10^{12} N-mm^2.

Ans. $\delta_{max} = $ **43.76 mm** at a point **5.4023 m from A**.

Q.10.14. A SS beam of 10m span carries an *udl* 3kN/m from C to D (CD =5m). AC is 2m, and EI is constant and equal to = 98 × 10^8 kN-mm^2. Determine the deflection at its midspan point.

Ans. y_{mid} ($x = 5$m) = **28.046 mm**

Q.10.15. A SS beam AB of 4m span is subjected to an anticlockwise moment couple of 60 kN-m at the RH support B. Determine the slopes at A and B, and maximum deflection. EI of the beam is constant and is equal to **7435.6 × 10^6 kN-mm^2**.

Ans. $\theta_A = $ **0.00538 radians**, $\theta_B = $ **– 0.01076 radians**, $y_{max} = $ **8.28 mm**

Q.10.16. A cantilever beam AB of span L and fixed at B. A couple of moment M acts at the free end A. Derive the deflection curve for the cantilever and find the slope and deflection at the free end. EI is constant.

Ans. $y_x = \dfrac{M(L-x)^2}{2EI}$, $y_A = \dfrac{ML^2}{2EI}$, $\theta_A = \dfrac{ML}{EI}$.

Q.10.17. A SS beam $ADCB$ of 8m span carries a pint load of 4kN at D ($AD = 3$m) and an udl of 1 kN/m from C to B ($CB = 4$m). E for the material = 200 kN/mm^2, and I (cross-section) = 2×10^7mm^4. Determine the deflection at the centre C and maximum deflection. Also find the slopes at the supports.

Ans. $y_c = 16.4$mm, y_{max} $(x = 3.96$m$) = 16.43$mm, $\theta_A = +0.003696$ radian, $\theta_B = -0.006438$ radians

Q.10.18. A cantilever beam AB of 4m span carries a load of 10 kN at a point 3m from the fixed end B. (i) Find deflection at the free end A when there is no prop at A, (ii) If the free end A is supported by a rigid prop at the same level as the fixed end, find the prop reaction, (iii) If the elastic prop at the free end yields by $\dfrac{55}{EI}$ m, find the elastic prop reaction and draw BM diagram.

Ans. (i) $\delta_A = \dfrac{135}{EI}$ m, (ii) $R_A = \dfrac{405}{64} = 6.328$ **kN**, (ii) $R_P = 3.75$ **kN**, BM (fixed) $= -15$**kN-m**

Q.10.19. A cantilever beam of 4m span carries udl of 5 kN/m on the entire span, determine (a) the deflection at the free end, (ii) If the free end is supported by a prop at the same level, find the prop reaction. Also determine the BM at the fixed end if $EI = 7436 \times 10^9$ **N-m^2**.

Ans. $y_A = 21.517$ mm, (ii) $R_{Prop} = 7.5$ kN, (ii) $M_B = -10$**kN-m**.

Q.10.20. A SS beam AB of 10m span carries an udl of 20 kN/m over the entire span. If the beam is propped at the centre by a rigid prop at the level of end supports, find the prop reaction and maximum deflection. $EI = 14666 \times 10^9$ N-mm^2.

Ans. $R_{Prop} = 125$ kN, $R_A = R_B = 37.5$ kN, BM at $C = -62.5$ kN-m, $y_{max} = 4.616$**mm** at $x = 2.1075$**m**

Q.10.21. A cantilever AB of span L carries a point load 'W' at its free end A. This cantilever AB rests on another cantilever CD of the same x-section but span L_1 from the fixed end D (Just below the fixed end B.) Prove the lower cantilever gets a pressure $P = \dfrac{3W}{4}\left(\dfrac{L}{L_1} - \dfrac{1}{3}\right)$.

Hint: Deflection curve are different and the contact point will be at C where prop reaction P on cantilever AB will be equal to pressure 'P' on CD. Deflection of point C of lower cantilever will be equal to net defletion of upper cantilever at C due to load W and prop reaction P.

Q.10.22. A beam ABC is supported at A and B ($AB = 8$m) and overhangs from B to C ($BC = 2$m). A point load 10kN acts at the free end C of the overhang. Determine maximum deflection between the supports A and B, deflection at C and slope at A. $EI = 75$ kN-m^2.

Ans. y_{max} (between A and B) $= -\dfrac{W \cdot a \cdot L^2}{9\sqrt{3}EI} = -10.9483$ **mm**, (upward),

$y_C = \dfrac{W \cdot a^2(a+L)}{3EI} = +17.78$ **mm**; $\theta_A = \dfrac{W \cdot a \cdot L}{6EI} = -0.00355$ radian

Q.10.23. A simply supported beam AB of span 10m carries a uniformly varying load with zero on the supports and 12 kN/m at the midspan point. Determine the maximum slope and deflection in the beam. $EI = 8000$ kN-m^2

Ans. $\theta_A = -\theta_B = \dfrac{5w \cdot L^3}{192EI} = 0.039063$ **radian**, $y_C = \dfrac{w \cdot L^4}{120EI} = 0.125$**m (125mm).**

Q.10.24. A cantilever AB of span 3m carries a uniformly varying load from zero at the free end A to 60 kN/m at the fixed end. If $EI = 10000$ kN-m^2 constant, find the maximum slope and deflection.

Ans. $\theta_A = 0.00675$ **radians**, $y_A = 0.01617$**m** (16.17mm)

Q.10.25. State Mohr's theorems of Area–moment related to slope and deflections of beams.

Q.10.26. Calculate the maximum slope and deflection for the cantilever AB of span 3m and carrying a point load 10 kN at 2m from the fixed end B using moment area method. $EI = 8000$ kN-m^2, constant.

Ans. $\theta_{max} = -0.0025$ **radian**, $y_{max} = 5.833$**mm**

Q.10.27. A cantilever AB of 4m span carries an udl of 2kN/m on the entire span. Calculate the maximum slope and deflection using Mohr's Area–moment method.

Ans. $\theta_{max} = 0.002667$ radian, $y_{max} = 0.008$m (8mm).

Q.10.28. A simply supported beam AB of span 10m carries an *udl* of 4kN/m on the entire span. Determine the slope at the supports, deflections at the centre and at 2m from the support using moment area method. $EI = 15000$ kN-m^2, constant throughout.

Ans. $\theta_A = -\theta_B = 0.01111$ **radian**, $y_C = 0.03472$**m** (34.72mm), $y_2 = 0.020089$**m** (20.089mm)

Q.10.29. A SS overhanging beam $ACDBE$ of 12m total length and supported at A and B with 10m supported span. The overhang BE is 2m. $AC = 2$m, $CD = 5$m, $DB = 3$m. Point loads of 10kN, 20kN, and 10kN are acting at C, D and E respectively. EI is constant and is equal to 15000 kN-m^2. Find the deflections at the load points and slope at A, unsing moment area method.

Ans. $\theta_A = 0.0070467$ **radian**, $y_C = 13.0267$ **mm**, $y_D = 17.4467$ **mm**, y_E (free end) = 9.4867 **mm**

Q.10.30. A SS beam AB of 5m span carries a couple of moment 300 kN-m (anticlockwise) at the point C at a distance of 3m from the left hand support. Calculate the slopes at supports A and B and deflection at the moment point C. EI = 10,000 kN-m^2.

Ans. $\theta_A = 0.013005$ **radian** (clockwise), $\theta_B = 0.001995$ **radian** (clockwise), $y_C = 12.0$mm (above)

Q.10.31. A simply supported beam AB of 6m span carries a uniformly varying distributed load from zero at the support A to 80 kN/m at the support B. Determine the slopes at supports A and B. Also find the deflection at the middle of the span. EI is constant and is equal to 15000 kN-m^2.

Ans. $\theta_A = 0.0224$ **radians (clockwise)**, $\theta_B = 0.0256$ **radians (anticlockwise)**, $y_C = 45.0$ **mm.** y_{max} $(x = 3.116) = 45.084$mm.

Q.10.32. A SS beam AB of 10m span carries an *udl* of 10 kN/m over the left hand side half span and a point load of 20 kN at the centre of RH side half span. $E = 200$ kN/mm^2, $I = 20 \times 10^7$mm^4. Find the slopes at supports and deflections at the midspan point and under the load at the point D. Use moment area method.

Ans. $\theta_A = 0.00558$ **radians (clockwise)**, $\theta_B = 0.00521$ **radians (clockwise)**, $y_{centre} = 16.74$ mm, $y_D = 11.65$ mm.

Q.10.33. (a) State two theorems of conjugate beam related to slope and deflections.

(b) Show a conjugate beam corresponding to a cantilever AB (span 4m) and carrying two point loads each 10 kN at the free end A and midspan point C.

Ans. (b) Equivalent conjugate beam :

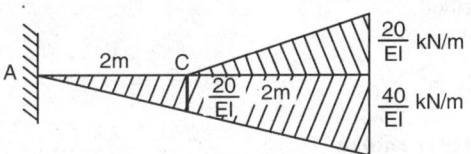

Fig. Q10.33b.

Q.10.34. In question 10.33b, if $EI = 8000$ kN-m^2, find the slope and deflections at the free end A and point C of the real beam ACB.

Ans. θ_A = SF at A = 0.0125 radian, θ_C = 0.0075 radian, y_A = BM at A = 35mm, y_C = 11.667mm.

Q.10.35. A SS beam AB of span 8m carries a point load of 15 kN at the midspan point. If EI is constant and equal to 4000 kN-m^2, find the slope and deflection at the centre by using principle of conjugate beam.

Ans. $\theta_A = -\theta_B = -0.015$ radians, $y_C = 0.04$m (40mm).

Q.10.36. A SS beam AB of span 10m carries a variable distributed load with zero intensity at both the supports A and B, while the intensity gradually increases to 15kN/m at the centre C. If $EI = 12000$ kN-m^2, find the slope at supports and deflection at the centre C using conjugate beam method.

Ans. $\theta_A = -\theta_B = 0.03255$ radians, $y_C = $ **0.1042 m** (104.2 mm).

Q.10.37. A wooden beam of 3.0m length is symmetrically supported over 2m central span AB with 0.50m overhang D and E on each side. The overhang carry a load of 5 kN on each free end D and E. If EI for the wooden beam section = 100 kN-m^2, find the slopes at the supports and deflections at the free ends and the mid point by conjugate beam method.

Ans. $\theta_A = -\theta_B = +0.025$ radians, $\theta_D = -\theta_E = 0.03125$ radians, $y_A = y_B = 0$, (Hinge),

$y_D = y_E = 0.014583$m (14.583 mm) downward, $y_C = -$ **0.0125m** (12.5 mm) upward.

Q.10.38. A SS beam ACB of span 10m carries a point load of 20 kN at the midspan pooint C. Moment of Inertia of the cross-section is $2I$ from A to C, and I from C to B. Determine deflection at the middle point C, and slopes at A and B by using conjugate beam method. EI of the material = 500 kN-m^2.

Ans. $\theta_A = F'_A = +$ **0.0041667 radian**, $\theta_B = F_B = -$ **0.0052083 radian**, $\theta_C = F'_C = +$ **0.0010417 radian**, $y_C = M_C = 0.015625$m (15.625 mm)

Q.10.39. A SS beam $ADCEB$ of 8m span has middle half span (DCE) cross-section of Moment of Inertia (I) of 6×10^7 mm^4 while end onefourth spans have cross-section with $M.I. = 3 \times 10^7$ mm^4. Determine the slopes at end supports and deflection as the middle of span AB if the beam carries a point load of 20kN at the midspan point and material of the beam has $E = 200$ kN/mm^2. Use conjugate beam method.

Ans. $\theta_A = SF_A$ of conjugate beam, $\theta_A = SF_B$ of conjugate beam, $y_C = $ BM of conjugate beam at C. $EI = 12000$ kN-m^2

$$\theta_A = -\theta_B = 0.008333 \text{ radians} \left(\frac{5WL^2}{64EI}\right), y_C = \frac{3}{128}\frac{WL^3}{EI} = \mathbf{0.020m} \text{ (20mm).}$$

Q.10.40. A cantilever ACB of 4m span carries concentrated load of 5kN at its free end A. The member is solid circular in section with diameter of 200mm for 2m length from the fixed end and then the diameter becomes 100mm for the next 2m length upto free end. Find the slope and deflection at the midspan point C and free end A. E for the material = 200 kN/mm². Use conjugate beam method.

Ans. $EI_{AC} = \dfrac{\pi}{32} \times 10^4$ kN-m², $EI_{CB} = \dfrac{\pi \times 10^4}{2}$ kN-m², $\theta_A = SF'_A = -\,0.012102$ **radian,**

$\theta_C = SF'_C = -\,0.001911$ **radian**

$y_A = M'_A = \dfrac{23WL^3}{384EI} = 0.019533\text{m} \ (\textbf{19.533mm}), \ y_C = M'_C = \dfrac{5WL^3}{768EI} = 0.00212314\text{m}$

(2.12314mm).

Q.10.41. State deflection and slope equations in terms of strain energy due to bending.

$$y_x = \frac{1}{W} \int_0^L \frac{M_x^2 \, dx}{EI}, \text{ where '}M_x\text{' is BM in terms of distance } x \text{ and } W \text{ is the load at } X-X,$$

$$\theta = \frac{1}{\mu} \int_0^L \frac{M_x^2 \cdot dx}{EI}, \text{ where '}\mu\text{' is the moment couple where the slope is desired.}$$

Q.10.42. A cantilever AB span 3m carries a point load of 2kN at the free end A. Determine the deflection at the free end by strain energy method. EI for the beam section = 4000 kN-m².

Ans. $y_A = \dfrac{2}{W} \int_0^L \dfrac{(W \cdot x)^2}{2EI} \cdot dx = \left(\dfrac{Wx^3}{3EI} \right)_0^L = 0.0045\text{m}$ **(4.5mm)**

Q.10.43. State *unit load* method of finding deflection.

$$y_C = \int_0^L \frac{M_x \cdot m_x \cdot dx}{EI}, \text{ Where } M_x \text{ is BM at any point } X-X \text{ due to loading and } m_x \text{ is } BM \text{ at}$$

$X - X$ due to unit load applied at C (point where delection is required).

Q.10.44. A cantilever of span 4m carries an udl of 2kN/m over the entire 4m span. Calculate the deflection at the free end A and 2m from the free end by unit load method. EI for the beam section = 8000 kN-m²

Ans. $y_A = \int_0^4 \left(-\dfrac{2 \times x^2}{2} \right)(-1 \cdot x) \dfrac{dx}{EI} = \dfrac{2 \times (4)^4}{8 \times 8000} = 0.0080\text{m}$ **(8.0mm)**

$y_2 = \int_0^2 \left(-\dfrac{2x^2}{2} \right)(0) \dfrac{dx}{EI} + \int_2^4 \dfrac{2(x^2)}{2}(x-2) \dfrac{dx}{EI} = \dfrac{2}{24EI} [3(4)^4 - 4(2)(4)^3 + (2)^4]$

$= 0.002833\text{m (2.833mm)}$

Q.10.45. A simply supported beam ACB of 10m span carries a point load of 30kN at a point 6m from the LHS A. EI is constant and equal to 29500 kN-m². Determine deflection at the midspan point 'C' by using unit load method.

Ans. Apply unit load at C and find m_x for various sections 0 to 4 and 4 to 5m.

Also apply a load of 30kN at 4m from the LHS for symmetry.

$$y_C = \frac{1}{2} \left[2 \int_0^5 M_x \cdot m_x \cdot \frac{dx}{EI} \right] = \frac{2}{2EI} \left[\int_0^4 M_x \cdot m_x \cdot dx + \int_4^5 M_x \cdot m_x \cdot dx = \frac{1}{29500} \left[\frac{960}{3} + 270 \right] \right.$$

$= 0.0200\text{m (20.0mm)}$

Q.10.46. State *castigliano's theorem* for displacement by the *partial derivative* of the total strain energy U with respect to any one of the forces or moments, taken individually (including fictitious load) in its directions of action.

Ans. $\delta_1 = \dfrac{\partial U}{\partial W_1} = \displaystyle\int_0^L M_x\left(\dfrac{\partial M_x}{\partial W_1}\right)\dfrac{dx}{EI}$, W_1 may be fictitious load at the point where deflection is required.

$\theta_1 = \dfrac{\partial U}{\partial M_0} = \displaystyle\int_0^L M_x\dfrac{\partial M_x}{\partial M_0}\cdot\dfrac{dx}{EI}$, M_0 may be fictitious moment at the point where the slope is desired.

Q.10.47. A simply supported beam of 6m span carries udl of 5kN/m on the entire span. Calculate the deflection at the centre of span and slopes at the supports by Castigliano's theorem. $EI = 6000$ kN-m^2, constant.

Ans. $y_C = \dfrac{2}{EI}\displaystyle\int_0^3\left\{\left(\dfrac{5\times6}{2}+\dfrac{W_0}{2}\right)\cdot x - \dfrac{5\cdot x^2}{2}\right\}\dfrac{x}{2}\cdot dx = \dfrac{5}{384}\times\dfrac{5\times(6)^4}{6000} = \mathbf{0.0140625m}$ (14.0625mm)

$\theta_A = \dfrac{1}{EI}\displaystyle\int_0^6\left\{\left(\dfrac{5\times6}{2}-\dfrac{M_0}{6}\right)\cdot x - \dfrac{5x^2}{2}+M_0\right\}\left(1-\dfrac{x}{6}\right)dx = \dfrac{5\times6^3}{24\times6000}(6-8+3) = \mathbf{0.0075\ radian}$

Q.10.48. An overhanging beam ABC of 8m length is hinged at A and supported at B ($AB = 6$m) carries a moment couple of 20 kN-m. EI is constant and equal to 12000 kN-m^2. Determine deflection and slope of the over hanging end C by *Castigliano Theorem*.

Ans. $\theta_C = \dfrac{1}{EI}\displaystyle\int_0^6\dfrac{M\cdot x}{6}\dfrac{\partial M_x}{\partial M_0}\cdot dx + \dfrac{1}{EI}\displaystyle\int_6^8 M\cdot\dfrac{\partial M_x}{\partial M}\cdot dx = \dfrac{1}{12000}\left[\dfrac{20}{36}\times\dfrac{216}{3}+20(8-6)\right]$

$=\dfrac{80}{12000} = 0.00667$ radian.

$y_C = \dfrac{1}{EI}\displaystyle\int_0^6\dfrac{(M+2W_0)}{6}\cdot x\cdot\dfrac{x}{3}\cdot dx + \dfrac{1}{EI}\displaystyle\int_6^8\{M+W_0(8-x)\}(8-x)dx = \dfrac{6M}{EI} = \mathbf{0.010m}$ (10mm)

Q.10.49. A cantilever AB of span 3m carries uniformly varying load with zero untensity at the free end 'A' and 2kN/m at the fixed end 'B'. Using castigliano's theorem determine slope and deflection at the free end 'A' if EI is constant and equal to 540 kN-m^2.

Ans. $y_A = \displaystyle\int_0^3 M_x\cdot\dfrac{\partial M_x}{\partial W_0}\cdot\dfrac{dx}{EI} = \displaystyle\int_0^3\dfrac{wx^4}{6\times3}\cdot\dfrac{dx}{540} = \dfrac{2\times3^4}{30\times540} = \dfrac{1}{100}\ m = 0.010\text{m (10mm)}$

$\theta_A = \displaystyle\int_0^3 M_x\dfrac{\partial M_x}{\partial M_0}\dfrac{dx}{EI} = \displaystyle\int_0^3\dfrac{2\cdot x^3}{6\times3}\cdot\dfrac{dx}{540} = \dfrac{2\times3^3}{24\times540} = \dfrac{1}{240} = \mathbf{0.004167\ radian}$

Q.10.50. A SS beam of span 8m carries a triangular distribyted load with 10kN/m intensity at the centre C and zero at the two supports A and B. If EI is constant and is equal to 51200 kN-m^2, find the slopes at the supports A and B and deflection at the midspan point C by using castigliano's theorem.

Ans. $\theta_A = -\theta_B = \dfrac{1}{EI}\displaystyle\int_0^4 M_x\cdot\dfrac{\partial M_x}{\partial M_0}\cdot dx + \dfrac{1}{EI}\displaystyle\int_4^8 M_x\cdot\dfrac{\partial M_x}{\partial M_0}\cdot dx = \dfrac{10\times8^3(6-1)}{51200\times192} = \mathbf{0.0026042\ radian}$

$$y_C = \frac{1}{EI} \int_0^8 M_x \cdot \frac{\partial M_x}{\partial W_0} \cdot dx = \frac{2}{EI} \int_0^4 \left\{ \frac{10 \times 8}{4} \cdot x + \frac{W_0}{2} \cdot x - \frac{10}{3 \times 8} \cdot x^3 \right\} \left(\frac{x}{2} \right) dx = \frac{2 \times 10 \times 8^4}{51200} \times \frac{1}{240}$$

$$= 0.00667m \ (6.667mm)$$

Q.10.51. A simply supported beam of 90mm × 200mm deep has span of 10m and carries a point load of 10kN at the midspan point. The beam material has modulus of rigidity $G = 80 \ kN/mm^2$. Determine the total strain energy in the beam due to shear stresses.

Ans. Shear stress $q = \dfrac{F}{2I} \left(\dfrac{d^2}{4} - y^2 \right) = \dfrac{3W}{bd^3} \left(\dfrac{d^2}{4} - y^2 \right),$

Total starain energy $U_s = 2 \displaystyle\int_0^{L/2} \int_{-d/2}^{+\frac{d}{2}} \frac{q^2}{2G} \cdot b \cdot dy \cdot dx = \frac{9W^2}{G \cdot b \cdot d^6} \int_0^{\frac{L}{2}} \left\{ \frac{d^5}{16} + \frac{d^5}{80} - \frac{d^5}{24} \right\} dx$

$$= \frac{3W^2 \cdot L}{20G \cdot b \cdot d} = 104.167 \ \text{N-mm}$$

Q.10.52. (a) State the equation of deflection due to shear.

(b) A SS beam AB of 10m span carries a point load of 10kN at the midspan point C. Cross-section of the beam is rectangular 90mm width × 200mm depth. The shear modulus of elasticity of the beam material $G = 80 \ kN/mm^2$. The modulus of elasticity $E = 200 \ kN/mm^2$. Calculate the *total deflection* at the centre due to **bending and shear**.

Ans. (a) $y_{shear} = \dfrac{2}{W} \displaystyle\int_0^L \int_{y_t}^{y_c} \frac{q^2}{2G} \cdot b \cdot dy \cdot dx$

(b) $y_{shear} = 0.020833$ mm, $y_{bending} = \dfrac{WL^3}{48EI} = 17.3611$ **mm**, $y_{total} = 17.38196$ **mm** (Deflection

due to shear is only **0.12%** and can be neglected for the purpose of practical designs)

Q.10.53. State implication of impact loading on deflections in comparison to gradual loading.

Ans. Deflection due to impact load (suddenly applied) will be *twice* **the deflection** caused by gradual loading ($\delta_{impact} = 2\delta_{static}$).

Deflection due to falling loads $\delta = \sqrt{2h \cdot \delta}$, when '$h$' is very large.

Deflection due to falling loads $\delta = \delta_s + \sqrt{\delta_s^2 + 2h\delta_s}$, δ_s = static deflection.

Q.10.54. A SS beam AB is of 10m span. A load of 10kN is dropped on to the beam at the midspan point from a height of 50mm. The cross-section of the beam is 80mm × 150mm rectangular. Determine the instantaneous deflection and instantaneous bending stress in the beam. E(material) = 200 kN/mm^2.

Ans. $I = 225 \times 10^5 mm^4$, $EI = 45 \times 10^{11}$ N-mm^2, $P.\delta = 2 \times 10^4$ ($\delta + 50$), $\delta = \dfrac{PL^3}{48EI}$

$\delta_{inst.} = $ **128.593mm**, $P_{inst.} = $ **2776N**, $\delta_{st.} = 46.296$ mm,

$f_{bending}$ (instantaneous) = 231.467 N/mm^2.

Unit V

Torsion in Shafts

11

Torsional Elements (Shafts)

11.1 INTRODUCTION

Shafts are structural components mainly subjected to torque (torsion) to transmit power. When a couple acts on a member perpendicular to its longitudinal axis, it tends to rotate the element about its longitudinal axis. The tendency to rotate is resisted by setting internal forces (opposite torque). Shafts are structural members used to transmit power in machines from one point to many other points.

Shafts are supported at various points and carry its own weight *in its axial plane* causing bending moment. Shafts are predominantly subjected to *torsion* and carries minor bending moment. Thus *Lateral Loads* acting along the axial plane causes bending moment while the loads acting in a perpendicular plane *not passing through* longitudinal axis causes *Torque* or *torsion* in the shaft (Fig. 11.1). If F is the *pair of* force acting at a distance 'd_2', then the torque $T = F.d_2$.

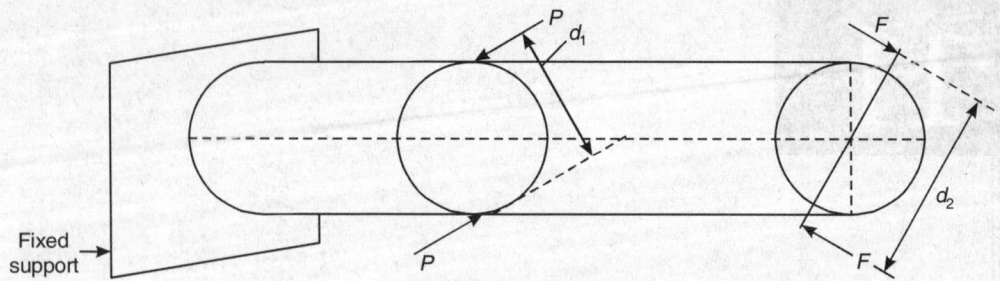

Fig. 11.1: Torsion in shaft

If resistive *pair of* force P acts at the surface of the shaft of diameter 'd_1', the resistive torque $T_1 = P.d_1$.

11.2 TORSION EQUATION

We have derived bending equation to relate bending stress, curvature, and bending moment of resistance, etc making certain assumptions. Torsion equation can also be derived to relate torsion, shear stress, and shear modulus of elasticity by making certain *assumptions* to simplify the derivation. These assumptions are :

i. Shafts are considered of *uniform circular section*;

ii. Cross-sections which are *plane before* torsion *remain plane after* application of torsion;

iii. Torque is *uniform along the longitudinal axis,* i.e. a same torque at the same radial distance from the axis on all sections;

iv. *Radial lines* of circular shafts *remain straight* even after application of torque;

v. *Shear stresses* in the shaft remains within the *elastic limit.*

vi. The material of the shaft is *homogeneous and isotropic.*

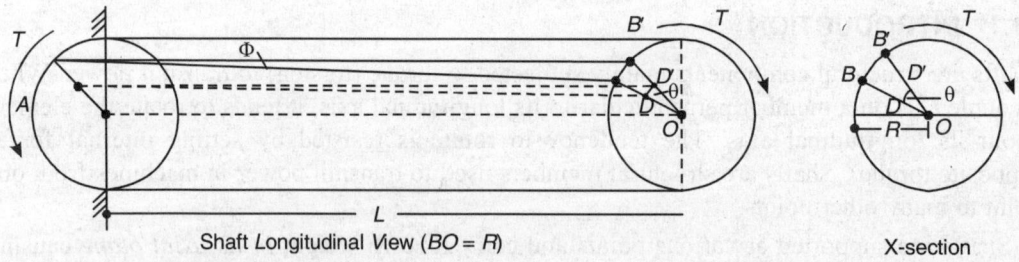

Shaft Longitudinal View (BO = R)   X-section

Fig. 11.2: Circular shaft (centre O, radius = R)

It may be noted that above assumptions hold good *more accurately* when the *angle of twist* θ *is small.* In case of small *angle of twist* θ, the stress obtained by the Torsion equation based on assumptions and the stress measured *practically* are *equal.* As the angle of twist θ increases, the stresses obtained practically and by torsion equation differs. For practical *design* purposes, the torsion equation provides fairly good results.

If a circular shaft rotates with *uniform speed* under a torque '*T*' or the shaft is *stationary* under the torque '*T*', the stresses and strains are the same in both the cases. For deriving torsion equation, assume one end of the shaft as fixed and the other end subjected to a torque '*T*' as shown in Fig. 11.2. The fixed end will offer a resisting torque '*T*' of opposite nature.

Consider a line *AB* on the surface of the shaft with A on the fixed end. With the application of twisting moment '*T*' the line *AB* distorts to *AB'* deforming radius *OB* on the free end section to *OB'* with the angle of twist θ (Fig. 11.2). If the shear stress on the surface *BB'* is say 'f_S' and the shear modulus of elasticity is *G*, then by basic principle

$$\phi = \frac{f_S \text{ (Shear stress)}}{G \text{ (Shear modulus)}} \qquad \qquad \text{... (i)}$$

or
$$\frac{BB'}{AB} = \frac{f_S}{G}$$

or
$$\frac{R \cdot \theta}{AB} = \frac{f_S}{G} \quad \text{(Since θ and φ are very small)}$$

or
$$\frac{f_S}{G} = \frac{R \cdot \theta}{L}$$

or
$$\frac{f_S}{R} = \frac{G\theta}{L} \qquad \qquad \text{... (ii)}$$

Similarly internal element *CD* located at a radius '*r*' will deform

$$\phi_1 = \frac{q \text{ (Shear stress at radius } r)}{G} \leftarrow \text{(Shear modulus)}$$

or
$$\frac{DD_1}{CD} = \phi_1 = \frac{q}{G} \quad \text{or} \quad \frac{r \cdot \theta}{L} = \phi_1 = \frac{q}{G}$$

or
$$\frac{q}{r} = \frac{G\theta}{L} \qquad \qquad \text{... (iii)}$$

From (ii) and (iii) we have $\dfrac{f_S}{R} = \dfrac{G\theta}{L} = \dfrac{q}{r}$ \qquad \qquad \text{... (iv)}

Thus, it can be stated that the shear stress '*q*' in any fibre in the cross–section is proportional to its radial distnace '*r*' from the central axis of the shaft. The shear stress 'f_S' at the surface is maximum while shear stress *q* at the central axis will be *zero*. Thus, the general equation is

$$\frac{q}{r} = \frac{G\theta}{L} = \frac{f_S}{R} \qquad \qquad \text{... (11.1)}$$

Twisting Moment Resistance (T_r)

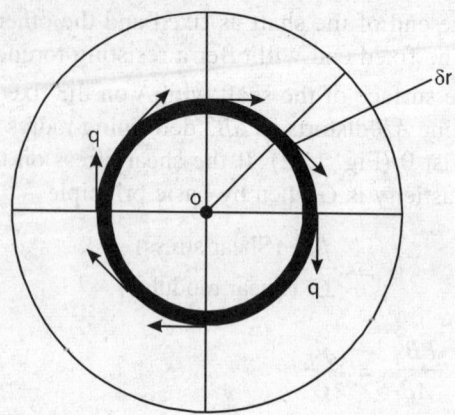

Fig. 11.3: Cross-section of shaft with shear resistance at 'r'

Consider a small elementary strip (ring) of thickness 'δ_r' at a radius 'r' from the central axis at 0 (Fig. 11.3). The shear stress at this strip will be

$$q = \frac{f_S}{R} \cdot r \qquad \qquad \text{... (v)}$$

Shear force on the strip = (area of strip) × (shear stress)

i.e. $$force = (2\pi r.\ \delta_r).q = (2\pi r.\ \delta_r)\ \frac{f_S}{R} \cdot r = \delta_a \cdot \frac{f_S}{R} \cdot r \qquad \text{... (vi)}$$

Elementary torque moment along the strip of area δ_a will be

$$\delta_T = \left(\delta_a \cdot \frac{f_S}{R} \cdot r\right) \cdot r = \frac{f_S}{R}(\delta_a \cdot r^2)$$

Total torque moment of the whole section = $\Sigma\ \delta_T = \int_0^R dT = \Sigma\ \frac{f_S}{R}(\delta_a \cdot r^2)$

or $$T_r = \sum_0^R \frac{f_S}{R}(\delta_a.r^2) = \int_0^R \frac{f_S}{R}(r^2)d_a = \frac{f_S}{R}\int_0^R r^2.d_a = \frac{f_S}{R}(I_p)$$

or $$\frac{T_r}{I_p} = \frac{f_S}{R} \qquad \qquad \text{... (vii)}$$

From (iv) and (vii) $$\frac{f_S}{R} = \frac{G\theta}{L} = \frac{T_r}{I_p} \qquad \qquad \text{... (11.2)}$$

$$I_p = \int_0^R r^2.d_a = \int_0^R r^2.(2\pi r.dr) = \int_0^R 2\pi r^3 \cdot dr$$

or
$$I_p = 2\pi \left(\frac{r^4}{4}\right)_0^R = \frac{\pi}{2}(R^4) = \frac{\pi D^4}{32} \qquad \text{... (viii)}$$

Thus I_p (Solid circular shaft) $= \dfrac{\pi D^4}{32} = \dfrac{\pi R^4}{2}$ ... (11.3)

$$\frac{T}{I_p} = \frac{f_S}{R} \quad \text{or} \quad T = \frac{f_S \cdot I_p}{R} = \frac{f_S \cdot \pi R^3}{2}$$

or
$$T = \frac{\pi R^3}{2} \cdot f_S = \frac{\pi D^3}{16} \cdot f_S \qquad \text{... (11.4)}$$

The term $\dfrac{I_p}{R}$ is known as *polar section modulus* of the shaft (Z_p)

Thus $\qquad\qquad\qquad T = Z_P \cdot f_S$ ... (11.5)

For a shaft of given material (f_S), the maximum permissible shear stress (f_S) is fixed and thus the maximum *twisting moment* which the shaft can withstand is proportional to the *polar modulus* of the shaft. Thus the **polar section modulus** of the shaft **represents the measure of the torsional strength** of the shaft.

The torsion equation $\qquad \dfrac{T}{I_p} = \dfrac{f_S}{R} = \dfrac{G\theta}{L}$, is quite similar to bending equation

$$\frac{M}{I} = \frac{f_b}{y} = \frac{E}{R},$$

Where : T = torsional moment, while M = bending moment,

$\qquad\quad I_p$ = Polar moment of inertia, while I = Moment of inertia about axis of *bending*.

$\qquad\quad f_S$ = Shear stress at the surface, while f_b = bending stress in the extreme fibre;

$\qquad\quad R$ = Radial distance of extreme fibre, while y = distance of extreme fibre from NA

$\qquad\quad G$ = Shear modulus, while E = Elastic modulus;

$\qquad\quad \dfrac{\theta}{L}$ = Torsional strain (twist) / unit length, while $\dfrac{1}{R}$ = Curvature due to bending.

Hollow Circular Shafts

The torsion equation equally holds good both for solid and hollow shafts. Consider a hollow circular shaft subjected to a torque T. The shaft has outer radius = R, inner radius = r_0

Shear stress at the surface at a radius $R = f_S$

consider a small ring strip of thickness = δ_r, at a radius r

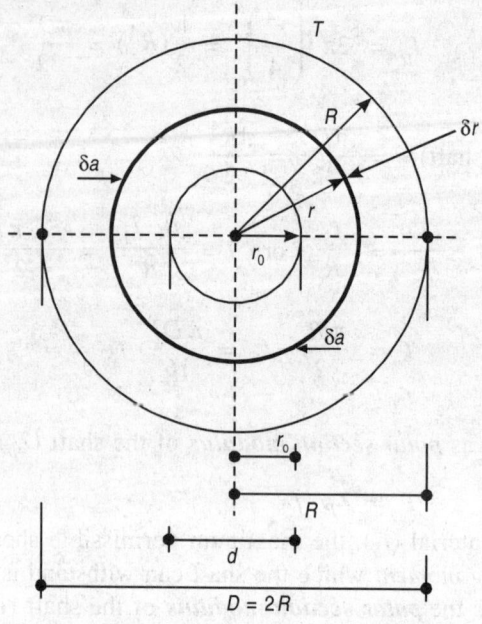

Fig. 11.4: Hollow shaft (outer R, inner r_0)

Turning moment of strip $\delta_T = q\ (2\pi r \cdot \delta r) \cdot r = 2\pi r^2 \cdot \dfrac{f_S}{R} \cdot r \delta_r$

Integrating : $\qquad \displaystyle\int d_T \ = \ \int_{r_0}^{R} 2\pi \cdot r^3 \cdot \frac{f_S}{R} \cdot dr$

or $\qquad T = \left.\dfrac{2\pi f_S}{R}\left(\dfrac{r^4}{4}\right)\right|_{r_0}^{R} = \dfrac{2\pi f_S}{4\ R}(R^4 - r_0^4) = \dfrac{\pi}{2} \cdot \dfrac{f_S}{R}(R^4 - r_0^4)$

$$R = \frac{D}{2},\ r_0 = \frac{d}{2}, \quad \therefore\ T = \frac{\pi f_S}{32\dfrac{D}{2}}(D^4 - d^4) = \frac{\pi f_S}{16D}(D^4 - d^4)$$

Torsional strength of hollow shafts,

$$T = \frac{\pi f_S}{16D}(D^4 - d^4) \qquad\qquad \text{... (11.6)}$$

The equation $\dfrac{T}{I_p} = \dfrac{f_S}{R} = \dfrac{G\theta}{L} = \dfrac{q}{r}$, also holds good for hollow shafts. $\qquad$ ... (11.2)

We have earlier derived the relationship of various modulii

$$E = 2G(1 + \mu) = 3K(1 - 2\mu),$$

where, E = Modulus of elasticity

G = Shear modulus or modulus of rigidity

K = Bulk modulus of elasticity

$\dfrac{1}{m}$ or μ = Poison's ratio

Thin Circular Tube Subjected to Torsion

(Thin circular Tube shafts)

Consider a thin circular tube of external diameter D and wall thickness 't', t being very small as compared to diameter D.

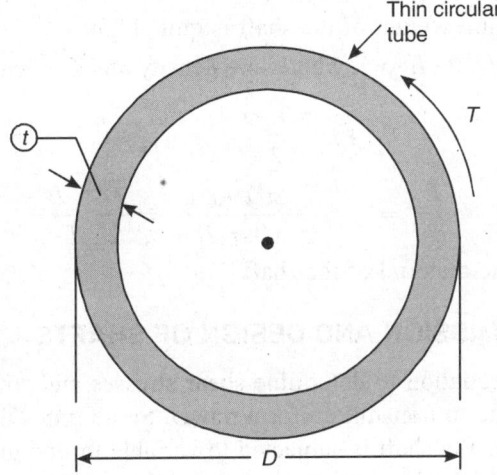

Thin circular tube

T

t

D

Fig. 11.5: Thin hollow circular shaft

Polar moment of Inertia of the tube section

$$I_p = \text{Area of the section} \times (\text{radial distance})^2$$

or

$$I_p = \pi D \cdot t \left(\frac{D}{2}\right)^2 = \frac{\pi}{4} D^3 \cdot t$$

Torsional Resistance :

$$\frac{T}{I_p} = \frac{f_S}{R}$$

or

$$T = f_S \cdot \frac{I_p}{R} = f_S \cdot \frac{\pi D^3 \cdot t}{4 \cdot D/2} = f_S \frac{\pi D^2 \cdot t}{2} \qquad \qquad \dots (11.7)$$

Angle of twist

$$\frac{T}{I_p} = \frac{G\theta}{L} \text{ or } \theta = \frac{TL}{G \cdot I_p} = \frac{T \cdot L}{G\left(\dfrac{\pi D^3 \cdot t}{4}\right)} = \frac{4T \cdot L}{\pi D^3 \cdot t G} \quad \text{... (11.8)}$$

Where I_p = Polar moment of inertia of the tube
  G = Modulus of rigidity
  T = Torsional moment of resistance
  D = Diameter of the tube (external)
  t = Thickness of the tube wall
  L = Length of the shaft
  f_S = Shear stress due to torsion (maximum limit).

Weight of tube shafts are quite less compared to solid shafts.

Torsional strength per unit weight of the shaft is quite high.

Weight of tube $W = (\pi D \cdot t \cdot L \cdot w)$, where w = density and L = length of the shaft

Torsion $\qquad\qquad T = f_S \cdot \dfrac{\pi D^2 \cdot t}{2}$

$$\therefore \qquad \frac{T}{W} = f_S \cdot \frac{\pi D^2 \cdot t}{2 \cdot \pi D \cdot t \cdot l \cdot w} = \frac{f_S}{2} \cdot \frac{D}{L \cdot w}, \qquad \text{... (11.9)}$$

where w = density of the material of the shaft.

11.3 POWER TRANSMISSION AND DESIGN OF SHAFTS

We have derived torsion equation to determine shear stresses and strains when the shafts are subjected to uniform torque to transmit uniform power. Some times the shafts transmit power to several points and hence the shaft is subjected to variable torque in different sections. Thus designing a shaft *uniformly* for the maximum torque will be *uneconomical*. The shaft is, thus, *designed for the maximum torque* in a particular section and the shaft of variable diameters may be provided. For this purpose let us establish *relationship* between *power* transmitted, *torque* and extreme *stresses*, etc.

Consider a shaft of radius R subjected to a pair of parallel and opposite forces 'P' tangentially along the surface of the shaft.

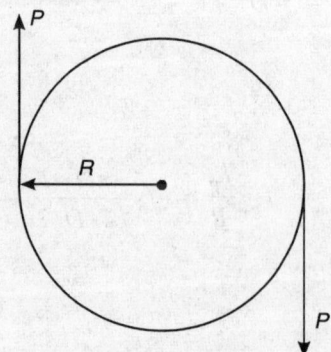

Fig. 11.6: Power transmission

The torque moment by these forces 'P' will be $T = P\ (2R)$ ... (i)

If the shaft rotates with a speed of 'n' rpm (rovolutions per minute).

Each force 'P' will travel (rotate) by a distance = $2\pi R \cdot n$ per minute. Work done by the two forces of P per minute will be

$P \times 2\ \pi R \cdot n \times 2$ per minute. ... (ii)

Since $2\ P \cdot R = T$, we have work done / minute $W = 2\pi \cdot T \cdot n$ ... (iii)

One $HP = 750$ N–m / Sec.

If Torque T is taken in N–m units, the shaft transmits power:

$$HP = \frac{T \cdot 2\pi\,n}{60 \times 750}, \ T \text{ in N–m}, n = RPM,$$ ... (11.10)

Similarly power transmitted in KW (Kilo–Watt) is obtained as

$$\text{Power } (KW) = \frac{T_{av}.2\pi.n}{60 \cdot 1000}, \text{ where } T \text{ is in N–m}, n = RPM, 2\pi \text{ radians (1 revolution)}$$

and $\dfrac{2\pi \cdot n}{60}$ radians / sec.

Power in Watts is work done in N–m /sec. Torque T is in N–m.

By using equation 11.2, the *shaft diameter* can be determined if the **permissible *maximum stresses* (f_S) and *power transmitted P or T and RPM (n)***, etc are given. In other cases if the shaft *diameter is given*, the *stress* or *angle of twist*, can be determined to transmit certain given *power*.

Whenever two set of data are available for the design of the shaft *greater of the two diameters* **shall** be adopted from safety point of view.

Considering a realistic generalised case of power transmitting shaft, at one end the shaft may be getting certain power P and transmitting this power to 2 – 3 points. The total power supplied :

$$P = P_1 + P_2 + P_3 + \textbf{Losses (if any)}$$ ... (11.11)

From power transmission, torque moments can be determined for various sections of the shaft. Torque moment diagram can be drawn for various sections of the shaft. Various sections of the shaft can be *designed* for the *maximum* power transmitted in the given section.

These principles and equations can be explained through solved examples, Relationship in various **units** are given as :

1 Metric Horse Power = 750 Watts = 750 N–m / sec, 1 Watt = 1 N–m / sec,

$GN = 10^9$ N, MN = 10^6 N, kN = 10^3 N

1 MN / m^2 = 1N / mm^2, 1GMP$_a$ = 10^9 N / m^2 = 1000 N/mm^2

EXAMPLE 11.1: A solid shaft of 80 mm diameter rotates at 150 rpm to transmit 100 Horse power. The shaft is 4m long and its shear modulus is 8×10^4 N/mm^2. Find the maximum shear stress and angle of twist.

Solution: $D = 80$mm, $n = 150$rpm, $G = 8 \times 10^4$ N/mm^2, $L = 4$m $= 4000$ mm

$$HP = 100 = 100 \times 750 \text{ N–m/s}, \quad I_p = \frac{\pi}{32}(80)^4 \text{ mm}^4$$

$$HP = \frac{T_{av.} \times 2\pi n}{60} \text{ or } 75000 \text{ N–m/s} = T_{av} \times 2\pi.\frac{150}{60}$$

or $\quad T_{av.} = \dfrac{75000 \times 60}{300\,\pi} = \dfrac{15000}{\pi}$ N–m $= 4772.72$ N–m

$$\frac{T}{I_p} = \frac{f_S}{R} = \frac{G\theta}{R} \text{ or } \frac{4772.72 \times 1000}{\dfrac{\pi}{32}(80)^4} = \frac{f_S}{40} = \frac{8 \times 10^4\theta}{4000}$$

$\therefore \quad f_S = \dfrac{4772.72 \times 1000 \times 40 \times 32}{\pi(8)^4 \times 10000} = \mathbf{47.5\ N/mm^2}$

$$\theta = \frac{4772.72 \times 32000}{\pi\ (8)^4 \times 10000} \times \frac{4000}{8 \times 10^4} = \frac{4772.72 \times 32 \times 4 \times 10^6}{\pi(8)^4 \times 8 \times 10^8}$$

$$= \mathbf{0.05937\ radian\ (3.4°)}$$

EXAMPLE 11.2: Determine HP transmitted by a 100mm solid shaft rotating at 150 rpm if the maximum permissible shear stress $= 60$ N/mm^2

Solution: $D = 100$ mm, $n = 150$ rpm, $f_S = 60$ N/mm^2

$$\frac{T'}{I_p} = \frac{f_S}{R},$$

$$T = \frac{\dfrac{\pi}{32}(100)^4}{50}.60 = 11.775 \times 10^6 \text{ N–mm.}$$

$$HP = \frac{T \times 2\pi \cdot n}{60 \times 750000} = \frac{11.775 \times 10^6 \times 2\pi \times 150}{60 \times 750000} = 10^2 \times \frac{11.775 \times 2\pi \times 15}{6 \times 75} = \mathbf{246.49}$$

Maximum horse power transmitted $= \mathbf{246.49\ HP}$.

EXAMPLE 11.3: A steel rod of 25mm diameter and gauge length of 200mm is subjected to a pull of 50kN. The gauge length is stretched 0.0975 mm. A torsion test piece from the same steel rod twists 0.025 radian over a length of 200mm when subjected to a torque of 0.4 kNm. Find the values of poisson's ratio and 3 elastic constants of the material.

Solution: $D = 25$ mm, $L = 200$ mm, $P = 50$ kN, $\delta_l = 0.0975$ mm.
torsion length $= 200$ mm, $\theta = 0.025$ rad, $T = 0.4$ kNm

$$E = \frac{P \cdot L}{A \cdot \delta_1} = \frac{50000 \times 200}{\frac{\pi}{4}(25)^2 \times 0.0975} = \frac{20 \times 10^6 \times 2}{\pi(625)(0.0975)}$$

$$= \mathbf{2.09 \times 10^5 \ N/mm^2} \ (209 \ GN/m^2)$$

$$\frac{T}{I_p} = \frac{G\theta}{L}$$

or $$G = \frac{T \cdot L}{I_p \cdot \theta} = \frac{400 \times 1000 \times 200}{\frac{\pi}{32}(D)^4 \cdot (0.0250)} = \frac{8 \times 32 \times 10^7}{\pi(25)^4(0.0250)} = \mathbf{0.8348 \times 10^5 \ N/mm^2}$$

$$G = \mathbf{0.8348 \times 10^5 \ N/mm^2} \ (83.48 \ GN/m^2)$$
$$E = 2G(1 + \mu) \ \text{or} \ 2.09 \times 10^5 = 2 \times 0.8348 \times 10^5 (1 + \mu)$$

or $$(1 + \mu) = \frac{2.09}{2 \times 0.8348} = 1.2517$$

∴ $$\mu = \mathbf{0.2517}$$

Also $$E = 3K(1 - 2\mu) \ \text{or} \ K = \frac{E}{3(1 - 2\mu)} = \frac{2.09 \times 10^5}{3(1 - 2 \times 0.2517)}$$

$$K = \frac{2.09 \times 10^5}{1.4898} = \mathbf{1.4029 \times 10^5 \ N/mm^2} \ (140.29 \ GN/m^2).$$

EXAMPLE 11.4: A solid circular shaft transmits 205 HP at 150 rpm. If the maximum torsion is 20% higher than the average and the permissible shear stress is 60 N/mm^2, determine the safe diameter of the shaft.

Solution: HP = 205, $n = 150$, $T_{max} = 1.20 \ T_{av.}$, $f_S = 60 \ N/mm^2$.

Power transmitted $= \dfrac{2\pi \, n.T_{av.}}{60}$, $T_{av.} = \dfrac{H.P. \times 750 \times 60}{2\pi n} = \dfrac{205 \times 750 \times 60}{2\pi(150)} = \mathbf{9793 \ N-m}$

$$T_{max} = 1.2 \ T_{av} = 1.2 \times 9793 \ N-m = 11751.6 \times 10^3 \ N-mm$$

$$\frac{T_m}{I_p} = \frac{f_S}{R} \ \text{or} \ \frac{T_m}{f_S} = \frac{I_p}{R} \ \text{or} \ \frac{\pi}{32} \frac{(D^4)2}{D} = \frac{11751.6 \times 10^3}{60}$$

or $$D^3 = \frac{11751.6 \times 100 \times 16}{6\pi} = 998012,$$

$$D = \mathbf{99.9336 \ mm} \ (\text{say} \ \mathbf{100 \ mm}).$$

EXAMPLE 11.5: A solid steel shaft transmits 300 HP at 200 rpm. If the maximum shear stress is 40 N/mm^2 and maximum angle of twist is *half* degree per metre length of the shaft, determine the safe diameter of the shaft. Modulus of rigidity of the shaft material $G = 8 \times 10^4$ N/mm^2.

Solution: $HP = 300$, $n = 200$ rpm, $f_S = 40$ N/mm^2, $G = 80000$ N/mm^2

$$HP = \frac{T_{av.} \times 2\pi n}{60 \times 750} \text{ or } \frac{T_{av.} 2\pi (200)}{60 \times 750} = 300 \qquad (\because 1\ HP = 750\ \text{N–m/s})$$

$$T_{av} = \frac{300 \times 60 \times 750}{400\,\pi} = 10748.4\ \text{N.m} = 10748400\ \text{N–mm}.$$

Also
$$f_S = \frac{T}{I_p} \cdot R = \frac{10748.40 \times 1000 \times 16}{\pi (D^3)} \text{ or } D^3 = \frac{10748.4 \times 16000}{\pi (40)}$$

$$= 1369.223 \times 10^3$$
$$D = \mathbf{111.04\ mm}.$$

Also
$$\frac{T}{I_p} = \frac{G\theta}{L} \text{ or } \frac{10748400 \times 32}{\pi (D^4)} = \frac{80000}{1000}\left(0.5\frac{\pi}{180}\right) = \frac{40\,\pi}{180}$$

$\therefore$
$$D^4 = \frac{10748400 \times 32}{\pi}\frac{18}{4\pi} = 15698.097 \times 10^4,\ D = \mathbf{111.93\ mm}$$

Thus safe diameter is greater of the two values, i.e. $D = \mathbf{111.93\ mm}$ (say 112 mm)

Safe diameter will be 112 mm where neither the stress will exceed 40 N/mm^2 nor the angle of twist will exceed 1/2°.

EXAMPLE 11.6: A hollow circular shaft transmits 300 HP at 200 rpm. Maximum shear stress is 70 N/mm^2 and maximum angle of twist is 0.5° per metre length and modulus of rigidity $G = 8 \times 10^4$ N/mm^2, determine the external and internal diameters.

Solution: $HP = 300$, $n = 200$ rpm, $f_S = 70$ N/mm^2, $\theta = \dfrac{0.5\pi}{180}$ rad, $G = 8 \times 10^4$N/mm^2

$$300 \times 750 = \frac{2\pi \times 200}{60} \cdot T_{av.},\ T_{av.} = \frac{300 \times 750 \times 60}{2\pi\,200} = 10748.4\ \text{N–m}$$

$$= 10748.4 \times 10^3\ \text{N–mm}$$

$$\frac{T}{I_p} \cdot R \leq 70.0$$

or
$$\frac{10748.4 \times 10^3}{\dfrac{\pi}{16D}(D^4 - d^4)} \leq 70.0$$

or
$$\frac{(D^4 - d^4)}{D} \geq \frac{10748.4 \times 10^3 \times 16}{\pi \times 70.0} = 781701.8 \qquad \text{... (i)}$$

$$\frac{T}{I_p} \leq \frac{G\theta}{L}$$

or
$$\frac{10748.4 \times 1000 \times 32\delta}{\pi \times (D^4 - d^4)} \leq \frac{8 \times 10^4}{1000} \times \frac{0.5\pi}{180}$$

or
$$(D^4 - d^4) \geq 156980972.7 \qquad \text{... (ii)}$$

Dividing (ii) by (i) $\dfrac{(D^4 - d^4)D}{(D^4 - d^4)} \geq \dfrac{156980972.7}{781701.8} = 200.8$

or
$$D \geq = \textbf{200.8 mm} \text{ (say } D = 200.8 \text{ mm)} \qquad \text{... (iii)}$$

From (ii) and (iii) $(200.8^4 - d^4) \leq 15698.09727$

or $(200.8^4 - 156980972.7) \geq d^4$

or $d^4 \leq (200.8^4 - 156980972.7)$

$$= (162575.4010 - 15698.0972.7)10^4 = 146877 \times 10^4$$

$$d \leq 195.767 \text{ (say } d = \textbf{195.7 mm)}.$$

EXAMPLE 11.7: A machine shaft is required to transmit 500 *HP* at 125 rpm. If the maximum shear stress is 60 N/mm^2, design a (i) solid (ii) hollow shaft with internal diameter 0.8 times external diameter. (iii) compare the weight of two shafts and percent saving.

Solution: $HP = 500$, $n = 125$ rpm, $f_S = 60$ N/mm^2, solid Dia. $= D_0$,

Hollow external diameter $= D$, internal dia. $d = 0.8D$.

$$\text{Power} = 500 \times 750 \text{ N–m/sec} = \frac{2\pi n \cdot T_{av.}}{60}$$

or
$$T_{av.} = \frac{500 \times 750 \times 60}{2\pi \times 125} = 28662.42 \text{ N–m}$$

$$= 2866.242 \times 10^4 \text{ N–mm}$$

i. Solid (D_0)

$$\frac{T}{I_p} = \frac{f_S}{R} \quad \text{or} \quad \frac{I_p}{R} = \frac{T}{f_S} \quad \text{or} \quad \frac{\pi}{16} . D_0^3 = \frac{2866.242 \times 10^4}{60}$$

or
$$D_0^3 = \frac{2866.242 \times 10^4 \times 16}{60\pi} = 2434.176 \times 10^3,$$

$$D_0 = \textbf{134.52 mm}$$

ii. Hollow shaft (D, $d = 0.8D$)

$$\frac{I_p}{R} = \frac{2866.242 \times 10^4}{60.0} \quad \text{or} \quad \frac{\pi}{16D}\left\{D^4 - (0.8D)^4\right\}$$

$$= \frac{2866.242 \times 10^3}{6} = 477.707 \times 10^3$$

or $\quad \dfrac{\pi}{16D}\left(D^4 - \dfrac{256}{625}D^4\right) = 477.707 \times 10^3$

or $\quad \dfrac{369}{625}\dfrac{\pi D^3}{16} = 477.707 \times 10^3$

or $\qquad D^3 = \dfrac{477.707 \times 10^3 \times 625 \times 16}{369\pi} = 4122.9265 \times 10^3,$

$$D = \textbf{160.35 mm}$$
$$d = 0.8 \times 160.35 = \textbf{128.28 mm}$$

iii. Weight of solid shaft $W_s = \dfrac{\pi}{4}(134.52)^2 \times \text{length} \times \text{density}$

$$W_s = \textbf{14205.07 L.w} \hspace{4cm} \text{... (i)}$$

Weight of hollow shaft $\quad W_h = \dfrac{\pi}{4}\{160.35^2 - 128.28^2\}\text{L.w}$

$$W_h = \dfrac{\pi}{4}\{25712.1225 - 16455.7584\} = \textbf{7266.2458 L.w}$$

$$\text{Ratio } W_s : W_h = \frac{14205.07 \; L.w}{7266.2458 \; LW} = 1.955$$

$$\text{Saving \%} = \frac{(W_s - W_h)}{W_s} \times 100 = \frac{(14205.07 - 7266.2458)}{14205.07} \times 100$$

$$= \textbf{48.85 \%}.$$

EXAMPLE 11.8: A solid steel shaft of 40 mm diameter and 2000 mm length is fixed at its two ends A and C. The shaft is subjected to a torque moment of 1000 N–m at a point B, 500mm away from the end A. (i) Determine fixed end moments at A and C. (ii) Find the maximum shear stress in the shaft. (iii) Find the angle of twist at the point of application of torque. Shear modulus $G = 8 \times 10^4$ N/mm^2.

Solution: $G = 80000$ N/mm^2, Dia. 'D' $= 40$ mm

$\quad L_1 = 500$ mm, $\; L_2 = 1500$ mm, $\; T = 1000$ N–m

$\quad f_S = ?$

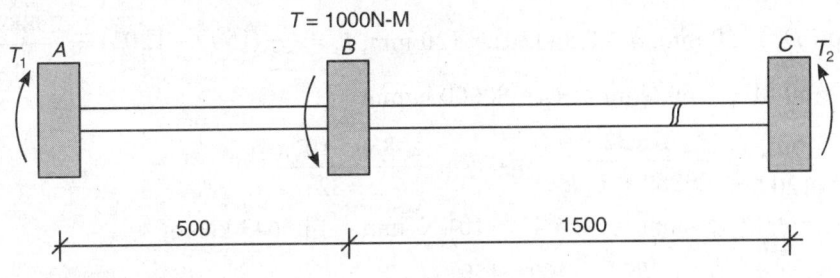

Fig. 11.7.

Torque of 1000 N–m at B is distributed to the shaft BA and BC such that

Total opposing torque $T_1 + T_2 = -T$ ... (i)

Also the angle of twist q at B will be same in BA and BC.

$$\frac{G\theta}{L_1} = \frac{f_{S_1}}{R} = \frac{T_1}{I_p} \text{ Also } \frac{G\theta}{L_2} = \frac{f_{S_2}}{R} = \frac{T_2}{I_p}$$

or $\quad \theta = \dfrac{T_1 L_1}{I_p G} = \dfrac{T_2 L_2}{I_p G}$

or $\quad \theta = \dfrac{T_1 L_1}{G I_p} = \dfrac{T_2 L_2}{G I_p}$, G and I_p are constant

$\therefore \quad T_1 L_1 = T_2 L_2$ or $T_1 = \dfrac{L_2}{L_1} \cdot T_2$

or $\quad T_1 = \dfrac{1500}{500} T_2$ or $T_1 = 3\,T_2$ ... (ii)

i. From (i) and (ii) $3\,T_2 + T_2 = 1000$, $T_2 = \dfrac{1000}{4} = 250$ N–m; $T_1 = 750$ N–m

ii. $\dfrac{f_{S_1}}{R} = \dfrac{T_1}{I_p}$ or $f_{S_1} = \dfrac{T_1 \cdot R}{I_p} = \dfrac{750 \times 1000}{\dfrac{\pi}{16}(D^3)} = \dfrac{75 \times 10^4 \times 16}{\pi (40)^3} = $ **59.71 N/mm^2**

iii. $\dfrac{G\theta}{L_1} = \dfrac{T_1}{I_p}$ or $\theta = \dfrac{T_1 L_1}{I_p \cdot G} = \dfrac{750000 \times 500 \times 32}{\pi (40)^4 \times 8 \times 10^4} = \dfrac{10^6 \times 75 \times 160}{10^8 \times \pi \times 256 \times 8}$

$\theta = $ **0.01866 radian** (1.07°).

EXAMPLE 11.9: A hollow circular shaft has an external diameter of 150 mm and the internal diameter is 80% of the external diameter. If the fibre stress at the inside surface is 60 MP$_a$ due to a torque T applied, find this torque, the maximum shear stress and the angle of twist per unit length. $G = 80$ GP$_a$.

Solution: $D = 150$ mm, $d = 0.8\,(150) = 120$ mm, $I_p = \dfrac{\pi}{32}(150^4 - 120^4) = \dfrac{29889 \times 10^4\,\pi}{32}$

$q = 60$ MP$_a$ = 60 N/mm^2, $G = 80000$ N/mm^2

or $\quad \dfrac{60}{120} = \dfrac{T \times 32}{29889 \times 10^4\,\pi}$, or $T = \dfrac{29889 \times 10^4\,\pi}{2 \times 32}$

$\qquad\qquad = \text{N–mm} = 1466.43 \times 10^4$ N–mm $= 14.6643$ kN–m

Also $\qquad \underset{\text{(max.)}}{f_S} = \dfrac{R}{r} \cdot q = \dfrac{60}{1.0} \times \dfrac{150}{120} = 75$ N/mm^2. (75 MP$_a$)

Angle of twist $\dfrac{G\theta}{L} = \dfrac{T}{I_p} = \dfrac{f_S}{R}$, $\theta = \dfrac{T\,L}{G \cdot I_p}$ or $\dfrac{f_S \cdot L}{R \cdot G}$

$$\theta = \dfrac{75 \times 1000}{150 \times 80,000} = \textbf{0.00625 radian} \ (0.3583°)$$

Also $\theta = \dfrac{1466.43 \times 10^4 \times 1000 \times 32}{80,000 \times 29889 \times 10^4\,\pi} = \textbf{0.00625 radian} \ (0.3583°).$

EXAMPLE 11.10: A solid shaft AD of 6m length and of 80 mm diameter is subjected to different torques as : AB ($L = 2$m), $T_B = -2000$ Nm (–) at B; BC ($L = 1.5$m), $T_C = +1500$ N–m at C, CD ($L = 2.5$m), $T_D = -1000$ Nm at D. Find the maximum shear stress and angular deformation of D with respect to A. $G = 80$ GP$_a$.

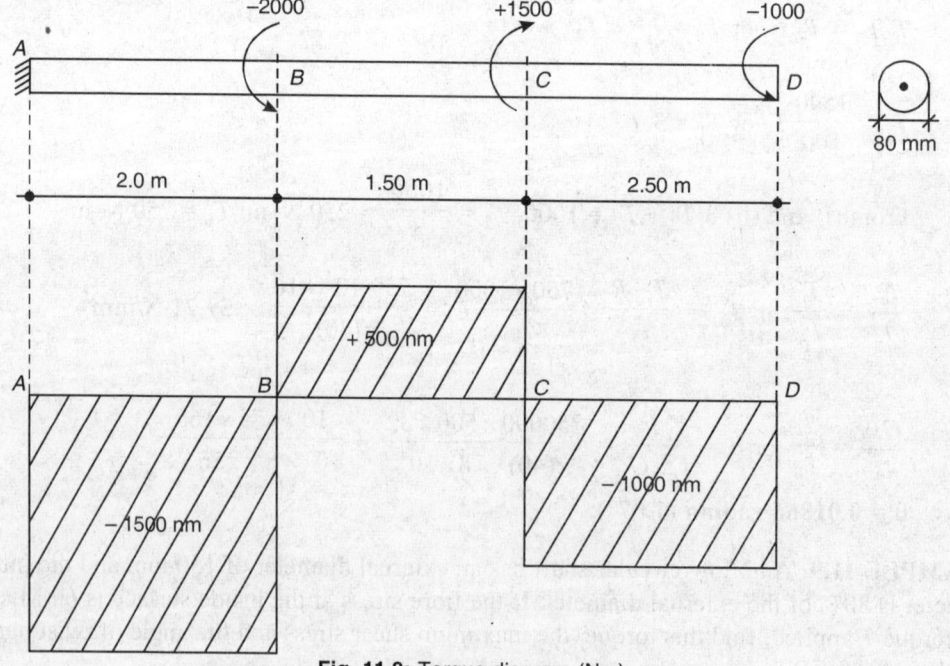

Fig. 11.8: Torque diagram (Nm)

Solution: $L_{AB} = 2m$, $L_{BC} = 1.5m$, $L_{CD} = 2.5m$, $G = 80000 \text{ N/mm}^2$, $D = 80 \text{ mm}$

$$I_p = \frac{\pi}{32}(80)^4 = \frac{4096\,\pi \times 10^4}{32} = (128\pi \times 10^4) \text{ mm}^4, \quad \frac{I_p}{R} = \frac{\pi}{16}(80)^3$$

$$Z_p = 32\pi \times 10^3 \text{ mm}^3.$$

From torque diagram shown in Fig. 11.7, we have : $T_{CD} = -1000$ Nm,

$T_{BC} = +1500 - 1000 = +500$ Nm, $T_{AB} = -2000 + 500 = -1500$ Nm

Since the shaft is of uniform diameter, the shear stress will be maximum wherever the torque applied is maximum. Thus the shaft section AB will have maximum shear stress.

$$\frac{f_S}{R} = \frac{T}{I_p} \quad \text{or} \quad f_S = \frac{T \cdot R}{I_p} = \frac{1500 \times 10^3}{32\,\pi \times 10^3} = \textbf{14.93 N/mm}^2$$

Angle of twist
$$\theta = \frac{T \cdot L}{G \cdot I_p},$$

Total deformation angle of D w.r.t A will be $\sum\left(\dfrac{TL}{GI_p}\right)$ ... (11.12)

$$\theta_{DC} = \frac{-1000 \times 10^3 . (2500)}{80{,}000 \times 128\pi\,10^4} = -0.00778 \text{ radian}$$

$$\theta_{BC} = \frac{+500 \times 10^3 \times 1500}{80{,}000 \times 128\pi\,10^4} = +0.00233 \text{ radian}$$

$$\theta_{AB} = \frac{-1500 \times 10^3 \times 2000}{80{,}000 \times 128\pi\,10^4} = -0.00933 \text{ radian}$$

$$\theta_{AD} = -0.00778 + 0.00233 - 0.00933$$
$$= \textbf{-0.01478 radian} \text{ (anticlockwise) } (0.8473°)$$

11.4 SHAFTS OF VARIABLE AND STEPPED SECTIONS

The shafts may be designed with uniformly varying diameter or stepped diameters according to power transmission or applied torque in a particular section. When the shaft diameters change in different steps, the total angle of twist is calculated as *algebraic sum of all the angle of twists for different sections.*

Angle of twist for *each section* is calculated by using the same general equation

$$\frac{T}{I_p} = \frac{f_S}{R} = \frac{G\theta}{L}, \text{ with usual notations explained earlier.} \qquad \text{... (11.13)}$$

In case of uniformly varying diameter with respect to its length 'x', the angle of twist $d\theta_x$ is first determined for a small strip of dx length by using the general equation

$$\frac{T}{I_{P_x}} = \frac{Gd\theta_x}{dx}.$$

For total angle of twist θ over the total length 'L', $d\theta_x$ is integrated within limits of length o to L.

i.e. $d\theta_x = \dfrac{T \cdot dx}{I_{P_x} G}$, $\qquad \theta = \displaystyle\int_0^L \dfrac{T \cdot dx}{I_{P_x} \cdot G}$, I_{P_x} is expressed in terms of x.

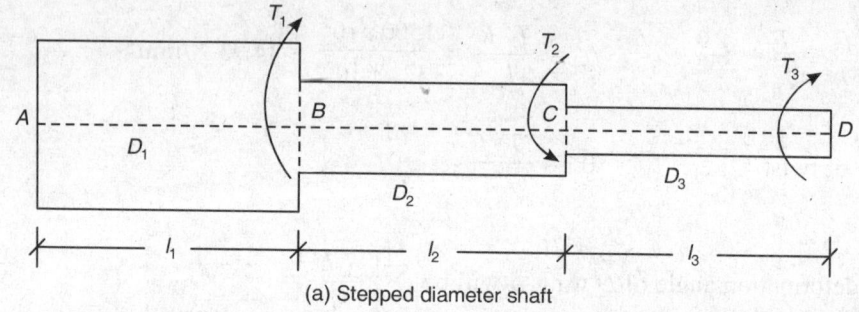

(a) Stepped diameter shaft

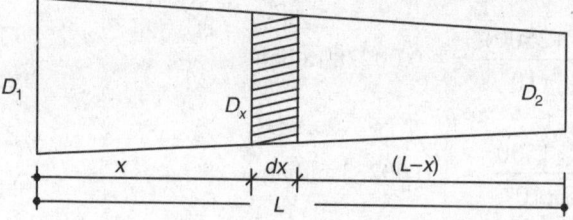

(b) Uniformly varying diameter shaft

Fig. 11.9: Variable diameter shafts

$$D_x = \left\{ D_1 - (D_1 - D_2)\frac{x}{L} \right\}, \quad I_{P_x} = \frac{\pi}{32} \cdot D_x^4$$

$$\therefore \quad I_{P_x} = \frac{\pi D_x^4}{32} = \frac{\pi}{32}\left\{ D_1 - \frac{(D_1 - D_2)}{L} x \right\}^4 = \frac{\pi}{32}\left\{ D_1 - \frac{x}{L}(D_1 - D_2) \right\}^4$$

$\delta\theta$ for the length dx will be

$$\frac{G \cdot d\theta_x}{dx} = \frac{T}{I_p}, \text{ or } d\theta = \frac{T \cdot dx}{I_p \cdot G} = \frac{T \cdot dx \cdot 32}{\pi\left\{ D_1 - \dfrac{(D_1 - D_2)}{L} x \right\}^4}$$

$$\theta = \int_0^L d\theta = \int_0^L \frac{32T\,dx}{\pi\left\{D_1 - \dfrac{(D_1 - D_2)}{L}x\right\}^4} = \frac{32T}{\pi}\int_0^L \frac{dx}{\left\{D_1 - \dfrac{x(D_1 - D_2)}{L}\right\}^4} \qquad \dots (11.14)$$

Thus θ can be obtained by integrating the expression within the given limits and given values. The process will be clarified by examples.

EXAMPLE 11.11: A solid circular shaft of 2.00 m length and diameters varying from 100 mm at one end to 50 mm at the other end. If the shaft is subjected to a torque of 1kN–m, find the maximum shear stress in the shaft and the angle of twist between the two ends. $G = 80\ GP_a$.

Solution: $D_1 = 50$ mm, $D_2 = 100$ mm, $L = 2000$ mm, $D_x = 50 + \left(\dfrac{100 - 50}{2000}\right)x = \left(50 + \dfrac{x}{40}\right)$,

$G = 80 \times 1000$ N/mm^2, $T = 1$ kN–m $= 10^6$ N–mm.

Shear stress will be maximum at the shaft of minimum section.

i.e. $\dfrac{f_S}{R} = \dfrac{T}{I_p}$ or $f_S = \dfrac{T}{I_p/R} = \dfrac{T}{\dfrac{\pi}{16}(D_1)^3} = \dfrac{10^6 \times 16}{\pi(50)^3} = \mathbf{40.764\ N/mm^2}$

$$\text{Minimum } f_S = \frac{16 \times 10^6}{\pi(100)^3} = \mathbf{5.0955\ N/mm^2}$$

Angle of twist $d\theta = \dfrac{T \cdot dx}{I_{P_x} G} = \dfrac{10^6 \times 32\,dx}{\pi\left(50 + \dfrac{x}{40}\right)^4 \times 80000} = \dfrac{400\,dx}{\pi\left(50 + \dfrac{x}{40}\right)^4}$ radian

$$\theta = \int_0^{2000} \frac{400\,dx}{\pi\left(50 + \dfrac{x}{40}\right)^4} = \frac{400}{\pi}\int_0^{2000}\frac{dx}{\left(50 + \dfrac{x}{40}\right)^4} = \frac{400}{\pi}\int_0^{2000}\left(50 + \frac{x}{40}\right)^{-4}dx$$

$$= \frac{400}{\pi}\left[\left(50 + \frac{x}{40}\right)^{-3}\right]_0^{2000}\left(\frac{1}{-3}\right)\left(\frac{40}{1}\right) = \frac{-400}{\pi}\left(\frac{40}{3}\right)\left\{\frac{1}{\left(50 + \dfrac{2000}{40}\right)^3} - \frac{1}{(50)^3}\right\}$$

$$= \frac{-400 \times 40}{3\pi}\left\{\frac{1}{(100)^3} - \frac{1}{50^3}\right\}$$

$$= \frac{-16000}{3\pi}\left(\frac{50^3 - 100^3}{100^3 \times 50^3}\right) = \frac{+16}{3\pi \times 50^3}(1000 - 125) = \mathbf{0.01189\ radian\ (0.6810°)}$$

11.5 FLANGED COUPLINGS AND KEYS

Power is transmitted from the rotating shaft through pulley on the shaft. The pulley is connected to the shaft with the help of a key. In this process, the key is also subjected to the twisting moment resulting in development of shear stress.

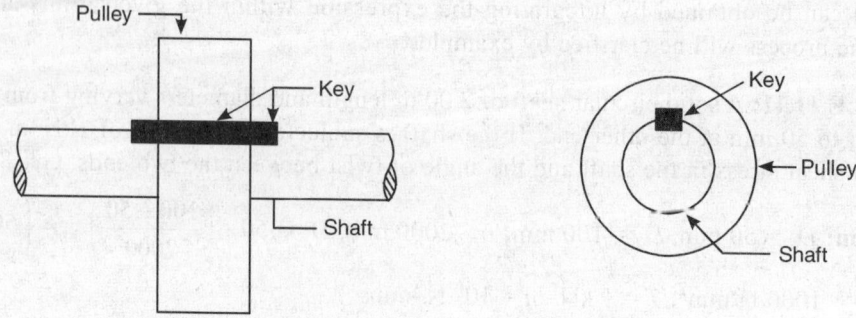

(a) Key for pulley and shaft connection

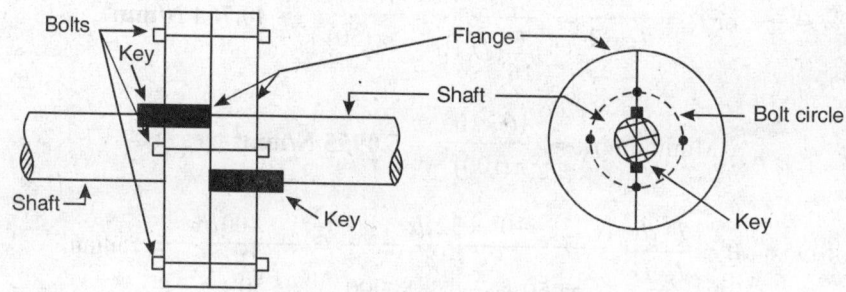

(b) Flange coupling connected with bolts

Fig. 11.10: Flange coupling and key

A rectangular **notch** is cut on the circumference of the shaft and a similar notch *is cut* in the inner side of the pulley or flange to house the key between the shaft and the pulley (or flange) as shown in Fig. 11.10. The pulley (or flange) is placed on the shaft in such a way that the two notches form a rectangular slot in which a rectangular key is inserted.

Let l_k = Length of the key

         b_k = width (thickness) of the key

         f_{sk} = shear stress in the key

         r_k = radial distance of the centre of the key from the shaft centre

$$= \frac{d}{2} \text{ (where } d \text{ is the shaft diameter).}$$

∴ The moment of resistance of the key = $f_{sk} \cdot l_k \cdot b_k \cdot r_k$

This must be equal to the torque T transmitted

$$\therefore \quad T = f_{sk}.\ l_k.\ b_k.\ r_k \ = \left(f_{sk}.l.b.\frac{d}{2} \right)$$

where T transmitted by the shaft $= \dfrac{\pi\, d^3}{16}.f_s$, (d is the diameter of the shaft)

Thus $\qquad\qquad \dfrac{\pi d^3}{16}.f_s \ = \ l \cdot b \cdot \dfrac{d}{2} \cdot f_{sk}$ ... (11.15)

Equation (11.15) provides f_{sk} if f_s, l, b and d are given.

Couplings

Two different sections of shaft are joined by a *flange* coupling. Flanges are fixed with the ends of the shaft sections to be joined. Flanges are connected with the shaft by fixing key in the **matching** notches cut in the shaft and the **flange or pulley**. Some times the pulley transmitting torque or power is used as flange. Matching holes are drilled in each flange. Holes in each of flanges of each section of the shaft are matched and conected with the help of bolts along certain circle. Bolt holes in the two flanges are cut along one circle or two circles symetrically. Torque is transmitted from one section of the shaft to the other section through a number of the bolts. Hence each bolt is subjected to a certain shear force.

If $\qquad\qquad\qquad n \ = $ no. of bolts along a circle of radius R_b ;

$\qquad\qquad\qquad d_b \ = $ diameter of each bolt;

$\qquad\qquad\qquad f_{sb} \ = $ shear stress in each bolt.

Total torque transmitted by n bolts placed along a circle of radius R_b will be

$$T \ = \ n\left(\frac{\pi}{4} d_b^2 \right) f_{sb}.R_b$$

T transmitted by the shaft $= \dfrac{\pi}{16} d^3 \cdot f_s$, where d is the diameter of the solid shaft.

Hence $\qquad\qquad T \ = \ \dfrac{\pi\, d^3}{16} f_s \ = \ n \cdot \left(\dfrac{\pi}{4} d_b^2 \cdot f_{sb} \right) R_b$

$$\therefore \qquad\qquad T \ = \ \frac{\pi d^3}{\ _s 16} \cdot f_s \ = \ n\left(\frac{\pi}{4} d_b^2 \cdot f_{sb} \right) \cdot R_b \ = \ f_{sk} \cdot l_k \cdot b_k \cdot \frac{d}{2} \qquad \text{... (11.16)}$$

From given values of d, f_s, d_b, R_b, n etc., f_{sb} can be found or from the given permissible stresses, bolts can be designed by using equation 11.16.

EXAMPLE 11.12: A shaft of 100mm diameter transmits 120 KWatt power at 200 rpm. A flanged coupling is keyed to the shaft by means of a key 80mm long and 20mm wide. The coupling has 8 bolts of 12mm diameter, symetrically arranged along a bolt circle of 280 mm diameter. Determine the shearing stresses in the shaft, the key, and the bolts of the coupling.

Solution: $d = 100$ mm, Power $= 120$ kW, rpm $= 200$, $l_k = 80$mm, $b_k = 20$mm,

$$d_b = 12 \text{ mm}, n = 8, R_b = \frac{280}{2} = 140 \text{ mm}$$

Torque $T_{av} = \dfrac{\text{Power}}{2\pi\left(\dfrac{200}{60}\right)} = \dfrac{120 \times 60}{2\pi(200)} = \textbf{5.7325 kN–m} = \textbf{5.7325} \times \textbf{10}^6 \textbf{ N–mm},$

$$\frac{T}{I_p} = \frac{f_S}{R}, \quad \therefore f_S = \frac{T}{I_p / R} = \frac{16T}{\pi(d)^3} = \frac{16 \times 5.7325 \times 10^6}{\pi(100)^3} = 29.21 \text{ N/mm}^2$$

$$f_{sk} \cdot l_k \cdot b_k \cdot \frac{d}{2} = T \text{ or } f_{sk}\left(80 \times 20 \times \frac{100}{2}\right) = 5.7325 \times 10^6 \text{ N–mm}, f_{sk} = \frac{5.7325 \times 10^6}{8 \times 10^4}$$

or $\qquad f_{sk} = \textbf{71.656 N/mm}^2.$

$$f_{sb} = \frac{T}{n\left(\dfrac{\pi}{4} d_b^2\right)(R_b)} = \frac{5.7325 \times 10^6 \times 4}{8(\pi 12^2)140} = 45.28 \text{ N/mm}^2.$$

EXAMPLE 11.13: A solid circular shaft of 80 mm diameter transmits 160 horse power at 200 rpm. Flange coupling is connected to the shaft with a key 150 mm long and 20 mm wide. Two flanges are connected with 5 bolts of 20mm diameter placed along a circle of 250 mm diameter.

Find the shear stresses in the (i) shaft, (ii) key, and (iii) bolts.

Solution: $HP = 160, \quad d = 80\text{mm}, \quad RPM = 200, \quad l_k = 150\text{mm}, \quad b_k = 20\text{mm},$

$$d_b = 20\text{mm}, n_b = 5, R_b = \frac{250}{2} \text{ mm}$$

$$H.P = \frac{2\pi \cdot N \cdot T_{av}}{60 \times 750}$$

or $\qquad T_{av} = \dfrac{160 \times 60 \times 750}{2\pi \times 200} \quad$ or $\quad T_{av} = 5730.0 \text{ N–m} = \textbf{573.0} \times \textbf{10}^4 \textbf{ N–mm}$

$$f_s \text{ (shaft)} = \frac{T}{I_p / R} = \frac{573.0 \times 10^4 \times 16}{\pi(80)^3} = \textbf{57.00 N/mm}^2$$

$$f_k = \frac{5730.0 \times 1000}{150 \times 20 \times 40} = \textbf{47.75 N/mm}^2$$

$$f_b = \frac{5730.0 \times 1000 \times 2 \times 4}{5 \times \pi(20)^2 \cdot 250} = \textbf{29.20 N/mm}^2.$$

EXAMPLE 11.14: A 100 mm diameter solid shaft transmits power at maximum shear stress of 70 N/mm², while the stress in key is 60 N/mm² and in coupling bolts it is 50 N/mm². If there are 4 bolts placed symmetrically at a bolt circle of 150 mm radius, design the bolt diameter and key length, if the width is 30 mm.

Solution: $d = 100$ mm, $f_s = 70$ N/mm², $f_{sk} = 60$ N/mm², $f_{sb} = 50$ N/mm², $n_b = 4$, $R_b = 150$ mm, $b_k = 30$ mm.

$$T = \frac{\pi}{16}(100)^3 \cdot f_s = \frac{\pi}{16}(10)^6 \times 70 = \textbf{13.75} \times \textbf{10}^6 \textbf{ N–mm}$$

$$L_k = \frac{T}{f_{sk} \cdot b_k \dfrac{d}{2}} = \frac{13.75 \times 10^6 \times 2}{60 \times 30 \times 100} = \textbf{152.78 mm}$$

$$\frac{\pi}{4}(d_b)^2 \cdot n_b \cdot f_{sb} \cdot R_b = T,$$

$$d_b^2 = \frac{T \times 4}{\pi \cdot n_b \cdot f_{sb} \cdot R_b} = \frac{13.75 \times 10^6 \times 4}{\pi(4)50 \times 150} = \textbf{583.864},$$

∴ $d_b = \textbf{24.16 mm} = \textbf{24.2 mm}$ (say).

EXAMPLE 11.15: A solid circular alloy shaft 50mm diameter is to be coupled in series with a hollow steel shaft of the same external diameter. The angle of twist per unit length of the steel shaft is to be 70 percent of the alloy shaft. Find the internal diameter of the hollow steel shaft. Also find the speed of the shaft so as to transmit 40 KW. if permissible shear stresses in the alloy and the steel are respectively 56 MN/m² and 80 MN/m². Modulus of rigidity for steel is 2.25 times the modulus of rigidity for the alloy.

Solution: $D_a = 50$ mm, $D_s = 50$ mm, $d_s = ?$ $f_a = 56$ MN/m² $= 56$ N/mm², $f_s = 80$ N/mm²

Power $= 40$ KW $= 40 \times 10^3$ N–m/s. ($\because 1$ watt $= 1$ N–m/s), $G_s = 2.25\ G_a$,

i. $\dfrac{T}{I_p} = \dfrac{G\theta}{L}$ or $\dfrac{\theta}{L} = \dfrac{T}{I_p \cdot G}$, $\therefore$ $\dfrac{T_s}{I_{P_s} \cdot G_s} = \dfrac{T_a \times 0.7}{I_{P_a} G_a}$

or $\dfrac{T_s}{\dfrac{\pi}{32}(D_s^4 - d_s^4)2.25 G_a} = \dfrac{T_a \times 0.7}{\dfrac{\pi}{32}(D_a^4) G_a}$

or $T_s = \dfrac{0.7 \times (D_s^4 - d_s^4)2.25 T_a}{D_a^4} = \dfrac{2.25 \times 0.7(50^4 - d_s^4)}{50^4} \cdot T_a = 1.575\dfrac{(50^4 - d_s^4)}{50^4} \cdot T_a$

or $T_s = 1.575\dfrac{(50^4 - d_s^4)}{50^4} \cdot T_a$ ... (i)

Torque transmitted will be the same since the shaft is continuous and moves at the same speed.

From (i) $\qquad \dfrac{T_s}{I_a} = 1.575\,\dfrac{(50^4 - d_s^4)}{50^4}$ or $\dfrac{50^4}{1.575} = (50^4 - d_s^4)$

or $\qquad d_s^4 = \dfrac{50^4\,(1.575 - 1)}{1.575}$

$\qquad d_s^4 = 50^4 \times \dfrac{0.575}{1.575}$, $\quad d = 50 \times 0.7773 = 38.866\ \text{mm} = \mathbf{38.86\ mm}$

(ii) Let the speed be n RPM

For the shaft : $\qquad \dfrac{f_s}{R_s} = \dfrac{G_s\,\theta_s}{L_s}$, $\quad \dfrac{f_a}{R_a} = \dfrac{G_a \cdot \theta_a}{L_a}$

Dividing the two expressions $\dfrac{f_s}{f_a} \cdot \dfrac{R_a}{R_s} = \dfrac{G_s}{G_a} \cdot \dfrac{\theta_s}{\theta_a} \cdot \dfrac{L_a}{L_s}$ or $\dfrac{f_s}{f_a} = \dfrac{G_s}{G_a} \cdot \dfrac{R_s}{R_a} \cdot \dfrac{L_a}{L_s} \cdot \dfrac{\theta_s}{\theta_a}$

or $\qquad \dfrac{f_s}{f_a} = 2.25 \times \dfrac{50}{50} \cdot \dfrac{\theta_s}{L_s} \cdot \dfrac{L_a}{\theta_a} = 2.25\,(0.70) = 1.575$ or $f_s = 1.575\,f_a$ $\qquad$ ... (ii)

$$f_s = \mathbf{80 N/mm^2}, f_a = \dfrac{80}{1.575} = 50.794\ \mathbf{N/mm^2} < 56\ \text{N/mm}^2,\ \text{permissible.}$$

Thus maximum permissible torque will be corresponding to $f_s = \mathbf{80\ N/mm^2}$.

$$\dfrac{80}{50/2} = \dfrac{T_s}{\dfrac{\pi}{32}\,(50^4 - 38.86^4)},$$

$$T_a = \dfrac{\pi}{16}\,(50)^3 \cdot (f_a) = \dfrac{\pi}{16}\,(50)^3 \times 50.794 = \mathbf{1246040\ N\text{--}mm} = \mathbf{1246.040\ N\text{--}m}$$

Power transmitted,

$$P = \dfrac{2\pi.\,nT}{60} = \dfrac{2\pi\,(1246.04)\,n}{60} = \text{N--m/s}$$

or $\quad 40 \times 10^3 = \dfrac{2\pi \times 1246.04\,n}{60}$, $\therefore\ n = \dfrac{40 \times 10^3 \times 60}{2\pi \times 1246.04} = \mathbf{306.7\ R.P.M.}$ (Say **306 RPM**).

EXAMPLE 11.16: A steel shaft is rigidity fixed within a sleeve of bronze metal. The compsite shaft is used to transmit power jointly. If the torque on the metal sleeve is 4 times the torque on the steel shaft, find the ratio of outer diameter of sleeve to the diameter of the steel shaft. Modulus of rigidity of steel is twice of metal.

Solution: $G_s = 2G_m$, Diameter of solid steel shaft $= d$, external dia. sleeve $= D$.

T_s = Torque of steel, T_m = torque shared by the metal sleeve = $4T_s$ (given),

$$\frac{G_s \theta}{L} = \frac{T_s}{I_s} \text{ or } \frac{\theta}{L} = \frac{T_s}{I_s \cdot G_s} = \frac{T_m}{I_m \cdot G_m},$$

Since θ/L is same for steel shaft and metal sleeve.

$$\therefore \quad \frac{T_s}{I_s \cdot G_s} = \frac{T_m}{I_m \cdot G_m} \text{ or } \frac{T_s}{T_m} = \frac{I_s \cdot G_s}{I_m \cdot G_m} = \frac{\dfrac{\pi}{32}(d)^4}{\dfrac{\pi}{32}(D^4 - d^4)} \cdot (2), \quad \frac{G_s}{G_m} = 2, \text{ given}$$

or $$\frac{2d^4}{(D^4 - d^4)} = \frac{T_s}{T_m} = \frac{1}{4}, \qquad \left(\frac{T_m}{T_s} = 4 \text{ given}\right)$$

or $(D^4 - d^4) = 8d^4$ or $D^4 = 9d^4$, $\left(\dfrac{D}{d}\right)^4 = 9$, $\therefore \dfrac{D}{d} = \mathbf{1.732}$.

Ratio of external diameter of sleeve to internal diameter of the sleeve is **1.732**.

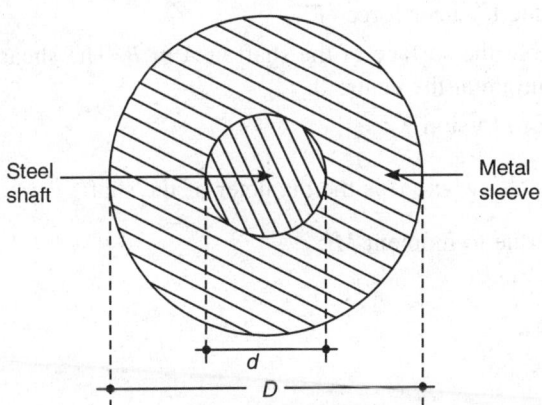

Fig. 11.11: Steel shaft with metallic sleeve

11.6 COMBINED BENDING AND TORSION

The shafts are generally used to transmit power but these are also subjected to transverse loading due to (i) self weight of shaft as uniformly distributed (ii) weight of pulleys (or flanges) as concentrated loads, and (iii) pull exerted by belt or rope drives. The combined

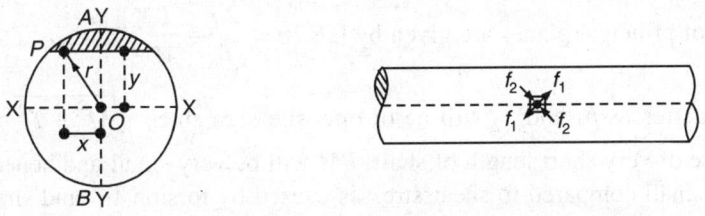

(a) Shaft X-section (b) Shaft-principal stresses

Fig. 11.12: Effect of combined stresses

effect of torsion and bending moments may result in *Principal* stresses which may be more than normally allowable stresses.

Consider a point '*P*', in the shaft, at a radial distance '*r*' from the shaft centre and at a distance '*y*' from the *NA* at the centre, shear stress (q) due to torsion T is given by

$$q = \frac{T \cdot r}{I_p} \qquad \text{... (i)}$$

The bending stress $f = \dfrac{M}{I_X} \cdot y$, *where 'M' is the BM at the section* ... (ii)

The shear stress caused from bending forces at the plane is given by

$$q' = \frac{F(A\bar{y})}{I_X Z}, \text{ where 'F' is the SF at the section} \qquad \text{... (iii)}$$

Thus, the component stresses at any point in the shaft are:

i. Shear stress q due to torsion T

ii. Bending stress 'f' (tensile or compressive) due to *BM* of *M*,

iii. Shear stress q' due to shear force '*F*'.

q and f are maximum at the surface of the shaft at *A* or *B*. The shear stress q' due to *SF* is zero at *A* or *B* and maximum at the centre 0.

(i) Shear stress q due to torsion T :

$$q = \frac{T}{I_P} \cdot R = \frac{16T}{\pi d^3}, \text{ where d is the diameter of the shaft} \qquad \text{... (iv)}$$

(ii) Bending stress f_b due to moment M :

$$f = \frac{M}{I_X} \cdot R = \frac{32M}{\pi d^3}, \qquad \text{... (v)}$$

The principal stresses are

$$p_{1,2} = \frac{f_b}{2} \pm \sqrt{\frac{f_b^2}{4} + q^2} = \frac{16M}{\pi d^3} \pm \sqrt{\left(\frac{16M}{\pi d^3}\right)^2 + \left(\frac{16T}{\pi d^3}\right)^2}$$

or $$p_{1,2} = \frac{16}{\pi d^3}\left[M \pm \sqrt{M^2 + T^2}\right] \qquad \text{... (11.17)}$$

The position of principal planes are given by $\tan 2\theta = \dfrac{2q}{f_b} = \dfrac{T}{M}$ ... (11.18)

The principal stresses p_1 and p_2 will be of opposite sign since $\sqrt{M^2 + T^2} > M$. It may be noted that in case of very short length of shaft, *BM* will be very small and hence bending stress f_b shall be very small compared to shear stresses caused by torsion '*T*' and shear force '*F*'. In such cases the maximum principal stresses may occur within the shaft. Usually, the maximum **principal stress occurs at the surface of the shaft.**

The design of the shaft is based on the following criterion:

- i. Maximum principal stress (tensile or compressive)
- ii. Maximum shear stress (positive or negative)
- iii. Maximum principal strain in the shaft (due to principal stresses)
- iv. Maximum elastic strain energy.
- v. Maximum elastic shear strain energy in the shaft (due to torsion)

i. *Maximum principal stress* $f = \dfrac{f_b}{2} + \sqrt{\dfrac{f_b^2}{4} + q^2} = \dfrac{16}{\pi d^3}\left[M + \sqrt{M^2 + T^2}\right]$ ... (11.19)

'f' must be within permissible limits

ii. *Maximum shear stress* $q_{max} = \dfrac{p_1 - p_2}{2} = \sqrt{\dfrac{f_b^2}{4} + q^2} = \dfrac{16}{\pi d^3}\left(\sqrt{M^2 + T^2}\right)$

$$= \dfrac{16}{\pi d^3} \cdot T_{eq} \qquad\qquad ... (11.20)$$

where $T_{eq} = \sqrt{M^2 + T^2}$ = Equivalent twisting moment.

iii. *Maximum principal strain criterion*

$f = (p_1 - \mu p_2) = \left(\dfrac{1}{2}f_b - \mu\dfrac{f_b}{2}\right) + (1+\mu)\sqrt{\dfrac{f_b^2}{4} + q^2}$, where p_1 and p_2 are principal

stresses due to f_b and q, μ = poisson's ratio.

$f = \dfrac{16}{\pi d^3}\left[M(1-\mu) + (1+\mu)\sqrt{M^2 + T^2}\right]$ ... (11.21)

iv. *Maximum elastic strain energy criterion*

equivalent stress $f^2 = p_1^2 + p_2^2 - 2p_1 p_2 \cdot \mu$

or $f^2 = f_b^2 + 2q^2(1+\mu) = \left(\dfrac{32M}{\pi d^3}\right)^2 + 2\left(\dfrac{16T}{\pi d^3}\right)^2 (1+\mu)$

or $f = \dfrac{16}{\pi d^3}\left[\sqrt{4M^2 + 2(1+\mu)T^2}\right]$ ... (11.22)

EXAMPLE 11.17: At a certain cross-section, a solid shaft of 120 mm diameter is subjected to a bending moment of 12 kNm, and a twisting moment of 18 kNm. Find the maximum principal stress in the section, indicating the position of the plane where it acts. Poisson's ratio $\mu = 0.25$, find the stress, which acting alone, may produce the same maximum (i) strain (ii) strain energy.

Solution: $d = 120$mm, $M = 12$ kNm $= 12 \times 10^6$ N–mm, $T = 18 \times 10^6$ N–mm

max. shear stress $\quad q = \dfrac{16T}{\pi d^3} = \dfrac{16 \times 18 \times 10^6}{\pi (120)^3} = \dfrac{16 \times 18 \times 10^3}{\pi \times 12 \times 12 \times 12} = \mathbf{53.08 \ N/mm^2}$

max. bending stress $\quad f_b = \dfrac{32M}{\pi d^3} = \dfrac{32 \times 12 \times 10^6}{\pi (120)^3} = \dfrac{32000}{\pi 144} = \mathbf{70.77 \ N/mm^2}$

$$\tan 2\theta = \frac{2q}{f_b} = \frac{T}{M} = \frac{18 \times 10^6}{12 \times 10^6} = \mathbf{1.5}$$

$2\theta = 56°18'$, $\theta_1 = 28°9'$ or $61°51'$ with the shaft axis.
$\theta_2 = 118°9'$ or $28°9'$

$$p_{1,2} = \frac{f_b}{2} \pm \sqrt{\frac{f_b^2}{4} + q^2} = \frac{70.77}{2} \pm \sqrt{\frac{70.77^2}{4} + 53.08^2}$$

$$= 35.385 \pm \sqrt{1252.1 + 2817.49}$$

$$= 35.385 \pm 63.793 = + \ \mathbf{99.18 - 28.41 \ N/mm^2}$$

i. For max. *strain*, we have $f = p_1 - \mu p_2 = 99.18 + 0.25\,(28.41) = \mathbf{106.28 \ N/mm^2}$

ii. For max. *strain energy* $f^2 = p_1^2 + p_2^2 - 2p_1 p_2 \mu$

$$= 99.18^2 + 28.41^2 + 2 \times 0.25 \ (99.18)(28.41) = 12052.6$$

$$f = \mathbf{109.78 \ N/mm^2}.$$

EXAMPLE 11.18: An electric motor rotates at 180 rpm and receives 520 KW power. The armature weight and magnetic pull of the poles exerts a force of 90 kN at the centre of the shaft supported between the two end bearing located 1200 mm apart centre to centre. The shaft has *ultimate tensile* and *shear strength* of 440 MN/m² and 400 MN/m² respectively. Determine the diameter of the shaft for a factor of safety of 5.

Solution: Max *BM* at the shaft centre

$$M = \frac{WL}{4} = \frac{90 \times 1200}{4} = 27000 \text{ kN–mm} = 27 \times 10^6 \text{ N–mm}$$

$$P = 520 \text{ kW, rpm} = 180, FS = 5,$$

max. tensile stress $\quad f_t = \dfrac{440}{5} = 88 \text{ MN/m}^2 = 88 \text{ N /mm}^2$

(f_s) max shear stress $\dfrac{400}{5} = 80 \text{ MN/m}^2 = 80 \text{ N/mm}^2 \ (f_s)$

$$\text{Power} = \frac{2\pi \cdot n}{60} \cdot T_{av}, \ T_{av} = \frac{520 \times 60}{2\pi \times 180} = 27.6 \text{ kNm} = \mathbf{27.6 \times 10^6 \ N\text{–}mm}$$

Equivalent T_e (combined moment and torsion) $= \sqrt{M^2 + T^2} = 10^6 \sqrt{27^2 + 27.6^2}$

$$T_e = 10^6 \,(38.61) \text{ N–mm.} = f_s \cdot \frac{\pi}{16}(d^3),$$

where f_s is max. permissible shear stress

or $\quad 80 \dfrac{\pi}{16}(d^3) = T_e = 38.61 \times 10^6$ or $d^3 = \dfrac{38.61 \times 10^6}{5\pi}$, $d_s = 100\,(2.45928)^{\frac{1}{3}}$

$\quad = 134.98 \; d_s = 135 \text{ mm (say)}$

Also $\quad M_e = \dfrac{M + \sqrt{M^2 + T^2}}{2} = \dfrac{27 \times 10^6 + 10^6 \sqrt{27^2 + 27.6^2}}{2} = \dfrac{10^6}{2}\left(27 + \sqrt{27^2 + 27.6^2}\right)$

$\qquad M_e = 32.805 \times 10^6 \text{ N–mm}$

or $\quad M_e = \dfrac{\pi}{32} d^3 \cdot f_t$, or $d^3 = \dfrac{M_e}{f_t} \cdot \dfrac{32}{\pi} = \dfrac{32.805 \times 10^6 \times 32}{88 \times 3.14} = 3.79907 \times 10^6$

$\therefore$ Safe 'd_b' $= 100\,(3.79907)^{\frac{1}{3}} = $ **156.04 mm**.

Safe diameter will be greater of d_s or $d_b =$ **156.04 mm**.

EXAMPLE 11.19: A hollow circular shaft is subjected to a torque of 80 kN–m and bending moment of 60 kN–m. The internal diameter is 3/4th the external diameter. If the maximum shear stress should not exceed 80 MN/m², find the diameter of the shaft.

Solution: $T = 80$ kN–m $= 80 \times 10^6$ N–mm, $BM = 60$ kN–m $= 60 \times 10^6$ N–mm, $d = \dfrac{3D}{4}$,

$\quad f_s$ (Permissible maximum shear stress) $= 80$ N/mm².

$\quad T_e$ (equivalent torsion) $= \sqrt{M^2 + T^2} = \sqrt{60^2 + 80^2} = $ **100 kN–m** $= 100 \times 10^6$ N–mm

$$T_e = f_s \cdot \frac{\pi}{16D}(D^4 - d^4) = 80 \frac{\pi}{16D}\left(D^4 - \frac{81}{256}D^4\right) = 5\pi\left(D^3 - \frac{81}{256}D^3\right)$$

or $\quad T_e = 100 \times 10^6 = 5\pi \dfrac{(256 - 81)}{256} D^3 = \dfrac{5 \times 175\,\pi}{256} D^3$.

$$D^3 = \frac{25600 \times 10^6}{5 \times 175\,\pi} = 9317561.42, \; D = \textbf{210.427 mm}, \; d = \textbf{157.82 mm}$$

The shaft with external diameter **210.5 mm** and internal diameter **157.8 mm**, will be ok.

11.7 SUMMARY–SHAFTS

Shafts are structural members used to *transmit power* in machines from one point to many other points and develop torsional stresses (shear) in itself. Torsion equation is

$$\frac{f_S}{R} = \frac{q}{r} = \frac{G\theta}{L} = \frac{T_{max}}{I_p}, \qquad \text{... (11.2)}$$

where $I_p = \dfrac{\pi}{32} D^4$ for solid and $I_p = \dfrac{\pi}{32}(D^4 - d^4)$ for hollow with usual notations.

Relations of elastic modulii $E = 2G(1 + \mu) = 3k(1 - 2\mu)$, with usual notations.

Power transmitted by rotating shafts

$$P \text{ (Watts)} = T_{av} \times \frac{2\pi n}{60}, \text{ where } T \text{ is in N–m and } n = \text{rpm} \qquad \text{... (11.10)}$$

$MN/m^2 = N/mm^2$, $GMP_a = 10^9 \ N/m^2 = 1000 \ N/mm^2$.

1 Metric Horse Power = 750 Watts = 750 N–m/sec, 1 KW = 1000 N–m/sec

Total power $\quad P = \Sigma\,(p_1 + p_2 + p_3 + ...)$ $\qquad \text{... (11.11)}$

Stepped section shafts consider each section, $\theta = \Sigma\,(\theta_1 + \theta_2 + ...) = \Sigma\,\dfrac{TL}{GI_p}$ $\qquad \text{... (11.13)}$

where $\theta_1 = \dfrac{T_1 L_1}{I_{P_1} \cdot G}$, G is shear modulus or modulus of rigidity.

Uniformly varying section $\theta = \displaystyle\int_0^L d\theta = \frac{32T}{\pi} \int_0^L \frac{dx}{\left\{ D_1 - \left(\dfrac{D_1 - D_2}{L} \cdot x \right) \right\}^4}$ $\qquad \text{... (11.14)}$

For keys and flange couplings, with usual notations, we have

$$T = f_s \frac{\pi d^3}{16} = f_{sk} \cdot l_k \cdot b_k \cdot \frac{d}{2} = n_b \left(\frac{\pi d_b^2}{4} \right) f_{sb} \cdot R_b \qquad \text{...(11.16)}$$

where d = diameter of shaft, d_b= diameter of the bolt, R_b = radius of bolt circle, etc.

Combined effect of bending and torsion plays important role in the design and failure of shafts under different modes.

i. Maximum principal criterion (direct) stress : $p_{1,2} = \dfrac{f_b}{2} \pm \sqrt{\dfrac{f_b^2}{4} + q^2}$

$$= \frac{16}{\pi d^3}\left[M \pm \sqrt{M^2 + T^2} \right] = \frac{16}{\pi d^3}\ M_e, \text{ where } M_e = \text{equivalent } BM \qquad \text{... (11.17)}$$

Principal plane tan $\quad 2\theta = \dfrac{2q}{f_b} = \dfrac{T}{M}$ $\qquad \text{... (11.18)}$

ii. Maximum shear stress criterion

$$q_{max} = \frac{p_1 - p_2}{2} = \sqrt{\frac{f_b^2}{4} + q^2} = \frac{16}{\pi d^3}\left(\sqrt{M^2 + T^2}\right) = \frac{16}{\pi d^3} \times T_e \quad \dots (11.20)$$

where T_e = Equivalent torsion = $\sqrt{M^2 + T^2}$,

iii. Maximum principal direct strains (μ = poisson's ratio)

$$f = (p_1 - \mu\, p_2) = \left[\left(\frac{f_b}{2} - \frac{\mu f_b}{2}\right) + (1+\mu)\sqrt{\frac{f_b^2}{4} + q^2}\right]$$

$$= \frac{16}{\pi d^3}\left[M(1+\mu) + (1+\mu)\sqrt{M^2 + T^2}\right] \quad \dots (11.21)$$

iv. Maximum elastic strain energy criterion :

$$f^2 = p_1^2 + p_2^2 - 2\mu\, p_1\, p_2 = f_b^2 + 2q^2(1+\mu)$$

$$f = \frac{16}{\pi d^3}\left[\sqrt{4M^2 + 2(1+\mu)T^2}\right], \text{ where } f = \text{equivalent stress} \quad \dots (11.22)$$

Thus for design appropriate condition of limiting stress (equivalent stress) is used.

EXERCISE 11

Q.11.1. Explain the functions and resistive forces developed in shafts.

Q.11.2. Derive the equation $\dfrac{f_s}{R} = \dfrac{q}{r} = \dfrac{G\theta}{L}$, with usual notations.

Q.11.3. Derive the equation $\dfrac{f_s}{R} = \dfrac{G\theta}{L} = \dfrac{T}{I_P}$, with usual notations.

Q.11.4. Prove that for a hollow circular shaft $T = \dfrac{\pi}{16D}(D^4 - d^4)f_s$, with usual notations.

Q.11.5. Prove that power transmitted in watts by a shaft rotating at 'n' revolutions per minute and subjected to a torque $T\,(N-m)$ will be

$$P = \frac{2\pi n}{60}\cdot T_{average},$$

Q.11.6. Determine the metric Horse Power transmitted by a 50 mm solid circular shaft rotating at 300 rpm, if the maximum permissible shear stress is 60 N/mm^2.

Ans. (61.62 H.P.)

Q.11.7. A solid circular shaft of 160 mm diameter rotates at 150 r.p.m. to transmit 200 H.P. The shaft is 4m long. If modulus of rigidity G of the shaft material is 80 GP_a, find the maximum shear stress and the angle of twist.

Ans. $f_s = 11.88$ N/mm^2, $\theta = 0.00743$ rad (0.426°).

Q.11.8. A solid circular shaft transmits 410 HP at 300 rpm. If the maximum torque is 20% higher than the average torque and the permissible shear stress is 120 N/mm^2, determine the safe diameter of the shaft.

Ans. $D = 79.52$ mm ≈ 80 mm say).

Q.11.9. A hollow circular shaft transmits 150 HP at 100 rpm. Maximum permissible shear stress is 80 N/mm^2 and the maximum angle of twist is limited to 1 degree per metre length. If the shear modulus $G = 80$ GP$_a$, determine the external and internal diameters of the shaft.

Ans. $D = 114.65$ mm (say 115 mm), d internal $= 99.5$ mm (say 99 mm)}.

Q.11.10. A machine shaft is required to transmit 400 HP at 100 rpm. If the maximum shear stress is limited to 80 MP$_a$, design a (i) solid and (ii) hollow shaft with internal diameter 3/4th of the external diameter (iii) find the % age saving in weight in case of hollow shaft.

Ans. $D_S = 122.22$ mm, $D_H = 138.74$ mm, $d_H = 104.0$ mm, % saving in weight $= 43.63$ %).

Q.11.11. A thin steel tube of 75 mm diameter and 3 mm thickness is subjected to torsion. The maximum shear stress is limited to 80 N/mm^2, $G = 80$ GP$_a$. Find the safe twisting moment that can be applied (ii) the twist angle in 600 mm length.

[**Hint:** $I_p = \pi Dt\ (R^2) = \dfrac{\pi D^3 \cdot t}{4}$, $T = \dfrac{f_s \cdot I_p}{R}$, $= \dfrac{f_s \pi D^2 \cdot t}{2}$, $\theta = \dfrac{4TL}{\pi D^3 \cdot t G}$,

i. $T = 2.12$ kN–m,   ii. $\theta = 0.92°$.

Q.11.12. A solid circular shaft of 50 mm diameter and 1500 mm length is fixed at its two ends A and C. The shaft is subjected to a torque moment of 1500 N–m at a point B 500 mm away from A. Determine the (i) fixed end moments at A and C, (ii) maximum and minimum shear stress in the shaft. (iii) Find the angle of twist at the point of application of torque. Shear modulus $G = 80$ GP$_a$.

Ans. $T_1 = 1000$ N-m, $T_2 = 500$ N-m, f_s (max) $= 40.74$ N/mm^2, $\theta = 0.01019$ radians $(0.584°)$.

Q.11.13. A propeller shaft of a ship has 200 mm external diameter and 160 mm inner diameter. The shaft rotates at 120 rpm and develops a stress 78 N/mm^2. Determine Horse Power transmitted and angle of twist in a length of 5m. $G = 84.4$ kN / mm^2.

Ans. $HP = 1213$, $\theta = 0.0462$ radians $= 2°39'$)

Q.11.14. A propeller shaft of a ship has 200 mm external diameter and 160 mm internal diameter and rotates at 120 rpm. If the angle of twist is 1° maximum in a length of 3m, determine the maximum shear stress in the shaft and the average HP transmitted if the maximum torque is 20% higher than average torque. $G = 80 \times 10^3$ N/mm^2.

Ans. $f_S = 46.56$ N/mm^2, $HP = 905$).

Q.11.15. A steel shaft transmits 190 HP at 200 rpm. If the angle of twist is not to exceed 1° in a length of 2m and maximum shear stress is not to exceed 50 N/mm^2, design the (i) solid shaft, (ii) hollow shaft with internal diameter 3/4th of external diameter. G (shear modulus) $= 80$ GP$_a$. Also find the ratio of weight of solid to hollow shaft.

Ans. $D_S = 99.8$ mm   (ii) $D_H = 110$ mm and $d_h = 82.5$ mm,   (iii) $W_S/W_H = 1.888$).

Q.11.16. Two solid shafts of 100 mm diameter are joined by a flange couplings to transmit 200 HP at 200 rpm. Flange coupling is connected to the shaft with the flat key of 100 mm × 30 mm. Two flanges are joined with 6 bolts of 20 mm diameter placed at a bolt circle of 300 mm diameter. Determine the shear stresses in key, shaft, and bolts.

Ans. $f_s = 36.46$ N/mm^2, $f_k = 47.73$ N/mm^2, $f_b = 25.35$ N/mm^2).

Q.11.17. A shaft *ABC* transmits torque of 1 kN–m at the pulley *A* (anticlock wise), 3 kN–m at the pulley *B* (clockwise) and 2 kN–m at the pulley *C* (anticlock wise) as shown in figure Q11.17. The shaft has diameters of 40 mm from *A* to *B* and 60 mm from *B* to *C* respectively. Determine maximum shear stresses in sections *AB* and *BC* and also the relative *angle of twist* between pulleys at *A* and *C*. *G* = 80 GP$_a$.

Ans. f_S (*AB*) 83.3 N/mm^2, f_S (*BC*) = 47.15 N/mm^2,   θ (*A* to *C*) = 0.01043 radian)

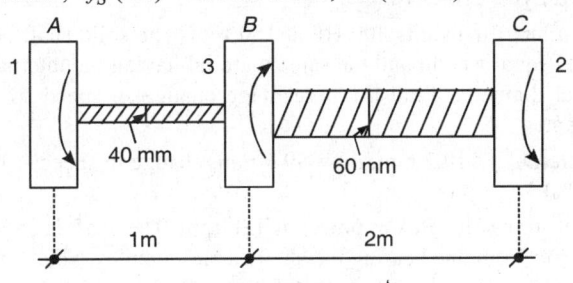

Fig. Q11.17.

Q.11.18. A shaft of 50 mm diameter and 3m length is fixed at both the ends. *A* torque of 3 kN–m acts at a distance of 1m from one end. Find the (i) torque at the two ends (ii) maximum shear stress, (iii) angle of twist at the point of application of torque.

Ans. *T* = 2 kN–m at one end, *T* = 1 kN–m at the other end,

(ii) f_S = 81.6 N/mm^2 ;   (iii) θ = 0.0398 radian)

Q.11.19. A solid shaft of 200 mm diameter is subjected to 48 kN–m torque and 32 kN–m bending moment at a certain point. Find the (i) maximum direct principal stress, (ii) maximum direct stress which produces maximum principal strains. Poisson's ratio μ = 0.3.

Ans. f_{max} = 57.1 N/mm^2, θ = $28°9'$ with the shaft axis,   (ii) f_{max} = 62.1 N/mm^2).

Q.11.20. A solid shaft of 100 mm is subjected to a torque of *T* and bending moment of *M* at certain point. The maximum principal stress is limited to 88 N/mm^2 and bending stress and shear stress are also of equal magnitude. Determine the values of *T* and *M*.

Ans. *T* = 10.668 kN–m, *M* = 5.334 kN–m).

Q.11.21. A propeller shaft of a ship has 160 mm external and 120 mm internal diameters respectively. The shaft transmits 605 HP at 105 rpm. The shaft is subjected to 4125 N–m bending moment and 44 kN axial thrust. Find the (i) Max. Principal stress and its direction (ii) direct stress which can produce equivalent maximum strain. Poisson's ratio μ = 0.3

[**Hint:** axial comp. stress = 5.0 N/mm^2, f_b = ± 15.0 N/mm^2, f_x (max) = 5 + 15 = 20 N/mm^2

$$T = \frac{605 \times 60 \times 750}{2\pi 105} = 41268 \text{ N–m}, \quad q = 75.0 \text{ N/mm}^2, \quad p_{1,2} = \frac{f_x}{2} \pm \sqrt{\frac{f_x^2}{4} + q^2}$$

$$= 85.664 \text{ N/mm}^2 \text{ (comp.)} - 65.66 \text{ N/mm}^2 \text{ (tensile)}$$

$$\theta_1 = 41.2°, q_{max} = \frac{p_1 - p_2}{2} = 75.66 \text{ N/mm}^2,$$

$$f_{(\text{max. strain})} = (p_1 - \mu p_2) = 105.36 \text{ N/mm}^2.$$

Q.11.22. A propeller shaft 160 mm external diameter and 80 mm internal diameter transmits 1000 kW of power at 100 rpm. It is also subjected to a *BM* of 8 kNm and a thrust of 120 kN. Find the (i) principal stresses (ii) maximum shear stress (iii) the direct stress which acting alone can produce the maximum strain. $\mu = 0.3$.

Ans. (i) $p_1 = 142.1$ N/mm^2 comp, $p_2 = 112.9$ N/mm^2 (tensile) (ii) $f_s = 127.5$ N/mm^2,

(iii) $f = (p_1 - \mu\, p_2) = 176.0$ N/mm^2.

Q.11.23. A solid circular shaft transmits 300 HP at 150 rpm. The solid shaft needs to be replaced by a hollow shaft of equal weight and the same material having the internal diameter equal to 50% of the external diameter. Find the percentage change in speed so that the HP transmitted remains the same.

Ans. $D_S = 103.3$ mm, $D_H = 119.2$ mm, $d_H = 59.6$ mm, change in speed = 103.56 rpm, % change in speed = 30.9 %).

Q.11.24. A circular shaft, transmits 50 kW power at 120 rpm. The shaft is supported in end bearings 4 m apart. At 1.5 m from one bearing the shaft carries a pulley which exerts a transverse load of 16 kN. Determine the suitable **diameter** of the shaft if (a) The maximum direct stress is not to exceed 120 N/mm^2, (b) The maximum shear stress is not to exceed 60 N/mm^2, (c) The stress which acting alone would produce the same maximum strain, and is not to exceed 120 N/mm^2 and (d) The direct stress which acting alone would produce the same *maximum strain energy* is not to exceed 120 N/mm^2. Take m = 0.30.

($T = 3.9789 \times 10^6$ N–mm, $M = 15 \times 10^6$ N–mm, d_1 (max. principal stress) = 109 mm, d_2 (max. shear stress) = 109.2 mm, (iii) d_3 (max. strain) = 109.2 mm, d_4 (max. strain energy) = 109.2 mm, **Adopt dia. of 110 mm**).

Q.11.25. A shaft of 150 mm diameter under combined bending and torsion develops a maximum direct stress and shear stress of 120 N/mm^2 and 80 N/mm^2 respectively. Find the bending moment (*M*) and torque (*T*). If the maximum shear stress limit is increased to 100 N/mm^2, find the increase in torsion if the bending remains constant.

[**Hint:** $f_{max} = \dfrac{16}{\pi d^3}\left[M + \sqrt{M^2 + T^2} \right]$, $q_{max} = \dfrac{16}{\pi d^3}\left[\sqrt{M^2 + T^2} \right]$, $M = 26.5 \times 10^6$ N–mm;

$T = 45.9 \times 10^6$ N–mm. Increase in torsion $T = 14.8 \times 10^6$ N–mm]

Q.11.26. A thin steel tube of 50 mm diameter and 2 mm thickness is subjected to torsion. The maximum shear stress is limited to **80 N/mm^2**, Modulus of rigidity $G = 80$GP$_a$. Find (i) shaft twisting moment that can be applied (ii) the twist angle in **400 mm length**.

[**Hint:** $T = f_s \dfrac{\pi D^2 \cdot t}{2}$, $\theta = \dfrac{4Tl}{\pi D^3 \cdot t \cdot G}$, (i) $T = 628.32$ N–m (ii) $\theta = 0.9170°$]

Q.11.27. A solid aluminium shaft of 100 mm diameter and 1 m length is coupled with hollow steel shaft of 100 mm external and 50 mm internal diameters and 2 m length end to end. The compound shaft is fixed at ends and a torque of 15 kN–m is applied at the coupling. Neglecting the effect of coupling find the

i. Maximum shear stresses in the two portions of the shaft.

ii. Angle of *twist* at the coupling.

$G_{steel} = 80$ GP$_a$, $G_{aluminium} = 25$ GP$_a$.

Ans. $f_{SS} = 48.8$ N/mm^2, $f_{sa} = 30.4$ N/mm^2, $\theta = 1.4°$]

Unit VI

Long and Short Columns: Load Capacity Analysis

12

Long and Short Columns in Structures

12.1 INTRODUCTION

Engineering Mechanics

Structural elements are subjected to loads in different ways. We have already studied structural elements having **lengths** much **larger** than the **cross-sectional** dimensions and subjected to **transverse loads** along the longitudinal axis. Such elements are called **beams** and offer mainly **shear and bending stress resistances** equivalent to shear force and bending moment respectively. If such structural elements are **subjected to axial compressive loads** instead of transverse loads, the elements are called **columns and struts**. Thus **columns and struts** are defined as structural elements mainly carrying **axial compressive forces**. Columns are called **short** if the **ratio of length** to its **minimum cross-sectional** dimension is not more than certain specified value for a given material and the element **basically fails by crushing** under **compressive axial loads**.

If the **ratio of length to its minimum cross-sectional dimension** is large and the element **fails by buckling** (bending) under **axial compressive loads**, the element is called **long column**.

The structural members, forming **part of a truss** and subjected to **compressive forces**, are called **struts**. When the structural elements are placed independently in **vertical direction** to carry compressive loads, such members are called **columns**. Generally the columns comprise of heavy sections carrying **heavy compressive loads**. **Struts** are of lighter sections and used for members of truss or frame structures. Struts can be **fixed** or **hinged** (or pinned) at **one or both ends**.

The members subjected to **axial compressive** forces also undergo lateral deflections by minor lateral excitation when the axial load exceeds certain limiting value. The compression members (column or strut) remain in stable equilibrium with its axis straight if the applied compressive loads are small. Loads beyond certain limit cause lateral deflection even by a minor disturbance. The **limiting** axial compressive **load**, upto which **no instability** of lateral deflection occurs is called **critical load** for specific ratio of length to minimum cross-sectional dimension.

The **failure** of compressive members can occur by:

i. **Pure compression (crushing)** in case of short column or strut.

ii. **Buckling** (bending) in case of very long column or strut.

iii. **Combination of compression and buckling** depending on the slenderness ratio of the member.

12.2   DEFINITIONS

Column is a long vertical member, subjected to an axial compressive load. Column is generally rigidly fixed at both ends.

Strut is a slender bar or member of a truss or frame in any position other than vertical and subjected to a compressive force and fixed rigidly or hinged or pinned at ends.

Slenderness ratio (K) is the ratio of unsupported length or effective length of the column (or strut) to the **minimum radius of gyration** of the cross-section of the compressive member. It has no units.

Buckling is the phenomenon representing lateral deflection of members under **axial compressive load**. If the loading is within certain limit, the lateral deflection disappears on removal of the axial compressive load. The lateral **deflection is elastic**. The equation of bending moment and slope deflection can be used to define geometric shape of the axis of column under load.

The structural member remains in of **stable** condition if the axial compressive load is within certain **limit** depending on the **material properties** and its **slenderness** ratio. When column member axis is straight under axial load P, a small lateral load W causes **deflection 'δ'**. The column axis returns to **original straight** position on removal of lateral load **W** if the axial compressive load **P** is small. This limiting **axial**

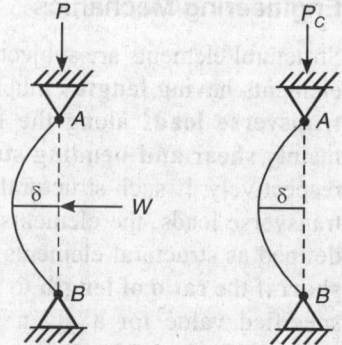

Fig. 12.1: Column stability

compressive load 'P' upto which the column axis retains its original position on removal of small lateral load W is called **critical load (P$_c$)**.

If this **limiting axial compressive load P$_c$** is increased by small amount, the lateral deflection **increases many fold** and ultimately the **column fails** by bending (also known as **buckling**). The column is subjected to **axial stresses** and large amount of **bending stresses** causing ultimately **failure of the column**. This axial **compressive load P$_c$** beyond which the column undergoes excessive lateral deflection and **fails primarily by bending stresses** is called **buckling load** P$_C$ or **crippling load**. For **design** of column sections this crippling or buckling load plays most important role.

12.3 END CONDITIONS

The long column axis deflects under axial compressive loads and the pattern and extent of deflection depends on the **end support condition**. Two ends of a column can be supported in any of the following conditions:

   i. Both ends **hinged** (or **pinned**).

   ii. One end **fixed** and other end **free**.

  iii. One end **fixed** and other end **hinged** (or **pinned**).

  iv. **Both ends fixed** rigidly.

Critical Euler's load for all these conditions shall be determined by studying slope-deflection patterns with respect to moment of resistance equation. As per moment-deflection

equation we have $\boxed{EI\,\dfrac{d^2y}{dx^2} = M_x}$.

Moment-deflection equation is formulated according to the **deflected shape** and **end support conditions**. Differential equation is solved by finding **constants of integration** using known **support conditions** in respect of **deflection 'y'** and **slope** $\dfrac{dy}{dx}$ at $x = 0$ and $x = l$.

Euler's critical loads (P_E) are determined by considering above four set of supports conditions.

For finding Euler's buckling load P$_E$, following assumptions are made:

   i. Column axis is **initially straight**.

   ii. Column cross-section is uniform from one end to the other end. i.e. moment of **inertia 'I' is uniform**.

  iii. Column material is **homogeneous and isotropic** having values of **E** same in all directions.

  iv. Column axis bends in weaker **direction only** at a time.

   v. **Change in length** due to bending is neglected and assumed the same 'L' as initial length.

  vi. For **slender** long columns, the **bending stresses** are considered **very high** compared to **direct compressive stress** which can be **neglected**.

vii. **Bending moment** due to deflection is considered **negative if the deflected shape is concave** towards **original axis** and **positive if the deflected shape is convex towards original axis.**

12.4 EULER'S BUCKLING LOAD

Case I: Both ends hinged (Pinned)

BM at any section at distance x from B

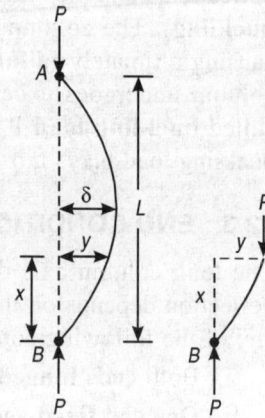

$$EI\frac{d^2y}{dx^2} = -P.y$$

$$\frac{d^2y}{dx^2} + \frac{P}{EI} \cdot y = 0$$

$$y = C_1 \cos\left(x\sqrt{\frac{P}{EI}}\right) + C_2 \sin\left(x\sqrt{\frac{P}{EI}}\right)$$

At $x = 0, y = 0, 0 = C_1 \cos 0, C_1 = 0,$

Fig. 12.2.

$$\therefore \quad y = C_2 \sin\left(x\sqrt{\frac{P}{EI}}\right)$$

At $x = l, y = 0, 0 = 0 + C_2 \sin\left(l\sqrt{\frac{P}{EI}}\right),$ i.e. $C_2 \sin\left(l\sqrt{\frac{P}{EI}}\right)$

$$\therefore \quad C_2 \neq 0, \quad \therefore \sin\left(l\sqrt{\frac{P}{EI}}\right) = 0, \qquad \text{i.e.} \quad l\sqrt{\frac{P}{EI}} = 0, \pi, 2\pi, 3\pi, ...$$

Consider first practical value

$$l\sqrt{\frac{P}{EI}} = \pi, \quad \text{or} \quad l^2P = EI\pi^2 \qquad \text{or} \quad P = \frac{\pi^2 EI}{l^2}$$

$$P = \frac{\pi^2 EAK^2}{l^2} = \frac{\pi^2 EAK^2}{l^2} = \frac{\pi^2 EA}{\left(\dfrac{l}{k}\right)^2} \qquad\qquad ... (12.1)$$

where K = radius of gyration   and $AK^2 = I.$

Case II: One end fixed and other free

BM at any section *X-X* at a distance x from the fixed end B

$$EI\frac{d^2y}{dx^2} = P(\delta - y) \quad \text{or} \quad EI\frac{d^2y}{dx^2} + P \cdot y = P \cdot \delta$$

$$\frac{d^2y}{dx^2} + \frac{P}{EI} \cdot y = \frac{P}{EI} \cdot \delta$$

$$y = C_1 \cos\left(x\sqrt{\frac{P}{EI}}\right) + C_2 \sin\left(x\sqrt{\frac{P}{EI}}\right) + \delta$$

At $\quad x = 0, \quad y = 0, \qquad C_1 + \delta = 0,$

$$C_1 = -\delta$$

At $\quad x = 0,$

$$\frac{dy}{dx} = 0 = -C_1\sqrt{\frac{P}{EI}} \cdot \sin\left(x\sqrt{\frac{P}{EI}}\right) + C_2\sqrt{\frac{P}{EI}}\cos\left(x\sqrt{\frac{P}{EI}}\right)$$

or $\quad 0 = -C_1\sqrt{\frac{P}{EI}} \cdot \sin 0 + C_2\sqrt{\frac{P}{EI}}\cos 0$

or $\quad C_2\sqrt{\frac{P}{EI}} \cdot \cos 0 = 0, \qquad \sqrt{\frac{P}{EI}} \neq 0, \qquad C_2 = 0$

$\therefore \quad y = C_1 \cos\left(x\sqrt{\frac{P}{EI}}\right) + \delta = (-\delta)\cos\left(x\sqrt{\frac{P}{EI}}\right) + \delta$

At $\quad x = l, \quad y = \delta = (-\delta)\cos\left(l\sqrt{\frac{P}{EI}}\right) + \delta$

$\therefore \quad \cos\left(l\sqrt{\frac{P}{EI}}\right) = 0, \qquad l\sqrt{\frac{P}{EI}} = \frac{\pi}{2}, \frac{3\pi}{2}, \frac{5\pi}{2}, ...,$

$$P = \frac{\pi^2 EI}{4l^2} = \frac{\pi^2 EA}{4l^2 \big/ K^2} = \frac{\pi^2 EA}{\left(2l^2 \big/ K\right)^2} = \frac{^2 EA}{\left(l_e \big/ K\right)^2}, \quad \text{where } l_e = 2l \qquad \qquad ... (12.2)$$

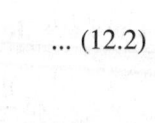

Fig. 12.3.

Case III: One end fixed and other end hinged

$$EI\frac{d^2y}{dx^2} = -P \cdot y + H(l - x)$$

$$\frac{d^2y}{dx^2} + \frac{P}{EI} \cdot y = \frac{H}{EI}(l - x)$$

$$y = C_1 \cos\left(x\sqrt{\frac{P}{EI}}\right) + C_2 \sin\left(x\sqrt{\frac{P}{EI}}\right) + \frac{H}{P}(l - x)$$

$$\frac{dy}{dx} = -C_1\sqrt{\frac{P}{EI}}\sin\left(x\sqrt{\frac{P}{EI}}\right) + C_2\sqrt{\frac{P}{EI}}\cos\left(x\sqrt{\frac{P}{EI}}\right) - \frac{H}{P}$$

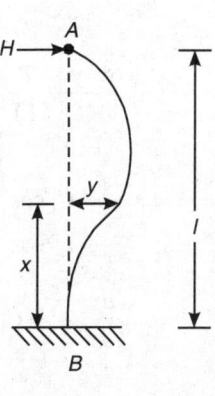

Fig. 12.4.

At $\quad x = 0, \quad y = 0, \qquad C_1 = -\dfrac{H}{P} \cdot 1$

$\therefore \quad y = -\dfrac{H \cdot 1}{P} \cos\left(x\sqrt{\dfrac{P}{EI}} \right) + C_2 \sin\left(x\sqrt{\dfrac{P}{EI}} \right) + \dfrac{H}{P}(1 - x)$

At $\quad x = 0, \quad \dfrac{dy}{dx} = 0, \qquad C_2 \cdot \sqrt{\dfrac{P}{EI}} - \dfrac{H}{P} = 0 \quad$ or $\quad C_2 = \dfrac{H}{P}\sqrt{\dfrac{EI}{P}}$

At $\quad x = 1, \quad y = 0, \qquad 0 = -\dfrac{H}{P} \cdot 1\cos\left(1\sqrt{\dfrac{P}{EI}} \right) + \dfrac{H}{P}\sqrt{\dfrac{EI}{P}} \sin\left(1\sqrt{\dfrac{P}{EI}} \right)$

Simplify we get: $\quad \tan\left(1\sqrt{\dfrac{P}{EI}} \right) = \left(1\sqrt{\dfrac{P}{EI}} \right)$

Solution is $\qquad 1\sqrt{\dfrac{P}{EI}} = 4.5$ radians $\quad$ or $\quad 1^2 \cdot \dfrac{P}{EI} = (4.5)^2,$

$$P = 20.25\dfrac{EI}{1^2} \approx \dfrac{2\pi^2 \cdot EI}{1^2}$$

$$\boxed{P = 2\pi^2 \cdot \dfrac{EI}{1^2} = \dfrac{\pi^2 EA}{\left(\dfrac{1}{\sqrt{2}k} \right)^2} = \dfrac{\pi^2 EA}{\left(\dfrac{1_e}{K} \right)^2}}, \; P\left(\text{where} 1_e = \dfrac{1}{\sqrt{2}} \right) \qquad \dots (12.3)$$

Case IV: Both ends fixed

$$EI\dfrac{d^2 y}{dx^2} = (M_0 - P \cdot y)$$

$$\dfrac{d^2 y}{dx^2} + \dfrac{P}{EI} \cdot y = \dfrac{M_0}{EI}$$

$$y = C_1 \cos\left(x\sqrt{\dfrac{P}{EI}} \right) + C_2 \sin\left(x\sqrt{\dfrac{P}{EI}} \right) + \dfrac{M_0}{P}$$

$$\dfrac{dy}{dx} = -C_1\sqrt{\dfrac{P}{EI}} \sin\left(x\sqrt{\dfrac{P}{EI}} \right) + C_2\sqrt{\dfrac{P}{EI}} \cos\left(x\sqrt{\dfrac{P}{EI}} \right)$$

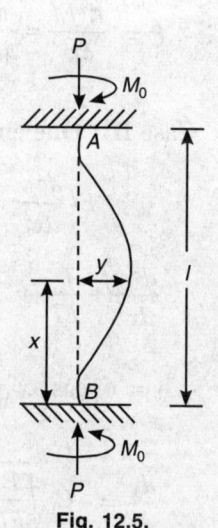

Fig. 12.5.

At $\quad x = 0, \quad y = 0, \qquad C_1 = -\dfrac{M_0}{P}$

At $\quad x = 0, \quad \dfrac{dy}{dx} = 0, \qquad 0 = C_2 \sqrt{\dfrac{P}{EI}}$ (1), $\quad C_2 = 0$, as $\sqrt{\dfrac{P}{EI}} \neq 0$

At $\quad x = l, \quad y = 0, \qquad 0 = -\dfrac{M_0}{P} \cos\left(l\sqrt{\dfrac{P}{EI}}\right) + 0 + \dfrac{M_0}{P}$

$\cos\left(l\sqrt{\dfrac{P}{EI}}\right) = 1$

$l\sqrt{\dfrac{P}{EI}} = 0, 2\pi, 4\pi, \ldots$

First Value: $\qquad l^2 \cdot \dfrac{P}{EI} = 4\pi^2,$

$$P = \frac{4\pi^2 \cdot EI}{l^2} = \frac{\pi^2 EAK^2}{l^2/4} = \frac{\pi^2 EA}{\left(^{1}/_{2}K\right)^2}$$

or $\quad P = \dfrac{\pi^2 E \cdot A}{\left(\dfrac{l_e}{K}\right)^2}, \quad \left(\text{where } l_e = \dfrac{l}{2}\right).$ $\qquad\qquad\qquad$ ... (12.4)

12.5 LIMITATIONS OF EULER'S FORMULA

Considering general formula of $P_E = \dfrac{\pi^2 EA}{\left(^{l_e}/_{K}\right)^2}$

$$f_{\text{critical}} = \frac{P_E}{A} = \frac{\pi^2 E}{\left(^{l_e}/_{K}\right)^2} = C \cdot \left(\frac{K}{l_e}\right)^2$$

where 'C' is a constant depending on the *material value of* 'E', l_e is *effective* length and 'K' is *minimum* radius of gyration.

If a graph is drawn between f_{critical} v/s slenderness ratio (le/K), we get the relationship as a *hyperbola* (as shown in Fig. 12.6).

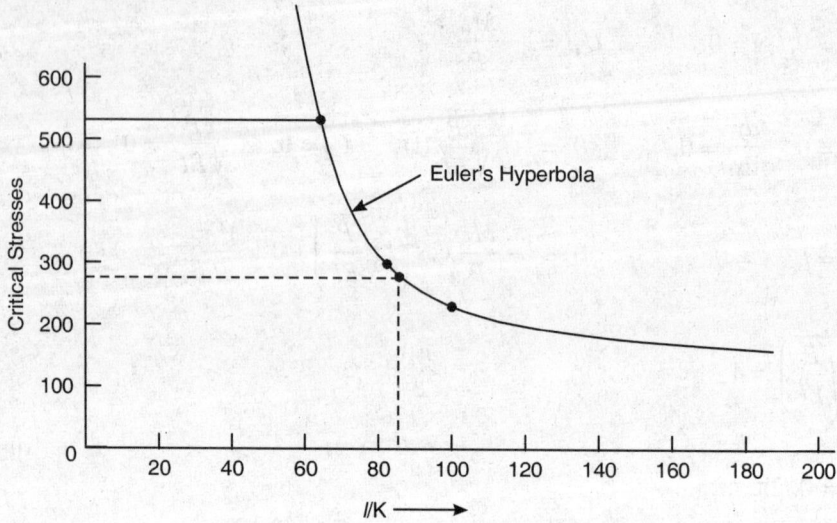

Fig.12.6: Euler's hyperbola

Generally for mild steel, value of $E \approx 2 \times 10^5$ N/mm^2, $f_{critical}$ is calculated for different **slenderness ratios** considering minimum value of K. Table 12.1 gives values of (l_e/K) and $f_{critical}$.

Table 12.1: $f_{critical}$ for different values of slenderness

Slenderness Ratio (le/K)	200	150	100	80	60
$f_{critical}$ (N/mm^2)	49.5	88.0	198	308	548.0

These values of f_c are valid if f_c *does not exceed proportional limit* of the material. Proportional limit of mild steel is approximately 300 N/mm^2 and hence (l_e/K) for $f_c \leq 300$ N/mm^2

will be $\dfrac{\pi^2 \times 2 \times 10^5}{\left(\frac{l_e}{K}\right)^2} \leq 300, \dfrac{l}{K} \geq \sqrt{\dfrac{\pi^2 \times 2 \times 10^5}{300}}$, say 82 for *mild steel* and similarly for

aluminium this limit ≥ 60. Thus columns of *mild steel* will be *long if slenderness ratio is more than* 82 and *aluminnium* column *more than* 60. Thus Euler's formula applies to *long* columns

for $\dfrac{l}{K} \geq 80$ in case of *mild steel* and $\dfrac{l}{K} \geq 60$ in case of aluminium.

Consider a high strength steel and mild steel columns having the same values of E, the actual critical loads will be different although from Euler's formula which is dependent on E *will give the same load*. Critical load v/s (l/K) curve becomes parallel to X-axis showing that slenderness ratio in this zone represents very long columns. Long columns may fail at very low critical stress, similarly very short columns may fail at very high critical stress. But actual failure does not occur according to Euler's equation of critical load for medium values of (l/K) which does not represent very long or short columns. These are the limitations of Euler's formula as based on values of 'E' and perfect elasticity.

EXAMPLE 12.1: A wooden rectangular section of 120 mm × 150 mm is used for a column of 5.50 m with both ends hinged. Determine the Euler's buckling load if $E_w = 10$ kN/mm^2.

Solution:

$$\text{Minimum } I = \frac{120^3 \times 150}{12} = 2160 \times 10^4 \text{ mm}^4$$

Effective length l_e (both ends hinged) = 5.50×10^3 mm

$$\text{PE (Euler's critical load)} = \frac{\pi^2 EI}{l_e^2} = \frac{\pi^2 \times 10 \times 2160 \times 10^4}{(5.50 \times 10^3)^2}$$

$$P_E = \textbf{70.5 kN.}$$

EXAMPLE 12.2: A steel column has 60 mm × 40 mm cross-section and both the ends are fixed. Determine the actual length of the column for which the Euler's formula for long columns can be applied if the limit of elastic stress is 210 N/mm^2 and the modulus of elasticity $E = 210$ kN/mm^2. Also find the Euler's load if the length is 4.0 m (both ends hinged).

Solution:

$$I_{\min} = \frac{60 \times 40^3}{12} = \textbf{32} \times \textbf{10}^4 \textbf{ mm}^4,$$

$$K^2 = \frac{I}{A} = \frac{32 \times 10^4}{60 \times 40} = \frac{400}{3} \text{ mm}^2$$

$$\text{Euler's load } P_E = \frac{\pi^2 \cdot E \cdot AK^2}{l_e^2} \quad \text{or} \quad \frac{P_E}{A} = \frac{\pi^2 \cdot E \cdot K^2}{l_e^2}$$

$$\frac{P_E}{A} \leq \text{Limiting stress}$$

$$\frac{P_E}{A} = \frac{\pi^2 \cdot E \cdot K^2}{l_e^2} \leq 210$$

or

$$l_e^2 \geq \frac{\pi^2 EK^2}{210} = \frac{21 \times 10^4 \pi^2}{210} \cdot \frac{400}{3}$$

or

$$l_e = \sqrt{\frac{1000\pi^2 \cdot 400}{3}} = 100 \text{ p } (3.6515) = \textbf{1146.56 mm} = \frac{L}{2},$$

$$L = \textbf{2293 mm}$$

Euler's formula is applicable for lengths more than **2.293 m**.

ii.
$$P_E = \frac{\pi^2 \cdot E \cdot I}{l_e^2} = \frac{\pi^2 \times E \cdot I}{(4000)^2} = \frac{\pi^2 \times 210000 \times 32 \times 10^4}{4 \times 4 \times 10^6} \, \text{N}$$

$$= \pi^2 \times 4200 \, \text{N} = \textbf{41.41 kN}.$$

EXAMPLE 12.3: A mild steel tube of 40 mm outer diameter and 5 mm wall thickness is used for 4 m long strut. Determine the critical load on the strut if (*i*) both the ends are hinged, (*ii*) both the ends are fixed, (*iii*) one end is fixed and the other is hinged.

Solution:

$$E = 210 \, \text{kN/mm}^2, \qquad D = 40 \, \text{mm}, \quad d = 40 - 2 \times 5 = 30 \, \text{mm}$$

$$I = \frac{\pi}{64}(D^4 - d^4) = \frac{\pi}{64}(40^4 - 30^4) = \frac{10^4 \pi}{64}(175)$$

(*i*) $l_e = 4000$ mm $\qquad$ (*ii*) $l_e = \dfrac{4000}{2} = 2000$ mm $\qquad$ (*iii*) $l_e = \dfrac{4000}{\sqrt{2}}$

(*i*) $\quad P = \dfrac{\pi^2 \cdot E \cdot I}{l_e^2} = \pi^2 \cdot \dfrac{175}{64} \dfrac{\pi \times 10^4 \times 210}{(4000)^2} = \textbf{11.125 kN}$

(*ii*) $\quad P_E = \dfrac{\pi^2 \cdot E \cdot I}{l_e^2} = \pi^2 \cdot \dfrac{175}{64} \dfrac{\pi \times 10^4 \times 210}{(2000)^2} = \textbf{44.50 kN}$

(*iii*) $\quad P_E = \dfrac{\pi^2 \cdot E \cdot I}{l_e^2} = \pi^2 \cdot \dfrac{175}{64} \dfrac{\pi \times 10^4 \times 210}{(2000\sqrt{2})^2} = \textbf{22.25 kN}.$

EXAMPLE 12.4: A circular rod of **20 mm** diameter and **2 m** length is simply supported horizontally over a span of 2.0 m. The rod undergoes a maximum deflection of 25 mm under a concentrated load of 250 N at midspan point. Determine **critical load** if this rod is used for vertical column to carry axial load with both ends hinged. Also find the ratio of maximum stresses in two situations.

Solution:

$$D = 20 \, \text{mm}, \qquad\qquad \text{Span } L = 2000 \, \text{mm}, \qquad\qquad l_e = 2000 \, \text{mm}$$

$$\delta_{max} = \frac{W \cdot L^3}{48EI} = 25 \, \text{mm}, \quad \text{or} \quad \frac{250 \, (2000)^3}{48E\left(\dfrac{\pi}{64} 20^4\right)} = 25$$

$$E = \frac{250 \times 8 \times 10^9 \times 64}{48 \times 25\pi \times 16 \times 10^4} = \frac{10^6 \times 2}{3\pi}$$

$$P_c = \frac{\pi^2 EI}{l_e^2} = \frac{\pi^2 \cdot \dfrac{2 \times 10^6}{3\pi} \cdot \dfrac{\pi}{64}(20)^4}{(2000)^2} = \frac{\pi^2 \times 32 \times 10^{10}}{3 \times 64 \times 4 \times 10^6} = \frac{\pi^2 \times 10^4}{24} = \textbf{4115 N}$$

$$f_c = \frac{P_c}{A} = \frac{4115}{\frac{\pi}{4}(40)^2} = \textbf{13.11 N/mm}^2,$$

$$f_b = \frac{M}{I} \cdot y = \frac{250 \times 2000 \times 32}{4 \times \pi \times (20)^3} = \textbf{159.1 N/mm}^2$$

Ratio of $f_b : f_c = \dfrac{159.1}{13.11} = \textbf{12.13}$.

EXAMPLE 12.5: A 4 m long hollow circular strut with both ends fixed carries an axial load of **50 kN**. If the permissible stress is **100 N/mm**^{2} and the strut is **long** for which critical load and crushing loads are equal, determine the external and internal diameters of the tube. $E = $ **200 kN/mm**^{2}.

Solution:

$$P_E = \frac{\pi^2 EI}{l_e^2} = \frac{\pi^2 \cdot 200 \times \dfrac{\pi}{64}(D^4 - d^4)}{(2000)^2} = 50$$

or $\qquad \dfrac{\pi}{64}(D^4 - d^4) = \dfrac{50 \times 20000}{\pi^2} = \dfrac{10^6}{\pi^2}$ ... (1)

Also crushing load $= 50{,}000 = 100\,\dfrac{\pi}{4}(D^2 - d^2)$

or $\qquad \dfrac{\pi}{4}(D^2 - d^2) = 500$ ... (2)

Divide (1) by (2)

$$\frac{(D^2 + d^2)}{16} = \frac{10^6}{\pi^2\,500} = \frac{2 \times 10^3}{\pi^2}$$

or $\qquad D^2 + d^2 = 3245.56$ ... (3)

Also from (2), $\qquad D^2 - d^2 = 636.364$

Adding $\qquad 2D^2 = 3876.04, \qquad D^2 = 1938.02, \qquad \textbf{\textit{D}} = \textbf{44.060 mm}.$ (44.10 mm), say

Also $\qquad 1938.02 + d^2 = 3245.56 \quad$ or $\quad d^2 = 1307.5, \textbf{\textit{d}} = \textbf{36.158 mm}.$ (36.16 mm), say

EXAMPLE 12.6: A circular hollow steel column section has external diameter 160 mm, wall thickness 20 mm and length 5 m with ends hinged. Determine the Euler's buckling load if $E = 210$ kN/mm^2.

Solution:

$l_e = 5000$ mm, $\qquad\qquad I = \dfrac{\pi}{64}(D^4 - d^4) = \dfrac{\pi}{64}(160^4 - 120^4)$

$$= \frac{\pi \times 10^4 \times 400 \times 112}{64} = 7 \times \pi \times 10^6 \text{ mm}^4$$

$$P_E = \frac{\pi^2 \cdot EI}{l_e^2} = \frac{\pi^2 \times 210 \times 7 \times \pi \times 10^6}{(5000)^2} = \frac{\pi^2 \times \pi \times 21 \times 10^7 \times 7}{25 \times 10^6}$$

$$= 1825.37 \text{ kN}.$$

12.6 RANKINE'S HYPOTHESIS

Rankine has stated a general theory that the *critical* or *crippling* load on any column can be determined by using a relation $\frac{1}{P_R} = \frac{1}{P_E} + \frac{1}{P_C}$. It takes into consideration the limitations of Euler's formulae which is applicable for very long columns having (l_e/K) more than certain values based on material crushing strength. Rankine's formula is applicable for all type of columns.

In this equation:

P_R is the *Rankine's crippling* or critical *failure load*,

P_C is the *crushing load* due to pure *axial loading,* and

P_E is the *Euler's buckling* load for long columns

P_C (Axial crushing load) $= f_c.A$ ... (1)

$$P_E \text{ (Buckling load)} = \frac{\pi^2 EA}{\left(\frac{l_e}{K}\right)^2}$$ ... (2)

Rankine's Theory

$$\frac{1}{P_R} = \frac{1}{P_E} + \frac{1}{P_C}$$

(*i*) If the column or strut is short, i.e. (l_e/K) is small and hence $P_E = \dfrac{\pi^2 EA}{\left(\dfrac{l_e}{K}\right)^2}$ will be very large, and $\dfrac{1}{P_E}$ will be *very small* and can be considered *negligible* in comparison to $\dfrac{1}{P_C}$.

$$\therefore \quad \frac{1}{P_R} = \frac{1}{P_C} \text{ for short columns, } \quad i.e. \quad \boldsymbol{P_R = P_C. \text{ For short columns.}}$$

(*ii*) If the columns or struts are long, i.e. the values of (l_e/K) will be large and the Euler's load $P_E = \dfrac{\pi^2 EA}{\left(\dfrac{l_e}{K}\right)^2}$ will be very small and $\dfrac{1}{P_E}$ will be substantially large compared to

$\dfrac{1}{P_C}$ and $\dfrac{1}{P_C}$ can be neglected as compared to $\dfrac{1}{P_E}$.

$$\therefore \qquad \frac{1}{P_R} = \frac{1}{P_E} \quad \text{or} \quad \boldsymbol{P_R = P_E}$$

Hence the *Rankine's theory* is applicable to both *short and long* columns. Thus it can be stated that Rankine's theory is applicable to all type of columns whether *short, long* or *medium*.

i.e.
$$\frac{1}{P_R} = \frac{1}{P_E} + \frac{1}{P_C} = \frac{\left(\frac{l_e}{K}\right)^2}{\pi^2 EA} + \frac{1}{f_c \cdot A} = \frac{\left\{ f_c \cdot A \left(\frac{l_e}{K}\right)^2 + \pi^2 \cdot EA \right\}}{(f_c \cdot A\pi^2 E \cdot A)}$$

or
$$\frac{1}{P_R} = \frac{f_c \cdot \left(\frac{l_e}{K}\right)^2 + \pi^2 E}{f_c \cdot \pi^2 EA},$$

$$\therefore \quad P_R = \frac{f_c \cdot A\pi^2 \cdot E}{f_c \cdot \left(\frac{l_e}{K}\right)^2 + \pi^2 E} = \frac{f_c \cdot A}{1 + \frac{f_c}{\pi^2 E}\left(\frac{l_e}{K}\right)^2} = \frac{f_c \cdot A}{1 + a\left(\frac{l_e}{K}\right)^2} \qquad \dots (12.6)$$

Rankine's formula for crippling load has constant $a = \dfrac{f_c}{\pi^2 E}$ which can be determined empirically by actual testing. Table 12.2 provides values of 'f_c' and 'a' for column or struts of different materials.

Table 12.2: Values of compressive stress f_c and rankine's constant 'a'

Material	Mild steel	Medium carbon steel	Cast iron	Wrought iron	Strong timber
Stress fc (MN/m^2)	320	500	550	250	50
Rankine's constant $a = \dfrac{f_c}{\pi^2 E}$	$\dfrac{1}{7500}$	$\dfrac{1}{5000}$	$\dfrac{1}{1600}$	$\dfrac{1}{9000}$	$\dfrac{1}{750}$

$$\textbf{Rankine's load } \boldsymbol{P_R} = \frac{f_c \cdot A}{1 + a\left(\frac{l_e}{K}\right)^2} = \frac{\text{Crushing Load}}{1 + a\left(\frac{l_e}{K}\right)^2}$$

The factor $\left\{ 1 + a\left(\frac{l_e}{K}\right)^2 \right\}$ takes into account effect of *buckling* and *crippling load*.

Rankine's formula takes into account both *crushing* and *buckling loads*.

EXAMPLE 12.7: A 5.0 m long column with both ends hinged comprises of Indian standard heavy beam section ISHB 300. Determine the safe axial load on the column if critical load stress $f_c = 315$ N/mm^2, Rankine's constant $a = \dfrac{1}{7500}$ and factor of safety = 3.

Properties of ISHB 300: $I_X = 12545.2 \times 10^4$ mm^4, $I_Y = 2193.6 \times 10^4$ mm^4, $A = 7485$ mm^2.

Solution:

$$l_e = 5000 \text{ mm}, f_c = 0.315 \text{ kN/mm}^2$$

$$P_R \text{ (Cripping load)} = \frac{f_c \cdot A}{1 + a\left(\dfrac{l_e}{K}\right)^2},$$

$$\text{Minimum } K^2 = \frac{1}{A} = \frac{2193.6 \times 10^4}{7485} = 29.307 \times 10^2 \text{ mm}^2$$

$$P_R = \frac{315 \times 7485}{\left\{1 + \dfrac{1}{7500}\left(\dfrac{5000 \times 5000}{2930.7}\right)\right\}} = \frac{2357.775}{(1 + 1.13738)} = \frac{2357.8}{2.13738} = \mathbf{1103.13 \text{ kN}}$$

$$\text{Safe load } P = \frac{P_R}{F_S} = \frac{1103.13}{3} = \mathbf{367.71 \text{ kN}.}$$

EXAMPLE 12.8: A cast iron column has 4 m length with one end fixed and other end hinged. The column is of hollow circular section with internal diameter 75% of external diameter. Determine the external diameter for an axial load of 250 kN if the factor of safety is 4, Rankine's stress $f_c = 567$ N/mm^2 and constant 'a' $= \dfrac{1}{1600}$.

Solution:

Rankine's critical load $\quad P_R = $ F.S. $\times$ Safe Load $= 4 \times 250 = 1000$ kN

Diameter D, $\quad d = \dfrac{3}{4} D$, $\quad A = \dfrac{\pi}{4}(D^2 - d^2) = \dfrac{7\pi}{64} D^2$, $\quad l_e = \dfrac{4000}{\sqrt{2}} = 2000\sqrt{2}$ mm

$$K^2 = \frac{I}{A} = \frac{\pi}{64} \times \frac{4(D^4 - d^4)}{\pi(D^2 - d^2)} = \frac{(D^2 + d^2)}{16} = \frac{25}{16} \cdot \frac{D^2}{16}$$

$$\left(\frac{l_e}{K}\right)^2 = \frac{2000\sqrt{2} \times 2000\sqrt{2}}{25 D^2} \times 256 = \frac{81.92 \times 10^6}{D^2}$$

$$P_R = 1000 = \frac{567 \times \dfrac{7\pi D^2}{64}}{\left\{1 + \dfrac{1}{1600}\left(\dfrac{81.92 \times 10^6}{D^2}\right)\right\}} = \frac{567 \times 7\pi D^4}{64 \{D^2 + 51200\}}$$

$$D^2 + 51200 = 0.1949 \, D^4, \quad \text{or} \quad D^4 - 5.13 \, D^2 - 262699 = 0$$

$$D^2 = \frac{5.13 \pm \sqrt{1050822}}{2} = \frac{5.13 \pm 1025.1}{2}$$

$$D = 22.70 \text{ mm}.$$

EXAMPLE 12.9: Two rolled steel IS channels of 250 mm × 100 mm are joined together by two flange plates each of 400 mm × 12.5 mm with channels separated by 150 mm back to back. Determine the safe load on the 8 m long column section if the factor of safety is 3 and one end is fixed and the other end is hinged. Properties of each channel section are given as under :

IS channel 250 mm × 100 mm,

$A = 3565 \text{ mm}^2$,

$I_X = 3687.9 \times 10^4 \text{ mm}^4$,

$I_Y = 298.4 \times 10^4 \text{ mm}^4$,

Cg from outer web face = 27.0 mm,

Rankine's Constant 'a' = $\dfrac{1}{7500}$,

Critical Stress = 315 N/mm²,

Rankine's Critical Stress = 315 N/mm².

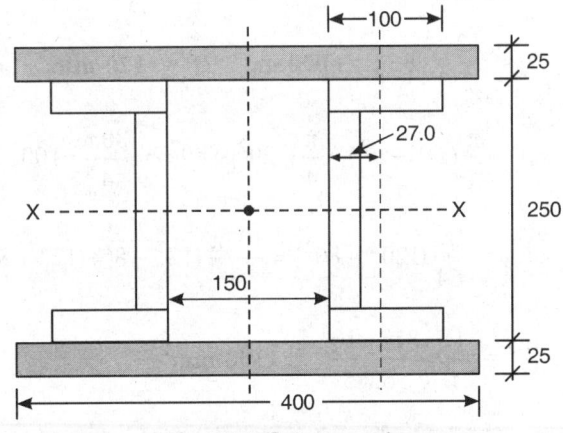

Fig. 12.7: Column section

Solution:

A (compound section) = $2 \times 3565 + 2 \times 400 \times 25 = 7130 + 2000 =$ **27130 mm²**

$$I_{XX} \text{ (compound section)} = 2 \times 3687.9 \times 10^4 + \frac{400}{12}(300^3 - 250^3)$$

$$I_{XX} = 7375.8 \times 10^4 + 10^3 (379167) = \mathbf{45292.5 \times 10^4 \text{ mm}^4}$$

$$I_{YY} = 2 \times 298.4 \times 10^4 + 2\left[3565 \, (75 + 27)^2 + \frac{2 \times 25 \times 400^3}{12}\right]$$

$$I_{YY} = 596.8 + 10^4 + 7418.0520 \times 10^4 + 26666.67 \times 10^4 = \mathbf{34681.52 \times 10^4 \, mm^4}$$

$$K \text{ (minimum)} = \sqrt{\frac{I}{A}} = \sqrt{\frac{34681.52 \times 10^4}{27130}} = 100 \times 1.1306 = \mathbf{113.06 \text{ mm}}$$

$$\text{Effective Length } l_e = \frac{8000}{\sqrt{2}} = 5656 \qquad \frac{l_e}{K} = \frac{5656.85}{113.06} = \mathbf{50.03}$$

$$P_R \text{ (crippling load)} = \frac{315 \times 27130}{1} = \frac{315 \times 27130}{(1 + 0.33376)} = \mathbf{6407412 \ N}$$

$$\text{Safe Load } P = \frac{6407412}{3} = \mathbf{2135804 \ N = 2135.804 \ kN}$$

EXAMPLE 12.10: A hollow circular column section with both ends fixed is 12 m long and has an outer diameter of 120 mm with 20 mm wall thickness. Compare the crippling load according to Euler's and Rankine's formulae. E for the material is 80 kN/mm^2 and critical stress $f_c = 550$ N/mm^2. Rankine's constant $a = \dfrac{1}{1600}$. Determine the length at which both the crippling loads are equal.

Solution:

$$l_e = \frac{12}{2} = 6 \text{ m} = \mathbf{6000 \ m}, \quad D = \mathbf{120 \ mm}, \quad d = 120 - 2 \times 20 = \mathbf{80 \ mm}$$

$$A = \frac{\pi}{4}(D^2 - d^2) = \frac{\pi}{4}(120^2 - 80^2) = \frac{80\pi}{4} \times 100 = \mathbf{6285 \ mm}$$

$$I_{XX} = \frac{\pi}{64}(120^2 - 80^2) = \frac{10^4 \pi}{4}(12^2 - 8^2)(12^2 + 8^2) = \mathbf{817 \times 10^4 \ mm^4}$$

$$K^2 = \frac{1}{A} = \frac{817 \times 10^4}{6285} = \mathbf{1300 \ mm^2}$$

(i) $\quad P_E = \dfrac{\pi^2 \cdot E(AK^2)}{l_e^2} = \dfrac{\pi^2 \times 80 \times 10^3 \times 6285 \times 1300}{(6000)^2} = \mathbf{179343 \ N}$

$$P_R = \frac{f_c \cdot A}{\left\{1 + a\left(\dfrac{l_e^2}{K^2}\right)\right\}} = \frac{550 \times 6285}{1 + \dfrac{1}{1600}\left(\dfrac{36 \times 10^6}{1300}\right)} = \frac{3456750}{(1 + 17.3077)} = \mathbf{188.814 \ kN}$$

PE : PR = 1 : 1.0528

(ii) $\quad PE = PR$

i.e. $\qquad \dfrac{\pi^2 EAK^2}{l_e^2} = \dfrac{550 \cdot A}{\left\{1 + \dfrac{1}{1600}\left(\dfrac{l_e^2}{K^2}\right)\right\}}$

or $\qquad \left\{ 1 + \dfrac{l_e^2}{1600 \times 1300} \right\} = \dfrac{550 \times l_e^2}{\pi^2 \times 80 \times 10^3 \times 1300} = \dfrac{l_e^2}{1.86776 \times 10^6}$

or $\qquad \left\{ 1 + \dfrac{l_e^2}{208 \times 10^4} \right\} = \dfrac{l_e^2}{186.776 \times 10^4} \qquad \text{or} \qquad 1 = \dfrac{l_e^2 \, (208 - 186.776)}{208 \times 186.776 \times 10^4}$

or $\qquad\qquad l_e^2 = \dfrac{208 \times 186.776 \times 10^4}{21.224} = 1830.447 \times 10^4,$

$$l_e = \textbf{4278.37 mm}$$

$\qquad L = 2l_e = \textbf{8.557 m}$ (Actual length).

EXAMPLE 12.11: Determine the ratio of the critical stresses using Rankine's and Euler's formulae for struts with slenderness ratios of 50, 80, 100, 150 and 200. Assume that both ends are hinged.

$E = 200$ kN/mm^2, Rankine's constant $a = \dfrac{1}{7500}$, $f_c = 300$ N/mm^2.

Solution:

$$P_E = \frac{\pi^2 EAK^2}{l^2} \qquad \text{or} \qquad \frac{P_E}{A} = \frac{\pi^2 \cdot E \cdot K^2}{l^2} = \frac{1975.51 \times 10^3}{\left(\dfrac{l}{K}\right)^2}$$

$$P_R = \frac{f_c \cdot A}{1 + a\left(\dfrac{l}{K}\right)^2} \qquad \text{or} \qquad \frac{P_R}{A} = \frac{f_c}{1 + a\left(\dfrac{l}{K}\right)^2} = \frac{300}{\left\{ 1 + \dfrac{1}{7500}\left(\dfrac{l}{K}\right)^2 \right\}}$$

Thus $\dfrac{P_E}{P_R} = \dfrac{1975.51 \times 10^3}{\left(\dfrac{l}{K}\right)^2} \cdot \dfrac{\left\{ 1 + \left(\dfrac{l}{K}\right)^2 \right\}}{300} = \dfrac{6585.034}{\left(\dfrac{l}{K}\right)^2} \cdot \left\{ \dfrac{7500 + \left(\dfrac{l}{K}\right)^2}{7500} \right\}$

$$= \frac{7500 + \left(\dfrac{l}{K}\right)^2}{\left(\dfrac{l}{K}\right)^2} \times 0.878$$

(i) $l/K = 50$; $p_E : p_R = 3.510$

(ii) $l/K = 80$; $p_E : p_R = 1.900$

(iii) $l/K = 100$; $p_E : p_R = 1.5365$

(iv) $l/K = 150$; $p_E : p_R = 1.1707$

(v) $l/K = 200$; $p_E : p_R = 1.0426$

This shows that Euler's formula always gives higher values specially for short columns.

EXAMPLE 12.12: An angle iron section of 100 mm × 100 mm size is used for a strut of 4 m length with both ends hinged. Determine Rankine's and Euler's critical load if yield stress $f_c = 335$ N/mm^2,

Rankine's constant $a = \dfrac{1}{7500}$, and $E = 210$ kN/mm^2.

Also find (l/K) for which Euler's formula can be for crippling loads.

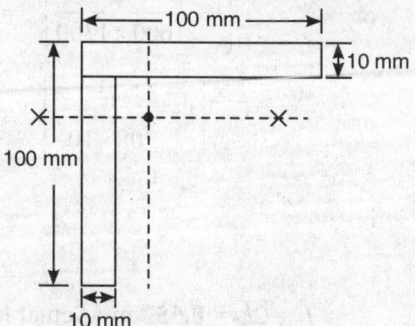

Fig. 12.8: Angle iron strut

Solution:

$$A = 100 \times 10 + 90 \times 10 = 1900 \text{ mm}^2,$$

$$\bar{x} = \frac{1000 \times 5 + 900 \times 55}{1900} = \mathbf{28.7 \text{ mm}}$$

$$I_X = I_Y = \left\{ \frac{90 \times 10^3}{3} + \frac{10 \times 100^3}{3} \right\} - 1900(28.7)^2$$

$$= \mathbf{180 \times 10^4 \text{ mm}^4}$$

$$K^2 = \frac{I}{A} = \frac{180 \times 10^4}{1900} = \mathbf{947 \text{ mm}^2}, \quad (K = 30.76 \text{ mm})$$

$$l_e = 4000 \text{ mm (both ends hinged)}, \quad \frac{l_e^2}{K^2} = \frac{(4000)^2}{947} = 16897$$

$$P_R = \frac{f_c \cdot A}{1 + a\left(\dfrac{l_e}{K}\right)^2} = \frac{335 \times 400}{1 + \dfrac{1}{7500}(16897)} = \frac{636500}{3.25293} = 195670 \text{ N} = \mathbf{195.67 \text{ kN}}$$

$$P_E = \frac{\pi^2 EI}{l_e^2} = \frac{\pi^2 \times 210 \times 180 \times 10^4}{(4000)^2} \text{ kN} = \mathbf{233.36 \text{ kN}}$$

Ratio $\dfrac{P_R}{P_E} = \dfrac{1956.70}{233.36} = \mathbf{0.839}$

For Euler's formula, critical stress shall not be more than yield stress i.e.

$$\frac{P_E}{A} \leq 335 \text{ N/mm}^2 \quad \text{or} \quad \frac{\pi^2 \cdot E \cdot AK^2}{Al_e^2} \leq 335$$

i.e. $\left(\dfrac{l}{K}\right)^2 \geq \dfrac{\pi^2 \cdot E}{335} = \dfrac{\pi^2 \times 210,000}{335}$

$$\left(\frac{l}{K}\right) = \pi \sqrt{\frac{210 \times 10^2}{335}} = \mathbf{78.6}, \text{ slenderness ratio to be more than 79, say.}$$

12.7 ECCENTRIC LOADING ON LONG COLUMNS AND SECANT FORMULA

If the column is *short* and the axial compressive load 'W' acts eccentrically, the axial stress is the algebraic *sum of direct* axial stress and *bending stress* due to eccentricity.

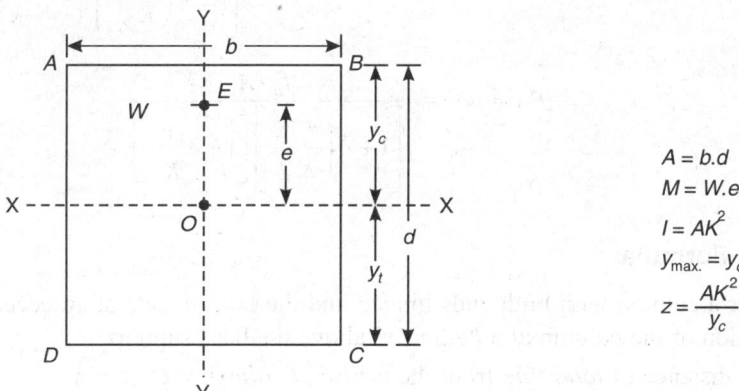

$$A = b.d$$
$$M = W.e$$
$$I = AK^2$$
$$y_{max.} = y_c$$
$$z = \frac{AK^2}{y_c}$$

Fig. 12.9: Column cross-section with ecentric load W acting at a distance 'e' from C_g 'O'

Maximum axial stress:

$$f_{max.} = f_0 + f_b = \frac{W}{A} + \frac{W \cdot e}{Z} = \frac{W}{A} + \frac{W \cdot e}{I} \cdot y$$

$$= \frac{W}{A} + \frac{W \cdot e}{AK^2} \cdot y_c$$

$$f_{max.} = \frac{W}{A}\left(1 + \frac{e}{K^2} \cdot y_c\right)$$

$$f_{min.} = \frac{W}{A}\left[1 - \frac{e}{K^2} \cdot y_t\right]$$

Critical load $W = \dfrac{f \cdot A}{\left(1 + e \cdot \dfrac{y_c}{K^2}\right)}$, for eccentric loading. Thus eccentricity factor for critical

load is $\dfrac{1}{\left(1 + e \cdot \dfrac{y_c}{K^2}\right)}$. If buckling effect is considered in case of long columns, the Rankine's

critical load is obtained multiplying the direct ($f \cdot A$) with Rankine's buckling factor

$$\frac{1}{\left\{1 + \left(\dfrac{l_e}{K}\right)^2\right\}}.$$

In case of eccentric loading, the Rankine's critical load can be obtained by multiplying with a combination factor of buckling and eccentricity : $\dfrac{1}{\left\{1+a\left(\dfrac{l_e}{K}\right)^2\right\}\left\{1+\dfrac{e \cdot y_c}{K^2}\right\}}$

i.e. $\qquad P_R = W' = \dfrac{f_c \cdot A}{\left[1+a\left(\dfrac{l_e}{K}\right)^2\right]\left[1+e\dfrac{y_c}{K^2}\right]}$ $\qquad$ ... (12.7)

Secant Formula

Consider a column with both ends hinged and the load W acts at an eccentricity 'e'. Consider any section of the column at a *height* 'x' above the base support.

Total distance of *load line* from the *centre of column* section $= y$

Moment about $\qquad C_g = W \cdot y$

$\therefore \qquad EI \dfrac{d^2 y}{dx^2} = -Wy$

or $\qquad EI \dfrac{d^2 y}{dx^2} + Wy = 0$

or $\qquad \dfrac{d^2 y}{dx^2} + \dfrac{W}{EI} \cdot y = 0$

Solving the differential equation

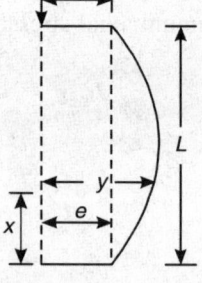

Fig. 12.10: Column with eccentric load

$$y = C_1 \sin\left(x\sqrt{\dfrac{W}{EI}}\right) + C_2 \cos\left(x\sqrt{\dfrac{W}{EI}}\right) \qquad ... (i)$$

At $\;x = 0, \; y = e, \; \therefore \qquad C_2 = y = e$ $\qquad$ ... (ii)

At mid height, i.e. $\qquad x = \dfrac{L}{2}, \quad \dfrac{dy}{dx} = 0$, (Symmetry).

i.e. $\qquad \dfrac{dy}{dx} = 0 = C_1 \sqrt{\dfrac{W}{EI}} \cos \dfrac{L}{2}\sqrt{\dfrac{W}{EI}} - C_2 \sqrt{\dfrac{W}{EI}} \sin \dfrac{L}{2}\sqrt{\dfrac{W}{EI}}$

Substituting $\qquad C_2 = e$

We have $\qquad C_1 = e \tan \dfrac{L}{2}\sqrt{\dfrac{W}{EI}}$ $\qquad$ ... (iii)

Substituting in equation (*i*)

$$y = e \tan \frac{L}{2}\sqrt{\frac{W}{EI}} \cdot \sin\left(x\sqrt{\frac{W}{EI}}\right) + e\cos\left(x\sqrt{\frac{W}{EI}}\right)$$

At $\quad x = L/2, \quad y = y_{max.}$

$$y_{max.}\left(x = \frac{L}{2}\right) = \frac{e\sin\left(\frac{L}{2}\sqrt{\frac{W}{EI}}\right)}{\cos\left(\frac{L}{2}\sqrt{\frac{W}{EI}}\right)} \cdot \sin\left(\frac{L}{2}\sqrt{\frac{W}{EI}}\right) + e\cos\left(\frac{L}{2}\sqrt{\frac{W}{EI}}\right).$$

or $\qquad y_{max.} = \dfrac{e\left[\sin^2\left(\dfrac{L}{2}\sqrt{\dfrac{W}{EI}}\right) + \cos^2\left(\dfrac{L}{2}\sqrt{\dfrac{W}{EI}}\right)\right]}{\cos\left(\dfrac{L}{2}\sqrt{\dfrac{W}{EI}}\right)}$

or $\qquad \boldsymbol{y_{max.} = e\sec\left(\dfrac{L}{2}\sqrt{\dfrac{W}{EI}}\right)}$ ... (*iv*)

Maximum bending moment will be developed at a section where y is maximum.

$$M_{max.} = W \cdot y_{max.} = W \cdot e \cdot \sec\left(\frac{L}{2}\sqrt{\frac{W}{EI}}\right) = W \cdot e \cdot \sec\left\{\left(\frac{L}{2}\sqrt{\frac{W}{E \cdot AK^2}}\right)\right\}$$

$$M_{max} = W \cdot e \cdot \sec\left\{\frac{L}{2K}\sqrt{\frac{W}{AE}}\right\}$$

Maximum stress development in column sectioin will be sum of direct and bending stresses.

i.e. $\quad f_{max.} = \dfrac{W}{A} + \dfrac{M}{Z} = \dfrac{W}{A} + \dfrac{W}{Z} \cdot e\sec\left(\dfrac{L}{2}\sqrt{\dfrac{W}{EI}}\right)$ ... (12.8 (*a*))

$$= \frac{W}{A} + \frac{W \cdot y_c}{I} \cdot e \cdot \sec\left(\frac{L}{2K}\sqrt{\frac{W}{EA}}\right), \qquad\qquad \left\{Z = \frac{I}{y_c} = \frac{AK^2}{y_c}\right\}$$

$$= \frac{W}{A}\left[1 + \frac{y_c \cdot e}{K^2}\sec\left(\frac{L}{2K}\sqrt{\frac{W}{EA}}\right)\right] \qquad\qquad\qquad ... (12.8 (b))$$

or $\quad \dfrac{W}{A} = \dfrac{f_{\text{max.}}}{1 + \dfrac{e \cdot y_c}{K^2} \sec\left\{\dfrac{l}{2K}\sqrt{\dfrac{W}{AE}}\right\}}$ $\qquad \qquad$ ... (12.8 (c))

This equation is for hinged ends and other conditions can be solved by replacing l_e for l. 'e' is eccentricity of load, y_c is the distance of extreme compression fibre, K is radius of gyration. The equation is an identity as the term $\dfrac{W}{A}$ occurs on both sides and hence it is solving by trial and error.

For various support conditions 'L' may be replaced with 'L_c'.

$$f_{\text{max.}} = \dfrac{W}{A}\left[1 + \dfrac{e \cdot y_c}{K^2} \sec\left(\dfrac{L_e}{2K}\sqrt{\dfrac{W}{AE}}\right)\right]$$

or $\quad \dfrac{W}{A} = \dfrac{f_{\text{max.}}}{1 + \dfrac{e \cdot y_c}{K^2} \sec\left(\dfrac{L_e}{2K}\sqrt{\dfrac{W}{AE}}\right)}$ $\qquad \qquad$ ... (12.8 (c))

where: $f_{\text{max.}}$ = Maximum compressive stress = yield stress

$\qquad \quad W$ = Eccentric load

$\qquad \quad A$ = Cross-sectional area

$\qquad \quad e$ = Eccentricity from the centroid

$\qquad \quad y_c$ = Extreme compression fibre distance from NA

$\qquad \quad K$ = Minimum radius of gyration

$\qquad \quad I$ = Minimum moment of inertia = AK^2

$\qquad \quad L_e$ = Effective length of the column as per conditions

$\qquad \quad E$ = Modulus of elasticity

In this formula the term $\text{secant}\left(\dfrac{L}{2K}\sqrt{\dfrac{W}{AE}}\right)$ occurs and therefore the formula is known as *secant formula* for critical load on columns with eccentric loading. Generally the loading may not be in perfect centric alignment and hence some initial eccentricity 'e' may be considered for the critical load assessment.

(a) Design Formula as per IS Code 800-1962

The Indian standards have developed a formula for standard steel sections assuming certain initial eccentricity 'e' and standard factor of safety (1.68) to develop a table of safe compressive stresses by trial for different slenderness ratios (upto 160 and for more than 160).

(i) Slenderness ratio less than 160

Permissible stress $\quad f_a\left(\dfrac{W}{A}\right) \;=\; f_a' = \dfrac{\dfrac{f_y}{m}}{\left\{1+0.20\sec\left(\dfrac{l_e}{K}\sqrt{\dfrac{mf_a'}{E}}\right)\right\}}$ $\qquad$ ... (12.9)

where, $\quad f_a$ = Average axial compressive permissible stress

$\qquad\; f_a'$ = Average permissible compressive stress as obtained by Indian standard trials

$\qquad\; f_y$ = Maximum *yield* strength of material

$\qquad\; m$ = Factor of safety (1.68)

$\qquad\; l_e$ = Effective length of column based on end conditions

$\qquad\; K$ = Minimum radius of gyration

$\qquad\; E$ = Modulus of elasticity

$\qquad\; W$ = Axial working load

$\qquad\; A$ = Area of cross-section

(ii) Slenderness Ratio more than 160

When the slenerness ratio is more than 160, the permissible **stress** f_a' obtained by Indian standard is modified by following equation.

$$f_a \;=\; f_a'\left(1.2-\frac{l_e}{800\,K}\right) \qquad\qquad ... (12.10)$$

where $\qquad\qquad f_a' \;=\; \dfrac{\dfrac{f_y}{m}}{\left\{1+0.20\sec\left(\dfrac{l_e}{K}\sqrt{\dfrac{mf_a'}{E}}\right)\right\}} \qquad\qquad ... (12.9)$

Indian standard formulae is quite complex and it considers initial eccentricity and curvature and the solution of such an equation is quite complex. Indian standard organization has provided a table of permissible compressive stresses for different values of slenderness ratios considering standard values of material properties specially for mild steel.

$E = 200\ \text{kN/mm}^2,\qquad\qquad f_y = 250\ \text{N/mm}^2.$

The values are given in Table 12.3.

Table 12.3: Approximate permissible compressive stresses (as per IS 800-1962 formula)

Slenderness ratio	Permissible stress 'f_a' (N/mm^2)
0	122.5
10	122.1
20	121.4
30	119.9
40	117.9
50	114.86
60	110.7
70	105.35
80	98.7
90	90.9
100	82.3
110	73.8
120	65.76
130	58.5
140	52.04
150	46.45
160	41.45
170	36.95
180	32.93
190	29.4
200	26.46
210	23.81
220	21.46
230	19.5
240	17.74
250	16.27
300	10.68

(b) Design Formula Revised as per IS Code 800-1984

IS 800-1984 has recommended the following *Merchant Rankine Formula* for determining the permissible stress:

$$f_{ac} = \frac{f_{cc} \cdot f_y}{m\left[(f_{cc})^n + (f_y)^n\right]^{\frac{1}{n}}} \qquad \ldots (12.11)$$

where, f_{ac} = Permissible axial compressive stress, N/mm^2 (or (MPa)

$$f_{cc} = \text{Elastic critical compressive stress } \left\{ \frac{\pi^2 E}{\left(\frac{l_e}{K} \right)^2} \right\}, \text{ N/mm}^2 \text{ (or MPa)}$$

f_y = Yield stress in steel, N/mm^2 (or MPa)

E = Modulus of elasticity of steel, 2×10^5 N/mm^2 (or MPa)

(l/K) = Slenderness ratio

m = Factor of safety $\left(\text{equal } \frac{10}{6} \right)$

n = An assumed factor as 1.40

hence
$$f_{ac} = \frac{0.60\, f_{cc} \cdot f_y}{\left\{ (f_{cc})^{1.40} + (f_y)^{1.40} \right\}^{\frac{1}{1.40}}} \qquad \qquad \text{... (12.12)}$$

According to IS: 800-1984, the direct compressive stress on the gross area of cross-section for axially loaded column *shall not exceed* 0.60 f_y, nor the permissible value f_{ac} calculated from the equation (12.11). The value of f_{ac} for Indian standard structural steels are given in Table: IS: 800-1984 formula:

$$f_{ac} = \frac{0.60\, f_{cc} \cdot f_y}{\left[(f_{cc})^n + (f_y)^n \right]^{\frac{1}{n}}}$$

$f_{cc} = \dfrac{\pi^2 EI}{l_e^2}$ (MPa), f_{yy} = Yield stress (MPa), f_{ac} = Permissible stress (MPa), $n = 1.40$

(c) Design of Compression Members as per IS: 800-2007

According to IS: 800-2007 the compression members may fail due to combination of factors such as crushing or flexural buckling. The flexural buckling depends on residual stresses in the material, initial curvature and accidental eccentricities of the load. To account for all these factors, the strength of compressive members in defined by buckling class a, b, c or d as given in Table 12.7 of IS: 800-2007. The actual 'P' on the compression member is given by:

$$P > Pd$$

where, $P_d = Ac \cdot f_{cd}$

A_c = Effective cross-sectional area

f_{cd} = Design compressive stress

Table 12.4: Permissible stress fac (N/mm^2) in axial compression

$f_y \rightarrow$ $(l_e/K) \downarrow$	220	230	240	250	260	280	300	320	340	360	380	400	420	450	480	510	550
10	132	138	144	150	156	168	180	192	204	215	227	239	251	269	287	305	323
20	131	137	142	148	154	166	177	189	201	212	224	235	246	263	280	297	314
30	128	134	140	145	151	162	172	183	194	204	215	225	236	251	266	280	295
40	124	129	134	139	145	154	164	174	183	192	201	210	218	231	243	255	267
50	118	123	127	132	136	145	153	161	168	176	183	190	197	207	216	225	233
60	111	115	118	122	126	133	139	146	152	158	163	168	173	180	187	193	199
70	102	106	109	112	115	120	125	130	135	139	142	147	150	155	160	164	168
80	93	96	98	101	103	107	111	115	118	121	124	127	129	133	136	139	141
90	85	87	88	90	92	95	98	101	103	105	108	109	111	114	116	118	119
100	76	78	79	80	82	84	86	88	90	92	93	94	96	97	99	100	101
110	68	69	71	72	73	74	76	77	79	80	81	82	83	84	85	86	87
120	61	62	63	64	64	66	67	67	69	70	71	71	72	73	73	74	75
130	55	55	56	57	57	58	59	60	61	61	62	62	63	63	64	64	65
140	49	50	50	51	51	52	53	53	54	54	54	55	55	56	56	56	57
150	44	45	45	45	46	46	47	47	48	48	48	49	49	49	49	50	50
160	40	40	41	41	41	42	42	42	43	43	43	43	43	44	44	44	44
170	36	36	37	37	37	37	38	38	38	38	39	39	39	39	39	39	39
180	33	33	33	33	33	34	34	34	34	35	35	35	35	35	35	35	35
190	30	30	30	30	30	30	31	31	31	31	31	31	32	32	32	32	32
200	27	27	28	28	28	28	28	28	28	28	28	28	28	28	28	28	28
210	25	25	25	25	25	25	26	26	26	26	26	26	26	26	26	26	26
220	23	23	23	23	23	23	23	24	24	24	24	24	24	24	24	24	24
230	21	21	21	21	21	21	21	21	22	22	22	22	22	22	22	22	22
240	20	20	20	20	20	20	20	20	20	20	20	20	20	20	20	20	20
250	18	18	18	18	18	18	18	18	18	19	19	19	19	19	19	19	19

f_{cd} = Design compressive stress

Design compressive stress is calculated using the following equation:

$$f_{cd} = \frac{\frac{f_y}{m}}{\phi + [\phi^2 - \lambda^2]^{0.5}} = R \cdot \frac{f_y}{\gamma_{m0}} \le \frac{f_y}{\gamma_{m0}} \qquad \dots (12.13)$$

where, $\phi = 0.50 [1 + \alpha (\lambda - 0.2) + \lambda^2]$

λ = Non-Dimensional effective slenderness ratio = $\sqrt{\dfrac{f_y}{f_c}} = \sqrt{\dfrac{f_y \cdot \left(\dfrac{l_e^2}{K^2}\right)}{\pi^2 \cdot E}}$

$$f_{cc} = \text{Euler's buckling stress} = \frac{\pi^2 \cdot E}{\left(\dfrac{l_e}{K}\right)^2}$$

l_e = Effective length based on support condition

K = Appropriate radius of gyration

(l_e / K) = Effective slenderness ratio (λ)

R = Stress reduction factor based on slenderness ratio yield stress as given in Table: 12.8 in IS: 800-2007.

$$= \frac{1}{[\phi + (\phi^2 - \lambda^2]^{0.5}}$$

α = Imperfection factor based on class of buckling as per IS: 800-2007.

Class of Buckling →	a	b	c	d
Factor α →	0.21	0.34	0.49	0.76

y_{m0} = Partial safety factor for material strength

IS: 800-2007 specifies calculated values of design compressive stress, f_{cd} for different buckling classes as given in Table: 12.9.

Tables: 12.8, 12.9, 12.10 and 12.11 of IS: 800-2007 are given annexture.

For solution of any problem on compression members design stresses can be adopted from the Tables 12.8 and 12.9 of IS: 800-2007. Effective lengths can be found from the Table: 12.11 and buckling type from the Table: 12.10 of IS: 800-2007.

12.8 EMPIRICAL FORMULA FOR CRITICAL LOADS ON COLUMN

For the design of column sections, it is ensured that the column does not fail even under the worst combination of loads. For safety *actual stresses* developed shall be kept below the *ultimate strength* of the material by designing an *appropriate column section*. Depending on *slenderness ratio*, the column section *fail* by either *yielding* or *buckling*. Design has to ensure safety against both the conditions. *Long columns* fail by *buckling* while *very short columns* fail by *crushing* when the *actual stresses* are higher than the *yield stress* of the material. When the columns are moderately long the failure may occur by initiating either *buckling* or *yielding*.

For safe design of the section, a factor of safety is taken on yield strength in case of short column and factor of safety is taken on *buckling* or *crippling strength* in case of long columns.

$$\text{Safe stress} = \frac{\text{Critical strength}}{\text{Factor of safety}}$$

Thus the load is equal to safe strength multiplied by the area of cross-section. From practical observations, it is generally found that the initial *eccentricity* on column increases with increase in *slenderness ratio*. The factor of safety must be increased for higher slenderness ratio (i.e. for longer columns take higher factor of safety).

For simplicity, the engineers have developed *emperical formulae* for safe or critical stress which is considered based on various practical factors related to *slenderness ratio* in one of the other way. Different countries adopt different emperical formula based on slenderness ratio. The equations are:

   i. Gorden-Rankine's Formula

   ii. Straight Line Formula

   iii. Indian Standard (IS Secant) Formula

   iv. Johnson Parabolic Formula

In earlier sections Gorden-Rankine and Indian standard formulae have already been explained. Straight line emperical formula is based on the critical stress of the material 'f_c' linearly related to (l_e/K) values as under:

$$\frac{P}{A} = f_c - z\left(\frac{l_e}{K}\right) \qquad \qquad \text{... (12.14)}$$

i.e. P (Critical load values) $= A\left[f_c - z\left(\frac{l_e}{K}\right)\right]$

where, z is constant and f_c is basic critical strength of the material.

The variation of load has straight line relation with the effective *slenderness ratio* l_e/K.

Indian standard secant formula has already been explained in section 12.7 which is based on the assumption of initial eccentricity in loading or initial curvature in the column axis. This initial eccentricity reduces the loading capacity of the column. This reduced capacity of critical load can be calculated by considering eccentricity 'e' and using the *secant formula* as explained in earlier section.

Johnson's Parabolic Formula

Johnson has suggested a formula based on critical load variation in proportion to square of the *slenderness ration* $\{(l_e/K)^2\}$.

Johnson has simplified Rankine's formula into a parabolic expression as below:

$$\frac{P}{A} = f_c\left\{1 - a\left(\frac{l_e}{K}\right)^2\right\}, \text{ approximately as subsequent terms will be small.}$$

or $$\frac{P}{A} = f_c - f_c \cdot a\left(\frac{l_e}{K}\right)^2 = f_c - b\left(\frac{l}{K}\right)^2 \qquad \qquad \text{... (12.15)}$$

where, $b = \dfrac{f_c^2}{4\pi^2 E}$ approximately, and Johnson accepted the value $\left(\dfrac{f_c^2}{64 E}\right)$ for pinned ends.

EXAMPLE 12.13: Determine an axial permissible load on a ISHB 150, 5 m long column with both ends hinged. ISHB 150 has following properties:

Area = 3898 mm^2,   Minimum radius of gyration K = 34.4 mm

Apply Indian standard secant formula.

Solution:

$$L_e = L = 5000 \text{ mm}, \quad K_{min} = 34.4, \quad \frac{l_e}{K} = \frac{5000}{34.4} = \textbf{145.35}$$

From Table 12.3: $\dfrac{l_e}{K} = 140, f_a = 52.04$ N/mm^2, For $l_e = 150, f_a = 46.45$ N/mm^2

For $\dfrac{l_e}{K} = 145.35, f_a = 52.04 - \dfrac{(52.04 - 46.45)}{10} 5.35 = 49.05$ N/mm^2

Safe axial load on the column $= f_a \cdot A = 49.05 \times 3898 = \textbf{191.197 kN.}$

EXAMPLE 12.14: Determine the length of a circular strut having 32 mm external and 24 mm internal diameter with both ends hihged. Material properties E = 200 kN/mm^2.

Rankines constant $a = \dfrac{1}{7500}$, critical stress $f_c = 315$ N/mm^2.

Solution:

$$A = \frac{\pi}{4}(D^2 - d^2) = \frac{\pi}{4}(32^2 - 24^2), \quad I = \frac{\pi}{64}(D^4 - d^4) = 35200 \text{ mm}^4$$

$$K^2 = \frac{I}{A} = \frac{35200}{352} = 100 \text{ mm}^2$$

Let the length $l_e = L$,

$$P_E = \frac{\pi^2 EI}{L^2} = \frac{\pi^2 \times 200,000 \times 35200}{L^2} = \frac{704\pi^2 \times 10^7}{L^2}$$

$$P_R = \frac{f_c \cdot A}{1 + a\left(\dfrac{L^2}{K^2}\right)} = \frac{315 \times 352}{1 + \dfrac{1}{7500}\left(\dfrac{L^2}{100}\right)}$$

$$P_E = P_R$$

i.e. $\left(1 + \dfrac{L^2}{75 \times 10^4}\right) = \dfrac{110880 L^2}{704 \times \pi^2 \times 10^7} = \dfrac{L^2}{627146}$ or $1 = \dfrac{L^2(750000 - 627146)}{750000 \times 627146}$

or $\qquad L^2 = \dfrac{750000 \times 627146}{122854} = 3828605, \quad L = \textbf{1956.63 mm = 1.95 m.}$

EXAMPLE 12.15: A circular piston rod of steam engine 480 mm long is subjected to a maximum compression of 60 kN. Determine the diameter of the rod if the permissible critical compressive stress is 100 N/mm². Piston rod may be assumed hinged. Rankine's constant $a = \dfrac{1}{7500}$ for hinged ends.

Solution:

$l_e = 400$ mm, $f_c = 100$ N/mm², Load $P = 60{,}000$ N

$$a = \frac{1}{7500}, \quad K^2 = \frac{I}{A} = \frac{\dfrac{\pi}{64} D^4}{\dfrac{\pi}{4} D^2} = \frac{D^2}{16}, \quad K = \frac{D}{4}$$

$$P = \frac{f_c \cdot A}{1 + a\left(\dfrac{l_e}{K}\right)^2}$$

or

$$60{,}000 = \frac{100\left(\dfrac{\pi}{4} D^2\right)}{1 + \dfrac{1}{7500}\left(\dfrac{480 \times 4}{D}\right)^2} = \frac{78.54\,D^2}{1 + \dfrac{491.52}{D^2}} = \frac{78.54\,D^4}{(D^2 + 491.52)}$$

$$78.54\,D^4 - 60{,}000\,D^2 - 29491200 = 0$$
$$D^4 - 763.94\,D^2 - 375492.74 = 0$$

Let $D^2 = x$, $\quad x^2 - 763.94\,x - 375492.74 = 0$

$$x = D^2 = \frac{+763.94 \pm \sqrt{(763.94)^2 - 4 \times 375492.74}}{2}$$

$$D^2 = \frac{+763.94 \pm 1444.15}{2} = 1104 \text{ mm}^2, \quad D = \mathbf{33.22\ mm.}$$

EXAMPLE 12.16: A column of length 'l' is fixed at one end and free at the other end. An eccentric load 'P' is applied at the free end. Determine the expression for the maximum length of the column from the first principle if the deflection of the free end does not exceed the eccentricity of loading.

Solution:

Figure 12.11 shows a column fixed at A and free end B with eccentric load 'P' at C ($BC = e$). Let y be the deflection at any section X-X at a distance of x.

The bending moment at X-X is $P\,(\delta + e - y)$.

$$EI\frac{d^2 y}{dx^2} = P\,(\delta + e - y)$$

or $\dfrac{d^2y}{dx^2} + \dfrac{P}{EI} \cdot y = \dfrac{P}{EI}(\delta + e)$

The solution of this differential equation is:

$$y = C_1 \cos\left(x\sqrt{\dfrac{P}{EI}}\right) + C_2 \sin\left(x\sqrt{\dfrac{P}{EI}}\right) + (\delta + e) \qquad \dots (2)$$

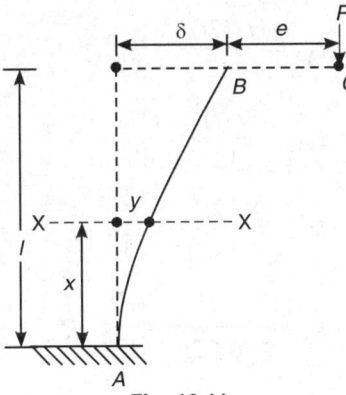

Fig. 12.11.

At fixed end A, $x = 0$, $y = 0$

$0 = C_1 \cos(0) + C_2 \sin(0) + (\delta + e)$, $\qquad C_1 = -(\delta + e)$

Differentiating equation (2)

$$\dfrac{dy}{dx} = -C_1\sqrt{\dfrac{P}{EI}}\sin\left(x\sqrt{\dfrac{P}{EI}}\right) + C_2\sqrt{\dfrac{P}{EI}}\cos\left(x\sqrt{\dfrac{P}{EI}}\right)$$

$$\dfrac{dy}{dx} = +(\delta + e)\sqrt{\dfrac{P}{EI}}\sin\left(x\sqrt{\dfrac{P}{EI}}\right) + C_2\sqrt{\dfrac{P}{EI}}\cos\left(x\sqrt{\dfrac{P}{EI}}\right) \qquad \dots (3)$$

At $x = 0$, $\dfrac{dy}{dx} = 0 = +(\delta + e)\sqrt{\dfrac{P}{EI}}\sin(0) + C_2\sqrt{\dfrac{P}{EI}}\cos(0)$, $\qquad C_2 = 0$

$\therefore$ $\quad y = +(\delta + e)\cos\left(x\sqrt{\dfrac{P}{EI}}\right) + (\delta + e)$

At $x = l$, $y = \delta = -(\delta + e)\cos\left(l\sqrt{\dfrac{P}{EI}}\right) + (\delta + e)$,

$\therefore$ $\quad \cos\left(l\sqrt{\dfrac{P}{EI}}\right) = \dfrac{e}{(\delta + e)}$. But $\delta = e$, $\qquad \dots (4)$

$\cos\left(l\sqrt{\dfrac{P}{EI}}\right) = \dfrac{e}{2e} = \dfrac{1}{2}$, $\quad \therefore \quad l\sqrt{\dfrac{P}{EI}} = \dfrac{\pi}{3}$ or $l = \dfrac{\pi}{3}\sqrt{\dfrac{EI}{P}}$.

EXAMPLE 12.17: A rectangular strut of link of 500 mm length and cross-section with width-twice the thickness. The strut is subjected to a maximum compressive load of 10 kN. Assuming a factor of safety of 3, find the thickness of the strut. The strut ends act as fixed in the direction of weaker axis and hinged in the other direction. Rankine's constant $a = \dfrac{1}{7500}$, $f_c = 325$ N/mm^2.

Solution:

Let the thickness be 't', width = $2t$ $\quad A = 2t^2$

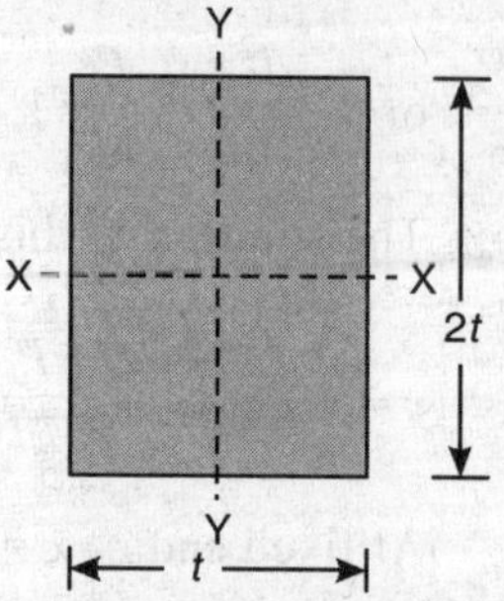

$$I_{XX} = \frac{t \cdot (2t)^3}{12} = \frac{2}{3} t^4,$$

$$I_{YY} = \frac{2t (t)^3}{12} = \frac{t^4}{6}$$

$$K^2_{xx} = \frac{2t^4}{3 \times 2t^2} = \frac{t^2}{3}, \quad K_{xx} = \frac{t}{\sqrt{3}} \qquad \text{... (1)}$$

$$K_{yy}^2 = \frac{t^4}{6 \times 2t^2} = \frac{t^2}{12}, \quad K_{yy} = \frac{t}{2\sqrt{3}} \qquad \text{... (2)}$$

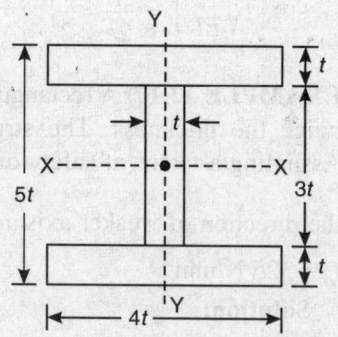

Fig. 12.12: Strut link

$$(l_e)x = 500 \text{ mm}, \qquad (l_e)_y = \frac{500}{2} = 250 \text{ mm}$$

$$\frac{(l_e)_x}{K_{xx}} = \frac{500\sqrt{3}}{t}, \qquad \frac{(l_e)_y}{K_{yy}} = \frac{250 \times 2\sqrt{3}}{t} = \frac{500\sqrt{3}}{t}$$

Thus (l_e / K) is same in both the directions represents equal strength in both the directions.

$$P_R = 10 \times 3 = 30 \text{ kN}$$

$$P_R = \frac{f_c \cdot A}{1 + a \left(\dfrac{l}{K} \right)} \quad \text{or} \quad 30 \times 1000 = \frac{325 \times 2t^2}{1 + \dfrac{1}{7500} \left(\dfrac{500\sqrt{3}}{t} \right)^2}$$

or $\quad 30000 = \dfrac{650t^2}{1 + \dfrac{250000 \times 3}{7500t^2}} = \dfrac{650t^4}{(t^2 + 100)}$

or $\quad 650\, t^4 - 30000\, t^2 - 3 \times 10^6 = 0, \quad$ or $\quad t^4 - 46.154\, t^2 - 4615.4 = 0$

or $\quad t^2 = \dfrac{46.154 \pm \sqrt{2130.9 + 18461.6}}{2} = \dfrac{46.154 \pm 143.495}{2} = 94.82.$

∴ $\quad$ **$t = 9.738$ mm.**

EXAMPLE 12.18: A forged steel connecting rod of an engine is 300 mm long and is subjected to an axial load of 28 kN. Using a factor of safety of 6, find the suitable value of 't' for the cross-section shown in Fig. 12.13. The section behaves as a strut fixed at ends for bending about the axis Y-Y and as a strut hinged at ends for bending about the axis X-X.

Rankine's constant $a = \dfrac{1}{7500}, \ = f_c = 325 \text{ N/mm}^2.$

Fig. 12.13: Connecting rod section

Solution:

$$l = 300 \text{ mm}, \quad (l_e)_x = l = 300 \text{ mm}, \quad (l_e)_y = l/2 = 150 \text{ mm}, \quad A = 11\,t^2$$

Crippling load $P_R = 28 \times 6 = 168 \text{ kN} = \textbf{168000 N}$

$$I_{XX} = \frac{4t(5t)^3}{12} - \frac{3t(3t)^3}{12} = \frac{t^4}{12}(500-81) = \frac{419t^4}{12}, \qquad K_{xx} = \sqrt{\frac{419t^4}{12 \times 11t^2}} = t\,(1.782)$$

$$I_{XX} = \frac{t \times (4t)^3}{12} \times 2 + \frac{3t(3t)^3}{212} = \frac{t^4}{12}(128+3) = \frac{131t^4}{12}, \qquad K_{yy} = \sqrt{\frac{131}{12 \times 11}} \times t = \textbf{0.996 } t$$

$$\frac{(l_e)_x}{K_{xx}} = \frac{300}{1.782t} = \frac{168.35}{t},$$

$$\frac{(l_e)_y}{K_{yy}} = \frac{150}{0.9961} = \frac{150.6}{t}$$

Larger slenderness will govern the design.

$$168000 = \frac{325 \times 11t^2}{1 + \dfrac{1}{7500}\left(\dfrac{168.35}{t}\right)^2} = \frac{3575t^4}{(t^2 + 3.779)}$$

or $\qquad t^4 - 46.993\,t^2 - 177.59 = 0$

$$t^2 = \frac{46.993 \pm \sqrt{2208.34 + 710.36}}{2} = \frac{46.993 \pm 54.025}{2} = 50.51$$

$$t = \textbf{7.107 mm.}$$

EXAMPLE 12.19: A rolled steel beam section IS WB 200 @ 288 N/m is used as a column and has an unsupported length of 5 m. It is hinged at both ends. Determine the axial load this column can support, if the yield stress (f_y) for steel is 250 N/mm^2 and modulus of elasticity $E = 200$ kN/mm^2. Use IS code formula (1984). For IS WB 200 @ 288 N/m, $K_{xx} = 84.6$ mm, $K_{yy} = 29.9$ mm and $A = 3671$ mm^2.

Solution:

$$f_y = 250 \text{ N/mm}^2, \quad l_e = l = 5000 \text{ mm}, \quad E = 200 \text{ kN/mm}^2 = 2 \times 10^5 \text{ N/mm}^2$$

$$K_{\min.} = K_{yy} = 29.9 \text{ mm}$$

$$\text{Maximum slenderness ratio} = \frac{l_e}{K_{\min}} = \frac{5000}{29.9} = 167.22$$

$$f_{cc} = \frac{\pi^2 \cdot E}{\left(\dfrac{l_e}{K_{\min}}\right)^2} = \frac{\pi^2 \times 2 \times 10^5}{(167.22)^2} = 70.65 \text{ N/mm}^2$$

Hence,
$$f_{ac} = \frac{0.6 \times 70.65 \times 250}{[(70.65)^{0.4} + (250)^{0.4}]^{\frac{1}{0.4}}} = 37.90 \text{ N/mm}^2$$

Alternatively from the Table 12.4, $(l/K) = 167.22$, $f_y = 250$ N/mm^2

$$f_{ac} = 41 - (41 - 37)\frac{7.22}{10} = 38.112 \text{ N/mm}^2$$

Safe load $= f_{ac} \cdot A = 37.9 \times 3671 = 139131 \text{ N} = \mathbf{139.131 \text{ kN}.}$

EXAMPLE 12.20: A solid circular rod of 2 m length (both ends hinged) and 100 mm diameter carries safely an axial compressive load of 250 kN. If the rod of the same material and length 3 m is used for the strut with both ends fixed, determine the safe load using the straight line formula, $f_y = 250$ N/mm^2.

Solution:

$$l = l_e = 2000 \text{ mm}, \quad d = 100 \text{ mm}, \quad A = (\pi/4) \times (100)^2 = 0.7857 \times 10^4 \text{ mm}^2$$

$$I_{min} = \frac{\pi}{64}(100)^4 = \frac{0.7857 \times 10^8}{16} \text{ mm}^4$$

$$K = \sqrt{\frac{I}{A}} = \sqrt{\frac{\pi}{64} \times \frac{4(100)^4}{\pi(100)^2}} = \frac{100}{4} = \mathbf{25 \text{ mm},}$$

$$\frac{l_e}{K} = \frac{2000}{K} = 80$$

Load $= \mathbf{250 \text{ kN} = 250000 \text{ N}}$

By straight line formula

$$\frac{P}{A} = \left[f_y - z\left(\frac{l_e}{K}\right)\right]$$

or $\quad \dfrac{250,000}{\dfrac{\pi}{4} \times 10^4} = [250 - z(80)] \quad$ or $\quad z = \dfrac{\left(250 - \dfrac{100}{\pi}\right)}{80} = \dfrac{218.17}{80} = 2.727$

$$l_e = \frac{1}{2} \times 3000 = 1500 \text{ mm}$$

$$\frac{l_e}{K} = \frac{1500}{25} = 60$$

∴ Load $P = A\left[f_y - z\left(\dfrac{l}{K}\right)\right] = \dfrac{\pi}{4} \times 10^4 [250 - 2.727(60)]$

$$P = \frac{\pi \times 10^4}{4}[250 - 163.63] = \mathbf{678350 \text{ N} = 678.35 \text{ kN}.}$$

12.9 SUMMARY

Columns and struts are structural elements carrying *axial compressive loads. Columns* are generally vertical members carrying large vertical axial compressive loads. *Struts* are light structural elements placed in a pinned truss or a framed structure carrying *axial compressive forces.* If the length of a compressive member (column or strut) is much larger than its minimum cross-section and the member *fails primarily by buckling,* the member is called *long.* The effective length of a column or strut depends on its end support conditions. For different end conditions, the effective lengths are given as under.

End Conditions →	Both Ends Hinged	Both Ends Fixed	One End Fixed and Other Hinged	One End Fixed and Other Free
Effective Length	$l_e = L$	$l_e = \dfrac{L}{2}$	$l_e = \dfrac{L}{\sqrt{2}}$	$l_e = 2L$

The *ratio of effective length* (l_e) to its minimum *radius of gyration* (K) of the cross-section is known as *slenderness ratio* (l_e / K). Columns or struts with *large slenderness* ratio fails by *buckling* and the Euler's general equation of load will be:

$$P_E = \frac{\pi^2 \cdot E \cdot I}{l_e^2} = \frac{\pi^2 \cdot E \cdot A}{\left(\dfrac{l_e}{K}\right)^2}$$

Rankine-Gorden has derived general equation of crippling load (P_R) for all type of columns (short or long) using the principle of

$$\frac{1}{P_R} = \frac{1}{P_C} + \frac{1}{P_E} \quad \text{i.e.} \quad \frac{1}{P_R} = \frac{1}{f_c \cdot A} + \frac{\left(\dfrac{l_e}{K}\right)^2}{\pi^2 EA}$$

or

$$P_R = \frac{f_c \cdot A}{1 + a \left(\dfrac{l_e}{K}\right)^2}$$

Where Rankine's constant $a = \dfrac{f_c}{\pi^2 E}$ depends on the *material* properties. When loads are eccentric, or the material are not perfect, Secant formulae or other Indian standard formulae or tables of stresses can be used for the design of compression members.

$$\frac{W}{A} = \frac{f_{max}}{1 + \dfrac{e \cdot y_c}{K^2} \sec\left(\dfrac{l_e}{2K}\sqrt{\dfrac{W}{AE}}\right)} \qquad \text{... (12.8 (d))}$$

(a) Indian Standard IS: 800-1962 Formula

(i)   for $(l_e/K) < 160$; $f_{ac}\left(\dfrac{W}{A}\right) = f_a' = \dfrac{\dfrac{f_y}{m}}{\left\{1 + 0.20 \sec\left(\dfrac{l_e}{2K}\sqrt{\dfrac{mf_a'}{E}}\right)\right\}}$ ... (12.9)

(ii) for $(l_e/K) > 160$; $f_a = f_a'\left(1.2 - \dfrac{l_e}{800\,K}\right)$ ... (12.10)

(b) Indian Standard IS: 800-1984 Formula

$$f_{ac} = \dfrac{f_{cc} \cdot f_y}{m\left[(f_{cc})^n + (f_y)^n\right]^{\frac{1}{n}}} \qquad \text{... (12.11)}$$

$$f_{ac} = \dfrac{0.60\, f_{cc} \cdot f_y}{\left[(f_{cc})^{1.4} + (f_y)^{1.4}\right]^{\frac{1}{1.4}}} \qquad \text{... (12.12)}$$

where, f_{ac} = Permissisble axial compressive stress (N/mm^2 or MPa)

$\qquad f_{cc}$ = Elastic critical compressive stress $\left\{\dfrac{\pi^2 E}{l_e/K^2}\right\}$ N/mm^2

$\qquad\quad m$ = Factor of safety $\left(m = \dfrac{10}{6}\right)$

$\qquad\quad E$ = Modulus of elasticity of material (steel) – 2×10^5 N/mm^2

$\quad (l_e/K)$ = Slenderness ratio

$\qquad\quad n$ = Assumed factor as 1.4

(c) Indian Standard IS: 800-2007 Formula

$$\text{Design compressive stress}\, f_{cd} = \dfrac{\dfrac{f_y}{\gamma_{m0}}}{\phi + [\phi^2 - \lambda^2]^{0.5}} = R \cdot \dfrac{f_y}{\gamma_{m0}} \le \dfrac{f_y}{\gamma_{m0}}$$

where, $\phi = 0.50\,[1 + \alpha\,(\lambda - 0.2) + \lambda^2]$

λ = Non-Dimensional effective slenderness ratio = $\sqrt{\dfrac{f_y}{f_c}} = \sqrt{\dfrac{f_y \cdot \left(\dfrac{l_e^2}{K^2}\right)}{\pi^2 \cdot E}}$

$$f_{cc} = \text{Euler's buckling stress} = \frac{\pi^2 \cdot E}{\left(\dfrac{l_e}{K}\right)^2}$$

l_e = Effective length based on support conditions

K = Appropriate radius of gyration

(l_e/K) = Effective slenderness ratio (λ)

R = Stress reduction factor based on slenderness ratio and yield stress as given in IS: 800-2007 Table: 12.8.

$$= \frac{1}{\left[\phi + (\phi^2 - \lambda^2)\right]^{0.5}}$$

α = Imperfection factor based on class of buckling as per IS: **800-2007** Table 12.7.

γ_{m_0} = Partial safety factor for material strength

Other empirical formulae are also based on slenderness ratio.

(*i*) Straight line formula

$$P = f_c \cdot A\left\{1 - b\left(\frac{l_e}{K}\right)\right\}, \text{ with usual notations}$$

(*ii*) Johnson's parabolic formula

$$P = f_c \cdot A\left\{1 - z\left(\frac{l_e}{K}\right)^2\right\}, \text{ with usual notations.}$$

EXERCISE 12

Q.12.1. Derive equation of Euler's buckling load in case of a long column section of length '*L*' with one end hinged and other fixed. The column element has modulus of elasticity '*E*' and minimum cross-sectional moment of inertia '*I*'.

Ans. $\dfrac{2\pi^2 EI}{l^2}$

Q.12.2. Compare the strength of a column of solid circular to that of hollow circular section of internal diameter of $3/4^{\text{th}}$ external diameter. Both the columns comprise of the same material weight per metre, length and end support conditions.

Ans. $P_{\text{solid}} : P_{\text{hollow}} = 7 : 25$

Q.12.3. Write general equations of buckling load for different support conditions of column of actual length '*L*', cross-sectional area '*A*', minimum moment of inertia '*I*', and modulus of elasticity '*E*'.

Ans. P_E (Both ends hinged) $= \dfrac{\pi^2 EI}{L^2}$, $\qquad P_E$ (Both ends fixed) $= \dfrac{4\pi^2 EI}{L^2}$

P_E (One end fixed and other hinged) $= \dfrac{2\pi^2 EI}{L^2}$, P_E (One end fixed and other free) $= \dfrac{\pi^2 EI}{4L^2}$

Q.12.4. Calculate the safe compressive load on a hollow cast iron column of 10 m length with one end fixed and other end hinged. Outer diameter of the column is **150 mm** and wall thickness is **25 mm**. Consider modulus of elasticity $E = 95000$ N/mm^2 and factor of safety of 3.

Ans. Safe Compressive Load = **125 kN**

Q.12.5. A 20 mm solid circular bar of 2 m length is simply supported. The bar deflects 25 mm under a concentrated mid span load of 250 N. If this bar is used to carry vertical compressive axial load, determine the buckling load with both ends hinged. Also determine the ratio of maximum stresses developed as beam and as column. Deflection under central load

$$W = \frac{WL^2}{48EI}.$$

Ans. From deflection $\dfrac{EI}{L^2} = \dfrac{1250}{3}$, $\qquad P_E = \dfrac{\pi^2 \cdot EI}{L^2} = \pi^2 \left(\dfrac{1250}{3}\right) = 4108.17$ N.

Q.12.6. A hollow alloy tube 5 m long has 40 mm (external) and 25 mm (internal) diameters. The tube extends 3.2 mm under a tensile load of 30 kN. Find the Euler's buckling load when the tube is used as strut with both ends pinned. Also find the safe load if the factor of safety is 5.

Ans. Buckling Load PE = 257 N, $\qquad\qquad$ Safe Load 514.2 N.

Q.12.7. Explain Rankine's theory of crippling load for columns and derive general equation of Rankine's load for column having both ends hinged and minimum radius of gyration for the cross-section 'K'.

Ans. $\dfrac{1}{P_R} = \dfrac{1}{P_E} + \dfrac{1}{P_C}$, $P_R = \dfrac{f_c \cdot A}{1 + a\left(\dfrac{l_e}{K}\right)}$

Q.12.8. Compare the Euler's Buckling load to Rankine's load for a tubular strut of 2.25 m effective length having 40 mm and 30 mm outer and inner diameters respectively.

Hints: $P_E = \dfrac{\pi^2 E \cdot AK^2}{l_e^2}$, $\qquad\qquad P_R = \dfrac{f_c \cdot A}{1 + \dfrac{1}{7500}\left(\dfrac{l_e^2}{K^2}\right)}$

$\qquad P_R = 3344.5 : 32536.2 = 1.028$

Q.12.9. A mild steel tubular strut of 2 m length and 80 mm external diameter and 10 mm thickness is welded rigidly at ends. Find the Rankines crippling load and safe load if factor of safety is 3, yield stress $f_c = 320$ N/mm^2, and constant $a = \dfrac{1}{7500}$.

Ans. $P_R = 580.22$ kN, $\qquad\qquad$ Safe Load = 193.41 kN.

Q.12.10. Fig. Q12.10 shows a compound stanchion made up of two channels ISJC 200 weighing 139 N per channel and two 250 mm × 10 mm plates riveted one of each flange. Calculate the safe load that can be carried by the column of 5.5 m length if both ends are fixed and factor of safety considered is 4. Properties of each ISJC 200 are: $A = 1777$ mm^2, $I_X = 1161.2 \times 10^4$ mm^4, $I_{YY} = 84.2 \times 10^4$ mm^4. Distance of centroid from the back of web = 19.7 mm, Rankine's constant = $\dfrac{1}{7500}$ and critical stress $f_c = 320$ N/mm^2.

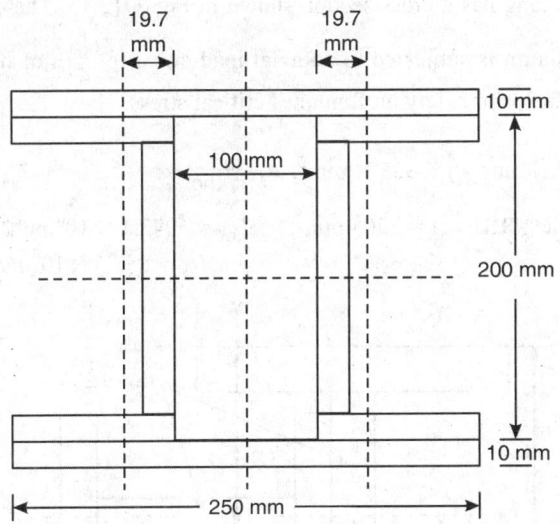

Fig. Q12.10.

Hints: Total $A = 2 \times 1777 + 250 \times 10 \times 2 = 8554$ mm^2

$$I_X = \left[1161.2 \times 2 + \frac{250 \times (220)^3 - 250 \times (200)^3}{12} \right] \times 10^4 = \mathbf{7839 \times 10^4 \ mm^4}$$

$I_Y = 4492.3 \times 10^4$ mm^4

Minimum $K^2 = \dfrac{I}{A} = \dfrac{4492.3 \times 10^4}{8554} = 5250$ mm^2

$l_e = \dfrac{5500}{2} = 2750$ mm

$$P_R = \frac{f_c \cdot A}{1 + a \left(\dfrac{l_e}{K} \right)^2} = \frac{320 \times 8554}{1 + \dfrac{2750 \times 2750}{5250} \times \dfrac{1}{7500}} = 2296 \times 10^6 \ N$$

Safe $P = \mathbf{574 \ kN}$

Q.12.11. A hollow cast iron column with outside diameter = 200 mm and t = 20 mm, L = 4.5 m fixed at both ends. Calculate safe load by Rankine-Gordon formula, factor of safety = 4, f_c = 550 MN/m^2, $a = \dfrac{1}{1600}$.

Ans. $D = 0.2$ m, $d = 0.16$ m, Safe Load $= \dfrac{3.51}{4} = 0.877$ MN, $l_e = \dfrac{4.5}{2} = 2.25$ m.

Q.12.12. A straight bar of alloy, 1 m long and 12.5 mm × 4.8 mm in section is loaded as strut with both ends hinged. It is loaded axially till buckling failure. Estimate the maximum central deflection before the material attains its yield stress of 280 N/mm^2. Take E = 72000 N/mm^2. Use Euler's formula for loading.

Ans. 163 mm.

Q.12.13. A column 8 m long has a cross-section shown in Fig. Q12.13. The column is hinged at both ends. If the column is subjected to an axial load equal to $\dfrac{1}{4}$th of its Euler's buckling load, determine the factor of safety on Rankines critical stress.

Take $E = 200$ kN/mm^2, $f_c = 325$ N/mm^2, $a = \dfrac{1}{7500}$.

Properties of each RSJ: $A = 5205$ mm^2, $I_{XX} = 5943.1 \times 10^4$ mm^4,

$$I_{YY} = 857.5 \times 10^4 \text{ mm}^4, \qquad t_{web} = 6.7 \text{ mm.}$$

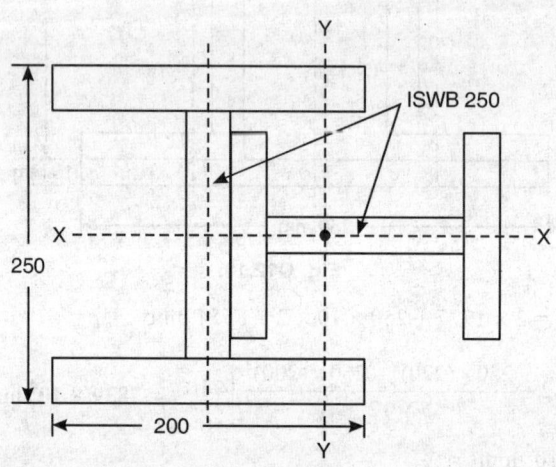

Fig. Q12.13.

Hints: Total $A = 2 \times 5205 = 10410$ mm^2,

C_g (YY of compound) $= 64.15$ mm

$I_{XX} < I_{YY}$ compound,

$I_{XX} = 5943.1 \times 10^4 + 857.5 \times 10^4 = 6800.6 \times 10^4$ mm^4

$l_e = 8000$ mm,

$K^2_{(min)} = 6532.76$ mm^2

$$P_E = \frac{\pi^2 EI}{l_e^2} = \frac{\pi^2 \times 200 \times 10^3 \times 6800.6 \times 10^4}{(8000)^2} = 2097 \times 10^3 \text{ N}$$

Actual Load $= \dfrac{1}{4}$ (2097) kN $= 524.25$ kN

$$\text{Rankine's Load} = \frac{f_c \cdot A}{1 + a\left(\dfrac{l}{K}\right)^2} = \frac{325 \times 10410}{\left\{1 + \dfrac{1}{7500}\left(\dfrac{8000^2}{6532.76}\right)\right\}} = \frac{325 \times 10410}{\{1 + 1.30624\}} = 1467 \text{ kN}$$

Ans. $\dfrac{1467}{524.25} = \mathbf{2.80}$

Q.12.14. A hollow circular steel pipe of 80 mm outside and 60 mm inside diameters, 2.0 m length with both ends fixed to prevent expansion in length. The pipe is unstressed at normal temperature. If the temperature rises by 40°C, using Rankine's formula calculate: (*i*) Temperature stresses in the pipe (*ii*) Factor of safety against failure as strut. Rankine's constant $a = \dfrac{1}{7500}$, $f_c = 330$ N/mm^2, $E = 200$ kN/mm^2, $\alpha = 12 \times 10^{-6}/°C$.

   Ans. (*i*)   Stress = 96.0 N/mm^2,   (*ii*) *FS* = 2.83

Q.12.15. A steel tube having 88 mm outer and 66 mm inner diameters respectively and 2800 mm length, is used as a compression member with both ends hinged. The load acting on the member is eccentric but parallel to the axis. Determine the maximum eccentricity '*e*' so that the *crippling load* on the strut is 60 % of the *Euler's buckling* load. Take $E = 210$ kN/mm^2 and yield strength = 320 N/mm^2.

   Hints: $A = 2662$ mm^2, $K^2 = 756.25$ mm^2, $l_e = 2800$ mm

   $P_E = 0.6 \times 523.63 = 319.58$ kN,   Direct $f_o = 120.05$ N/mm^2

   $f_b = f_{max.} - f_o = 320.0 - 120.05 = 199.95$ N/mm^2

$$f_b = \frac{M}{Z} = \frac{y}{AK^2} P \cdot e \cdot \sec\left\{ \frac{l_e}{2}\sqrt{\frac{P}{EI}} \right\} = \frac{P}{A} \cdot \frac{ey}{K^2} \sec\left\{ \frac{l_e}{2}\sqrt{\frac{P}{EI}} \right\}$$

$$199.95 = 120.05 \left(\frac{e \cdot \dfrac{88}{2}}{756.25} \right) \sec\left\{ \frac{l_e}{2}\sqrt{\frac{P}{E}} \right\}$$

$$\sec\left\{ \frac{l_e}{2}\sqrt{\frac{P}{EI}} \right\} = \left\{ \frac{\pi}{2}\sqrt{\frac{P}{P_E}} \right\} = \sec\left\{ \frac{\pi}{2}\sqrt{0.6} \right\} = \sec (1.216 \text{ rad.}) = 2.882$$

   Ans. $e = \dfrac{199.95}{120.05} \times \dfrac{756.25}{44} \cdot \dfrac{1}{\sec\left\{ \dfrac{l_e}{2}\sqrt{\dfrac{P}{EI}} \right\}} = $ **9.933 mm.**

Q.12.16. Compare the ratio of compressive strengths of a solid and hollow steel circular column sections having the same length, support conditions and cross-sectional areas. Internal diameter may be taken as $\dfrac{3}{4}$th of the external diameter.

   Ans. Strength of solid to hollow = **7 : 25.**

Q.12.17. A steel stanchion is build with 2 *RS* joists 300 mm × 150 mm × 12.5 mm connected with flanges by 20 mm thick and 400 mm wide plates keeping the ends of plates and flanges flush. Using Rankine's formula for columns, calculate the safe load for the stanchion if it is **8 m long** with both ends hinged, $f_c = 330$ N/mm^2 and $a = \dfrac{1}{7500}$, factor of safety = **4.**

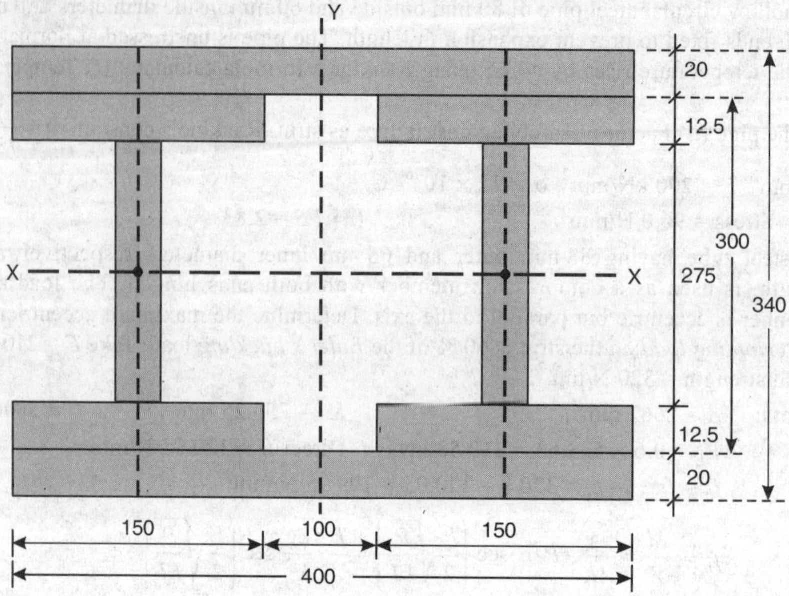

Fig. Q12.17.

Hints: Total $A = 30375$ mm², $l_e = 8000$ mm,

$I_{min} = I_{XX} = 21070.9 \times 10^4$ mm⁴, $K^2 = 6936.92$ mm²

Ans. $P_R = 4494.7$ kN,   Safe Load = 1123.67 kN.

Q.12.18. A mild steel bar of 2 m length and 8 mm × 12 mm section is tested as a strut and loaded axially till it buckles. Assuming Euler's formula for hinged ends to apply, estimate the maximum central deflection before the material attains its yield stress of 330 N/mm². Take $E = 210$ kN/mm².

Hints: $P_E = \dfrac{\pi^2 EI}{l_e^2} = 265.3$ N, $M = P_E \cdot \delta_{max} = 265.3 \times \delta_{max}$, $I_{min} = 512$ mm⁴

$$\text{Maximum stress} = 330 = p_0 + p_b = \frac{265.3}{(8 \times 12)} + \frac{265.3 \times \delta}{\dfrac{b}{6} d^2} = 2.7635 + 2.0726\,\delta$$

$$2.0726\,\delta = (330 - 2.7636)$$

Ans. $\delta = \dfrac{327.2365}{2.0726} = \mathbf{157.89}$ **mm.**

Q.12.19. Calculate the permissible load on a solid circular rod of 40 mm diameter and 5 m height with both ends fixed. $E = 2 \times 10^5$ N/mm², $f_c = 320$ N/mm². Refer Table of stresses using Indian standard formula (IS: 800-1984).

Hints: $\left(\dfrac{l_e}{K}\right) = \dfrac{2500}{\left(\dfrac{d}{4}\right)} = 250$

From IS: 800-1984 (for $f_y = 320$, $\lambda = 250$): $f_{permissible} = 18$ N/mm²

Ans. Permissible load $= 18 \times (\pi/4)(40)^2 = \mathbf{22608}$ **N = 22.608 kN**

Q.12.20. A T-section 200 mm × 200 mm × 12 mm is used as a strut of 2 m length with both ends hinged. (i) Determine Rankine's safe load if factor of safety = 3.0 and Rankine's constant $a = \dfrac{1}{7500}$ and critical stress $f_y = 320$ N/mm^2, (ii) Also calculate the design load for the strut from IS: 800-2007 if the column buckling class is 'c' on the basis of imperfections, initial curvatures, etc.

Hints: $A = 4656$ mm^2, $I_{XX} = 1830.234 \times 10^4$ mm^4, $I_{YY} = 802.707 \times 10^4$ mm^4

   $K_{min} = 41.52$ mm, $l_e = 2000$ mm, $(l_e/K) = 48.2 = \lambda$

   $f_y = 320$ N/mm^2,

Ans. (i) $P_R = 1138$ kN,   Safe $P_R = 379.3$ kN

   (ii) f_{cd} (from tables 9 'c' of IS: 800-2007) = **225.96 N/mm^2**,

   Design load = **1052.07 kN**.

Q.12.21. An *I*-joist ISMB 250 mm × 150 mm @ 37.3 kg/m has an effective length of 5 m. This joist is used as stanchion by welding flange plates of 250 mm × 10 mm symmetrically to each flange. Calculate the load carrying capacity using IS: 800-2007 formula (or tables) for buckling class 'a' and $f_y = 300$ N/mm^2. Each joist has area = 4755 mm^2, $I_{XX} = 513.6 \times 10^4$ mm^4, $I_{YY} = 334.5 \times 10^4$ mm^4.

Hints: $I_{XX} = $ **13586** $\times 10^4$ mm^4, $I_{YY} = 2938.7 \times 10^4$ mm^4,

   $A = 9755$ mm^2, $l_e = 5000$ mm

$$K_{min} = 54.89 \text{ mm}, \quad (l_e/K) = \textbf{91.1}, \quad f_{cd} \text{ (Table 9 '}a\text{')} = \left\{ 161 - \frac{1.1}{10}(161-139) \right\} = 158.6$$

Ans. Design load capacity of column = **1547 kN**.

Combined Stresses and Principal Stresses

13

Principal Planes and Principal Stresses

13.1 INTRODUCTION

We have already studied separately stresses developed in structural elements subjected to *axial loads*, *shear forces* or *bending moments*. If a structural element is subjected to *all these forces at the same time*, the structural behaviour is quite complicated. In all such cases, we need to find the *worst case of normal* or shear stress to *assess safety* under any of the extreme conditions of stresses. Whenever a structural element is subjected to forces in a plane, *two dimensional* normal (f_x and f_y) and shear (q_{xy}, q_{yx}) stresses are setup in the structural element. *Worst combination* occurs along certain plane inclined at 'θ_x' with the plane *X-X*. It may be noted that *normal stresses* are maximum along some plane while *shear stresses* are maximum along some other plane. The structural elements which are *safe separately* under normal or shear stresses (f_x, f_y or q_{xy}) may suffer *failure* under *combination of stresses* (f_θ or q_θ). Thus the *design of structural elements* under combination of plane stresses is based on these *maximum normal* (f_θ) and maximum shear stresses (q_θ).

In real life, the structural elements are generally subjected to *three dimensional* normal forces (**F_x**, **F_y** or **F_z**) and shear forces (Q_x, Q_y, Q_z), but for simplicity in this chapter, only *two dimensional* stresses (f_x, f_y, q_x, q_y) caused by two dimensional plane forces, shall be studied. Three dimensional stress system shall be studied in advance mechanics of structures.

13.2 TWO DIMENSIONAL PLANE STRESS SYSTEM

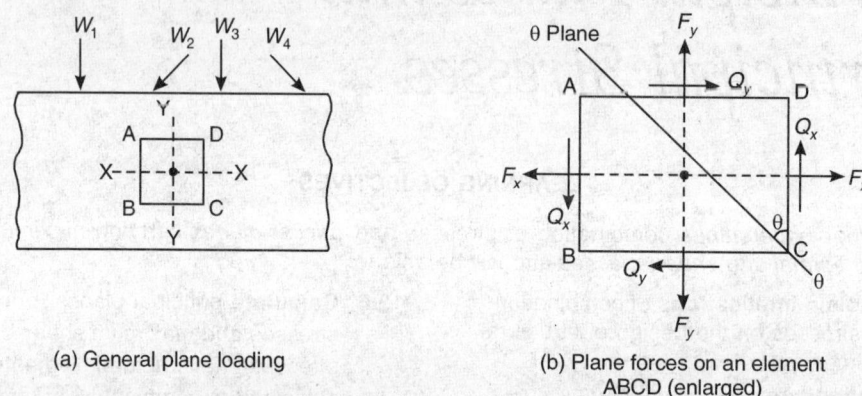

(a) General plane loading

(b) Plane forces on an element
ABCD (enlarged)

Fig. 13.1: Two dimensional plane stress system

Consider a structural beam element subjected to transverse plane loads W_1, W_2, W_3, W_4, ... R_1, R_2, ... etc. causing certain bending, axial and shear stresses at some small element **ABCD** at some point in the structural member (Fig. 13.1a). Indicate various internal forces developed in the small element **ABCD** (shown enlarged sin Fig. 13.1b). Consider any *plane θ-θ* inclined at an angle $θ_x$ with the *plane X-X*.

We shall consider plane stress-system (set of *two perpendicular* **axes** *X-X* and *Y-Y*) in the plane of loading (as shown in Fig. 13.2).

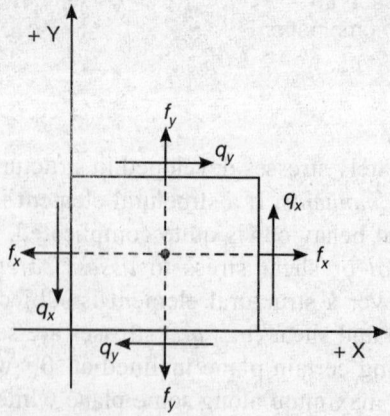

Fig. 13.2: Plane stress system (f_x, f_y, q)

13.3 SIGN CONVENTION

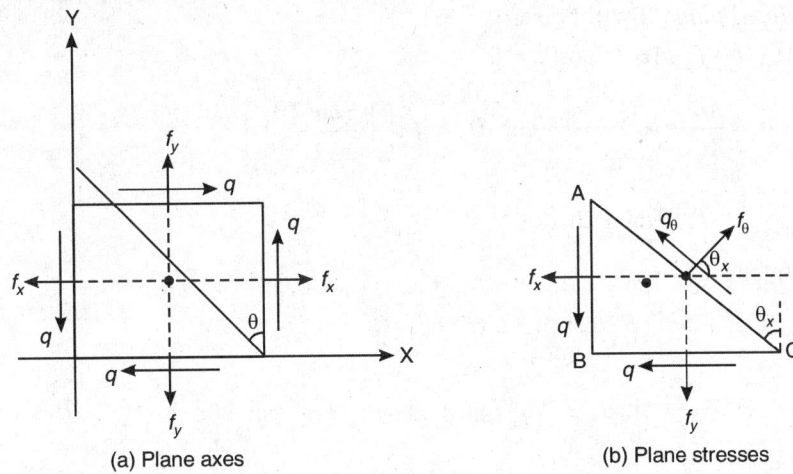

(a) Plane axes   (b) Plane stresses

Fig. 13.3: Axes and plane stresses on small element

Consider horizontal axis as X-X and vertical axis as Y-Y. Normal stresses are represented by 'f' and suffix indicates direction (e.g. 'f_x' – normal stress in X-direction). Shear stresses are represented by 'q' and suffix indicates the plane on which it acts (e.g. 'q_x' – shear stress on the plane X). Sign conventions for various stresses and axes shall as given below:

i. Normal stresses f_x, f_y, f_n, f_θ .... etc. *positive* when *tensile* and *negative* when *compressive*.

ii. Shear stresses q_x, q_y, q_θ, q_{xy}, ... etc. *positive* if it develops *anticlockwise rotation* of the element and *negative if clockwise rotation* of the element.

iii. Inclination of any plane 'θ' with respect to plane X will be *positive in anticlockwise* and *negative in clockwise* direction.

iv. In graphical method: *Normal stresses* (f_x, f_y, f_θ, f_n, ... etc.) are measured along X-*axis* and *shear stresses* (q_x, q_y, q_θ, ... etc.) are measured along Y-*axis*.

v. Normal stresses are *positive in positive direction* of X-*axis* and *negative in negative direction* of X-*axis*.

vi. Shear stresses are considered along Y direction and measured *positive downwards* and *negative upwards*.

13.4 STRESSES ALONG INCLINED PLANE

Case I: Element subjected to unidirectional normal stress

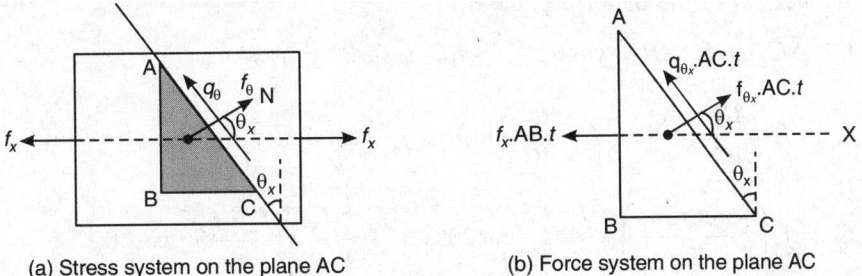

(a) Stress system on the plane AC   (b) Force system on the plane AC

Fig. 13.4: Unilateral normal stress

Since the element is in equilibrium, the algebraic sum of all forces along the *normal* and along the *inclined plane* θ will be zero.

i.e. $f_\theta \cdot AC \cdot t - f_x \cdot AB \cdot t \cos\theta_x = 0$

or $f_{\theta_x} = f_x \cdot \dfrac{AB}{AC} \cdot \cos\theta_x = f_x \cdot \cos^2\theta_x = f_x \left(\dfrac{1+\cos 2\theta_x}{2}\right)$ ... (13.1)

$\left(f_{\theta_x}\right)_{max} = f_x$, when $2\theta_x = 0$

Along the plane

$q_{\theta_x} \cdot AC \cdot t + f_x \cdot AB \cdot t \cdot \sin\theta_x = 0$

$q_{\theta_x} = -f_x \cdot \dfrac{AB}{AC} \cdot \sin\theta_x = -f_x \cdot \sin\theta_x \cdot \cos\theta_x = \dfrac{f_x}{2}\sin 2\theta_x$ ... (13.2)

$\left(q_{\theta_x}\right)_{max} = -\dfrac{f_x}{2}$, when $2\theta_x = 90°$, $\theta_x = 45°$

Case II: Element subjected to two mutually perpendicular normal stresses (f_x, f_y)

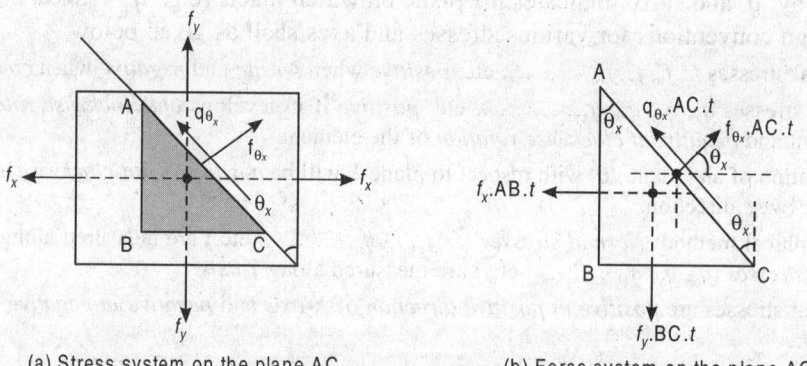

(a) Stress system on the plane AC   (b) Force system on the plane AC

Fig. 13.5: Element subjected to two mutually perpendicular stresses (f_x, f_y)

Since the element is in *equilibrium*, the algebraic sum of resolved components of all stresses in normal and tangential direction will be zero. i.e.

Along the normal of the inclined plane θ_x:

$f_{\theta_x} \cdot AC \cdot t - f_x \cdot AB \cdot t \cdot \cos\theta_x - f_y \cdot BC \cdot t \cdot \sin\theta_x = 0$

or $f_{\theta_x} = f_x \cdot \dfrac{AB}{AC} \cdot \cos\theta_x + f_y \cdot \dfrac{BC}{AC} \cdot \sin\theta_x = f_x \cdot \cos^2\theta_x + f_y \sin^2\theta_x$

or $f_{\theta_x} = f_x\left(\dfrac{1+\cos 2\theta_x}{2}\right) + \dfrac{f_y}{2}(1-\cos 2\theta_x) = \dfrac{f_x + f_y}{2} + \dfrac{f_x - f_y}{2}\cos 2\theta_x$ ... (13.3)

Along the surface of the inclined plane θ_x:

$$q_{\theta_x} \cdot AC \cdot t + f_x \cdot AB \cdot t \cdot \sin\theta_x - f_y \cdot BC \cdot t \cdot \cos\theta_x = 0$$

or $\quad q_{\theta_x} = -(f_x - f_y)\sin\theta_x \cdot \cos\theta_x = -\dfrac{(f_x - f_y)}{2}\sin 2\theta_x$

or $\quad q_{\theta_x} = -\dfrac{(f_x - f_y)}{2}\sin 2\theta_x$

Maximum values of $f_{\theta_x} = \left(\dfrac{f_x + f_y}{2}\right) + \left(\dfrac{f_x - f_y}{2}\right) \cdot 1$, when $\cos 2\theta_x = 1$, $2\theta_x = 0$

$$(f_{\theta_x})_{\text{max.}} = f_x$$

$$(q_{\theta_x})_{\text{max.}} = -\dfrac{f_x - f_y}{2}, \text{ when } 2\theta_x = 90°, \text{ or } \theta_x = 45°$$

Case III: Element subjected to pure shear stresses

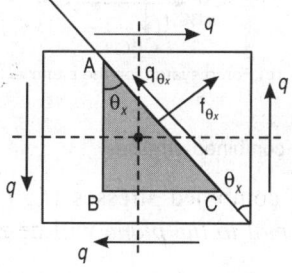   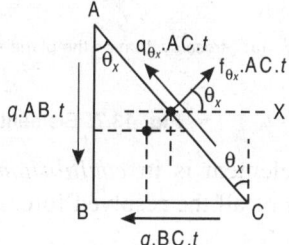

(a) Stress system on the plane AC   (b) Force system on the plane AC

Fig. 13.6: Element subjected pure shear 'q'

Since the element is in *equilibrium* under pure shear stresses 'q', the *algebraic sum* of all resolved forces along the normal to the plane and along the plane will be zero. i.e.

Along the normal:

$$f_{\theta_x} \cdot AC \cdot t - q \cdot AB \cdot t \cdot \sin\theta_x - q \cdot BC \cdot t \cdot \cos\theta_x = 0$$

or $\quad f_{\theta_x} = q \cdot \dfrac{AB}{AC} \cdot \sin\theta_x + q \cdot \dfrac{BC}{AC} \cdot \cos\theta_x = q\,(\sin\theta_x \cdot \cos\theta_x + \sin\theta_x \cdot \cos\theta_x)$

$$f_{\theta_x} = q \cdot \sin 2\theta_x \qquad\qquad\qquad\qquad \dots (13.5)$$

$$(f_{\theta_x})_{\text{max}} = \boldsymbol{q}, \text{ when } \sin 2\theta = 1, \; \boldsymbol{2\theta = 90°}, \; \theta = 45°, \; 135°$$

Along the plane:

$$q_{\theta_x} \cdot AC \cdot t - q \cdot AB \cdot t \cdot \cos\theta_x + q \cdot BC \cdot t \cdot \sin\theta_x = 0$$

or $\quad q_{\theta_x} = q \cdot \dfrac{AB}{AC} \cdot \cos\theta_x - q \cdot \dfrac{BC}{AC} \cdot \sin\theta_x = q \, (\cos^2\theta_x - \sin^2\theta_x)$

or $\quad q_{\theta_x} = q \dfrac{(1 + \cos 2\theta_x)}{2} - q \dfrac{(1 - \cos 2\theta_x)}{2} = q \, \cos 2\theta_x$ $\qquad \dots (13.6)$

$\quad (q_{\theta_x})_{max} \cdot = q$, when $2\theta_x = 0, 180°$ i.e. $\theta_x = 0°, 90°$

Case IV: Element subjected to two normal stresses combined with pure shear

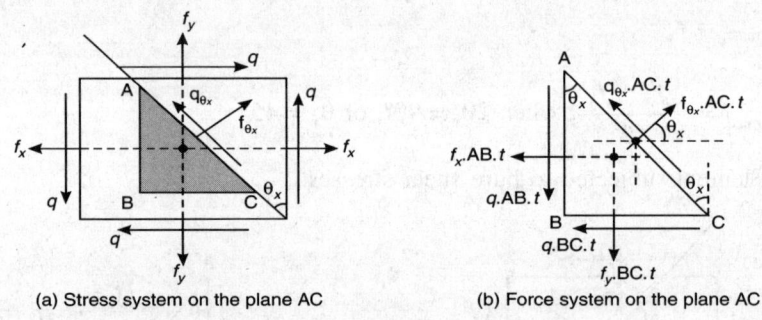

(a) Stress system on the plane AC   (b) Force system on the plane AC

Fig. 13.7: Element subjected to the combined stresses

Since the element is in *equilibrium* under all the combined stresses (f_x, f_y, q, etc.), the *algebraic sum* of all the resolved forces *along* and *normal to the plane* will be zero, i.e.

$$f_{\theta_x} \cdot AC \cdot t - f_x \cdot AB \cdot t \cdot \cos\theta_x - f_y \cdot BC \cdot t \cdot \sin\theta_x - q \cdot AB \cdot t \cdot \sin\theta_x - q \cdot BC \cdot t \cdot \cos\theta_x = 0$$

or $\quad f_{\theta_x} = \dfrac{f_x + f_y}{2} + \dfrac{f_x - f_y}{2} \cdot \cos 2\theta_x + q \, \sin 2\theta_x$ $\qquad \dots (13.7)$

$$q_{\theta_x} \cdot AC \cdot t + f_x \cdot AB \cdot t \cdot \sin\theta_x - f_y \cdot BC \cdot t \cdot \cos\theta_x - q \cdot AB \cdot t \cdot \cos\theta_x + q \cdot BC \cdot t \cdot \sin\theta_x = 0$$

or $\quad q_{\theta_x} = -\dfrac{(f_x - f_y)}{2} \sin 2\theta_x + q \, \cos 2\theta_x$ $\qquad \dots (13.8)$

Rewriting (13.7) and (13.8) and squaring

$$\left\{ f_{\theta_x} - \frac{f_x + f_y}{2} \right\}^2 = \left\{ \frac{f_x - f_y}{2} \cos 2\theta_x + q \sin 2\theta_x \right\}^2$$

Also $\left\{ q_{\theta_x} - 0 \right\}^2 = \left\{ -\dfrac{f_x - f_y}{2} \sin 2\theta_x + q \cos 2\theta_x \right\}^2$

Adding and simplifying

$$\left\{ f_{\theta_x} - \frac{f_x + f_y}{2} \right\}^2 + \{q_{\theta_x}\}^2 = \left\{ \frac{f_x - f_y}{2} \right\}^2 (1) + q^2 \cdot (1) \; (\because \; \sin^2 2\theta_x + \cos^2 2\theta_x = 1) \; ...(13.9)$$

This is equation of a circle in f_θ and q_θ with *centre* $\left(\dfrac{f_x + f_y}{2}, 0 \right)$ and *radius*

$$'r' = \sqrt{ q^2 + \left(\frac{f_x + f_y}{2} \right)^2 }.$$

13.5 TWO DIMENSIONAL PRINCIPAL STRESSES AND PLANES

Principal planes are those planes which *carries **only normal stresses*** and *shear stresses* are zero along these planes. *Normal stress* along the *principal plane* will be *maximum* or *minimum* and *shear stress zero*. The *maximum normal stress* across the principal plane is known as *major principal stress* and the *minimum normal stress* across the principal plane is known as *minor principal stress*. Principal planes (both major or minor) have *zero shear stress* across them.

We have already derived equations for *normal and shear stresses* across any plane inclined at θ_x with the plane *X-X*. These general equations are:

$$f_{\theta_x} = \frac{f_x + f_y}{2} + \frac{f_x - f_y}{2} \cdot \cos 2\theta_x + q \, \sin 2\theta_x \qquad \qquad ... (13.7)$$

and $\quad q_{\theta_x} = -\dfrac{(f_x + f_y)}{2} \sin 2\theta_x + q \, \cos 2\theta_x \qquad \qquad ... (13.8)$

For maximum values of $f_{\theta_x}, \dfrac{df_{\theta_x}}{d\theta} = 0$

i.e. $\quad \dfrac{d}{d\theta} \left\{ \dfrac{f_x + f_y}{2} + \dfrac{f_x - f_y}{2} \cos 2\theta_x + q \sin 2\theta_x \right\} = 0$

or $\quad 0 + -\left(\dfrac{f_x - f_y}{2} \sin 2\theta_x \times 2 \right) + 2q \cos 2\theta_x = 0, \;$ or $\; \dfrac{\sin 2\theta_x}{\cos 2\theta_x} = \dfrac{2q}{(f_x - f_y)}$

or $\quad$ **tan2θ_x** $= \dfrac{2q}{(f_x - f_y)} \qquad \qquad ... (13.10)$

Also by definition of principal stress we have, $\; q_{\theta_x} = 0$

$$q_{\theta_x} = 0 = -\left(\frac{f_x - f_y}{2}\right)\sin 2\theta_x + q\cos 2\theta_x \qquad \ldots (13.8)$$

or $\quad \tan 2\theta_x = \dfrac{2q}{(f_x - f_y)} \qquad \ldots (13.10)$

From equation (13.10) principal planes are given by:

$$\tan 2\theta_x = \frac{2q}{(f_x - f_y)}, \quad \text{i.e. } 2\theta_x = \tan^{-1}\left(\frac{2q}{f_x - f_y}\right) \qquad \ldots (13.10)$$

$$\sin 2\theta_x - \pm\frac{2q}{\sqrt{4q^2 + (f_x - f_y)^2}}, \quad \cos 2\theta_x = \pm\frac{(f_x - f_y)}{\sqrt{4q^2 + (f_x - f_y)^2}}$$

Substituting these values of $2\theta_x$, principal stresses are f_{n_1} or f_{n_2}

$$f_{n_{1,2}} = \left(\frac{f_x + f_y}{2}\right) \pm \left(\frac{f_x - f_y}{2}\right)\frac{(f_x - f_y)}{\sqrt{4q^2 + (f_x - f_y)^2}} \pm q \cdot \frac{2q}{\sqrt{4q^2 + (f_x - f_y)^2}}$$

or $\quad f_{n_{1,2}} = \dfrac{f_x + f_y}{2} \pm \dfrac{1}{2}\dfrac{4q^2 + (f_x - f_y)^2}{\sqrt{4q^2 + (f_x - f_y)^2}}$

or $\quad f_{n_{1,2}} = \dfrac{f_x + f_y}{2} \pm \dfrac{1}{2}\sqrt{4q^2 + (f_x - f_y)^2} \qquad \ldots (13.12)$

Thus *principal planes* $\qquad \theta_x = \dfrac{1}{2}\tan^{-1}\left(\dfrac{2q}{f_x - f_y}\right) \qquad \ldots (13.11)$

Principal stresses $\qquad f_{n_{1,2}} = \dfrac{1}{2}(f_x + f_y) \pm \sqrt{q^2 + \left(\dfrac{f_x - f_y}{2}\right)^2} \qquad \ldots (13.12)$

As indicated principal stresses lie on the circle having no ordinate for shear stress ($q = 0$).

i.e. $\left\{f_n - \dfrac{f_x + f_y}{2}\right\}^2 + q_n^2 = \left\{\sqrt{q^2 + \left(\dfrac{f_x - f_y}{2}\right)^2}\right\}^2$, $q_n = 0$, for principal stresses, and centre of

the circle coordinates $\left(\dfrac{f_x + f_y}{2}, 0\right)$ i.e. the centre lies on the horizontal axis X-X. Radius of the

circle $\quad r = \sqrt{q^2 + \left(\dfrac{f_x - f_y}{2}\right)^2}$, $f_{n_1} = \left(\dfrac{f_x + f_y}{2}\right) + \sqrt{q^2 + \left(\dfrac{f_x - f_y}{2}\right)^2}$

$$\text{and } f_{n_2} = \left(\frac{f_x + f_y}{2}\right) - \sqrt{q^2 + \left(\frac{f_x - f_y}{2}\right)^2}.$$

The concept of stress circle will be used in *graphical* determination of principal or normal stresses and shear stresses across any plane inclined at any angle θ_x with the X-X axis.

$$(q_\theta)_{max.} = \left(\frac{f_{n_1} - f_{n_2}}{2}\right) = \frac{1}{2}\sqrt{(f_x - f_y)^2 + 4q^2} \text{ , on a plane at } 45° \text{ to the principal planes}$$

$$\qquad ... (13.13)$$

Principal stresses f_{n_1}, f_{n_2} also by substituting the values of $\sin 2\theta_x$ and $\cos 2\theta_x$ will be:

$$f_{n_{1,2}} = \frac{f_x + f_y}{2} + \frac{f_x - f_y}{2}\cos 2\theta_x + q \sin 2\theta_x$$

$$f_{n_{1,2}} = \frac{f_x + f_y}{2} \pm \left(\frac{f_x - f_y}{2}\right)\frac{(f_x - f_y)}{\sqrt{(f_x - f_y)^2 + 4q^2}} \pm q \cdot \frac{2q}{\sqrt{(f_x - f_y)^2 + 4q^2}}$$

$$f_{n_{1,2}} = \frac{f_x + f_y}{2} \pm \tfrac{1}{2}\sqrt{(f_x - f_y)^2 + 4q^2} \qquad\qquad ... (13.12)$$

Values of maximum $q_\theta = \dfrac{f_{n_1} - f_{n_2}}{2} = \tfrac{1}{2}\sqrt{(f_x - f_y)^2 + 4q^2}$ $\qquad\qquad ... (13.13)$

EXAMPLE 13.1: A body is subjected to 50 N/mm^2 tensile and 40 N/mm^2 (compressive) stresses along two perpendicular planes. Determine normal stress, shear stress and resultant stress along a plane inclined at 30° with the plane of 50 N/mm^2 stress.

Solution:

$$f_{\theta x} = \frac{f_x + f_y}{2} + \frac{f_x - f_y}{2}\cos 2\theta = \frac{50 - 40}{2} + \frac{50 - (-40)}{2}\cos(2 \times 30°)$$

$$= \frac{10}{2} + 45 \times \cos 60° = 5 + 22.5 = \mathbf{27.5 \text{ N/mm}^2} \text{ tensile}$$

Shear $\quad q_{\theta x} = -\dfrac{f_x - f_y}{2}\sin 2\theta = -\dfrac{50 - 40}{2}\sin 60° = -\dfrac{45\sqrt{3}}{2} \text{ N/mm}^2$

Maximum shear stress $\quad = \mathbf{-45 \text{ N/mm}^2}$

Resultant stress $\qquad f_r = \sqrt{(27.5)^2 + \dfrac{45^2 \times 3}{4}} = 47.7 \text{ N/mm}^2$

$$\tan \phi = \frac{225\sqrt{3}}{275} , \quad \phi = 54°48'$$

EXAMPLE 13.2: A stressed body is subjected to 500 N/mm² (tensile) and 300 N/mm² (compressive) along two perpendicular planes. The plane carrying 500 N/mm² (tensile) stress has shear stress + 200 N/mm² (anticlockwise). Determine normal and shear stresses along a plane inclined at 50° with the 500 N/mm² plane.

Solution:

(i) $f_{\theta x} = \dfrac{f_x + f_y}{2} + \dfrac{f_x - f_y}{2}\cos 2\theta + q\sin 2\theta$

$\qquad = \dfrac{500 - 300}{2} + \dfrac{500 - (-300)}{2}\cos 100 + 200\sin 100$

$\qquad = 100 + 400\cos 100 + 200\sin 100$

$\qquad = 100 - 69.46 + 196.96 = 227.5$ N/mm² **(tensile)**

(ii) $q_{\theta x} = -\dfrac{f_x - f_y}{2}\sin 2\theta_x + q\cos 2\theta_x$

$\qquad = -\dfrac{500 + 300}{2}\sin 100 + 200(\cos 100)$

$\qquad = -400(0.98481) + 200\,(-0.17365)$

$\qquad = \mathbf{-428.65}$ **N/mm²** (clockwise)

13.6   MOHR CIRCLE (GRAPHICAL) APPROACH

Sign conventions

- Normal stresses are taken along *X-axis*. *Tensile stresses* are considered +ve, while compressive stresses are considered –ve.
- All shear stresses are taken along *Y-axis*. Shear stress causing *anticlockwise moment* or rotation will be considered +ve.
- *Positive shear* stresses are drawn *downward* while –*ve shear stress* shall be drawn *upwards*.
- Substitution of all the values in the formula will be with proper *signs* as given above.
- All the stresses will be drawn by adopting *suitable linear scale* for all type of stresses.
- *Units* of all the stresses must be *consistent and uniform*.
- *Accuracy of results* of graphical analysis will greatly depend on the *adopted scale* and *skill* of drawing.
- Graphical analysis may be carried out on graphs with suitable scale.

Graphical Method (Mohr Circle Approach)

$$\left\{f_{\theta x} - \frac{f_x + f_y}{2}\right\}^2 + q_{\theta x}^2 = \left(\sqrt{\left(\frac{f_x - f_y}{2}\right)^2 + q^2}\right)^2 \qquad \text{... Equation of Circle}$$

If f_θ (normal stress) is drawn along X-axis and shear stress q_θ along Y-axis above equation represents circle. *Shear* stress is considered *+ve* when on X-plane it tends to rotate the body *anticlockwise*. In graphical method *+ve shear* stress is drawn *below*.

$$\text{Radius of the circle} = \sqrt{\left(\frac{f_x - f_y}{2}\right)^2 + q^2}, \; r = \frac{f_x - f_y}{2}, \text{ when } q = 0$$

$$\text{Centre of the circle} = \left(\frac{f_x + f_y}{2}, 0\right) \text{ lies on X-X axis.}$$

$$\text{Maximum shear stress} = \text{Radius} = \sqrt{\left(\frac{f_x - f_y}{2}\right)^2 + q^2}$$

EXAMPLE 13.3: A body is subjected to a tensile stress of 500 N/mm² on a plane and a compressive stress of 400 N/mm² along a perpendicular plane. Determine normal shear stress along plane at 30° with the plane of 500 N/mm².

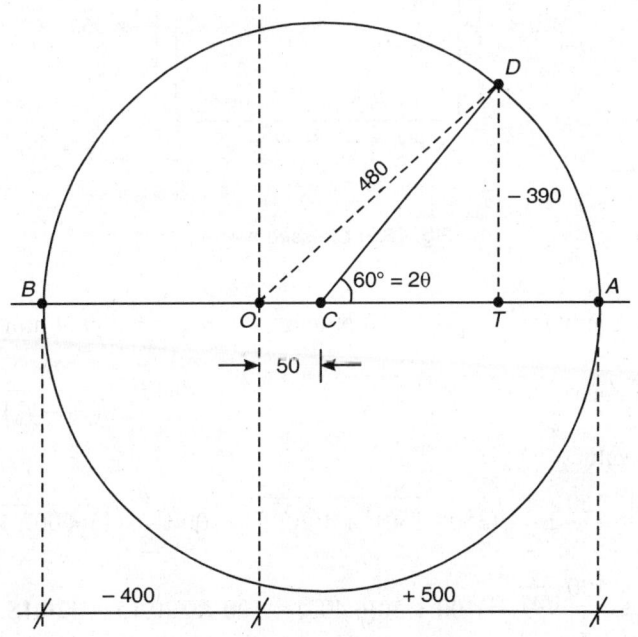

Fig. 13.8: Mohr circle of stresses

Solution:

$f_x = +500 \text{ N/mm}^2,$ $\qquad f_y = -400 \text{ N/mm}^2,$ $\qquad q = 0,$ $\qquad \theta = 30°$

Assume a scale $\qquad 100 \text{ N/mm}^2 = 10 \text{mm}$

Choose the centre and radius for $q = 0$, $r = \dfrac{f_x - f_y}{2}$

$$R = \frac{f_x - f_y}{2} = \frac{500 - (-400)}{2} = 450 \ (45 \text{ mm}), \qquad \text{Centre} = \left(\frac{500 - 400}{2}, 0\right)$$

$f_{30} = OT = 28$ mm (+ 280 N/mm^2)

$q_{30} = DT = 39$ mm (– 390 N/mm^2), $\qquad\qquad\qquad (q_{max} = CD = 450$ N/mm^2)

$f_r = OD = 48$ mm (480 N/mm^2)

EXAMPLE 13.4: A stressed body is subjected to a tensile stress of 350 N/mm^2 and a compressive stress of 150 N/mm^2 on two mutually perpendicular planes. These planes are also subjected to shear stress of 200 N/mm^2 anticlockwise on the first plane. Determine analytically and graphically the *principal planes* and *principal stresses*. Also find the *maximum shear stress*.

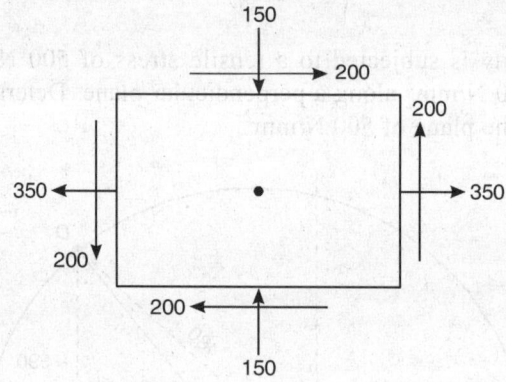

Fig. 13.9: Stressed element

Solution:

$$f_x = + 350 \text{ N/mm}^2, \qquad f_y = -150 \text{ N/mm}^2, \qquad q = + 200 \text{ N/mm}^2$$

$$f_{m_{1,2}} = \frac{f_x + f_y}{2} \pm \frac{1}{2}\sqrt{(f_x - f_y)^2 + 4q^2}, \qquad\qquad \tan \beta = \frac{2q}{(f_x - f_y)}$$

Substituting the values

$$f_{m_{1,2}} = \frac{350 - 150}{2} \pm \frac{1}{2}\sqrt{(350 + 150)^2 + 4(200)^2} = 100 \pm \frac{1}{2}\sqrt{250000 + 160000}$$

$$= 100 \pm \frac{100}{2}\sqrt{41} = 100 \pm 50(6.403) = 100 \pm 320.15 = \mathbf{420.15, -220.15 \ N/mm^2}$$

$$2\theta = \beta = \tan^{-1}\frac{2 \times 200}{500} = \tan^{-1}(0.8), \quad \beta = 38.66°(38° \ 40'), \ 218° \ 40'$$

$\theta_{1,2} = \mathbf{19° \ 20', \ 109° \ 20'}$ *(Principal planes* with the plane of 350 N/mm^2)

Principal stresses

$\qquad f_{m_1} = 420.15$ N/mm^2 **(tensile)**, $\qquad\qquad \theta_1 = \mathbf{19° \ 20'}$

$\qquad f_{m_2} = \mathbf{-220.15 \ N/mm^2}$ (compressive), $\qquad \theta_2 = \mathbf{109° \ 20'}$

Mohr Circle

Scale $10 \text{ N/mm}^2 = 1$ mm

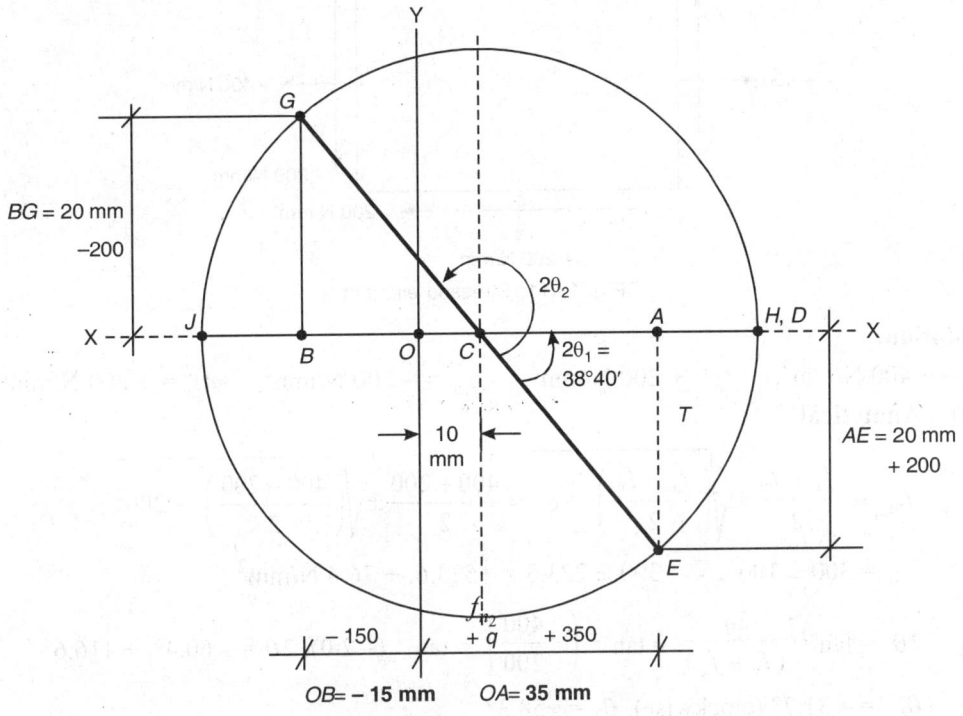

Fig. 13.10: Mohr circle of stresses

Origin – O

X-axis: $OA = f_x = \dfrac{350}{10} = \mathbf{35 \ mm,}$ $\qquad\qquad OB = f_y = \dfrac{150}{10} = \mathbf{15 \ mm}$

Y-axis: $AE = q = +\dfrac{200}{10} = 20$ mm (downward), $\quad BG = -\dfrac{200}{10} = -20$ mm (upward)

Join GE, GE cuts X-axis at C. Take C as centre and CE = CG radius and draw the circle.
Measure angle $ECH = 38°\ 40'\ (2\theta_1)$, $\ \theta_1 = 19°\ 20'$

Measure $f_{n_1} = OH = +42$ mm (tensile), $\qquad\qquad \theta_1 = 19°\ 20'$

$\qquad f_{n_2} = OJ = -22$ mm (compressive), $\qquad\qquad \theta_2 = 109°\ 20'$

EXAMPLE 13.5: An element in a stressed body is shown in Fig. 13.11. Determine principal stresses and the principal planes.

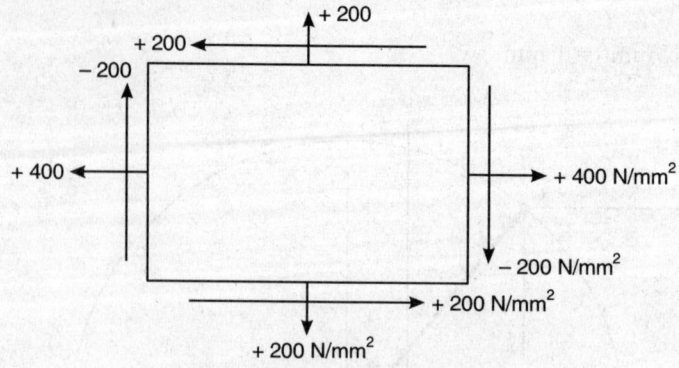

Fig. 13.11: Stressed element

Solution:

$f_x = + 400 \text{ N/mm}^2$, $\quad f_y = + 200 \text{ N/mm}^2$, $\quad q_{xy} = -200 \text{ N/mm}^2$, $\quad q_{yx} = +200 \text{ N/mm}^2$

(a) Analytical

$$f_{n_{1,2}} = \frac{f_x + f_y}{2} \pm \sqrt{\left(\frac{f_x - f_y}{2}\right)^2 + q^2} = \frac{400 + 200}{2} \pm \sqrt{\left(\frac{400 - 200}{2}\right)^2 + 200^2}$$

$$= 300 \pm 100 \sqrt{5} = 300 \pm 223.5 = +523.6, +76.4 \text{ N/mm}^2$$

$$2\theta = \tan^{-1} \frac{2q}{(f_x - f_y)} = \tan^{-1}\left(-\frac{400}{200}\right) = \tan^{-1}(-2.0), \, 2\theta = -60.4°, +116.6°$$

$$\theta_1 = -31.7° \text{ (clockwise)}, \, \theta_2 = +58.3°$$

(b) Mohr Circle

Scale 10 N/mm^2 = 1 mm

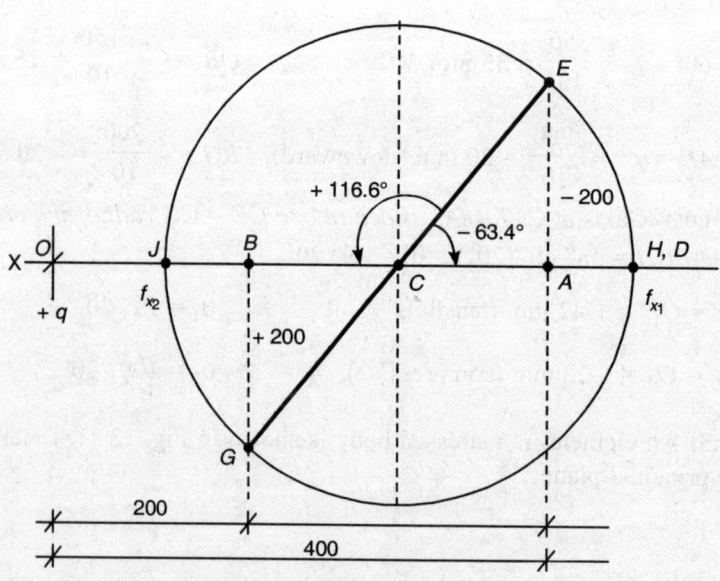

Fig. 13.12: Mohr circle of stresses

OA = + 400 (40 mm), OB = + 200 (20 mm)

AE = – 200 (– 20 mm) up, BG = + 200 (+ 20 mm) down

CE = radius

Draw circle with centre at C and radius equal to CE or CG.

Measure

$$2\theta_1 = -63.4^\circ \ (\theta_1 = -31.7^\circ), \text{ clockwise}$$

$$f_{n_1} = OH = 52.4 \text{ mm } (+ 524 \text{ N/mm}^2)$$

$$f_{n_2} = OJ = 7.7 \text{ mm} = 77 \text{ N/mm}^2$$

$$q_{max} = \frac{524 - 77}{2} = 223.5 \text{ N/mm}^2$$

EXAMPLE 13.6: A structural element is subjected to a tensile stress of 70 N/mm² (MPa) along any direction and a compressive stress of 30 N/mm² (MPa) along the perpendicular direction. The element also carries a shear stress of 40 N/mm² along the first plane and has tendency to rotate the element anticlockwise. (a) Calculate the principal stresses and the planes where these principal stresses exist (b) Also determine principal planes and principal stresses by Mohr circle approach.

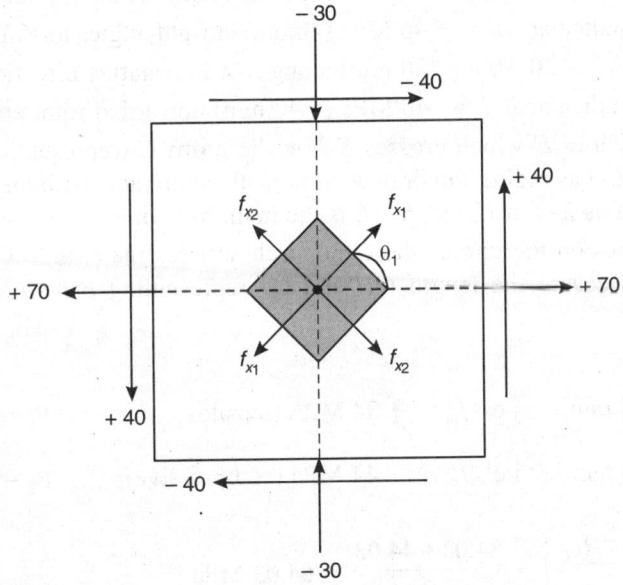

Fig. 13.13: Stressed element (stress in MPa)

Solution:

Given that

$$f_x = + 70 \text{ MPa,} \qquad f_y = -30 \text{ Mpa,} \qquad q_x = + 40 \text{ MPa,}$$

$$f_{n_1} = ? \qquad\qquad f_{n_2} = ? \qquad\qquad \theta_1 = ? \qquad\qquad \theta_2 = ?$$

a. Analytical Approach

$$\tan 2\theta_x = \frac{2q}{(f_x - f_y)} = \frac{2 \times 40}{(70 + 30)}$$

$$2\theta_x = \tan^{-1}(0.80) = 38°40', 218°40'$$

$\theta_{x_1} = 17°20', \theta_{x_2} = 109°20'$, with the plane of 70 MPa tensile stress.

$$f_{m_{1,2}} = \left(\frac{f_x + f_y}{2}\right) \pm \sqrt{q^2 + \left(\frac{f_x - f_y}{2}\right)^2} = \frac{70 - 30}{2} \pm \sqrt{(40)^2 + \left(\frac{70 + 30}{2}\right)^2}$$

$$= 20 \pm 64.03 = +84.03, -44.03 \text{ MPa}$$

i.e. *major* principal stress + 84.03 MPa (tensile) along a plane inclined at 19°20'. Minor principal stress – 44.03 MPa (compressive) along a plane inclined at 109°20' with plane of 70 MPa (tensile) stress (Fig. 13.13).

b. Graphical Approach

Assume force (stress) scale of 1 MPa = 1 mm.

Draw axes *X-X* and *Y-Y* at right angles with origin *O*.

Measure *OA* = + 70 MPa (70 mm) with scale along the horizontal axis *X-X*.

Draw perpendicular *AE* = + 40 MPa (40mm) at right angles to *X-X*. (+ve downwards).

Measure *OB* = – 30 MPa (– 30 mm) along *X-X* in negative direction.

Draw perpendicular *BG* = – 40 MPa (– 40 mm) upward at right angles to *X-X*.

Join the points **G and E** which crosses *X-X* at the **point C** (representing the centre of the circle). Take *CE* or *CG* as radius and draw a circle with centre at *C* with the help of a compass. The circle cuts the axis *X-X* at *H* and *J*. *GE* is the main base line.

Points *H* and *J* lies on the circle where the circle crosses the axis *X-X* i.e. ordinate along *Y-Y* will be zero. Measure the length *OH* and *OJ* representing *principal stresses* with the assumed scale.

By measuring:

$OH = + 84$ mm,   i.e. $f_{m_1} = +84$ MPA (tensile), $\theta_1 = 19°15'$

$OJ = -44$ mm,   i.e. $f_{n_2} = -44$ MPa (compressive), $\theta_2 = 109°15'$

$$q_{max.} = \left(\frac{f_{m_1} - f_{n_2}}{2}\right) = \frac{84.03 + 44.03}{2} = 64.03 \text{ MPa}$$

By Mohr Circle $q_{max.} = $ radius $= \pm 64.0$ mm $= \pm 64.0$ MPa

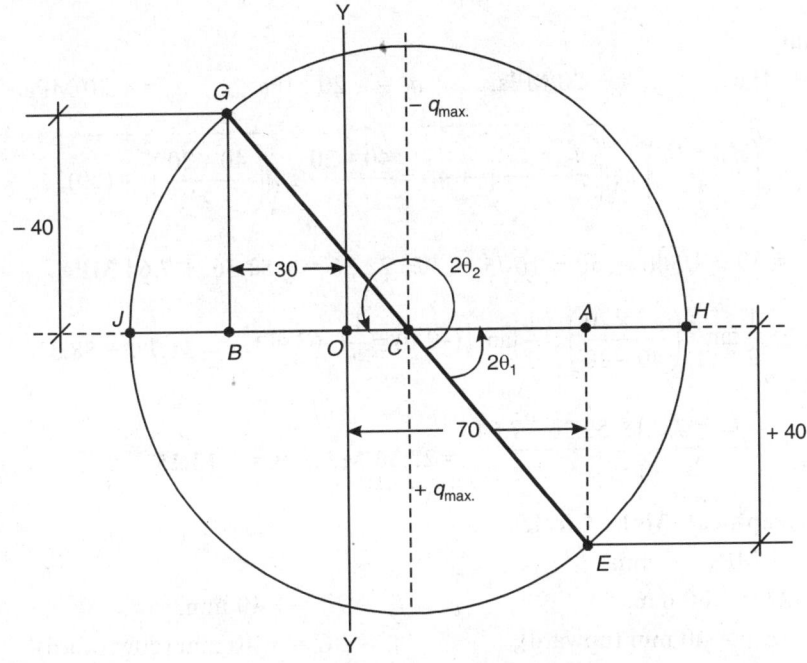

Fig. 13.14: Mohr circle of stresses

O – Origin

C – Centre

$OA = f_x$, $OB = f_y$

$AE = + q$, $BG = -q$

$OH = f_{n_1}$, $OJ = f_{n_2}$

• All quantities in MPa

• All linear dimensions in mm

$\angle ECH = 2\theta_1 = 38.5°$

$\angle ECJ = 2\theta_2 = 218.5°$

EXAMPLE 13.7: A structural element is subjected to different stresses as shown in Fig. 13.15a Calculate the principal stresses and planes and maximum shear stress and its plane. (b) Determine by graphical method: principal planes inclination, principal stresses and maximum shear stresses and their planes.

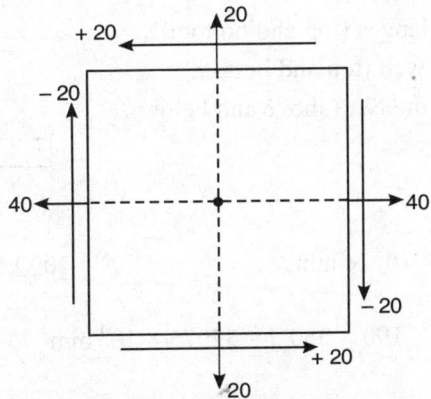

Fig. 13.15: Stressed element

Solution:

Given that

$$f_x = +40 \text{ MPa}, \qquad f_y = +20 \text{ MPa}, \qquad q_x = -20 \text{ MPa}, \qquad q_y = +20 \text{ MPa}$$

a. $\quad f_{n_{1,2}} = \left(\dfrac{f_x + f_y}{2}\right) \pm \sqrt{\left(\dfrac{f_x - f_y}{2}\right)^2 + q^2} = \dfrac{40 + 20}{2} \pm \sqrt{\left(\dfrac{40 - 20}{2}\right)^2 + (20)^2}$

$$= 30 \pm \sqrt{500} = 30 \pm 10\sqrt{5} = 30 \pm 22.36 = +\mathbf{52.36}, +\mathbf{7.64 \text{ MPa}}$$

$$\theta_{1,2} = \frac{1}{2}\tan^{-1}\left(\frac{-2 \times 20}{40 - 20}\right) = \frac{1}{2}\tan^{-1}(-2.0) = \frac{1}{2}(-63.4°) = -\mathbf{31.7°}, +\mathbf{58.3°}$$

$$q_{max} = \left(\frac{f_{n_1} - f_{n_1}}{2}\right) = \frac{52.36 - 7.64}{2} = 22.36 \text{ MPa}, \quad \theta = +\mathbf{13.25°}$$

b. **By Graphical (Mohr Circle)**

Scale 1 MPa = 2 mm

$f_x = OA = +80 \text{ mm}, \qquad\qquad\qquad f_y = OB = +40 \text{ mm},$

$q_x = AE = -40 \text{ mm (upward)}, \qquad\quad q_y = BG = +40 \text{ mm (downward)}$

Join EG as main base line which cuts X-X axis in C (centre). Draw circle with radius $r = CE = CG$. The circle cuts X-X axis at H and J. $OH = f_{n_1}$ and $OJ = f_{n_2}$. Measure lengths OH and OJ. From actual graph we have: $OH = 105 \text{ mm} = +52.5 \text{ MPa}$, $OJ = 15.3 \text{ mm} = +7.65 \text{ MPa}$, $\theta_1 = -31.5°$, $\theta_2 = +58.5°$, $q_{max.} = 44.8 \text{ mm} = 22.4 \text{ MPa}$.

EXAMPLE 13.8: An Indian rolled steel joist 400 mm × 200 mm overall has flanges 200 mm × 20 mm and web 360 mm × 10 mm. The RSJ is used for a simply supported beam having bending moment of 200 kN-m and shear force of 400 kN at certain section. Determine principal stresses and maximum shear stresses across the section under consideration at various points:

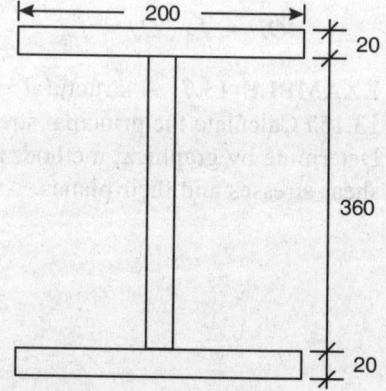

Fig. 13.16-I: Section 400×200

i.   Outermost fibre of flanges (top and bottom).

ii.   Joints of flange and web (top and bottom).

iii. 100 mm away from the NA (above and below).

iv.   NA

Solution:

Given

$$M = 200 \text{ kN-m} = 2 \times 10^8 \text{ N-mm}, \qquad\qquad F = 400 \text{ kN} = 4 \times 10^5 \text{ N}$$

$$I_x = \frac{1}{12}[200 \times 400^3 - 190 \times 360^3] = 32975 \times 10^4 \text{ mm}^4$$

Bending stress at a distance y from the N.A.

$$f_y = \pm \frac{M}{I} \cdot y = \pm \frac{2 \times 10^8}{32795 \times 10^4} \cdot y = \pm 0.60985y \ (y \text{ in mm})$$

Bending stress above NA in case of simply supported beam will be compressive (−) while below NA it will be tensile (+).

Shear stress 'q' will be +ve at all the points as the SF is +ve.

We shall calculate bending stresses at various points across the section.

f_1 (top outer fibre) = − 0.60985 (200) = − 122 N/mm^2

f'_1 (bottom outer fibre-flange) = + 122 N/mm^2

f_2 (junction flange-web top) = − 0.60985 × 180 = − 110 N/mm^2

f'_2 (junction flange-web bottom) = + 110 N/mm^2

f_3 (100 mm above NA in web top) = − 0.60985 × 100 = − 61.0 N/mm^2

f'_3 (100 mm below NA in web bottom) = + 61 N/mm^2

f_4 (at NA, $y = 0$) = ± 0.0

Shear stress at various points at a distance from NA: $q = \dfrac{FA\bar{y}}{I \cdot b}$

(i)   Outermost fibres of flanges: top $q_1 = 0$

                                         bottom $q'_1 = 0$

(ii) Junction of flanges and web in *flanges*:

$$q_{2f} = \frac{F(D^2 - d^2)}{8I} = \frac{4 \times 10^5 \times (400^2 - 360^2)}{8 \times 32795 \times 10^4} = \textbf{4.635 N/mm}^2$$

Junction of flanges and web **in web:**

$$q_{2w} = \frac{B}{d} \frac{F(D^2 - d^2)}{8I} = \textbf{+ 92.7 N/mm}^2$$

(iii) Web 100 mm above and below NA:

$$q_3 = \frac{B}{d} \frac{F(D^2 - d^2)}{8I} + \frac{F}{2I}\left(\frac{d^2}{4} - y^2\right) = 92.7 + 13.7 = \textbf{+ 106.4 N/mm}^2$$

(iv) Shear stress at **NA:**

$$q_4 = 92.7 + \frac{4 \times 10^5 (180)^2}{2 \times 32795 \times 10^4} = 92.7 + 19.8 = \textbf{112.5 N/mm}^2$$

Principal stresses

$$f_{m_{1,2}} = \frac{f_x}{2} \pm \sqrt{q^2 + \left(\frac{f_x}{2}\right)^2} \ , \quad \tan 2\theta = \frac{2q}{f_x} , \text{ as } f_y = 0$$

Thus principal stresses at various points are shown in Table 13.1

Table 13.1: Principal stresses, bending and shear stresses at various points

Points		Bending stress (f N/mm^2)	Shear stress (q N/mm^2)	Principal stresses N/mm^2		Slope θ_1, θ_2	Maximum shear N/mm^2
				Major	Minor		
i	Flanges top	− 122	0.0	− 122.0	0.0	0, π/2	− 61.0
	outer bottom	+122	0.0	+122.0	0.0	0, π/2	+ 61.0
ii	junction flange top	− 110.0	+ 4.64	− 110.20	+ 0.020	− 2°25′,87°35′	− 55.2
	junction flange Bottm	+ 110.0	+ 4.64	+ 110.20	− 0.20	− 87°35′, 2°25′	+ 55.2
	web top	−110.0	+ 92.7	− 162.80	+ 52.80	− 29°42′, 60°18′	− 107.8
	web bottom	+110.0	+ 92.7	+ 162.80	− 52.80	− 60°18′, 29°42′	+ 107.8
iii	web 100mm above	− 61.0	+ 106.4	− 141.2	+ 80.2	− 37°, 53°	− 110.7
	100mm below	+ 61.0	+ 106.4	+ 141.2	− 80.2	− 53°, 37°	+ 110.7
iv	NA	0.0	+ 112.5	+112.5	−112.5	45°, − 45°	112.5

EXAMPLE 13.9: A steel plate has circular hole of 200 mm diameter in the centre of the plate. The steel plate is subjected to two perpendicular normal plane stresses of 80 N/mm^2 (tensile) and 20 N/mm^2 (tensile and accompanied by shear stresses of + 20 N/mm^2 (anticlockwise about centre on 80 N/mm^2 plane). Determine the major and minor diameters of the circular hole and their direction. The steel material has modulus of elasticity (E) = **200 kN/mm^2** and Poisson's ratio (μ) = **0.25.**

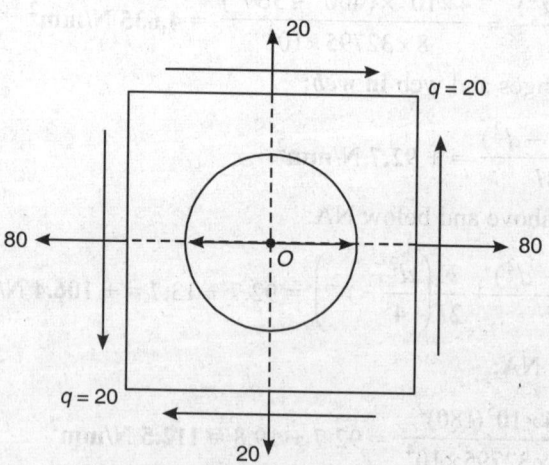

Fig. 13.17: Stress system (N/mm^2)

Solution:

Given that

$$f_x = + 80 \text{ N/mm}^2, \qquad f_y = + 20 \text{ N/mm}^2, \qquad q = + 20 \text{ N/mm}^2$$
$$f_{m,2} = ?, d_1, d_2, \qquad d_0 = d = 200 \text{ mm}$$

$$f_{m_{1,2}} = \frac{f_x + f_y}{2} \pm \sqrt{\left(\frac{f_x - f_y}{2}\right)^2 + q^2} = \frac{80.0 + 20.0}{2} \pm \sqrt{\left(\frac{80.0 - 20.0}{2}\right)^2 + (20.0)^2}$$

$$= 50 \pm 10\sqrt{13} = 50 \pm 36.08 = +\textbf{86.08}, +\textbf{13.92 N/mm}^2$$

$$\tan 2\theta = \frac{2q}{f_x - f_y} = \frac{2 \times 20}{(80 - 20)} = 0.6667$$

$$2\theta_1 = 33°42', \qquad \theta_1 = 16°51', 106°51'$$

Major principal strain:

$$e_1 = \left(\frac{f_{n_1}}{E} - \frac{\mu f_{n_2}}{E}\right) = \frac{86.08 - (0.25)(13.92)}{2 \times 10^5} = +4.13 \times 10^{-4} \text{ (elongation)}$$

Minor principal strain:

$$e_2 = \left(\frac{f_{n_2}}{E} - \frac{\mu f_{n_1}}{E}\right) = \frac{13.92 - 0.25(86.08)}{2 \times 10^5} = -0.38 \times 10^{-4} \text{ (shortening)}$$

Major diameter $= 200 (1 + 4.13 \times 10^{-4}) = 200.083$ mm

Minor diameter $= 200 (1 - 0.38 \times 10^{-4}) = 199.9924$ mm

EXAMPLE 13.10: An element is subjected to a resultant stress of 120 N/mm^2 (tensile) at a point and inclined at 30° to the normal of the plane AB. The element is also subjected to another tensile stress of 80 N/mm^2 (tensile) along a plane AC at right angles to the plane AB at the same point. Determine resultant stress on the plane AC, principal stresses and maximum shear stress at the same point analytically and also graphically.

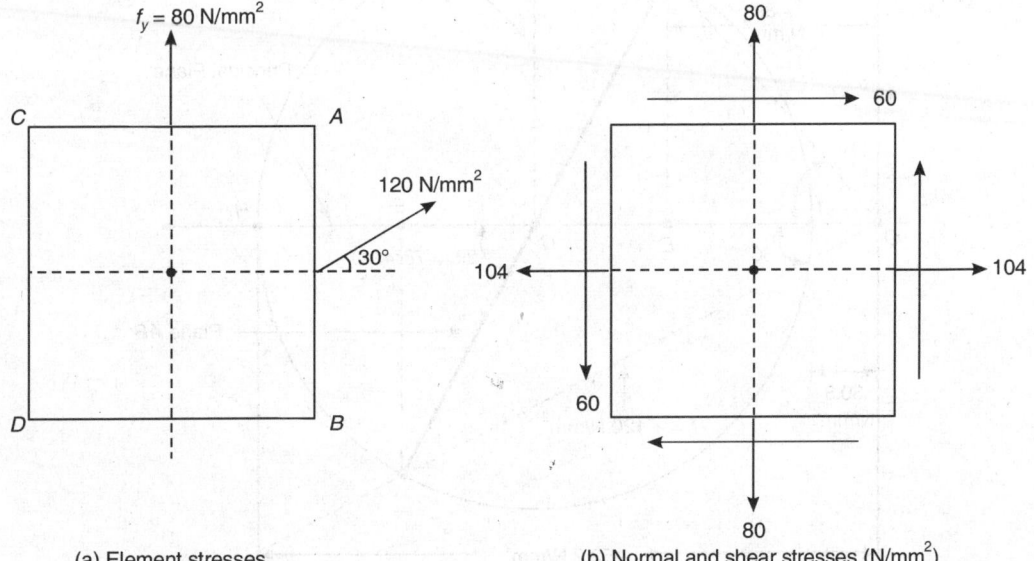

(a) Element stresses

(b) Normal and shear stresses (N/mm^2)

Fig. 13.18: Stress system

Solution:

Given that

$f_r = +120$ N/mm² at 30°, $f_x = 120 \cos 30° = +104$ N/mm²

$q = +120 \sin 30° = +60$ N/mm² (anticlockwise)

The plane AC will also have *complimentary shear* $q' = q = 60$ N/mm² (clockwise). Normal and shear stresses are shown in Fig. 13.18b.

a. Analytic method

$$f_{n_{1,2}} = \frac{f_x + f_y}{2} \pm \sqrt{\left(\frac{f_x - f_y}{2}\right)^2 + q^2} = \frac{104 - 80}{2} \pm \sqrt{\left(\frac{104 - 80}{2}\right)^2 + 60^2}$$

$$= +92 \pm 61.2 = \mathbf{+153.2, +30.8} \text{ N/mm}^2$$

$$\tan 2\theta = \frac{2q}{(f_x - f_y)} = \frac{2 \times 60}{(104 - 80)} = \frac{120}{24} = 5.0, \quad 2\theta = 78°42'$$

$$\theta_1 = 39°21', \qquad\qquad \theta_2 = 129°21'$$

$$q_{max.} = \frac{f_{n_1} - f_{n_2}}{2} = \frac{153.2 - 30.8}{2} = 61.2 \text{ N/mm}^2$$

Plane of $q_{max.} = 39°31' + 45° = \mathbf{84°21'}$

b. Graphical solution

Assume certain scale and draw Mohr circle. Assume scale say   1N/mm² = 1 mm

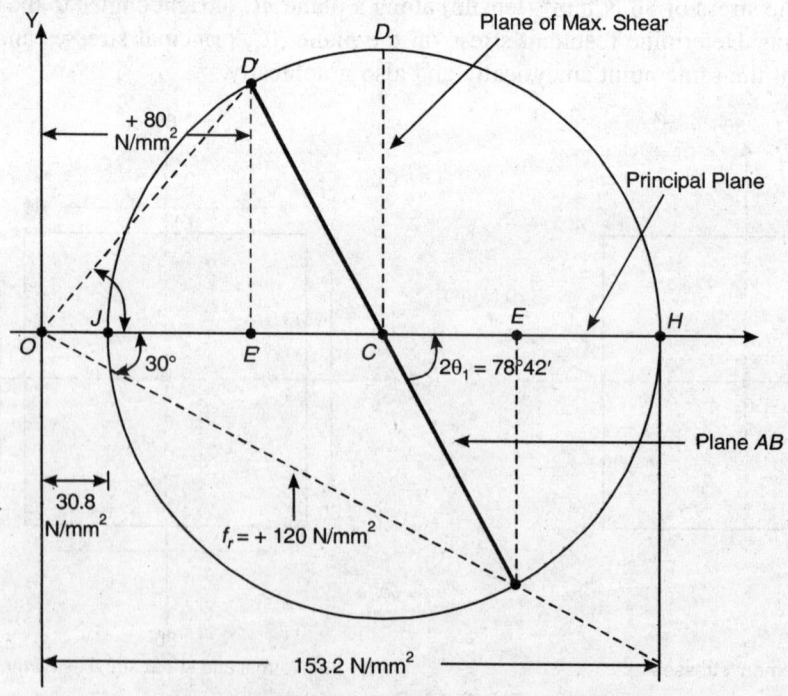

Fig. 13.19: Mohr circle of stresses (not to scale)

Draw resultant stress of 120 N/mm^2 (OD = 120 mm) at 30° with the X-axis as shown in positive direction as the stress is tensile. Draw perpendicular DE from the point D cutting on the axis X-X at E. Measure f_y = + 80 N/mm^2 (OE' = 80 mm). Take centre C of EE' and draw a circle with radius CD. Draw $E'D'$ perpendicular to X-X axis from E'. Points D and D' lie on the circle. The circle cuts the axis X-X at the points H and J. OH (153.2 mm) represents *major principal stress* (f_{n_1} = + 153.2 N/mm^2) and OJ (30.8 mm) represents *minor principal stress* (f_{n_2} = + 30.8 N/mm^2). CD_1 (61.2 mm) represents *maximum shear stress* ($q_{max.}$ = 61.2 N/mm^2). Angle of major principal plane is ½ ($\angle HCD$) = ½ (78°42′) = 39°21′.

OD' is the resultant stress on the plane AC. Angle $\angle COD'$ = 47°.

Shear stress on the plane AB = ED = 60 mm = 60 N/mm^2 (positive) i.e. causing anticlockwise rotation about the centre. Similarly shear stress on the plane AC = $E'D'$ = 60 N/mm^2 (negative) causing clockwise rotation of the centre.

EXAMPLE 13.11: An elastic material is subjected to a normal tensile stress of 140 N/mm^2 and normal compressive stress of 100 N/mm^2 *mutually at right angles* at a point. If the maximum principal stress is limited to 150 N/mm^2, determine the allowable shear stress on the planes at the point. Also determine the magnitude and direction of the minor principal stress and maximum shear stress.

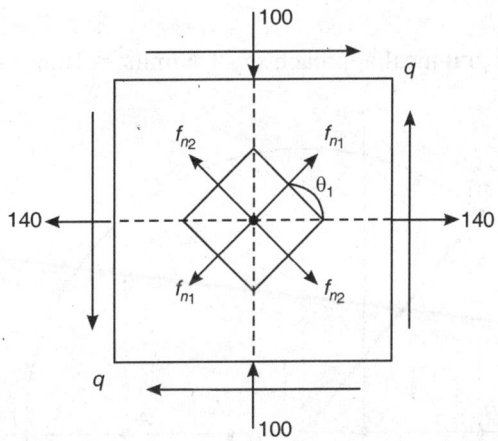

Fig. 13.20: Stress system at a point in an elastic material (N/mm^2)

Solution:

Given that

$$f_x = + 140 \text{ N/mm}^2, \qquad f_y = -100 \text{ N/mm}^2,$$
$$f_{n_1} = 150 \text{ N/mm}^2, \qquad f_{n_2} = ?$$

a. Analytical Method

$$f_{n_1} = \frac{f_x + f_y}{2} + \sqrt{\left(\frac{f_x - f_y}{2}\right)^2 + q^2}$$

$$150 = \frac{140 - 100}{2} + \sqrt{\left\{\frac{140 - (-100)}{2}\right\}^2 + q^2}$$

or $\quad 150 = 20 + \sqrt{(120)^2 + q^2}$

or $\quad (150 - 20)^2 = (120)^2 + q^2, \quad$ or $q^2 = 130^2 - 120^2 = 16900 - 14400 = 2500$

$\quad\quad q = \pm\, \mathbf{50\ N/mm^2}$

$$f_{n_2} = \frac{f_x + f_y}{2} - \sqrt{\left(\frac{f_x - f_y}{2}\right)^2 + q^2} = \frac{140 - 100}{2} - \sqrt{120^2 + 50^2} = 20 - 130 = -\,\mathbf{110\ N/mm^2}$$

$$q_{max.} = \frac{f_{n_1} - f_{n_2}}{2} = \frac{150 - (-110)}{2} = 130\ N/mm^2$$

$$\tan 2\theta = \frac{2q}{(f_x - f_y)} = \frac{2 \times 50}{\{140 - (100)\}} = \frac{100}{240} = 0.4167$$

$2\theta = \mathbf{22.62°} \qquad\qquad \theta_1 = 11.31° = \mathbf{11°20'} \qquad\qquad \theta_2 = 101°20'$

Plane of maximum shear $45° + 11°20' = \mathbf{56°20'}$

b. Graphical approach

Assume scale for graphical approach say $1\ N/mm^2 = 1mm$

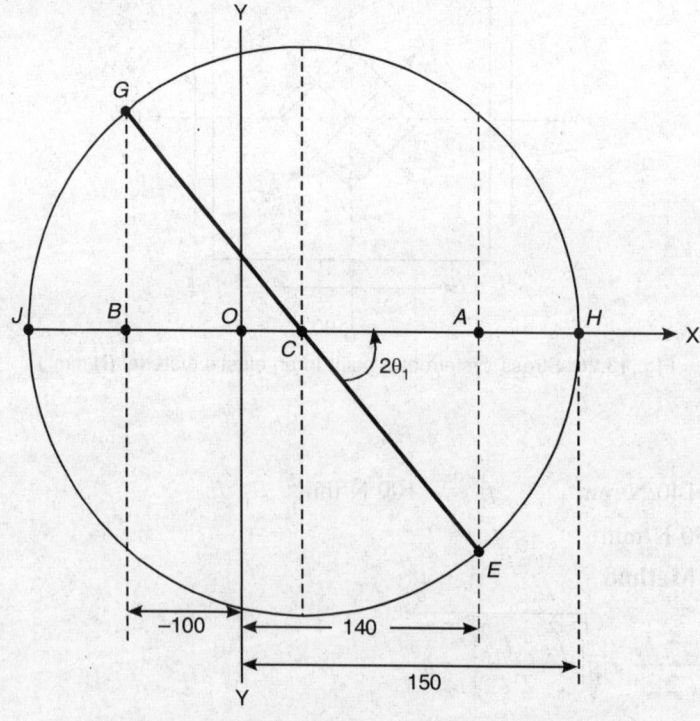

Fig. 13.21: Mohr circle of stresses (not to scale)

$f_x = + 140$ N/mm$^2 = + 140$ mm (OA), $f_y = -100$ N/mm$^2 = -100$ mm (OB)

Find centre of AB at C. Mark the point H on X-axis representing major principal stress $f_{n_1} = + 150$ N/mm^2 (taking OH $= + 150$ mm).

Draw a circle with 'C' as the centre and CH as radius cutting the base line at H and J. OH and OJ representing two principal stresses. Measure OJ (110 mm) representing minor principal stress ($f_{n_2} = -110$ N/mm^2). Measure the length AE and BF representing q_x ($AE = + 50$ mm) and q_y respectively. Draw perpendicular AE and BF. Join E and F. Measure $\angle ECH = 2\theta_1 = + 22.62°$, $\theta_1 = 11°20'$. $\theta_2 = 11°20' + 90° = + 101°20'$.

$q_x = AE = + 50$ mm $= + 50$ N/mm^2 (anticlockwise), $q_y = -50$ N/mm^2 (clockwise)

13.6   SUMMARY (PRINCIPAL PLANES, STRESSES AND STRAINS)

Individual effect of different type of loads on structural elements have been *studied separately*. In real life situations different loads *act simultaneously* and the structural elements offer resistance to the *combination of forces acting* at the same time. For safety it is therefore, essential to study *worst combination* of loads by finding *principal normal* stresses and *maximum shear* stresses. In this context study is limited to *two dimensional* plane stresses and strains on structural elements. Effect of loading is first converted into *two dimensional normal* and shear stresses. The effect of these two dimensional stress-system f_x, f_y and q_{xy} in X-Y plane is studied along any plane inclined at 'θ' with the *plane-X*.

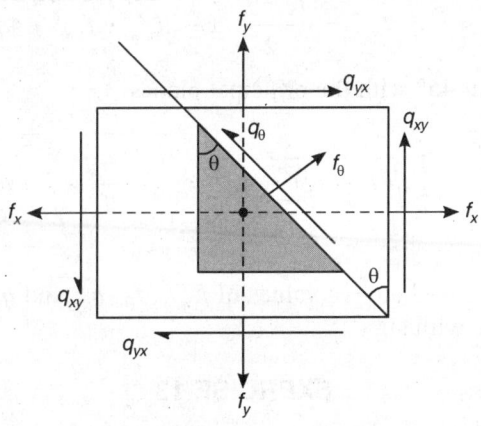

Fig. 13.22: Stresses at a point

f_x, f_y are normal stresses in X and Y directions acting at a point along with shear stresses q_{xy} and q_{yx} in *X-Y plane*.

$$f_\theta = \frac{f_x + f_y}{2} + \frac{f_x - f_y}{2} \cos 2\theta + q_{xy} \sin 2\theta \qquad \ldots (13.7)$$

$$q_\theta = -\left(\frac{f_x - f_y}{2}\right) \sin 2\theta + q_{xy} \cos 2\theta \qquad \ldots (13.8)$$

Mohr circle equation

$$\left\{ f_\theta - \frac{f_x + f_y}{2} \right\}^2 + q_\theta^2 = \left\{ \sqrt{\left(\frac{f_x - f_y}{2} \right)^2 + q^2} \right\}^2$$

Coordinates of centre $\left[\dfrac{f_x + f_y}{2}, 0 \right]$

Radius of circle $r = \sqrt{\left(\dfrac{f_x - f_y}{2} \right)^2 + q^2}$

Principal planes are those planes where only maximum or minimum normal stresses exist and *shear stresses are zero*. Principal stresses are *maximum* or *minimum normal stresses* and *not accompanied by shear stress*.

Principal planes: $\qquad \theta_x = \dfrac{1}{2} \tan^{-1} \left(\dfrac{2q}{f_x - f_y} \right)$ ... (13.11)

Principal stresses: $\qquad f_{m_{1,2}} = \dfrac{f_x + f_y}{2} \pm \dfrac{1}{2} \sqrt{(f_x - f_y)^2 + 4q^2}$ ... (13.12)

Maximum shear stress at 45° with the principal planes:

$$(q_\theta)_{max.} = \frac{f_{m_1} - f_{m_2}}{2} = \frac{1}{2}\sqrt{(f_x - f_y)^2 + 4q^2} = \sqrt{\left(\frac{f_x - f_y}{2} \right)^2 + q^2} \qquad \text{... (13.13)}$$

Note: All equations are based on +ve values of $f_x, f_y, f_\theta, q_{xy}$ and q_θ. Any −ve value shall be substituted in the formulae with sign.

EXERCISE 13

Q.13.1. Define principal stress, principal plane, complementary shear stress, maximum shear stress, and normal stress.

Q.13.2. Explain sign convention for normal plane stresses and shear stresses along an inclined plane in a stressed element.

Q.13.3. An element *ABCD* is subjected to a system of plane stresses, derive the equations for normal and shear stresses along an inclined plane at an angle 'θ' with the plane of first normal stress (i.e. f_x). Consider all stresses to be positive.

Ans. $f_\theta = \dfrac{f_x + f_y}{2} + \dfrac{f_x - f_y}{2} \cos 2\theta + q \sin 2\theta$, $\quad q_\theta = -\left(\dfrac{f_x - f_y}{2} \right)$ with usual notations of f_x, f_y and q.

Q.13.4. Prove that normal and shear stresses on any plane can be represented by a circle. Indicate the coordinate of the centre and radius of the stress circle.

Ans. $\left\{f_\theta - \dfrac{f_x + f_y}{2}\right\}^2 + q_\theta^2 = \left\{\sqrt{\left(\dfrac{f_x - f_y}{2}\right)^2 + q^2}\right\}^2$

Centre $\left\{\dfrac{f_x + f_y}{2}, 0\right\}$, Radius $r = \sqrt{\left(\dfrac{f_x - f_y}{2}\right)^2 + q^2}$

Q.13.5. Derive the equations of principal planes and principal stresses in a stressed element subjected to f_x and f_y normal stresses along mutually perpendicular axes X-X and Y-Y with shear stress of q.

Ans. $\theta_{1,2} = \dfrac{1}{2}\tan^{-1}\left\{\dfrac{2q}{f_x - f_y}\right\}$, $f_{m,2,} = \dfrac{1}{2}(f_x + f_y) \pm \sqrt{\left(\dfrac{f_x - f_y}{2}\right)^2 + q^2}$ with usual notations.

Q.13.6. Draw Mohr Circle of stresses and indicate principal stresses and maximum shear stress for a stressed element showing critical details.

Q.13.7. An element is subjected to plane stresses of $f_x = 240$ N/mm^2 (tensile) and $f_y = 80$ N/mm^2 (compressive) and $q_{xy} = 90$ N/mm^2 (clockwise rotation) as shown in Fig. Q7. Determine analytically or graphically (a) Principal stresses and planes (b) Maximum shear stresses and its planes.

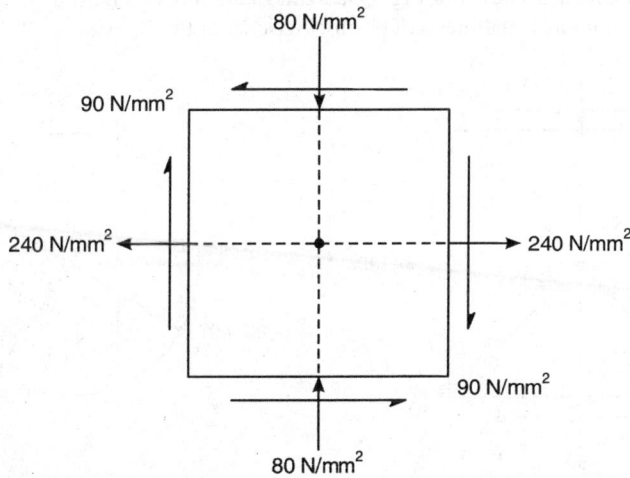

Fig. Q13.7: Stressed element

Ans. (a) $f_{n_1} = +263.6$ N/mm^2 (b) $q_{max.} = +183.6$ N/mm^2

$f_{n_2} = -103.6$ N/mm^2 (comp.) $\theta_{q_{max}} = 120.3°$

$\theta_1 = 165.3°$, $\theta_2 = 75.3°$

Q.13.8. A circle of 200 mm radius is scribed on a mild steel plate before it is subjected to stresses as shown in Fig. Q8. After stressing the circle deforms to an ellipse, calculate the lengths of the major and minor axes of the ellipse and their directions.

Take $\mu = 0.286$, modulus of elasticity $E = 205$ kN/mm^2.

Ans. $f_{n_1} = +220$ N/mm^2, $\theta_1 = -18.5°$; $f_{n_2} = +20$ N/mm^2, $\theta_2 = +71.5°$

major axis $= 400 (1 + 10.450 \times 10^{-4}) = 400.418$ mm

minor axis $= 400 (1 - 2.09 \times 10^{-4}) = 399.916$ mm

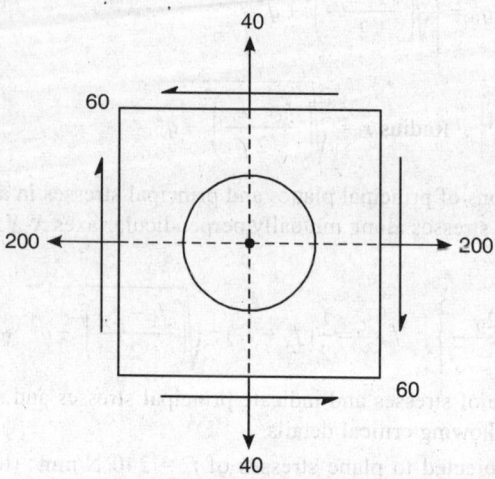

Fig. Q13.8: Stressed element (N/mm^2)

Q.13.9. A stressed element is shown in Fig. Q9a With Plane Stresses (N/mm^2). Find the stresses on a plane whose normal is inclined at 20° clockwise from the X-axis.

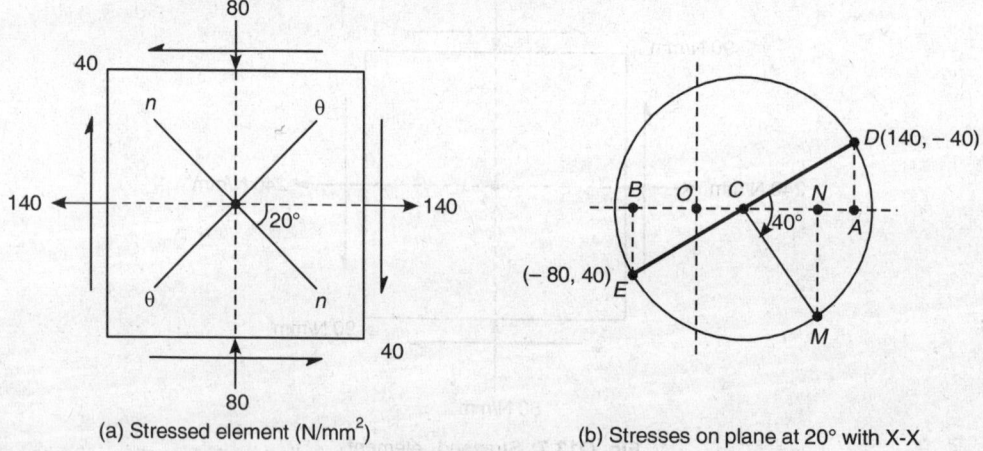

(a) Stressed element (N/mm^2)   (b) Stresses on plane at 20° with X-X

Fig. Q13.9.

Ans. $F_{20} = +88.55$ N/mm^2 (ON), $q_{20} = +101.34$ N/mm^2 (MN) {Fig. Q9b}

Q.13.10. Orthogonal plane stress system at a point in an element is shown in Fig. Q10a. Find the stresses on a plane inclined at 30° (anticlockwise) to the plane of 200 N/mm^2 tensile stress on X-X axis.

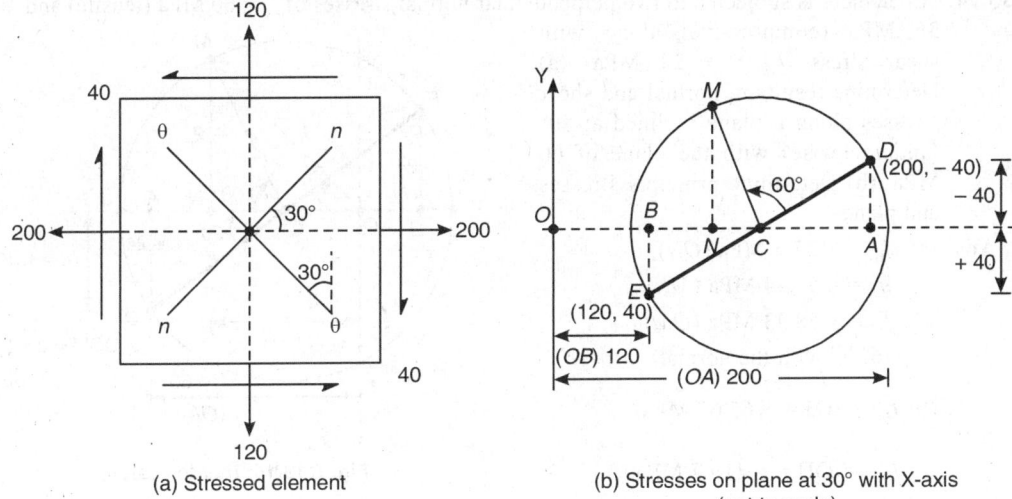

(a) Stressed element

(b) Stresses on plane at 30° with X-axis
(not to scale)

Fig. Q13.10.

Ans. $f_{30°} (ON) = \dfrac{f_x + f_y}{2} + \dfrac{f_x - f_y}{2} \cos 60° + q \sin 60° = + 145.4 \text{ N/mm}^2$

$q_{30°} (MN) = -\dfrac{f_x - f_y}{2} \sin 60° + q \cos 60° = -54.63 \text{ N/mm}^2$

Q.13.11. A circular rod of 80 mm diameter is pulled with a force of 400 kN. Determine resultant, normal and shear stresses along a plane inclined at 15° with the axis of the rod.

Ans. $f_r = 80.5 \text{ N/mm}^2$, $f_{15} = 77.8 \text{ N/mm}^2$, $q_{15} = 20.85 \text{ N/mm}^2$

Q.13.12. A stressed element is subjected to a normal stress f_x of 80 N/mm² (compressive) and a stress of 'f_y' on perpendicular plane equal to 40 N/mm² (compressive). Determine resultant normal and shear stress along a plane inclined at 40° with 80 N/mm² direction.

Ans. $f_\theta = -63.47 \text{ N/mm}^2$ (compressive), $q_\theta = 19.7 \text{ N/mm}^2$

$f_{r\theta} = 66.46 \text{ N/mm}^2$ (22°48′ with the direction of 80 N/mm²)

Q.13.13. An element is subjected to two mutually perpendicular normal stresses of $f_x = 60$ MPa (tensile) and $f_y = 80$ MPa (tensile) along with shear stresses $q_{xy} = + 15$ MPa. Determine resultant, normal and shear stresses along a plane inclined at 30° with the plane of 60 MPa by using analytical and graphical method.

Ans. $f_r = 65.7$ MPa along a direction at 5° with the normal of the plane (OM)

$f_\theta = 65.4$ MPa (ON), $q_\theta = -4.8$ MPa (MN)

Hints: $OA = 60$ MPa (60 mm),

$AD = + 15$ MPa (15 mm), $OB = 30$ MPa (30 mm),

$BE = -15$ MPa (15 mm), $DE = 2r$ and C as centre. $\angle DCM = 2 \times 30° = 60°$,

$ON = f_\theta$, $MN = q_\theta$, $OM = f_r$, $\angle MON = 5°$ (resultant with f_θ)

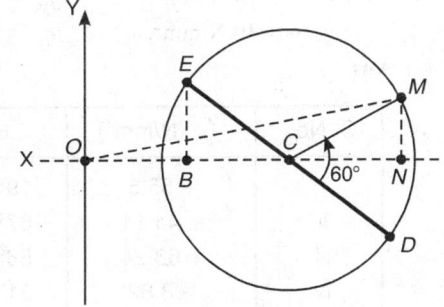

Fig. Q13.13: Mohr circle
(not to scale)

Q.13.14. An element is subjected to two perpendicular normal stresses of $f_x = 60$ MPa (tensile) and $f_y = 36$ MPa (compressive) along with shear stress $q_x = +24$ MPa. (a) Determine resultant, normal and shear stresses along a plane inclined at $50°$ (anticlockwise) with the plane of 60 MPa. (b) Determine principal stresses and planes.

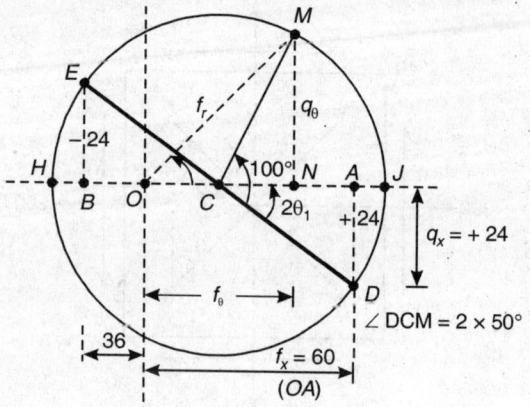

Ans. (a) $f_\theta = +27.3$ MPa (ON),

$q_\theta = -51.44$ MPa (MN),

$f_r = +58.23$ MPa (OM)

$(62°6'$ with the normal$)$

(b) $f_{n_1} = OJ = +65.67$ MPa,

$f_{n_2} = OH = -41.67$ MPa

$\theta_1 = 13.45°$, $\theta_2 = 103.45°$

Fig. Q13.14: (not to scale)

Q.13.15. An element is subjected to stresses as shown in Fig. Q15. Determine the principal stresses and maximum shear stress by graphical method and verify the results by analytic method in various cases when values are shown.

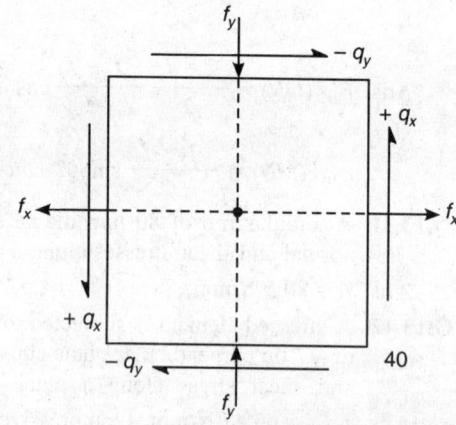

Fig. Q13.15.

Values are:

(i) $f_x = -30$ N/mm^2, $f_y = +45$ N/mm^2,

$q_x = -30$ N/mm^2, $q_y = +30$ N/mm^2

(ii) $f_x = +40$ N/mm^2, $f_y = +20$ N/mm^2,

$q_x = -10$ N/mm^2, $q_y = +10$ N/mm^2

(iii) $f_x = +33$ N/mm^2, $f_y = +50$ N/mm^2,

$q_x = +20$ N/mm^2, $q_y = -20$ N/mm^2

(iv) $f_x = -10$ N/mm^2, $f_y = -20$ N/mm^2,

$q_x = +10$ N/mm^2, $q_y = -10$ N/mm^2

Ans.

S. No.	f_{n_1} (N/mm^2)	θ_1	f_{n_2} (N/mm^2)	θ_2	$q_{max.}$ (N/mm^2)
i	+ 55.5	19°21′	− 40.5	109°21′	± 48.0
ii	+ 44.14	67°30′	15.86	157°30′	± 14.14
iii	+ 63.24	56°29′	19.76	146°29′	± 21.74
iv	− 3.82	31°43′	− 26.18	121°43′	± 11.18

Sample Mohr Circle

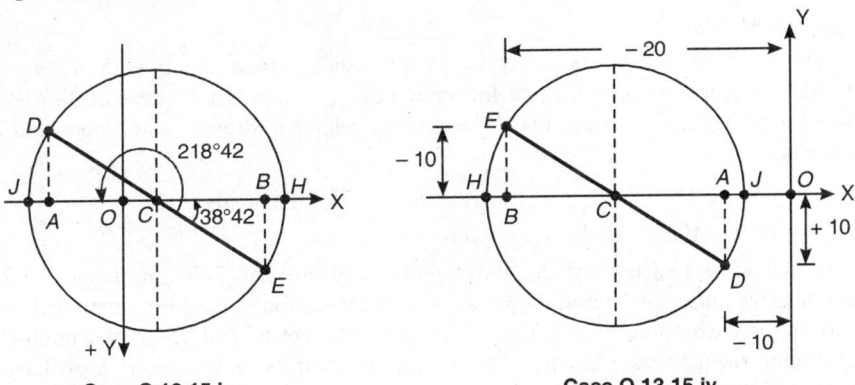

Case Q.13.15 i	Case Q.13.15 iv

Q.13.16. A bolt 20 mm diameter carries an axial load of 20 kN and 5 kN shear force. Determine principal stresses and principal planes, maximum shear stress, and stress causing maximum linear strain if $\mu = 0.25$.

Ans. $f_{n_1} = +67.5$ N/mm^2, $\theta_1 = 13°17'$ inclination of normal with the axis of the bolt.

$$f_{n_2} = -3.77 \text{ N/mm}^2, \quad \theta_2 = 103°17', \quad q_{max} = \frac{f_{n_1} - f_{n_2}}{2} = 35.63 \text{ N/mm}^2.$$

f (causing equivalent linear strain) $= 68.45$ N/mm^2.

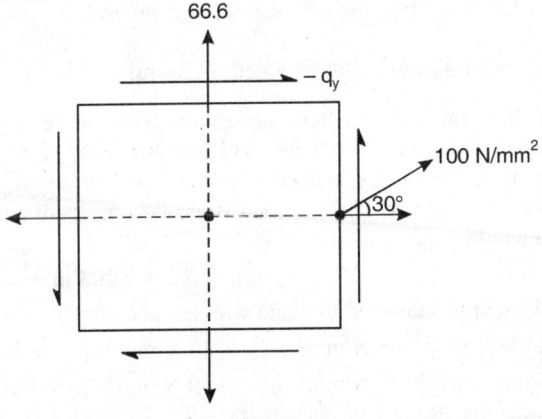

Fig. Q13.16.

Q.13.17. A stressed element has *resultant* stress of 100 N/mm^2 (tensile) along 30° with the normal of the plane. The element is subjected to a normal stress of $\dfrac{200}{3}$ N/mm^2 (tensile) along perpendicular plane with respect to the first plane. Determine principal planes and stresses, maximum shear stress with the help of *Mohr Circle* and confirm the results analytically.

Hints: $f_x = 100 \cos 30° = +86.6$ **N/mm^2** (tensile), $q_x = 100 \sin 30° = +50$ **N/mm^2**,

$f_y = +66.66$ **N/mm^2**

$f_{n_1} = +127.7$ **N/mm^2** (Normal $\theta_1 = 39°21'$ with the X-axis)

$f_{n_2} = + 25.7 \text{ N/mm}^2$ ($\theta_2 = 129°21'$ with the X-axis)

$q_{max.} = \pm 51 \text{ N/mm}^2$

Q.13.18. A stressed element is subjected to 54 MPa shear stress along with a tensile stress of 72 MPa on certain plane X-X. At the same point a compressive stress of 48 MPa also acts along a perpendicular plane (Y-Y). Determine principal stresses and planes and maximum shear stress.

Ans. $f_{n_1} = + 92.72 \text{ MPa}$, $\quad \theta_1 = 21°$; $\quad f_{n_2} = - 68.72 \text{ MPa}$, $\quad \theta_2 = 111°$

$q_{max.} = 80.72 \text{ MPa}$, $\quad$ (θ for $q_{max.}$) = 66°

Q.13.19. A rolled steel I-section of 200 mm × 400 mm (deep) has 200 mm flanges of 20 mm and 360 mm web plate of 10 mm thickness. The beam section is used for simply supported beam carrying a shear force of 200 kN and sagging moment of 100 kN-m at a point. Determine maximum shear stress, principal stresses and planes at the (i) top junction of flange and web (ii) point 100 mm below the NA if the moment of inertia of the I-section = $3.28 \times 10^8 \text{ mm}^4$.

Hints: $f_x = \dfrac{M}{I} \cdot y$, $\qquad q_x = \dfrac{F \cdot A \cdot \bar{y}}{I \cdot b}$, $\qquad f_{n_{1,2}} = \dfrac{f_x}{2} \pm \sqrt{\left(\dfrac{f_x}{2}\right)^2 + q^2}$

Ans. Junction flange: $f_{n_{1,2}} = + 0.60 \text{ N/mm}^2$, $- 55.5 \text{ N/mm}^2$, $\theta_1 = 87°35'$, $q_{max.} = 28.05 \text{ N/mm}^2$

Junction web: $f_{n_{1,2}} = + 26.43 \text{ N/mm}^2$, $- 81.33 \text{ N/mm}^2$, $\theta_1 = 60°18'$, $q_{max.} = 53.9 \text{ N/mm}^2$

100 mm below NA: $f_{n_{1,2}} = + 70.57 \text{ N/mm}^2$, $- 40.07 \text{ N/mm}^2$, $\theta_1 = 37°$, $q_{max.} = + 55.32 \text{ N/mm}^2$

Q.13.20. A circular rod carries an axial tensile load of 99 kN. If the maximum permissible shear stress on any plane is 70 N/mm², find the safe diameter of the rod.

Hint: $q_{max.} = \dfrac{f_x}{2} = 70 \text{ N/mm}^2$, Safe diameter = **30 mm**

Q.13.21. A steel flat of 200 mm *gauge length* undergoes 0.16 mm elongation when subjected to a longitudinal tensile stress of 160 N/mm². (i) If another tensile stress of **160 N/mm²** also acts at perpendicular direction of first tensile stress of 160 N/mm², find the change in 200 mm gauge length when $\mu = 0.3$. (ii) If the second stress is compressive instead of tensile, find the change in gauge length.

Ans. i. $\delta L = 0.112 \text{ mm}$, $\qquad\qquad$ ii. $\delta L = 0.208 \text{ mm}$

Q.13.22. A rectangular element is subjected to plane stresses as under:

i. $f_x = - 30 \text{ N/mm}^2, f_y = + 45 \text{ N/mm}^2, q_x = - 30 \text{ N/mm}^2, q_y = + 30 \text{ N/mm}^2$,

ii. $f_x = + 40 \text{ N/mm}^2, f_y = + 20 \text{ N/mm}^2, q_x = - 10 \text{ N/mm}^2, q_y = + 10 \text{ N/mm}^2$.

If $E = 200 \text{ kN/mm}^2$, $\mu = 0.30$, find the principal strain and strain energy.

Ans. i. $e_1 = 0.0003382$, $\qquad e_2 = - 0.002857$, $\qquad U = 0.01517 \text{ N-mm/mm}^3$ vol.

ii. $e_1 = + 0.0001969$, $\qquad e_2 = 0.0000131$, $\qquad U = 0.00445 \text{ N-mm/mm}^3$ vol.

Q.13.23. The principal tensile stresses at a point across two perpendicular plane are **80 N/mm²** and **40 N/mm²**. Find the normal, tangential and resultant stresses on a plane at 20° with the major principal plane. Find also the intensity of single stress which alone can produce the same maximum strain. μ (Poisson's ratio) = 0.25.

Hints: $f_r = \sqrt{f_\theta^2 + q_\theta^2}$, $\qquad \tan\phi = \dfrac{f_\theta}{q_\theta}$

Ans. $f_\theta = + 75.32 \text{ N/mm}^2$, $\qquad q_\theta = - 12.86 \text{ N/mm}^2$, $\qquad f_\theta = 76.02 \text{ N/mm}^2$

$f = 70.00 \text{ N/mm}^2$ producing the same maximum strain.

14

Unsymmetrical Bending

LEARNING OBJECTIVES

After studying this chapter, the learner **understands** unsymmetrical bending and will be able to:

14.1 **Describe** unsysmmetrical bending.

14.2 **Calculate** moment of Inertia about principal axes of a section from the given cross-sectional moment of inertia about normal axes X-X and Y-Y.

14.3 **Determine** the moment of inertia about principal axes of a section from the given cross-sectional moment of inertia about normal axes **X-X** and **Y-Y** by **Mohr's graphical** approach.

14.4 **Calculate** bending stresses in a beam section subjected to unsymmetrical bending.

14.5 **Calculate** Deflection of a beam subjected to unsymmetrical bending.

14.6 **Explain** the concept of shear centre.

14.7 **Calculate** the location of shear centre in **symmetrical** and **unsymmetrical** beam cross-sections.

14.1 INTRODUCTION

In simple theory of bending, it was assumed that the neutral axis of the corss-section is at right angles to the plane of loading, i.e. the plane of loading is perpendicular to the plane of neutral axis of the cross-section. According to this assumption the simple equation of bending $\dfrac{M}{I} = \dfrac{f}{y}$, was derived.

This assumption implies that the *plane of loading* or the plane of bending, is *coincident* with or *parallel to*, a plane containing the *principal centroidal* axis of the cross-section of the beam. In many cases, the *plane of loading* or the *plane of bending* does not lie in a plane of the principal centroidal axis of the cross-section. Thus the bending is *not symmetrical*. For example, the *purlin* in a system of trusses, is generally placed on the inclined rafter and the plane of bending (or loading) *does not coincide* with the centroidal axis (Fig. 14.1).

The *plane of loading is vertical*, while the centroidal axes plane $V - V$ and $U - U$ are not vertical and hence bending is unsymmetrical. *NA* will not be perpendicular to the plane of bending.

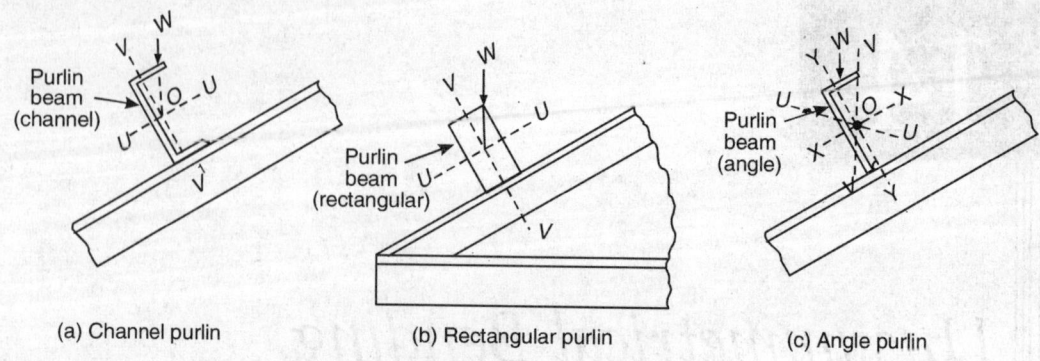

(a) Channel purlin   (b) Rectangular purlin   (c) Angle purlin

Fig. 14.1: Unsymmetrical bending in purlins

14.2   PRINCIPAL AXES OF A SECTION

The *principal centroidal* axes of a section are a set of perpendicular axes through the centroid of the sectional area for which the product of inertia will be zero, i.e.

$I_{UV} = 0 = \Sigma u.v.\delta_a$ or $I_{XY} = 0 = \Sigma x.y.\delta_a$, in case of principal axes.

X–X, Y–Y, are any pair of rectangular axes.

U–U, V–V, are pair of *principal* centroidal axes.

α, is the angle between principal centroidal axis U–U and X–X. (Fig. 14.2).

Let U–U and V–V are the set of principal axes for which the product of inertial $\Sigma\ u.v.\delta_a = 0$, where δ_a is an elementary small area with coordinates 'u' and 'v' with reference to the principal axes U–U and V–V. The axis of symmetry is also the principal axis since it satisfies the condition $\Sigma\ u.v.\delta_a = 0$. Some times, a plane area *may not have* any axis of symmetry, but the principal axes may be found by *known properties* about any pair of rectangular axes X–X and Y–Y.

Consider an elementary area δ_a shown in Fig. 14.2 having rectangular co-ordinates x, y with reference to X–X and Y–Y. Let the coordinates with reference to the principal axes U–U and V–V be, u,v. By definition we have $I_{XX} = \Sigma\ y^2.\delta_a$; $\qquad I_{YY} = \Sigma x^2\delta_a$; $\qquad Ixy = \Sigma\ x.y.\delta_a$

Similary : $\qquad\qquad\qquad\qquad I_{UU} = \Sigma v^2.\delta_a$; $\qquad I_{V-V} = \Sigma u^2.\delta_a$; $\quad I_{u-v} = \Sigma u.v.\delta_a$

From the Fig. 14.2, we have : $u = x\cos\alpha + y\sin\alpha$;

$$v = y\cos\alpha - x\sin\alpha;$$

$\therefore\quad I_{U-U} = \Sigma\ (y\cos\alpha - x\sin\alpha)^2.\delta_a = \cos^2\alpha\ \Sigma y^2\delta_a + \sin^2\alpha\ \Sigma x^2\delta_a - 2\sin\alpha\ \cos\alpha\ \Sigma x.y.\delta_a.$

or $\quad\mathbf{I_{U-U} = I_{XX}\cos^2\alpha + I_{YY}\sin^2\alpha - I_{XY}\sin 2\alpha}$ $\qquad\qquad\qquad$ ...(14.1)

similary $I_{V-V} = \Sigma\ (x\cos\alpha + y\sin\alpha)^2\delta_a = \cos^2\alpha\ \Sigma x^2\delta_a + \sin^2\alpha\ \Sigma y^2\delta_a + 2\sin\alpha.\cos\alpha\ \Sigma x.y.\delta_a$

or $\quad\mathbf{I_{V-V} = I_{XX}\sin^2\alpha + I_{YY}\cos^2\alpha + I_{X-Y}\sin 2\alpha}$ $\qquad\qquad\qquad$ ...(14.2)

$\qquad I_{U-V} = \Sigma\ u.v.\delta_a = \Sigma\ (x\cos\alpha + y\sin\alpha)\ (y\cos\alpha - x\sin\alpha)\delta_a$

or $\quad\mathbf{I_{U-V} = \cos^2\alpha\ \Sigma\ x.y.\delta_a - \sin^2\alpha\ \Sigma x.y.\delta_a + \sin\alpha\ \cos\alpha\ (\Sigma\ y^2\delta_a - \Sigma\ x^2\delta_a)}$

or $\quad I_{U-V} = I_{X-Y}(\cos^2\alpha - \sin^2\alpha) + \dfrac{\sin 2\alpha}{2}(I_{XX} - I_{YY}) = \left(\dfrac{I_{XX} - I_{YY}}{2}\right)\sin 2\alpha + I_{XY}\cos 2\alpha$

$$...(14.3)$$

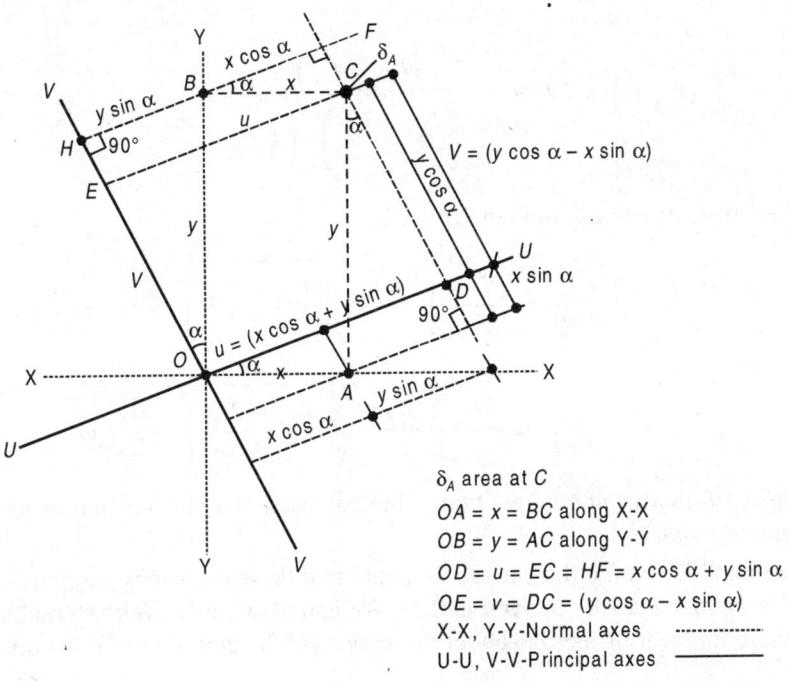

δ_A area at C

$OA = x = BC$ along X-X

$OB = y = AC$ along Y-Y

$OD = u = EC = HF = x\cos\alpha + y\sin\alpha$

$OE = v = DC = (y\cos\alpha - x\sin\alpha)$

X-X, Y-Y-Normal axes ------------

U-U, V-V-Principal axes ————

Fig. 14.2: Unsymmetrical bending about U–V axes

By definition $I_{U-V} = 0$, since $U-U$ and $V-V$ are the principal axes.

i.e. $\quad \left\{(I_{XX} - I_{YY})\dfrac{\sin 2\alpha}{2} + I_{xy}\cos 2\alpha\right\} = 0$

or $\qquad\qquad \mathbf{tan2\alpha} = \dfrac{2I_{XY}}{I_{XX} - I_{YY}} \qquad\qquad ...(14.4)$

Thus 'α' can be calculated from the known values of I_{XX}, I_{YY} and I_{XY}.

Now substituting the values of 'α' in equation 14.1 and 14.2 we can determine the values of I_{UU}, I_{VV}. i.e.

$$I_{U-U} = \dfrac{I_{XX} + I_{YY}}{2} + \dfrac{I_{XX} - I_{YY}}{2}\cos 2\alpha - I_{xy}\sin 2\alpha \qquad ...(14.1)$$

$$I_{VV} = \dfrac{I_{XX} + I_{YY}}{2} - \dfrac{I_{XX}\ I_{YY}}{2}\cos 2\alpha + I_{XY}\sin 2\alpha \qquad ...(14.2)$$

From equation (14.4): $\sin 2\alpha = \dfrac{-I_{XY}}{\sqrt{\left(\dfrac{I_{XX} - I_{YY}}{2}\right)^2 + I^2_{XY}}}$ ... (i)

$\cos 2\alpha = \dfrac{(I_{XX} - I_{YY})}{2\sqrt{\left(\dfrac{I_{XX} - I_{YY}}{2}\right)^2 + I^2_{XY}}}$ ... (ii)

Putting the values of $\cos 2\alpha$ and $\sin 2\alpha$, we have

$$I_{UU} \atop (Max) = \dfrac{I_{XX} + I_{YY}}{2} + \sqrt{\left(\dfrac{I_{XX} - I_{YY}}{2}\right)^2 + I^2_{XY}}$$ ... (14.5)

$$I_{VV} \atop (Min) = \dfrac{I_{XX} + I_{YY}}{2} - \sqrt{\left(\dfrac{I_{XX} - I_{YY}}{2}\right)^2 + I^2_{XY}}$$ ... (14.6)

The moment of inertia about the two principal axes will be *maximum about one* and *minimum about the other.*

From equation (14.5) and (14.6), it can be noted that these equations are quite *similar* to the equations of *principal stresses* derived earlier. We can also apply *Mohr's* graphical approach for finding these moment of inertias about the *major (U–U)* and *minor (V–V)* principal axes.

14.3 GRAPHICAL METHOD FOR PRINCIPAL AXES

(A) Mohr circle equations of moment of inertia about *principal axes U–U* and *V–V* are similar to the equations of *principal stresses*. These equations can be solved by Mohr's graphical method in a similar manner by drawing *Mohr's circle* by adopting suitable scale and similar conventions.

$$\tan 2\alpha = \dfrac{-2I_{XY}}{I_{XX} - I_{YY}}$$ ... (14.4)

$$I_{UU} = \dfrac{I_{XX} + I_{YY}}{2} + \sqrt{\left(\dfrac{I_{XX} - I_{YY}}{2}\right)^2 + I^2_{XY}}$$ ... (14.5)

$$I_{VV} = \dfrac{I_{XX} + I_{YY}}{2} - \sqrt{\left(\dfrac{I_{XX} - I_{YY}}{2}\right)^2 + I^2_{XY}}$$ ... (14.6)

I_{XX} is similar to σ_x, I_{YY} similar to σ_y and I_{XY} similar to τ_{xy} or q_{xy}
I_{UU} – Major principal MI, I_{VV} – minor principal MI,
α – Similar to the location of principal plane.

Thus Mohr circle can be used to find MI about the two principal axes and to locate the principal axes.

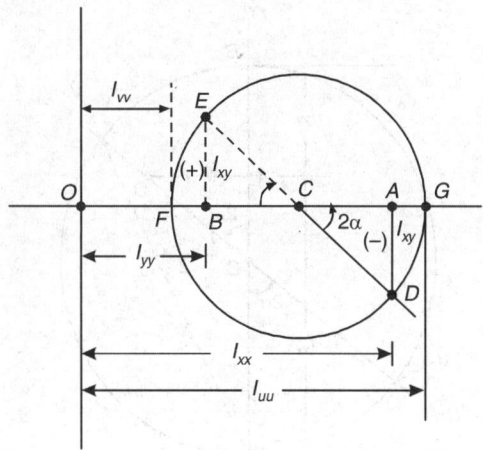

Fig. 14.3: Mohr circle of inertia

Draw :

$OA = I_{XX}$ to scale, $OB = I_{YY}$ to scale

$AD = I_{XY}$ to scale, $BE = I_{XY}$ to scale

DCE = Diameter of the circle

Draw circle with 'C' as centre and $CD = CE$ as radius.

Measure :

OG represents I_{UU} and OF represents I_{VV} to the same scale. $\angle DCA$ represents 2α, where α being angle between UU and X–X.

$$OC = \frac{I_{XX} + I_{YY}}{2}, \qquad\qquad CA = \frac{I_{XX} - I_{YY}}{2}$$

$$OG = I_{UU} = OC + CG = \frac{I_{XX} + I_{YY}}{2} + \sqrt{\left(\frac{I_{XX} - I_{YY}}{2}\right)^2 + I^2_{XY}}$$

$$OF = I_{VV} = OC - CF = \frac{I_{XX} + I_{YY}}{2} - \sqrt{\left(\frac{I_{XX} - I_{YY}}{2}\right)^2 + I^2_{XY}}$$

It may be noted that I_{XY} is plotted below the line OA (AD) if it is negative and plotted above the line OA if it is positive (BE). The principal axis UU is inclined to axis X–$X = 2\alpha = \angle ACD$ in anti clockwise if I_{XY} is negative. Similary UU axis makes 2α clockwise with X–X axis if I_{XY} is positive $\angle BCE$.

(*B*) *Circle of Inertia* (Mohr – Land construction)

The values of I_{uu} and I_{vv} can also be determined by drawing a circle of inertia (also known as Mohr – Land construction) as shown in Fig. 14.4. Let X–O–X and YOY represent any set of rectangular axes through the centroid 'O'. Draw $OA = I_{XX}$ and $AB = I_{YY}$ along Y – Y (OAB).

$$OA = I_{XX} \text{ (To Scale)}, \quad AB = I_{YY} \text{ (To Scale)}, \quad AD = I_{XY} \text{ (To Scale)},$$
$$QD = I_{UU} \text{ (To Scale)}, \quad DP = I_{VV} \text{ (To Scale)}, \quad I_{UV} = O, \text{ (Principal axes)}$$

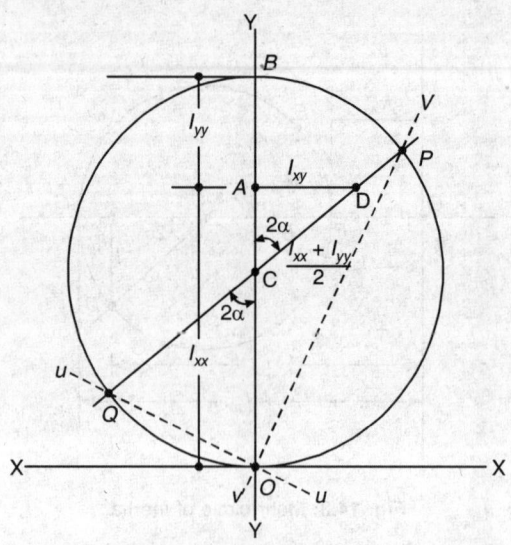

Fig. 14.4: Circle of inertia

Draw a circle with OB *as* diameter and C as the centre. $AD = I_{XY}$ perpendicular to OAB. Draw AD to right if I_{XY} is positive and AD to left side if I_{XY} is negative. Join C to D and extend the line to meet the circle in P and Q (Fig. 14.4, Circle of Inertia). Join the points P to O and Q to O. The line QO and PO will be at right angles and PCQ will be diameter of the circle. The line QO represents U–U (Major Principal axis) and PO represents $V - V$ (Minor Principal axis) and are at right angles to each other as the $\angle POQ = 90°$. QD represents I_{UU} and DP represents I_{VV}.

From Fig. 14.4: $\quad CA = (CB - AB) = \dfrac{1}{2}\left(I_{XX} + I_{YY}\right) - I_{YY} = \dfrac{1}{2}\left(I_{XX} - I_{YY}\right)$

$$AD = I_{XY}, \quad \tan 2\alpha \atop (\angle DCA) = \frac{AD}{AC} = \frac{2I_{XY}}{(I_{XX} - I_{YY})}, \text{ numerically} \qquad \dots (1)$$

Also
$$CD = \sqrt{AC^2 + AD^2} = \sqrt{\left(\frac{I_{XX} - I_{YY}}{2}\right)^2 + I^2_{XY}}$$

$\therefore \qquad$
$$QD = \left(\frac{I_{XX} + I_{YY}}{2}\right) + \sqrt{\left(\frac{I_{xx} - I_{YY}}{2}\right)^2 + I^2_{XY}} \qquad \dots (2)$$

$$DP = CP - CD = \left(\frac{I_{XX} + I_{YY}}{2}\right) - \sqrt{\left(\frac{I_{XX} - I_{YY}}{2}\right)^2 + I^2_{XY}} \qquad \dots (3)$$

EXAMPLE 14.1: Find the product of inertia for the section shown in Fig. 14.5 about axes $X' - X'$ and $Y' - Y'$ along edges and parallel axes X–X and Y–Y through the centroid O of the total section.

Solution: Portion (1) 200×20 mm^2, (2) 100×20 mm^2

Combined centroid of (1) and (2)

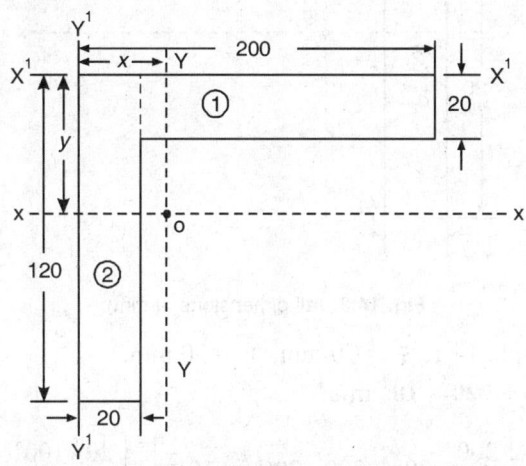

Fig. 14.5: (all dimensions in mm)

$$\bar{x} = \frac{200 \times 20 \times 100 + 100 \times 20 \times 10}{300 \times 20} = \frac{420}{6.0} = 70 \text{ mm}$$

$$\bar{y} = \frac{200 \times 20 \times 10 + 100 \times 20 \times (20 + 50)}{300 \times 20} = 30 \text{ mm}$$

$$I_{X'Y'} = \text{Total product of inertia about } X' - X' \text{ and } Y' - Y' \text{ of } A_1$$
$$= 0 + 4000 \,(100 \times -10) = -4 \times 10^6 \text{ mm}^4$$
$$I_{X'-Y'} \,(A_2) = 0 + 2000 \,(10 \times -70) = -14 \times 10^5 \text{ mm}^4$$
$$I_{X'Y'} \,(\text{total}) = -54 \times 10^5 \text{ mm}^4.$$
$$I_{XY} = I_{X'Y'} - A.\bar{x}.\bar{y} = -54 \times 10^5 - (6000)\,(+70)\,(-30)$$
$$= -54 \times 10^5 + 126 \times 10^5$$
$$= +72 \times 10^5 \text{ mm}^4.$$

EXAMPLE 14.2: Find the *principal moments* of inertia and directions of principal axes for the angle section shown in example 14.1 (Fig. 14.5).

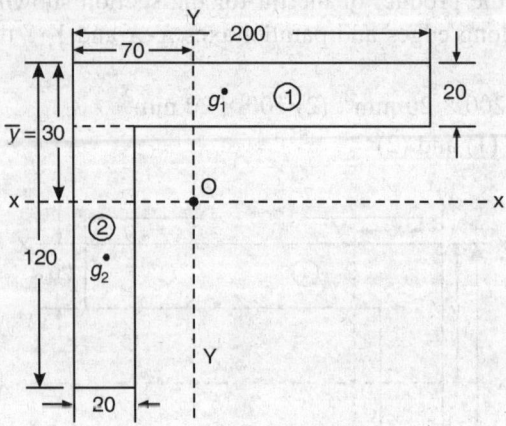

Fig. 14.6: (all dimensions in mm)

Solution: From example 14.1, $\bar{x} = 70$ mm, $\bar{y} = 30$ mm.

$$I_{XY} = + 720 \times 10^4 \text{ mm}^4.$$

$$I_{XX} = \left[\frac{200}{12} \times 20^3 + 200 \times 20(30 - 10)^2 \right] + \left[\frac{20 \times 100^3}{12} + 100 \times 20(40)^2 \right]$$

$$= 10^5 \left[\frac{16}{12} + 16 + \frac{200}{12} + 32 \right] = 10^5 \left(\frac{216}{12} + 48 \right) = 660 \times 10^4 \text{ mm}^4$$

$$I_{YY} = \left[\frac{20 \times 200^3}{12} + 200 \times 20(30)^2 \right] + \left[\frac{100 \times 20^3}{12} + 20 \times 100 \ (60)^2 \right]$$

$$= 10^5 \left[\frac{1600}{12} + 36 + \frac{8}{12} + 72 \right] = 10^5 \left[\frac{1608}{12} + 108 \right] = 2420 \times 10^4 \text{ mm}^4$$

Direction of Principal axes:

$$\tan 2\theta = \frac{2I_{XY}}{I_{YY} - I_{XX}} = \frac{2 \times 720}{(2420 - 660)} = \frac{2 \times 720}{1760} = 0.8182,$$

$$2\theta = 39.3°, \ \theta_1 = \mathbf{19.65°}, \ \theta_2 = \mathbf{109.65°}$$

Principal moment of inertia:

$$I_{UU} = \left(\frac{I_{YY} + I_{XX}}{2} \right) + \sqrt{\left(\frac{I_{YY} - I_{XX}}{2} \right)^2 + I^2{}_{XY}}$$

$$= \left(\frac{2420 + 660}{2} \right) 10^4 + \left(\sqrt{\left(\frac{2420 - 660}{2} \right)^2 + 720^2} \right) 10^4$$

$$= \left(1540 + \sqrt{1292800}\right)10^4$$

$$= (1137.014 + 1540)10^4 = 2677.014 \times 10^4 \text{ mm}^4.$$

$$I_{VV} \text{ (Minor principal axis)} = \frac{I_{YY} + I_{XX}}{2} - \sqrt{\left(\frac{I_{YY} - I_{XX}}{2}\right)^2 + I^2_{XY}}$$

$$= \left[\frac{2420 + 660}{2} - \sqrt{\left(\frac{2420 - 660}{2}\right)^2 + 720^2}\right]10^4$$

$$= (1540 - 1137.014)10^4 = \textbf{402.986} \times \textbf{10}^4 \textbf{ mm}^4.$$

Substituting θ to find major or minor moment of inertia in the general equation.

$$I_\theta = \frac{I_{YY} + I_{XX}}{2} - \frac{I_{YY} - I_{XX}}{2}\cos 2\theta - I_{XY}\sin 2\theta$$

$$= 10^4\left[\frac{2420 + 660}{2} - \frac{(2420 - 660)}{2}\cos(39.3°) - 720\sin(39.3°)\right]$$

$$= 10^4\left[1540 - 880 \times 0.77384 - 720 \times 0.63338\right]$$

$$= 10^4\left[1540 - 680.98 - 456.034\right]$$

$$= 402.987 \times 10^4 \text{ mm}^4 \text{ (This is minor principal } MI).$$

Thus Major principal moment of inertia $I_{UU} = 2677.014 \times 10^4$ mm^4 is at 109.65° and the minor principal moment of Inertia $I_{VV} = 402.986 \times 10^4$ mm^4 is at 19.65° with X–X axis. This can also be determined by drawing *Mohr Circle* for *MI* as shown in Fig. 14.7.

Draw Mohr circle adopting a linear scale for *MI*. Say 1 mm = 50 × 10^4 mm^4.

Draw two perpendicular axes with origin at 0 as *ON* and *OM*.

$$I_{XX} = OP = \frac{660}{50} = \textbf{13.2 mm}, I_{YY} = OQ = \frac{2420}{5} = 48.4 \text{ mm}$$

Draw *PR* (perpendicular to *ON*) at $P = I_{XY} = \frac{720}{50} = \textbf{14.4 mm}$

Bisect *PQ* at *C*. Take *C* as centre to draw a circle with *radius CR*. Circle cuts the axis *ON* at *T* and *W*.

$$OW = I_{UU} = 53.50 \text{ mm} = 2675 \times 10^4 \text{ mm}^4$$

$$OT = I_{VV} = 8.10 \text{ mm} = 405 \times 10^4 \text{ mm}^4$$

Minor principal *MI* = 405 × 10^4 mm^4 at 19.65° plane and major principal *MI* = 2675 × 10^4 mm^4 at 109.65° with X–X axis.

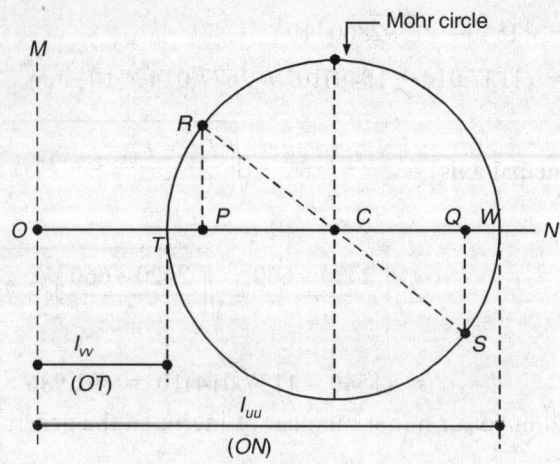

Fig. 14.7: Mohr circle of MI

EXAMPLE 14.3: An equal angle section 80 mm × 80 mm × 10 mm is used as a horizontal simply supported beam over a span of 4.0 m. It carries a vertical point load of 800 N at the mid span point along the vertical line passing through the centroid of the section. Find

(i) Stresses at points A, B, and C of the section at the mid span points as shown in Fig. 14.8.

(ii) Deflection at the, midspan point. $E = 200\text{GN/m}^2$ (200 kN / mm²).

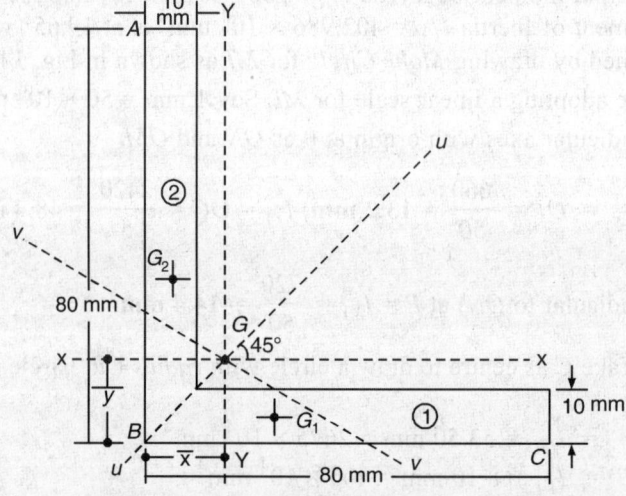

Fig. 14.8.

Solution: $E = 200 \times 10^3 \text{ N / mm}^2$, span $L = 4\text{m}$, Load $W = 800 \text{ N}$

G is the c.g. of the section, G_1 is the cg of part (1) and G_2 is the cg of part (2).

CG 'G' at $\bar{x}, \bar{y}$.

$$\text{Equal angle so } \bar{x} \ = \ \bar{y} = \frac{80\times10\times40+70\times10\times5}{(80\times10+70\times10)} = \textbf{23.666 mm}$$

$$MI \ I_{XX}= \left[\frac{1}{12}\times80\times10^3+800(23.666-5.0)^2\right]+\left[\frac{10\times70^3}{12}+700(45-23.666)^2\right]$$

$I_{XX} \ = I_{YY} = 6666.67 + 27875.7 + 285833.3 + 318586.8 = \textbf{889842.45 mm}^4$
$\ \ \ \ \ = 88.9843 \times 10^4 \text{ mm}^4.$

Coordinates of G_1, w.r.t. X–X, Y–Y $(40 - 23.666) \ (- 23.666 + 5.0) = (16.334, - 18.666)$ and G_2 $(-23.666 + 5.0), (45 - 23.666) = (-18.66, 21.334)$

Product of inertia, $I_{XY} \ = \ 800 (16.334 \times -18.666) + 700 (-18.666 \times 21.334)$
$$= \ \textbf{--52.267} \times \textbf{10}^4 \textbf{ mm}^4$$

$$\tan 2\theta \ = \ \frac{2I_{XY}}{I_{YY}-I_{XX}} = \frac{2\times52.267\times10^4}{0} = \infty, \ 2\theta = 90°, \ \theta_1 = 45°,$$

$$\theta_2 \ = \ 90 + 45 = 135° \text{ with } X\text{–}X.$$

$$\text{Principal } MI \ = \ I_{UU} = \frac{1}{2}(I_{XX}+I_{YY})+\frac{1}{2}(I_{XX}-I_{YY})\cos 2\theta-I_{XY}\sin 2\theta$$

$$I_{UU} \ = \ 88.9813 \times 10^4 + 0 - (-52.267 \times 10^4 \text{ mm}^4)$$
$$= \ 141.250 \times 10^4 \text{ mm}^4 = \textbf{141.25} \times \textbf{10}^4 \textbf{ mm}^4.$$
$$I_{VV} + I_{UU} \ = \ I_{XX} + I_{YY}, \ I_{VV} = 2 \times 88.9813 \times 10^4 - 141.25 \times 10^4$$
$$= \ \textbf{36.720} \times \textbf{10}^4 \textbf{ mm}^4$$

BM : Max. BM at the midspan point

$$= \ \frac{WL}{4} = \frac{800\times4}{4} \text{ N–m} = 800 \times 1000 = \textbf{8} \times \textbf{10}^5\textbf{N–mm}$$

(i) Components of BM about U–U and V–V axes.
$$M_u \ = \ M_x \sin\theta = 80 \times 10^4 \sin 45° = \textbf{56.5685} \times \textbf{10}^4 \text{ N–mm}.$$
$$M_v \ = \ M_x \cos\theta = 80 \times 10^4 \cos 45° = \textbf{56.5685} \times \textbf{10}^4 \text{ N–mm}.$$

Co–ordinates of A, B, C with respect to U–U, V–V :

Point A : $x \ = \ -23.66 \text{ mm}, \ y = 80 - 23.66 = \textbf{56.34 mm}$
$u \ = \ x \cos\theta + y \sin\theta = -23.66 \textbf{ cos45°} + \textbf{56.34 sin45°} = + \textbf{23.1012 mm}$
$v \ = \ y \cos\theta - x \sin\theta = 56.34 \cos45° - (-23.66) \sin45° = + 56.57 \text{ mm}$

Point B : $x \ = \ \textbf{--23.666 mm}, \ y = \textbf{--23.666 mm}$
$u \ = \ \textbf{--23.666 cos45°} + (\textbf{--23.666 sin45°}) = - \textbf{33.46 mm}$
$v \ = \ -23.666 \text{ cost45°} - (-23.666 \sin45°) = 0$

Point C : $x \ = \ 80 - 23.666 = 56.334 \text{ mm}, \ y = - 23.666 \text{ mm}$
$u \ = \ 56.334 \cos45° + (23.666 \sin45°) = + 23.10 \text{ mm}$

$$v = (-23.666 \cos 45°) - (56.334 \sin 45°) = -56.56 \text{ mm}$$

$$f_A = \frac{M_u \cdot u}{I_{VV}} + \frac{M_v \cdot v}{I_{uu}} = \frac{56.5685 \times 10^4 \times 23.1012}{36.7143 \times 104} + \frac{56.5685 \times 10^4 \times 56.57}{141.25 \times 10^4}$$

$$= 35.594 + 22.655 = \mathbf{58.25 \ N/mm^2}$$

$$f_B = \frac{56.5685 \times 10^4 \times (-33.46)}{10^4 \times 36.7143} + \frac{56.5685 \times 10^4}{141.25 \times 10^4}(0) = \mathbf{-51.554 \ N/mm^2}$$

$$f_C = \frac{56.5685 \times 10^4 (23.10)}{36.7143 \times 10^4} + \frac{56.5685 \times 10^4 (-56.56)}{141.25 \times 10^4} = 35.592 - 22.651$$

$$= \mathbf{12.94 \ N/mm^2}$$

(ii) Deflection δ is given by :

$$\delta = \frac{K \cdot WL^3}{E} \sqrt{\frac{\sin^2 \theta}{I^2_{UU}} + \frac{\cos^2 \theta}{I^2_{vv}}}, \ K = \frac{1}{48} \text{ For point load}$$

$$W = 800 \text{N}, \ L = 4.0 \text{ m} = 4000 \text{ mm}, \ E = 2 \times 10^5 \text{ N/mm}^2,$$

$$I_{UU} = 141.25 \times 10^4 \text{ mm}^4, \ I_{VV} = 36.7143 \times 10^4 \text{ mm}^4,$$

$$\delta = \frac{800}{48} \times \frac{(4000)^3}{2 \times 10^5} \sqrt{\frac{\sin^2 45}{36.7143^2 \times 10^8} + \frac{\cos^2 45}{(141.25)^2 10^8}}$$

$$= \frac{10^6}{10^4} \times \frac{16}{3} \sqrt{\frac{1}{2 \times 36.7143^2} + \frac{1}{2 \times 141.25^2}} = \frac{16000}{30} \left(\sqrt{\frac{1}{2695.9} + \frac{1}{39903.1}} \right)$$

$$= \frac{16000}{30} \sqrt{3.9599 \times 10^{-4}} = \frac{16000 \times 10^{-2}}{30} \times 1.99 = \mathbf{10.61 \ mm}.$$

EXAMPLE 14.4: A cantilever, of I section, is subjected to a load of 400 N at the free end and is at 30° with the vertical. The span of the cantilever is 3.0 m. Determine the bending stresses at corners A, B and C near the fixed support of the cantilever (Fig. 14.9).

Solution: $L = 3$ m, $W = 400$ N, $\theta = 30°$ with the vertical

The load is inclined at 30°, $\therefore$ vertical load 400 cos30°, horizontal load 400 sin30°.

Since the section is symmetrical about both axes (X–X and Y–Y), the principal axes will be X–X and Y–Y.

Max. BM near the fixed end = $W.L.$ = 400 × 3 = 1200 N–m = 12 × 10^5 N – mm

$$M^{11} = M \cos 30° = 12 \times 10^5 \times \frac{\sqrt{3}}{2} = \mathbf{10.3923 \times 10^5 \ N\text{–}mm}$$

$$M = M \sin 30° = 12 \times 10^5 \times \frac{1}{2} = \mathbf{6.0 \times 10^5 \ N\text{–}mm}$$

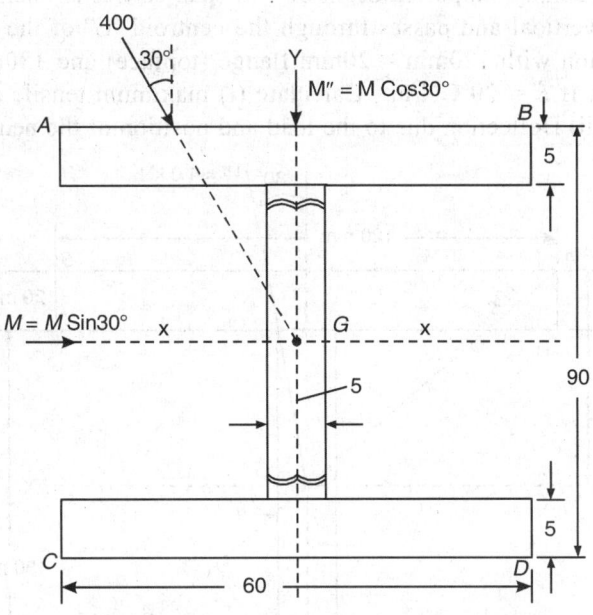

Fig. 14.9: Section of cantilever

MI about principal axes

$$I_{XX} = \frac{1}{12}(60 \times 90^3 - 55 \times 80^3) = \frac{10^4}{12}(4374 - 2816) = \mathbf{129.833 \times 10^4 \ mm^4}.$$

$$I_{YY} = \frac{1}{12}(5 \times 60^3 \times 2 + 80 \times 5^3) = \frac{1}{12}(2160000 + 10000) = \mathbf{18.08333 \times 10^4 \ mm^4}$$

$$\sigma_A = \frac{M^{11} \times 45}{I_{XX}} + \frac{M^1 \times 30}{I_{YY}} = \frac{10.3923 \times 10^5 \times 45}{129.833 \times 10^4} + \frac{6.0 \times 10^5 \times 30}{18.08333 \times 10^4}$$

$$= \mathbf{36.02 + 99.54 = 135.56 \ N/mm^2 \ (tensile)}$$

$$\sigma_B = \frac{M^{11} \times 45}{I_{XX}} - \frac{M^1 \times 30}{I_{YY}} = \frac{10.3923 \times 10^5 \times 45}{129.833 \times 10^4} - \frac{6.0 \times 10^5 \times 30}{18.0833 \times 10^4}$$

$$= 36.02 - 99.54 = -63.52 \ N/mm^2 \ (compressive)$$

$$\sigma_C = -\frac{M^{11} \times 45}{I_{XX}} + \frac{M^1 \times 30}{I_{YY}} = -36.02 + 99.54 = 63.52 \ N/mm^2 \ (tensile).$$
$$\quad\ \ \text{(Comp.)} \quad \text{(tensile)}$$

EXAMPLE 14.5: A simply supported beam of 4 m span carries an inclined point load of 40 kN with 30° to the vertical and passes through the centroid 'G' of the section. The section comprises of T–section with 120mm × 20mm flange (topface) and 180mm × 20mm web as shown in Fig. 14.10. If E = 20 GN/m², Calculate (i) maximum tensile stress and maximum compressive stress, (ii) Deflection due to the load and position of the neutral axis.

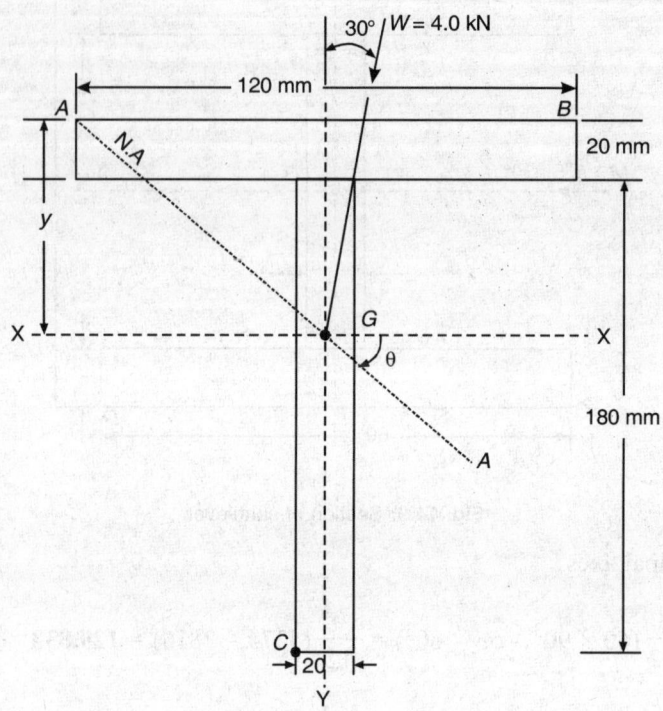

Fig. 14.10: T–section

Solution: *SS* span L = 4.0m, W = 4.0 kN at 30° with vertical,
E = 200 GN/m² = 200000 N/mm².

Centroid G from top horizontal flange

$$\bar{y} = \frac{120 \times 20 \times 10 + 180 \times 20 \times (20 + 90)}{(120 \times 20 + 180 \times 20)} = 70 \text{ mm.}$$

The section is symmetrical about vertical axis and hence the principal axes pass through the centroid 'G'. The principal axes UU and VV will be along XX and YY only.

$$I_{XX} = I_{UU} = \left[\frac{120 \times 20^3}{12} + 120 \times 20(70-10)^2\right] + \left[\frac{20 \times 180^3}{12} + 180 \times 20(130-90)^2\right]$$

$$= 10^4[8 + 864 + 972 + 576] = 2420 \times 10^4 \text{ mm}^4$$

$$I_{YY} = I_{VV} = \left[\frac{20 \times 120^3}{12} + \frac{180 \times 20^3}{12}\right] = 10^4(288 + 12) = 300 \times 10^4 \text{ mm}^4,$$

Load along V–V : $W \cos 30° = 4.0 \times \dfrac{\sqrt{3}}{2} = 2\sqrt{3} = 3.464$ kN.

Load along U–U : $W \sin 30° = 4.0 \times \dfrac{1}{2} = 2.0$ kN.

$$M_{VV} = \frac{W_V \cdot L}{4} = \frac{3.464 \times 4}{4} = 3.464 \text{ kN–m} = 3.464 \times 10^6 \text{ N–mm}$$

$$M_{UU} = \frac{M_u \cdot L}{4} = \frac{2.0 \times 4}{4} = 2.0 \text{ kN–m} = 2 \times 10^6 \text{ N-mm}$$

Max. comp. stress will be at B due to M_u and M_v. Similarly max. tensile stress will be at C.

$$f_b = \frac{M_u}{I_{VV}} \cdot 60 + \frac{M_V}{I_{uu}} \cdot 70 = \frac{2 \times 10^6}{3 \times 10^6} \times 60 + \frac{3.464 \times 10^6}{2420 \times 10^4} \times 70 = 40 + 10.019$$

$$= \textbf{50.02 N/mm}^2 \textbf{ (comp.)}$$

$$f_c = \frac{M_V}{I_{uu}} \times (200 - 70) + \frac{M_u}{I_{VV}} \times 10 = \frac{3.464 \times 10^6}{2420 \times 10^4} \times 130 + \frac{2 \times 10^6}{300 \times 10^4} \times 10$$

$$= \textbf{18.61 + 6.66 = 25.27 N/mm}^2 \text{ (tensile)}$$

(ii) Deflection due to the load :

$$\delta = \frac{K W L^3}{E} \sqrt{\frac{\sin^2 \theta}{I^2_{VV}} + \frac{\cos^2 \theta}{I^2_{uu}}} = \frac{K W L^3}{E I_{uu}} \sqrt{\sin^2 \theta \left(\frac{I_{uu}}{I_{VV}}\right)^2 + \cos^2 \theta}$$

$$= \frac{1}{48} \times \frac{4 \times 10^3 \times (4000)^3}{2 \times 10^5 \times 2420 \times 10^4} \sqrt{\frac{1}{4}(8.067)^2 + \frac{3}{4}} = \frac{8 \times 100}{3242} \sqrt{16.2691 + 0.75}$$

$$= 1.10193 \times 4.12524 = \textbf{4.546 mm (Deflection)}$$

Position of NA

$$\tan \beta = \frac{I_{uu}}{I_{VV}} \tan \theta = \frac{2420}{300} \tan 30° = 4.6573, \ \beta = 77.88° \text{ with } X–X.$$

14.4 THE SHEAR CENTRE

The concept of shear centre plays important role in study of bending of members with sections having only one or no axis of symmetry. Beam sections with axes of symmetry and with no axes of symmetry are shown in Fig. 14.11.

The *shear centre* in any transverse section of the beam is the point of *intersection* of the *bending axis* and the plane of the transverse section. The *shear centre* of a section can be defined as a *point about which the applied force is balanced* by the *set of internal shear forces* obtained by *summing the shear stresses over the section* (For unsymmetrical sections such as

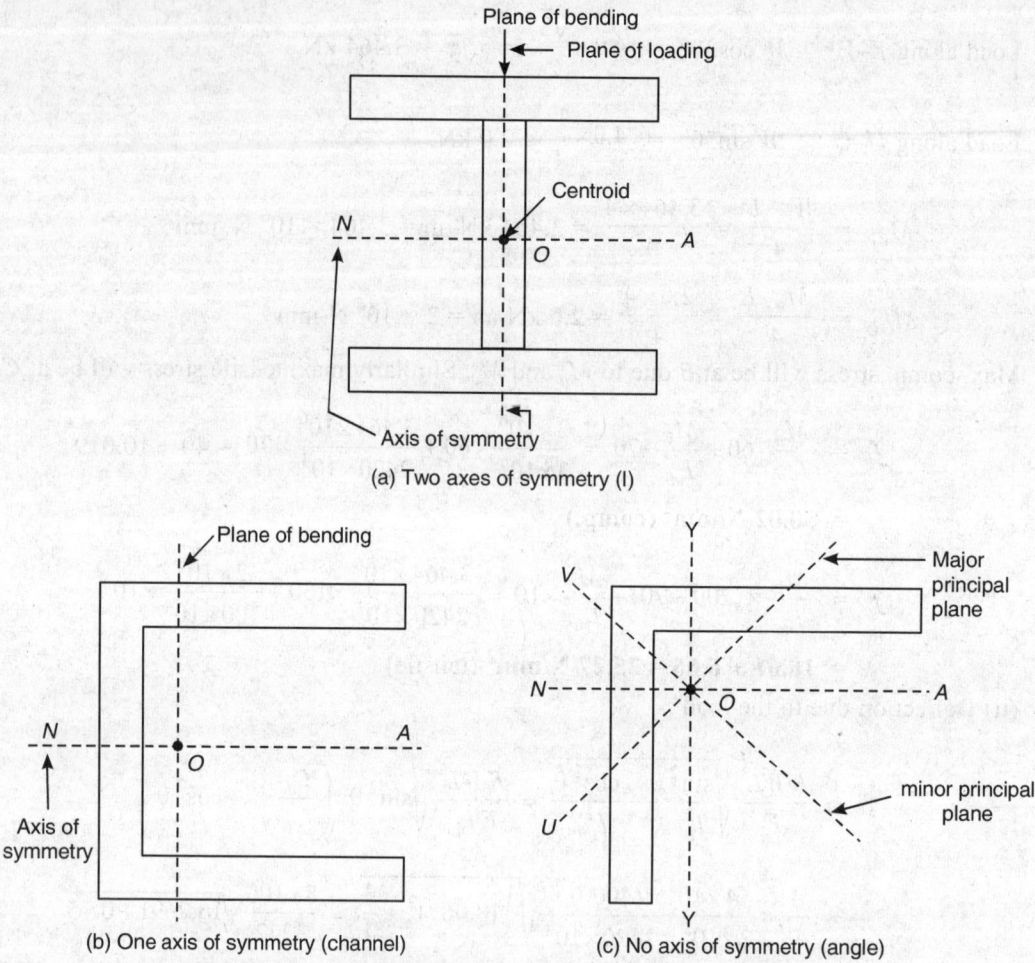

Fig. 14.11: Different sections with axes of symmetry or no symmetry

channels and angles sections, *Summation of shear stresses in each leg* gives a set of forces which should be in equilibrium with the applied shear force). The *resultant of internal shear forces* must balance with the external shear force and these must pass through the *shear centre*. The shear centre is also known as the *"centre of twist"*.

In case of a beam having two axes of symmetry, the *shear centre* coincides with the *centroid*. In case of sections having *one axis of symmetry*, the *shear centre* does not coincide with the centroid but *lies on the axis of symmetry*.

When the load *passes through the shear centre* then there will be *only bending* in the cross-section and *no twisting*.

The principle involved in locating the shear centre for a cross-section of a beam is that the loads acting on the beam must lie in a plane which contains the resultant shear force on each cross-section of the beam as calculated from the shear stresses produced in the beam when it is loaded so that it *does not twist* at its ends.

In real beams both *BM* and *SF* exist at a section. When the beam section has two axes of symmetry, as in the case of the *I*–section shown in Fig. 14.12.

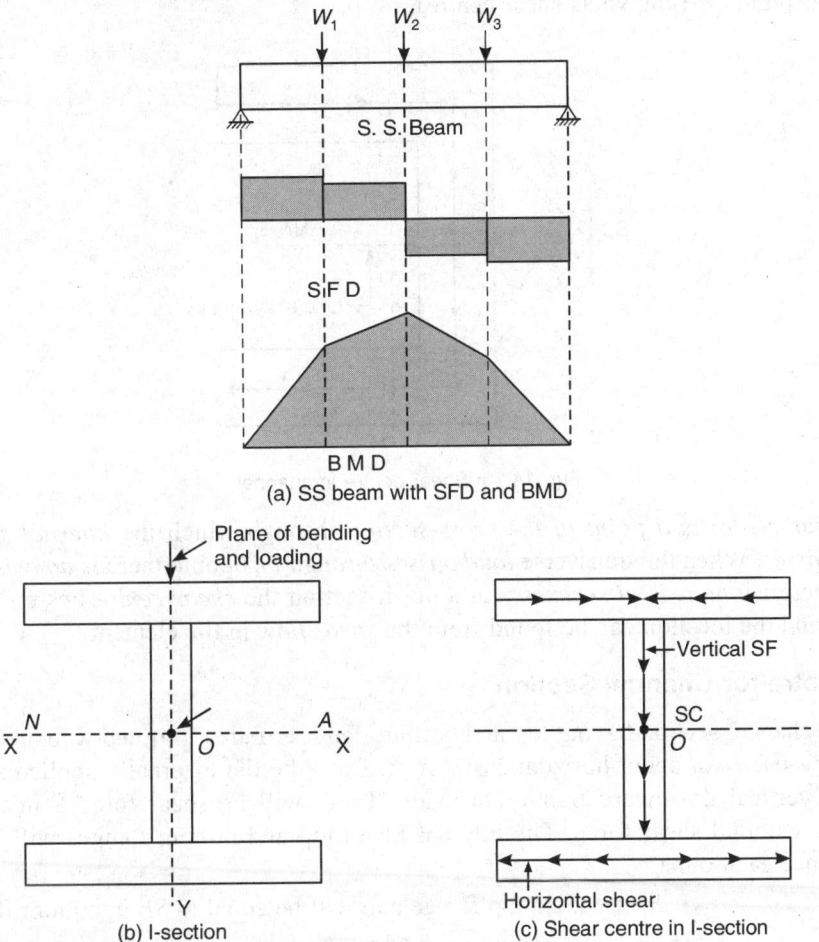

(a) SS beam with SFD and BMD

(b) I-section

(c) Shear centre in I-section

Fig. 14.12: SS beam with I–section

One of the axes of symmetry (*X–X*) is the neutral Axis (NA) and the *plane of loading* and that of bending contains the other axis of symmetry (*Y–Y*). This avoids torsion in the section. If the loading is inclined but the line of action passes through the shear centre (centroid in case of symmetrical *I*–section), it does not cause any torsion. The bending moment can be resolved in along the two axes of symmetry to find the bending stresses in two directions. In case of *I*–section, the *centroid* is also the *shear centre*.

In a channel section with one axis of symmetry, which becomes the NA if used as beam with plane of loading and bending the same. The Shear Flow in the flanges and web will be as shown in Fig. 14.13. The *resultant* of the forces *H* and *V* can be passing at the point *SC* lying on the line of symmetry. If the *applied external SF* also passes through this point *SC* on the line of symmetry, there will be *no twisting* of the section. If the external *SF* '*V*' does not pass through this point *SC*, then the external *SF* of *V* will form a couple with the *resultant internal*

SF of V. This moment of couple is about the *axis–Z* (perpendicular to the plane of the paper), and is called *twisting moment*. Thus the *twisting moment* can be avoided if the load line passes through this point *SC* (known as shear centre).

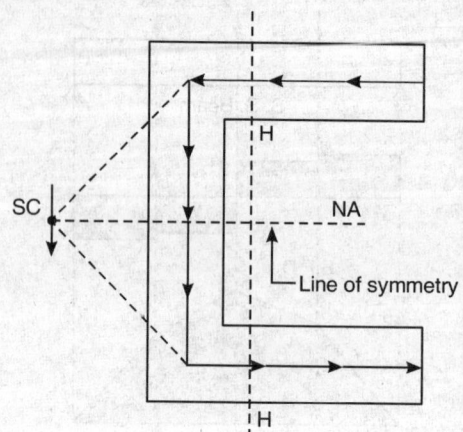

Fig. 14.13: Shear centre in channel

The *shear centre* is *a point in the cross-section* through which the *internal shear force resultant passes*. When the transverse *load passes through this* point, there is *no twisting* of the beam. If there is *one axis of symmetry* in a beam section the *shear centre* lies on the axis of symmetry and the location can be found from the *shear flow* in the element.

Shear Centre for Channel Section

Consider a channel section having top and bottom flanges of area $b. t_1$ and web area $h. t_2$, The section is *symmetrical* about horizontal *axis X–X.* Let *V* be the externally applied shear force caused by vertical downward transverse loads. There will be shear force *S* in the web to balance the external shear force. The internal SF in top and bottom flanges will be S_1 each which balance each other.

$$S_1 = SF \text{ in top flange and will be equal to } SF \text{ in bottom flange.}$$

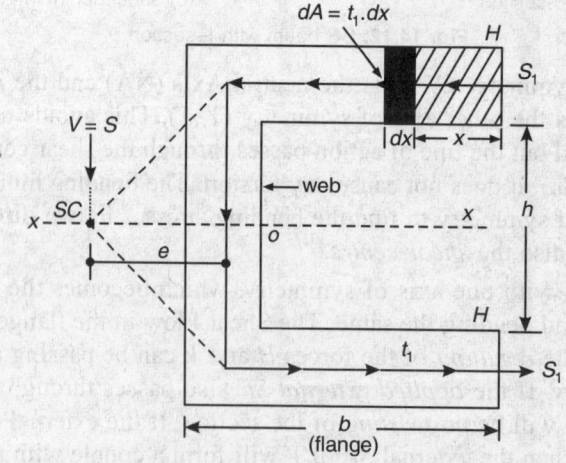

Fig. 14.14: Shear centre in channel

Consider top flange and a strip of width dx at a distance of x from the extreme *RH* edge.

Shear stress q (at any point) $= \dfrac{S}{I_{XX}} \cdot \dfrac{A\overline{y}}{t_1}$, where $A\overline{y} = (x \cdot t_1)\dfrac{h}{2}$, and $\underline{h}$ is the height of the web up to bottom of the flange.

or
$$q = \frac{S}{I_{XX}} \cdot \frac{x \cdot t_1}{t_1} \cdot \frac{h}{2} = \frac{S \cdot x \cdot h}{2I_{XX}} \qquad \text{... (i)}$$

Thus *SF* in *elementary strip* of area dA $(t_1 \cdot dx) = q \cdot dA = q \cdot t_1 \cdot dx$

Total *SF* in top flange at the junction with the web $= \displaystyle\int_0^b q \cdot t_1 \cdot dx$, where b = flange width

or
$$S_1 = \int_0^b q \cdot t_1 \cdot dx = \int_0^b \frac{S \cdot x \cdot h}{2I_{xx}} \cdot t_1 \cdot dx, \left(\text{since } q = \frac{S \cdot x \cdot h}{2I_{xx}} \right)$$

or
$$S_1 = \frac{S \cdot h \cdot t_1}{2I_{XX}} \int_0^b x \cdot dx = \frac{S \cdot h \cdot t_1}{2I_{XX}} \left(\frac{x^2}{2} \right)_0^b = \frac{S \cdot h \cdot t_1 \cdot b^2}{4I_{XX}} \qquad \text{... (ii)}$$

The couple formed by S_1 will be equal to couple formed by vertical SF V and shear force S (equal to V) in the web. If the shear centre (*SC*) is located at a distance of 'e' from the mid point of the web, we get.

$$S \cdot e = S_1 \cdot h = \frac{S \cdot h \cdot t_1 \cdot b^2}{4I_{XX}} \cdot h = \frac{S \cdot h^2 \cdot b^2 \cdot t_1}{4I_{XX}}$$

$\therefore$
$$e = \frac{h^2 \cdot b^2 \cdot t_1}{4I_{XX}} \qquad \text{... (14.7)}$$

(Distance of shear 'e' centre from the web centre along the axis of symmetry).

$$I_{XX} = 2\left[\frac{b \cdot t_1^3}{12} + b \cdot t_1 \left(\frac{h}{2} + \frac{t_1}{2} \right)^2 \right] + \frac{t_2 \cdot h^3}{12} = 2\left[\frac{b \cdot t_1}{4} \left(h^2 + t_1^2 + 2t_1 h \right) \right] + \frac{t_2 \cdot h^3}{12},$$

Neglecting higher powers of t_1 as compared to h.

$$I_{XX} \text{(Approx.)} = \frac{b \cdot t_1 h^2}{2} + \frac{t_2 \cdot h^3}{12} + bh \cdot t_1^2 = \frac{h^2}{12} \left(6b \cdot t_1 + t_2 \cdot h + 12 \frac{b \cdot t_1^2}{h} \right),$$

where $t_1 b$ = Area Af, $t_2 h = Aw$

$\therefore$
$$I_{XX} = \frac{h^2}{12} \left(6A_f + A_w + \frac{12b \cdot t_1^2}{h} \right)$$

$$e = \frac{h^2 \cdot b^2 \cdot t_1}{\dfrac{4h^2}{12} \left(6A_f + A_w + \dfrac{12b \cdot t_1^2}{h} \right)} = \frac{3b \cdot A_f}{\left(6A_f + A_w + \dfrac{12A_f \cdot t_1}{h} \right)}$$

$$e = \frac{3b}{\left(6 + \dfrac{A_w}{A_f} + \dfrac{12t_1}{h}\right)}, \qquad \qquad \dots (14.8)$$

where A_f = area of flange (bt_1) and A_w = Area of web, b = breadth of flange.

EXAMPLE 14.6: Calculate the location of *shear centre* for the channel section (220 × 100 × 10 mm) shown in Fig. 14.15. Also show the shear flow due to a vertical shear force of 5 kN.

Solution: $SF = V = 5000N$, X–X is NA

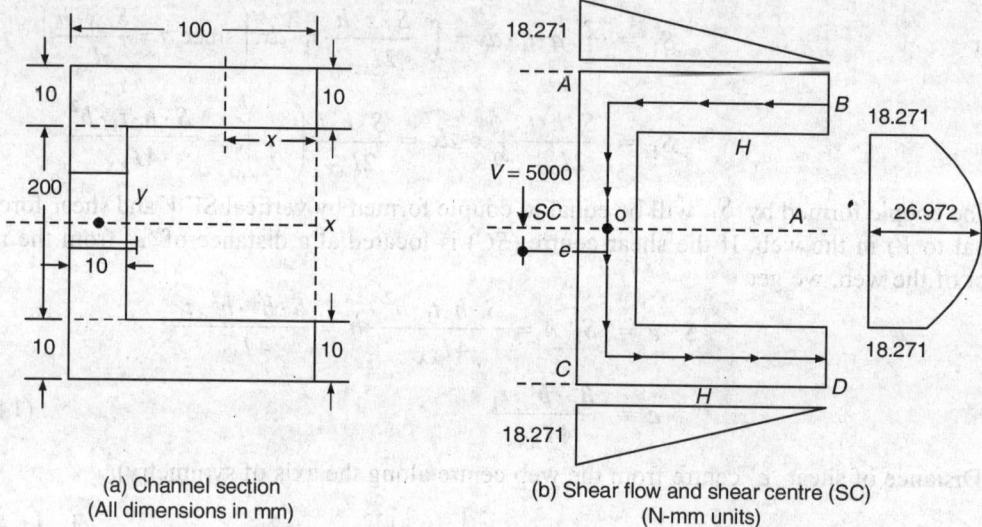

(a) Channel section
(All dimensions in mm)

(b) Shear flow and shear centre (SC)
(N-mm units)

Fig. 14.15: SC in channel (not to scale)

$$I_{XX} = 2\left(\frac{100 \times 10^3}{12} + 100 \times 10 \times 105^2\right) + \frac{10 \times 200^3}{12}$$

$$= 10^5\,[0.16667 + 220.50 + 66.6667] = \mathbf{287.334 \times 10^5 \ mm^4}$$

Consider section in the flange at x

Shear stress flow $\quad q = \dfrac{V}{I} \cdot A\bar{y} = \dfrac{5000}{287.334 \times 10^5} \dfrac{(10 \cdot x \cdot 105)}{}$

or $\qquad \qquad q = \dfrac{8750\,x}{47889}, \quad q_B\,(x = 0) = 0, \quad q_A\,(x = 100) = \mathbf{18.271 \ N/mm}$

Consider a section in web 'Y' above NA

$$q = \frac{V}{I} \cdot (A\bar{y}) = \frac{5000}{287.334 \times 10^5}\left[100 \times 10 \times 105 + (100 - y)10\left(y + \frac{100 - y}{2}\right)\right]$$

$$= \frac{5}{28733.4}[105000 + 5\,(100^2 - y^2)]$$

$$q_{A'}\,(y = 100) = \textbf{18.271 N/mm.}$$

$$q_{NA}\,(y = 0) = \textbf{26.972 N/mm}$$

Let the shear centre be at a distance of 'e' from the point 0 (centre of web).

Moment of forces about 0 :

$$5000.e = \text{Horizontal shear } H \times 200, \text{ where } H = \frac{100}{2} \times 18.271 = \textbf{913.55 N}$$

$$\therefore \qquad e = \frac{H \times 200}{5000} = \frac{913.55 \times 200}{5000} = \textbf{36.542 mm}$$

The *shear centre* is 36.542 mm from the centre of web (31.542 mm from the outer face of the web).

EXAMPLE 14.7: A symmetrical channel section shown in Fig. 14.16 is used as a beam carrying a maximum vertical shear force of 10 kN at a particular section. Determine the shear centre.

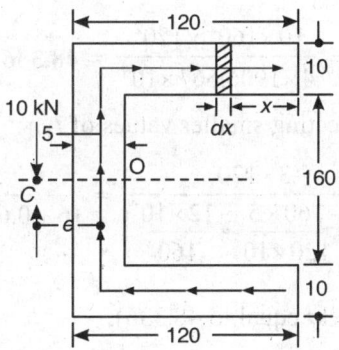

Fig. 14.16: Channel section (all dimensions in mm)

Solution: $SF = S = 10000\text{N}$, flange $b = 120\text{mm}$, $t_1 = 10\text{mm}$, web $h = 160\text{mm}$, $t_2(\text{web}) = 5\text{mm}$.

$$I_{XX} = 2\left\{\frac{120 \times 10^3}{12} + 120 \times 10(85)^2\right\} + \frac{5 \times 160^3}{12}$$

$$= 2 \times 10^4\{1 + 867\} + 10^4(170.668) = \textbf{1906.667} \times \textbf{10}^4 \textbf{ mm}^4$$

$$q \text{ flange} = \frac{S \cdot A\bar{y}}{I_{xx} \cdot t_1}, \quad A\bar{y} = x \cdot t_1 \cdot \frac{h}{2}$$

or

$$q = \frac{S}{I_{XX}} \cdot x \cdot \frac{h}{2},$$

SF in elementary strip $d_A = q.d_A = q \cdot t, \cdot dx,$

Total SF in top flange $S_1 = \int_0^b q \cdot t_1 \, dx$

or $\quad S = \int_0^b \left(\dfrac{S}{I_{XX}} \cdot x \cdot \dfrac{h}{2} \right) \cdot t_1 \, dx = \dfrac{S \cdot h \cdot t_1}{2 I_{XX}} \int_0^b x \cdot dx = \dfrac{S \cdot h \cdot t_1}{2 I_{XX}} \left(\dfrac{x^2}{2} \right)_0^b = \dfrac{S \cdot h \cdot t_1 \cdot b^2}{4 I_{XX}}$

$\quad S_1 = \dfrac{S \cdot h \cdot t_1 \cdot b^2}{4 I_{XX}} = $ SF in bottom flange also. .

Taking moment about mid point of web (0):

$$10000 \times e = S_1 . h = \left(\dfrac{S \cdot h \cdot t_1 \cdot b^2}{4 I_{XX}} \right) h = \dfrac{S \cdot h^2 \cdot t_1 \cdot b^2}{4 I_{XX}} = \dfrac{10000(160)^2 \cdot 10 \cdot (120)^2}{4 \times 1906.667 \times 10^4}$$

or $\quad e = \dfrac{(160)^2 \times 10 \times (120)^2}{4 \times 1906.667 \times 10^4} = \mathbf{48.336\ mm}$

Directly by formula

$$e = \dfrac{t_1 \cdot h^2 \cdot b^2}{4 I_{XX}} = \dfrac{10 \times 160^2 \times 120^2}{4 \times 1906.667 \times 10^4} = \mathbf{48.336\ mm}$$

By approximate formula neglecting smaller values of t_1

$$e' = \dfrac{3b}{\left(6 + \dfrac{A_w}{A_f} + \dfrac{12 t_1}{h} \right)} = \dfrac{3 \times 120}{\left(6 + \dfrac{160 \times 5}{120 \times 10} + \dfrac{12 \times 10}{160} \right)} = \dfrac{360}{(6 + 0.667 + 0.75)} = \dfrac{360}{7.417}$$

$= \mathbf{48.537\ mm}$ (Approximately equal to 48.336).

EXAMPLE 14.8: Determine the position of the shear centre of the unsymmetrical beam section shown in Fig. 14.17. The section is subjected to 20 kN vertical SF

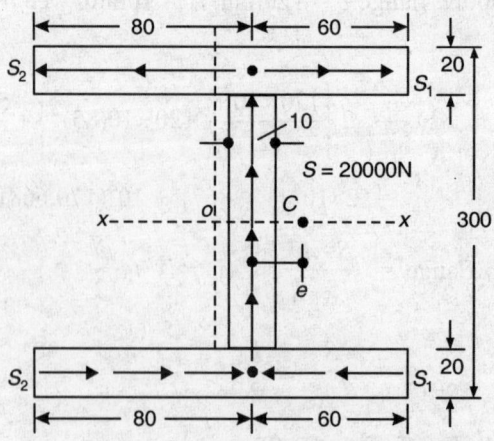

Fig. 14.17: Unsymmetrical I–section

Solution: $b_1 = 60$mm, $t_1 = 20$mm, $h_w = 300 - 20 \times 2 = 260$ mm, $b_2 = 80$ mm, $b_1 = 60$ mm $t_2 = t_1 = 20$mm, $t_3 = 10$mm, $V = 20,000$ N.

The section is symmetrical about X–X axis.

Shear stress in any layer $q = \dfrac{S\,A\bar{y}}{I_{XX} \cdot t}$

$$I_{XX} = 2\left[(80+60)\frac{20^3}{12} + (80+60)20 \times (130+10)^2\right] + \frac{10 \times 260^3}{12}$$

$$= 2[93.333 \times 10^3 + 5488 \times 10^4] + 1464.667 \times 10^4$$

$$= 10494.6667 \times 10^4 + 1464.6667 \times 10^4$$

$$= 12459.33 \times 10^4 \text{ mm}^4 = 12459.33 \times 10^4 \text{ mm}^4.$$

Shear force in 60 mm part

$q = \dfrac{S}{I_{XX}}\dfrac{A\bar{y}}{t}$, Small strip $d_A = 20 \cdot dx$, $A\bar{y} = 20 \cdot x \cdot (130+10)$

$$S_1 = \int_0^{60} q \cdot d_A = \int_0^{60} \frac{S \cdot A\bar{y}}{I_{XX} \cdot 20} \cdot d_A = \int_0^{60} \frac{S}{I_{XX}} \cdot \frac{20.x\ (140)}{20} \cdot (20dx) = \int_0^{60} \frac{S}{I_{XX}} \cdot 2800\,x\,dx$$

$$S_1 = \frac{S \cdot 2800}{I_{XX}}\left(\frac{x^2}{2}\right)_0^{60} = \frac{S}{I_{XX}}\left(\frac{2800}{2} \times 3600\right) = \frac{S}{I_{XX}}(1400 \times 3600)$$

$$= \frac{20000 \times 10^4 \times 504}{12459.33 \times 10^4}$$

$S_1 = $ **809.032 N**

Similarly $S_2 = $ **1438.28 N**, Let the shear centre 'C' be at 'e' from the web cetre '0'. Taking moment of the shear forces about the centre of the web '0', we get

$$S \times e = S_2 \times 280 - S_1 \times 280, \quad \{S_3 = \text{SF in web} = S \text{ for equilibrium.}$$

$$e = \frac{1438.28 \times 280 - 809.032 \times 280}{20,000} = \textbf{8.8095 mm (from the centre of the web)}$$

Shear centre lies at **8.809** mm from the mid point of the web towards the R_H side of the section.

EXAMPLE 14.9: Find the location of the shear centre in case of thin semicircular ring shown in Fig. 14.18 and subjected to a SF of S.

Let $\qquad\qquad\qquad\qquad CO = e$

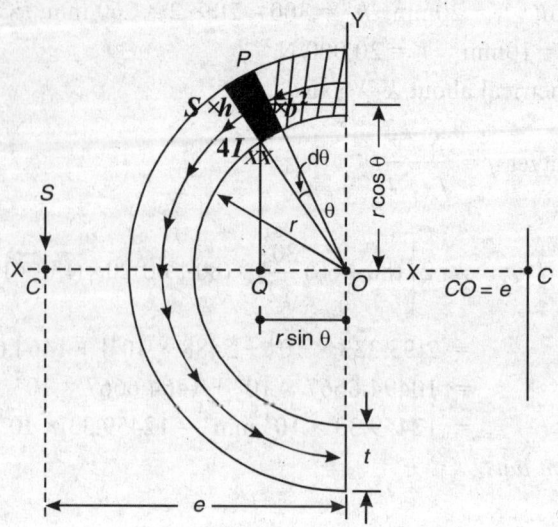

Fig. 14.18: Semicircular ring

Solution: Let the axis of symmetry be X–X and also the NA for the section.

$$I_{XX} = \frac{\pi}{8}[(r + t)^4 - r^4] = \frac{\pi}{8}[4r^3.t], \text{ neglecting higher powers of '}t\text{'}$$

$$I_{XX} = \frac{\pi r^3 \cdot t}{2}, \textbf{ approx.}$$

Consider small sector $rd\theta$ at an angle of θ with the vertical line Y–Y, say.

Shear flow 'q' $= S \cdot \dfrac{A\bar{y}}{I} = \dfrac{S}{I} \cdot (A\bar{y})$

$$A\bar{y} = \text{moment of shaded area about } X\text{–}X = \int_0^\theta (r \cdot d\theta) t \cdot r \cos\theta$$

$$\therefore \qquad q = \frac{S}{I} \int_0^\theta r^2 \cdot t \cos\theta \cdot d\theta = \frac{S}{I} r^2 \cdot t (\sin\theta)_0^\theta = \frac{S r^2 t \sin\theta}{I}$$

This indicates that $q = 0$, when $\theta = 0$ at the top and $q = \dfrac{S \cdot r^2 \cdot t}{I}$, when $\theta = 90°$ along X–X.

If the shear centre is at C, the moment of shear force S about '0' will be equal to the moment of internal shear force resultant from $\theta = 0$ to $180°$, i.e.

$$S.e = \int_0^{180} \left(\frac{S \cdot r^2 t \sin\theta}{I} \right) (r \cdot d\theta) \, r = \frac{S r^4 \cdot t}{I} \int_0^{180} \sin\theta \cdot d\theta$$

$$= \frac{S \cdot r^4 \cdot t}{I} (-\cos\theta)_0^{180} = \frac{S \cdot r^4 \cdot t}{I} (2)$$

$$S.e = \frac{2S \cdot r^4 \cdot t}{I}, \quad \therefore \quad e = \frac{2r^4 \cdot t}{I} = \frac{2r^4 \cdot t}{\dfrac{\pi r^3 t}{2}} = \frac{4 \cdot r}{\pi}$$

Thus the shear centre is located at a distance of $\dfrac{4r}{\pi}$ from the centre '0' of the semicircular thin ring of radius 'r'.

EXAMPLE 14.10: Determine the shear centre of the box type section shown in Fig. 14.19 having uniform thickness 't' and horizontal flanges of breadth b_2. Web projections on RHS are each 'b_1' as shown. Web height = h.

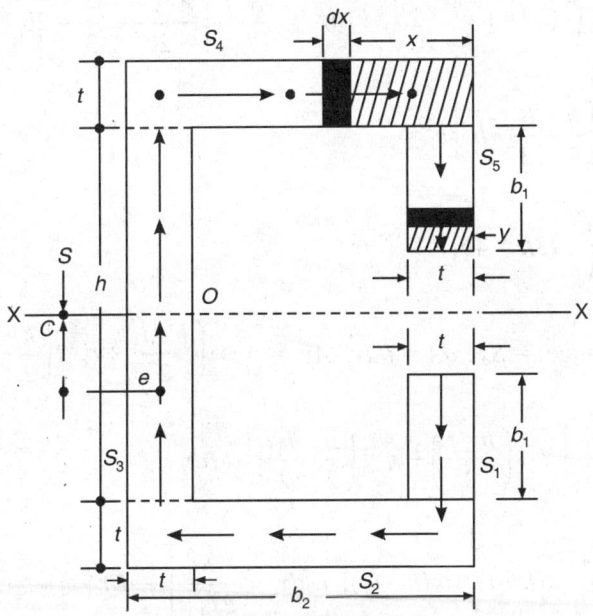

Fig. 14.19: Box type section

Solution: Uniform thickness = t, web ht. = h. Projections on RHS = $b_1 t$ (each), top and bottom flanges = $b_2.t$ (each).

The section will be symmetrical and the axis of symmetry will be X–X. Let the shear centre 'C' lies on X–X at a distance of 'e' from the web centre '0'.

The external SF = S, and internal SFs in various elements are S_1, S_2, S_3, S_4 and S_5 as shown.

$$S_1 = S_5, \quad S_2 = S_4, \text{ due to symmetry.}$$

Shear stress in any layer 'q' $= \dfrac{S}{I_{XX}} \cdot \dfrac{A\overline{y}}{t}$

Shear force $S_1 = S_5$ (web projections RHS) : Consider layer 'dy' at y from the edge

$$dA = t \cdot dy \text{ and } A = y \cdot t, \text{ c.g. } \quad \bar{y} = \left(\frac{h}{2} - b_1\right) + \frac{y}{2} = \frac{1}{2}(h + y - 2b_1)$$

$$S_5 \text{ (Shear force)} = \int_0^{b_1} \cdot q \cdot dA = \int_0^{b_1} \frac{S \cdot y \cdot t \, (h + y - 2b_1)}{I_{XX} \, t} \frac{1}{2} \, dA$$

$$= \int_0^{b_1} \frac{S}{2I_{XX}} y \, (h + y - 2b_1) \, t \cdot dy$$

$$\text{or} \quad S_5 = \frac{S \cdot t}{2I_{XX}} \int_0^{b_1} (h \cdot y + y^2 - 2b_1 \cdot y) dy = \frac{S \cdot t}{2I_{XX}} \left[\frac{h \cdot y^2}{2} + \frac{y^3}{3} - 2b_1 \cdot \frac{y^2}{2} \right]_0^{b_1}$$

$$= \frac{S \cdot t}{2I_{XX}} \left[\frac{h}{2} + \frac{b_1}{3} - b_1 \right] b_1^2$$

$$S_5 = \frac{S \cdot t \cdot b_1^2}{12 I_{XX}} \, (3h - 4b_1) \qquad \qquad \dots \text{(i)}$$

Shear force S_2 (Flange) $= S_4 : dA = t.dx, \quad A\bar{y} = x \cdot t \cdot \left(\frac{h+t}{2}\right) + b_1 \cdot t\left(\frac{h}{2} - b_1 + \frac{b_1}{2}\right)$

$$S_4 = \int_0^{b_2} \frac{S}{I_{XX} \cdot t} \left[x \cdot t\left(\frac{h+t}{2}\right) + b_1 \cdot t\left(\frac{h}{2} - \frac{b_1}{2}\right) \right] \cdot t \cdot dx$$

$$\text{or} \quad S_4 = \frac{S}{2I_{XX}} \int_0^{b_2} [x(h+t) + b_1(h - b_1)] \cdot t \cdot dx = \frac{S}{2I_{XX}} \left[\frac{x^2}{2}(h+t) + b_1(h - b_1) \cdot x \right]_0^{b_2} \cdot t$$

$$S_4 = \frac{S \cdot t}{4I_{XX}} \left[b_2^2(h+t) + 2b_1 \cdot b_2(h - b_1) \right] = \frac{S \cdot t}{4I_{XX}} \left[h(b_2^2 + 2b_1 b_2) + t b_2^2 - 2b_1^2 b_2 \right] \quad \dots \text{(ii)}$$

S_3 (SF in web) passes through the mid point '0'.

Taking moment about the mid point of web '0', we get.

$$S \cdot e = 2S_5 (b_2 - t) + 2S_4 \cdot \left(\frac{h+t}{2}\right) + S_3 \times 0$$

$$\text{or} \quad S \cdot e = 2(b_2 - t) \frac{S t \cdot b_1^2}{12 \ I_{XX}}(3h - 4b_1) + (h+t) \cdot \frac{St}{4I_{XX}}\left[h\left(b_2^2 + 2b_1 b_2\right) + t b_2^2 - 2b_1^2 b_2 \right]$$

$$e = \frac{2(b_2 - t) \cdot t \cdot b_1^2 (3h - 4b_1)}{12 I_{XX}} + \frac{(h+t)t}{4I_{XX}}\left[h\left(b_2^2 + 2b_1 b_2\right) + t b_2^2 - 2b_1^2 b_2 \right]$$

or $\quad e = \dfrac{t}{12I_{XX}}\left[2b_1^2(b_2-t)(3h-4b_1)+3(h+t)\left\{h\left(b_2^2+2b_1b_2\right)+tb_2^2-2b_1^2b_2\right\}\right]$

$$I_{XX} = \dfrac{t\cdot h^3}{12}+2\left\{\dfrac{b_2\cdot t^3}{12}+b_2\cdot t\left(\dfrac{h+t}{2}\right)^2\right\}+2\left\{\dfrac{t\cdot b_1^3}{12}+t\cdot b_1\left(\dfrac{h}{2}-b_1+\dfrac{b_1}{2}\right)^2\right\}$$

$$= \dfrac{t\cdot h^3}{12}+2\left[\dfrac{b_2\cdot t^3}{12}+\dfrac{b_2\ t}{4}(h+t)^2+\dfrac{t\,b_1^3}{12}+\dfrac{t\ b_1}{4}(h-b_1)^2\right]$$

$$I_{XX} = \dfrac{1}{12}\left[t\cdot h^3+2t\cdot b_1^3+2b_2\cdot t^3+6b_2\cdot t(h+t)^2+6t\cdot b_1(h-b_1)^2\right]$$

e (distance of shear centre) can be found by substituting the values of I_{XX} and various dimensions.

14.5 SUMMARY (UNSYMMETRICAL BENDING AND SHEAR CENTRE)

We have studied beam elements carrying transverse loading along the plane of bending in earlier units. In *simple* theory of bending we calculate bending stresses due to *pure bending moment* by using the basic bending equation: $\dfrac{M}{I}=\dfrac{f}{y}=\dfrac{E}{R}$, which is based on certain assumptions. The *important one is plane sections* before loading remain *plane after bending*, which means the *bending deformations* across the section *remain linear*. Here the effect of *shear stresses is neglected* in calculating bending stresses. In real situations the loading may not result in pure bending and also the *plane of loads* may not lie along the plane of bending which may cause *unsymmetrical* bending. Further, if the load does not pass through the *shear centre* of the section, it may also cause *twisting* of the section. Different sections may develop shear stresses to resist external shear forces. Both *bending* and *shear* stresses are determined *independently* for simplicity. The relation $\dfrac{M}{I}=\dfrac{f}{y}$, relates to the stress distribution across the section and calculation of maximum bending stress to resist external BM. The equation $\dfrac{M}{I}=\dfrac{E}{R}$, relates to deformations caused by the applied BM.

The shear stress 'q' is tangential to the section and the distribution is found by $q=\dfrac{S\cdot A\overline{y}}{I\cdot b}$, with usual notations. Beam sections are *designed separately for safety against BM and SF and stronger of the two designs adopted*.

Thin walled sections, such as I, T, channels, angles, etc, the shear flow can be determined by assuming the *shear stress constant* across the thickness which is small. Shear flow $q=\tau\ t$. The concept of *shear centre* is based on the shear flow. **Shear centre is defined as a point in the cross-section through which the internal shear force resultant passes.**

If the transverse loading passes through *shear centre*, there will be *no twisting* of the beam. If there is one axis of symmetry, the shear centre lies on this axis of symmetry. *Shear centre* coincides with the *centroid* when both the axes of the section are symmetrical. If the external load does not pass through the shear centre, there will be twisting moment in the section in addition to bending moment and shear force.

Unsymmetrical bending occurs, if the beam sections are *unsymmetrical* or the *loading* does not lie in the symmetrical *plane of bending*. For *no twisting* the load must pass through the *shear centre*. In case of unsymmetrical sections, the bending occurs about the *Principal axes* of the section. Bending stress at any point can be calculated by super imposition of bending stresses about two principal axes. The neutral axis may be oblique line representing zero stress.

EXERCISE 14

Q.14.1. Explain the structural behaviour of a beam element carrying transverse loads.

Q.14.2. Explain the significance of various assumptions in simple bending theory.

Q.14.3. Explain the bending equation with meaning and units of each term.

Q.14.4. Explain the shear stress equation with meaning and units of each term.

Q.14.5. Indicate *shear flow* and location of shear centre in cases of

(a) I–section (b) Channel (c) Angle (d) *T*–sections with thin wall thickness.

Q.14.6. Define *shear centre* and explain with sketch.

Q.14.7. Show approximately the location of shear centre for sections shown in Fig. Q14.7.

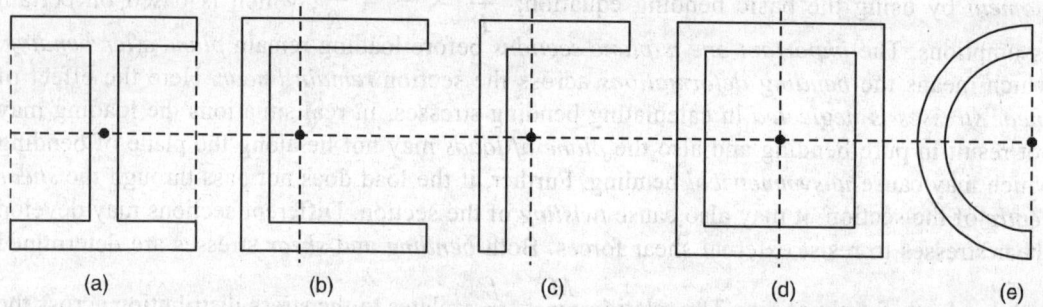

(a)   (b)   (c)   (d)   (e)

Fig. Q14.7: Different sections

Q.14.8. An angle section of 100mm × 80mm × 10mm, is used in a *S.S.* beam of 5m span. Find the point load *W* acting at the midspan point and passes through the shear centre of the section. Maximum permissible stress both in tension and compression is 200 MPa.

Ans. *W* = 1350 N.

Q.14.9. A cantilever beam of 3m span comprises of a channel section 250mm × 100mm × 10mm. Find the maximum point load *W* acting at the free end and inclined at 30° to the vertical if the maximum permissible stress in the material is 75 MPa in tension and 40 MPa in compression. The load passes through the shear centre of the section.

Ans. *W* = 2796 N.

Q.14.10. A simply supported beam of 6m span carries udl along a loading palne inclined at an angle of 40° to the vertical plane and passes through the shear centre of the section. Find the rate of loading if the permissible stress in the material is 200 MPa both in comperssion and tension and the section consists of 150 mm × 200 mm × 10 mm *T*–section. 150 mm flange is horizontal.

Ans. $w = 2183$ N/m.

Q.14.11. Determine the location of *shear centre* in a channel section of 100mm × 10mm horizontal flanges and 200mm × 10mm internal vertical web. The beam carries 8000 N vertical shear force. Determine the shear centre and show the shear flow due to SF.

Ans. Shear centre '*e*' from the centre of web = **36.5 mm**.

Shear flow at the flange junction = **29230 N/mm** (max.) and **431.50 N/mm at the *NA* (max.)**.

Q.14.12. An equal angel ABC of 80mm × 80mm × 10mm is used for a simply supported beam over a span of 3.0 m. The beam carries a vertical point load of 800 N and passes through the centroid the section, calculate the stresses at *A*, *B* and *C* representing outer corner points (ii). Also calculate the deflection (max.) if $E = 200$ GN/m².

Ans. $I_{UU} = 141.25 \times 10^4$ mm⁴, $I_{VV} = 36.7 \times 10^4$ mm⁴, A (23.1, 56.56), B (–33.45, 0),

C (23.1, –56.56), $f_A = $ **43.68 N/mm²**, $f_B = $ **– 38.68 N/mm²**, $f_C = $ **9.7 N/mm²**.

ii. Deflection $\delta = \dfrac{K.\ WL^3}{E} \sqrt{\dfrac{\sin^2\theta}{I_{VV}^2} + \dfrac{\cos^2\theta}{I_{UU}^2}}$, $\theta = 45°$, $\delta = $ **4.479 mm**.

Q.14.13. A cantilever of span 2m comprises of *I*–section 3 mm × 2.5 mm horizontal flanges and 45 mm × 2.0 mm internal web. The cantilever carries an inclined load of 100N at the free end inclined at 20° with the vertical. Calculate the maximum bending stresses near the fixed end in top flange left hand corner *A* and *RH* corner *B*.

Ans. $I_{XX} = 9.9875 \times 10^4$ mm⁴, $I_{YY} = 1.128 \times 10^4$ mm⁴, $f_A = $ **138 N/mm²** (tensile),

$f_B = $ **43.94 N/mm²** (comp.)

Q.14.14. Determine the location of the *shear centre* for the section shown in Fig. Q14.14. Also sketch the shear flow in the section with uniform thickness (10mm).

Ans. **30.372 mm** from the centre of the web.

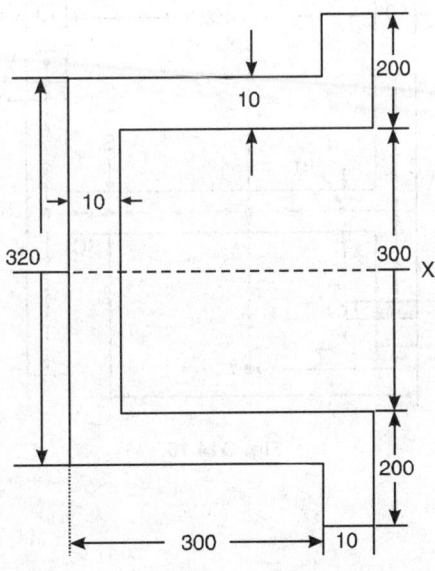

Fig. Q14.14.

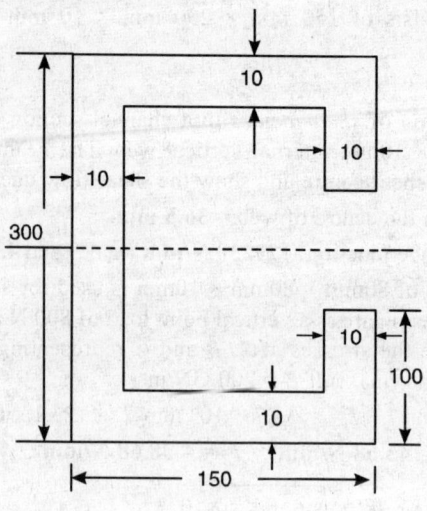

Fig. Q14.15.

Q.14.15. Determine the location of the *shear centre* for the section shown in Fig. Q14.15. Also sketch the shear flow in the section having uniform thickness of 10 mm.

Ans. Shear centre = **65.775 mm** from the centre of the web.

Q.14.16. A beam section is shown in Fig. Q14.16. Find the locations of the centroid and the shear centre.

Ans. Centroid 0, $\bar{x}$ = **251.0 mm** from LH edge.

Shear centre C, e = **369.38 mm** from LH along X–X axis outer ege.

I_{XX} = **6003.333 × 10⁴ mm⁴**.

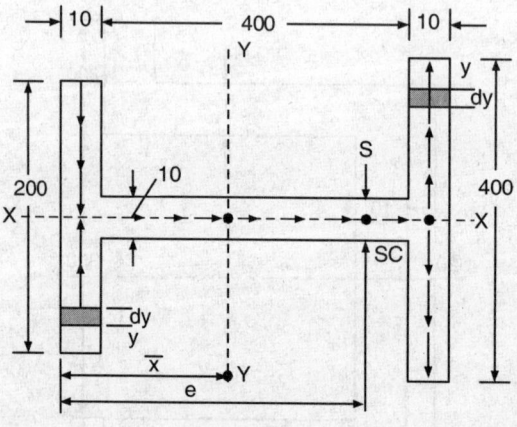

Fig. Q14.16.

15

Combined Axial and Bending Stresses (Combined Direct and Eccentric Forces)

LEARNING OBJECTIVES

After studying this chapter the learner, understands the behaviour of structures or structural elements subjected to axial loads and moments and will be able to:

15.1 **Differentiate** between **direct axial** and **eccentric** forces.

15.2 **Explain eccentricity** of loading and its effect on stresses.

15.3 **Explain general conditions** of stability and safety of masonry structures.

15.4 **Explain** the effect of **lateral pressures** on stresses on foundation bases.

15.5 Determine **eccentricity** in bases of dams or retaining walls subjected to lateral earth pressures or hydraulic pressures.

15.6 Calculate the stresses on foundation bases due to **lateral pressures** on the retaining walls or dams.

15.7 **Design** gravity retaining walls and dams with reference to
(a) overturning, (b) sliding, (c) crushing, (d) tension.

15.8 **Analyse** masonry chimney shafts, etc.

15.1 INTRODUCTION

Any short structural member loaded axially causes direct axial stress which can be determined

by stress $= \dfrac{\text{Load}}{\text{Area of the cross-section}}$, i.e. $p = \dfrac{W}{A}$ ... (15.1)

This equation 15.1 holds good both for compressive and tensile loads if the length of the member is *short*. When the axial loads are compressive, the members are called *struts*. When the axial loads are tensile, the members are called *ties*. We have already studied centric axial loads and calculated *direct stresses* by equation $p = \dfrac{W}{A}$ (equation 15.1).

This equation is valid only when the loading is *perfectly centric* and the struts are *short*. In case of *long columns and struts* we have used *Euler's, Rankine's* and various other formulae viz *IS code, Perry*, etc. *Columns* are generally vertical members carrying vertical compressive loads and when the column comprises of rolled steel sections, these are known as *stanchions*. Various members used to carry compressive forces in trusses are called *struts*. Long columns and struts mainly fail due to *buckling*. We have already studied these cases of long and short columns under axial loading.

We shall now study *axial elements* subjected to eccentric loads. The stresses in axial members shall comprise of two components, i.e. *direct* axial and axial *bending* stresses caused by *eccentricity*.

Combined stress $\qquad p_{1,2} = p_0 \pm p_b,$ $\qquad\qquad\qquad$ ...(15.2)

Where $\qquad\qquad\qquad p_0 = \dfrac{W}{A}, p_b = \dfrac{W \cdot e_x}{I_{x-x}} y$

W = Axial eccentric load, e_x = eccentricity from the centroid of the section.

I_{x-x} = Moment of Inertia of the X-section about the axis of bending $x - x$,

y = Distance of the fibre where stress is calculated from the neutral axis $x - x$.

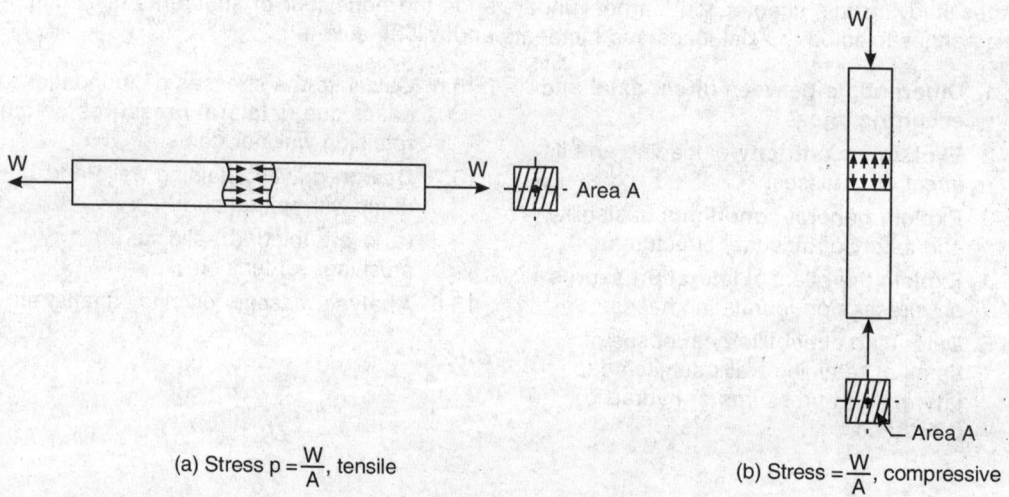

(a) Stress $p = \dfrac{W}{A}$, tensile $\qquad\qquad\qquad\qquad\qquad$ (b) Stress $= \dfrac{W}{A}$, compressive

Fig. 15.1: Direct stresses in axial force elements

15.2 COMBINED DIRECT AND BENDING STRESSES

Sometimes the structural members are simultaneously subjected to axial loads and bending moments. For example if a beam is subjected to inclined loads, it will result in axial force as well as transverse force. Axial force will cause axial stress and transverse force will result in shear stresses and bending stresses. In axial elements if the force is eccentric, it results in bending moment equal to (load) × (eccentricity), where eccentricity is measured from the centroid of the section.

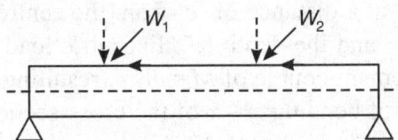

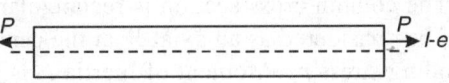

Horizontal component of resultant of W_1 and W_2 along axis results in direct stress $p0 = \dfrac{W_{axial}}{A}$,

BM at any section causes bending stress $p_b = \dfrac{M}{I} \cdot y$

SF at any section causes shear stress $q = \dfrac{F.A\bar{y}}{I.b.}$

(a) Transverse loads

Eccentric force P, Eccentricity = e
Bending moment $P.e$
Direct stress $p_0 = P/A$

Bending stress $pb = \dfrac{P.e}{I_{x-x}} , y$

(b) Eccentric axial loads

Fig. 15.2: Direct and bending stresses

15.3 ECCENTRIC LOADING AND NATURE OF STRESS

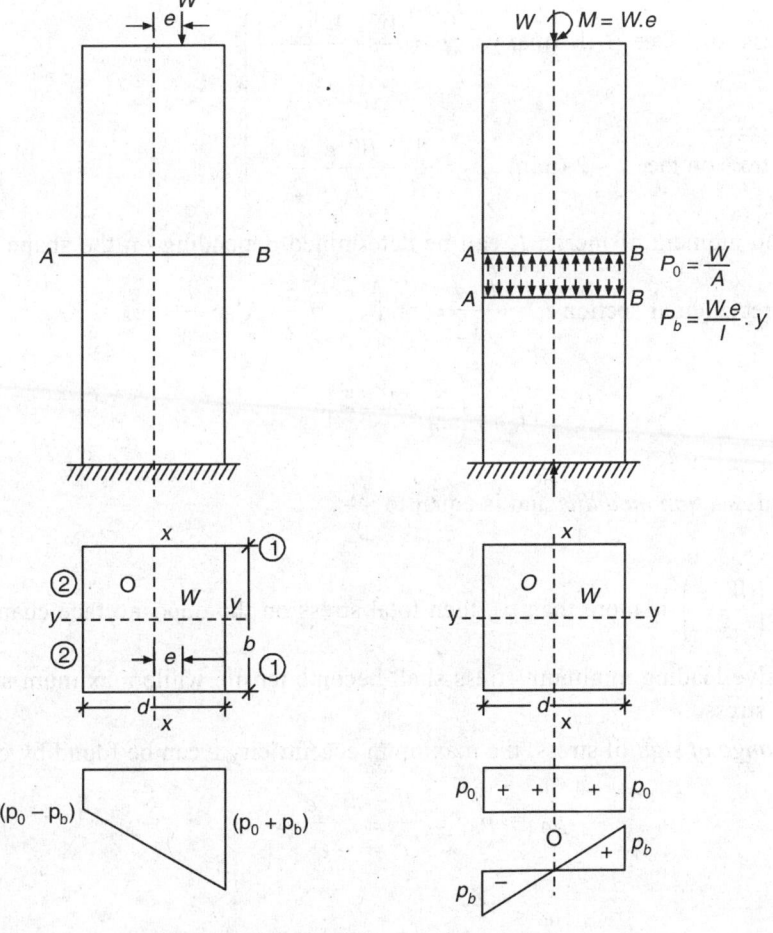

Fig. 15.3: Combined direct and bending stresses

Consider a short column carrying an eccentric load W at a distance of 'e' from the centroid 0, The column cross-section is rectangular with width 'b' and the depth 'd'. Eccentric load W can be considered as an axial W at the centroid *plus* a moment couple of $M = W. e$, resulting in bending stress p_b. Moment of Inertia I_x is about the axis of bending $x - x$ of the cross-section.

Bending stress $p_b = \dfrac{M}{I}(\pm y)$

i.e. $\qquad\qquad\qquad p_{b(max)} = \pm\dfrac{W \cdot e}{I_x}\cdot\dfrac{d}{2}$ $\qquad\qquad$ ... (i)

Direct axial stress $\qquad\qquad p_0 = \dfrac{W}{A}$ $\qquad\qquad\qquad\qquad\qquad$ ... (ii)

Total stress $p = p_0 \pm p_b = \dfrac{W}{A}\pm\dfrac{W \cdot e}{I_x}\cdot\dfrac{d}{2}$ $\qquad\qquad\qquad$ ... (15.3)

Total stress on face 1–1 (max) : $p_1 = \dfrac{W}{A}+\dfrac{W \cdot e}{I_x}\cdot\dfrac{d}{2}$

$\qquad\qquad\qquad\qquad\qquad\qquad\qquad\qquad\qquad\qquad\qquad$ ... (15.3 a)

Total stress on face 2 – 2 (min) : $p_2 = \dfrac{W}{A}-\dfrac{W \cdot e}{I_x}\cdot\dfrac{d}{2}$

Area A and moment of inertia I_x can be determined depending on the shape of the cross-section. For rectangular section $I_x = \dfrac{b \cdot d^3}{12}$, and $y = \pm\dfrac{d}{2}$. Also $\dfrac{I_x}{y} = z_x$

Hence $\qquad\qquad\qquad p = \left(\dfrac{W}{A}\pm\dfrac{W \cdot e}{z_x}\right)$ $\qquad\qquad$ ... (15.3 b)

Where, z_x is *section modulus* and is equal to $\dfrac{I_x}{y}$.

In case $p_b\left(\dfrac{W \cdot e}{z_x}\right)$ is more than p_0, then total stress on the opposite face changes sign, i.e.

for compressive loading minimum stress shall become tensile while maximum stress shall be compressive stress.

For no *change of sign* of stress, the maximum eccentricity e can be found by equating

$$p_0 = p_b \text{ i.e. } \dfrac{W}{A}\geq\dfrac{W \cdot e}{z_x}$$

or $\qquad\qquad\qquad\qquad e \leq \dfrac{Z_x}{A}$ $\qquad\qquad\qquad\qquad\qquad$ ... (15.4)

For *rectangular* section $A = b.\,d., \; z_x = \dfrac{1}{6} b \cdot d^2$

$$\therefore \quad \underset{(max.)}{e} \; \leq \; \frac{1}{6}\frac{bd^2}{bd} \leq \frac{d}{6}, \text{ for rectangular section} \qquad \dots (15.5\ a)$$

For solid circular section $\quad A = \dfrac{\pi}{4} d^2, \; z_x = \dfrac{\pi d^3}{32}.$

$$\therefore \quad \underset{(max.)}{e} \; \leq \; \frac{\pi d^3}{32\dfrac{\pi d^2}{4}} \leq \frac{d}{8}, \text{ for circular section} \qquad \dots (15.5\ b)$$

Similarly maximum eccentricity for *no change of sign* of stress can be calculated by equating p_0 to p_b. Thus there is no change of nature of stress if the eccentric load remains within e_{max} distance from the centroid of the section on either side.

(i) *Rectangular section–Middle third Rule*

$$\text{or} \qquad p_0 \leq p_b \quad \text{or} \quad \frac{W}{A} \geq \frac{W \cdot e}{I} \cdot \frac{d}{2} \geq \frac{W \cdot e \cdot d}{2 \cdot (Ak^2)}$$

$$\text{or} \qquad e \leq \frac{2k^2}{d}, \qquad \dots (15.6)$$

where, k = radius of gyration of the section with respect to the *NA*

d = depth of the section.

For rectangular section $\quad k^2 = \dfrac{I}{A} = \dfrac{d^2}{12},$

$$\therefore \quad \underset{(max.)}{e} \; \leq \; \frac{2k^2}{d} \leq \frac{2d^2}{12 \cdot d} \leq \frac{d}{6}, \text{ same as in equation 15.5(a)}$$

Total distance on either side of the centroid $= \dfrac{d}{6} + \dfrac{d}{6} = \dfrac{d}{3}.$

Therefore if the load lies within this *middle third* $\left(\dfrac{d}{3}\right)$, there is no change of nature of stress and the total stress remains of the same sign as the direct stress $p_0 \left(= \dfrac{W}{A} \right)$. Thus if the load remains within middle one third width, total stress does not change the sign and this is known as **middle one third rule** for rectangular sections.

(ii) *Solid circular section – Middle one fourth rule*

For no change in nature of the stress, $p_0 \geq p_b$. or $p_b \leq p_0$

i.e.
$$\frac{W \cdot e}{\frac{\pi}{64} \cdot d^4} \cdot \frac{d}{2} \le \frac{W}{\frac{\pi}{4}(d^2)}$$

or $\quad e \le \frac{d}{8}$, i.e. $e_{max} \le \frac{d}{8}$, on either side of the centroid $\qquad$ ...(15.7)

i.e. total portion for *no change* in nature of the stress due to eccentricity 'e' $\le \frac{d}{8} + \frac{d}{8} \le \frac{d}{4}$.

Thus, if the load lies within *middle fourth* of the section, the nature of the stress *remains unchanged*. This rule is called **middle one fourth rule** for solid circular sections.

(iii) *Hollow Circular Section*

$$p_0 = \frac{W}{\frac{\pi}{4}(D^2 - d^2)}, \quad p_b = \frac{W \cdot e}{\frac{\pi}{64}(D^4 - d^4)} \cdot \frac{D}{2}$$

For no change in the nature of the stress

$$p_0 \ge p_b \quad \text{or} \quad p_b \le p_0, \quad \text{i.e.} \quad \frac{W \cdot e}{\frac{\pi}{64}(D^4 - d^4)} \cdot \frac{D}{2} \le \frac{W}{\frac{\pi}{4}(D^2 - d^2)}$$

or $\quad \dfrac{32 D \cdot e}{(D^4 - d^4)} \le \dfrac{4}{(D^2 - d^2)}, \quad \text{or} \quad e \le \dfrac{(D^2 + d^2)}{8D}$

or $\quad e_{max.} \le \dfrac{(D^2 + d^2)}{8D}$, (either side) $\qquad$ ...(15.8)

for no change in nature of stress in hollow circular sections.

15.4 THE CENTRAL CORE OF A SECTION

We have seen that bending stresses are developed in the cross-section in addition to direct stress if the loading is not centric. The nature of the total stress remains the same as the direct stress, if the eccentricity remains within certain limits in certain sections as determined in earlier section. *Core of the section* is defined as the area of the section in which if the **line of action of load lies, the total stress remains of the same nature throughout**.

Each section has two rectangular axes (i.e. x–x and y–y axes). If the line of action of the load is on neither of the axes of the section, the *bending is unsymmetrical*. If the total stress in the section is *not to change the nature of stress*, the line of action of the load W must *cut the cross – section* within certain *limits of area*. This area is called the **core or kern**. The shape of the **kern of a section** depends on the shape of the section.

i. *Rectangular section* ($b \times d$)

If the load W acts at P with coordinates x, y with reference to axes X–X and Y–Y. Let both x and y are positive. The bending moment about X–X and Y–Y Shall be

$$M_x = W \cdot e_x = W \cdot y, \quad M_y = W \cdot e_y = W \cdot x.$$

$$I_{xx} = \frac{d \cdot b^3}{12}, \; I_{yy} = \frac{b \cdot d^3}{12}$$

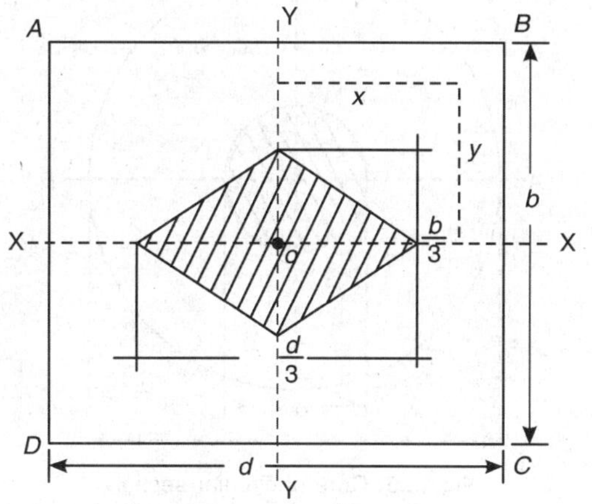

Fig. 15.4: Core of the section

Stress at any point Q (x', y') shall be p

$$p = \frac{W}{A} + \frac{M_x}{I_{XX}} \cdot y' + \frac{M_y}{I_{yy}} \cdot x' = \frac{W}{b \cdot d} + \frac{W \cdot y \cdot 12 y'}{d \cdot b^3} + \frac{W \cdot x \cdot 12 x'}{b \cdot d^3}$$

or $\quad p = \dfrac{12W}{b \cdot d} \left[\dfrac{1}{12} + \dfrac{y \cdot y'}{b^2} + \dfrac{x \cdot x'}{d^2} \right]$

At point D, stress 'p' shall be minimum and if no change of nature of stress occurs,

$$p_{\min} \geq O, \text{ at } D, x' = \frac{d}{2}, y' = \frac{b}{2}.$$

$\therefore \qquad p_{\min} = \dfrac{12W}{b \cdot d} \left[\dfrac{1}{12} - \dfrac{y}{2b} - \dfrac{x}{2d} \right] \quad$ or $\quad O = \left[\dfrac{1}{6} - \dfrac{y}{b} - \dfrac{x}{d} \right] \dfrac{6W}{bd}$

or $\quad \dfrac{x}{d} + \dfrac{y}{b} = \dfrac{1}{6} \quad$ or $\quad \left(\dfrac{6x}{d} + \dfrac{6y}{b} \right) = 1$ (Equation of straight line) $\qquad$ ... (15.9)

Hence for *no change in nature* of stress due to eccentricity, the eccentricity is governed in such a way that the load W lies at a distance (x, y) from the two reference axes so that

$\dfrac{6x}{d} + \dfrac{6y}{b} = 1$ and points on the reference axes are $\dfrac{d}{6}$ and $\dfrac{b}{6}$ from the origin on either side.

Thus core lies within $\dfrac{d}{3}$ and $\dfrac{b}{3}$ in the middle.

ii. *Solid circular section*

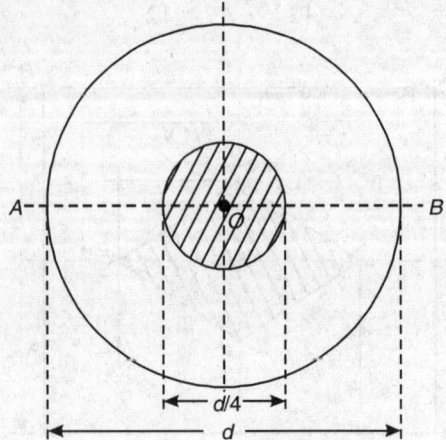

Fig.15.5: Core of circular section

We have seen that e_{max} for no change in nature of stress is $\dfrac{d}{8}$, where d is diameter of the

circular section. Therefore central core on either side of the centre O will be $= \dfrac{d}{8} + \dfrac{d}{8} = \dfrac{d}{4}$. i.e.

the diameter of the *core* $= \dfrac{d}{4}$, for *no change in the nature* of the stress.

Diameter of *core* $= \dfrac{d}{4}$, in case of solid circular section. ... (15.10)

(iii) *Hollow circular section*

We have seen that for no change in the nature of the stress $e_{max} = \dfrac{D^2 + d^2}{8D}$, where

D = outer diameter and d = inner bore diameter.

Hence the core for the hollow circular section shall be a concentric circle of diameter

$$d' = \frac{D^2 + d^2}{8D} + \frac{D^2 + d^2}{8D} = \left(\frac{D^2 + d^2}{4D} \right), \text{ for hollow circular section ... (15.11)}$$

(iv) *I – Section*

We have seen that in rectangular section *at the point* (x', y') stress

$$p = \frac{W}{A} \pm \frac{W \cdot e_x}{I_X} \cdot y' \pm \frac{W \cdot e_y}{I_y} \cdot x'$$

or $$p = \frac{W}{A} \left[1 \pm \frac{e_X \cdot y'}{k_x^2} \pm \frac{e_Y \cdot x'}{k_y^2} \right]$$

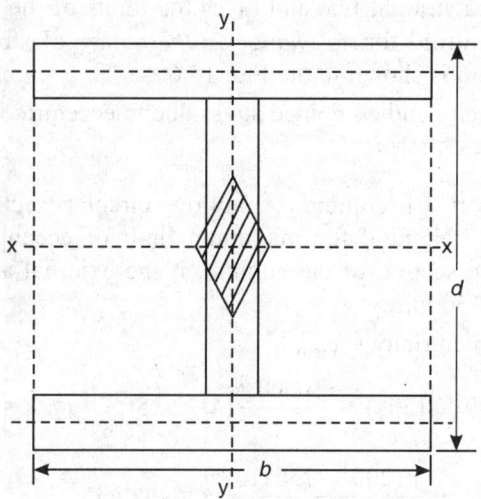

Fig. 15.6: I-section core

If coordinates of the load point are (x, y) we have $p = \dfrac{W}{A}\left[1 \pm \dfrac{y \cdot y'}{k_x^2} \pm \dfrac{x \cdot x'}{k_y^2}\right]$ ... (15.12)

Where k_x = radius of gyration of the section about X–X axis

k_y = radius of gyration of the section about Y–Y axis

(x', y') is the point where stresses are determined

(x, y) is the point where eccentric load W acts $(e_x = y, e_y = x)$.

For stress p to be zero, we have $\dfrac{W}{A}\left[1 - \dfrac{yy'}{k_x^2} - \dfrac{xx'}{k_y^2}\right] = 0$

or $\qquad \dfrac{y \cdot y'}{k_x^2} + \dfrac{x \cdot x'}{k_y^2} = 1$ ... (15.13)

This is the general equation for no change in the nature of stress in the extreme cases. In case of corners of the I–section, we have $x' = \dfrac{b}{2}, y' = \dfrac{d}{2}$, therefore the equation 15.13 can be rewritten as

$$\dfrac{y \cdot \dfrac{d}{2}}{k_x^2} + \dfrac{x \cdot \dfrac{b}{2}}{k_y^2} = +1, \quad \text{or} \quad \dfrac{y \cdot d}{2k_x^2} + \dfrac{x \cdot b}{2k_y^2} = 1,$$

or $\qquad\qquad y = -\dfrac{k_x^2}{k_y^2} \cdot \dfrac{b}{d} \cdot x + \dfrac{2k_x^2}{d}$ ...(15.14)

This is the equation of a straight line and gives the limits of the deviation of the load line (*eccentricity*) from the centroid for *no change in the nature* of stress for an *I*–section. The central core for an *I*–section is shown in the Fig. 15.6.

The concept of eccentricity and combined stress due to eccentric loading shall be explained through variety of solved examples.

EXAMPLE 15.1: A short CI column of hollow circular section carries an eccentric compressive load of 200 kN. Find the maximum limit of eccentricity so that no tension develops any where in the section of the column, if the external and internal diameters are respectively 200 mm and 150 mm.

Solution: Maximum eccentricity = e_{max}

$$W = 200 \text{ kN}, A = \frac{\pi}{4}(200^2 - 150^2) = \frac{10000\pi}{4}(4 - 2.25) = 13744.5 \text{ mm}^2$$

$$I_x = \frac{\pi}{64}(200^4 - 150^4), Z_x = \frac{\pi(200^4 - 150^4)}{32 \times 200} = 536893 \text{ mm}^3.$$

$$p_0 = \frac{200 \times 1000}{13744.5} = 14.55 \text{ N/mm}^2 \text{ (comp)}.$$

$$p_b = \frac{W \cdot e}{Z} = \frac{2 \times 10^5 \cdot e}{536893} \text{ N/mm}^2$$

For no tension

$$p_0 - p_b \geq 0, \text{ i.e. } \frac{2 \times 10^5 \cdot e}{536893} \leq \frac{200 \times 1000}{13744.5}$$

$$e_{max} \leq \frac{2 \times 10^5 \times 536893}{2 \times 10^5 \times 13744.5} = \textbf{39.06 mm.}$$

EXAMPLE 15.2: A strut 1m long and 150 mm × 200 mm in cross-section is subjected to an eccentric load of 120 kN at an eccentricity of 40mm in a plane bisecting the thickness. Find the maximum and minimum stresses and its nature.

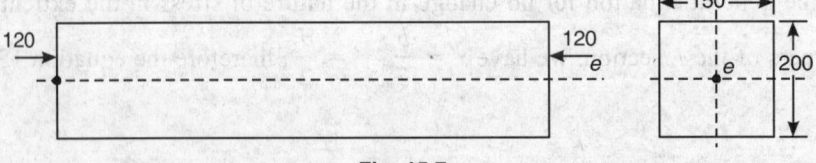

Fig. 15.7.

Solution: $W = 120000 \text{ N}$

$A = 150 \times 200 = 30000 \text{ mm}^2$

$M = W \cdot e = 120000 \times 40 = 48 \times 10^5 \text{ N–mm}$

$$I_X = \frac{150 \times 200^3}{12}, \quad Z_X = \frac{15 \times 8 \times 10^7}{12 \times 100} = 1 \times 10^6 \text{ mm}^3$$

$$p_0 = \frac{W}{A} = \frac{120000}{30000} = 4.0 \text{ N/mm}^2 \text{ (comp)}$$

$$p_b = \pm \frac{W \cdot e}{Z_x} = \frac{48 \times 10^5}{1 \times 10^6} = \pm \textbf{4.8 N/mm}^2 \text{ (comp. in top and tension in bottom fibres)}$$

p_{max} (top) = $p_0 + p_b$ = 4.0 + 4.8 = 8.8 N/mm^2 (comp.)

p_{min}(bottom) = $p_0 - p_b$ = 4.0 - 4.8 = -0.8 N/mm^2 (-ve means tensile).

EXAMPLE 15.3: A short column of 200 mm external diameter and 150 mm internal diameter when subjected to a load, stresses of 300 N/mm^2 (compressive) and 50 N/mm^2 (tensile) are developed across the two ends of the diameter. Determine the load and eccentricity of the load from the centroid.

Solution: D = 200mm, d = 150mm. $A = \dfrac{\pi}{4}(200^2 - 150^2) = \textbf{13744.47 mm}^2$

p_{max} = 300 N/mm^2 (comp), p_{min} = 50 N/mm^2 (tensile),

Load = W, eccentricity = e

$$Z = \frac{I}{y} = \frac{\pi \cdot 2}{64 \cdot 200}(200^4 - 150^4) = \frac{\pi \times 10^8}{6400}(16 - 5.0625) = 536893 \text{ mm}^3.$$

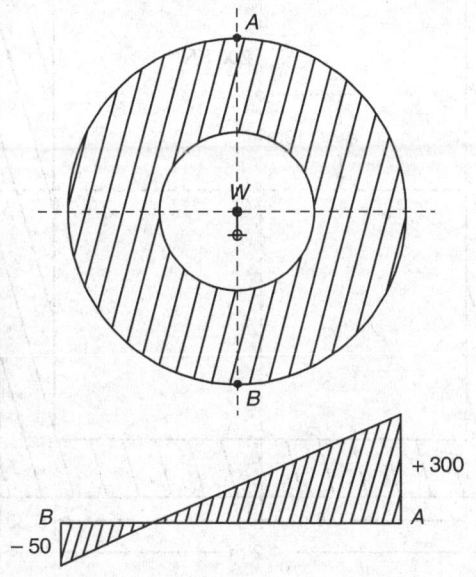

Fig. 15.8.

$$p_{max} = \frac{W}{A} + \frac{W \cdot e}{Z} = 300 \text{ N/mm}^2 \qquad \qquad \ldots \text{(i)}$$

$$p_{min} = \frac{W}{A} - \frac{W \cdot e}{Z} = -50 \text{ N/mm}^2 \qquad \qquad \ldots \text{(ii)}$$

Adding (i) and (ii) $\dfrac{2W}{A} = 250 \text{ N/mm}^2$.

$\therefore \qquad W = \dfrac{250 \times 13744.47}{2} \text{N} = 1718059\text{N} = 1718.059 \text{ kN}$

Subtracting (ii) from (i) we have

$$\frac{2W \cdot e}{Z} = 350, \text{ or } e = \frac{350Z}{2W} = \frac{350 \times 536893}{2 \times 1718059} = \textbf{54.687 mm}$$

EXAMPLE 15.4: A hollow rectangular masonry pier is 1.5m × 1.2m external and having wall thickness of 0.30m throughout. A vertical load 200 kN is transmitted in the vertical plane bisecting 1.2m side at an eccentricity of 0.10m from the geometric axis of the section.

Calculate the maximum and minimum stresses in the section.

Also check whether the pier is safe if maximum crushing stress is 200 kN/m².

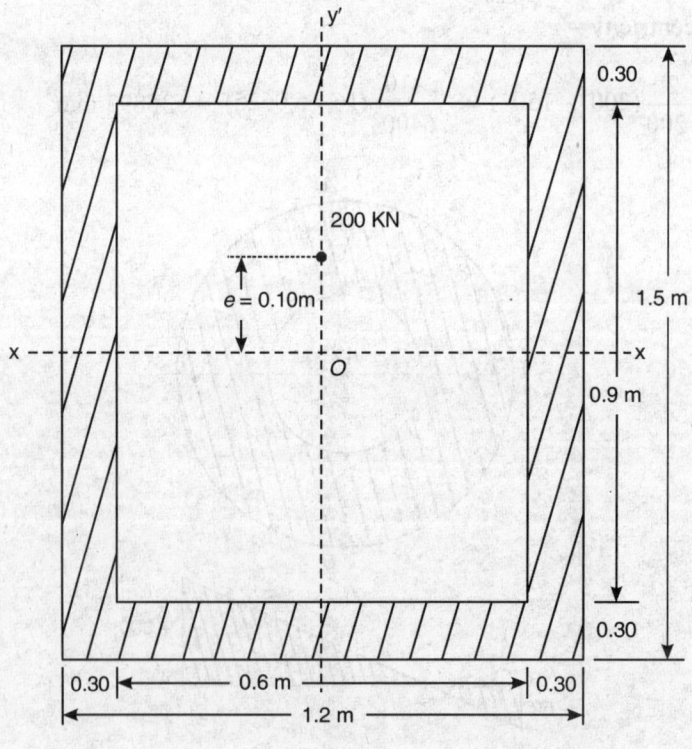

Fig. 15.9.

Solution:

$$A = [1.5 \times 1.2 - 0.90 \times 0.60] = 1.26 \text{m}^2, \ e = 0.10 \text{ m}$$

$$Z_x = \frac{I_x}{y} = \frac{1}{12} \frac{[1. \times 1.5^3 - 0.60 \times 0.90^3]}{1.5/2} = (0.3375 - 0.03645)\frac{2}{1.5} = 0.4014 \text{ m}^3.$$

$$p_o = \frac{200}{1.26} = 158.73 \text{ kN/m}^2$$

$$p_b = \pm\frac{W \cdot e}{Z_x} = \pm\frac{200 \times 0.10}{0.4014} = 49.826 \text{ kN/m}^2$$

$$p_{max} = 158.73 + 49.826 = 208.556 \text{ kN/m}^2, \text{ (comp)}$$
$$p_{min} = 158.73 - 49.826 = 108.904 \text{ kN/m}^2, \text{ (comp)}$$

Since crushing stress permitted is 200 kN/m² and the actual maximum compressive stress is 208.556 kN/m², the pier is not safe.

EXAMPLE 15.5: A steel bar of rectangular section 150 mm × 20 mm carries a tensile load (pull) of 210 kN. The load acts on a plane bisecting thickness and at a distance of 3.0 mm from the centroid of the section. Determine the maximum and minimum stresses.

Solution:

$$A = 150 \times 20 = 3000 \text{ mm}^2, \ e = 3 \text{ mm}, \ W = 210 \text{ kN}$$

$$I_{yy} = \frac{20}{12} \times 150^3 = 5625000 \text{ mm}^4, \ x = 75 \text{ mm}$$

$$Z_y = \frac{I}{X} = \frac{5625 \times 10^3}{75} = 75 \times 10^3 \text{ mm}^3.$$

$$p_o = \frac{210 \times 1000}{3000} = 70 \text{ N/mm}^2 \text{ (tensile)}$$

$$p_b = \frac{W \cdot e}{Z} = \frac{210 \times 3 \times 10^3}{75 \times 10^3} = 8.4 \text{ N/mm}^2 \text{ (tensile and comp.)}$$

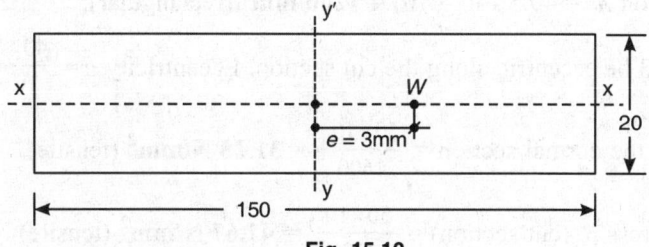

Fig. 15.10.

$$p_{max} = 70 + 8.4 = 78.4 \text{ N/mm}^2 \text{ (tensile)}$$
$$p_{min} = 70 - 8.4 = 61.6 \text{ N/mm}^2 \text{ (tensile).}$$

EXAMPLE 15.6: A rod of square cross-section carries an axial pull of 50 kN. The cross-section has 40 mm side and has a cut 10 mm deep and 40 mm wide at a section PQ as shown in Fig. 15.11. Find the maximum and minimum stresses.

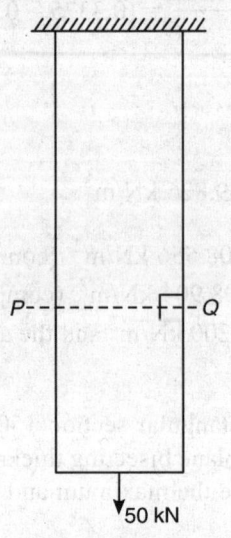

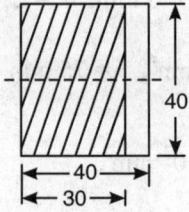

Fig. 15.11.

Solution:

W = 50 kN, along the axis 20 mm from the face

Normal cross – section, $A = 40 \times 40 = \mathbf{1600\ mm^2}$ (square).

Cut section $A' = 40 \times (40 - 10) = \mathbf{1200\ mm^2}$ (rectangular).

Load will be eccentric along the cut section. Eccentricity $e = \dfrac{40}{2} - \dfrac{30}{2} = \mathbf{5\ mm}$

Stress at the normal section $= \dfrac{50 \times 10^3}{1600} = \mathbf{31.25\ N/mm^2}$ (tensile)

Direct stress p_o (cut section) $= \dfrac{50 \times 10^3}{1200} = \mathbf{41.67\ N/mm^2}$ (tensile)

I_y of the cut section $\dfrac{40 \times 30^3}{12} = \mathbf{9 \times 10^4\ mm^4}$.

p_b (bending stress along the edges) $= \dfrac{W \cdot e}{I_y} \cdot x = \dfrac{50 \times 10^3 \times 5}{9 \times 10} \times \dfrac{30}{2}$ N/m^2

$p_b = \pm$ **41.67 N/mm^2**

$p_{max} = 41.67 + 41.67 =$ **83.33 N/mm^2** (tensile)

$p_{min} = 41.67 - 41.67 = 0$.

EXAMPLE 15.7: A rod is formed by connecting two flats each 120 mm $\times$ 20 mm with a T – section 120 mm $\times$ 120 mm $\times$ 20 mm. The rod carries a pull of 24200 N along the centroid of the flat as shown in Fig. 15.12. Determine the maximum stress any where in the rod.

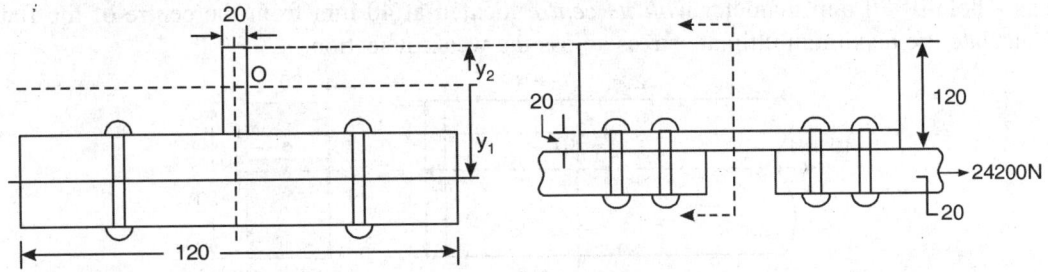

Fig. 15.12: (not to scale)

Solution:

W = 24200 N

A, $(T) = 120 \times 20 + 100 \times 20 = 4400$ mm^2.

c g o of T–section

$\bar{y} = \dfrac{2400 \times 60 + 2000 \times 10}{4400} = \dfrac{410}{11}$ mm (from flange edge y_1)

y_2 (from web edge) $= 120 - \dfrac{410}{11} = \dfrac{910}{11}$ mm.

Eccentricity of load from the cg of T–section to cg of flat

$e = \dfrac{410}{11} + \dfrac{20}{2} = \dfrac{520}{11}$ mm.

BM about NA $= 24200 \times \dfrac{520}{11} = 10^3 \times 1144$ N–mm.

MI about the NA $= \dfrac{100 \times 20^3}{3} + \dfrac{120^3 \times 20}{3} - 4400 \left(\dfrac{410}{11} \right)^2$

$I_X = 10^5 \left(\dfrac{8}{3} + 115.2 \right) - 611.27273 \times 10^4 = 567.3943 \times 10^4$ mm^4.

p_o (direct stress) $= \dfrac{24200}{4400} = 5.5$ N/mm^2 (tensile)

$$p_{b_1} = +\frac{1144 \times 10^3}{567.394 \times 10^4} \times \frac{410}{11} = +7.515 \text{ N/mm}^2 \text{ (tensile)}$$

$$p_{b_2} = -\frac{1144 \times 10^3}{567.394 \times 10^4} \times \frac{910}{11} = -16.68 \text{ N/mm}^2 \text{ (comp)}$$

$$p_{max} = p_o + p_{b_1} = 5.5 + 7.515 = 13.015 \text{ N/mm}^2 \text{ (tensile)},$$

$$p_{min} = 5.5 - 16.68 = 11.18 \text{ (comp) N/mm}^2.$$

EXAMPLE 15.8: A steel flat of 120 mm × 20 mm carries a tensile force of 100 kN. The flat has a hole of 20 mm diameter *with its centre* located at 40 mm from the centre of the flat. Calculate the maximum ultimate stress across the weakest section.

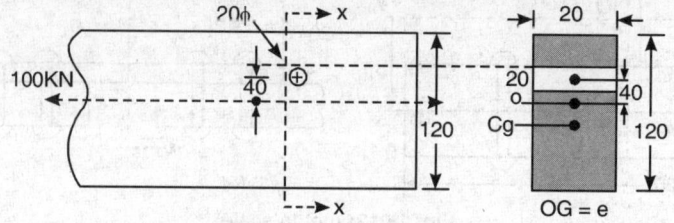

Fig. 15.13: (not to scale).

Solution:

Load = 100 kN,

Minimum 'A' = 120 × 20 – 20 × 20 = 2000 mm^2

cg of the section from the bottom at $X-X$

$$\bar{y} = \frac{2400 \times 60 - 20 \times 20 \ (60 + 40)}{2000} = 52 \text{ mm}$$

$y_1 = 68$ mm, $y_2 = 52$ mm, $e = 60 - 52 = 8$mm

$$p_o \text{ (direct)} = \frac{100 \times 1000}{2000} = 50 \text{ N/mm}^2 \text{ (tensile)}$$

$$p_{b_1} = \frac{100 \times 1000 \times 8}{I_X} \cdot y_1, \ p_{b_2} = \frac{100 \times 1000 \times 8}{I_X} y_2$$

$$I_X \text{(centroid)} = \left\{ \frac{1}{12} \times 20 \times 120^3 + 2400 \ (8)^2 \right\} - \left\{ \frac{20 \times 20^3}{12} + 20 \times 20(48)^2 \right\}$$

$$= 10^4[288 + 15.36 - (1.333 + 92.16)]$$

$I_X = \textbf{209.87} \times \textbf{10}^4 \textbf{ mm}^4$

$$p_{b_1} = \frac{10^5 \times 8 \times 68}{209.87 \times 10^4} = +25.92 \text{ N/mm}^2, \ p_{b_2} = -\frac{10^5 \times 8 \times 52}{209.87 \times 10^4} = \textbf{–19.822 N/mm}^2.$$

$p_{max} = 50 + 25.92 = \textbf{75.92 N/mm}^2 \text{ (tensile)}.$

$p_{min} = 50 - 19.822 = \textbf{30.178 N/mm}^2 \text{ (tensile)}.$

EXAMPLE 15.9: A 2m long horizontal strut carrying an axial load of 10 kN in addition to a transverse load of 20 kN/m over the entire span is simply supported. The strut comprises of rolled steel angle section of 120 mm × 120 mm × 20 mm with 120 mm leg placed horizontally in top face.

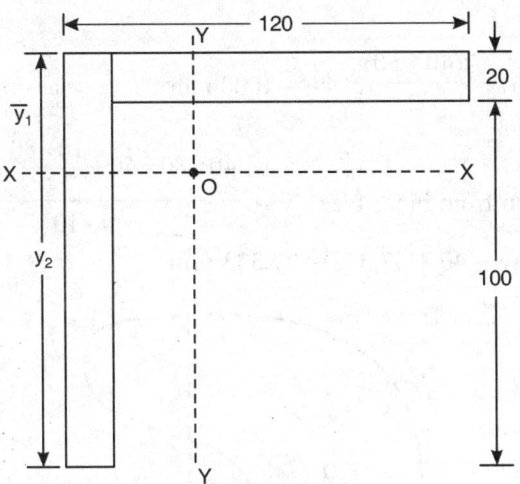

Fig. 15.14.

$A = 120 \times 20 + 100 \times 20 = 4400 \text{ mm}^2$.

$$\bar{y} = \frac{120 \times 20 \times 60 + 100 \times 20 \times 10}{4400} = \frac{410}{11} \text{ mm}$$

$$I_{XX} = \frac{20}{3} \times 120^3 + \frac{100 \times 20^3}{3} - 4400 \left(\frac{410}{11}\right)^2 = 567.4 \times 10^4 \text{ mm}^4$$

Max. mom. due to transverse loading at midsqan point.

$$M = \frac{20 \times 2 \times 2}{8} = \text{kN–m} = \mathbf{10 \times 10^6 \text{ N–mm}.}$$

$P = \mathbf{10 \times 10^3 \text{ N}.}$

$$p_o \text{ (Direct)} = \frac{10^4}{4400} = \mathbf{2.273 \text{ N/mm}^2 \text{ (comp.)}}$$

$$p_{b_1} \text{ (comp – top)} = \frac{10^7}{567.4 \times 10^4} \times \frac{410}{11} = \mathbf{+ 65.69 \text{ N/mm}^2 \text{ (comp)}.}$$

$$p_{b_2} \text{ (bottom)} = \frac{10^7}{567.4 \times 10^4} \times \left(120 - \frac{410}{11}\right) = \frac{10^7 \times 910}{567.4 \times 10^4 \times 11} = \mathbf{-145.80 \text{ N/mm}^2}$$

Max. stress (top) = 2.273 + 65.69 = **67.963 N/mm²** (comp).

Min. stress (bottom) = 2.273 – 145.80 = **–143.53 N/mm²** (tensile)

EXAMPLE 15.10: A crane hook cross-section is trapezoidal with internal breadth 60 mm and external breadth 40 mm and depth 80 mm. If the line of action of the load is 40 mm from the inner face, find the maximum load which can be safely lifted. The permissible tensile stress of the material is 50 N/mm^2.

Solution:

$$A \text{ (Hook section)} = \frac{(40+60)}{2} \times 80 = 4000 \text{ mm}^3.$$

$$cg \text{ (Hook section from inner face) } \bar{y} = \frac{40 \times 80 \times 40 + \dfrac{10 \times 80}{2} \times \dfrac{80}{3} \times 2}{4 \times 10^3} = 37.333 \text{ mm}$$

Total eccentricity = 40 + 37.333 = **77.333 mm**

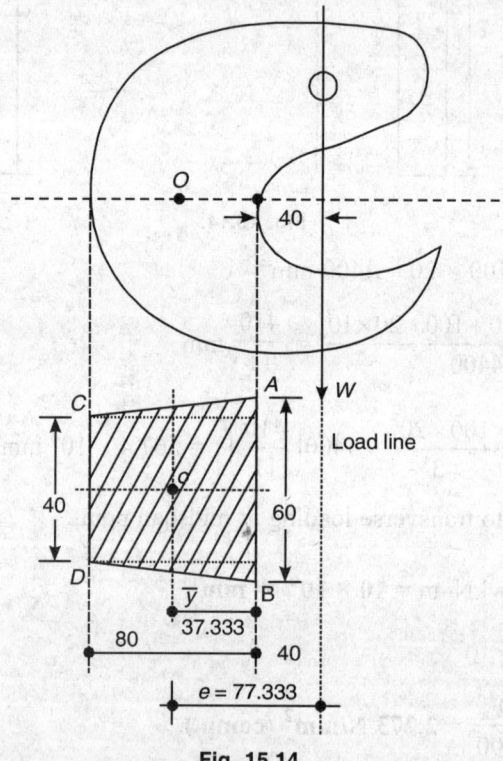

Fig. 15.14.

$$I_X \text{(centroid)} = \frac{40 \times 80^3}{3} + \frac{20 \times 80^3}{12} - 4000(37.333)^2 = \mathbf{210.5 \times 10^4 \text{ mm}^4}.$$

Bending stress p_{b_1} (inner) $= \dfrac{W. \ e. \ y_1}{I_X}$

$$p_{b_1} = \frac{W.77.333}{210.5 \times 10^4} \times 37.333 = \frac{13.715 \ W}{10^4} \text{ N/mm}^2 \textbf{(tensile)}$$

$$p_{b_2} = \frac{W \times 77.333}{210.5 \times 10^4} \times (80 - 37.333) = \frac{15.675}{10^4} \frac{W}{} \text{ N/mm}^2 \text{ (comp).}$$

$$p_0 \text{ (Direct)} = \frac{W}{A} = \frac{W}{4000} = \frac{2.5}{10^4} \frac{W}{} \text{ N/mm}^2 \text{ (tensile)}$$

$$\text{Max. tensile} = \frac{2.5}{10^4} \frac{W}{} + \frac{13.715}{10^4} \frac{W}{} = \frac{16.215}{10^4} \frac{W}{} \text{ N/mm}^2$$

$$\text{Thus max. permissible stress } \frac{16.215}{10^4} \frac{W}{} \leq 50$$

$$\therefore \quad W \leq \frac{50 \times 10^4}{16.215} N \leq \textbf{30.835 kN}$$

EXAMPLE 15.11: A hollow cast iron circular column section has 400 mm outer diameter and 300 mm inner diameter. The inner bore is eccentric which gives maximum *wall thickness* of 60 mm and minimum wall thickness of 40 mm. If a load of 1540 kN acts along the central axis of bore, find maximum and minimum stresses in the column section.

Solution:

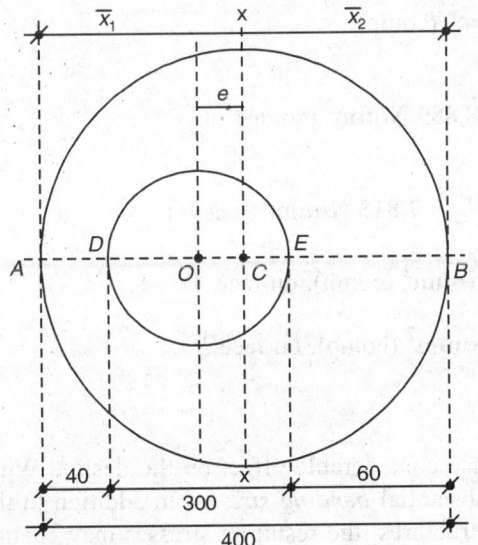

$CO = e$
C–Centroid of the Area
O–Centre of bore and loading point
AB–Diameter of outer circle (400 mm)
DE–Diameter of bore (300 mm)
AD–40 mm
EB–60 mm

Fig. 15.16: (not to scale)

$$A \text{ (Solid)} = \frac{\pi}{4}(400^2 - 300^2) = 5.5 \times 10^4 \text{ mm}^2$$

Let cg be $\bar{x}_1$ from the thinner section side, A

$$\bar{x}_1 = \frac{\frac{\pi}{4}(400)^2 \times 200 - \frac{\pi}{4}(300^2)\left(40 + \frac{300}{2}\right)}{\frac{\pi}{4}(400^2 - 300^2)} = \frac{10^5(320 - 171)}{10^4(7)} = \frac{1490}{7}$$

$$x_2 = \left(400 - \frac{1490}{7}\right) = \frac{1310}{7} \text{ mm (from } B)$$

Load is at the centre of bore O and hence

$$e = \left(40 + \frac{300}{2}\right) - \frac{1490}{7} = -\frac{160}{7} = 22.857 \text{ mm i.e. load is towards } A.$$

I_x (about centroide C)

$$= \frac{\pi}{64}(400)^4 + \frac{\pi}{4}(400)^2\left(\frac{1490}{7} - 200\right)^2 - \left[\frac{\pi}{64}(300)^4 + \frac{\pi}{4}(300)^2\left(\frac{1490}{7} - 190\right)^2\right]$$

$$I_x = \frac{\pi \times 10^4}{4}\left[\frac{256000}{16} + \frac{16(90)^2}{49} - \frac{81 \times 10^4}{16} - 9 \times \frac{160 \times 160}{7 \times 7}\right] = 84288 \times 10^4 \text{ mm}^4$$

$$p_0 \text{ (Direct)} = \frac{W}{A} = \frac{1540 \times 1000}{5.5 \times 10^4} = 28 \text{ N/mm}^2 \text{ (comp)}.$$

$$p_{b_1} \text{ (face } A) = \frac{1540 \times 160 \times 10^3}{84288 \times 10^4 \times 7} \times \frac{1490}{7} = \textbf{8.889 N/mm}^2 \text{ (comp)}$$

$$p_{b_2} \text{ (face } B) = \frac{1540 \times 160}{84288 \times 10^4 \times 7} \times \frac{1310}{7} \times 10^3 = \textbf{7.815 N/mm}^2 \text{ (tensile)}$$

$$p_{max} = p_0 + p_{b_1} = 28 + 8.889 = \textbf{36.889 N/mm}^2 \text{ (comp), on face } A$$

$$p_{min} = p_0 - p_{b_2} = 28 - 7.815 = \textbf{20.185 N/mm}^2 \text{ (comp), on face } B$$

15.5 WIND LOADS

In case of *tall structures*, the *wind pressures* cause considerable effect on the design. Wind pressures cause transverse loading resulting in substantial *bending stresses* in addition to the *axial gravitational* stresses. In case of very tall structures, the resulting stresses may change even the nature of stresses in extreme fibres. Chimney stack, walls, water tanks shafts, etc. fall under tall structures and subjected to primarily wind loading.

"Consider a chimney stack of height 'h' cross-sectional breadth 'B', depth 'D' and cross-sectional area 'A'. If density of the stack material is 'g', weight of stack 'W' = g. Ah on the base. Axial gravitational stress 'p_0' =

$$\frac{\text{Total load}}{\text{Area}} = \frac{\gamma \cdot A \cdot h}{A} = \gamma h \qquad \qquad \dots \text{(i)}$$

Consider breadth '*B*' facing the wind pressure '*p*',
Total wind pressure on the height '*h*' of the stack

will be $P = p.(B.h)$. cg of wind load will be at $\dfrac{h}{2}$,

and hence maximum *BM* at the base of stack =

$$p.(B .h).\frac{h}{2} = \frac{pBh^2}{2}$$

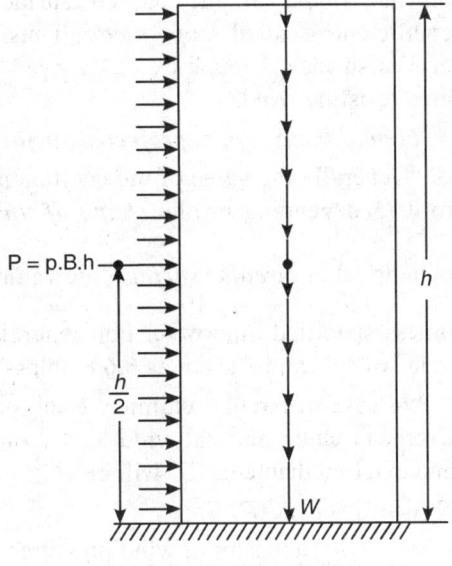

BM due to wind pressure causes bending about
$Y - Y$. Moment of inertia about $Y - Y$ for rectangular
section *B.D*:

$$I_{YY} = \frac{1}{12} B.D^3$$

Bending stress $p_b = \dfrac{M}{I} \cdot \dfrac{D}{2}$

or $\quad p_b = \pm \dfrac{p \cdot Bh^2 \cdot D/2}{2 \times \dfrac{B}{12}D^3}$

$$= \pm \frac{p \cdot 12 \cdot h^2}{4 \times D^2} = \pm \frac{3 p \cdot h^2}{D^2} \qquad \text{... (ii)}$$

Max. stress at the base

$$p_{max} = p_0 + p_b = \left(\gamma h + \frac{3 \; p \cdot h^2}{D^2} \right) \qquad \text{... (iii)}$$

$$p_{min} = p_0 - p_b = \left(\gamma h - \frac{3 ph^2}{D^2} \right) \qquad \text{... (iv)}$$

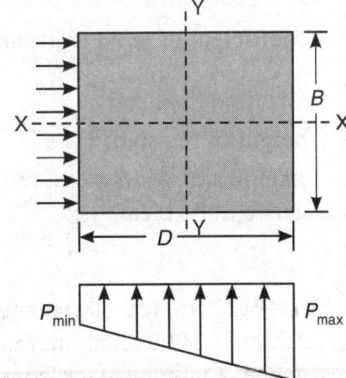

Fig. 15.17: Chimney stack

The minimum stress can be of opposite nature (i.e. *tensile*) if the bending stress $(3\,ph^2/D^2)$
is greater than direct gravitational stress (γh). In design of masonry stacks, tensile stresses are
avoided for safety.

15.6 COEFFICIENT OF WIND RESISTANCE

The wind pressure acts normally to the surface on which it acts. When the stack surface is
curved (or circular), the *projected vertical surface* area is considered for the calculation of the
total wind pressure. The *total wind* load on a *curved surface* is actually *lessthan* the total wind
load on the *flat surface*. The *ratio of total wind resistance* on a *curved surface* of the same
projected surface area as the flat surface to the *total wind resistance on the flat surface* is
known as the *coefficient of wind* resistance (c). This coefficient (c) of wind resistance depends

on the *shape* of the curved surface. Some times this coefficient is called *shape factor*. Thus, the *wind load* on the curved surface is found by $P_{wind} = p \times c \times$ projected surface area resisting wind.

P(wind load) = *p. c. projected surface area* ... (15.15)

"Generally the value of the coefficient '*c*' varies from **0.50** to **0.75** depending on the **shape of the curved** surface. For cylindrical or **circular surface**, the value of '*c*' is taken as $\dfrac{2}{3}$, unless specified otherwise. For general curved surfaces, the value of '*c*' can be taken as **0.60**, unless specified otherwise.

"In case of circular chimney stack or cylindrical shaft for overhead tank, the total wind load on the shaft of height '*h*' and external diameter '*D*' will be.

P (wind) = *p. c.* (*D. h*) ... (15.16)

where p = intensity of wind pressure

c = coefficient of wind resistance = $\dfrac{2}{3}$

D = external diameter

h = height of the shaft

Examples of wind pressure will make the concept of wind resistance quite clear.

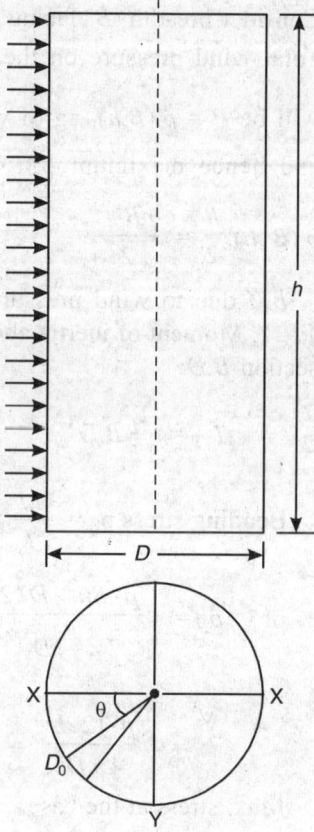

Fig. 15.18: Circular stack

EXAMPLE 15.12: A rectangular chimney stack 3000 mm × 2000 mm externally and 2000 mm × 1000 mm internally is 20 m high. Wind blows with 3000 mm side offering resistance. Chimney stack consists of brick masonry having unit weight of 21 KN/m³. If the wind pressure is 1200 N/m², find the maximum and minimum stresses in the brick masonry. Also find the maximum and minimum stresses if the wind starts blowing at 90°.

Solution:

$A = (3000 \times 2000 - 2000 \times 1000) = 4 \times 10^6$ mm²

Direct stress due to weight of chimney $= \dfrac{21000 \times A \times h}{A} = 42 \times 10^4$ N/m² (420 kN/m²)

Wind load $P_W = 1200 \times 3 \times 20 = \mathbf{72 \times 10^3}$ N,

BM due to wind load $= 72 \times 10^3 \times \dfrac{20}{2} = 72 \times 10^4$ N–m

I (About the axis of bending) $= \dfrac{1}{12}(3000 \times 2000^3 - 2000 \times 1000^3) = \dfrac{22}{12} \times 10^{12}$ mm^4

$$= \dfrac{22}{12} \text{ m}^4$$

Max. bending stress $f_b = \dfrac{72 \times 10^4}{\dfrac{22}{12}} \times \dfrac{2.0}{2} = \pm \dfrac{72 \times 12 \times 10^4}{22}$ N/m$^2 = \pm$ 392.73 kN/m^2.

Max. stress $= f_0 + f_b = 420 + 392.73 =$ **812.73 kN/m^2** (comp).
Min. stress $= f_0 - f_b = 420 - 392.73 =$ **27.27 kN/m^2** (comp).

EXAMPLE 15.13: A circular chimney stack has 3m external diameter and 20m height and is subjected to a wind pressure of 1200 N/m^2. Find the internal diameter of the chimney stack if *no tension* has to develop in the base. Unit weight of masonry = 21 kN/m^3. The coefficient of wind resistance may be taken as 0.60.

Solution:
Axial stress due to weight $f_0 = 21 \times 20 = 420$ kN/m^2.

If internal diameter is 'd', $Z = \dfrac{\pi}{32D}(D^4 - d^4) = \dfrac{\pi}{32 \times 3}(3^4 - d^4) = \dfrac{(81 - d^4)}{96} \pi$ m^4

Wind load $P_w = 0.60 \times 3 \times 20 \times 1200$N = **43.2 kN**, B.M. $= 43.2 \times \dfrac{20}{2} =$ **432 kN–m**

For no tension $f_b \le f_0$ or $\dfrac{432 \times 96}{(81 - d^4)\pi} \le 420$

or $\quad (81 - d^4) \ge \dfrac{432 \times 96}{\pi \times 420} \ge 31.4308$, or d$^4 \le (81 - 31.4308) = 49.5692$

$d \le 2.6534$ m, say $d = 2.65$ m.

EXAMPLE 15.14: 4 conical chimney of 25 m *height* has base diameter of 3 m and top diameter of 1.50 m. The wind pressure at any height is proportional to *square root* of height from the base. If the wind pressure is 1200 N/m^2 at a height of 16m from the base, find the bending moment at the base due to the wind pressure on the whole chimney. The coefficient of wind resistance $c = 0.60$.

Solution:

$$H = 25 \text{ m}, D_1 = 3.0 \text{ m}, D_2 = 1.50 \text{ m}, D_x = 1.5 + \dfrac{1.5 \ (25 - x)}{25}$$

$$p_x = k\sqrt{x} \text{ , at } x = 16 \text{ m}, p_{16} = 1200 \text{ N/m}^2$$

$$p_{16} = 1200 = k\sqrt{16} = k(4), k = 300$$

$\therefore \quad p_x = 300\sqrt{x}$ $\hfill$ ... (i)

$$Dx = 1.5 + \frac{1.5}{25}(25 - x) = \frac{75 + 3\ (25 - x)}{50} = \frac{150 - 3x}{50}$$

$$Dx = \frac{3}{50}(50 - x) \qquad \qquad \text{... (ii)}$$

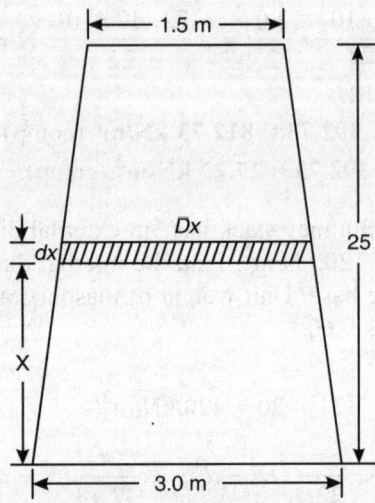

Fig. 15.19: Conical stack

Consider a small strip of dx height.

Total wind force on dx:

$$dp = \frac{3}{50}(50 - x) \cdot c \cdot dx \cdot p_x = 0.60 \times 300\sqrt{x} \cdot \frac{3(50 - x)}{50} \cdot dx$$

Moment about base $dM_x = x \cdot dp = 0.60 \times 300\sqrt{x} \cdot \frac{3(50 - x)}{50} \cdot dx \cdot x$

$$dM_x = \frac{180 \times 3}{50}(50\sqrt{x} - x\sqrt{x}) \cdot x \ dx = \frac{54}{5}\left(50 \cdot x^{\frac{3}{2}} - x^{5/2}\right) \cdot dx$$

Total moment due to total wind pressure on the whole chimney stack

$$M = \int_0^{25} dM_x = \int_0^{25} \frac{54}{5}\left(50.x^{\frac{3}{2}} - x^{5/2}\right) \cdot dx = \frac{54}{5}\left[\frac{50 \cdot x^{\frac{5}{2}}}{5/2} - \frac{x^{\frac{7}{2}}}{7/2}\right]_0^{25}$$

or $\quad M = \frac{54}{5}\left[20\ \left(25^{\frac{5}{2}}\right) - \frac{2}{7}\ \left(25^{\frac{7}{2}}\right)\right] = \frac{54}{5}\left[20\ (3125) - \frac{2}{7}(78125)\right]$

$$= \frac{54}{5}[62500 - 22321.43] = 433928.5 \text{ N-m}$$

Moment at the base due to wind pressure on the chimney stack = **433.9285 kN–m**.

15.7 WATER AND EARTH RETAINING STRUCTURES

Dams and retaining walls are subjected to lateral pressures caused by water or earth. Lateral pressures cause bending moment about the base in addition to direct stresses caused by the weight of the structure. Thus dams and retaining walls are examples of combined bending and direct stresses.

Criteria of safe design of dams or retaining walls are based on safety against:

(i)   **Overturning** due to lateral forces (water pressure or earth pressure),

(ii) **Sliding due to horizontal forces** and frictional resistance at the base,

(iii) **Crushing** due to excessive **compressive stresses** caused by combined forces,

(iv) **Excessive tension** due to combined effect of forces.

The safety against *overturning* is dependent on *resisting moment* and overturning moment about the base.

$$FS = \frac{\sum \text{Resisting moment about toe}}{\sum \text{Overturnig moment about toe}}$$

The safety against sliding is dependent on the resultant *weight of* the structure, coefficient of friction and horizontal thrust caused by water pressure or soil pressure. Frictional resistance can be increased by benching the foundation base on the soil. The coefficient of static friction μ generally ranges from 0.65 to 0.75.

The safety against crushing *at the toe* depends on the maximum compressive stress caused by the combined forces of *thrust* (or pressure) and gravitational forces (weight of components) and eccentricity caused due to bending of the structure.

The safety against development of tension at the heel depends on the ratio of the eccentricity (e) and the base width of the structure (b). For *no tension* to develop in a rectangular base, eccentricity 'e' should not be more than $\frac{b}{6}$. i.e. for **no tension** $e \leq \frac{b}{6}$.

All these safety considerations will be explained by examples.

EXAMPLE 15.15: A triangular dam section has base width 'b' and top width *zero*, and height H. If water is filled upto top find the safe base width 'b' when the uplift is considered and notconsidered. Uplift intensity coefficient is 'k' and friction coefficient is 'μ'. Density of Dam structure $= \gamma \text{ kN/m}^3$, water density $= \omega \text{ kN/m}^3$.

Solution:

(i)   Total weight of dam/m $= \dfrac{b \cdot H}{2} \times 1 \times \gamma$

i.e. $W = \dfrac{1}{2}\, \gamma b H$, acting at $\dfrac{b}{3}$ from A.

(ii) Horizontal pressure

$$P = \frac{\omega H \times H}{2} = \frac{\omega H^2}{2}, \text{ acting at } \frac{H}{3} \text{ above (base)}$$

(iii) Uplift pressure on the base/m

$$U = \frac{k \cdot \omega H (b \times 1)}{2} = \frac{k}{2} \omega H b \text{ (average)}$$

cg at $\frac{b}{3}$ from the inner face A.

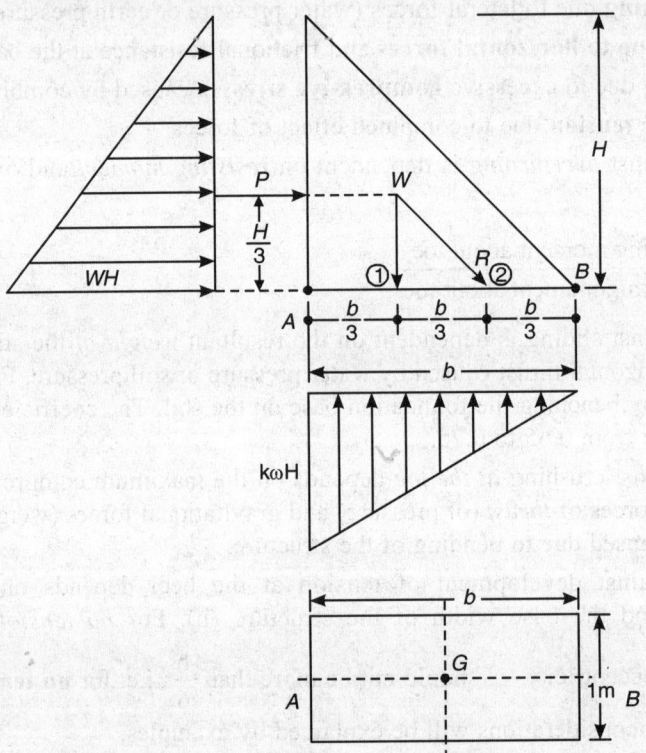

Fig. 15.20: Dam section, base and pressure diagram

(A) When dam is fully empty

i.e. $P = 0$

Weight of the dam $W = \dfrac{\gamma b \cdot H}{2}$ acts at $\dfrac{b}{3}$ from the inner face A. The load is eccentric on the

base and $e = \dfrac{b}{2} - \dfrac{b}{3} = \dfrac{b}{6}$.

Uplift $U = 0$.

$$p_0 = \text{(Direct stress) on the base} = \frac{\gamma}{2} \cdot \frac{b \ H}{(b \times 1)} = \frac{\gamma}{2} H \qquad \text{... (i)}$$

$$p_b = \text{(bending stress) on the base} = \frac{\frac{1}{2} \gamma b H}{\frac{1}{6} \cdot b^2} \times \frac{b}{6} = \frac{\gamma}{2} H \qquad \text{... (ii)}$$

$$p_{max} = \frac{\gamma H}{2} + \frac{\gamma}{2} \cdot H = \boldsymbol{\gamma H} \le \text{crushing strength.} \qquad \text{... (iii)}$$

$$p_{min} = \frac{\gamma H}{2} - \frac{\gamma H}{2} = \boldsymbol{0}, \text{ No tension} \qquad \text{... (iv)}$$

Since no lateral pressure acts, the dam is *safe in sliding* and *overturning*.

(B) When dam is full

i.e. $P = \dfrac{\omega H^2}{2}$, acting at $\dfrac{H}{3}$ above the base.

Uplift $U = \dfrac{k \cdot \omega H \cdot b \times 1}{2} = \dfrac{k b H \omega}{2}$ upward at $\dfrac{b}{3}$ from the inner face

Sliding Safety

Resistance to sliding $= \mu \ (W - U)$, where W = weight of dam

$$U = \text{uplift}$$

or $R = \mu \left(\dfrac{\gamma b H}{2} - \dfrac{k \omega b H}{2} \right) = \dfrac{\mu b H}{2} (\gamma - k \omega)$, where γ = density of dam material

ω = density of water, k = coeff. of uplift

For **safety against sliding**

$$\frac{\mu b H}{2} (\gamma - k \omega) \ge = \frac{\omega H^2}{2}$$

or $\mu b (\gamma - k \omega) \ge \omega H$

or $b \ge \dfrac{\omega H}{\mu (\gamma - k \omega)},$

neglecting uplift $b \ge \dfrac{\omega . H}{\mu \gamma} = \dfrac{H}{\mu \left(\dfrac{\gamma}{\omega} \right)} = \dfrac{H}{\mu \rho} \qquad \text{... (v)}$

where $\dfrac{\gamma}{\omega} = \rho$ = specific gravity of dam material (viz concrete or masonry)

For safety against overturning

Moment of horizontal pressure about *toe* of base should be *less than the moment caused by the weight and uplift* component of the resultant.

i.e. $\dfrac{\omega H^2}{2} \cdot \dfrac{H}{3} \le (W - U)\dfrac{b}{3}$

or $\dfrac{\omega H^3}{6} \le \left(\dfrac{\gamma b H}{2} \times 1 - \dfrac{k\omega b H}{2} \right)\dfrac{b}{3}$

or $\dfrac{\omega H^2}{6} \le \dfrac{b^2}{6}(\gamma - k\omega)$

or $b^2 \ge \dfrac{\omega H^2}{(\gamma - k\omega)}$ or $b^2 \ge \dfrac{H^2}{\left(\dfrac{\gamma}{\omega} - k \right)}$

or $b \ge \dfrac{H}{\sqrt{\dfrac{\gamma}{\omega} - k}}$, where $\dfrac{\gamma}{\omega}$ = specific gravity of dam material 'ρ'

or $\boldsymbol{b \ge \dfrac{H}{\sqrt{\rho - k}}}$

Neglecting uplift : $b \ge \dfrac{H}{\sqrt{\rho}}$ ... (vi)

Maximum and minimum stresses:

p_{max} to be less than crushing strength,

$p_{max/min} = \dfrac{\Sigma W}{b \times 1}\left(1 \pm \dfrac{6e}{b} \right)$, where e = eccentricity = $\dfrac{b}{6}$, max. and $\Sigma W = (W - U)$

$p_{min} = 0, p_{max} = \dfrac{2\Sigma W}{b \times 1}$, neglecting uplift, $\boldsymbol{p_{max} = \dfrac{2W}{b}}$ ... (vii)

Generally the profile of dam section is *trapezoidal* for providing roads on the top surface of the dam and for *resisting extra loads* of the traffic and machinery moved on the top surface. We have dealt with theoretical dam section as *triangular*. Rectangular dam section may be provided for very small heights, otherwise it becomes *uneconomical*. Trapezoidal sections also cover free board for retaining extra water heights during floods.

EXAMPLE 15.16: A rectangular masonry dam has base width of 4.5 m. Calculate the maximum height H when (*a*) *no tension* is allowed, and (*b*) the factor of safety against sliding is 2. Coefficient of friction μ = .6 sp, gravity of dam masonry = 2.4, density of water 10 kN/m^3 and coefficient of uplift c = 1. Also find the height if the *uplift is ignored*.

Solution:

$w = 10$ kN/m^3, Density of masonry = $2.4 \times 10 = 24$ kN/m^3.

Weight of the dam structure = $4.50 \times H \times 1 \times 24 = 108\,H$ kN

Uplift $U = c \cdot b \cdot \times \dfrac{10H}{2} = 5 \times 1 \times 4.5\,H = 22.5\,H$ kN

Water pressure (total) $P = \dfrac{\omega H^2}{2} = \dfrac{10H^2}{2} = 5\,H^2$ kN.

$\Sigma W = (108H - 22.5H) = 85.5\,H$ kN.

Moment about the toe $\Sigma M = 108H \times \dfrac{4.5}{2} - 22.5H \times \dfrac{2}{3} \times 4.5 - 5H^2 \times \dfrac{H}{3}$

$= 243H - 67.5H - 1.667H^3 = 175.5H - 1.667H^3$

Distance of resultant from the toe $= \dfrac{\Sigma M}{\Sigma W} = \dfrac{175.5H - 1.667H^3}{85.5H}$

$= (2.05263 - 0.01949H^2)$

eccentricity 'e' $= \dfrac{4.5}{2} - (2.05263 - 0.01949H^2) = (0.19737 + 0.01949H^2)$

For no tension $e \leq \dfrac{b}{6}$ or $e = \dfrac{4.5}{6} = 0.75$ m

i.e. $0.19737 + 0.01949H^2 = 0.75$, or $H^2 = 28.354541$, $H =$ **5.3249 m**

Max. Height $H =$ **5.32 m for no tension.**

For no sliding : $\mu\,(W - U) = 2P$ (factor of safety in sliding = 2)

or $0.60(108H - 22.5H) = 2 \times 5H^2$ or $51.3H = 10H^2$, $H =$ **5.13 m**

Thus safer H will be lesser of **5.13 m** and **5.32 m.**

∴ $H =$ **5.13 M**

Thus safe H for safety against sliding and no tension $H = 5.13$ m

(b) Uplift ignored

Moment about toe: $\Sigma M = 108H \times \dfrac{4.5}{2} - \dfrac{5H^3}{3} = H\left(243 - \dfrac{5H^2}{3}\right)$

$\Sigma W = 108H, \quad \bar{x} = \dfrac{\Sigma M}{\Sigma W} = \dfrac{H\left(243 - \dfrac{5H^2}{3}\right)}{108H} = \left(2.25 - \dfrac{5H^2}{324}\right)$

$$\text{Eccentricity } e = \frac{4.5}{2} - \overline{x} = 0 + \frac{5H^2}{324} = \frac{5H^2}{324}$$

$$\text{For no tension : } e \leq \frac{b}{6} \text{ i.e. } \frac{5H^2}{324} \leq \frac{4.5}{6}$$

or $\quad H^2 \leq \dfrac{4.5}{6} \times \dfrac{324}{5} = 48.6, \; \therefore \; \textbf{\textit{H}} \boldsymbol{\leq} \textbf{6.9714m}$

For stability in sliding

$\quad \mu \, W \geq 2P, \qquad\qquad \text{or } 0.60 \times 108H \geq 2 \times 5H^2$

or $\quad H \leq \dfrac{0.60 \times 108}{10} = \textbf{6.48m}$

Thus safer H will be lesser of 6.48 and 6.9714, i.e. $\textbf{\textit{H}} \boldsymbol{=} \textbf{6.48m}$

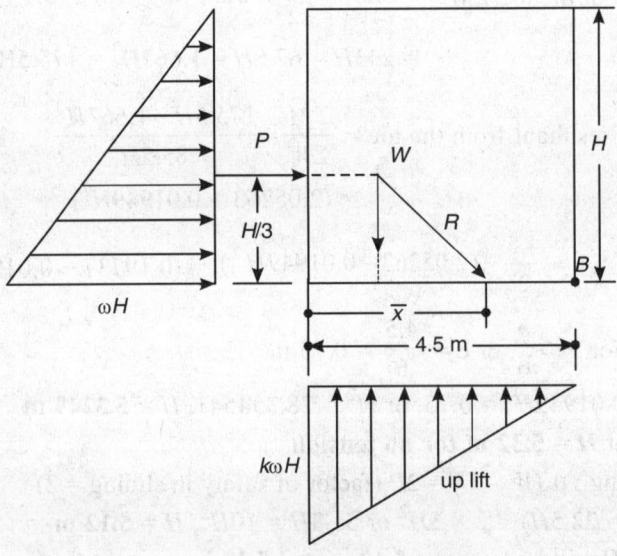

Fig. 15.21: Rectangular dam section

EXAMPLE 15.17: A trapezoidal dam section has 10 m height, top width 3 m and bottom width 6 m. Water face of the dam is vertical and masonry density is 21 kN/m³ while density of water is 10 kN/m³. Determine the *maximum* and *minimum* stresses on the base when the dam is *filled up to* 9 m *height* and when the dam has *no water*. Neglect uplift pressure.

Solution:

$\quad H = 10$ m, h(water $= 9$ m), $b_1 = 3$ m, $b_2 = 6$ m, γ (masonry) $= 21$ kN/m³, $w = 10$ kN/m³, *water filled for* 9 m (consider 1m length of the dam).

$$P = \frac{1}{2}wH^2 = \frac{10}{2}(9)^2 \times 1 = 405 \text{ kN acting at 3 m from the base.}$$

$$W = \left(\frac{3+6}{2}\right)10 \times 1 \times 21 = 945 \text{ kN}$$

$$\bar{x}_1 \text{ (from the vertical water face)} = \frac{3 \times 10 \frac{3}{2} + \frac{(6-3)10}{2}\left(3 + \frac{3}{3}\right)}{\left\{3 \times 10 + \frac{3}{2} \times 10\right\}} = \frac{105}{45} = \frac{7}{3}\text{ m}$$

Let the resultant of weight and water pressure (P) strikes at $\bar{x}_r$ from the vertical face, taking moment at the point of strike on the base,

we have $P.\dfrac{h}{3} = W(x_r - \bar{x}_1)$

or $405 \times \dfrac{9}{3} = 945\left(x_r - \dfrac{7}{3}\right) = x_r = \dfrac{405 \times 3}{945} + \dfrac{7}{3} = \dfrac{9}{7} + \dfrac{7}{3} = \dfrac{76}{21} = 3.619$ m

Eccentricity 'e' on the base $= 3.619 - 3.0 = 0.619$m
Thus moment due to eccentricity 'e' (0.619m) $= 945 \times 0.619 = 584.955$ kN–m

$$I_y = (MI) \text{ about the axis of bending} = \frac{1}{12} \times 1 \times 6^3 = 18\text{m}^4$$

$$p_b = \text{(bending stresses)} = \frac{M}{I} \cdot x = \pm\frac{584.955}{18} \times \frac{6}{2} = \textbf{97.4925 kN/m}^2$$

$$p_0 = \text{(Direct axial)} = \frac{W}{1 \times b} = \frac{945}{1 \times 6} = 157.5 \text{ kN/m}^2 \text{ (comp)}$$

$p_{max} = 157.5 + 97.4925 = \textbf{254.9925 kN/m}^2$ (comp) toe end
$p_{min} = 157.5 - 97.4925 = \textbf{60.0075 kN/m}^2$ (comp) at heel end.

(b) When the dam is totally empty

$$W = 945 \text{ kN}, \bar{x} \text{ (from vertical face)} = \frac{7}{3}m, e = 3 - \frac{7}{3} = \frac{2}{3}m, P = 0$$

$$p_b = \pm\frac{M}{I_y} \cdot x = \frac{945 \times \frac{2}{3} \cdot \left(\frac{6}{2}\right)}{18} = \textbf{105 kN/m}^2, p_0 = \textbf{157.5 kN/m}^2.$$

$p_{max} = 157.5 + 105 = 262.5 \text{ kN/m}^2$ (comp) vertical face side (Heel)
$p_{min} = 157.5 - 105 = 52.5 \text{ kN/m}^2$ (comp) at toe end.

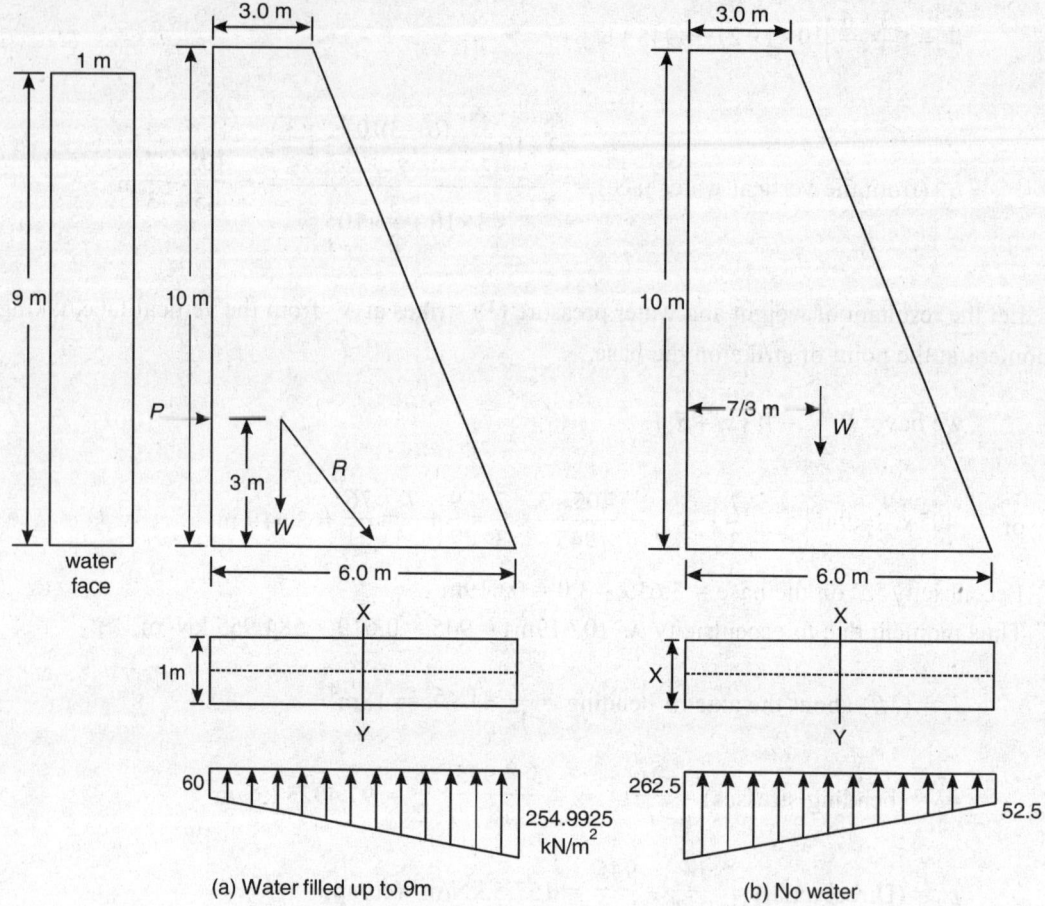

Fig. 15.21: Dam section and stresses on the base (not to scale).

EXAMPLE 15.18: A retaining wall is trapezoidal in cross-section and 10 m high. One side of the wall is vertical and has top width 3 m, *while bottom width 6 m. The wall retains earth upto the top of vertical face. The angle of repose of earth is 30° and the density of earth is* 16.2 kN/m³. The masonry has density of 21 kN/m³. Determine the maximum and minimum stresses on the base.

Solution:

$H = 10$ m, $b_1 = 3$ m, $b_2 = 6$ m, $\theta = 30°$, density, of earth $= 16.2$ kN/m³. density of masonry $= 21.0$ kN/m³.

Weight of retaining wall W/m Length $= \dfrac{(3+6)}{2} 10 \times 1 \times 21 = 945$ kN

cg of W from the vertical face $= \dfrac{b_1^2 + b_2^2 + b_1 b_2}{3(b_1 + b_2)} = \dfrac{(3^2 + 6^2 + 3 \times 6)}{3(3+6)} = \dfrac{7}{3}$ **m.**

$$c_p = \text{(coeff. of earth pressure)} = \frac{1-\sin\theta}{1+\sin\theta} = \frac{(1-\sin 30°)}{(1+\sin 30°)} = \frac{1-0.50}{1+0.50} = \frac{1}{3}$$

Total earth pressure $P = \dfrac{1}{2}c_p wh^2 = \dfrac{1}{2\times 3}\times 16.2\times 10^2 = \textbf{270 kN}$

cg of total earth pressure $= \dfrac{h}{3}$ from the base $= \dfrac{10}{3}$ m from the base.

Let the resultant of P and W strikes the base at a distance of $\bar{x}_2$

$$\bar{x}_2 = \frac{7}{3} + \frac{P\cdot h}{3W} = \frac{7}{3} + \frac{270\times 10}{3\times 945} = \frac{7}{3} + \frac{20}{21} = \textbf{3.2857 m}$$

$\therefore$ Eccentricity of vertical load $W = 3.2857 - \dfrac{6}{2} = 0.2857$ m

BM due to eccentricity $(M) = W\cdot e = 945\times 0.2857 = 270$ kN–m

MI of base for 1m length $I_y = \dfrac{1\times 6^3}{12} = 18\text{m}^4$

Bending stress at the base, $f_b\cdot\left(x = \dfrac{6}{2}m\right) = \dfrac{M}{I}\times 3 = \pm\dfrac{270\times 3}{18} = \pm\,\textbf{45 kN/m}^2$

Direct stress on the base, $f_0 = \dfrac{W}{6\times 1} = \dfrac{945}{6} = \textbf{157.5 kN/m}^2$.

Max. stress (toe) $= f_0 + f_b = 157.5 + 45 = 202.5$ kN/m^2 **(comp).**
Min. stress (heel) $= f_0 - f_b = 157.5 - 45 = 112.5$ kN/m^2 **(comp).**

EXAMPLE 15.19: A rectangular weir has 4.5 m height. Top 3 m height has thickness of 2.0 m. The thickness is increased by 't' on either side. Water can be filled on either side of wall. Find the thickness 't' so that no tension develops in the base of the wall. Also find the maximum stress in the base if Masonry density is 22.5 kN/m^3 and water density is 10.0 kN/m^3.

Solution:

Base width $(2 + 2t)$ m. Worst condtion of stress will be when one side water is full and other side is completely *empty*. Consider 1 m length of weir

Weight of weir

$$W_1 = (2\times 4.5\times 1 + 2t\times 1.5\times 1)22.5$$
$$= (9 + 3t)22.5$$
$$= \textbf{(202.5 + 67.5}t\textbf{)kN}$$

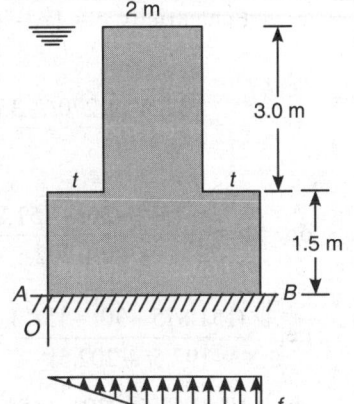

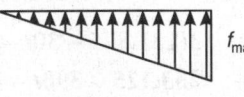

Fig. 15.22.

cg of $W_1 = \left(t + \dfrac{2}{2}\right) = (1 + t)$m from the face A.

Weight of water over the projection on water side

$W_2 = 1 \times t \times 3 \times 10 = 30t$ kN, c.g. of W_2 (from A) $= \dfrac{t}{2}$ m

Water pressure (horizontal) $= \dfrac{1}{2} w \cdot h^2 \times 1 = \dfrac{10}{2} \times 4.5^2 = 101.25$ kN

cg of pressure P above base $= \dfrac{h}{3} = \dfrac{4.5}{3} = 1.50$m.

Take moment of all the forces about the point where the resultant strikes the base.

i.e. $101.25 \times 1.5 - W_2 \cdot \left(x - \dfrac{t}{2}\right) - W_1(x_r - 1 - t) = 0$

or $151.875 - 30t\left(x_r - \dfrac{t}{2}\right) - (202.5 + 67.5t)(x_r - t - 1) = 0$

or $151.875 - x_r\{(30t + 67.5t) + 202.5\} + 202.5 + 67.5t^2 + 202.5t + 15t^2 + 67.5t = 0$

or $354.375 - x_r(97.5t + 202.5) + 82.5t^2 + 270.0t = 0$

or $x_r(97.5t + 202.5) = 354.375 + 270t + 82.5t^2$

$$x_r = \dfrac{(82.5t^2 + 270t + 354.375)}{(97.5t + 202.5)}$$

$\therefore$ Eccentricity $e = \{x_r - (1 + t)\} = \dfrac{(82.5t^2 + 270t + 354.375)}{(97.5t + 202.5)} - (1 + t)$

or $e = \dfrac{(82.5t^2 + 270t + 354.375) - (97.5t + 202.5 + 97.5t^2 + 202.5t)}{(97.5t + 202.5)}$

or $e = \dfrac{-15t^2 - 30t + 151.875}{97.5t + 202.5}$, For no tension on the base $e \leq \dfrac{2 + 2t}{6}$

i.e. $\dfrac{(151.875 - 30t - 15t^2)}{(97.5t + 202.5)} \leq \dfrac{2 + 2t}{6}$

or $3(151.875 - 30t - 15t^2) \leq (97.5t + 202.5 + 97.5t^2 + 202.5t)$

or $\mathbf{253.125 - 390t - 142.5t^2 \leq 0}$

or $142.5t^2 + 390t - 253.125 \geq 0$, $t = -\dfrac{390 \pm \sqrt{390^2 + 4 \times 253.125 \times 142.5}}{2 \times 142.5}$

or $t \geq 0.5418\text{m}$, say $t = \mathbf{0.55m}$

Assuming $t = 0.55$m , $f_{\max}$ will be calculated.

Base width = $2 + 0.55 \times 2 = \mathbf{3.10m}$.

$\therefore$ $e = \dfrac{\{151.875 - 30(0.55) - 15(0.55)^2\}}{97.5(0.55) + 202.5} = \mathbf{0.51\ m}$

Total vertical load $W_1 + W_2 = 202.5 + 62.5 \times 0.55 + 30 \times 0.55 = \mathbf{253.375\ kN}$

Section modulus $Z_y = \dfrac{1}{6} \times 1 \times (3.1)^2 = \mathbf{1.6017\ m^3}$

Direct stress on the base $f_0 = \dfrac{253.375}{1 \times 3.1} = \mathbf{81.734\ kN/m^2}$.

Bending stress $= \pm\dfrac{253.375 \times 0.51}{1.6017} = \mathbf{80.678\ kN/m^2}$.

Max. stress = $81.734 + 80.678 = \mathbf{162.412\ kN/m^2\ (comp.)}$

Min. stress = $81.734 - 80.678 = \mathbf{1.056\ kN/m^2\ (comp.)}$

Max. stress will be $2 \times 81.734 = \mathbf{163.468\ kN/m^2}$, if the min. stress = 0 by taking exact value of t.

EXAMPLE 15.22: A trapezoidal weir section has 4.50 m height and 2 m top width. The weir wall tapers on both the sides to make bottom width of 3 m symmetrically. The density of masonry = 24 kN/m^3, while density of water is 10 kN/m^3. Calculate the maximum and minimum stresses in the base.

   Solution:

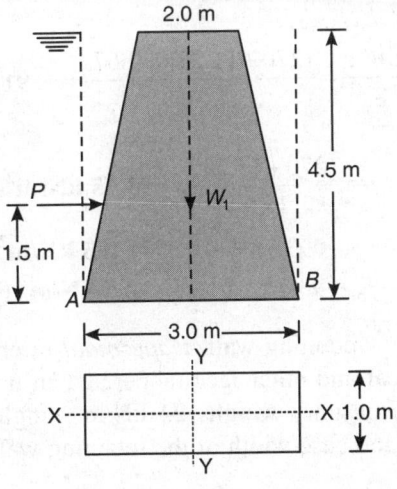

Fig. 15.23: Weir section

Consider 1m length of weir

$b_1 = 2.0$ m, $b_2 = 3.0$ m, Taper on each side = 0.5 m

$H = 4.50$ m, Masonry density = 24 kN/m³

Water density = 10.0 kN/m³.

Weir will have worst condition when water is on one side only.

$$W_1 \text{ (}w_i \text{ of weir)} = \frac{(2+3)}{2} 4.5 \times 1 \times 24 = 270 \text{ kN}$$

$$W_2 = (w_i \text{ of water}) = \left(\frac{0.5 \times 4.5}{2}\right) \times 10 = 11.25 \text{ kN}$$

at $\dfrac{0.50 \times 1}{3} = \mathbf{0.166\ m}$ from A (Heel)

$$P\text{(Water Pressure)} = \frac{10 \times 4.5 \times 4.5}{2} = 101.25 \text{ kN at 1.5 m above } A.$$

Let the resultant strikes the base at $\bar{x}_r$, take moment of all forces about this point.

$\Sigma M = 0$,

$$101.25 \times 1.5 - 11.25 \times (\bar{x}_r - 0.1667) - 270(\bar{x}_r - 1.5) = 0$$

or $151.875 - 11.25\bar{x}_r + 1.875 - 270\bar{x}_r + 405 = 0$

or $(11.25 + 270)\bar{x}_r = 558.75$, $\bar{x}_r = 1.9867$ m.

Eccentricity 'e' on the base = $1.9867 - 1.5 = \mathbf{0.4867\ m}$

$$Z_y = \frac{1 \times 3^2}{6} = \mathbf{1.50\ m^3}$$

Bending stress $f_b = \dfrac{\Sigma W.e}{Z_y} = \dfrac{(270 + 11.25)0.4867}{1.50} = \mathbf{\pm 91.25\ kN/m^2}$

Direct stress on base $f_0 = \dfrac{(270 + 11.25)}{1 \times 3} = \mathbf{93.75\ kN/m^2}$

Maximum stress = $f_0 + f_b = 93.75 + 91.25 = \mathbf{185\ kN/m^2}$ (comp.)

Minimum stress = $f_0 - f_b = 93.75 - 91.25 = \mathbf{2.5\ kN/m^2}$ (comp.)

EXAMPLE 15.23: A masonry retaining wall, *trapezoidal* in cross-section is 12 m high and has earth retaining face vertical and other face tappered 1 in 6. It retains earth upto the top. Earth weighs 16 kN/m³ and masonry weighs 24 kN/m³, angle of repose of earth may be considered as 30°. Determine the base width of the retaining wall.

Solution:

Let the top width $= b_1$, bottom width $b_2 = b_1 + 12 \times \dfrac{1}{6} = (b_1 + 2)$m.

$\theta = 30°$, coeff. of earth resistance $K = \dfrac{1 - \sin 30°}{1 + \sin 30°} = \dfrac{1}{3}$, w_e(earth) $= 16$ kN/m³

$w_m = 24$ kN/m³.

Total earth pressure $/_m$: $P = K \cdot w_e \dfrac{H^2}{2} = \dfrac{16}{3} \times \dfrac{12^2}{2} = $ **384 kN**

Weight of masonry/m length : $W = \left(\dfrac{b_1 + b_2}{2} \right) h \times 1 \times w_m = \left(\dfrac{b_1 + b_1 + 2}{2} \right) 12 \times 24$

$= (b_1 + 1) 288$ kN

cg masonry weight x_m

$= \dfrac{b_1^2 + b_1 b_2 + b_2^2}{3(b_1 + b_2)} = \dfrac{b_1^2 + b_1(b_1 + 2) + (b_1 + 2)^2}{3(2b_1 + 2)} = \dfrac{b_1^2 + b_1^2 + 2b_1 + b_1^2 + 4b_1 + 4}{6(b_1 + 1)}$

$x_m = \dfrac{(3b_1^2 + 6b_1 + 4)}{6(b_1 + 1)}$ ... (i)

Moment of forces about A

$P \times \dfrac{12}{3} + W(x_m) = 384 \times 4 + (b_1 + 1) 288 \dfrac{(3b_1^2 + 6b_1 + 4)}{6 \ (b_1 + 1)}$

$\Sigma M = 1536 + 48(3b_1^2 + 6b_1 + 4) = 1728 + 48 \times 3b_1 \ (b_1 + 2)$

Resisting force ΣW at C must balance

i.e. $288(b_1 + 1) \ x_r = 1728 + 144b_1 \ (b_1 + 2)$

$x_r = \dfrac{1728 + 144b_1(b_1 + 2)}{288(b_1 + 1)} = \dfrac{144 \ [12 + b_1(b_1 + 2)]}{288(b_1 + 1)}$

$x_r = \dfrac{[12 + b_1(b_1 + 2)]}{2(b_1 + 1)}$

Eccentricity $e = \left(x_r - \dfrac{b_2}{2} \right)$

or $e = \dfrac{12 + b_1^2 + 2b_1}{2(b_1 + 1)} - \dfrac{b_1 + 2}{2} = \dfrac{1}{2} \dfrac{[12 + b_1^2 + 2b_1 - (b_1^2 + 3b_1 + 2)]}{(b_1 + 1)}$

$$e = \frac{(10 - b_1)}{2(b_1 + 1)}$$

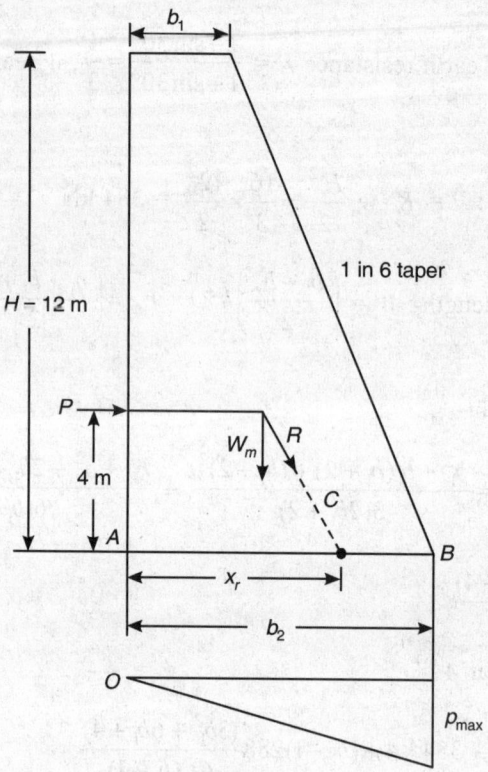

Fig. 15.24.

For no tension $e \leq \dfrac{b_2}{6}$.

i.e.　$\dfrac{(10 - b_1)}{2(b_1 + 1)} = \dfrac{b_1 + 2}{6}$ or $6 (10 - b_1) = 2 (b_1 + 1) (b_1 + 2)$

or　$30 - 3b_1 = b_1^2 + 3b_1 + 2$

or　$b_1^2 + 6b_1 - 28 = 0, \; b_1 = \dfrac{-6 \pm \sqrt{6^2 + 4 \times 28}}{2} = \dfrac{-6 \pm 2 \; \sqrt{37}}{2}$

or　$b_1 = (\sqrt{37} - 3) = (6.0828 - 3) = \mathbf{3.0828 \ m}$

∴　$b_2 = \mathbf{5.0828 \ m}$

EXAMPLE 15.24: A hollow cast iron short column has external diameter of 600 mm and internal diameter of 400 mm. The casting was found to be eccentric so that minimum thickness is 60 mm while the maximum thickness 140 mm. The column carries a load of 2000 kN along the axis of the bore. Determine the maximum and minimum stresses in the section.

Solution:

External diameter D = 600 mm, internal diameter = 400 mm.

Maximum thickness t_1 = 140 mm, minimum thickness t_2 = 60 mm.

Net area of cross-section = $\dfrac{\pi}{4}(600^2 - 400^2) = 10^4(15.708)$ mm^2

Load W = 2000 kN = 2 × 10^6 N (excluding self weight being neglected)

cg of solid portion x_0 = $\dfrac{\dfrac{\pi}{4}600^2 \times 300 - \dfrac{\pi}{4}(400)^2(140 + 200)}{\dfrac{\pi}{4}(600^2 - 400^2)} = \dfrac{(10800 - 5440)}{20}$

= 268 mm (332 mm from other end)

Distance of external centre = 300 − 268 = 32 mm, load on bore centre

Distance of cg from bore centre = $\left(140 + \dfrac{400}{2}\right) - 268 = $ **72 mm**,

e = 340 − 268 = 72 mm

Moment of inertia about cg of the section

$$I_y = \left[\frac{\pi}{64}(600)^4 + \frac{\pi}{4}(600)^2 \cdot (32)^2\right] - \left[\frac{\pi}{64}(400)^4 + \frac{\pi}{4}(400)^2(72)^2\right]$$

or $I_y = \dfrac{\pi}{4}\dfrac{10^4}{} \left[\dfrac{36 \times 36 \times 10^4}{16} + \dfrac{36 \times 32 \times 32}{1}\right] - \dfrac{\pi}{4}\dfrac{10^4}{}[16 \times 10^4 + 16 \times 72 \times 72]$

$$= \frac{10^4 \pi}{4}[810000 + 36864 - 160000 - 82944] = 474317 \times 10^4$$

or x_1 = 268 mm, x_2 = 332 mm

f_0 (direct stress) = $\dfrac{2 \times 10^6}{15.708 \times 10^4}$ = **12.73 N/mm^2** (comp.)

Bending stress due to eccentricity 'e' = $\dfrac{W \cdot e}{I_y} \cdot x_1$ and $\dfrac{W \cdot e}{I_y} \cdot x_2$

$$f_{b_1} = \frac{2 \times 10^6 \times 72 \times 268}{474317 \times 10^4} = \textbf{8.136 N/mm}^2 \text{ (tensile)}$$

$$f_{b_2} = \frac{2 \times 10^6 \times 72 \times 332}{474317 \times 10^4} = \textbf{10.079 N/mm}^2 \text{ (comp.)}$$

Net $f_{max} = 12.73 + 10.079 = \textbf{22.809 N/mm}^2$ (comp.)

$f_{min} = 12.73 - 8.136 = \textbf{4.594 N/mm}^2$ (comp.)

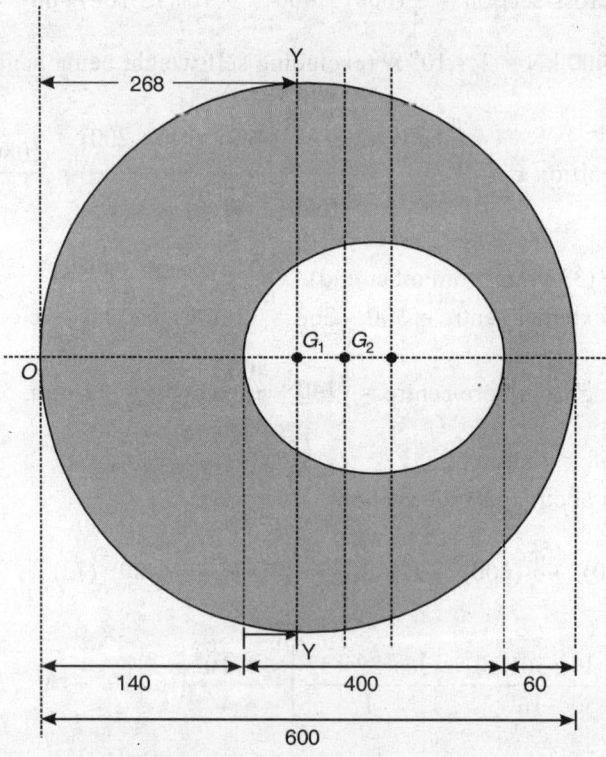

Fig. 15.25: Eccentric bore in casting

EXAMPLE 15.25: A retaining wall is 3 m wide at the top and 6 m wide at the bottom and is 15 m high. Earth retaining face is vertical. Earth weighs 18 kN/m³ and masonry weighs 24 kN/m³. The angle of repose of earth is 30° and earth is filled upto top level. Find the maximum and minimum stresses on the base and also cheek if the retaining wall is safe. Maximum permissible stress in compression is 600 kN/m² and $\mu = 0.45$. Maximum tensile stress permitted = 40 kN/m².

Solution:

Top width $b_1 = 3$m, bottom width = 6m

$H = 15$m, $\theta = 30°$, $\mu = 0.45$

$w_e = 18$ kN/m³, $w_m = 24$ kN/m³,

c_p (coeff. of Earth resistance) $= \dfrac{1-\sin 30°}{1+\sin 30°} = \dfrac{1}{3}$.

W (Masonry wall) $= \left(\dfrac{3+6}{2}\right) 15 \times 24 = 1620$ kN

P (Earth pressure) $= \dfrac{w_e \cdot H^2 c_p}{2} = \dfrac{18 \times 15^2}{3 \times 2} = 675$ kN acting at 5 m above the base

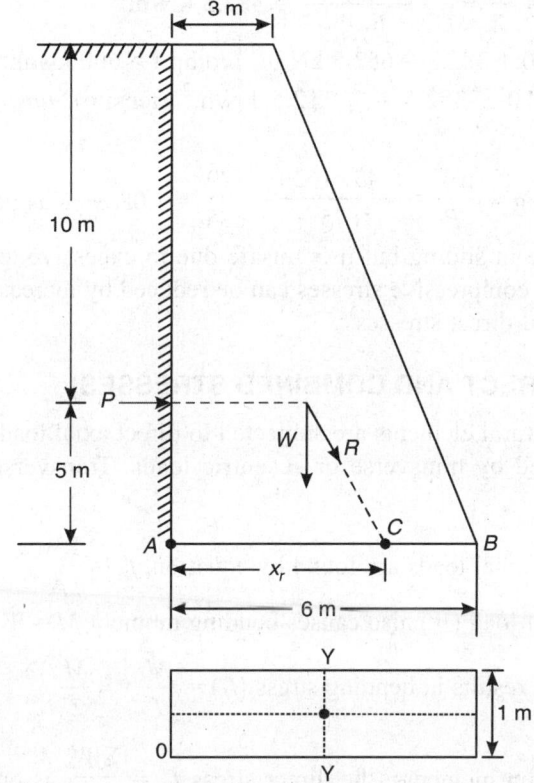

Fig. 15.26: Retaining wall

cg of weight $W : x_w = \dfrac{b_1^2 + b_1 b_2 + b_2^2}{3(b_1 + b_2)} = \dfrac{3^2 + 3 \times 6 + 6^2}{3(3+6)} = \dfrac{7}{3}$ m

$x_r = \dfrac{\Sigma M \text{ about } A}{\Sigma W} = \dfrac{675 \times 5 + 1620 \times \dfrac{7}{3}}{1620} = 4.4167$ m

Eccentricity 'e' = $4.4167 - \dfrac{6}{2} = 1.4167$ m

BM = $W \cdot e = 1620 \times 1.4167$ kNm.

$I_y = \dfrac{1 \times 6^3}{12}$, $Z_y = \dfrac{36 \times 6}{12 \times 3} = 6$ m^3.

Direct stress $f_0 = \dfrac{W}{A} = \dfrac{1620}{6 \times 1} = \textbf{270 kN/m}^2$

Bending stress = $\dfrac{M}{Z} = \dfrac{1620 \times 1.4167}{6} = \textbf{382.5 kN/m}^2$

f_{max} (base) = $270 + 382.5 = 652.5$ kN/m^2 (comp.) $>$ 600 kN/m^2, *unsafe*

f_{min} (base) = $270 - 382.5 = -112.5$ kN/m^2 (tensile). *unsafe* as tension is more than 40 kN/m^2

FS against sliding = $\dfrac{\mu W}{P} = \dfrac{0.45 \times 1620}{675} = \dfrac{729}{675} = 1.08$, *safe* as FS $>$ 1.

Retaining wall is safe in sliding but it is unsafe due to excessive tensile and compressive stresses. The tensile and compressive stresses can be reduced by increasing the base width and reducing the bending and direct stresses.

15.8 SUMMARY (DIRECT AND COMBINED STRESSES)

Many structures or structural elements are subjected to direct axial loads in addition to certain bending moments caused by transverse or eccentric loads. Transverse loads cause bending moments (M).

Stresses due to direct axial loads are found by equation, $f_0 = \dfrac{\Sigma W}{A}$ ... (i)

The eccentricity 'e' of load (W) also causes bending moment $M = W \cdot e$ ... (ii)

Bending moment 'M' results in bending stress (f_b) = $\dfrac{M}{I} \cdot y = \dfrac{M}{Z}$ ... (iii)

In case of compression members, the direct stress $f_0 = \dfrac{\Sigma W}{A}$, is only applicable in *short columns* because *long columns* develop buckling effect and the simple formula for direct stress $\left(f_0 = \dfrac{\Sigma W}{A} \right)$ does not apply.

The *resultant stress* at any point in the section will be due to direct stress (f_0) and bending stress (f_b), i.e. $f_r = (f_0 \pm f_b)$ ... (iv)

"For rectangular sections, the resultant stress due to axial loads ΣW,

shall be $f_r = \dfrac{W}{A} \pm \dfrac{W \cdot e}{I} \cdot y = \left(\dfrac{W}{b \cdot d} \pm \dfrac{6W \cdot e}{bd^2} \right)$ ... (v)

If the eccentricity '*e*' remains within limits of $\leq \dfrac{b}{6}$ or $\dfrac{d}{6}$, the resultant stress does not change its nature in rectangular section. For $e_x = \dfrac{b}{6}$ or $e_y = \dfrac{d}{6}$, the maximum stress $f_{max} = \dfrac{2W}{b \cdot d}$ and minimum stress = 0 in case of rectangular section. In case of biaxial eccentricity e_x and e_y, the resultant stress at any point will be given by

$$f_r = \frac{W}{A} + \frac{W \cdot e_x \cdot x}{I_y} + \frac{W \cdot e_y \cdot y}{I_x} \qquad \qquad \text{... (vi)}$$

The *central area* in which the resultant stress does not change sign is called *core* (or *kern*). In case of rectangular sections the *kern* is diamond shaped and its conditions are given by :

$$\frac{6e_x}{d} + \frac{6e_y}{b} = 1 \qquad \qquad \text{... (vii)}$$

Thus central diamond with diagonals along two axes with diagonals $\dfrac{b}{3}$ and $\dfrac{d}{3}$ is called *core (kern)*. The equation of kern shall be $\dfrac{6x}{b} + \dfrac{6y}{d} = 1$.

"In circular solid section, the *kern or core* is formed by the circle with its centre at the centre of the section and diameter $= \dfrac{D}{4}$ $\qquad \qquad \text{... (viii)}$

For no tension in hollow circular section due to direct weight and lateral loads, the weight must lie within radius $\left(\dfrac{D^2 + d^2}{8D} \right)$ of the centre line. $\qquad \qquad \text{... (ix)}$

"In case of earth pressure $P = \left(\dfrac{1 - \sin\theta}{1 + \sin\theta} \right) \dfrac{wH^2}{2}$, where '$\theta$' = angle of repose and '*w*' is weight of earth / unit volume. $\qquad \qquad \text{... (x)}$

Total wind pressure on circular or curved surface will be

$P = k \cdot p \cdot A_p$, where k = coefficient of wind resistance, p wind pressure, A_p projected area of surface. $\qquad \qquad \text{... (xi)}$

EXERCISES 15

Q.15.1. A 20 mm diameter rod is pulled by a 4.4 kN load along a axis at a distance of 0.50 mm from the central axis. Determine the maximum and minimum stresses across the section.

Ans. p(minimum) = 11.2 N/mm^2 (tensile), p(maximum) = 16.8 N/mm^2 (tensile)

Q.15.2. A cast iron column of 160mm external diameter and 120mm internal diameter carries an axial load W at an eccentricity of 50mm. Determine the load W if the permissible stresses are 104.0 N/mm^2 (maximum compressive) and 30 N/mm^2 (max. tensile). Also show stress variation for the suggested load.

Ans. Safe W = 352 kN, f_{max} = 104 N/mm^2, f_{min} = 24 N/mm^2

Q.15.3. A drilling machine is shown in figure Q15.3. The drilling machine drill A is 300 mm away from the centre of the circular body B. Find the drill body element B diameter if maximum tensile stress permitted in the body B is 30 N/mm^2.

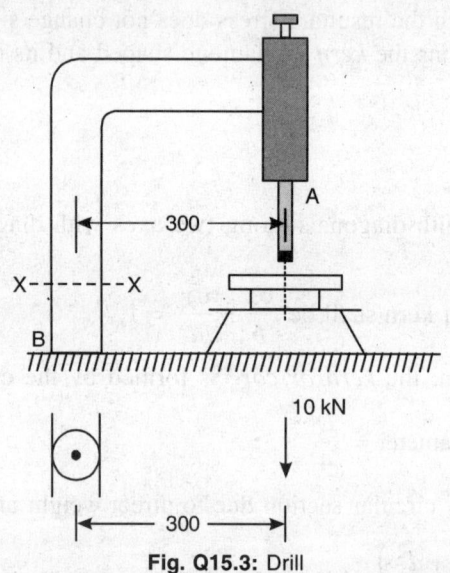

Fig. Q15.3: Drill

Ans. Element diameter at B = 102 mm

Q.15.4. A hollow pier of 3m × 2m external and 1m × 1m internal carries a load of 100 kN at a distance of 600 mm from the c.g. of the section along the bisector of 2m side. The self weight of the pier is 50 kN.

Ans. f_{max} = **50.378 kN/m^2**, (comp).

f_{min} = **9.622 kN/m^2** (comp).

Q.15.5. Vertical column of a crane comprises of 550 mm × 190 mm I – section. Crane lifts 40 kN load at a distance of 2.0 m from the cg of the I – section. Determine the maximum and minimum stresses in the column I – section. Consider Area of I – section = **13200 mm^2** and its section modulus z = **2360 × 10^3 mm^3**. Neglect self weight of the column.

Ans. f_{max} = 36.93 N/mm^2 (comp), f_{min} = –30.87 N/mm^2 (tensile).

Q.15.6. A crane hook is trapezoidal in section with inner width = 60 mm and outer width = 40 mm, and depth = B mm. If the load line is symmetrical and located at a distance of 50 mm from the inner face. Find the safe load if the permissible stress in the hook = 100 N/mm^2. (Also tensil and comp.) Find the minimum stress in the hook.

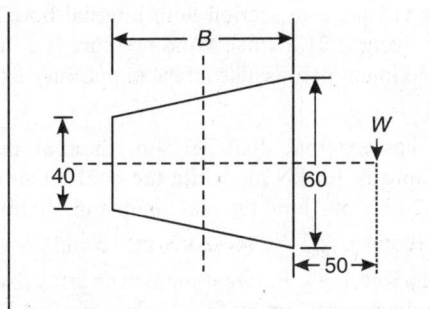

Fig. Q15.6: Section of crane hook

Ans. B = 25 mm, Safe W = 8.333 kN, p_{max} = 100 tensile N/mm^2, p_{min} = 100 N/mm^2 comp.

Q.15.7. A cast iron circular column has 200 mm external and 150 mm internal diameters. The column carries an axial load of 400 kN in addition to 80kN eccentric load at an eccentricity of 250 mm. Determine the maximum and minimum stresses in the column.

Ans. f_{max} = **72.15 N/mm^2** (comp), f_{min} = **2.33 N/mm^2** (tensile)

Q.15.8. A cast iron hollow circular column has 200 mm external diameter and 150 mm internal diameter. Error in casting resulted in wall thickness variation of maximum 40 mm to minimum 10 mm at the other end. A load of 400 kN acts at the centre of the bore. Determine the maximum and minimum stresses in the column section.

Ans. f_{max} = **65.77 N/mm^2** (comp), f_{min} = **4.20 N/mm^2** (comp).

Q.15.9. A tie rod 200 mm wide and 15 mm thick transmits an axial pull of 200 kN. At certain section a hole of **40mm** diameter is drilled with its centre 50 mm from the mid point of the rod. Determine the maximum and minimum stresses at the weakest section.

Ans. f_{max} = **118.3 N/mm^2** (tensile), f_{min} = **56.14 N/mm^2** (tensile)

Q.15.10. A vertical short column is formed by channels and plates of 300 mm × 25 mm on each flange. If the column carries a load of **1500 kN** at an eccentricity of **40 mm** along Y–Y axis. Determine the maximum and minimum stresses on the column section. Each channel section I_{xx} = 3816.8 × 10^4 mm^4 and A = 3867 mm^2.

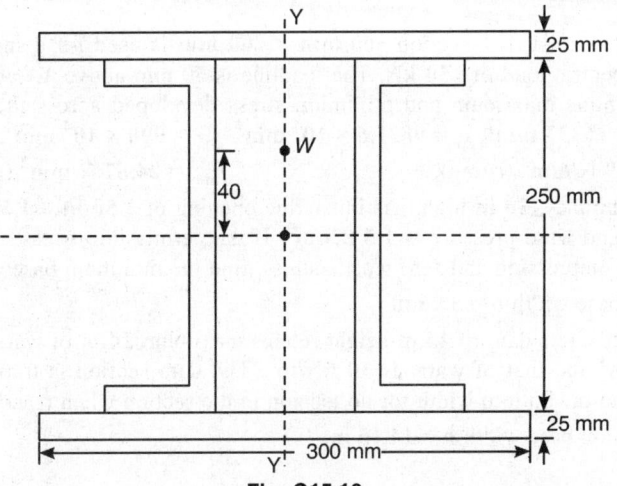

Fig. Q15.10

Ans. f_{max} = 90.93 N/mm^2 (top) comp. and f_{min} = 41.03 N/mm^2 (bottom) comp.

Q.15.11. A 20m high chimney is square in section with internal bore 1000mm × 1000mm and wall thickness 't'. Masonry weighs 21 kN/m^3. Wind pressure is 1.20 kN/m^2. Find the thickness of masonry wall if the maximum permissible stress in masonry is **780 kN/m^2**.

Ans. $t = 525$ mm (0.525m).

Q.15.12. A circular chimney has external diameter 4m, internal diameter 2m and height 30m. Horizontal wind pressure is 1.5 kN/m^2 while the coefficient of wind resistance = 2/3. The density of masonry = 20 kN/m^3. Find the maximum and minimum stresses on the base.

Ans. f_{max} = **905.58 kN/m^2** (comp), f_{min} = **294.42 kN/m^2** (comp).

Q.15.13. A tappered chimney is 30m high. External diameter varies from 4m at the base to 2m at the top. The internal bore diameter is 2m uniformly through out. The total weight of the chimney is 2400 kN. Wind pressure is 2.40 kN/m^2. Coefficient of wind resistance may be taken as $\dfrac{2}{3}$. Find the bending moment due to wind pressure at the base and also calculate maximum and minimum stresses at the base.

Ans. BM at the base = **1920 kNm**, f_{max} = **580.6 kN/m^2** (comp). f_{min} = **71.3 kN/m^2** (i.e. tensile)

Q.15.14. Find the height of a rectangular masonry retaining wall if *no tension* is to be developed in the wall and the density of earth is 20 kN/m^3, angle of repose $\theta = 30°$. Density of masonry = 25 kN/m^3.

Ans. Maximum height h = **3.87 m.**

Q.15.15. A trapezoidal dam section has top width 2 m, bottom width 4 m and height 6 m. Water face of the wall is vertical. Water rises upto the top level. If the masonry weighs 24 kN/m^3. Determine the maximum and minimum stresses on the base when the reservior is full and when the reservior is totally empty.

Ans. Reservoir full : f_{max} = **171 kN/m^2**, (comp.), f_{min} = **45.0 kN/m^2** (comp).
Reservoir empty : f_{max} = **180 kN/m^2** (comp), f_{min} = **36 kN/m^2** (comp)

Q.15.16. A chimney of 1500 kN total weight tilts and the centre of gravity shifts by **0.15 m** at the base. The chimney section at the base has **3.0 m** external and **1.80 m** internal diameter. Find the horizontal load acting at **12 m** above the base so that there is no tension developed in the chimney.

Ans. P = **45 kN.**

Q.15.17. A rolled steel joist of I–section **300 mm** × **200 mm** is used for a short column carrying a vertical eccentric load of **150 kN**. The loadline is 50 mm above XX and **25 mm** right of YY axis. Determine maximum and minimum stress developed across the cross-secton. For I–section : A = 6133 mm^2, I_x = 9821.6 × 10^4 mm^4, I_{yy} = 990 × 10^4 mm^4.

Ans. f_{max} = **73.79 N/mm^2** (comp), f_{min} = **24.87 N/mm^2** (tensile).

Q.15.18. A square chimney, **30 m** high, has inner flue opening of **1.50 m** × **1.50 m**. Masonry weighs **20 kN/m^3** and wind pressure is **1.5 kN/m^2**. If the permissible stress in the masonry is **1000 KN/m^2** in compression and zero tensile stress, find the minimum base width.

Ans. Minimum base width b = **3.26 m.**

Q.15.19. A cement concrete dam of **25 m** height retains maximum **24 m** of water. Density of concrete is **24 kN/m^3** and that of water is **10 kN/m^3**. The dam section is trapezoidal with top with **2.50 m**. Find the bottom width for no tension in the section when reservoir is full or empty.

Ans. For no tension base width b = **14.15 m.**

Q.15.20. A tie rod of 100mm square section is cut from one side so that the remaining section is **100 mm × 50 mm** rectangular as shown in Fig. Q15.20. The remaining cut section carries an axial load of **15 kN**. Find the maximum and minimum stresses in both the sections.

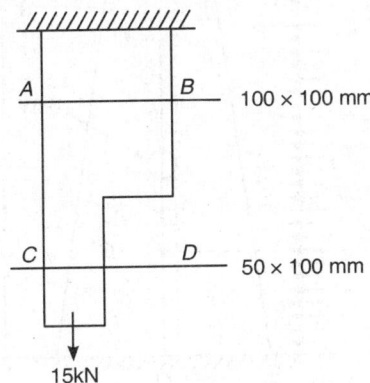

Fig. Q15.20.

Ans. f_{max} (CD) = 3 N/mm² (tensile uniform) $\quad$ f_{max} (AB) = **3.75 N/mm²** (tensile)
$\quad$ f_{min} (AB) = **0.75 N/mm²** (comp.)

Q.15.21. A solid circular short CI column carries a load of 100 kN at the top. The column is bent as shown in the Fig. Q15.21. The load line is $\dfrac{600}{7}$ mm away from the centre of the base of the column. Find the diameter of the base so that the maximum stress does not exceed **100 N/mm²**.

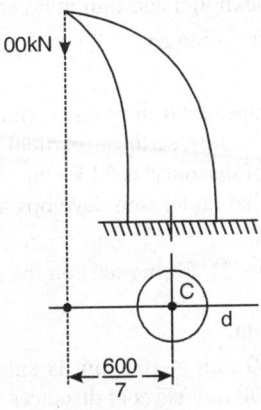

Fig. Q15.21.

Ans. d reqd = **100 mm**.

Q.15.22. A tapering chimney of circular bore of **0.80 m** diameter is **30 m** high. Diameter (external) varies from **2.40 m** at the base to **1.6 m** at the top. Wind pressure of **2.2 kN/m²** acts on the projected area of chimney. Weight of chimney is **4000 kN**. Determine the maximum and the minimum stresses at the base (Fig. Q15.22).

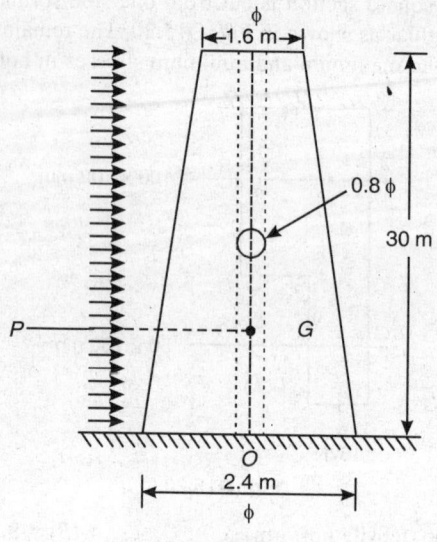

Fig. Q15.22.

Ans. $P = 2.2 \times 60 = \mathbf{132\ kN}$, centroid from base = **14 m**

$f_{max} = \mathbf{2374.1\ kN/m^2}$, $\qquad\qquad\qquad f_{min} = \mathbf{-\ 384.1\ kN/m^2}$ (tensile).

Q.15.23. A short hollow cast iron column having an external diameter of 600 mm and internal diameter of 400 mm but the bore is eccentrically cast with wall thickness varying from 140 mm at one end to 60 mm at the other opposite end. The column carries a load of 2000 kN along the axis of the bore. Determine the maximum and minimum stresses in the section.

Ans. Direct stress = 12.7334 MN/m² (comp), $\qquad f_{max} = \mathbf{22.81\ MN/m^2}$ (comp),

$f_{min} = \mathbf{4.6\ MN/m^2}$ (comp).

Q.15.24. A masonry retaining wall, trapezoidal in cross-section is 12 m high and has one face vertical and other tapered 1 in 6. It retains earth on vertical face upto top. Weight of earth may be taken as **16 kN/m³** and that of masonry as 24 kN/m³. The angle of repose for the earth is 30°. Calculate the base width so that no tension develops any where in the base.

Ans. $p_e = 384$ kN, top width = $(b - 2)$, For no tension the resultant strikes base at $\dfrac{2}{3}b$.

$b = \mathbf{5.08m}$, top width = **3.08 m**.

Q.15.25. A rectangular section of 200 mm × 100 mm is subjected to an eccentric load of 160 kN parallel to the 200 mm and 100 mm sides at distances of 40 mm and 20 mm from the centroid of the section parallel to 200 mm and 100 mm sides respectively. Find the net stresses at the four corners. Also find the additional axial compressive load at the centroid so as to make net tensile *stress zero* any where in the section.

Ans. f_A (load side corner) = **27.2 N/mm²** (comp); $\qquad f_B = \mathbf{8\ N/mm^2}$ (comp);

$f_C = \mathbf{-\ 11.2\ N/mm^2}$ (tensile); $\qquad\qquad f_D = \mathbf{8\ N/mm^2}$ (comp).

Additional axial load at the centroid = **224 kN**, for zero tension.

Q.15.26. A hollow circular masonry chimney of **20 m** height has 2 m internal diameter. Find the **external** *diameter* if *no* tension has to be developed in the base. Wind pressure is **3 kN/m²** of the projected area. Coefficient of wind resistance may be taken as **2/3** and density of masonry may be assumed as **20 kN/m³**.

Ans. External Diameter $D = 3.42$ **m.**

Q.15.27. A trapezoidal dam section of **12 m** *height* with **3 m** top width and **9 m** *bottom* width. Unit weight of dam masonry may be taken as **21 kN/m³**. It is proposed to raise the storage capacity of reservoir by erecting steel plate shuttering above the top on vertical water face of the dam. Determine the maximum height of shuttering so that no tension develops at the base under reservoir full condition. Ignore the effect of uplift due to seepage.

Ans. Height of steel shuttering not to exceed **1.65 m.**

Index

Reader's Notes

Reader's Notes